Principles of Modern Agriculture

Principles of Modern Agriculture

Editor

Richard Renneboog, MSc

SALEM PRESS

A Division of EBSCO Information Services
Ipswich, Massachusetts

GREY HOUSE PUBLISHING

Cover Image: Growing young maize seedling in cultivated agricultural farm field with modern technology concepts. Photography by lamyai (iStock).

Principles of Modern Agriculture, published by Grey House Publishing, Inc., Amenia, NY, under exclusive license from EBSCO Information Services, Inc.

For information, contact Grey House Publishing/Salem Press, 4919 Route 22, PO Box 56, Amenia, NY 12501.

∞ The paper used in these volumes conforms to the American National Standard for Permanence of Paper for Printed Library Materials, Z39.48 1992 (R2009).

Publisher's Cataloging-In-Publication Data
(Prepared by The Donohue Group, Inc.)

Names: Renneboog, Richard, editor.
Title: Principles of modern agriculture / editor, Richard Renneboog, MSc.
Description: [First edition]. | Ipswich, Massachusetts : Salem Press, a division of EBSCO Information Services, Inc. ; Amenia, NY : Grey House Publishing, [2019] | Series: Principles of | Includes bibliographical references and index.
Identifiers: ISBN 9781642652635
Subjects: LCSH: Agriculture.
Classification: LCC S493 .P75 2019 | DDC 630--dc23

First Printing
Printed in the United States of America

Contents

PUBLISHER'S NOTE

Salem Press is pleased to add *Principles of Modern Agriculture* as the thirteenth title in the *Principles of* series that includes *Chemistry, Physics, Astronomy, Computer Science, Physical Science, Biology, Scientific Research, Sustainability, Biotechnology, Programming & Coding, Climatology,* and *Robotics & Artificial Intelligence.* This new resource introduces students and researchers to the fundamentals of modern agriculture using easy-to-understand language for a solid background and a deeper understanding and appreciation of this important and evolving subject. All of the entries are arranged in an A to Z order, making it easy to find the topic of interest.

Entries related to basic principles and concepts include the following:

- Fields of Study related to the topic;
- Principal Terms and definitions;
- A Summary that provides brief, concrete summary of the topic and how the entry is organized;
- Text that gives an explanation of the background and significance of the topic to modern agriculture by describing developments such as Apiology, Animal Husbandry, Organic Farming, Seed Banks, Strip Farming, Sustainable Agriculture, and Water Conservation;
- Illustrations that clarify difficult concepts via models, diagrams, and charts of such key topics as the Hydrologic Cycle, Horizons, Hybridization, Habitats & Biomes; and
- Further Reading lists that relate to the entry.

This reference work begins with a comprehensive introduction to climatology, written by volume editor Richard Renneboog, MSc.

The book includes helpful appendixes as another valuable resource, including the following:

- Time Line of Major Events in Agriculture, tracing the field back to the earliest advances in agriculture;
- Glossary;
- General Bibliography; and
- Subject Index.

Salem Press and Grey House Publishing extend their appreciation to all involved in the development and production of this work. The entries have been written by experts in the field. Their names and affiliations follow the Editor's Introduction.

Principles of Modern Agriculture, as well as all Salem Press reference books, is available in print and as an e-book. Please visit www.salempress.com for more information.

Editor's Introduction

Modern agriculture is a complex, multi-faceted activity, global in scale and increasingly dependent on advanced technology. Yet for all that, the fundamental principle of agriculture is exactly the same today as it was ten thousand and more years in the past and will remain for years into the future. At its most basic, the raison *d'être* of agriculture is to produce food. It does not matter how well the agricultural producers themselves understood the biochemical and biophysical aspects of photosynthesis and energy transport. Agriculture boils down to the simple act of putting a seed into fertile soil and reaping the fruits of the plant that grew from that seed. It is simplicity itself... in principle. As the saying goes, however, the devil is in the details.

Agriculture is one of the oldest social exercises of the human race. Many consider that agriculture is what made human civilization possible. Our hunter-gatherer forbearers undoubtedly saw that the wild plants they collected for food produced new plants where seeds from those plants had fallen to the ground. It would have been a natural step to intentionally collect the seeds and put them into the ground on purpose. As wandering peoples, they would have found those new food plants ready and waiting for them when they returned to that same location. After that would have come the realization that roaming about the countryside in a wide-ranging search to find edible plants was no longer necessary. They could instead remain in one location and the edible plants would in effect come to them as they grew in place. At that point, some nomadic hunter-gatherers became settled in a specific area, and thus agriculture began. Those who farm vegetable crops today, more than ten thousand years later, carry out exactly the same function as did their ancient ancestors.

The hunter-gatherer tradition didn't die out with the emergence of crop agriculture, though. The same process that resulted in the domestication of certain plants would also have pointed the way to the domestication of certain animals. Just as plants could be made to grow in the areas where the people settled, some migrating animals could be prevented from wandering and made to remain where the people lived. This marks the beginning of animal husbandry, the second primary aspect of agriculture that persists into the present day and beyond.

From these beginnings—both of immeasurable value to the human race—arose the multi-millennia long traditions of agriculture. This simplified description of how agriculture got its start understates the innately complex nature of agriculture. As a practice, agriculture represents the most intimate relationship the human species has with the planet it calls home. Agriculture depends on the favorable interplay of every aspect of the natural environment of Earth, from the functional structure of the soil below the surface, to the behavior of the farthest reaches of the atmosphere, and everything in between. It is worth keeping in mind that agriculture, despite its grounding in the soil, sky, and water of the planet, agriculture is an artificial construct. Despite its organic nature and basis in the natural rhythms of the plants and animals, agriculture is a sort of parasite. To exist at all, agriculture must maintain a symbiotic relationship with the planet's total ecology in order to coexist and integrate itself as a functioning component of the larger natural environment.

Earth's great cycles and seasonal changes establish an effective self-balancing steady state subject to the laws that govern the physical world. The climate of the planet goes through cycles of seasonal change. The water cycle describes the continuous movement of water on Earth, rising as water vapor from the oceans and condensing back into liquid and solid forms to fall as precipitation, and then making its way back to the oceans to start the cycle once more. The carbon-oxygen cycle describes how carbon and oxygen are in a constant state of motion, as atmospheric gases are changed into carbohydrates through photosynthesis and back again into gases as the end products of respiration and energy transport in living things. Gasses are also captured as carbonate minerals which are, in their turn, broken down to their component parts in the complex chemical environment that is Earth. Predator and prey species exist in cycles of life and death within the natural environment. Earth and the environment it supports is a huge, complex and dynamic system. The human endeavor of agriculture impacts the cycles and rhythms that make up this complex milieu, inserting new segments into each natural cycle as it draws inputs from each one and delivers outputs back to each one.

Lands that have been domesticated for agricultural uses have always coexisted with lands that are conserved to be wild. One unavoidable consequence of this is that predatory animals in the natural environment have an interest in the descendants of the prey animals that were domesticated thousands of years ago. This quite naturally brings some agriculturalists into conflict with their wild neighbors. In the first quarter of the twentieth century, for example, the gray wolf population in the wilds of Yellowstone National Park was eradicated, primarily to placate ranchers who claimed the wolves were killing their livestock. The reintroduction of gray wolves to Yellowstone National Park in 1995 was met with resistance from local ranchers who feared the wolves would soon be killing their livestock as they roamed outside of the boundaries of the park. As observation has amply revealed, the removal of those apex predators did severe harm to every aspect of the natural environment of the park, including its geography, because it disrupted the natural order. Their reintroduction initiated a trophic cascade that has restored some of that balance and gone some way toward mending the damage, while predation by gray wolves outside of the park boundaries has not been observed, although other wild predators such as cougars, coyotes, and grizzly bears can and do prey on livestock.

Plant crops are also not immune to competition with the wild. A huge part growing crops of any sort is keeping the wild at bay. Weeds are anathema to plant crops. They can take up a significant amount of the water and soil nutrients needed for the proper growth of the desired crop plants. Agricultural operations meant to enhance crop production such as irrigation and fertilization, extensive soil preparation, and field maintenance carry large costs and demand careful attention to assure their proper application as well as appropriate conservation of resources, even before considerations related to pest and weed control. A whole new set of costs and cautions are involved in the constant battle to eliminate undesirable weeds from the desirable crop plant. Other threats to crops include funguses, viruses, bacterial infections, insects, birds, and animals, forcing the farmer to fight the battle on several fronts at the same time. Physical deterrents, pesticides, and herbicides to counter these intruders add huge costs for farmers. Such is the nature of the artifice of agriculture in the milieu of the natural environment.

As much as the natural environment impacts how agriculture functions, it is equally true that agriculture affects the natural environment. Since agriculture is by nature an activity by and for the support of humans, any consideration of how agriculture affects the natural environment has to begin with an examination of how the human population affects the natural environment. In that long ago time when agriculture was first evolving, it would have noted that the seeds grew best when they were planted in cleared soils where they didn't have competition. Land areas could be cleared to make plots of ground where plantings could be concentrated. That meant that the yield—grains, seeds, fruits, nuts, leaves, shoots, stems, tubers—from those plants could be recovered with much less effort than wild harvesting. Now that humans could produce amounts of food that exceeded their basic needs, they could trade the excess food for material goods and services. In this way, agriculture became the foundation for commerce and the driving force for the establishing of villages, towns, and eventually cities, nations, and civilizations. All of these human establishments, from small cleared plots of land to sprawling cities, have one thing in common: the conversion of the corresponding and ancillary land areas, such as roads, from their natural state. This effectively removes those areas from the natural environment, reducing the area of the natural environment, fragmenting the layout of the natural environment, and drastically altering the role of those areas in natural cycles. Rainfall that had for millions of years percolated through soil to replenish underground aquifers, for example, is instead directed as surface run-off by the impermeable surfaces and artificial channels in the terrain modern man's environment of paved roads, buildings, airports, parking lots, and industrial complexes.

As agriculture became a more efficient means to feed ourselves and there was more food available, there was a simultaneous increase in the human population. As the population increased, the need for food from the practice of agriculture also increased. Greater need for more agricultural production means there is a concurrent the need for more lands devoted to agriculture, again to the detriment of the natural environment and the wild. Modern agriculture must address this problem by working to develop methods of producing more food in the same amount of space without depleting resources or ruining the long-term viability of the soil. Agricultural practices that eliminate the wild and convert it to agricultural land by clear-cutting or slash-and-burn techniques are no longer viable, although they are still widely practiced. In the Amazon basin, for example, huge tracts of ancient forests there are being clear-cut and converted to agricultural crop land and cattle pasture. Research has shown that the soil structure in those areas will not support these agricultural activities reliably in the long term. The soils that have been cleared are now exposed to erosive forces of wind and tropical rains without benefit of the protective cover that kept them in place. Flooding and desertification inevitably follow such practices, yet the clearing continues at an estimated twenty thousand hectares (fifty thousand acres) per day. Species extinctions, which occur at a normal background rate of between one and five species per year, are estimated to be occurring at an actual rate of between one thousand and ten thousand times the normal background rate as a direct result of agricultural and other human activities in 2018. Some may be tempted write the losses off as having no commercial use, but it is worth remembering that each and every one of the species, both plant and animal, that the human race relies on for its food was once also just a wild species with no perceived value.

Reliance on a relatively small number of species has created a type of farming known as monoculture, wherein the majority of agricultural activities are based on just a single species on the local or individual farm level or on just a small number of species at a regional level. Monoculture is a natural consequence of human nature. Once a given species—whether soybeans or Holstein cows or bananas—shows a desirable return on the investment of effort and expenditure required to bring it to market, it will be used by more and more farmers in an area. At some point, unless new species or genetic variants of those species are brought in to displace them, every farmer will end up raising those same species over a very large area. Husbandry would be geared only to obtaining better individual Holstein cows and more productive soybean plants within the same genetic variety. This is a normal practice in modern agriculture, and produces equitable results in terms of herd improvement and crop yields. But it is a practice that also opens the door for some very serious problems.

Individuals of the same species are all susceptible to the same diseases and abnormal genetic conditions. Without genetic variability within a species, entire populations can be wiped out by a single disease. This is true of both domestic and wild populations. The *Brucellosis* bacterium, for example, will affect every animal in a herd of the same species, whether it is a herd of dairy cattle or a herd of wild deer. The *Fusarium* virus that affects any plant of a particular plant species like tomatoes can infect every plant in a crop of that species. In 2018, corn crops in North America have been deemed unusable for animal and human food due to the presence of unacceptable amounts of the compound deoxynivelenol (also known as vomitoxin) that they contain. That compound is produced by a mold that flourished during warm, moist fall weather of that year.

Nowhere is the danger of monoculture more apparent than in crops in which every plant is genetically identical, a clone of some corresponding parent plant. This is a danger that the commercial banana industry knows all too well. First introduced to the United States at the Chicago World's Fair at the end of the nineteenth century, bananas quickly became the most popular fruit in the country. Huge plantations of banana trees were established in tropical countries, leading to the so-called "banana republics," to provide enough bananas to meet the consumer demand. Every tree in those plantations was the clone of a banana species known as the *Gros Michel* (French for Big Mike). A fungal infestation struck, and due to the extreme monoculture of the plantations, every single tree was susceptible to the fungus and the entire species was wiped out. A new variety was found that could be grown to replace the *Gros Michel* banana, known as the Cavendish banana. The Cavendish is smaller than and not as sweet as the *Gros Michel*, but it was resistant to the fungus that eradicated the *Gros Michel*. The Cavendish is the variety currently sold in grocery stores around the world. The story of *Gros Michel* is set to repeat itself once again, however, as banana plantations relied on extreme monoculture for rapid production, and the Cavendish variety now faces eradication due to its susceptibility to a different fungus species. That fungus has already infested and destroyed banana plantations in several parts of the world. Researchers are scrambling to find a new variety of banana that can be an acceptable replacement for the Cavendish banana. While some comb through tropical forests and fruit markets in the hope of finding a suitable type of banana, others, are pursuing another avenue available to modern agriculture: gene modification.

Gene modification, practiced since the earliest days of agriculture and animal husbandry, has, until quite recently, relied on the traditional practices of plant cross-breeding and of animal husbandry. Plants that exhibited desirable traits such as large, robust seed heads in comparison to others of the same species would either be grown preferentially or used to transfer the desirable traits to other individuals by cross-pollination. This type of modification took successive generations and might require many years to produce a variety that reliably exhibited the desirable traits. In animal husbandry, the same principles applied, and successive generations of animals were selectively bred to have specific breed characteristics. This type of breeding continues to the point at which genetic lines would "breed true." For example, all dogs can interbreed to produce "mutt" offspring, yet German shepherd dogs produce German shepherd puppies, poodles produce poodle puppies, and so on. It is interesting top note that all modern dogs are believed to have arisen through the likely domestication of wild wolves more than forty thousand years ago (the age of the oldest known domestic dog remains). Domestic dogs are interfertile with wolves like those roaming Yellowstone National Park, a fact that underscores the ancient link between modern agriculture and the natural environment.

In modern agriculture, genetic modification generally refers to the direct manipulation of the DNA molecule by scientific methods. In this process, a specific segment of a DNA molecule called a gene can be activated, deactivated, inserted, or removed to produce a genetic variant having an innate property that the organism would not normally have. It is a controversial practice, but one with great economic and social potential. One example was the so-called Flavr Savr tomato, a variety modified to deactivate the gene that controlled the ripening of the fruit. The goal was to produce a tomato that would ripen very slowly when grown to maturity, enabling it to resist bruising and damage during harvesting and shipping. The Flavr Savr could be harvested while still very green and hard and then allowed to ripen slowly in storage or as they sat on store shelves. Unfortunately, the Flavr Savr tomato failed because it could not ripen reliably to the state that consumers desired and it lacked the flavor of a properly ripened tomato. More successful examples of genetically modified species are found in corn and soybeans. Genetic modification of corn produced the 'Bt' corn varieties that are resistant to insect infestations. Other modified varieties of corn are given the 'ss' designation, indicating they are supersweet varieties that have higher levels of sugar content and sweetness compared unmodified varieties of sweet corn. Soybeans are a crop that is sensitive to the presence of weeds, and yields are significantly affected by the presence of weeds in soybean fields. For soybean growers, the herbicide of choice is Monsanto's Round-Up product. Unfortunately, Round-Up is a post-emergence herbicide that must be applied after weeds have begun to grow, and soybean germination is also occurring. Glyphosate, the active ingredient in Round-Up, does not discriminate between weeds and soybeans. This made its use in weed-infested soybean fields problematic. To get around this, genetic modification was used to produce so-called Round-Up Ready soybean varieties that are resistant to glyphosate and are not be killed off by the herbicide. The result has been strong, viable soybean production.

The controversy regarding genetically-modified organisms (GMOs) like Round-Up Ready soybeans is rooted in understanding what happens when GMOs cross-pollinate with traditional plant species. That is a sound concern, but it has been subsequently blown out of proportion by talk of "Frankenfoods" and other such fears. A related and more realistic concern is the genetic modification of animal species. Because humans and other animals consume the flesh of food animals, there is a real potential for the transfer of potentially harmful proteins called prions to the consumer. This is an effect that has actually been observed in domestic animals that have been fed with waste animals proteins as part of their usual feed regimen. Accordingly, there are many controls in place to limit the introduction of GMOs into the consumer market or to improve labeling requirements so that consumers can make more informed choices.

Increasingly, modern agriculture depends on advanced and highly technological equipment. Digital control of agricultural equipment operations is a particularly broad field of application in modern agriculture. A modern tractor or combine, for example, is not much like the tractors and combines of a mere twenty or so years ago. While a tractor remains essentially a powerhouse on wheels and combines recover ripened grain in the same way that they always have, these machines are now controlled through highly sophisticated electronics. Before these advances, land preparation, seeding, drilling, harvesting, and other operations depended solely on the skill of the human operator. The operator's goal was to manipulate the machines as efficiently as possible to produce straight rows with no overlap or broadened spaces between them, thus minimizing operating costs. Nowadays, machinery movement across fields, speed, seeding rate, fertilizer application, and all of the other aspects of machine control that were once the province of the human operator can now be completely controlled electronically and governed by guidance from GPS satellites. A human operator is only needed for the handling of random events, and can just sit back in the comfort of the climate-controlled cab and enjoy the ride.

Technology has made its way into essentially every aspect of modern agriculture, as more and more farm operations become fully automated. Every aspect of the feeding, hygiene, and milking of the dairy herd animals is carried out by computer-controlled automation in some large dairy farm operations. Similar automation is in place in many other large commercial operations such as sheep, pig, chicken, goat and turkey farms. Drones can monitor crops and help determine which fields need specific amendments or irrigation. A human presence is there essentially only to monitor the operations and to ensure that the systems are functioning as they are intended to function. It is a far cry from the days when one rose before the sun to milk a few cows by hand and gather eggs from a dozen chickens. It is an even farther cry from those long ago days when people followed herds of wild animals and gathered what wild plants would provide. But even though the methods have changed so much, the basic principles of agriculture have not. As they always have and always will, farmers do agriculture within the natural environment of Earth, to produce the food that feeds the world.

—Richard M. Renneboog, MSc

Contributor's List

Ronju Ahammad
Craig W. Allin
Richard W. Arnseth
Michael P. Auerbach
Alexandra M. Avila
Clara B.
Anita Baker-Blocker
Melissa A. Barton
Hannah Bement
Cynthia A. Bily
Tyler Biscontini
Rachel Leah Blumenthal
Sarah Boslaugh
Judith J Bradshaw-Rouse
Alan Brown
Michael A. Buratovich, PhD
Scott F. Burns
William E. Burns
Mary E. Carey
Robert E. Carver
Diana M. Casey
Zhiqiang Chen
Sneed B. Collard
David R. Coyle
Mark S. Coyne
Charles R. Crumly
William A. Dando
Peter L. deFur
Rene M. Descartes
Gordon Neal Diem
John P. DiVincenzo
Sally Driscoll
Maria Droujkova
Corinne Elliott
Jessica O. Ellison
Jack B. Evett
Jim Fowler
Lenela Glass-Godwin
José F. González–Maya
Nancy M. Gordon
D. R. Gossett
Gina Hagler
Angela Harmon
Jasper L. Harris
Peter B. Heller
Thomas E. Hemmerly
Virginia L. Hodges
Robert M. Hordon
Jason A. Hubbart
Lawrence E. Hurd
Mary Hurd
Micah L Issitt, BS
Jerome A. Jackson, PhD
Domingo M. Jariel Jr., PhD
Kamal J. K. Gandhi
Karen N. Kähler
Jamie Michael Kass
Diann S. Kiesel
Michael M. Kimberley
Robert T. Klose
Denise Knotwell
James Knotwell
Grove Koger
Dawn A. Laney, MS
Eugene Larson
Jeffrey Larson
Kari L. Lavalli
M. Lee
J. Lipman-Boon
Robert Lovely
Spencer G. Lucas
Nancy Farm Männikkö
Erika Marín-Spiotta
Jordan M. Marshall, PhD
Michael W. Mayfield
Laurence W. Mazzeno
John S. Mecham
Nathan H. Meleen
Roman J. Miller
Jacqueline E. Mohan
Thomas C. Moon
David C. Morley
M. Marian Mustoe, PhD
Bryan Ness
Oghenekome U. Onokpise
George R. Plitnik
Eduardo Ponce
P. S. Ramsey
C. Nicholas Raphael
C. Mervyn Rasmussen
Alexis S. Reed
Gregory J. Retallack
Mariana Rhoades
James L. Robinson, PhD
Lindsay Rohland
Krishna Roka
Carol A. Rolf, JD
Julia A. Rosenthal, BA
Lisa Jo Rudy
Panagiotis D. Scarlatos
Elizabeth D. Schafer, PhD
Martha A. Sherwood, PhD
John F. Shroder Jr.
Sanford S. Singer
Robert A. Sinnott
Amy Sisson
Elizabeth Slocum
Adam B. Smith
Philip A. Smith
Kymberley A. Snarr
Richard S. Spira
Dion Stewart
Toby Stewart
Bruce L. Stinchcomb
Frederick M. Surowiec
Janine Ungvarsky
Krishna Prasad Vadrevu
Megan E. Watson
Christine Watts, PhD
Jeffrey Wielgus
George Wrisley
Lisa A. Wroble
Robin L. Wulffson
Ming Y. Zheng

A

Agriculture & Agricultural Land

—Ming Y. Zheng

FIELDS OF STUDY

Agronomy; Environmental Studies; Soil Mechanics.

PRINCIPAL TERMS

carbon cycle: processes through which carbon atoms circulate among Earth's atmosphere, terrestrial biosphere, oceans, and sediments—including fossil fuels

climate change: a statistically significant variation in either the mean state of the climate or its variability

energy-intensive agriculture: a method of farming that involves working on a large scale, utilizing significant resources, energy, and mechanization; often also referred to as industrialized agriculture

greenhouse effect: phenomenon in which certain gases in a planet's atmosphere trap heat that would otherwise escape into outer space, thereby increasing the planet's surface temperature

SUMMARY

Agriculture is a practice that utilizes the intimate structure of the soil for the purposeful growth of animal and vegetable crops. The methods used in agriculture have significant effects on the nature of the soils being used, requiring that methods must also be utilized that maintain the long-term health and viability of those soils. This has not always been the case, and remediation of agricultural lands that have been mismanaged is difficult.

BACKGROUND

Modern, large-scale agriculture has led to increased greenhouse gas (GHG) emissions, primarily resulting from high energy inputs, land clearing, soil degradation, and overgrazing by livestock. The massive conversion of forests into farms has reduced the land's capacity to function as a carbon sink. As a result, more GHGs are emitted into the atmosphere, contributing to global warming and climate change.

There is now a consensus that the mean temperature of Earth will increase by an average of 2.0° to 5.8° Celsius in the twenty-first century. Some environmental models, taking into account likely changes in vegetation cover, predict an even higher rise of 8° Celsius during the century. The resulting elevated temperatures (even at the lower end of the estimates) will have significant effects on Earth's biosphere, including human life. Many factors are associated with this predicted temperature rise, but agriculture is among the major contributors.

DIRECT IMPACTS ON CLIMATE CHANGE

Agricultural activity is a significant source of greenhouse gases (GHGs). GHG levels are affected by land clearing, high energy inputs, soil degradation, and intensive animal husbandry. Based on current estimates, agriculture contributes to 25 percent of the world's carbon dioxide (CO_2) emissions, 60 percent of methane gas emissions, and 80 percent of nitrous oxide emissions. Agriculture's high energy input results primarily from manufacturing chemical fertilizers, herbicides, and pesticides; operating farm machinery; irrigating farmland using pumps and other machines; and transporting products over long distances. Collectively, these activities account for more than 90 percent of the total energy expenditure in agriculture.

The burning of fossil fuels releases CO_2 into the atmosphere. The CO_2 concentration in the atmosphere has increased from 277 parts per million to 382 parts per million since the beginning of the Industrial Revolution in the mid-eighteenth century. Industrialized agriculture is believed to have contributed to 25 percent of that increase.

Overuse of fertilizers, in addition to energy inputs in fertilizer manufacturing, contributes significantly to climate change. More than half of all synthetic fertilizers applied to the soil either end up in local waterways or emit to the atmosphere. A portion of the excess nitrogen fertilizers in the soil is converted into nitrous oxide, which is 296 times more potent than CO_2 in trapping heat and which has a long atmospheric lifetime of 114 years. Each year, nitrous oxide emissions alone account for the equivalent of 1.9 billion metric tons of CO_2 emissions.

China has the largest agricultural output of any country. Image by Anna Frodesiak [CC0].

The second greatest GHG emission by agriculture is methane, released in small amounts by rice paddies and in much larger amounts by livestock. As the demand for meat increases, more livestock are raised and are fed higher-protein diets. Both the number of livestock and their protein-rich diets increase the amount of methane they emit. Methane gas is fourteen times more potent than CO_2 in trapping heat. Its concentration has almost tripled since the Industrial Revolution, from 600 parts per billion to 1,728 parts per billion.

INDIRECT IMPACTS ON CLIMATE CHANGE

Agriculture also contributes indirectly to climate change. Clearing trees and other natural stands to make land suitable for agricultural uses removes important carbon sinks, so less carbon is returned to the terrestrial biosphere and more CO_2 finds its way into the atmosphere, where it contributes to climate change.

The effect of land clearing on climate change is evident from the consequences of the destruction of tropical rain forests. For instance, large areas in Brazil have been cleared to facilitate soybean production. This clearing disrupts the local water cycle, which in turn alters Brazil's climate. In rain forests, water circulates as a result of evaporation, which greatly increases humidity. Natural tree stands act to buffer extremes of heat, cold, and drought. When the trees are removed, the buffer disappears. Moreover, the amount of water vapor in the air decreases, causing shifts in rainfall patterns, moisture levels, air temperature, and weather patterns generally.

The conversion of forests into agricultural lands has significantly altered Earth's vegetation cover. Such changes in the land surface affect Earth's albedo—that is, the proportion of incident radiation reflected by the planet's surface. Changes in the albedo in turn can affect the surface energy budget, which affects local, regional, and global climates. Changes in vegetation also produce changes in the global atmospheric concentration of CO_2. Agricultural landscape ranks among the lowest in carbon sequestration. Thus, as more land is devoted to agricultural uses, more of Earth's carbon is converted to CO_2 and emitted to the atmosphere, contributing to global warming.

CLIMATE CHANGE & AGRICULTURE

Potential climate changes associated with elevated GHGs and an altered surface energy budget include an increased incidence of heat waves, severe storms, and floods, as well as elevated sea levels. Some 30 percent of the agricultural lands worldwide could be affected by these changes. Global warming alone is projected to have considerable effects on agriculture.

A warming of 2° Celsius or more could reduce global food supplies and aggravate world hunger. The impact on crop yields will vary considerably across different agricultural regions. Warm regions, such as tropics and subtropics, will be threatened by climate change, while cooler regions, mainly in temperate or higher latitudes, may benefit from warming.

Global climate change may have significant effects upon livestock systems as well. First, the productivity and quality of rangelands may be adversely affected. This in turn will affect the quality and productivity of livestock. Second, higher grain prices resulting from the disruption of crop production will lead to higher costs for livestock products. Third, increased severity and frequency of storms may intensify soil erosion and decrease the productivity of rangelands. Fourth, global warming could result in changes in the distribution and severity of livestock diseases and parasites, which may threaten the health of animals, especially those in intensively managed livestock systems.

POSSIBLE SOLUTIONS

Unlike any other industrial GHG emitters, agriculture has the potential to change from being one of the largest GHG sources to being a net carbon sink, reversing its role in climate change. Several practical measures can be taken to mitigate the climate change caused by intensive agriculture. These include the reduced and more efficient use of chemical fertilizers, protection of soil, improvement of paddy rice production, and reduction of demand for meat.

Precision farming can reduce the need for chemical fertilizers. In precision farming, fertilizers and other agrochemicals are applied based on crops' needs, in precise amounts and on a carefully managed schedule. The reduced application of these chemicals not only cuts GHG emissions but also alleviates other environmental problems such as water pollution and eutrophication of waterways.

As a result of intensive farming, agricultural soils have some of the lowest carbon contents of all land types. If these soils can be modified to absorb more of Earth's carbon, the result will be a net reduction in atmospheric carbon. Low soil carbon content can be reversed through a number of measures, including planting cover crops, fallowing, and engaging in conservation tillage. These practices will increase the amount of organic matter (and thus the carbon content) in the soil. They will also reduce soil erosion and surface runoff, thereby reducing the need for chemical fertilizers. Collectively, these measures can turn agricultural soils into carbon sinks, changing the nature of their impact on climate change.

To reduce methane emissions from rice production, better cultivation techniques will need to be adapted. For example, rather than continuously flooding rice paddies, farmers could supply water to the paddies only when it is needed during the growing season and keep the paddies dry during the nongrowing season. Such measures could reduce methane emission from rice fields significantly.

Livestock raising is the second largest source of GHGs in agriculture. The most efficient way to cut methane emission due to livestock is simply to reduce the number of farm animals. As an ever-increasing demand for meat and dairy products drives increasing animal husbandry, one effective approach to cut methane emission is to reduce the demand for meat, especially in developed countries where consumers have tremendous buying power. Reduced meat and dairy consumption would go a long way toward curbing methane emissions.

CONTEXT

Agriculture and climate change are interlocked processes, in that each exerts effects on the other in a complex fashion. Climate changes, especially shifts in precipitation and temperature, are widely believed to have significant effects on agriculture, because these two factors determine the carrying capacity of any ecosystem. At the same time, modern agriculture is a major contributing factor to global warming, as altered land cover and the emission of CO_2, methane gas, and nitrous oxide from intensive farming increase the GHG content of the atmosphere. However, it remains possible to transform industrialized agriculture, using techniques that could render it more sustainable and mitigate its effects upon global and local climates.

Further Reading

Bazzaz, Fakhri, and Wim Sombroek, eds. *Global Climate Change and Agricultural Production: Direct and Indirect Effects of Changing Hydrological, Pedological, and Plant Physiological Processes.* New York: Wiley, 1996. Assesses the effects of global climate change on the agricultural production of crops and livestock. Covers issues such as the CO_2 fertilization effect, the adverse effects of elevated levels of ultraviolet-B (UVB) radiation and ozone on plant growth and productivity, and the environmental effects of livestock.

Marland, Gregg, et al. "The Climatic Impacts of Land Surface Change and Carbon Management, and the Implications for Climate-Change Mitigation Policy." *Climate Policy* 3 (2003): 149–157. Reviews evidence on how changes in land use and altered vegetation cover affect surface energy budget in addition to GHG emission, which in turn affects local, regional, and global climate.

Palo, Matti, and Heidi Vanhanen, eds. *World Forests from deforestation to Transition?* Boston: Kluwer Academic, 2000. Addresses global and subnational issues concerning the world's forests, societies, and environment from an independent and nonpolitical point of view. Emphasizes the importance of developing a scientific understanding of the interconnectedness between forests, human activity, and the environment, and of the consequences of environmental change for societies' development and growth.

Paustian, Keith, et al. *Agriculture's Role in Greenhouse Gas Mitigation.* Arlington, Va.: Pew Center on Global Climate Change, 2006. Well-researched and clearly written paper. Offers reasonable measures and reachable goals for mitigating GHG emission by agriculture.

AGRICULTURE & ANIMAL HUSBANDRY IN THE ANCIENT WORLD

—Rene M. Descartes

FIELDS OF STUDY

Anthropology; Archaeology; History.

PRINCIPAL TERMS

husbandry: the breeding and interbreeding of animals for the maintenance and improvement of existing breed stocks

irrigation: the natural and artificial or forced application of water to agricultural crops, typically to counteract a scarcity of natural rainfall

SUMMARY

Modern agriculture is the current state of development in a process that is believed to have begun approximately 10,000 years ago when human societies in all different parts of the world began to cultivate plants and domesticate various animal species. The degree of technological sophistication of the time determined the nature of agricultural processes, just as it does in the present day. However, the fundamental principles of agriculture and animal husbandry are exactly the same in the present day as they were a hundred centuries in the past.

INTRODUCTION

Plant cultivation and controlled stock breeding served to increase human population densities and resulted in the emergence of urbanism, occupational specialization, social stratification, writing, and long-distance trade. Agriculture also necessitated the development of metallurgy, engineering, astronomy, and mathematics and transformed both religious and political systems as humans altered the environment through deforestation, terracing, and irrigation.

Archaeological evidence that attends humankind's great transformation from hunter-gatherer to agriculturalist includes pollen samples, vegetal remains, and animal bones. Art, writing, agricultural artifacts, and topographical modifications provide information on early farming systems.

Although the precise mechanism that led to domestication remains unknown, models proposed to explain the transition from hunting and foraging to incipient agriculture emphasize climatic change, population growth, and accidental domestication. The domestication process began independently in the Near East, Mexico, sub-Saharan Africa, eastern North America, South America, and eastern Asia. From these regions, techniques, cultigens, and animals diffused throughout the world.

Burial chamber of Sennedjem, Scene: Plowing farmer. Image by Deutsch: Maler der Grabkammer des Sennudem.

THE NEAR EAST & EGYPT

Domestication began in an environmentally diverse region, the Fertile Crescent, a rugged, naturally rich habitat. Distribution of wild cereals, such as emmer and einkorn wheat and barley, and wild legumes, such as peas and lentils, paralleled the areas of earliest cultivation. Farming began in the Jordan Valley and spread into adjacent lands. By the eighth millennium BCE, autonomous villages had been established in areas occupied by modern Israel, Jordan, Syria, and Turkey.

By 8000 BCE, at Zawi Chemi Shanidar in modern Iraq, goats and sheep had been domesticated, and at Aswad in Syria and Cayonu in Turkey, wheat and barley had been cultivated. Peas and lentils may have been grown as early as the ninth millennium BCE at Aswad. At Cayonu, pigs had been domesticated by 7000 BCE. Evidence of domesticated cattle has been secured from sites in Turkey and in southwestern Europe dating to 6500 BCE. Initially, cattle were used for meat and draft purposes. Agriculture was a powerful subsistence strategy within the Fertile Crescent by 7000 BCE. Gradually, knowledge and the artifacts of agriculture diffused into Crete, Thessaly, Egypt, and lowland Mesopotamia.

Early tools were simple. The digging stick, which has been used in hunting and foraging communities from time immemorial, was a primary tool. Sickles for harvesting cereals were constructed of chipped flint. Food-processing tools such as mortars and pestles were made of ground stone.

The Tigris and Euphrates floodplains were farmed after 6000 BCE. By 5500 BCE, irrigation canals had been constructed to move water from the rivers to field systems and to facilitate field drainage following spring inundations. Construction and maintenance of large irrigation works required a centralized political authority for effectively coordinating labor.

Intensive irrigation, along with the plow, allowed population to increase in the developing urban centers on the Mesopotamian plains. The plow, or ard, probably evolved from a digging stick into a wooden device harnessed to oxen and designed to scratch the surface of the soil. Such plows were in use by the late fourth millennium BCE. Figs, olives, and grapes were grown in the eastern Mediterranean lands during the fourth millennium BCE. The date palm was cultivated for fruit and wood on the Mesopotamian floodplains.

Farming variations emerged by 3000 BCE in the Near East and eastern Mediterranean. One pattern emphasized dependence on rainfall for cereal cultivation and livestock raising. Another variant, centered in the river valleys, relied on irrigation. By the end of the first millennium BCE, canalization was accompanied by lifting devices such as the waterwheel. Such devices and practices persisted into the twentieth century CE.

Egyptian food production practices were derived from the cereal and livestock economies of the Mesopotamian floodplains. Early farming communities were evident in Egypt by the sixth millennium BCE, and by 4000 BCE, agricultural practices had spread across the Nile Valley. In contrast with Mesopotamian irrigation techniques, Egyptian farmers employed a system of basins and sluices to regulate and divert floodwater to growing crops. The annual deposits of silt that remained after the floodwaters receded in the Nile and Mesopotamian floodplains enhanced the soil's fertility. However, the Mesopotamian hydraulic system carried an overabundance of salts, ultimately retarding the soil's fertility there. In contrast, in Egypt, the receding floodwaters cleansed the system, creating and maintaining high fertility levels for millennia.

EUROPE

The crops, animals, and culture of early farming diffused from the Near East into Anatolia and the Aegean region. By the mid-seventh millennium BCE, cattle had been domesticated in Greece. Between 6000 and 4500 BCE, permanent farming villages were located over wide areas in the Balkans. By 6000 BCE, the cereal and livestock complex of southeastern Europe had spread up the Danube and into central Europe.

Agricultural technology changed and adapted as it spread and moved into the European forests. The diffusion of agriculture along the Mediterranean and Atlantic coasts, as well as the penetration of northern Europe, required varied subsistence and technological strategies. Climate and altitude were crucial considerations. In some locales, hunting, fishing, and shellfish collecting remained important.

Cultivation within the forest zones required fire and stone adzes to clear dense wooded tracts. After an area was burned, seeds were sown in the nutrient-rich ash using simple tools. Slash-and-burn farming was a key technique in the evolution of European agriculture. By the fourth millennium BCE, people had harnessed the power of oxen, which combined with the simple ard (plow) to enhance efficiency. Early European farmers planted barley, wheat, vetch, flax, peas, beans, and lentils. Oats and rye were probably domesticated in Europe.

By 4500 BCE, the rising productivity of agriculture, coupled with the evolving social organization of populations, had resulted in the construction of elaborate tombs in Western Europe. The most enduring achievement of these agrarian societies, however, was their megalithic edifices, the greatest of which was Stonehenge (3100–1550 BCE), located on Salisbury Plain, England.

Agriculture shaped economic foundations long before the appearance of the classical world. In Greece, implements remained simple and the general complex of cereals, fruits, and vine crops were cultivated. By the late fifth century BCE, mills had been constructed for pressing olives, and animals treaded stone floors to thresh grain. The Greeks wrote little on agriculture, but the Romans produced a broad literature on agricultural topics, covering topics such as soil fertility, farm management, and field procedures. To feed its great urban centers, Rome used slave power in tandem with numerous simple and mechanical wooden and iron tools. Cereals, olives, grapes, fruits, and vegetables moved from diverse environments along Roman roads and seaways.

European society in the early Middle Ages rested on an agrarian base. Manorialism evolved following

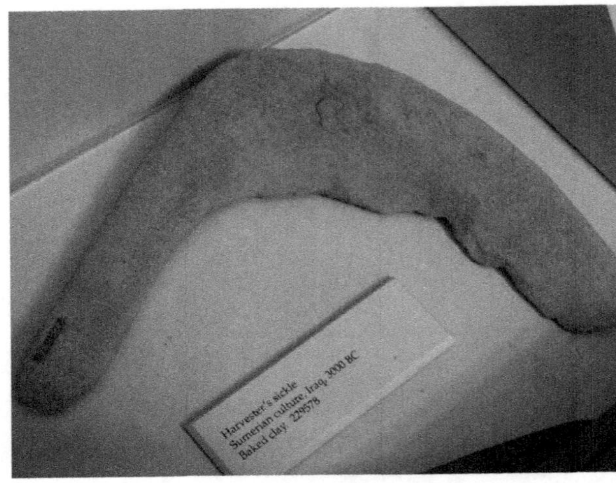

Sumerian Harvester's sickle, 3000 bce. Baked clay. Field Museum.

the collapse of Roman administration in the Western Empire (476 CE). Feudalism, the larger political and military framework that included manorialism, was based on a system of personal loyalties and territorial arrangements. Local landlords allowed peasants, or serfs, to reside in manorial villages and perform agricultural and other tasks in exchange for various services. Peasant agriculture in the post-Roman world was labor intensive, and physical mobility was circumscribed. Peasant life and work revolved around a religious and agricultural calendar.

Plowing and general hauling was accomplished with oxen. Northern European soils demanded a plow that would cut through the heavy soil and turn it over to facilitate drying. During the sixth century CE, German and Slavic populations developed such a plow. Conservation concerns were manifested in the two-field system, which allowed one of a pair of fields to lie fallow and replenish its fertility. The two-field system, which by the mid-eighth century CE had become the three-field system, was based on growing winter cereals while allotting equal fallow ground. It was used in southern Mediterranean Europe, where inadequate moisture meant poor spring seedings.

In contrast to the market-oriented large Roman estates, the manors during the early Middle Ages strived for self-sufficiency. Perhaps 90 percent of the European population was directly involved with agriculture and manorial labor.

AFRICA

In contrast to other regions of the Old World, the evidence for the development and spread of agricultural practices in Africa is poor. Before the Saharan region was largely abandoned because of desiccation (c. 2500–1000 BCE), millet and sorghum may have been domesticated between 4000 and 2000 BCE.

Vegeculture (the cultivation of multiple species within a single field) may have been practiced in West Africa by the fifth millennium BCE. Coffee was domesticated in Ethiopia, and millet, yams, sorghum, okra, African rice, and groundnuts were all cultivated in West Africa with stone tool technology. Plow agriculture was unknown until the nineteenth century CE. Sub-Saharan Africa posed many ecological obstacles including diseases and poor soils, and tropical cultivators adapted in numerous ways over the millennia.

The general cereal crop complex diffused into North Africa from the Middle East. In the forest zones, slash-and-burn was the foremost technique before tillage. As soil fertility ebbed, populations shifted and opened new forest tracts. Indigenous domesticated animals were few: guinea fowls, cattle, and perhaps donkeys. Evidence suggests cattle were domesticated in the Sahara region, perhaps by 4500 BCE. East African cattle culture is ancient, and pastoralism remained a major subsistence pursuit among tribes into the twentieth century CE.

SOUTH & EAST ASIA

Early food production in India is linked to the Near Eastern and East Asian patterns of domestication. The site of Mehrgarh (c. 7000 BCE) in modern Pakistan revealed evidence of the cultivation of wheat and barley. By 5500 BCE, the population in India and other parts of South Asia was extensively exploiting wild cattle, sheep, and goats, suggesting a predomestication period similar to that of the Near East. By the end of the third millennium BCE, domesticated water buffalo, pigs, cattle, goats, and sheep were mainstays of village life. A millennium later, donkeys, camels, and horses, used in transport, were to be found over much of South Asia.

Few crops and animals were domesticated independently in India. The exceptions are chickens and cotton, which date from as far back as 5000 BCE. Dates, grapes, and seed plants such as sesame were probably grown by the third millennium BCE. Agricultural hand tools, like the adze and ax, were simple and constructed of stone.

The dense urban populations of Harapp and Mohenjo-Daro, cities of the Indus civilization (c. 2500–2000 BCE), were supported through farming. The supporting technology included engineering projects to control flooding of the Indus River. The plow is believed to have been introduced into India following the collapse of the Indus civilization. The livestock, crops, and technological complex of village India has persisted into the modern world.

Early northern Chinese agriculture focused on the cultivation of millet, whereas in southern regions, rice became a staple. By 6000 BCE, foxtail and broomcorn millet as well as Chinese cabbage had been cultivated at the northern site of Peiligang, and rice had been cultivated in southern China at Pengtoushan. Additional plants included fruit trees (peach, pear, and apricot), soybeans, and the bottle gourd. The mulberry tree was grown for silkworm raising. At Hemudu (c. 5000 BCE) in the Yangtze Valley, rice was cultivated, and stone, bone, and wooden agricultural tools were discovered. Livestock included pigs, chickens, cattle, goats, and sheep. Cattle were used for traction and probably pulled wooden plows for rice cultivation. The principal tools remained the hoe and spade. Fishing provided an important supplement for agricultural communities.

From these peasant villages evolved monumental Chinese civilizations. During the third millennium BCE, the Longshan culture in north China began to shift from extensive to intensive cultivation. Social differentiation, copper metallurgy, and specialized labor intensified, increasing the social complexity. In the second millennium BCE, the traditional agricultural economy enabled the Shang Dynasty (1600–1066 BCE) to emerge, and people began to use the water buffalo. During the first millennium BCE, higher levels of efficiency were introduced into Chinese agriculture through the use of extensive irrigation works and iron implements, including the plow.

OCEANIA

Kuk swamp, Papua, in the New Guinea highlands provides the earliest evidence for agricultural activity: drainage systems, perhaps for sugarcane and taro cultivation, dating to 7000 BCE. This date suggests the possibility of another independent center for plant domestication. By 4000 BCE, tree foods, coconut and perhaps sago, were used in the lowland Sepik Basin. Independently of these precocious developments, migrations from Southeast Asia, beginning about 1500 BCE, initiated colonization through the remainder of Melanesia, as well as Micronesia and ultimately Polynesia, including New Zealand. These migrations diffused the primary domesticated fauna, dogs, pigs, and chickens, in addition to the greater yam, sugarcane, and new varieties of bananas throughout Oceania.

Successful settlement of the Pacific Basin was based on continuous adaptation to environmental diversity. The mainstay of life, vegeculture, persisted into the historic period, accompanied by the tree crops of breadfruit, coconuts, and bananas. Ecological variability caused prehistoric populations to evolve varied exploitative systems: raised bed cultivation in swampy locales for yams; sunken fields for taro, designed to benefit from subsurface moisture; and hillside terracing and irrigation. Swidden, or slash-and-burn technology, remained the premier cultivation technique into historic times. Tools were simple: the digging stick, a mainstay of tuber horticulture, and adzes of shell and stone. In contrast to mainland Asia, Melanesian chiefdoms and the Polynesian civilizations followed a different evolutionary course based on sophisticated vegeculture, tree food cultivation, and the resources of the sea.

NORTH AMERICA

The Native Americans lacked the technology of the Old World civilizations, which included large animals for traction, metallurgy, and the wheel. A powerful energy base and a formidable subsistence technology evolved, however, permitting sophisticated adaptations, and in some environments, spectacular urban civilizations.

In the woodland regions of eastern North America, native populations domesticated a number of seed plants: lamb's quarter, goosefoot, and the sunflower. Sunflowers and squash were domesticated by the mid-third millennium BCE. Hunting, fishing, and shellfish collecting supplemented wild food gathering and cultivation. Digging sticks and stone hoes were used in seed-bed preparation. Plants of ritual and economic importance diffused from Mexico: Tobacco and maize arrived in the Midwest by the second century CE. Maize was a minor cultigen until the ninth century CE.

An entirely different pattern evolved in the Salt and Gila River Valleys in the American Southwest. Rooted in the Mesoamerican tradition, the Hohokam (1–1400 CE) irrigated their fields in the hot Sonoran desert. Canalization, dated circa 300–200 BCE, included segments more than 15 miles (24 kilometers) long and irrigated thousands of acres. Hohokam agriculture included floodwater farming: These techniques allowed population expansion and political evolution. Large hydraulic works at the Snaketown site implied labor coordination, not only for construction but also for annual maintenance.

MESOAMERICA

In the regions of modern Mexico, Guatemala, and adjacent Central America, independent domestication took place. Plant species were numerous, the foremost being maize, beans, squash, and chili peppers. Animal domesticates were few, consisting of the turkey and hairless dog. At Tehuac n in Mexico, maize was cultivated by 5000 BCE, chili peppers by 4000 BCE, and beans during the late first millennium BCE. Evidence of squash from Guila Naquitz in Oaxaca may date to 6000 BCE. By the fourth millennium BCE, villages in the Tehuacan Valley employed small-scale canalization. The shift to complete reliance on domesticates encompassed five millennia, a period in which hunting and foraging remained important. In the lowlands, the slash-and-burn technique was used to clear heavy vegetation; in the highlands, the digging stick remained the primary tool. Ground stone tools (the mano-metate) processed corn into meal.

Domesticates (particularly the triad of maize, beans, and squash) provided the nutritional base for Mesoamerican civilizations. The Olmec (c. 1200–400 BCE) of the humid Gulf Coast farmed the rich soils adjacent to the lowland rivers. Teotihuacan (1–750 CE) in the central Mexican plateau became an urban center boasting a population estimated between 125,000 and 200,000 people. Irrigation was used selectively, as was *chinampas*, a form of raised field farming in which decomposing materials and mud are piled in a watercourse and secured by stakes and branches. After taking root, the branches and saplings held the "floating garden" firm.

In lowland Central America, a variant of shifting cultivation, *conuco*, was developed to grow such crops as sweet potatoes and manioc. The primary tool remained the digging stick, or *coa*.

SOUTH AMERICA

In pre-Columbian lowland and highland South America, regions with extreme ecological variability, emphasis was placed on root crops. Potatoes, manioc, quinoa, beans, peanuts, cotton, and oca are among the many plants that facilitated the growth of civilization. Domesticated fauna were limited: the duck, alpaca, llama, and guinea pig. Llamas and guinea pigs were domesticated by the fourth millennium BCE. Domestication enjoyed a significant time span in Andean South America: Potatoes, for example, were domesticated by the late third millennium BCE, cotton between 4000 and 1200 BCE, and beans by 5800 BCE.

Irrigation was widely used. In the Moche Valley of north coastal Peru (first to seventh centuries CE), water was diverted for cotton, peanuts, beans, and other crops. Cultivation was by hand. Canals were constructed of mud and earth: Engineering was sophisticated to ensure the correct gradient of water flow down hillsides. Soil fertility was replenished through guano gathered from offshore islands. Moche intensive agriculture may have supported a population of 50,000. The Moche state was a heavily stratified, politically complex society, as evidenced by monumental building projects and an elaborate mortuary ritual. An equally complex agricultural infrastructure supported the sociopolitical hierarchy.

Further Reading

Cowan, C. Wesley, and Patty Jo Watson, eds. *The Origins of Agriculture: An International Perspective.* Smithsonian Institution Press, 1992.

Grigg, D. B. *The Agricultural Systems of the World: An Evolutionary Approach.* Cambridge University Press, 1974.

Harris, David R., ed. *The Origins and Spread of Agriculture and Pastoralism in Eurasia.* Smithsonian Institution Press, 1996.

Isager, Signe, and Jens Erik Skydsgaard. *Ancient Greek Agriculture: An Introduction.* Routledge, 1995.

Agriculture & Mathematics

—Maria Droujkova

FIELDS OF STUDY

Accounting; Business; Mathematics.

PRINCIPAL TERMS

bioengineering: the application of methods to enhance and maintain the viability of individual animals and plants, typically in regard to the efficiency of their productivity

organic: referring to agricultural practices that rigorously exclude the use of fertilizers, pesticides, drugs and any other additives that are not native to the process

sexagesimal: the counting system based on the number 60; not to be confused with sexadecimal or more commonly hexadecimal, which is based on the number 16

SUMMARY

As a fundamentally important human activity, agriculture has long been a motivator for mathematical and statistics research. At minimum, mathematics is a tool for tracking the effectiveness and efficiency of agricultural activities with regard to input versus yield, or 'bean counting'. Increasingly sophisticated applications of mathematics to agriculture are the

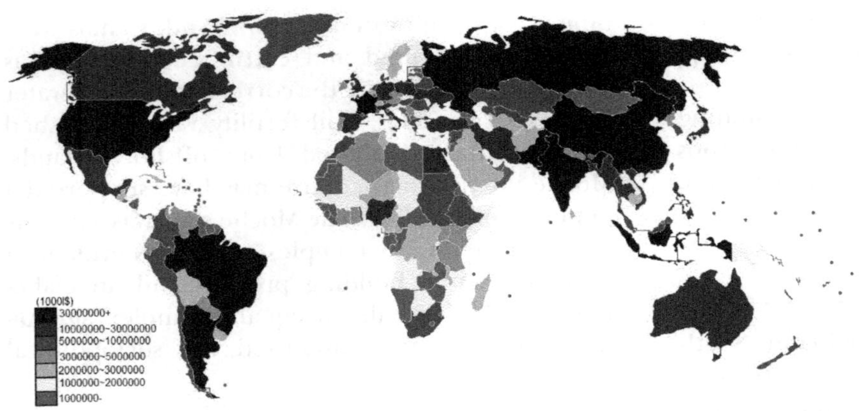

The agricultural value of each country.

basis of trade and the conduct of business on a global scale.

INTRODUCTION

Farming, also called "agriculture," is the production and distribution of plant and animal products. Farming methods range from organic farming to industrial agriculture. Farming operations are also categorized by their products, including foods, pets, decorative plants, pharmaceuticals, building materials, fibers, resins, and bioplastics. Agriculture has long been a motivator for mathematical and statistic research. Mathematical concepts and models have helped advance many agricultural methods beyond simple arithmetic calculations of quantities of seed and fertilizer. Many consider Ronald Fisher to be the father of modern statistics. Much of his research in statistical methods originated from his work with more than 60 years of agricultural data at Rothamsted Experimental Station, one of the oldest agricultural research institutions in the world. Methods pioneered by Fisher are still widely used in the twenty-first century, including hypothesis testing, analysis of variance, maximum likelihood estimation, and factorial experimental design. Mathematician Michael Weiss has worked in several mathematical areas with applications in agriculture, including nonlinear and chaotic dynamics, fuzzy set theory, and topological and algebraic entropy. Some applications of his work include a model of crop yields as a two-dimensional stochastic process, called "random surfaces," and assessing revenue risk as a probabilistic function of foodborne disease outbreaks. Precision farming models spatial variability in farmland and the resulting changes in yields as geometric surfaces.

Numerical characteristics of the farmland, such as fertilizer needs, are assigned to surfaces by functions and mapped to other surfaces by operators using modeling software. The so-called cobweb theorem relates price and production for situations in which there is a time lag between the marketing of a product and initially obtaining price information to determine production. This is common in agricultural markets, since prices in one year tend to influence planting in subsequent years.

THE HISTORY OF MATHEMATICS & SCIENCE

Agricultural development shaped the history of humankind, including the growth of science in mathematics. This impact is acknowledged in the historical tradition of naming major farming breakthroughs "revolutions," since the changes they produced in society were large and relatively fast. The neolithic revolution started circa 8000 BCE and included the development of permanent settlements. The resulting architecture and centralized management systems required abstract thought and systems of knowledge, including writing, mathematics, and science. The Arab agricultural revolution took place in the eighth through the thirteenth centuries CE and included the development and distribution of international knowledge exchange, sophisticated algebra and geometry, and astronomy for farming and navigation, as well as the scientific method and the modern number and computational system in mathematics. The British agricultural revolution started in the seventeenth century. It codeveloped with the Industrial Revolution and included the heavy use of mechanical tools and developments in the natural sciences, including chemistry and biology. This industrialization of agriculture continued into the twentieth century, driving the research in organic chemistry and genetics known as the green revolution.

Domestication of local crops, such as rice in China around 8000 BCE, allowed for both population growth and population concentration in villages and, later, towns. Planting, harvesting, and other timed activities required relatively exact time and weather observation, which in turn led to the development of astronomy and the development of sophisticated time measurement tools and calendars. Circa 5000 BCE, the people of Mesopotamia employed intensive farming methods, including monocrop fields, aggregation of crops for trade, and complex irrigation. Such methods called for and enabled major technological developments, such as better plows. It is hypothesized that the complex division of labor, distribution, and observation of water levels and calendars required for this type of agriculture led to the development and relatively widespread use of writing.

Mesopotamian clay tables show that quadratic and cubic equations, the Pythagorean theorem, and other topics currently found in algebra, geometry, and calculus were already widely used circa 2000 BCE in problems related to agriculture, such as astronomy-based calendars to time flooding and harvesting or the distribution of products. Some of this knowledge later was lost and then rediscovered by other cultures, and some continued to be used in the original form. For example, the practice of measuring time based on 60 minutes in an hour and 60 seconds in a minute comes from the Babylonian sexagesimal (base 60) number system. The number “60” was a convenient one for the Babylonians being highly composite (with more divisors than any number less than 60).

Agriculture promoted the development and spread of increasingly complex mechanisms, such as waterwheels in China. Excess crops supported the development of trade and transportation, from the domestication of draft and pack animals in ancient times to sophisticated spice trade fleets circa the sixteenth century. Starting in the eighth century CE, Muslim traders established an extensive network of trade routes among Asia, Europe, and Africa, enabling the diffusion of agricultural techniques and crops beyond their places of origin. This Arab agricultural revolution led to the development and distribution of science and mathematics, including the Arabic numerals used around the world in the twenty-first century. For example, one of the first documented uses of the scientific method comes from thirteenth-century work on medicinal plants and agronomy (the farming of plants).

The Industrial Revolution, starting in the eighteenth century, included the increasing mechanization of agriculture. Agricultural machines, such as the tractor, both decreased the number of people required for farming and increased productivity. The scientific advances associated with these developments primarily took place in engineering and chemistry. The green agricultural revolution of the second half of the twentieth century promoted advances in chemistry, genetics, and bioengineering, which led to high-yield, disease- and pest-resistant cultivation of major crops. The sustainability of these practices is not yet clear as the twenty-first century approaches its second quarter and ongoing climate change affects the global environment.

MEASUREMENTS IN AGRICULTURE

Metrics used in farming focus on average production of different cultivars of plants, breeds of animals, or farming methods; resource intensity of practices; efficiency of distribution; nutritional value of food products and industry-specific values of fibers, fuels, and lumber; environmental impact and sustainability; and the role of agriculture in local and global economy.

The global production levels, by crop type, are measured in tons per year. For example, cereals was the number one category of agricultural product, with worldwide production at around 2 billion tons per year in the early twenty-first century, while meat production at this same time was around 250 million tons per year. The total and per capita rates of production are frequently compared between years. For example, the total agricultural production grew by a factor of 16 between the early 1800s and 1970,

Man working in a ricefield.

while the world population grew by a factor of seven. This means that per capita consumption of agricultural products more than doubled during that period but not necessarily because of food items. Fiber or farmed trees for paper and construction are also included.

Farm yields are measured in crop weight per area for plants; in the ratio of seed input to seed output for grains; or in meat, fiber, or egg production per animal for animals. The yields are estimated using statistical methods of random sampling, or total outputs of a farm. In the United States, for example, corn yields averaged about 30 bushels per acre in the early 1900s and around 130 bushels per acre in the early 2000s. Food anthropologists estimate the minimal ratio of grain input to output necessary for sustaining farming as the main source of food as 1:3. For each grain planted, farmers get three grains, one of which is planted and two of which are either eaten by people or fed to farm animals. Yield metrics can be used to compare different methods of farming. For example, irrigation can raise corn yields by a factor of four or five. Industrial farming in developed countries produces yields that are about 10% greater than organic farming in nondrought years and about 70% less in drought years, netting about the same average yields over decades.

Resource intensity is measured by the outside input required per area of crops, per individual animal in meat or egg farms, or per unit of farm product output. For example, it takes about 1000 liters of water to produce 1 liter of corn-based ethanol. Resource intensity is one of many sustainability metrics used in farming. Other mathematical metrics of sustainability include nutrient leaching into water systems, which may cause proliferation of algae; biodiversity of farms; and pollution of soil, water, and air with herbicide and pesticide residues; as well as the carbon footprint of farming practices. For example, livestock production is currently responsible for about one-fifth of the total carbon footprint of humanity.

FARMING & THE ECONOMY

Agricultural systems include production, processing, packaging, distribution, marketing, and consumption. The proportion of resources and energy required for these activities varies with farming practices. For example, eating local foods reduces the resources expended in transportation; operating monocrop farms reduces labor per unit of production; eating processed foods increases packaging costs.

Agricultural economics is the study of resource allocation and distribution related to agriculture. It uses mathematical statistics for data analysis and trend prediction and mathematical modeling for research and development. Many general economic mathematical models were first developed in agricultural economics, for example, the cobweb model, which explains the cycles of price fluctuations through analyzing lags within the production chains, such as planting and harvesting.

Factory farming uses economies of scale by raising livestock in confinement and with high population densities. The calculations involved in factory farming include cost-output analysis and bioengineering of animals to optimize product output as well as the logistics of supplying food in to each animal in place and disease prevention through administering antibiotics. There are several measurements of factory farming impacts. For example, there are metrics involved with animal welfare, such as the degree of confinement, measured in area of pen per animal. Human health impact measures and research include studies of pesticide, antibiotic, and growth hormone levels in farm products and statistical studies of the impact of food on human health. Environmental impact measures are standard for all operations and include levels of specific air, water, and soil pollutants produced by the farm and its carbon footprint. Capital redistribution is the measure of movement of money among communities, which is relatively high for factory farming because of its centralized nature.

Industrial marketing and distribution models do not work well for organic farming because most organic products are not scalable. In the early 2000s, organic farmers developed a variety of peer-to-peer credence and distribution models, network marketing models, and sharing economy (mesh) models. Such modern models support decentralized production and disintermediated distribution. Some organic farmers join together in cooperatives and use economies of scale. Community-supported agriculture (CSA) is an economic model that provides a way to share the benefits and risks of farming. In a typical CSA, consumers buy farm shares and receive a weekly delivery of farm outputs.

FURTHER READING

Alspaugh, Shawn. "Farmer Ted Goes 3D." *Mathematics Magazine* 78, no. 3 (2005).

Glen, John. "Mathematical Models in Farm Planning: A Survey." *Operations Research* 35, no. 5 (1987).

Street, Deborah. "Fisher's Contributions to Agricultural Statistics." *Biometrics* 46 (1990).

Weiss, Michael. "Precision Farming and Spatial Economic Analysis: Research Challenges and Opportunities." *American Journal of Agricultural Economics* 78, no. 5 (1996).

AGRICULTURE & THE ENVIRONMENT

—Ming Y. Zheng

FIELDS OF STUDY

Ecology; Environmental Studies; Agronomy.

PRINCIPAL TERMS

carbon cycle: processes through which carbon atoms circulate among Earth's atmosphere, terrestrial biosphere, oceans, and sediments—including fossil fuels

climate change: a statistically significant variation in either the mean state of the climate or its variability

energy-intensive agriculture: a method of farming that involves working on a large scale, utilizing significant resources, energy, and mechanization; often also referred to as industrialized agriculture

greenhouse effect: phenomenon in which certain gases in a planet's atmosphere trap heat that would otherwise escape into outer space, thereby increasing the planet's surface temperature

SUMMARY

There is now a consensus that the mean temperature of Earth will increase by an average of 2.0° to 11.5° Celsius in the twenty-first century. (The higher end accounts for likely changes in vegetation cover.) The resulting elevated temperatures (even at the lower end of the estimates) will have significant effects on Earth's biosphere, including human life. Many factors are associated with this predicted temperature rise, but agriculture is among the major contributors.

Modern, large-scale agriculture has led to increased GHG emissions, primarily resulting from high energy inputs, land clearing, soil degradation, and overgrazing by livestock. The massive conversion of forests into farms has reduced the land's capacity to function as a carbon sink. As a result, more GHGs are emitted into the atmosphere, contributing to global warming and climate change.

BACKGROUND

There is now a consensus that the mean temperature of Earth will increase by an average of 2.0° to 11.5° Celsius in the twenty-first century. (The higher end accounts for likely changes in vegetation cover.) The resulting elevated temperatures (even at the lower end of the estimates) will have significant effects on Earth's biosphere, including human life. Many factors are associated with this predicted temperature rise, but agriculture is among the major contributors.

DIRECT IMPACTS ON CLIMATE CHANGE

Agricultural activity is a significant source of greenhouse gases (GHGs). GHG levels are affected by land clearing, high energy inputs, soil degradation, and intensive animal husbandry. Agriculture contributes to 25 percent of the world's carbon dioxide (CO_2) emissions, 60 percent of methane gas emissions, and 80 percent of nitrous oxide emissions. Agriculture's high energy input results primarily from manufacturing chemical fertilizers, herbicides, and pesticides; operating farm machinery; irrigating farmland using pumps and other machines; and transporting products over long distances. Collectively, these activities account for more than 90 percent of the total energy expenditure in agriculture.

The burning of fossil fuels releases CO_2 into the atmosphere. The CO_2 concentration in the atmosphere has increased from 277 parts per million to 382 parts per million since the beginning of the Industrial Revolution in the mid-eighteenth century. Industrialized agriculture is believed to have contributed to 25 percent of that increase.

Overuse of fertilizers, in addition to energy inputs in fertilizer manufacturing, contributes significantly to climate change. More than half of all synthetic fertilizers applied to the soil either end up in local waterways or emit to the atmosphere. A portion of the excess nitrogen fertilizers in the soil is converted into nitrous oxide, which is 296 times more potent than CO_2 in trapping heat and which has a long atmospheric lifetime of 114 years. Each year, nitrous oxide emissions alone account for the equivalent of 1.9 billion metric tons of CO_2 emissions.

The second greatest GHG emission by agriculture is methane, released in small amounts by rice paddies and in much larger amounts by livestock. As the demand for meat increases, more livestock are raised and are fed higher-protein diets. Both the number of livestock and their protein-rich diets increase the amount of methane they emit. Methane gas is fourteen times more potent than CO_2 in trapping heat. Its concentration has almost tripled since the Industrial Revolution, from 600 parts per billion to 1,728 parts per billion.

INDIRECT IMPACTS ON CLIMATE CHANGE

Agriculture also contributes indirectly to climate change. Clearing trees and other natural stands to make land suitable for agricultural uses removes important carbon sinks, so less carbon is returned to the terrestrial biosphere and more CO_2 finds its way into the atmosphere, where it contributes to climate change.

The effect of land clearing on climate change is evident from the consequences of the destruction of tropical rain forests. For instance, large areas in Brazil have been cleared to facilitate soybean production. This clearing disrupts the local water cycle, which in turn alters Brazil's climate. In rain forests, water circulates as a result of evaporation, which greatly increases humidity. Natural tree stands act to buffer extremes of heat, cold, and drought. When the trees are removed, the buffer disappears. Moreover, the amount of water vapor in the air decreases, causing shifts in rainfall patterns, moisture levels, air temperature, and weather patterns generally.

The conversion of forests into agricultural lands has significantly altered Earth's vegetation cover. Such changes in the land surface affect Earth's albedo—that is, the proportion of incident radiation reflected by the planet's surface. Changes in the albedo in turn can affect the surface energy budget, which affects local, regional, and global climates. Changes in vegetation also produce changes in the global atmospheric concentration of CO_2. Agricultural landscape ranks among the lowest in carbon sequestration. Thus, as more land is devoted to agricultural uses, more of Earth's carbon is converted to CO_2 and emitted to the atmosphere, contributing to global warming.

EFFECTS OF CLIMATE CHANGE ON AGRICULTURE

Potential climate changes associated with elevated GHGs and an altered surface energy budget include an increased incidence of heat waves, severe storms, and floods, as well as elevated sea levels. Some 30 percent of the agricultural lands worldwide could be

Soybean field in Buenos Aires, Argentina. Image by Alfonso. GNU Free Documentation License. Image by Alfonso. GNU Free Documentation License.

affected by these changes. Global warming alone is projected to have considerable effects on agriculture. A warming of 2° Celsius or more could reduce global food supplies and aggravate world hunger. The impact on crop yields will vary considerably across different agricultural regions. Warm regions, such as tropics and subtropics, will be threatened by climate change, while cooler regions, mainly in temperate or higher latitudes, may benefit from warming.

Global climate change may have significant effects upon livestock systems as well. First, the productivity and quality of rangelands may be adversely affected. This in turn will affect the quality and productivity of livestock. Second, higher grain prices resulting from the disruption of crop production will lead to higher costs for livestock products. Third, increased severity and frequency of storms may intensify soil erosion and decrease the productivity of rangelands. Fourth, global warming could result in changes in the distribution and severity of livestock diseases and parasites, which may threaten the health of animals, especially those in intensively managed livestock systems.

POSSIBLE SOLUTIONS

Unlike any other industrial GHG emitters, agriculture has the potential to change from being one of the largest GHG sources to being a net carbon sink, reversing its role in climate change. Several practical measures can be taken to mitigate the climate change caused by intensive agriculture. These include the reduced and more efficient use of chemical fertilizers, protection of soil, improvement of paddy rice production, and reduction of demand for meat.

Precision farming can reduce the need for chemical fertilizers. In precision farming, fertilizers and other agrochemicals are applied based on crops' needs, in precise amounts and on a carefully managed schedule. The reduced application of these chemicals not only cuts GHG emissions but also alleviates other environmental problems such as water pollution and eutrophication of waterways.

As a result of intensive farming, agricultural soils have some of the lowest carbon contents of all land types. If these soils can be modified to absorb more of Earth's carbon, the result will be a net reduction in atmospheric carbon. Low soil carbon content can be reversed through a number of measures, including planting cover crops, fallowing, and engaging in conservation tillage. These practices will increase the amount of organic matter (and thus the carbon content) in the soil. They will also reduce soil erosion and surface runoff, thereby reducing the need for chemical fertilizers. Collectively, these measures can turn agricultural soils into carbon sinks, changing the nature of their impact on climate change.

To reduce methane emissions from rice production, better cultivation techniques will need to be adapted. For example, rather than continuously flooding rice paddies, farmers could supply water to the paddies only when it is needed during the growing season and keep the paddies dry during the nongrowing season. Such measures could reduce methane emission from rice fields significantly.

Livestock raising is the second largest source of GHGs in agriculture. The most efficient way to cut methane emission due to livestock is simply to reduce the number of farm animals. As an ever-increasing demand for meat and dairy products drives increasing animal husbandry, one effective approach to cut methane emission is to reduce the demand for meat, especially in developed countries where consumers have tremendous buying power. Reduced meat and dairy consumption would go a long way toward curbing methane emissions.

CONTEXT

Agriculture and climate change are interlocked processes, in that each exerts effects on the other in a complex fashion. Climate changes, especially shifts in precipitation and temperature, are widely believed to have significant effects on agriculture, because these two factors determine the carrying capacity of any ecosystem. At the same time, modern agriculture is a major contributing factor to global warming, as altered land cover and the emission of CO_2, methane gas, and nitrous oxide from intensive farming increase the GHG content of the atmosphere. However, it remains possible to transform industrialized agriculture, using techniques that could render it more sustainable and mitigate its effects upon global and local climates.

Further Reading

Bazzaz, Fakhri, and Wim Sombroek, eds. *Global Climate Change and Agricultural Production: Direct and Indirect Effects of Changing Hydrological, Pedological, and Plant Physiological Processes.* Wiley, 1996.

Marland, Gregg, et al. "The Climatic Impacts of Land Surface Change and Carbon Management, and the Implications for Climate-Change Mitigation Policy." *Climate Policy* 3 (2003): 149–57.

Palo, Matti, and Heidi Vanhanen, eds. *World Forests from Deforestation to Transition?* Kluwer Academic, 2000.

Paustian, Keith, et al. *Agriculture's Role in Greenhouse Gas Mitigation.* Pew Center on Global Climate Change, 2006.

Sengar, Rakesh S., and Kalpana Sengar. *Climate Change Effect on Crop Productivity.* CRC, 2015.

Wollenberg, Eva. *Climate Change Mitigation and Agriculture.* Earthscan, 2012.

Agriculture as Energy Producer

—Sarah Boslaugh

FIELDS OF STUDY

Plant Science; Crop Science; Botany; Agronomy; Agriculture; Biochemistry; Ecology; Entomology; Soils Engineering.

ENERGY USE IN AGRICULTURE

Agriculture is both a consumer and a producer of energy, using it for heat, light, motor fuel, and other energy required for agricultural production as well as the production of materials such as fertilizer and pesticides. It produces crops, which can act as energy inputs.

In preindustrial Europe, as in many developing countries today, agriculture production was labor-intensive and governed largely by the availability of natural resources, including sunlight, water, and regionally available fuels. The primary source of energy was human and animal labor; fuels such as wood were often in short supply, and their use was strictly controlled by law. For instance, in 16th-century England timberlands were protected from poaching by stiff penalties, forcing many farmers to rely on inferior fuels such as peat for their own use. During the "agricultural revolution" of the seventeenth to nineteenth centuries, technological advances in agriculture, such as the introduction of the threshing machine, began the process of replacing human and animal labor with machines. However, the widespread use of fossil fuels in agricultural production is a fairly recent phenomenon, with rapid increases seen after 1940. Today, about 17 percent of the petroleum energy used annually in the United States is devoted to the process of food production, although less than half of that is devoted to direct production. In fact, only 6 percent of total energy consumed in the United States is used directly in crop and livestock production, an additional 6 percent is used in processing and packing food products, and 5 percent is used in food transportation and preparation.

The shift in types of energy used in American agriculture can be seen by examining trends in the production of corn. Corn yields increased about fourfold from 1945 to 2000, thanks to hybrids, irrigation, commercial fertilizers and insecticides, and other improved technologies, and total energy input has also increased about fourfold. In 1945, the average yield for the U.S. corn crop was 2,132 kilograms per hectare (kg/ha), whereas in 2000, it was 8,590 kg/ha. Labor inputs during that time decreased from 57 hours per hectare to 6.2 hours per hectare, whereas the amount of machinery used increased from 22 kg/ha to 55 kg/ha. The amount of fuel used in corn production has decreased slightly, from about 880 liters per hectare in 1975 to 850 liters per hectare in 2000, despite the large increase in corn produced per hectare, because farm machinery has become more efficient. The type of fuel used has also changed: In 1945, tractors ran primarily on gasoline, whereas today diesel fuel is more common. As a diesel-fueled engine generally performs better in terms of miles per gallon, this switch has contributed to increased energy efficiency in farming.

Changes in the use of fertilizer, pesticides, and irrigation have also increased agricultural productivity while representing another source of energy consumption. Early agriculture depended primarily on nutrients present in the soil, including those from wild vegetation growing on land allowed to lie fallow

Farmyard anaerobic digester converts waste plant material and manure from livestock into biogas fuel. Image by Volker Thies (Asdrubal) CC-BY-SA-3.0.

or from crop rotation. Commercial fertilizers came into use around 1945, and the average use in that year in the United States was 8 kg/ha of nitrogen and phosphorus and 6 kg/ha of potassium. By 2000, this had grown considerably, with an average use of 148 kg/ha of nitrogen, 53 kg/ha of phosphorus, and 57 kg/ha of potassium.

Pesticides also came into widespread agricultural use in the United States in this period. The average quantity of pesticides used in corn production grew from 0.1 kilogram per hectare in 1950 to 1.5 kilograms per hectare in 2000. Irrigation also increased over this period: In 1945, less than 1 percent of the corn crop in the United States was irrigated, versus 16 percent per day. This increase in the use of irrigation in 2000 represents an increased outlay of energy (for example, to pump the water and apply it to the corn) of more than 7.5 times that used in 1945.

Changes in methods of corn harvesting have changed the demands for energy as well. In 1945, most corn was harvested on the cob, but currently it is commonly harvested directly as grain. Because corn can have only 13–15 percent moisture when it is placed in storage, and corn harvested as grain may contain 25–30 percent moisture, corn harvested as grain must be dried before being stored. This process is estimated to require one-third more energy than harvesting the corn on the cob and drying it in corn cribs (in which case wind and solar energy do the drying). However, other agricultural practices have led to reductions in energy use. These include adopting conservation tillage practices and switching to more efficient methods of irrigation.

ENERGY & AGRICULTURE PRODUCTION TODAY

Although the direct use of energy in agricultural production represents a small percentage of the total energy use in the United States, agriculture is dependent on the timely availability of energy, and direct and indirect energy costs are a significant component of total production expenses. Examples of direct energy use include the uses of diesel fuel to operate farm machinery, gasoline to operate smaller farm vehicles, and electricity to produce light and power appliances. Examples of indirect energy use include the uses of natural gas used to produce fertilizer and petroleum or natural gas to produce pesticides; for example, 75–90 percent of the cost of anhydrous ammonia, the principal ingredient in nitrogen-based fertilizers, is due to the natural gas needed to produce it. Overall, the largest categories of farm energy use in the United States in 2002 were fertilizers (29.0 percent), diesel fuel (27.3 percent), electricity (20.7 percent), and gasoline (8.5 percent).

In the years 2000–03, total energy expenditures constituted 14.7 percent of U.S. farm production expenditures, with about two-thirds of this going to indirect expenses (primarily chemicals and fertilizers) and the remainder (5.2 percent of the total) to direct expenses such as fuel and electricity. Energy was the second-largest category of farm production expenses, after rent, interest, and taxes. Energy expenditures vary widely by crop. For instance, the cost of energy to produce sugar beets was $153.85 per acre (17.6 percent of production costs), compared to $148.32 per acre (21.5 percent of production costs) for peanuts, $113.33 per acre (18.4 percent of production costs) for rice, $91.85 per acre (16.8 percent of production costs) for cotton, $69.61 per acre (19.9 percent of production costs) for corn, and $30.09 per acre (15.7 percent of production costs) for wheat.

There has been some interest in using alternative energy sources in agriculture, although such applications constitute a small percentage of total energy used for agriculture in the United States, and the usefulness of each alternative depends partly on local conditions. Solar energy is currently used on some farms to create electricity (through a photovoltaic system), to heat greenhouses, and for space heating, crop drying, and water heating. Wind energy is also used to create electricity for farms, and both solar and wind systems may also create surplus energy that can be sold back to the local utility provider. Ethanol and biodiesel are used to operate some vehicles and other farm machinery. Geothermal heating is used to heat greenhouses, in aquaculture, and to heat homes and barns.

LIVESTOCK PRODUCTION

One trend seen in many countries as they become more prosperous is a change in diet to include more animal products, including meat. This is a cause for concern, not only because of the health consequences involved as diets shift away from plant products but also because livestock production as compared to grain production requires far more energy inputs to produce the same amount of food.

Overall, 5.9 percent of livestock production expenditures in the United States in 2002 were for energy, with the percentage varying by the type of livestock. For beef cattle ranching, 11.6 percent of the total production costs were energy expenditures; for sheep and goat farming, 8.5 percent of the costs; for aquaculture, 7.9 percent; for dairy cattle and milk production, 6.7 percent; for hog and pig farming, 4.6 percent; for poultry and egg production, 3.0 percent; and for cattle feedlots, 2.6 percent.

In the United States, livestock are annually fed about 45 million tons of plant protein—28 million tons of which is grain and 17 million tons of which is forage—producing about 7.5 million tons of animal protein for human consumption.

In terms of energy use, an average of 25 kilocalories is required to produce 1 kilocalorie of animal protein, about 10 times that required to produce 1 kilocalorie of plant protein (although it is worth noting that animal protein is about 1.4 times as nutritious as plant protein).

AQUACULTURE

Aquaculture—the farming of organisms such as fish, mollusks, and crustaceans—is one of the fastest-growing food production sectors in the world, accounting for about 30 percent of total global fish food production, with more than 220 species currently being produced. The amount of energy required varies widely, depending on the type of aquaculture practiced, and is usually expressed as the ratio of industrial energy input in joules to protein energy output in joules.

Intensive production in tanks—characterized by high stocking density, high dependence on exogenous energy inputs such as feed and chemicals, low inputs of labor, and production of high-value species—tends to have the highest ratio of energy input to protein output, whereas extensive production in a more natural environment tends to have a lower ratio. For instance, tilapia grown in extensive ponds in Indonesia have a ratio of 8, whereas semi-intensive tilapia culture in Africa has a ratio of 18 and intensive shrimp culture in Thailand has a ratio of 67. The highest ratios are for oysters and lobsters grown in tanks in the United States (136 and 480, respectively).

PRODUCTION OF ENERGY CROPS

The largest energy crop in the United States is corn, which is used to produce ethanol. Ethanol has been used as a source of energy in the United States since the nineteenth century and was also used as a fuel source for early automobiles. However, it was eclipsed by gasoline and did not again become popular as an automotive fuel source and gasoline additive until the late twentieth century.

Most ethanol in the United States is created from corn, with one bushel of corn yielding about 2.5 gallons of ethanol. Ethanol production in the United States has shown steady growth from 1980 (175 million gallons produced) to 2000 (1,630 million gallons produced), and by 2010 U.S. ethanol production had increased to 13 billion gallons. About 12 percent of all U.S. gasoline contains ethanol as a fuel additive. Corn stover, the wastes left in the field when corn is mechanically harvested, and crops such as switchgrass also show promise as sources of ethanol. One ton of corn stover yields about 80–90 gallons of ethanol, and a ton of switchgrass will yield about 75–100 gallons of ethanol. Using corn stover for this purpose has two advantages—disposing of waste while creating energy—and using switchgrass has the advantage of not using food crops for energy production.

Biodiesel is created from fats and oils, including those from crops such as soybeans grown specifically for biodiesel production. The U.S. biodiesel market is smaller than the ethanol market but growing. In the United States, about 2 million gallons of biodiesel were produced in 2000 and about 25 million gallons in 2004. Growth since then has been much more rapid, with 75 million gallons produced in 2005, 250 million in 2006, 450 million in 2007, and 700 million in 2008. Because biodiesel is created from renewable resources and offers less risk to the environment (it results in lower emissions and is less harmful if spilled), some states and cities have passed ordinances requiring vehicle fleets to use biodiesel or other biofuels or biofuel blends, further increasing demand.

Worldwide, about 10 percent of energy was produced from biomass and waste in 2007. This figure includes wood and animal dung burned as fuel, while liquid biofuels represented only 1.9 percent of the total. Use of biofuels is growing rapidly, however; for instance, use of biofuels for transportation more than doubled globally between 1990 and 2005 and is expected to almost triple between 2005 and 2015. In 2007, the United States was the leading producer of ethanol, with 26,500 million liters, followed by Brazil (19,000 million liters), the European Union (2,253 million liters), and China (1,840 million liters). The crops most commonly used to create ethanol worldwide are sugarcane, sugar beet, and cassava. In 2007, the European Union was the leading producer of biodiesel (6,109 million liters), followed by the United States (1,688 million liters), Indonesia (409 million liters), and Malaysia (330 million liters). Crops used for biodiesel production include the oil palm, soybeans, and rapeseed.

FURTHER READING

Food and Agriculture Organization (FAO) of the United Nations. *The State of Food and Agriculture 2008.* Rome: FAO, 2008. http://www.fao.org/docrep/011/i0100e/i0100e00.htm.

Renewable Fuels Association. "Ethanol Facts: Agriculture." http://www.ethanolrfa.org/pages/ethanol-facts-agriculture.

Schnepf, Randall D. *Energy Use in Agriculture: Background and Issues.* Washington, D.C.: BiblioGov Project, 2013. Print.

Svejkovsky, Cathy. "Renewable Energy Opportunities on the Farm." 2006. http://www.attra.org/attra-pub/pdf/energyopp.pdf.

U.S. Agriculture and Forestry Greenhouse Gas Inventory: 1990-2013. , 2016. Print.

U.S. Department of Agriculture, Alternative Farming Systems Information Center. "What Alternative Energy Options Are Available for Farms?" http://www.nal.usda.gov/afsic/pubs/altenergy.shtml.

U.S. Department of Energy. "Biofuels and Agriculture: A Factsheet for Farmers." September 2001. ORNL 2001–03624/tcc. http://www1.eere.energy.gov/biomass/pdfs/farmerfactsheet.pdf.

AGRICULTURE INDUSTRY

—D. R. Gossett

FIELDS OF STUDY

Agribusiness; Agronomy; Mathematics.

PRINCIPAL TERMS

climate change: a statistically significant variation in either the mean state of the climate or its variability

energy-intensive agriculture: a method of farming that involves working on a large scale, utilizing significant resources, energy, and mechanization; often also referred to as industrialized agriculture

monoculture: agriculture based on the cultivation of a single crop species rather than a number of different crop species

SUMMARY

The practice of agriculture has progressed over thousands of years from simple sow-and-reap and basic animal herding to the present-day methods of large-scale monoculture and animal husbandry operations. The basic goal of all agricultural practices has always been the production of food and usable material by-products, and as human population demands have increased over time, agriculture has become increasingly reliant on the application of technology.

BACKGROUND

The ability to produce sufficient food and fiber to feed and clothe its population is the most important natural resource a nation can have. In modern urban societies, it is also the natural resource that is most often taken for granted.

The beginnings of agriculture predate the written history of humankind. No one knows when the first crop was cultivated, but at some time in the distant past humans discovered that seeds from certain wild grasses could be collected and planted in land that could be controlled, the end product of which could later be gathered for food. Most authorities believe this occurred at about the same time in both the Old World and New World, some eight thousand to ten thousand years ago. The earliest attempts at growing crops were primarily to supplement the food supply provided by hunting and gathering. However, as the ability to produce crops increased, people began to domesticate animals, and their reliance on hunting and gathering decreased, allowing the development of permanent human settlements.

As far back as six thousand years ago, agriculture was firmly established in Asia, India, Mesopotamia, Egypt, Mexico, Central America, and South America. The earliest agricultural centers were located near large rivers that helped maintain soil fertility by the deposition of new topsoil with each annual flooding cycle. As agriculture moved into regions that lacked the annual flooding of the large rivers, people began to utilize a technique known as slash-and-burn agriculture. In this type of agriculture, a farmer clears a field, burns the trees and brush, and farms the field. After a few years, soil nutrients become depleted, so the farmer must repeat the process. This type of agriculture is still practiced in some developing countries and is one reason that the tropical rain forests are disappearing at such a fast rate.

Until the nineteenth century, most farms and ranches were family-owned, and most farmers practiced sustenance agriculture: Each farmer or rancher produced a variety of crops sufficient to feed himself and his family as well as a small excess to be sold for cash or bartered for other goods or services. Agricultural tools such as plows were made of wood, and almost all agricultural activities required human or animal labor. This situation placed a premium on large families to provide the help needed to tend the fields. The Industrial Revolution changed agriculture just as it did almost all other industries. Eli Whitney invented the cotton gin in 1793. The mechanical reaper was invented by Cyrus McCormick, and John Lane and John Deere began the commercial manufacture of the steel plow in 1833 and 1837, respectively. These inventions led the way for the development of the many different types of agricultural machinery that resulted in the mechanization of most farms and ranches. By the early part of the twentieth century, most agricultural enterprises in the United States were mechanized. The Industrial Revolution produced a significant change in society. The industrialized nations were gradually transformed from agrarian societies into urban societies. People involved in agricultural production left the farms to go to the city to work in the factories. At the same time, there was no longer a need for large numbers of people to produce crops. As a result, fewer people were required to produce more agricultural produce for an increasing number of consumers. This trend continues in developing nations with more rural dwellers, adding to the overcrowding in urban centers.

MODERN AGRICULTURE

As populations continued to grow, there was a need to select and produce crops with higher yields. The Green Revolution of the twentieth century helped to make these higher yields possible. Basic information supplied by biological scientists allowed agricultural scientists to develop new, higher-yielding varieties of numerous crops, particularly the seed grains that supply most of the calories necessary for maintenance of the world's population. These higher-yielding crop varieties, along with improved farming methods, resulted in tremendous increases in the world's food supply. The new crop varieties also led to an increased reliance on monoculture. While the practice of growing only one crop over a vast number of hectares has resulted in much higher yields than planting multiple crops, it has also decreased the genetic variability of many agricultural

An agronomist cradles a box of broccoli on a farm in Salinas, California. Image by United States Department of Agriculture.

plants, increased the need for commercial fertilizers, and produced an increased susceptibility to damage from a host of biotic and abiotic factors. These latter two developments have resulted in a tremendous growth in the agricultural chemical industry. Today's modern agricultural unit requires relatively few employees, is highly mechanized, devotes large amounts of land to the production of only one crop, and is highly reliant on agricultural chemicals such as fertilizers and pesticides.

AGRICULTURAL DIVERSITY

With all the diversity that has occurred in modern agriculture, the industry is subdivided into many different specialties. On the animal side of the industry, there is the beef industry, which deals with the production of beef cattle; the dairy industry, which focuses on the production of dairy cattle, milk, and milk products; the horse industry, which produces horses for work, sport, or pleasure; the sheep and goat industry; the swine industry, which deals with the production of pigs and hogs; and the poultry industry, which is concerned with the production of commercial birds and bird products such as eggs. Those agricultural industries that deal with plants include agronomy, the production of field crops (wheat, cotton, and so on); forestry, the growth and production of trees; and horticulture. Horticulture is subdivided into pomology, the growth and production of fruit crops (oranges, apples, and so on); olericulture, the growth and production of vegetable crops (tomatoes, lettuce); landscape horticulture, the growth and production of trees and shrubs that are used in landscape design; and floriculture, the growth and production of flowering plants used in the floral industry.

The various agriculture industries produce a tremendous number of different agricultural products. Those agricultural products that are derived from plants can be subdivided into timber products (such as lumber, furniture), grain products (wheat, oats), fiber products (cotton, flax), fruit products (grapes, peaches), nut crops (pecans, hazelnuts), vegetable products (lettuce, cabbage), beverage products (tea, coffee), spice and drug crops (garlic, mustard, opium, quinine), ornamental crops (carnation, chrysanthemum), forage crops (alfalfa, clover), and other cash crops, such as sugarcane, tobacco, artichoke, and rubber.

The animal industries provide products such as red meats from cattle (beef and veal), swine (pork), sheep (lamb and mutton), and goats (chevon). Milk and milk products, also referred to as dairy products, include milk, ice cream, and cheeses. Broiler chickens provide most of the world's poultry meat, but turkeys, roaster chickens, mature laying hens (fowl), ducks, geese, pigeons, and guinea hens are also consumed and may be more important than chickens in some parts of the world. Other than meat, poultry also provides eggs. The hair covering the skin of some farm animals is also considered an agricultural product. The two most important of these are wool from sheep and mohair from angora goats.

IMPACT ON OTHER NATURAL RESOURCES

While there have been tremendous increases in agricultural productivity through the use of modern agricultural practices, these same practices have had a significant impact on some other natural resources. Soil is one of the most overlooked and misunderstood resources. Most people think of soil as an inert medium from which plants grow. In reality, topsoil—that upper 15 to 25 centimeters of the Earth's terrestrial surface in which nearly all plants grow—is a complex mixture of weathered mineral materials from rocks, partially decomposed organic molecules, and a large number of living organisms. The process of soil formation is very slow. Under ideal conditions, topsoil can form at a rate sufficient to produce a layer of about 1 millimeter thick when spread over 1 hectare per year. Under less favorable conditions, it can take thousands of years to produce this small amount of soil. With proper management, topsoil can be kept fertile and productive indefinitely. However, many agricultural techniques lead to the removal of trees and shrubs, which provide windbreaks, or to the depletion of soil fertility, which reduces the plant cover over the field. These practices have exposed the soil to increased erosion from wind and moving water, and as a result, as much as one-third of the world's current croplands are losing topsoil faster than they can be replaced.

Because plants require water in order to grow, agriculture represents the largest single global user of water. Worldwide, about 70 percent of all fresh water withdrawn from groundwater supplies, rivers, and lakes is used to irrigate crops, and almost 15 percent of the world's croplands are irrigated. Water usage varies among countries. Some countries have abundant water supplies and irrigate liberally, while water is scarce in other countries and must be used carefully. Because as much as 60 percent of the water intended for irrigation is lost through old pipes and canals or to evaporation before the water reaches the field, the efficiency of water use in some countries can be very low. There is no doubt that irrigation has dramatically increased crop production in many areas, but some irrigation practices have been detrimental. Overwatering can lead to a waterlogging of the soil. Waterlogging cuts off the supply of oxygen to the roots, and the plants die. Irrigation of crops in dry climates can often result in salinization of the soil. In these climates, the irrigation water rapidly evaporates from the soil, leaving behind the mineral salts that were dissolved in the water. As the salts accumulate, they become lethal to most plants. Some experts estimate that as much as one-third of the world's agricultural soil has been damaged by salinization. There is also an argument as to whether or not the increased usage of water for agriculture has decreased the supply of potable water fit for other human uses.

A farm manager practices weed control in a soybean field as an aspect of the farm's weed-tillage system. Image by United States Department of Agriculture/Keith Weller.

Plants require sunshine, water either from rainfall or irrigation, carbon dioxide from the atmosphere, and thirteen mineral nutrients from the soil. Of these, calcium, magnesium, nitrogen, phosphorus, and potassium are required in the greatest amounts. Calcium and magnesium are plentiful in soils located in dry climates, but in wetter climates, these nutrients are often leached through the soil. In these regions, calcium and magnesium are returned to the soil in the form of lime, which is primarily used to raise the soil pH. Nitrogen, phosphorus, and potassium are the nutrients most often depleted from agricultural soils, and these nutrients are often referred to as the fertilizer elements. Because these nutrients stimulate plant growth and usually greatly increase crop yields, it is necessary to apply them to the soil regularly in order to maintain fertility.

The amount of fertilizer applied to the soil has increased more than twentyfold in the past fifty years. While this increase in the use of fertilizers has more than tripled the worldwide crop production, it has also caused some problems. The increased production of fertilizers has required the use of energy and mineral resources that could have been used elsewhere. In many cases, farmers tend to overfertilize. Overfertilization not only wastes money but also contributes to environmental degradation. Fertilizer elements, particularly nitrogen and phosphorus, are carried away by water runoff and are eventually deposited in the rivers and lakes, where they contribute to pollution of aquatic ecosystems. In addition, nitrates can accumulate in underground water supplies. These nitrates can be harmful if ingested by newborns.

Modern agriculture, as it is practiced in the industrialized nations, consumes large amounts of energy. Farm machinery utilized in planting, cultivating, harvesting, and transporting crops to market consumes the largest supplies of liquid fossil fuels such as gasoline or diesel. The energy required to produce fertilizers, pesticides, and other agricultural chemicals is the second largest energy cost associated with agriculture. The use of fuel-requiring pumps to irrigate crops is also a major energy consumer. Additional energy is used in food processing, distribution, storage, and cooking after the crop leaves the farm. The energy used for these activities may be five times as much as that used to produce the crop.

CURRENT TRENDS IN AGRICULTURE

The development of biofuels, fuels produced from plants, such as corn and soy ethanol and cellulosic ethanol (produced from inedible portions of plants), has been encouraged by the need to find a substitute for expensive and environmentally harmful fossil fuels. However, the fluctuating price of oil has caused this industry to advance in fits and starts. Critics point out that biofuels use cropland that otherwise would be producing food, and the rise of the electric car could speed the decline in the use of fossil fuels, making biofuels obsolete. The US government and private corporations have tried to address the first criticism by putting money into improving the process of distilling cellulosic ethanol, since it uses otherwise useless parts of food crops rather than requiring crops to be grown for the sole purpose of being turned into biofuels; however, producing cellulosic ethanol remains more expensive and difficult than distilling it from the edible parts of plants.

The next major development in agriculture will be the biotechnical revolution, in which scientists will be able to use molecular biological techniques to produce exotic new crop varieties. In the future, perhaps agricultural scientists will be able to use these techniques to develop crop plants that can be produced, processed, and distributed with less impact on other resources. Many scientists feel nanotechnology, the ability to restructure matter at the level of molecules and atoms, could meet the need for growth in agriculture through improving the production of both plants and animals and improving both the safety and quality of food. A wide range of developed and developing countries, from the United Kingdom to Iran to India, are providing funding to scientific laboratories to develop nanotechnology products. The potential products range from antibacterial agents to technology that signals when a product is near the end of its shelf life. There remains concern that including nanoparticles in food may pose a health risk, and consumer advocates are encouraging more research, consumer awareness, and governance.

The trend toward globalization in agriculture has been good for the developed countries, but it poses a threat to developing nations. For example, countries in Africa do not benefit from the advances in global agriculture. Rural dwellers have neither the money nor the natural resources to take advantage of modern agricultural methods. At the same time, the agricultural practices in the developed world bring with them many negative consequences for the environment. Water pollution from fertilizers and pesticides; global warming from increasing land under cultivation and decreasing forests; and decreased diversity of agricultural products in specific regions, which results in increased energy use to get these products to their global markets. Interest in organic farming, which is practiced in more than one hundred countries, offers opportunities for organic farmers from developing countries. However, if organic farming follows the pattern of commercial agriculture, with the growth of large farms, specialized products, and need for increasing capital, the benefit to the small, local farmer will disappear and the environmental impact will turn negative.

COMMERCIAL IMPACT OF THE AGRICULTURE INDUSTRY

Worldwide, some 45 percent of the population makes a living through agriculture, both subsistence and commercial. This also includes those people hired by the agriculture chemical companies, those companies that produce or sell agriculture implements and machinery, processing and canning plants, and wholesale and retail marketing firms, such as grocery stores. There are some eight thousand different agricultural products on the market, and while agriculture is big business, it amounts to less than 5 percent of the gross domestic product of all nations. Approximately one-third of the land worldwide is used for agriculture.

Further Reading

Akinyemi, Okoro M. *Agricultural Production: Organic and Conventional Systems.* Science Publishers, 2007.

Bickis, Ian. "Agriculture Industry Betting the Farm on Innovation to Boost Yields, Profits." CBC News, CBC/Radio-Canada, 10 Feb. 2016, www.cbc.ca/news/technology/farming-technology-1.3442023. Accessed 1 Nov. 2016.

Brody, Aaron L., and John B. Lord, eds. *Developing New Food Products for a Changing Marketplace.* 2nd ed. CRC Press/Taylor & Francis, 2008.

Field, Thomas G., and Robert E. Taylor. *Scientific Farm Animal Production: An Introduction to Animal Science.* 9th ed. Prentice Hall, 2008.

Janick, Jules. *Horticultural Science.* 4th ed. W. H. Freeman, 1986.

Kipps, M. S. *Production of Field Crops: A Textbook of Agronomy.* 6th ed. McGraw-Hill, 1970.

Metcalfe, Darrel S., and Donald M. Elkins. *Crop Production: Principles and Practices.* 4th ed. Macmillan, 1980.

Phillips, Denise, and Sharon Kingsland, eds. *New Perspectives on the History of Life Sciences and Agriculture.* Springer, 2015.

Southgate, Douglas, Douglas H. Graham, and Luther Tweeten. *The World Food Economy.* Blackwell, 2007.

Toovey, Michelle. "The Industry Handbook: The Agriculture Industry (COW, LSTK)." *Investopedia,* 10 Mar. 2016, www.investopedia.com/articles/investing/031016/industry-handbook-agriculture-industry-cow-lstk.asp. Accessed 1 Nov. 2016.

Weis, Tony. *The Global Food Economy: The Battle for the Future of Farming.* Zed Books, 2007.

Wojtkowski, Paul A. *Agroecological Economics: Sustainability and Biodiversity.* Elsevier/Academic Press, 2008.

Agriculture: History & Overview

—D. R. Gossett

FIELDS OF STUDY

Agronomy; Environmental Studies; History; Soil Mechanics.

SUMMARY

Agriculture is the ability to produce sufficient food and fiber to feed and shelter the population, the most important natural resource a nation can have. In modern urban societies, it is also the natural resource that is most often taken for granted.

The beginnings of agriculture predate written history. No one knows when the first crop was cultivated, but at some time in the distant past humans discovered that seeds from certain wild grasses could be collected and planted in land that could be controlled and the grasses later gathered for food. Most scholars believe this occurred at about the same time in both the Eastern and Western Hemispheres, some eight thousand to ten thousand years ago.

EARLY AGRICULTURE

The earliest attempts to grow crops were primarily to supplement the food supply provided by hunting and gathering. However, as the ability to produce crops increased, people began to *domesticate* plants and animals, and their reliance on hunting and gathering decreased, allowing the development of permanent settlements in which humans could live. As far back as six thousand years ago, agriculture was firmly established in Asia, India, Mesopotamia, Egypt, Mexico, Central America, and South America.

The earliest agricultural centers were located near large rivers that helped maintain soil fertility by the deposition of new *topsoil* with each annual flooding cycle. As agriculture moved into regions that lacked the annual flooding of the large rivers, people began to utilize a technique known as *slash-and-burn agriculture.* In this type of agriculture, a farmer clears a field, burns the tress and brush, and farms the field. After a few years soil nutrients become depleted, so

The earliest agricultural centers were located near large rivers that helped maintain soil fertility with the deposition of new topsoil during each annual flooding cycle. Image by Ephemeral Archives.

the farmer must repeat the process at a new location. This type of agriculture is still practiced in some developing countries and is one reason tropical rain forests are disappearing at a fast rate.

Until the nineteenth century, most farms and ranches were family-owned, and most farmers practiced *sustenance agriculture*. Each farmer produced a variety of crops sufficient to feed his or her family as well as a small excess which was sold for cash or bartered for other goods or services. Agricultural tools such as plows were made of wood, and almost all agricultural activities required human or animal labor. This situation placed a premium on large families to provide the help needed in the fields.

The arrival of the Industrial Revolution changed agriculture, just as it did almost all other industries. Eli Whitney invented the cotton gin in 1793. The mechanical reaper was invented by Cyrus McCormick, and John Lane and John Deere began the commercial manufacture of the steel plow in 1833 and 1837, respectively. These inventions led the way to the development of the many different types of agricultural machinery that resulted in the *mechanization* of most farms and ranches. By the early part of the twentieth century, most agricultural enterprises in the United States were mechanized. American society was transformed from an agrarian society into an urban society. People involved in agricultural production left farms to go to cities to work in factories. At the same time, there was no longer a need for large numbers of people to produce crops. As a result, fewer people were required to produce the growing amounts of agricultural products that supplied an increasing number of consumers.

MODERN AGRICULTURE

As populations continued to grow, there was a need to select and produce crops with higher yields. The *Green Revolution* of the twentieth century helped to make these higher yields possible. Basic information supplied by biological scientists allowed agricultural scientists to develop new, higher-yielding varieties of numerous crops, particularly the *seed grains* which supply most of the calories necessary for maintenance of the world's population. These higher-yielding crop varieties, along with improved farming methods, resulted in tremendous increases in the world's food supply.

The new crop varieties also led to an increased reliance on *monoculture*. While the practice of growing only one crop over a vast number of acres has resulted in much higher yields, it has also decreased the genetic variability of many agricultural plants, increased the need for commercial fertilizers, and produced an increased susceptibility to damage from a host of biotic and abiotic factors. These latter two developments have resulted in a tremendous growth in the *agricultural chemical industry*. Today's modern agricultural unit requires relatively few employees, is highly mechanized, devotes large amounts of land to the production of only one crop, and is highly reliant on agricultural chemicals such as fertilizers and pesticides.

AGRICULTURAL DIVERSITY

Modern agriculture is subdivided into many different specialties. Those agricultural industries that deal with plants include *agronomy*, the production of field crops; *forestry*, the growth and production of trees; and *horticulture*. Horticulture is subdivided into *pomology*, the growth and production of fruit crops such as oranges and apples; *olericulture*, the growth and production of vegetable crops (tomatoes, lettuce); *landscape horticulture*, the growth and production of trees and plants that are used in landscape design; and *floriculture*, the growth and production of flowering plants used in the floral industry.

The various agriculture industries produce a tremendous number of agricultural products. Agricultural products that are derived from plants can be subdivided into timber products (lumber, furniture), grain products (wheat, oats), fiber products (cotton, flax), fruit products (grapes, peaches), nut crops (pecans, hazelnuts), vegetable products (lettuce, cabbage), beverage products (tea, coffee), spice and drug crops (garlic, mustard, opium, quinine), ornamental crops (carnations, chrysanthemums), forage crops (alfalfa, clover), and other cash crops such as sugarcane, tobacco, artichokes, and rubber.

IMPACT ON SOIL RESOURCES

While there have been tremendous increases in agricultural productivity through the use of modern agricultural practices, these practices have had a significant impact on some other natural resources. *Soil* is one of the most overlooked and misunderstood resources. Most people think of soil as an inert medium from which plants grow. In reality, *topsoil*—that upper 15 to 25 centimeters (6 to 10 inches) of the earth's terrestrial surface in which nearly all plants grow—is a complex mixture of weathered mineral materials from rocks, partially decomposed organic molecules, and a large number of living organisms.

The process of soil formation is very slow. Under ideal conditions, enough topsoil can form in one year to produce a layer of about 1 millimeter (0.04 inch) thick when spread over 1 hectare (2.5 acres). With proper management, topsoil can be kept fertile and productive indefinitely. Unfortunately, many agricultural techniques lead to the removal of trees and shrubs, which provide windbreaks, or to the depletion of soil fertility, which reduces the plant cover over the field. These practices expose the soil to increased erosion from wind and moving water. As a result, as much as one-third of the world's current croplands are losing topsoil faster than it can be replaced.

WATER & IRRIGATION

Because plants require water in order to grow, agriculture represents the largest single use of global water. About 73 percent of all fresh water withdrawn from groundwater supplies, rivers, and lakes is used in the irrigation of crops. Almost 15 percent of the world's croplands are irrigated. Water use varies among countries. Some countries have abundant water supplies and irrigate liberally, while water is very scarce in other countries and must be used very carefully. Because as much as 80 percent of the water intended for irrigation is lost to evaporation before reaching the plants, the efficiency of water use in some countries can be very low.

There is no doubt that irrigation has dramatically increased crop production in many areas, but some irrigation practices have been detrimental. Overwatering can lead to a *waterlogging* of the soil. Waterlogging cuts off the supply of oxygen to the roots, and the plants die. Irrigation of crops in dry climates can often result in *salinization* of the soil. In these climates, the irrigation water rapidly evaporates from the soil, leaving behind mineral salts that were dissolved in the water. As the salts accumulate, they become lethal to most plants. Some experts estimate that as much as one-third of the world's agricultural soil has been damaged by salinization. There is also an argument as to whether the increased usage of water for agriculture has decreased the supply of potable water fit for other uses.

FERTILIZERS

Plants require sunshine, water either from rainfall or irrigation, carbon dioxide from the atmosphere, and thirteen mineral nutrients from the soil. Of these, calcium, magnesium, nitrogen, phosphorus, and potassium are required in the greatest amounts. Calcium and magnesium are plentiful in soils located in dry climates, but in wetter climates these nutrients are often *leached* through the soil. In these regions, calcium and magnesium are returned to the soil in the form of lime, which is also sometimes added to soil to raise its pH (increase its alkalinity). Nitrogen, phosphorus, and potassium are the nutrients which are most often depleted from agricultural soils, and these nutrients are often referred to as the *fertilizer elements*. Because these nutrients stimulate plant growth and can greatly increase crop yields, it is necessary to apply them to the soil regularly in order to maintain fertility.

The amount of fertilizer applied to the soil increased more than 450 percent in the second half of the twentieth century. While this increase in the use of fertilizers has more than doubled worldwide crop production, it has also caused some problems. The increased production of fertilizers has required the use of energy and mineral resources that could have been used elsewhere. In many cases, farmers tend to overfertilize. Overfertilization not only wastes money but also contributes to environmental degradation. Fertilizer elements, particularly nitrogen and phosphorus, are carried away by water runoff and are eventually deposited in the rivers and lakes, where they contribute to pollution of aquatic ecosystems. In addition, nitrates can accumulate in underground water supplies. These nitrates can be harmful if ingested by newborns.

OTHER RESOURCES

Modern agriculture, as it is practiced in the United States, consumes large amounts of energy. Farm machinery used in planting, cultivating, harvesting, and transporting crops to processing plants or to market consumes large supplies of liquid fossil fuels such as gasoline or diesel. The energy required to produce fertilizers, pesticides, and other agricultural chemicals is the second largest energy cost associated with agriculture. The use of fuel required by pumps to irrigate crops is also a major energy consumer. Additional energy is used in food processing, distribution, storage, and cooking after the crop leaves the farm. The energy used for these activities may be five times as much as that used to produce the crop.

About 16 percent of the total energy used in the United States is consumed by systems devoted to feeding people, and most of the foods consumed in the United States require more calories of energy to produce, process, and distribute to the market than they provide when they are eaten. The next major development in agriculture will be the *biotechnical revolution*, in which scientists will be able to use molecular biological techniques to produce new crop varieties. In the future, agricultural scientists may be able to develop crop plants that can be produced, processed, and distributed with less impact on other resources.

COMMERCIAL IMPACT

While fewer than 1 percent of Americans are directly involved in agricultural production, agriculture in the United States employs about seventeen million people in some phase of the industry, from production to retail sales. This includes workers hired by agricultural chemical companies which produce or sell agricultural implements and machinery, processing and canning plants, and wholesale and retail marketing firms such as grocery stores. In 1997 there were six thousand to eight thousand different agricultural products on the market. Agriculture is also big business, with assets of about one hundred billion dollars. This figure equals 88 percent of the combined capital assets of all the major U.S. corporations. American farmers now produce 76 percent more crops than the previous generation did on the same amount of land, and one-third of all American agricultural products are exported. This makes the United States the world's most agriculturally productive country and the largest exporter of agricultural products.

Further Reading

Acquaah, George. *Principles of Crop Production: Theory, Techniques, and Technology.* Englewood, N.J.: Prentice Hall, 2001. Comprehensive coverage of U.S. crop production issues, including plant anatomy, climate and weather, soil and land, radiation, temperature and air, and photosynthesis. Well illustrated.

Grigg, David. *An Introduction to Agricultural Geography.* London: Routledge, 2005. Print. Provides a comprehensive introduction to agriculture in developing and developed nations, describing both human and environmental issues. Covers the physical environment, economic behavior and demands, institutional, social and cultural influences, and the impact of farming upon the environment.

Janick, Jules. *Horticultural Science.* 4th ed. New York: W. H. Freeman, 1986. Contains sections on the major horticultural crops.

Smith, Bruce D. *The Emergence of Agriculture.* New York: W. H. Freeman, 1994. A lucid synthesis of many disciplines, explaining the origins of agriculture and the transition from wild to domesticated crops.

Zohary, Daniel, and Maria Hopf. *Domestication of Plants in the Old World: The Origin and Spread of Cultivated*

Plants in West Asia, Europe, and the Nile Valley. 3d ed. New York: Oxford University Press, 2001. A thorough review of information on the beginnings of agriculture, particularly utilizing new molecular biology findings on the genetic relations between wild and domesticated plant species.

MAJOR CROPS & PLACES OF ORIGINAL CULTIVATION

Crop	*Region*
Apples	Central Asia and the Middle East
Apricots	China
Asparagus	Mediterranean
Avocados	Central and South America
Bananas	India and Southeast Asia
Beans	Ethiopia
Bell peppers	Central and South America
Cantaloupes	Central Asia and the Middle East
Carrots	Central Asia and the Middle East
Cashews	Central and South America
Celery	Mediterranean
Cherries	China
Coconuts	India and Southeast Asia
Coffee	Ethiopia
Corn	Central and South America
Cotton	Central Asia and the Middle East
Cucumbers	China
Eggplant	China
Figs	Central Asia and the Middle East
Garlic	Central Asia and the Middle East
Ginger	India and Southeast Asia
Grapes	Central Asia and the Middle East
Leeks	Central Asia and the Middle East
Lettuce	Mediterranean
Lima beans	Central and South America
Mangoes	India and Southeast Asia
Mustard	India and Southeast Asia
Oats	Central Asia and the Middle East
Okra	Ethiopia
Olives	Mediterranean
Onions	Mediterranean
Oranges	China, India, and Southeast Asia
Papayas	Central and South America
Parsnips	Mediterranean
Peaches	China
Peanuts	Central and South America
Pears	Central Asia and the Middle East
Peas	Ethiopia
Pineapples	Central and South America
Potatoes	Central and South America
Pumpkins	Central and South America
Radishes	India and Southeast Asia
Rhubarb	Mediterranean
Rice	India and Southeast Asia
Rye	Central Asia and the Middle East
Soybeans	China
Spinach	Central Asia and the Middle East
Sugarcane	China
Sweet potatoes	Central and South America
Tangerines	India and Southeast Asia
Tea	China
Tomatoes	Central and South America
Turnips	Central Asia, the Mediterranean, and the Middle East
Walnuts	China
Wheat	Central Asia and the Middle East
Yams	India and Southeast Asia

Source: Data are from Brian Capon, *Botany for Gardeners: An Introduction and Guide* (Portland, Oreg.: Timber Press, 1990), p. 81.

AGRICULTURE: MARINE

—Virginia L. Hodges

FIELDS OF STUDY

Ecology; Environmental Studies; Agronomy.

SUMMARY

Marine agriculture uses techniques of artificial cultivation, such as growing, managing, and harvesting, and applies them to marine plants and animals. The products are then used for human consumption.

Marine agriculture is also known as *mariculture* or *aquaculture*, although aquaculture is a more general term referring to both freshwater and marine farming of organisms. The world's oceans cover approximately three-fourths of the globe, including vast regions of unexplored life and landforms. The potential for exploiting the oceans agriculturally is great but currently meets significant obstacles. Because of the expense of equipment and personnel involved, most marine species are not cultivated. Coastal pollution, habitat destruction, competition for land use, and economics all limit mariculture programs. Nevertheless, mariculture does offer several food, medical, and other products that are currently being marketed.

FOOD

Seaweeds are edible, especially the red and brown algae. The three most common types of seaweeds are known by their Japanese names: *nori* (*Porphyra*), a red seaweed high in vitamin C and digestible protein; *kombu* (*Laminaria*); and *wakame* (*Undaria*), high in calcium. They are eaten raw, cooked, or dried and have several vitamins and minerals as well as protein. Seaweeds are low in fats, and 35 to 50 percent of the dry weight of red seaweeds is protein. Seaweeds can be used to add taste and variety to foods. They are used as a hot vegetable, boiled and formed into cakes and fried, in salads, and in preparing desserts, breads, soups, casseroles, sandwiches, teas, and candy.

The world's yearly harvest of seaweeds is approximately 8.4 million tons of green seaweed, 2.8 million tons of brown seaweed, and 1.2 million tons of red seaweed. The total seaweed market in 1998 was worth more than $5 billion, with $600 million deriving from food additives alone. China is the leading harvester and the world's biggest seaweed consumer. Japan is the leading seaweed importer and, at the end of the twentieth century, employed more than thirty-five thousand people in the industry. Harvesting and marketing edible seaweed is a growing business in the United States, especially on the West Coast.

Seaweeds produce several types of *phycocolloids*, starchlike chemicals used in food processing and manufacturing. An important type called *algin*, which makes up alginic acid and alginates, is used in manufacturing dairy products such as ice cream, cheese, and toppings as well as to prevent frostings and pies from desiccation. Another extract is *agar*, used to form jellies and protect fish and meats during canning. Agar is also used in low-calorie foods and as a thickener. Red algae is a source of the agglutinant *carrageenan*, which is used in many food products as an emulsifier to give body to dairy products and other processed foods, including instant puddings. Additionally, seaweed-based food additives are common in prepared and fast foods, including hamburgers and yogurt.

Kelp farming is a major livelihood in the eastern Pacific, with approximately 140,000 tons harvested each year for the extraction of alginates used in

food and food additives. Kelp is a good source of calcium, potassium, iron, iodine, bromine, and zinc. It is also low-fat, has some protein, and is a natural tenderizer. Kelp flakes are used as a low-sodium salt substitute.

MEDICINE

The use of marine plants in medicine is still in the early stages of exploration and faces many challenges, including identification of useful chemicals and the cultivation of significant quantities. Dinoflagellates and other microalgae are being investigated for compounds that might fight cancerous tumors. Diluted algae toxins from red tides can be used to inhibit the growth of most bacteria. Green algae has halosphaerin, a strong antibiotic. Seaweed is used in wound dressings in hospitals and as a source of iodine, A, B, D, and E vitamins, calcium, magnesium, potassium, sodium, sulfur, and trace antioxidants such as selenium and zinc. The seaweed extract agar is used in laxatives and as a medium to grow bacteria and molds.

Kelp is rich in *chlorophyll*, which can help detoxify the body, fight inflammations, and increase the formation of oxygen-carrying red blood cells. Chlorophyll is also used to fight bad breath and as an ingredient in deodorants. Kelp is used to reduce cholesterol, treat gastrointestinal, respiratory, and genitourinary disorders, and lower blood pressure. The alginic acid produced by kelp can rid the body of radioactive strontium, the most dangerous to humans of all components in the fallout from atomic explosions.

OTHER USES

Marine plants are used for a variety of other purposes. Seaweed is used as a component of many fertilizers, as a food additive in animal feed, and to reduce soil acidity. Research on cattle and swine has revealed that the addition of seaweed to animal feed can enhance the immune system and makes the meat a more desirable color. It can also save cattle from the effects of fungus-infected grass.

Seaweed is used as an ingredient in cosmetics as well as to nourish, revitalize, condition, and improve the skin, hair, and body. It is used in cleansers, toners, moisturizers, scrubs, body lotions, and hair and bath products. The giant kelp (*Macrocystis*) is a major source of algin for commercial uses, as is the brown algae *Laminaria*, which is harvested in the north Atlantic. Algin is used in shampoos, shaving cream, plastics, pesticides, rubber products, paper, paints, and cosmetics. Additionally, kelp is used in emulsifiers for toothpastes and printing inks. Kelp has even been used to make fishing lines. Some research has been done on using kelp as a fuel to produce a clean-burning methane gas. Kelp can be used to ferment human waste and garbage, which can then be sold as fertilizers.

Further Reading

Castro, Peter, and Michael E. Huber. *Marine Biology.*, 2019. Print. Well-organized, basic textbook with summary and bibliography with each chapter. Good section on marine agriculture. Includes illustrations, glossary, and index.

Cousteau, Jacques. *The Ocean World.* New York: Harry N. Abrams, 1985. Discusses all marine life. Includes color photographs, glossary, bibliography, and index.

Levinton, Jeffrey S. *Marine Biology: Function, Biodiversity, Ecology.* Oxford University Press, 2018. Print. Well-organized, advanced textbook. Further Reading and review questions with each chapter. Includes glossary, index.

Morrissey, John F, James L. Sumich, and Deanna R. Pinkard-Meier. *Introduction to the Biology of Marine Life.* Jones and Bartlett Learning, 2018. Print. Textbook format with basic explanations, review and discussion questions, and bibliography with each chapter. Includes glossary and taxonomic and subject indexes.

Agriculture: Modern Problems

—*James Knotwell, Denise Knotwell,*
Bryan Ness, and Elizabeth Slocum

FIELDS OF STUDY

Ecology; Environmental Studies; Agronomy.

SUMMARY

Many modern problems in agriculture are not new. Erosion and pollution, for example, have been around as long as agriculture. However, agriculture has changed drastically within its ten-thousand-year history, especially since the dawn of the Industrial Revolution in the eighteenth century. Erosion and pollution are bigger problems than before and have been joined by a host of other issues that are equally critical—not all related to physical deterioration.

MONOCULTURE

Modern agriculture emphasizes crop specialization, also known as monoculture. Farmers, especially in industrialized regions, often grow a single crop on much of their land. Problems associated with this practice are exacerbated when a single variety or cultivar of a species is grown. Such a strategy allows the farmer to reduce costs, but it also makes the crop, and thus the farm and community, susceptible to widespread crop failure. The corn blight of 1970 devastated more than 15 percent of the North American corn crop. The corn was particularly susceptible to harmful organisms because 70 percent of the crop being grown was of the same high-yield variety. Chemical antidotes can

Monoculture This single barley field seems to cover all the land on the west side of Govel Hill. Image by Anne Burgess CC-BY-SA-2.0.

fight pests, but they increase pollution. Maintaining species diversity or varietal diversity—growing several different crops instead of one or two—allows for crop failures without jeopardizing the entire economy of a farm or region that specializes in a particular monoculture, such as tobacco, coffee, or bananas.

GENETIC ENGINEERING

Growing genetically modified (GM) crops is one potential way to replace post-infestation chemical treatments. Recombinant technologies used to splice genes into varieties of rice or potatoes from other organisms are becoming increasingly common. The benefits of such GM crops include more pest-resistant plants and higher crop yields. However, environmentalists fear new genes could trigger unknown side effects with more serious, long-term environmental and economic consequences than the problems they were used to solve. GM plants designed to resist herbicide applications could potentially pass the resistant gene to closely related wild weed species that would then become "super weeds." Also, just as pests can develop resistance to pesticides, they may also become resistant to defenses engineered into GM plants.

EROSION

An age-old problem, soil loss from erosion occurs all over the world. As soil becomes unproductive or erodes away, more land is plowed. The newly plowed lands usually are considered marginal, meaning they are too steep, nonporous or too sandy, or deficient in some other way. When natural vegetative cover blankets these soils, it protects them from erosive agents: water, wind, ice, or gravity. Plant cover "catches" rainwater that seeps downward into the soil rather than running off into rivers. As marginal land is plowed or cleared to grow crops, erosion increases.

Expansion of land under cultivation has not been the only factor contributing to erosion. Fragile grasslands in dry areas have also been used more intensively. Grazing more livestock than these pastures can handle decreases the amount of grass in the pasture and exposes more of the soil to wind, the primary erosive agent in dry regions. Overgrazing can affect pastureland in tropical regions, too. Thousands of acres of tropical forest have been cleared to establish cattle-grazing ranges in Latin America. Tropical soils, although thick, are not very fertile. After one or two growing seasons, crops grown in these soils will yield substantially less than before.

Tropical fields require fallow periods of about ten years to restore the soil after it is depleted. That is why tropical farmers using slash-and-burn agriculture move to new fields every few years in a cycle that returns them to the same place years later, after their particular lands have regenerated. Where there is heavy forest cover, soils are protected from exposure to the massive amounts of rainfall. Organic material for crops is present as long as the forest remains in place. When the forest is cleared, however, the resulting grassland cannot provide adequate protection, and erosion accelerates. Lands that are heavily grazed provide even less protection from heavy rains, and erosion accelerates even more.

The use of machines also promotes erosion, and modern agriculture relies on machinery such as tractors, harvesters, trucks, balers, and ditchers. Machinery is used intensely in industrialized nations, and its use has been on the rise in developing countries such as India, China, Mexico, and Indonesia, where traditional, nonmechanized farming methods were previously the norm. Farming machines, in gaining traction, loosen topsoil and inhibit vegetative cover growth, especially when farm implements designed to rid the soil of weeds are attached. The soil is then more prone to erosion.

Eco-fallow farming has become more popular in the United States and Europe as a way to reduce erosion. This method of agriculture, which leaves the crop residue in place over the fallow (nongrowing) season, does not root the soil in place as well as living plants do, so some erosion continues. Additionally, eco-fallow methods require the heavy use of chemicals, such as herbicides, to "burn down" weed growth at the start of the growing season. This contributes to increased erosion and pollution.

POLLUTION & SILT

Besides causing resistance among harmful bacteria, insects, and weeds, pesticides inevitably wash into surface and groundwater supplies, contaminating them. Pesticides are potentially harmful to human health; there has been concern that their seepage into land and water is linked to cancer.

An increasingly heavy silt load has been choking the life out of streams and rivers. Accelerated erosion from water runoff carries silt particles into streams, where they remain suspended and inhibit the growth of many forms of plant and animal life. The silt load in American streams has become so heavy that the Mississippi River delta has been growing faster than it once did. Heavy silt loads, combined with chemical residues, have been creating an expanded dead zone. By taxing the capabilities of ecosystems around the delta, sediments have been filtered out slowly, plant absorption of nutrients has decreased, and salinity levels for aquatic life have been unable to be stabilized. Most of the world's population lives in coastal zones, and 80 percent of the world's fish catch comes from coastal waters over continental shelves that are most susceptible to this form of pollution.

PESTICIDE RESISTANCE

With the onset of the Green Revolution of the mid-twentieth century, the use of herbicides, insecticides, and other pesticides increased dramatically all over the world. An increasing awareness of problems caused by the overuse of pesticides followed, extending even to household antibacterial cleaning agents and other products. Mutations among the genes of bacteria and plants have allowed these organisms to resist the effects of chemicals that were toxic to their ancestors. The use of pesticides leads to a cycle wherein more or different combinations of chemicals are used, and more pests develop resistance to these toxins. Additionally, the development of herbicide-resistant crop plants enables greater use of herbicides to kill undesirable weeds on croplands.

Increasing interest in biopesticides (biological pesticides) may slow the cycle of pesticide resistance. Types of biopesticides include beneficial microbes, fungi, and insects such as ladybugs that can be released in infested areas to prey upon specific pests. Biopesticides may be naturally occurring or genetically modified organisms. Their use also avoids excessive reliance on chemical pesticides.

FERTILIZERS & EUTROPHICATION

Increased use of fertilizers was another result of the Green Revolution. Particulate amounts of most fertilizers enter the hydrologic cycle through runoff. As a result, bodies of water become enriched by dissolved nutrients, such as nitrates and phosphates. The growth of aquatic plants in rivers and lakes is overstimulated, which results in the depletion of dissolved oxygen. This process of eutrophication can harm all aquatic life in these ecosystems.

WATER DEPLETION

An increasing reliance on irrigation has contributed to the mismanagement and overtapping of groundwater resources. The rate of groundwater recharge is slow, usually between 0.1 and 0.3 percent per year. When the amount of water pumped out of the ground exceeds the recharge rate, it is referred to as aquifer overdraft. An aquifer is a water-bearing stratum of permeable rock, sand, or gravel.

In Tamil Nadu, India, groundwater levels dropped twenty-five to thirty meters during the 1970s due to excessive pumping for irrigation. In Tianjin, China, the groundwater level has declined 4.4 meters per year. In the United States, aquifer overdraft has averaged 25 percent over the replacement rate. The Ogallala aquifer—located under parts of South Dakota, Wyoming, Nebraska, Colorado, Kansas, Oklahoma, Texas, and New Mexico—represents an extreme example of overdraft: the rate of depletion has annually been three times faster than its rate of recharge. The capacity of the aquifer decreased by an estimated 33 percent between 1950 and 2004. At this rate, the Ogallala aquifer, which supplies water to countless communities and farms, has been projected to become nonproductive by 2030.

SOIL SALINIZATION

In addition, continued irrigation of arid regions can lead to soil problems. Soil salinization has been widespread in the small-grained soils of these regions, which have a high water absorption capacity and a low infiltration rate. Some irrigation practices add large amounts of salts into the soil, increasing its natural rate of salinization. This can also occur at the base of a hill slope. Soil salinization has been recognized as a major process of land degradation.

Although surface and groundwater resources cannot be enriched by technology, conservation and improved environmental management can make the use of precious freshwater more efficient. In agriculture, for example, drip irrigation can reduce water use by nearly 50 percent. In developing countries, though, equipment and installation costs often limit the availability of these more efficient technologies.

URBAN SPRAWL

With the increasing mechanization of farms, the need for farmers and farm workers has been drastically reduced. From a peak in 1935 of about 6.8 million farmers farming 1.1 billion acres in the United States, for example, the country at the end of the twentieth century counted fewer than two million farmers farming 950 million acres. In 2012, the number of farm operations was at 2.17 million, according to the U.S. Department of Agriculture, with only 914 million acres of land in use.

Urban sprawl converts a tremendous amount of cropland into parking lots, shopping malls, industrial parks, and suburban neighborhoods. If cities were located in marginal areas, then concern about the loss of farmland to commercial development would be nominal. However, the cities attracting the greatest numbers of people have too often replaced the best cropland. Taking the best cropland out of primary production imposes a severe economic penalty.

ENERGY & TECHNOLOGICAL EFFICIENCY

The increasing mechanization of farms has led to major increases in the amounts of energy consumed by these farms, particularly those in industrialized nations. Farms use large quantities of energy for irrigation, to operate machinery, to heat and cool buildings, for food processing and shipment, to spray pesticides, and to fertilize crops. The latter two are products of fossil fuels. Raising livestock on grain also consumes large quantities of fossil fuel. Large-scale livestock farmers often feed their animals grains and protein byproducts rather than employing traditional methods of foraging and consuming crop waste. Grain and protein byproduct feeding requires less land and allows for the animals to grow to market weight quickly. However, this method of feeding can be inefficient, as animals convert only a fraction of their food energy into growth; for example, it has been estimated that seven kilograms of grain are needed to produce only one kilogram of beef.

Practices such as conservation tilling, which requires less working of the soil, have helped reduce energy use on farms. Another practice is drip irrigation, in which water drips slowly to the roots of plants, thus saving on both water and fertilizer. Farmers have also begun to plant genetically modified crops that do not need pesticide, which itself has become more sophisticated and therefore used in smaller quantities than before. Other farms have opted to grow organic (pesticide-free) food and to raise animals that are not given growth hormones and that are free-range, or not always confined in tight quarters. While these practices can be expensive, they can also save on energy costs and alleviate consumer concerns about ingesting potentially harmful chemicals.

Another way farmers have learned to save money on energy is to use renewable energy sources such as wind power, solar power, and biomass products (also called biofuel). By having electric wind turbines built on their farms, farmers can produce their own energy. Wind power can be used to power an entire farm or to power a specific area of the farm, such as pumping water for cattle. Solar energy can power a farm's lighting and heating (e.g., in greenhouses), pump water, and produce electricity. It can even be used to dry crops faster and more evenly than crops left prone to damage in the fields. Many farmers grow corn to make ethanol. Other crops have begun to be used for fuel as well, since there is virtually no limit to the type of plant and organic waste that can be used to produce energy. Agriculture creates a lot of waste; there is the potential for taking that waste and converting it into energy, thus saving on energy production costs, disposal costs, and pollution. Crops grown specifically for biofuel—for farms and other consumers—can be produced in large quantities and thus become profitable when sold.

Other technologies that have made farming become more efficient—and that have saved and made farmers money—are broadband Internet access, smartphones, and Global Positioning System (GPS) technology. Use of the Internet has helped farmers quickly exchange important data with each other and has helped farmers connect directly to their markets and consumers. GPS technology has helped farmers navigate their fields in a fraction of the time it took before this invention; equipment can be guided through fields, with no overlap or gaps, so that seeds can be planted and pesticides sprayed evenly. Smartphone applications (apps) for crop scouting can help farmers identify a problem and its specific source immediately, eliminating the need to apply pesticide, for example, to an entire field in the hopes of rectifying that one problem. There are numerous other mobile applications useful to farmers as well.

FURTHER READING

Campbell, Elizabeth. "Number of US Farms Fell to Six-Year Low in 2012, USDA Says." *Bloomberg*, 19 Feb. 2013, www.bloomberg.com/news/articles/2013-02-19/number-of-u-s-farms-fell-to-six-year-low-in-2012-usda-says-1-. Accessed 7 Apr. 2017.

Cockrall-King, Jennifer. *Food and the City: Urban Agriculture and the New Food Revolution.* Prometheus Books, 2012.

Hoag, Dana L. *Agricultural Crisis in America: A Reference Handbook.* ABC-CLIO, 1999.

Jackson, Wes. *New Roots for Agriculture.* 1980. U of Nebraska P, 1985.

Lægreid, M., et al. *Agriculture, Fertilizers, and the Environment.* Cabi, 1999.

Magdoff, Fred, and Brian Tokar, editors. *Agriculture and Food in Crisis: Conflict, Resistance, and Renewal.* Monthly Review Press, 2010.

McKee, Gregory J., et al., editors. *Pesticide Resistance, Population Dynamics, and Invasive Species Management.* Nova Science Publishers, 2010.

Paarlberg, Don, and Philip Paarlberg. *The Agricultural Revolution of the 20th Century.* Iowa State UP, 2000.

Pfeiffer, Dale A. *Eating Fossil Fuels: Oil, Food and the Coming Crisis in Agriculture.* Gabriola Island (BC: New Society Publishers, 2008. Print.

Pimentel, David, and Anne Wilson. "World Population, Agriculture, and Malnutrition." *World Watch*, Sept.–Oct. 2004, pp. 22–25. Accessed 7 Apr. 2017.

"Renewable Energy and Agriculture: A Natural Fit." *Union of Concerned Scientists*, 2003, www.ucsusa.org/clean_energy/smart-energy-solutions/increase-renewables/renewable-energy-and.html. Accessed 7 Apr. 2017.

Rissler, Jane, and Margaret Mellon. *The Ecological Risks of Engineered Crops.* MIT P, 1996.

"Unit 7: Agriculture // Section 8: Agriculture and Energy." *The Habitable Planet*, Annenberg Learner, www.learner.org/courses/envsci/unit/text.php?unit=7&secNum=8. Accessed 7 Apr. 2017.

AGRICULTURE: TRADITIONAL

—Nancy Farm Männikkö

FIELDS OF STUDY

Ecology; Environmental Studies; Agronomy.

SUMMARY

Two agricultural practices that are widespread among the world's traditional cultures, slash-and-burn agriculture and nomadism, share several features. Both are ancient forms of agriculture, both involve farmers not remaining in a fixed location, and both can pose serious environmental threats if practiced in a nonsustainable fashion. The most significant difference between the two is that slash-and-burn is associated with raising field crops, while nomadism usually involves herding livestock.

SLASH-AND-BURN AGRICULTURE

Farmers have practiced *slash-and-burn agriculture*, which is also referred to as *shifting cultivation* or *swidden agriculture*, in almost every region of the world where farming is possible. Although at the end of the twentieth century slash-and-burn agriculture was most commonly found in tropical areas such as the Amazon River basin in South America, swidden agriculture once dominated agriculture in more temperate regions, such as northern Europe. Swidden agriculture was, in fact, common in Finland and northern Russia well into the early decades of the twentieth century.

Slash-and-burn acquired its name from the practice of farmers who cleared land for planting crops by cutting down the trees or brush on the land and then burning the fallen timber on the site. The farmers literally slash and burn. The ashes of the burnt wood add minerals to the soil, which temporarily improves its fertility. Crops the first year following clearing and burning are generally the best crops the site will provide. Each year after that, the yield diminishes slightly as the fertility of the soil is depleted.

Farmers who practice slash-and-burn do not attempt to improve fertility by adding fertilizers such as animal manure to the soil. They instead rely on the soil to replenish itself over time. When the yield from one site drops below acceptable levels, farmers then clear another piece of land, burn the brush and other vegetation, and cultivate that site while leaving their previous field to lie fallow and its natural vegetation to return. This cycle will be repeated over and over, with some sites being allowed to lie fallow indefinitely, while others may be revisited and farmed again in five, ten, or twenty years.

Traditional farming methods had a low carbon footprint. Image by Augustus Binu CC BY-SA 3.0.

Farmers who practice slash-and-burn do not always move their dwelling places as they cultivate different fields. In some geographic regions, farmers live in a central village and farm cooperatively, with fields being alternately allowed to remain fallow and farmed, making a gradual circuit around the central village. In other cases, the village itself may move as new fields are cultivated. Anthropologists studying indigenous peoples in Amazonia, for example, discovered that village garden sites were on a hundred-year cycle. Villagers farmed cooperatively to clear a garden site. That garden would be used for about five years; then a new site was cleared. When the fields in use became an inconvenient distance from the village—about once every twenty years—the entire village would move to be closer to the new fields. Over a period of approximately one hundred years, a village would make a circle through the forest, eventually ending up close to where it had been located long before any of the present villagers had been born.

In more temperate climates, farmers often owned and lived on the land on which they practiced swidden agriculture. Farmers in Finland, for example, would clear a portion of their land, burn the covering vegetation, grow grains for several years, and then allow that land to remain fallow for five to twenty years. The individual farmer rotated cultivation around the land in a fashion similar to that practiced by whole villages in other areas but did so as an individual rather than as part of a communal society.

Although slash-and-burn is frequently denounced as a cause of environmental degradation in tropical areas, the problem with it is not the practice itself but the length of the cycle. If the cycle of shifting cultivation is long enough, forests will grow back, the soil will regain its fertility, and minimal adverse effects will occur. In some regions, a piece of land may require as little as five years to regain its maximum fertility; in others, it may take one hundred years. Problems arise when growing populations put pressure on traditional farmers to return to fallow land too soon. Crops are smaller than needed, leading to a vicious cycle in which the next strip of land is also farmed too soon, and each site yields less and less. As a result, more and more land must be cleared.

NOMADISM

Nomadic peoples have no permanent homes. They earn their living by raising herd animals, such as sheep, horses, or other cattle, and they spend their lives following their herds from pasture to pasture with the seasons, going wherever there is sufficient food for their animals. Most nomadic animals tend to be hardy breeds of goats, sheep, or cattle that can withstand hardship and live on marginal lands. Traditional nomads rely on natural pasturage to support their herds and grow no grains or hay for themselves. If a drought occurs or a traditional pasturing site is unavailable, they can lose most of their herds to starvation.

In many nomadic societies, the herd animal is almost the entire basis for sustaining the people. The animals are slaughtered for food, clothing is woven from the fibers of their hair, and cheese and yogurt may be made from milk. The animals may also be used for sustenance without being slaughtered. Nomads in Mongolia, for example, occasionally drink horses' blood, removing only a cup or two at a time from the animal.

In mountainous regions, nomads often spend the summers high up on mountain meadows, returning to lower altitudes in the autumn when snow begins to fall. In desert regions, they move from oasis to oasis, going to the places where sufficient natural water exists to allow brush and grass to grow, allowing their animals to graze for a few days, weeks, or months, then moving on. In some cases, the pressure to move on comes not from the depletion of food for the animals but from the depletion of a water source, such as a spring or well. At many desert oases, a natural water seep or spring provides only enough water to support a nomadic group for a few days at a time.

In addition to true nomads—people who never live in one place permanently—a number of cultures have practiced *seminomadic farming*: The temperate months of the year, spring through fall, are spent following the herds on a long loop, sometimes hundreds of miles long, through traditional grazing areas, then the winter is spent in a permanent village.

Nomadism has been practiced for millennia, but there is strong pressure from several sources to eliminate it. Pressures generated by industrialized society are increasingly threatening the traditional cultures of nomadic societies, such as the Bedouin of the Arabian Peninsula. Traditional grazing areas are being fenced off or developed for other purposes.

Environmentalists are also concerned about the ecological damage caused by nomadism. Nomads generally measure their wealth by the number of animals they own and will try to develop their herds to as large a size as possible, well beyond the numbers required for simple sustainability. The herd animals eat increasingly large amounts of vegetation, which then has no opportunity to regenerate. Desertification may occur as a result. Nomadism based on herding goats and sheep, for example, has been blamed for the expansion of the Sahara Desert in Africa. For this reason, many environmental policymakers have been attempting to persuade nomads to give up their traditional lifestyle and become sedentary farmers.

Further Reading

Beckwith, Carol, and Marion Van Offelen. *Nomads of Niger.* New York: Academic Press, 1987. Well-written depiction of the life of the ancient nomadic Wodaabe people, who have roamed for centuries across the sub-Saharan bushland. Photographer Beckwith spent eighteen months traveling with one particular band of Wodaabe.

Colfer, Carol J., with Nancy Peluso and Chin See Chung. *Beyond Slash and Burn: Building on Indigenous Management of Borneo's Tropical Rain Forest.* New York: New York Botanical Garden, 1994. From the Advances in Economic Botany series; contains bibliographical resources.

Goldstein, Melvyn C., and Cynthia M. Beall. *The Changing World of Mongolia's Nomads.* Berkeley: University of California Press, 1994. This anthropological study of a Mongolian herding community presents an intimate portrait of life on the steppes.

Grigg, David. *An Introduction to Agricultural Geography.* London: Routledge, 2005. Print. Provides a comprehensive introduction to agriculture in developing and developed nations, describing both human and environmental issues. Covers the physical environment, economic behavior and demands, institutional, social and cultural influences, and the impact of farming upon the environment.

Mortimore, Michael. *Roots in the African Dust: Sustaining the Sub-Saharan Drylands.* New York: Cambridge University Press, 1998. This volume, accessible to students, policymakers, and scholars, includes chapters on the extent of soil degradation, merits of agricultural intensification, and conservation of biotic resources. Offers a breadth of analysis in a holistic, wide-ranging view of rural livelihoods and landscapes. Mortimore argues that desertification is reversible.

Schmink, Marianne, and Charles H. Wood. *Contested Frontiers in Amazonia.* New York: Columbia University Press, 1992. This interdisciplinary analysis of the process of frontier change in one region of the Brazilian Amazon shows how deforestation and settlement patterns were outcomes of the competition for resources among various groups.

Agriculture: World Food Supplies

—William A. Dando

FIELDS OF STUDY

Ecology; Environmental Studies; Agronomy.

SUMMARY

Soil types, topography, climate, socioeconomics, dietary preferences, stages in agricultural development, and governmental policies combine to give a distinctive personality to regional agricultural characteristics and, hence, food supplies in various areas of the world.

All living things need food to live, grow, work, and survive. Almost all foods that humans consume come from plants and animals. Not all of earth's people eat the same foods, however. The types, combinations, and amounts of food consumed by different peoples depend upon historic, socioeconomic, and environmental factors.

HISTORY OF FOOD CONSUMPTION

Early in human history, people ate what they could gather or scavenge. Later, people ate what they could plant and harvest and the products of animals they could domesticate. Modern people eat what they can grow, raise, or purchase. Their diets or food composition is determined by income, local customs, religion or food biases, and advertising. There is a global food market, and many people can select what they want to eat and when they eat it according to the prices they can pay and what is available.

Historically, in places where food was plentiful, accessible, and inexpensive, humans devoted less time to basic survival needs and more time to activities that led to human progress and enjoyment of leisure. Despite a modern global food system, instant telecommunications, the United Nations, and food surpluses in some places, however, the problem of providing food for everyone on earth has not been solved.

In 1996 leaders from 186 countries gathered in Rome and agreed to reduce by half the number of hungry people in the world by the year 2015. United Nations data for 1998 revealed that more than 790 million people in the developing parts of the world did not have enough food to eat. This is more people than the total population of North America and Europe at that time. The number of undernourished people has been decreasing since 1990. Still, at the current pace of hunger reduction in the world, 600 million people will suffer from "acute food insecurity" and go to sleep hungry in 2015. Despite efforts being made to feed the world, outbreaks of

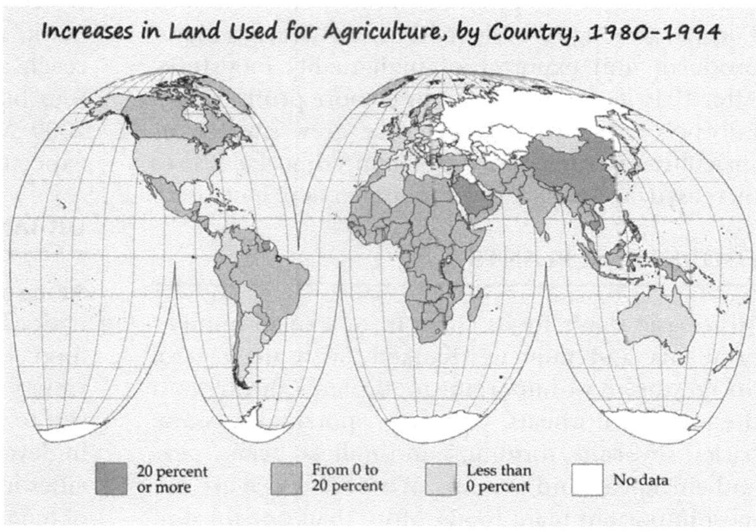

food deficiencies, mass starvation, and famine are a certainty in the twenty-first century.

WORLD FOOD SOURCE REGIONS

Agriculture and related primary food production activities, such as fishing, hunting, and gathering, continue to employ more than one-third of the world's labor force. Agriculture's relative importance in the world economic system has declined with urbanization and industrialization, but it still plays a vital role in human survival and general economic growth. Demands on agriculture in the twenty-first century include supplying food to an increasing world population of nonfood producers as well as producing food and nonfood crude materials for industry.

Soil types, topography, weather, climate, socioeconomic history, location, population pressures, dietary preferences, stages in modern agricultural development, and governmental policies combine to give a distinctive personality to regional agricultural characteristics. Two of the most productive food-producing regions of the world are North America and Europe. Countries in these regions export large amounts of food to other parts of the world.

North America is one of the primary food-producing and food-exporting continents. After 1940 food output generally increased as cultivated zacreage declined. Progress in improving the quantity and quality of food production is related to mechanization, chemicalization, improved breeding, and hybridization. Food output is limited more by market demands than by production obstacles. Western Europe, although a basic food-deficit area, is a major producer and exporter of high-quality foodstuffs. After 1946 its agriculture became more profit-driven. Europe's agricultural labor force grew smaller, its agriculture became more mechanized, its farm sizes increased, and capital investment per acre increased.

FOODS FROM PLANTS

Most basic staple foods come from a small number of plants and animals. Ranked by tonnage produced, the most important food plants throughout the world are wheats, corn, rice, potatoes, cassava, barley, soybeans, sorghums and millets, beans, peas and chickpeas, and peanuts. Wheat and rice are the most important plant foods. More than one-third of the world's cultivated land is planted with these two crops. Wheat is the dominant food staple in North America, Western and Eastern Europe, northern China, the Middle East, and North Africa. Rice is the dominant food staple in southern and eastern Asia.

Corn, used primarily as animal food in developed nations, is a staple food in Latin America and southeast Africa. Potatoes are a basic food in the highlands of South America and in Central and Eastern Europe. Cassava is a tropical starch-producing root crop of special dietary importance in portions of lowland South America, the west coast countries of Africa, and sections of South Asia. Barley is an important component of diets in North African, Middle Eastern, and Eastern European countries. Soybeans are an integral part of the diets of those who live in eastern, southeastern, and southern Asia. Sorghums and millets are staple subsistence foods in the savanna regions of Africa and South Asia, while peanuts are a facet of dietary mixes in tropical Africa, Southeast Asia, and South America.

THE WORLD'S GROWING POPULATION

The problem of feeding the world is compounded by the fact that population was increasing at a rate of nearly 80 million persons per year at the end of the twentieth century. That rate of increase is roughly equivalent to adding a country the size of Germany to the world every year. Compounding the problem of feeding the world are population redistribution patterns and changing food consumption standards. By 2001, the world had exceeded the six billion mark, and the world population was projected to reach approximately ten billion people by 2050—four billion people more than were on the earth in 2000. Most of the increase in world population was expected to occur within the developing nations.

URBANIZATION

Along with an increase in population in developing nations is massive urbanization. City dwellers are food consumers, not food producers. The exodus of young men and women from rural areas has given rise to a new series of megacities, most of which are in developing countries. By the year 2015, twenty-six cities in the world are expected to have populations of ten million people or more.

When rural dwellers move to cities, they tend to change their dietary composition and food-consumption patterns. Qualitative changes in dietary consumption standards are positive, for the most part, and are a result of educational efforts of modern nutritional scientists working in developing countries. During the last four decades of the twentieth century, a tremendous shift took place in overall dietary habits. Dietary changes and consumption trends have contributed to a decrease in child mortality, an increase in longevity, and a greater resistance to disease. This globalization of people's diets has resulted in increased demands for higher quality, greater quantity, and more nutritious basic foods.

PERSPECTIVES

Humanity is entering a time of volatility in food production and distribution. The world will produce enough food to meet the demands of those who can afford to buy food. In many countries, however, food production is unlikely to keep pace with increases in the demand for food by growing populations. The food gap—the difference between production and demand—could more than double in the first three decades of the twenty-first century. Such a development would increase the dependence of developing countries on food imports. About 90 percent of the rate of increase in aggregate food demand in the early twenty-first century is expected to be the result of population increases. Factors that could lead to larger fluctuations in food availability include weather variations, such as those induced by El Niño and climatic change, the growing scarcity of water, civil strife and political instability, and declining food aid.

FURTHER READING

Conway, Gordon, and Vernon W. Ruttan. *The Doubly Green Revolution: Food for All in the Twenty-First Century*. Ithaca, N.Y.: Comstock, 1998. Discusses the need for equitable food distribution and the need to pay more attention to the environment than did the Green Revolution of the 1960's-1980's.

DeRose, Laurie Fields, Ellen Messer, and Sara Millman. *Who's Hungry? and How Do We Know? Food Shortage, Poverty, and Deprivation*. New York: United Nations University, 1998. Differentiates between food shortage (regional food scarcity), food poverty (inadequate household food supplies), and food deprivation (malnutrition) in order to identify and respond to the causes of hunger.

Lappé, Frances Moore, Joseph Collins, and Peter Rosset. *World Hunger: Twelve Myths*. 2d ed. New York: Grove Press, 1998. Three experts on food and agriculture expose the flaws within many tenaciously held beliefs, which prevent effective treatment of the problems of world hunger and global food distribution.

Shiva, Vandana. *Stolen Harvest: The Highjacking of the Global Food Supply*. Lexington, KY: University Press Of Kentucky, 2016. Print. Charts the impacts of globalized, corporate agriculture on small-scale farmers, the environment, and food quality. Covers genetically modified foods, seed monopoly, and the need for sound environmental thinking throughout the world.

AGROECOSYSTEMS

—Krishna Prasad Vadrevu

FIELDS OF STUDY

Agronomy; Environmental Studies; Soil Mechanics.

SUMMARY

The term *ecosystem* was first coined by a British ecologist, Sir Arthur George Tansley, to denote complex systems comprising both abiotic and biotic components. Agroecosystems, or agricultural systems, are specialized, human-managed ecosystems in which communities of plants and animals interact with their physical and chemical environments modified for the production of food, fuel, and fiber for human consumption and processing. Agroecosystems are more open than natural systems (forests).

The term *ecosystem* was first coined by a British ecologist, Sir Arthur George Tansley, to denote complex systems comprising both abiotic and biotic components. Agroecosystems, or agricultural systems, are specialized, human-managed ecosystems in which communities of plants and animals interact with their physical and chemical environments modified for the production of food, fuel, and fiber for human consumption and processing. Agroecosystems are more open than natural systems (forests).

They rely on the external inputs of energy and nutrients, and most of the production is removed out of the system. In general, agroecosystems are not self-sustaining and are driven by population, market, and policy needs and regulated by environmental feedback mechanisms. Agroecosystems are the basic unit of agroecology studies. Specifically, agroecology refers to the science of understanding the ecology of agroecosystems. However, in recent times agroecology has denoted an integrated discipline that includes elements of agronomy, ecology, sociology, and economics.

Agroecosystem dynamics can be studied using key properties such as productivity, sustainability, stability, equitability, and autonomy. These are also called *system properties* or *emergent properties*. The term *productivity* relates to the quantity of food, fuel, or fiber that an agroecosystem can produce for human use. The term *stability* refers to consistency of production, while *sustainability* refers to the maintenance of a specified level of production over the long term. *Equitability* refers to sharing agricultural production fairly, whereas *autonomy* refers to agroecosystem self-sufficiency. These properties can be used to compare the performance of agroecosystems in diverse landscapes and at varied spatial scales.

However, assessing agroecosystem properties is more complicated as the properties themselves have multiple dimensions and meanings. The definitions also vary based on the scale at which the properties are analyzed. For example, the productivity of a shifting cultivation field may be high per unit area of land on the cultivated field itself, but the productivity may be low in terms of the total land area occupied by the entire shifting cultivation cycle including forest fallow.

STRUCTURE & FUNCTION

Significant structural and functional differences in agroecosystems versus natural ecosystems have been noted by several researchers. The structural composition of an agroecosystem varies based on the location and scale at which it is studied. Agroecosystem structure can also change over a period of time due to management practices. The structure of agroecosystems can be perceived in terms of hierarchy theory.

An entomologist and farm manager work together to eradicate pests on this agroecosystem, an apple orchard in Washington State. Image by USDA Agricultural Research Service.

Agroecosystems may be thought of as occurring in nested hierarchies, that is, agricultural farms nest in communities, which are nested in watersheds, which are further nested in larger regions, and so on, up to the globe. Each level in the nested hierarchy has its own unique properties (ecological, cultural, social, or economic features) and contributes to the nature of, and is affected by, levels above and below it. Across each level, agroecosystems may exhibit different interactions. For example, at a farm level, the input elements of agroecosystems include land, labor, and capital, which are interlinked together and interact with external attributes such as markets, policies, etc. To understand the complex interactions among different levels of agroecosystems, researchers have identified three different dimensions, that is, environmental (for example, soil and other biophysical attributes), economic (for example, market) and human dimensions (for example, rural community). The division of agroecosystems into individual dimensions such as those described above facilitates understanding in a more systematic manner.

In comparison to natural ecosystems such as forests, conventional agricultural ecosystems exhibit a simplified, trophic structure due to few selected crops or animal types. Agroecosystems consist of three intermingled and strongly interacting subsystems: the managed fields, referred to as the productive subsystem; the seminatural or natural habitats surrounding them; and the human subsystem composed of settlements and infrastructures. Agroecosystem function is a consequence of agroecosystem structure, which can vary widely based on location and management practices. Agroecosystem functions involve movement of material, energy, and information from one agroecosystem to another or movement in and out of the agroecosystem. Compared to natural ecosystems, agroecosystems are characterized by relatively high net productivity, simple trophic chains, low genetic and species diversity, low habitat heterogeneity, etc. Further, the multiple facets of agroecosystems and the services they offer have been well recognized. Agroecosystems offer a variety of functions. The functions refer to how agroecosystems operate in generating different services, which include production services, ecosystem regulation, and cultural services. Furthermore, these ecosystem services are strongly dependent upon the management practices.

The production services of agroecosystems include food, fuel, and fiber. Agroecosystems also serve as repositories of agrobiodiversity and genetic resources. The ecosystem services include providing habitat for biodiversity, primary production, biogeochemical cycling, soil formation and retention, carbon sequestration, water cycling, energy flow, etc. Regulation services include soil erosion control, climate mitigation, water purification, pollinating mechanisms, etc. The cultural services include agroecosystems serving as knowledge systems for education and inspiration, spiritual and religious values, recreation/aesthetic importance, etc.

In contrast to the above-mentioned positive ecosystem services, agroecosystems can also impact the environment in a negative way depending on management practices such as nutrient runoff due to excess nutrient application, greenhouse gas emissions, sedimentation, pesticide poisoning of humans, etc.

LIVESTOCK

Livestock are an important component of traditional agroecosystems and are considered a secondary food production system. Livestock provide draught power for farm operations including plough operations. Livestock are also the source of organic manure for crops and the main source of protein for human beings through the supply of meat and milk. In addition, livestock provide a source of income through the sale of animals. Compared to the specialized agroecosystems, such as single-cropping production systems or livestock-alone systems, integrated crop-livestock agroecosystems are considered highly beneficial in terms of agricultural productivity, environmental quality, operational efficiency, and economic performance. The synergies brought about by integrating crop and livestock systems can result in a positive feedback. For example, the crop residues that are left after harvesting can serve as a feed, while the livestock excreta can serve as manure to soils.

Integrated crop-livestock agroecosystems have also been associated with positive environmental effects. Increased organic carbon and nitrogen in soils of crop-livestock systems due to livestock manure have been associated with greater aggregate stability of soils, thereby conferring resistance to soil erosion. Another improvement obtained by diversifying solely from cropping systems or livestock systems and choosing integrated cropping-livestock systems is mitigation of losses from disturbances in either of the individual systems. In essence, crop-livestock agroecosystems can achieve both profitability and environmental benefits.

AGROECOSYSTEM HEALTH

The concept of agroecosystem health has been thoroughly debated in literature. Some researchers focus on symptoms of ecosystem stress, whereas others focus on ecosystem organization and change in relation to stress. In recent times, the health concept is generally judged to be well suited to describe the state of agroecosystems given the many ecosystem services that agroecosystems provide and their importance to human livelihood. In general, agroecosystem health has been closely linked to agroecosystem sustainability. Researchers preferring the agroecosystem health concept argue that sustainability is a broader concept while agroecosystem health is a more focused term. To describe agroecosystem health, different indicators have been proposed by earlier researchers, which vary from the farm to the landscape level. For example, agroecosystem health might be evaluated on the basis of structural criteria (resource availability, accessibility, diversity, equitability, and equity), functional criteria (productivity, efficiency, effectiveness), organizational criteria (integrity/coherence, self-organization, autonomy, self-dependence/self-reliance), or dynamic criteria (stability, resilience, capacity, and time to respond).

Alternatively, agroecosystem health evaluation could integrate different ecological processes such as cybernetics, water cycle, mineral cycle, and community dynamics in addition to human and biophysical phenomena and processes at different spatial scales. Several authors tried to quantify agroecosystem health, some using a conceptual framework, others using driving forces, and others by integrating biophysical and socioeconomic datasets using spatial tools and a multicriteria decision-making framework.

SCALE

There is a significant need to understand scaling aspects in agroecosystems. Agroecosystems exist and function at plot level (less than a hectare) to landscape scale (many square kilometers). The information that is generated at the plot or field level cannot be generalized to the regional, national, or global level. While most of the decisions made by policy makers are at a regional or national scale, those decisions may not be congenial to implement at the plot or farm level. Thus, interdisciplinary approaches are needed to link plot level data and extrapolate it to landscape scale. To implement any such approaches, the data requirements for analyzing agroecosystem properties and functions at multiple scales can vary. Extensive georeferenced data might be needed to quantify some of the agroecosystem properties, and such data are rapidly becoming available. With advances in remote sensing technology and availability of improved satellite data, crop-type mapping, including field size, is plausible. Further, geospatial information technologies facilitate storage analysis, retrieval, and display of spatial and nonspatial information useful for agroecosystem research. Both local and federal government agencies are rapidly developing geospatial technologies to establish databases and make them available to the scientific community. Therefore, characterizing agroecosystem properties and functions at different scales will become more feasible over time.

SUSTAINABLE AGROECOSYSTEMS

There is a growing recognition in the scientific community of the importance of adopting more sustainable and integrated practices for agricultural production in diverse landscapes that depend less on chemical and energy-based inputs. Relating to the same, the term *sustainable agriculture* is most commonly used to synthesize a variety of concepts and perspectives associated with agricultural practices that differ from conventional production. Conventional agricultural production relies heavily on external inputs. For several years the conventional practices emphasized short-term economic and production gains at the expense of long-term economic, environmental, and community interests. For improved agricultural management, the adoption of more holistic systems-based approaches, such as permaculture, have been proposed as an alternate path. The goal is to develop sustainable agroecosystems that have high production efficiency, economic viability, environmental compatibility, and social acceptability with less impact to the environment. Although designing such agroecosystems is highly challenging, efforts are underway through farming systems research.

Studies suggest organic farming as highly sustainable in the long term. Organic farming relies on management practices that enhance organic matter in the soil and limits the use of synthetic fertilizers, pesticides, plant growth regulators, livestock antibiotics, or genetically modified organisms. It also relies on improved crop rotations such as involving legumes, green manure, compost, and biological pest control. As defined by N. H. Lampkin in 1994, the aim of organic farming is "to create integrated, humane, environmentally and economically sustainable production systems, which maximize reliance on farm-derived renewable resources and the management of ecological and biological processes and interactions, so as to provide acceptable levels of crop, livestock and human nutrition, protection from pests and disease, and an appropriate return to the human and other resources."

Though organic agriculture is closely linked to sustainability, it can also have negative impacts to the environment such as the leaching of nitrates through legumes, or volatilization of ammonia from livestock waste. However, several researchers agree that the negative impacts caused by organic farming are low compared to conventional systems.

Four different attributes are most commonly used to assess agroecosystem sustainability in farming systems. They include diversity, cycling, stability, and capacity. Maintaining diversity in agroecosystems is important for risk minimization, genetic conservation, efficient resource use, and biological pest control. *Cycling* refers to effective flows of nutrients. Open nutrient cycles lead to losses, whereas closed cycles ensure that nutrients remain in the system. Thus, agroecosystems having closed or tighter nutrient cycles are better performing systems and are much more sustainable. *Stability* is also called *homeostasis* or *resilience* by different researchers. These terms refer to low variability or resistance to change and the ability of a system to maintain its productivity when subject to disturbances. *Capacity* refers to quality of the soil and water resource base and their ability to produce and sustain biomass. These attributes together with surrogate variables can be used to quantify the sustainability of farming systems.

Economic development and sustainability of agroecosystems will depend largely on our ability to manage these agroecosystems in a systematic manner. Maintaining crop diversity in agroecosystems is considered one of the important factors contributing to ecological sustainability. However, crop diversity can contribute to ecological sustainability only if the different crops fill various functions (for example, nitrogen fixation, production of organic matter to maintain soil quality, and provision of ground cover to prevent erosion) necessary for maintaining a productive agroecosystem. In addition, a diverse cropping system can reinforce monetary productivity and stability by allowing farmers to adjust the areas they plant to different crops in response to market opportunities.

AGROECOSYSTEMS & ROLE IN CLIMATE CHANGE

Climate change is mostly attributed to the increase in greenhouse gases in the atmosphere. The changes in greenhouse gas concentrations are projected to lead to regional and global changes in climate and related parameters such as temperature, precipitation, or soil moisture. These changes can also affect agroecosystems and their productivity. Further, agroecosystems act as both sources as well as sinks for greenhouse gases.

As of 2013, agriculture itself accounted for more than one-tenth of greenhouse gas emissions through the release of carbon dioxide (CO_2), methane (CH_4), and nitrous oxide (N_2O). Thus, agroecosystems have a major role to play in the mitigation of these gases. CO_2 is mostly released through deforestation for agriculture expansion. CO_2 is also released during the burning of agricultural crop waste, for example, rice or wheat residues. In Asia, agricultural residue burning is common as it is an easy way to dispose of residues. Fossil fuel combustion during agricultural management practices also releases CO_2. One of the largest sources of CH_4 from agriculture is through paddy fields. Because the paddy fields during most of their cropping period are submerged in water, CH_4 is released due to anaerobic decomposition. Other sources of CH_4 include animal husbandry (particularly of cattle and other ruminants), crop residue burning, and food waste. CH_4 persists for a shorter time in the atmosphere than CO_2 but traps at least twenty-eight times as much heat over a hundred years. Most of the N_2O emissions from agricultural systems are from nitrogen fertilizers, leguminous crops, animal waste, and agricultural residue burning. N_2O traps more than 250 times as much heat in the atmosphere as CO_2 does. Most of the above-mentioned greenhouse gas emissions can be mitigated through management practices both in the crop and livestock systems, such as breeding low-emission ruminants, changing animal diets to reduce flatulence, and shifting patterns of flooding and drying for rice paddies.

Agricultural soils can also act as a CO_2 sink mechanism. "Carbon sequestration" refers to removal of CO_2 from the atmosphere into a long-lived stable form, which can be beneath the soils. Soil carbon sequestration is considered an effective tool for mitigating CO_2. The most common management practices that are followed in the cropping systems for soil carbon sequestration include improved tillage practices, reduced cropping intensity, and incorporating organic inputs through fertilization. Tillage and soil carbon are negatively related as soil tillage accelerates organic carbon oxidation releasing high amounts of CO_2 to the atmosphere. Thus, management practices that reduce tillage result in increased soil carbon. Cropping intensity may positively enhance soil carbon as more biomass is incorporated into the soil through crop residues.

Carbon sequestration can also be maximized through fertilization options such as organic compost, livestock manure, etc., and other organic amendments.

One of the most important challenges for agroecosystems will be adapting to future climate change and unfavorable weather conditions. Developing cultivars that require either longer or shorter growing seasons based on location, that can tolerate drought conditions and heat resistance, and, most important, that can produce sufficient yields for a growing global population will be the key adaptation strategy.

Greater emphasis may also be placed on the moisture-conserving managing practices such as minimum tillage, conservation tillage, or no tillage as climate change predictions in different regions suggest moisture shortage. Practices such as intercropping, multicropping, and relay cropping may be beneficial in terms of achieving overall production per unit area and soil moisture conservation. The importance of no-till agriculture is gaining significance. It is a way of growing crops without disturbing the soil through a tillage mechanism. This practice has been shown to increase the amount of water and organic matter in the soil, increase nutrient retention, and reduce soil erosion. Cover-cropping is another option that has been shown to enhance soil fertility and soil quality, suppress weeds, and help control pests and diseases. In essence, the key for the future will be to design and manage agroecosystems that have high production efficiency as well as being economically and environmentally efficient over long-term periods.

FURTHER READING

Agroecology Knowledge Hub, Food and Agriculture Organization of the United Nations, 2018, www.fao.org/agroecology/en. Accessed 9 July 2018.

Ferguson, Rafter Sass, Sarah Taylor Lovell. "Permaculture for Agroecology: Design, Movement, Practice, and Worldview: A Review." *Agronomy for Sustainable Development*, vol. 34, no. 2, 2014, pp. 251–74, doi.org/10.1007/s13593-013-0181-6. Accessed 9 July 2018.

Frank, Stefan, et al. "Reducing Greenhouse Gas Emissions in Agriculture without Compromising Food Security?" *Environmental Research Letters*, vol. 12, no. 10, 2017, pp. 1–14. *IOP Publishing*, doi:10.1088/1748-9326/aa8c83. Accessed 9 July 2018.

Gewin, Virginia, and Patrick Monahan. "As U.S. Moves to Cut Greenhouse Emissions from Farms, New Study Finds Big Global Challenge." *Science*, 19 May 2016, doi:10.1126/science.aag0007. Accessed 9 July 2018.

"Global Emissions." *C2ES*, Center for Climate and Energy Solutions, www.c2es.org/content/international-emissions. Accessed 9 July 2018.

Pearce, Fred. "The Greenhouse Gas Riddle: What Is Causing the Recent Rise in Methane Emissions?" *Yale Environment 360*, Yale School of Forestry and Environmental Studies, 25 Oct. 2016, e360.yale.edu/features/methane_riddle_what_is_causing_the_rise_in_emissions. Accessed 9 July 2018.

AGROFORESTRY

—Jordan M. Marshall, PhD

FIELDS OF STUDY

Agronomy; Forestry; Environmental Studies; Soil Mechanics.

SUMMARY

Agroforestry is an interdisciplinary scientific field focused on combining and applying forestry and agricultural principles and techniques to create a sustainable land-use system. Agroforestry provides a wide range of ecological benefits, including improvements to water quality and wildlife habitat and reductions in soil erosion. By combining trees and shrubs with crops and livestock, agroforestry systems attempt to optimize the benefits of short-term crop and livestock rotations and long-term forest rotations. Blending of production practices, which has occurred for centuries, allows a landowner to reap the economic benefits of annual crops while waiting for forest products. In addition to economic benefits, agroforestry results in ecological benefits that may not occur in a traditional agricultural system, including the formation of windbreaks, wildlife corridors, and riparian buffers.

DEFINITION & BASIC PRINCIPLES

In many situations, land is managed using a single-approach (or monoculture) system, such as a farm devoted solely to raising crops or livestock or a forest that is cultivated for pulp or lumber production. These separate approaches can be managed successfully to produce a sustainable product. However, through the application of key forestry and agricultural techniques, agroforestry practitioners attempt to develop a sustainable land-use practice producing both short-term and long-term benefits. Although many combinations of forest- and crop-livestock-management techniques exist, agroforestry typically takes the forms categorized as tree-crop, tree-animal, shelterbelt, riparian forest, and natural-specialty crop systems.

Agroforestry provides the framework for a systems approach to land management, allowing for understanding how the physical structure, ecological influences, and economic outputs of trees and forests are interconnected with those same characteristics in crops and livestock. In many climatic regions, from tropical to semiarid zones, constant cropping can place a drain on soil nutrients, eventually creating infertile land. By incorporating fallow periods in the crop rotation, farmers can replenish these soil nutrients. However, during the fallow period no crops are produced, and depending on local policies and food-production demand, the fallow period may be excessively shortened and no longer serve its purpose. The goal of agroforestry is to incorporate deeper-rooted perennials (trees and shrubs) in order to maintain nutrient cycling and soil fertility so agriculture can continue for longer periods of time.

Parkland in Burkina Faso: sorghum grown under Faidherbia albida and Borassus akeassii near Banfora. Image by Marco Schmidt (Own work) CC-BY-SA-2.5.

BACKGROUND & HISTORY

Agroforestry as a technique for land management has been used for centuries, but it was not until the mid-1970s that the term was coined and defined. Early adoptions of what would later be defined as agroforestry probably incorporated the production of crops, animals, and building materials. Although there was a subsequent shift to farming and forestry with a focus on monocultures, agroforestry continued to exist. In later decades, agroforestry has become an important approach to land management in developing regions of the world. Focusing on sustainable land-management practices that allow for increases in economic and ecological diversity has facilitated survival and development of regions where heavy forestry or heavy agriculture could have destroyed resources. Because of the large economic and ecological impact on local communities, government and nongovernmental organizations have expanded research since the 1990s to refine agroforestry practices and disseminate information to the public.

HOW IT WORKS

Practitioners cannot assume that all forms of agroforestry will succeed in all environments. The surrounding ecological and economic environment, previous land-use practices, and goals of the manager, owner, and community all play into the success of any one of the numerous and overlapping agroforestry strategies. Although the techniques used may be structurally different, the outputs are the same: Establish an ecologically and economically sustainable system that provides short- and long-term products in the forms of agricultural crops and livestock, as well as tree and non-tree forestry products. The five common agroforestry systems (tree-crop, tree-animal, shelterbelt, riparian forests, and natural-specialty crops) are broad, and there is overlap between each one. The application of a specific system is typically related to the ecological and economic environment surrounding an individual owner and piece of land, which includes the selection of particular crop and tree species. Because of the broad nature of these systems, the individual goals defined by the practitioner may separate or merge the agroforestry practice being conducted, resulting in several agroforestry systems existing on a single piece of land.

Adding Trees to Agriculture. Two important decisions a practitioner must make regarding the incorporation of trees into an agricultural operation concern are spacing and timing of tree growth. First, trees can be grown in zones or scattered across the land. Zones can be in lines, plots, blocks, or any other systematic arrangement on the land. Spacing between trees is a powerful management tool, in both zoned and scattered arrangements, because it determines the level and intensity of competition and facilitates management. If the trees are growing too close together, growth may be reduced as a result of competition for moisture and nutrients, as well as issues related to crop planting, fertilizing, and harvesting by machinery.

Second, timing can influence competition between trees and crops, as well as labor costs for planting and harvesting. If trees and crops are grown sequentially, direct competition between the two groups of plants can be greatly reduced. However, there may be increases to required labor and associated costs by staggering the land-management techniques. If trees and crops are grown simultaneously, then direct competition between the two groups of plants may be increased, but labor costs may be reduced because of the combined management.

Adding Crops and Livestock to Silviculture. As with adding trees to establish agriculture, the same issues of spacing and timing exist when adding crops to a forest. With application of agroforestry techniques to land under more forest-focused management, the overall goals shift from long-term forest management to include short-term crop rotations. One different strategy of adding crops to silviculture is called *taungya*, which is a Burmese term for growing crops during the first few years of tree-plantation establishment. The goal of taungya is to gain short-term crop benefits during the early stages of tree establishment. After one to three years, the crops are no longer sown and the focus is turned to the management of the trees. Whether cropping ends after a few years or is continued indefinitely, there is a need to space trees effectively so that the future plantation can successfully reach the long-term goals defined at the outset.

APPLICATIONS & PRODUCTS

Tree-Crop Systems. Planting trees in rows, or any spacing configuration that allows for sowing of crops between the trees, is known as intercropping, alley-cropping, and agrosilviculture. Often, the spacing is arranged to allow modern agricultural machinery use and limit the need for physical labor. Two main approaches exist in tree-crop systems, which include planting trees on cropped land and planting crops on forested land. The former typically results in planting straight rows of trees while the latter may result in establishing tree plantations or utilizing natural forests, which may add difficulty in properly spacing trees. In either approach, there is a need for practitioners to make management decisions related to the pattern of trees on the land (zoned or scattered), spacing (between zones, between individual trees), and diversity of tree species. Crops may be temporarily sown during the establishment phase of a forest, which can have an added complication of multiple owners between trees and crops. Tree-crop systems are common in both temperate and tropical zones and have been used for fruit, nut, olive, and grape production for centuries in Europe and North and South America.

Tree-Animal Systems. Silvopasture is the management of livestock, their forage, and trees. If a pasture has trees, but those trees are not part of the active land management, then it is not agroforestry. The application of tree-animal agroforestry may occur as an agriculture-dominated, integrated, or forestry-dominated approach. In an agriculture-dominated approach, the focus is placed on maximizing the livestock and food-product outputs. This approach may occur when trees are planted in a pasture to provide shelter for livestock or when livestock are allowed to graze in an orchard. With an integrated approach, the focus is balanced between the management of livestock outputs and tree-product outputs. In temperate areas, such as the southeastern United States, this approach is used for the management of pine plantations with grazing for cattle.

Forestry-dominated approaches place management emphasis on tree outputs and utilizing livestock to maximize those tree products. In tropical areas where tree growth can be rapid, introduction of livestock may be beneficial to the trees. However, in temperate areas, tree growth may be slower. If livestock are introduced too early, there may be grazing and damage to the trees, reducing survival and success of the tree-animal system.

Shelterbelt Systems. When planted upwind from croplands, windbreaks provide linear shelter that reduces soil erosion. These windbreaks may include mixtures of trees and shrubs. A single row of trees does not provide a solid wall of protection, so it is necessary for multiple rows to be planted in order to create a stratified, continuous canopy to block wind. To reduce excess runoff and soil erosion related to water flow, contour hedgerows are planted along contours of moderate to steep slopes and can slow water movement. Contour hedgerows can be a less costly and less time-consuming strategy for reducing water erosion than the construction of terraces along a slope. Depending on strategies and goals for the trees, shelterbelts can function merely as filter strips, reducing the volume of soil erosion, or they can function as alley crops, providing mulch or tree crops mixed in with the annual row crops.

Riparian Forest Systems. Riparian forests are stands of trees adjacent to a linear, flowing body of water (a river or stream). The soils of these forests are either continuously saturated or repeatedly inundated with water. Hydrology and soil types play very important roles in determining which trees can survive and be productive. Riparian forest systems can benefit more than just the immediate community of landowners, as compared with tree-crop or tree-animal systems. Forested zones adjacent to rivers and streams can maintain water quality by filtering agricultural chemicals in runoff through infiltration and immobilization by plants and soil microbes; stabilize banks and reduce erosion by creating a soil-retaining structure with roots and coarse woody debris, thereby improving water infiltration, which reduces the quantity of runoff and the erosive energy; and reduce average and fluctuations of water temperature, maintaining fish and other aquatic-organism populations. Riparian forest systems benefit both the land adjacent to the river and adjacent land downstream.

Natural-Specialty Crop Systems. Forest farming takes advantage of the shade produced by a forest canopy to produce special, high-value non-timber forest products. In temperate agroforestry, these products include fruits, nuts, and mushrooms as food crops; ginseng, catnip, and echinacea as herbal medicines; and ferns and flowers for decorative and ornamental uses. Often, these specialty crops are added as products to forests where high-value timber is being produced in order to diversify the economic outputs.

SOCIAL CONTEXT & FUTURE PROSPECTS

Implementation of agroforestry techniques has improved the monetary, social, and ecological economies of local villages and regions in many tropical and temperate areas. However, in order to garner those benefits at the national level there is a need to scale up the techniques and practices used in agroforestry. To expand agroforestry and its benefits, national governmental agencies, as well as multinational non-government organizations, must take an active role in the dissemination of information and support to farmers. This has been demonstrated in several countries, including Cameroon, Indonesia, and Uganda, where increasing involvement of farmers in agroforestry programs has significantly improved the quality of products and the return on farmer investments. In each case, cooperative efforts by governmental agencies, as well as international relief organizations and other nongovernmental organizations, has expanded the application of agroforestry techniques to magnify the geographic scope of monetary, social, and ecological benefits.

FURTHER READING

Ashton, Mark S., and Matthew J. Kelty. *The Practice of Silviculture: Applied Forest Ecology*. 10th ed. Hoboken: Wiley, 2017.

Gordon, Andrew M., and Steven M. Newman, eds. *Temperate Agroforestry Systems*. 2nd ed. Wallingford, England: CAB International, 2015.

Huxley, Peter A. *Tropical Agroforestry*. Hoboken, N.J.: Wiley-Blackwell, 1999.

Jose, Shibu. "Agroforestry for Ecosystem Services and Environmental Benefits: An Overview." *Agroforestry Systems* 76, no. 1 (May, 2009): 1–10.

Shepard, Mark. *Restoration Agriculture: Real-World Permaculture for Farmers*. Austin: Acres USA, 2013.

SIDEBAR: AGROFORESTRY: FASCINATING FACTS ABOUT AGROFORESTRY

- More than 1 billion hectares of the world's agricultural land (about 46 percent) has greater than 10 percent tree cover.
- Researchers estimate that 17 percent of agricultural land managed using agroforestry methods involves tree cover great then 30 percent. It is also estimated that 46 percent of agricultural land has tree cover of 10 percent or more.
- Agroforestry is a major land-management strategy in South America and Southeast Asia, where it accounts for more than 80 percent of the agricultural land in each of these two regions.
- Planting of leguminous trees with maize provides necessary fertilization of soils in sub-Saharan Africa resulting in tripling of yields. The trees fix atmospheric nitrogen, making it available in the soil for maize. The trees' limbs are pruned creating mulch for soil protection, and the wood is used as fuel. Without the use of trees, typically a third of the area planted in maize is abandoned.

Agronomy

—Julia A. Rosenthal, BA

FIELDS OF STUDY

Agronomy; Environmental Studies; Soil Mechanics.

SUMMARY

Agronomy is the interdisciplinary field in which plant and soil sciences are applied to the production of crops. Agronomists develop ways in which crop yields can be increased and their quality improved. Some agronomists specialize in soil management and land use, which seeks to protect existing farmland and reclaim land for future use in growing crops. Other specialties cover areas such as weed and pest management, meteorology, and the impact of climate change on crop production. The growing importance of biofuels such as ethanol has increased interest in agronomy as a scientific and professional field.

DEFINITION & BASIC PRINCIPLES

Agronomy is the study of plants grown as crops for food, animal feed, and nonfood uses such as energy. In the United States, these crops include wheat, corn, soybeans, grasses, cotton, and a wide variety of fruits and vegetables. Leading crops in other countries vary widely, depending on the nature of the local soil, geography, and growing season.

Plant science is a major component of agronomy. Many agronomists look for ways to grow stronger, hardier plants with higher yields. New types of plants are bred by agronomists to contain specific improvements, such as increased nutrient levels or resistance to pests, over earlier breeds. An area of strong interest is the development of plant types that require fewer inputs such as fertilizers and insecticides to perform well.

The field of agronomy also covers the many factors in the environment that play a role in whether a crop succeeds or fails. The chemical makeup and water balance within a crop's soil are leading factors. Weather and climate patterns, both within a single season and over many years, affect the quantity and quality of crop yields. Technology and economics influence demand for certain types of crops, which in turn pushes market prices up and down. Agronomists help producers respond to these factors.

BACKGROUND & HISTORY

Agronomy is nearly as old as human civilization. According to archaeological findings, people have been growing plants for food for more than 10,000 years, starting in the western Asian regions of what was Mesopotamia and the Levant.

An agronomist field sampling a trial plot of flax. Image by Iowa State Department of Agronomy CC-BY-SA-3.0.

Many historians believe that plant cultivation, the earliest form of farming, led to a major change in human culture. The growing season required people to live in one place for long periods of time. Permanent settlements near fields most likely evolved into some of the first villages. These settlements were often near water sources such as rivers, which were needed to irrigate field crops. Some of the first developments in agronomy involved the design and building of water-delivery systems.

The Industrial Revolution brought widespread change to the field of agronomy. Steam-powered farming equipment replaced draft animals such as horses and mules. Plant scientists developed and standardized new breeds of field crops, which increased yields. By the mid-twentieth century, nearly all corn grown in the United States was from hybrid stock.

The use of inputs such as fertilizers and pesticides also increased but, in some cases, caused significant environmental harm. Since the 1990s, agronomists have focused more closely on ways to improve crops without damaging local ecosystems.

HOW IT WORKS

Field crops require the right plant type and breed, healthy soil, adequate water and nutrients, appropriate growing temperatures and rainfall, and the control of disease and pests in order to succeed.

Plant Breeding and Genetics. When choosing a type of field crop to plant, farmers and growers consider factors such as the hardiness of certain breeds and their expected yields at the end of the harvest season. Buyers of agricultural products such as food-manufacturing companies look for products that are high in quality and contain specific nutritional or chemical properties. Agronomists who specialize in plant breeding and genetics support the needs of both farms and buyers.

Multiple methods are used to create hybrid plants. Some hybrid strains are created by planting one breed next to another and allowing the two breeds to cross-pollinate. Plant scientists also use in vitro techniques, in which plant tissues are combined in a laboratory setting to create strains that would not occur in nature. One technique that has received significant media attention is genetic modification. Genetically modified plants contain genes introduced directly from other sources that create changes in the plant much more quickly than could be generated through traditional breeding.

Soil Health. To support a crop with the highest possible yields, the soil in which the seeds are planted must be in good condition and must match the needs of the particular plant breed. The health of soil can be measured on the basis of its physical properties, its chemical makeup, and the biological material it contains. These qualities are tracked by soil surveys that are conducted and published regularly. Farmers may switch the types of crops they grow in a particular field every few years in order to keep certain soil nutrients from being depleted. This practice is known as crop rotation.

Hydration and Irrigation. Field crops require vast amounts of water. When sources such as rainfall do not provide enough water for healthy plant growth, hydration must be supplemented by irrigation. Irrigation systems are often based on networks of pipes or hoses connected to sprinklers or drip mechanisms that can supply water to an entire field.

Weather and Climate. Even when a hardy breed of plant is grown in healthy soil and receives enough water, an entire crop can be damaged or destroyed by unexpected weather patterns. Many farmers and growers protect themselves against weather-related risks by purchasing crop insurance, which covers losses in situations such as storms or early freezes. While climate has less variance for individual farmers on a season-by-season basis, changes in climate over the long term can affect the types of plants that grow successfully in a given area. In some cases, climate change can increase or decrease the amount of land suited for growing crops.

Pathology and Pest Control. Like any living organism, field crops are susceptible to natural threats such as disease, predators, and competition from other plants. Agronomists specializing in plant pathology look for ways to fight disease through direct treatment as well as the breeding of new, hardier strains for future crops. Pests such as insects are controlled through the application of inputs such as insecticides. Pest control also influences plant genetics, as in the case of cotton bred to contain a natural compound toxic to boll weevils. Weeds are managed through a combination of herbicides, adjustments to soil properties such as adding or removing water, and the development of plant breeds resistant to weeds.

APPLICATIONS & PRODUCTS

Agronomic crops can be broken down into categories in a number of different ways, such as plant type or climate in which the crop is most likely to be found. One of the most common ways in which field crops are grouped is by the end use of the raw material. Nearly all crops in the world can be considered a form of food, animal feed, fiber, energy, or tools for environmental preservation. Some major crops such as corn can be classified in multiple ways, as corn is used for feeding both people and animals and is also refined into ethanol.

Food. Food represents one of the most diverse categories of agronomic crops in the United States and worldwide. When people think of field crops and food, the types of plants that come to mind first are grains such as corn, wheat, and rice. When measured by acres of land planted, grains make up the largest share of agronomic crops grown throughout the world. This category also includes fruits, tree nuts, vegetables, plants grown to be refined into sugar and sweeteners (such as beets and sugarcane), and plants from which oil is made (such as soybeans). While tobacco is not considered a food product, tobacco crops are often included in this category because cigarettes and other items made from tobacco leaves can be consumed only once.

The demand for food crops worldwide grows only as fast as the global population. Individual types of food crops may face sharp increases and drops in demand, however. These changes are influenced by factors such as weather patterns and crop failures, prices set by the commodities markets (which, in turn, make the prices of consumer items rise and fall), and the changing tastes of food buyers. The spike in popular interest in low-carbohydrate diets in the late 1990s had an impact on crops used to make products such as flours and sugars. Similarly, populations in developing countries where incomes are rising often change their food-buying habits and choose items more prominent in North American and European diets than in local ones. This pattern can push up prices for crops such as corn and lower demand for locally grown fruits and vegetables. Interestingly, this trend also works in reverse. Consumers in affluent economies such as the United States have become more interested in buying produce from crops grown locally in an effort to reduce the overall impact on the environment. These kinds of changes can lead to rapid shifts in demand for individual types of agronomic crops.

Animal Feed. As with food, the category of agronomic crops grown for animal feed is dominated by grains. In the United States, the leading feed grains tracked by the U.S. Department of Agriculture (USDA) are corn, sorghum, barley, and oats. Corn makes up the largest share of this category by volume. The USDA World Agricultural Outlook Board estimates that the 2013–14 growing season yielded a total of nearly 13.8 billion bushels of corn, much of which was used for animal feed. The combined yield for sorghum, barley, and oats was estimated at fewer than 700 million bushels in total.

A second type of animal-feed crop is hay. Much of the hay grown and harvested in the United States comes from alfalfa plants or a mixture of grass types. Demand for hay is influenced in part by the weather. Farmers feed more hay to their livestock—particularly cattle—in drier conditions. Hay is often grouped by the USDA with a type of crop known as silage. Silage is not a unique plant, but rather the plant stalks and leaves left after the harvesting and processing of grains such as corn and sorghum. Hay and silage together belong to a category of crops known as forage. While forage was once defined as plant matter eaten by livestock grazing in fields, it has been expanded to include plants that are cut, dried, and brought to the animals.

Fiber. Plants grown for nonfood use often fall into the category of fiber. Fiber crops are processed for use in making cloth, rope, paper and packaging, and composite materials, such as insulation for homes. In the United States, most of the yearly agronomic fiber crop is made up of cotton plants. In 2017, the United States was the third-largest grower of cotton in the world, according to the USDA. Other plants in this category are jute, sisal, and flax. Jute and sisal are frequently used to make rope, burlap, and rugs. Flax is refined into linen and used in a wide variety of applications ranging from fine clothing and home-decorating products to high-grade papers. Flax fiber is also used in making rope and burlap.

Energy and Environmental Preservation. Most of the agronomic crops in these categories also appear in one of the three categories above. Of the crops grown in the United States as sources of bioenergy, corn tops the list as a source of ethanol. Ethanol sources in other countries include crops such as sugarcane and grasses. Vegetable-based oils made from soybeans are blended into diesel fuel to make a composite known as biodiesel. The refining processes for turning crops into bioenergy sources are not yet as cost-effective as fossil fuels such as petroleum and coal, but this situation may change as technologies improve.

Environmental preservation from the standpoint of agronomic crops is a broad and developing category. It includes the strategic planting and rotating of crops to return depleted nutrients to the soil. It also includes the growth of plants near fields to minimize soil runoff and to protect endangered areas such as wetlands.

SOCIAL CONTEXT & FUTURE PROSPECTS

One of the turning points in public consciousness about agronomy was the release of Rachel Carson's book *Silent Spring* in 1962. Carson's book linked a number of ecological problems, particularly the deaths of wild plants and birds, to the widespread use of pesticides such as DDT. Many of these pesticides were used on field crops. The book led to the United States ban on DDT in 1972 and increased public awareness of the potential environmental harm in certain farming practices.

Consumer interest in the quality of food sources has been on the rise since the 1990s. While there has been demand for products such as organic foods for much of the twentieth century, the category has grown most quickly at the beginning of the twenty-first century. Some consumers participate in community-supported agriculture (CSA) programs in which fruits, vegetables, dairy, and other items are delivered directly from local farmers. Gourmet and chain restaurants are more likely to advertise their use of locally grown and environmentally sound food ingredients. This demand extends to nonfood items ranging from organic cotton fibers in clothing and linens to plant-based, biodegradable home products. It has also influenced the growth of plant-based, nonpetroleum energy sources such as ethanol.

Agronomists are well positioned to benefit professionally from these trends. Upcoming issues of interest for agronomists are likely to include the impact of changing weather and climate patterns and the ways in which crop yields can be increased without causing environmental harm.

FURTHER READING

"About Agronomy." *American Society of Agronomy*, 2018, www.agronomy.org/about-agronomy. Accessed 15 Mar. 2018.

Carson, Rachel. *Silent Spring*. 1962. Reprint. New York: Houghton Mifflin, 2002.

Fageria, Nand Kumar, Virupax C. Baligar, and Charles Allan Jones. *Growth and Mineral Nutrition of Field Crops*. 3rd ed. Boca Raton, Fla.: CRC Press, 2011.

Kingsbury, Noel. *Hybrid: The History and Science of Plant Breeding*. Chicago: University of Chicago Press, 2009.

Miller, Fred P. "After 10,000 Years of Agriculture, Whither Agronomy?" *Agronomy Journal* 100, no. 1 (2007): 22–34. Available at https://www.agronomy.org/files/about-agronomy/future-of-agronomy.pdf.

Reed, Matthew. *Rebels for the Soil: The Rise of the Global Organic Food and Farming Movement*. London: Earthscan, 2010.

Vandermeer, John H. *The Ecology of Agroecosystems*. Sudbury, Mass.: Jones and Bartlett, 2011.

AGRO-TERRORISM

FIELDS OF STUDY

Agronomy; Politics; Environmental Studies; Soil Mechanics.

SUMMARY

Agro-terrorism is a term used to describe attacks on crops or livestock, rather than directly on human beings, that are designed to induce uncertainty about food safety and/or to cause economic damage to the agriculture industry. While closely related to some forms of biological warfare, agro-terrorism has been treated as a separate threat. As of June 2008 no confirmed cases of agro-terrorism had been demonstrated. Nevertheless the U.S. Agriculture Department and the Customs Service both identified the prospect of spreading plant and animal diseases as a distinct threat that could easily be carried out by terrorists from Al Qaeda or other groups. Government officials warned that a successful agro-terrorist attack could create havoc in the agriculture industry far beyond, and long after, the exact site of an assault because of damaged public confidence in the safety of meat and plants.

"Agro-terrorism" describes potential efforts to poison or destroy the food supply or agricultural production by introducing diseases affecting livestock or plants. The term gained currency after the Al Qaeda attacks of September 11, 2001, which caused the Department of Agriculture and the Customs Service to increase inspections and record-keeping to protect livestock and edible plants in the United States, as well as agricultural imports, from an assault by terrorists.

In agro-terrorism the immediate aim is to infect animals or plants, which in turn would then harm human beings as well as undermining public confidence in the food supply. Agro-terrorism differs from biological warfare where the aim is to make a direct assault on human beings, such as spreading anthrax spores among human beings. Under most theoretical scenarios for an agro-terrorist attack, perpetrators introduce an infectious disease into the food chain, particularly livestock, with the result that disease spreads rapidly. Analysts from the Agriculture Department and elsewhere have theorized that attacks might be directed at concentrations of animals, such as cattle or swine in feedlots or concentrations of poultry in a handful of states. Such an attack could be devastating since many infectious diseases, such as hoof-and-mouth disease, have been eradicated in the United States, leaving animals without antibodies to fight against an infection.

The prospect of agro-terrorism, and proposals to guard against attacks, gained particular political currency in agricultural regions following the 2001 attacks on the World Trade Center and the Pentagon.

In addressing the issue the government lists both food safety and the adverse economic impact as possible results. Economic impacts include both the value of destroyed crops or livestock, and the future loss of confidence in the safety of food, which might divert consumers to other foods (for example, away from beef products in cases of infected cattle). Some studies have included lost tourism earnings in the potential cost of agro-terrorism.

Anti-terrorism officials site as precedents naturally-occurring outbreaks of disease. In 1983–1984, for example, an outbreak of avian influenza in Pennsylvania and surrounding states cost $63 million for federal efforts to control the disease. The incident also drove up poultry prices for consumers by an estimated $350 million. After a case of mad cow disease was found in a single cow in the United States in 2003, American meat products were banned by dozens of countries. Five years later, long after any incidence of mad cow disease in the United States, proposals to ease such a ban were an issue in a mass anti-government demonstration in Seoul. Contamination with dioxin found in chickens and animal feed in Belgium in 1999 forced the resignation of the prime minister and the banning of most Belgian livestock in Europe and Asia at a cost to the economy of $1 billion.

Agro-terrorism is also relatively easy to carry out undetected. Pathogens, such as hoof-and-mouth disease, could easily be smuggled into the United States and spread by a terrorist by simply approaching animals grazing near a roadside fence. The natural spread of such diseases among herds-or even just a few animals—could have a severe impact on whole industries even without a single human being affected. Even if terrorists were apprehended, the penalties for killing animals are far less severe than for attacking human beings directly.

Countermeasures. Agro-terrorist acts are extremely difficult to prove since most outbreaks could also occur naturally. As of June 2008 only tiny number of attacks had been identified. Security experts have focused on four main areas to defend against agro-terrorism: (1) developing resistance by plants and animals to known diseases; (2) monitoring farms, including feedlots, and animals to enable authorities to track immediately any infected animal to its origins; (3) wholesale monitoring and prevention techniques applied to whole sectors of the agriculture industry; and (4) policies designed to minimize the impact of an agro-terrorist attack.

In 2005 the Agriculture Department proposed the National Animal Identification System as a means of tracing a sick animal to the farm where it originated within 48 hours. The system assigned a unique 15-digit identification for every head of cattle-or, in the cases of pigs and chickens, an ID for groups of animals-as well as an identification number for each farm where they are raised. The tags were able to be scanned so that animals could be traced each step of the way from farm to slaughterhouse. In cases of outbreaks of a contagious disease, such as "mad cow disease," the government could quickly trace a stricken animal to its origins. The program was initially proposed to be mandatory when it was to be inaugurated in 2005. The proposal engendered significant opposition in the livestock industry, partly on grounds of cost and partly on grounds that it seemed to intrude on privacy. The Agriculture Department retreated and made it voluntary.

BACKGROUND

Although agro-terrorism became an issue in law enforcement and in the food industry after the 2001 attacks on the World Trade Center and the Pentagon. The term had been used before then, notably in cases of extortion or actions by groups opposed to genetically-modified plants. The term is also used to describe government (including the United States, Britain, Germany, and Japan) research and use of biological agents dating from World War I. The United States carried out research into biological weapons as recently as 1969. In 1972 the United States, Soviet Union, Britain, and Canada agreed to a Biological Weapons Convention that barred acquisition or use of biological weapons. The United States destroyed its stockpiles of such weapons by 1975.

During World War II the United States, Britain, Canada, the Soviet Union, and Japan all studied use of animal and plant diseases for offensive purposes. Since World War I, only ten instances of "agro-terrorism" have been recorded:

- 1915–1917: German spies tried to infect draft animals intended for use by Allied military with anthrax.
- 1943: A British naturalist alleged that Germany dropped cardboard bombs containing Colorado potato beetles. Other experts have challenged the assertion, claiming instead that the insect's appearance in England at the time was from natural causes.
- 1950: The East German government accused the United States of spreading Colorado potato beetles over its potato crops.
- 1952: Kenyan nationalists, the Mau Mau, poisoned 33 steers using a plant called African milk brush.
- 1962–1966: Cuba accused the United States of attacking its animal and plant populations on several occasions by spreading such diseases as Newcastle Disease (poultry), African swine fever (pigs), tobacco blue mold disease, and sugarcane rust disease. Other experts said outbreaks of these diseases were caused by natural events.
- 1982–1984: The Soviet Union was accused of spreading pathogens to kill horses used by anti-Soviet Mujahideen.
- 1983–1987: A Tamil rebel group threatened to attack crops of Sinhalese farmers in Sri Lanka; there was no evidence that the attack was carried out.
- 1984: The premier of Queensland state in Australia received two threats to infect wild pigs with foot-and-mouth disease unless prison reforms were implemented. One writer was later identified as a convicted murderer in prison; the second threat was not identified.
- 1996–1997: Cuba alleged that the United States used a crop duster airplane to spread the insect thrips palmi on Cuban fields. An investigation by the Biological and Toxin Weapons Convention into the complaint found the evidence to be inconclusive.
- 2000: Palestinians accused Israeli settlers of releasing sewer water into agricultural fields as a means of forcing Palestinian farmers off their land on the West Bank.

U.S. federal government responsibility for inspections to assure food safety dates from the early twentieth century. During the administration of President Theodore Roosevelt the Pure Food and Drug Act and the Meat Inspection Act established the principle of government oversight. Until 2001 the principal focus of government food inspection programs was detection of naturally occurring diseases. After the Al Qaeda attacks on New York and Washington in 2001, the focus came to include the deliberate introduction of chemical or biological agents into the food supply as a weapon of terror.

More information

Elisburg, Donald. "Review of Agroterrorism." (Book review of Agroterrorism; A Guide for First Responders; By Jason B. Moats, 2007.) *Journal of Homeland Security & Emergency Management* 5:1 (2008) 2p.

Foxell, Joseph W., Jr. "Current Trends in Agroterrorism (Antilivestock, Anticrop, and Antisoil Bioagricultural Terrorism) and Their Potential Impact on Food Security." *Studies in Conflict & Terrorism* 24:2 (April 2001) 23p.

Kohnen, Anne. "Responding to the Threat of Agroterrorism: Specific Recommendations for the United States Department of Agriculture." October 2000. PDF file online at http://ianrhome.unl.edu/inthenews/resources/bcsia%5Frecommendations.pdf

Monke, Jim. "Agroterrorism: Threats and Preparedness: RL32521." *Congressional Research Service*. Report. February 4, 2005. 51p.

Shea, Dana A. Jim Monke, and Frank Grottron. "The National Bio- and Agro-Defense Facility: Issues for Congress: RL34160." Congressional Research Service September 4, 2007. 16p.

Alternative Grains

—Thomas E. Hemmerly

FIELDS OF STUDY

Culinary Science; Chemistry; Bacteriology; Botany; Horticulture; Agriculture; Genomics; Geology; Microbiology.

DEFINITION

Various minor cereals, new cereals, and pseudocereals can be cultivated in climates that are not conducive to the growing of other high-yield grain crops, providing alternative food sources and avoiding the environmental damage caused by the fertilizers and pesticides used in large-scale intensive farming.

More than one-half of the calories consumed daily by the world's human population come from grains. Most of these grains are produced by plants of the grass family, Poaceae. Major cereal plants domesticated many centuries ago include rice (*Oryza sativa*), wheat (*Triticum aestivum*), and maize or corn (*Zea mays*). Other important grain crops, also plants of the grass family, include barley (originating in Asia), millets and sorghum (originating in Africa), and oats and rye (originating in Europe).

Since the early twentieth century, the scientific principles of genetics have been applied to improvements of crop plants, with some of the most notable improvements occurring between 1940 and 1970. As a result of irrigation, improved genetic varieties, and the use of large amounts of fertilizers and pesticides, yields of major crops greatly increased. Norman Borlaug received

Sorghum is a type of alternative grain. Image by Larry Rana, USDA.

the Nobel Peace Prize in 1970 for his contributions to these developments, which came to be called the Green Revolution. However, it soon became apparent that the Green Revolution was not the boon first envisioned. For maximum yield, large-scale farming involving huge investments of capital is required. Also, environmentalists became concerned over the resulting erosion and the environmental damage caused by the use of large amounts of fertilizer and pesticides.

Various alternatives to such farming have been proposed. In the case of grain crops, several approaches offer promise, including more widespread use of minor cereals, especially those tolerant of unfavorable growing conditions; development of new cereal plants through hybridization or other genetic manipulations; and utilization of pseudocereals, nongrass crop plants that produce fruits (grains) similar to those of cereal plants.

MINOR CEREALS & NEW CEREALS

Most sorghum (*Sorghum bicolor*) grown in the United States is used for silage (milo) or molasses (sweet sorghum). In Africa and India, various grain sorghums are grown in regions where rainfall is insufficient for most other grain crops. Well adapted to hot, dry climates, these grains are used to make a pancakelike bread. "Millet" refers to several grasses that are useful cereal plants because they also tolerate drought well. In Africa the most important are pearl millet (*Pennisetum glaucum*) and finger millet (*Eleusine coracana*). Grains of both species can be stored for long periods and are used to make breads and other foods. Other, perhaps less important, grain plants also called millet include foxtail millet (*Setaria italica*), native to India but now grown in China; proso millet (*Pamicum milaeceum*), native to China but grown in Russia and central Asia; sanwa millet (*Echinochloa frumentacea*), cultivated in East Asia; and teff (*Eragrostis teff*), an important food and forage plant of Ethiopia. Such grain sorghums and millets have the potential to grow in areas with hot, dry climates far beyond the regions where they are now being utilized.

In a distinct category is wild rice (*Zizania aquatica*). Native to the Great Lakes region of the United States and Canada, it has been, and still is, harvested by American Indians. Like the common (but unrelated) rice, it grows in flooded fields. Attempts to cultivate wild rice since the 1950's have been somewhat successful as the result of the development of nonshattering varieties. However, it remains an expensive, gourmet item.

Two cereal plants have promise because of the high protein content of their grains. Wild oat (*Avena sterilis*) is a disease-resistant plant with large grains. Job's tears (*Coix lachryma-jobi*), native to Asia, is now planted throughout the Tropics. Research on these and related species continues.

Although all important cereal plants have been improved, either by genetic engineering or by more conventional genetic techniques, the most notable new alternative grain plant is triticale (*Triticosecale* sp.). The first human-made cereal, it is the result of crossing wheat with rye. The sterile hybrid from such a cross was made fertile through the doubling of its chromosomes; thus triticale varieties produce viable seeds. Triticale combines the superior traits of each its parents: the cold tolerance of rye and the higher yield of wheat. The protein content of triticale compares favorably with that of wheat, and its quality, as measured by lysine content, is higher. However, flour made from triticale is inferior for making bread unless mixed with wheat flour.

PSEUDOCEREALS

Pseudocereals are plants that are not in the grass family but that produce nutritious hard, grainlike fruits that can be stored, processed, and prepared for food much like grains. They belong to several plant families. Many grow under conditions not suitable for the major cereal crops. Buckwheat (*Fagopyrum esculentum*), of the buckwheat family, Polygonaceae, probably originated in China. It tolerates cool conditions and is adapted to short growing seasons, thus permitting it to be grown in the temperate regions of North America and Europe. In the United States it is often associated with pancakes, but it is used in larger quantities for livestock feed. In Eastern Europe, the milled grain is used for soups.

Quinoa (*Chenopodium quinoa*), of the goosefoot family, Chenopodiaceae, has been cultivated by Indians of the Andes mountains for centuries. The leafy annual produces grainlike fruits (actually achenes) with a high protein content and exceptional quality (high in lysine and other essential amino acids). After its bitter saponins have been removed, quinoa can be cooked and eaten like rice or made into a flour. Quinoa has been cultivated in the Rocky Mountains of Colorado since the 1980's and has become a gourmet food in the United States.

Most amaranth (*Amaranthus* sp.) plants are New World weeds. They belong to the amaranth family, Amaranthaceae. A few species were used by Aztecs and other native peoples, but their use was banned by the Spanish. Since the late 1970's, plant breeders have targeted several species for improvement. The results are highly nutritious grains rich in lysine that are suitable for making flour. Research in Pennsylvania and California has resulted in improved varieties.

Further Reading

Carver, Brett F., ed. *Wheat: Science and Trade.* Ames, Iowa: Wiley-Blackwell, 2009.

Levetin, Estelle, and Karen McMahon. *Plants and Society.* 5th ed. Boston: McGraw-Hill Higher Education, 2008.

Schery, Robert W. *Plants for Man.* 2d ed. Englewood Cliffs, N.J.: Prentice Hall, 1972.

Simpson, Beryl B, and Molly Conner-Ogorzaly. *Plants in Our World: Economic Botany.* Boston, Mass: McGraw-Hill, 2014. Print.

Animal Domestication

—*James L. Robinson, PhD*

FIELDS OF STUDY

Agriculture; Agricultural Geography; Biology; Zoology; Biochemistry; Chemistry; Physical Science; Statistics; Botany; Insect And Pest Management; Food Production; Animal Science; Ecology.

PRINCIPAL TERM

dominance social behavior: organization around a dominant leader, whom the rest of the group follows

SUMMARY

Animal domestication is the process of bringing an animal species under the control of humans and gradually changing it, through careful selection, mating, and handling, so that it becomes more useful to humans. Some animals are more amenable to being domesticated than others. Domestication is a symbiotic relationship between humans and the animals involved, and has been integral to the development of human societies.

Although taming a wild animal, that is, acclimating it to the presence of people, is invariably a first step in domestication, the critical aspect is directing its breeding toward a functional goal. Genetic selection for desirable traits and against undesirable ones is a crucial feature of domestication. Captive wild animals in zoos or circuses cannot be considered domesticated, although they may be quite tame and their breeding is controlled by humans. However, this breeding is not directed toward some functional goal. Mere propagation of a species or maintaining genetic diversity is insufficient to constitute domestication. Domestication is a synergistic relationship between humans and the animals involved. Humans benefit from the resources and services that the animals provide, such as food, fiber, shelter, clothing, work, sentinel duty, and companionship. Animals benefit by having humans protect them from harm, provide for their needs, and increase their numbers and range.

CHARACTERISTICS OF ANIMALS FAVORING DOMESTICATION

In 1865, the English naturalist Francis Galton suggested the following six physiological and behavioral characteristics that make some animals good candidates for domestication: hardiness, dominance social behavior, herd behavior, utility to humans, facile (easy) reproduction, and facile husbandry. First, by hardiness, he refers to the ability of the young to be removed from its mother and to be around humans. The guinea pig is perhaps an extreme example of tolerating removal from its mother, as it is born ready to eat solid food. Most mammals, on the other hand, initially depend on their mother's milk. Primates are poor subjects for domestication because of their helplessness at birth and their relatively long dependence on their mothers for food and nurturing.

Second, dominance social behavior is in contrast to territorial behavior and refers to one animal assuming leadership, with the rest of the group acquiescing to him or her in the hierarchy. In domestication, humans co-opt the function of the leader, and animals remain submissive even as adults. Third, herd animals are contrasted to solitary animals, or ones that disperse in response to danger. Domesticated animals are penned or otherwise restricted at various times. If they remain together in herds, they are easier to manage. Fourth, utility to humans includes their use for food, fiber, work, companionship, and even worship. Humans would not make the effort to domesticate an animal unless it had some perceived value. The purpose for domestication may change with time, however. It is likely that the initial motivation for domesticating cattle was for worship—to capture the strength and aura of these animals, which were revered and used in religious ceremonies. Work, such as pulling carts, packing, and riding, became a subsequent goal, while contemporary utility in Western societies involves meat and milk production.

The fifth characteristic is facile reproduction under confined conditions; animals with finicky reproductive behaviors and/or elaborate courtship rituals make poor candidates for domes tication. Sixth, facile husbandry refers to placid disposition and versatility in terms of nutrition. Animals that are high-strung or dependent on unique foodstuffs would be weak prospects for domestication. Koalas, which eat leaves from only certain eucalyptus trees, are poor candidates. On the other hand, pigs and goats are excellent choices because they are not very fastidious in their eating habits. These six characteristics, enunciated more than a century ago, apply strongly to livestock species, but less well to dogs and cats; with regard to the latter, it has been argued that cats are not so much domesticated as merely tolerant of humans.

Grace Goodhue Coolidge (1879-1957) with two dogs, Sept., 1924. Image by Library of Congress.

HISTORY OF ANIMAL DOMESTICATION

Archaeological evidence suggests that agriculture developed about ten thousand years ago, after the last ice age, at a time when the climate became warmer and more stable. Predictability of the weather is particularly crucial for plant domestication, which apparently developed synergistically with animal domestication, leading to agriculture. Domesticated animals simplified the acquisition of food and provided food storage, in the form of "walking larders"; larger animals, such as cattle, buffalo, and horses, permitted heavier work to be done and larger distances to be covered. Farming the land made it possible for humans to abandon the nomadic lifestyle of hunter-gatherers and to adopt a more sedentary lifestyle. This allowed for development of new technologies, professional specializations, and new forms of social organization.

Domestication of any animal did not occur at once, but rather over a substantial period of time, perhaps hundreds of years. Furthermore, estimating the dates for domestication is subject to considerable uncertainty and may need to be modified as new information becomes available. For some species, domestication occurred independently at more than one location. The process may have begun almost accidentally, as by raising a captured young animal after its mother had been killed and observing its behavior and response to various treatments. The domestication of an animal subsequently spread from the site of origin through trade or war.

Animal domestication occurred in various parts of the world. In the Middle East, the Fertile Crescent, stretching from Palestine to southern Turkey and down the valley of the Tigris and Euphrates Rivers, was an important site. Sheep, goats, cattle, and pigs were domesticated there by around 6000 BCE. The Indian subcontinent and east Asia were independent sites for domesticating cattle and pigs, respectively. Llamas, alpacas, and guinea pigs were domesticated in the Andes Mountains of South America. Domestication of cats occurred in Egypt and of rabbits in Europe. No native animals were domesticated in Australia, likely because none of them were suitable for domestication. It is worthy of note that few domestications have occurred in the past thousand years. It is also of interest that domesticatable species were not evenly distributed over the globe, which probably has had a lasting effect on the differential development of various cultures.

Archaeology, coupled with the natural history of domesticated animals and their wild relatives, has been essential in reconstructing the history of domestication. Examining skeletal remains at archaeological sites for changes in morphology and distribution by age and sex has helped scientists to deduce the extent of domestication. Lately, traditional archaeology has been supplemented by the methods of molecular biology. Examining extant breeds for their genetic relatedness has been particularly useful in distinguishing single versus dual sites of domestication.

EARLY DOMESTICATES: DOGS & REINDEER

Dogs (*Canis familiaris*) are generally recognized as the earliest known domesticated animals. They were widespread across the Northern Hemisphere before other animals were domesticated. They were derived from wolves (*Canis lupus*), with whom they are fully interfertile. The earliest known dog is in a burial site in Northern Iraq that dates from 12,000 to 10,000 BCE. Other sites, dating from 10,000 to 7000 BCE, have been documented in England, Palestine, Japan, and Idaho. While it may have first occurred in China (a Chinese wolf has some of the detailed physical features of dogs), domestication probably occurred at a number of separate sites. Dogs accompanied the American Indians when they occupied the Americas in several waves prior to the end of the last ice age. Dingoes were brought to Australia by trade from Asia long after the Aborigines settled that continent 40,000 years ago. While dogs were considered a food animal, they have long been used for guarding, hunting, and companionship. Subsequently, they were developed for herding.

Reindeer (*Rangifer tarandus*) were another early domesticate, dating from around 12,000 BCE in northern Scandinavia and Russia. The reindeer has been little changed by domestication and its range has not been extended by the process. They are well-suited to their environment, but attempts to establish reindeer industries in Canada and Alaska have not been successful. Herding reindeer continues as a principal occupation of the Laplanders of Finland, Sweden, and Norway. Reindeer are used for draft (pulling loads), clothing and shelter (skins), tools (antlers), and food (meat and milk). Farming several other deer species (such as *Cervus dama* and *Cervus elaphus*) has recently gained in importance in New Zealand and western Europe, where they are raised for meat (venison) and "velvet," the new growth of antlers, the basis for traditional medicines in Asia. However, because these latter species have had little opportunity to be changed, they cannot be considered domesticated.

SHEEP, PIGS & CATTLE

Sheep (*Ovis aries*) were the first of the common food animals to be domesticated. They were derived from wild sheep (*Ovis orientalis*) and were first domesticated in the western Fertile Crescent around 7000 BCE. Goats (*Capra hircus*), derived from Persian wild goats (*Capra aegarus*), were first domesticated in the central Fertile Crescent slightly later, between 6000 and 7000 BCE. Sheep and goats were used for food, skins, and fiber (wool or hair). Both were later selected for milk production.

Pigs (*Sus domesticus*) probably originated at two separate sites, the central Fertile Crescent around 6000 BCE and in eastern Asia around 5000 BCE. Derived from wild pigs, they were primarily raised for meat. Despite restrictions against eating pork by Muslims and Jews, it has long been the principal meat consumed in the world. The most populous country, China, has nearly 50 percent of the world's pigs.

Cattle (*Bos taurus* and *Bos indicus*) are derived from now-extinct wild cattle (aurochs, *Bos pri migenius*) that ranged over much of Europe and Asia. They were probably domesticated independently at two locations, the western Fertile Crescent around 6000 BCE for *Bos taurus* and the Indian subcontinent around 5000 BCE for *Bos indicus*. Initially, the animals were worshiped and used in religious ceremonies. Reverence for cattle is still practiced by Hindus in India. Subsequently, they were developed for work, meat, and milk. Their hides are made into leather. Traditional cattle in Africa are derived from initial importations of *Bos taurus* and subsequent importations of male *Bos indicus*.

OTHER DOMESTICATED ANIMALS

Asiatic buffaloes (*Bubalus bubalis*) were domesticated as the water buffalo in India (3000 BCE) and as the swamp buffalo in east or southeast Asia (2000 BCE). While both were developed as draft animals, the water buffalo has also been selected as a dairy animal. Fully half of the milk production in India comes from buffaloes. In spite of its tropical origin, the Asiatic buffalo is not very heat tolerant and compensates by immersing in water or mud. Neither the African buffalo (*Syncerus caffer*) nor the American buffalo (more properly, bison, *Bison bison*) have been domesticated. Yaks (*Poephagus[Bos] grunniens*) were domesticated around 3000 BCE in Tibet or surrounding areas, where they are used as pack animals and as a source of milk, hair, hides, and, usually after an otherwise productive life, meat.

Horses (*Equus caballus*) originated from wild horses in the Caucasus Mountains around 4000 BCE. Originally used for food and skins, they were also developed for draft and, much later, for riding. Because they came to the Middle East after the development of written language, their arrival is documented, so scholars do not need to depend solely on the archaeological record. Donkeys (*Equus asinus*) were domesticated in the Middle East or Northern Africa (3000 BCE). They are used for carrying material or people, as is the mule, an infertile cross between a horse and a donkey.

Llamas (*Lama glama*) and alpacas (*Lama pacos*) were domesticated in Peru by Incas around 4000 BCE. Llamas are from wild guanacos and alpacas from wild vicuñas, found at higher elevations. Llamas are used as pack animals, alpacas are valued for their fine wool, and both serve as sources of meat. Camels, the one-humped dromedary (*Camelus dromedarius*) and the two-humped Bac trian (*Camelus bactrianus*) were domesticated in Arabia (2000 BCE) and Central Asia (1500 BCE), respectively. Both are pack animals, with the dromedary also used for meat.

Guinea pigs (*Cavia porcellus*) were domesticated in Peru around 3000 BCE. They continue to be used as a meat animal in parts of South America. Rabbits (*Oryctlagus cuniculus*) were domesticated between 600 and 1000 CE in France. They are primarily raised for meat and fur, with angora rabbits producing a fine wool.

Cats (*Felis catus*) are the animals least changed, morphologically, by domestication. In addition, they are quite capable of surviving without human intervention. Their domestication occurred relatively late, around 2000 BCE, in Egypt, the home of the African wild cat (*Felis catus libyca*), which resembles domestic tabby cats. The early Egyptians adopted cats enthusiastically, deifying them and prohibiting their export. After conversion to Christianity, Egyptians ceased worshiping cats, which were carried to all parts of the Roman Empire and thence to the rest of the world. Cats have been used for companionship and for rodent control.

Chickens (*Gallus gallus*), along with ducks and geese (from China, the Middle East and Europe), turkeys (from North America), Muscovy ducks (from South America), and guinea fowl (from Africa) are avian species that have been domesticated. Chickens were probably derived from wild red junglefowl in southeast Asia before 2000 BCE. Cockfighting was an initial purpose for their domestication. They acquired religious significance and were also used for meat and feathers. Their selection for egg production has been a recent development. Because of improvements in breeding, feeding, and management, poultry meat production is increasing rapidly and is second to pork worldwide. In the past few centuries, two other avian species, ostriches (from Africa) and emus (from Australia), have been farmed, but it is probably incorrect to call them domesticated, as they have so far been little changed from their wild relatives.

Two insects have also been domesticated: honeybees and silkworms. Honeybees were domesticated shortly after the last ice age and were the primary source of dietary sweetener until two hundred years ago. They were also valuable for wax and venom, the latter for medicinal purposes. Silkworms, one of the ten varieties of silk-producing insects, were domesticated around 3000 BCE in China, producing fiber used in apparel.

FURTHER READING

Clutton-Brock, Juliet. *A Natural History of Domesticated Mammals*. 2d ed. New York: Cambridge University Press, 1999. By an expert in the field of animal domestication, this book is well written and well photographed. It includes chapters on all domesticated mammals.

Diamond, Jared. *Guns, Germs, and Steel: The Fates of Human Societies.* New York: W. W. Norton, 1997. This well-written book examines the domestication of animals and argues that the location of animals that could be domesticated has had a lasting influence on the course of human history.

Harris, David R., ed. *The Origins and Spread of Agriculture and Pastoralism in Eurasia.* Washington, D.C.: Smithsonian Institution Press, 1996. This scholarly monograph includes chapters on domestication of animals in general and selected animals in detail, including the goat and camel.

Ryden, Hope. *Out of the Wild: The Story of Domesticated Animals.* New York: Lodestar Books, 1995. Written for adolescents, this book includes pictures of most domesticated animals and a description of their history, biology, and usefulness to humans.

Smith, Bruce D. *The Emergence of Agriculture.* New York: Scientific American Library, 1995. While emphasizing the domestication of plants, this book also details the archaeological evidence for domestication of animal species in various parts of the world.

ANIMAL GROWTH

—Roman J. Miller

FIELDS OF STUDY

Biology; Zoology; Biochemistry; Chemistry; Physical Science; Statistics; Botany; Insect And Pest Management; Food Production; Animal Science; Ecology.

PRINCIPAL TERMS

differentiation: the process during development by which cells obtain their unique structure and function

fertilization: the union of two gametes (egg and sperm) to form a zygote

gamete: a functional reproductive cell (egg or sperm) produced by the adult male or female

growth: the increased body mass of an organism that results primarily from an increase in the number of body cells and secondarily from the increase in the size of individual cells

mitosis: the process of cellular division in which the nuclear material, including the genes, is distributed equally to two identical daughter cells

zygote: a fertilized egg

SUMMARY

Animal development begins with a single cell, the fertilized egg. Continued growth and formation through numerous cellular replications result in the formation of the complex animal body.

Animal development has been a source of wonder for centuries. Development involves the slow, progressive changes that occur when a single cell—the zygote, or fertilized egg—undergoes mitosis. Mitosis is the process by which a cell divides into identical daughter cells. During development, mitosis occurs repeatedly, forming multiple generations of daughter cells. These cells increase in number and ultimately form all the cells in the body of a multicellular animal, such as a frog, mouse, or elephant. The simple experiment of opening fertile chicken eggs to observe the embryos on successive days of their three-week incubation period illustrates the process of embryonic development. A narrow band of cells can be seen increasing in number and complexity until the body of an entire, but immature, chick is seen.

ANIMAL GROWTH & DEVELOPMENT

An organism's growth occurs because of the increasing number of cells that form as well as because of the increasing size of individual cells. For example, a mouse increases from a single cell, the zygote, to about three billion cells during the period from fertilization to birth. Embryology is the study of the growth and development of an organism occurring before birth. Growth and development, however, continue after birth and throughout adulthood. Growth ceases only at death, when the life of the individual organism is ended. The bone marrow of human adults initiates the formation and development of millions of red blood cells every minute of life. About one gram of old skin cells is lost and replaced by new cells each day.

Development produces two major results: the formation of cellular diversity and the continuity of life. Cellular diversity, or differentiation, is the process that produces and organizes the numerous kinds of body cells. The first cell that determines an individual's unique identity, the zygote, ultimately gives rise to varying types of cells having diverse appearances and functions. Muscle cells, red blood cells, skin cells, neurons, osteocytes (bone cells), and liver cells are all examples of cells that have differentiated from a single zygote.

REPRODUCTION

Morphogenesis is the process by which differentiated cells are organized into tissues and organs. The continued formation of new individual organisms is called reproduction. The major stages of animal development include fertilization, embryology, birth, youth, adulthood—when fertilization of the next generation occurs—and death. A new individual animal is begun by the process of fertilization, when the genetic material from the sperm, produced by the father, and the egg, produced by the mother, are merged into a single cell, the zygote. Fertilization may be external, occurring in freshwater or the sea, or internal, occurring within the female's reproductive tract. While fertilization marks the beginning of a new individual, it is not literally the beginning of life, since both the sperm and egg are already alive. Rather, fertilization ensures the continuation of life through the formation of new individuals. This guarantees that the species of the organism will continue to survive in the future.

Following fertilization, the newly formed zygote undergoes embryological development consisting of cleavage, gastrulation, and organogenesis. Cleavage is a period of rapid mitotic divisions with little individual cell growth. A ball of small cells, called the morula, forms. As mitosis continues, this ball of cells hollows in the middle, forming an internal cavity called the blastocoel. Gastrulation immediately follows cleavage. During gastrulation, individual cell growth as well as initial cell differentiation occur. During this time, three distinct types of cells are formed: an internal layer called the endoderm, a middle layer, the mesoderm, and an external layer, the ectoderm. These cell types, or germ-cell layers, are the parental cells of all future cells of the body.

Cells from the ectoderm form the cells of the nervous system and skin. The mesoderm forms the cells of muscle, bone, connective tissue, and blood. The endoderm forms cells that line the inside of the digestive tract as well as the liver, pancreas, lungs, and thyroid gland. The transformation of these single germ layers into functional organs is called organogenesis. Organogenesis is an extremely complex period of embryological development. During this time, specific cells interact and respond to one another to induce growth, movement, or further differentiation; this cell-to-cell interaction is called induction. Each induction event requires an inducing cell and a responding cell.

In the formation of the brain and spinal cord, selected cells from the ectoderm form a long, thickened plate at the midline of the developing embryo. Through changes in cell shape, the outer edges of this plate fold up and fuse with each other in the middle, forming a tubular structure (a neural tube). This tubelike structure then separates from the remaining ectoderm. At the head region of the embryo, the neural tube enlarges into pockets that ultimately form brain regions.

For differentiation and development to occur, cells must be responsive to regulatory signals. Some of these signals originate within the responding cell; these signals are based in the genetic code found in the cell's own nucleus. Other signals originate outside the cell; they may include physical contact with overlying or underlying cells, specific signal molecules, such as hormones, from distant cells, or specialized structural molecules secreted by neighboring cells that map out the pathway along which a responding cell will migrate.

POSTNATAL DEVELOPMENT

Embryological development climaxes in the formation of functional organs and body systems. This period is concluded by birth (or hatching, in the case of some animals). Following birth, development normally continues. In some animals, such as frogs, newly hatched individuals undergo metamorphosis during which their body structures are dramatically altered. Newly hatched frogs (tadpoles), for example, are transformed from aquatic, legless, fishlike creatures into mature adults with legs that allow them to move freely on land.

In mammals, development and growth occur primarily after birth, as the individual progresses through the stages of infancy, childhood, adolescence, and adulthood. Mature adulthood is attained when the individual can produce his or her own gametes and participate in mating behavior.

Embryonic growth is especially impressive because the rate of cellular mitosis is so enormous. In the case of the mouse embryo, thirty-one cell generations occur during embryonic development. Thus, the zygote divides into two cells, then four, then eight, sixteen, thirty-two, and so on. This results in a newborn mouse consisting of billions of cells—produced in a period of only twenty-one days. When the newborn passes through its life stages to adulthood, its body cells may number more than sixty billion. One marine mammal, the blue whale, begins as a single zygote that is less than one millimeter in diameter and weighs only a small fraction of a gram. The resulting newborn whale (the calf) is about seven meters long and weighs two thousand kilograms: The embryonic growth represents a 200-millionfold increase in weight. Yet, for some animals, impressive growth periods also occur in the juvenile and adolescent stages of life.

In many cases, once an individual animal reaches its typical adult size, the rate of mitosis slows so that the number of new cells simply replaces the number of older, dying cells. At this maintenance stage, the individual no longer grows in overall size even though it continuously produces new cells. Since most of the cells in the mature adult have reached a final differentiated state, the function of mitosis is simply to replace the degenerating, aging cells. The slowing of the rate of cellular mitosis during this time may be attributable to the presence of specialized cell products called chalones. Chalones are thought to be local products of mature cells that inhibit further growth or mitosis.

STUDYING GROWTH

Historically, much study of animal development and growth was performed by simple observation. Aristotle, perhaps the first known embryologist, opened chick eggs during varying developmental periods. He observed and sketched what appeared to be the formation of the chick's body from a nondescript substance. With the invention of lenses and microscopes, growth and development could be studied on a cellular level. The concept of cellular differentiation arose, since investigators could see that embryonic muscle cells, for example, looked different from embryonic nerve cells. Again, much of the investigative information was descriptive in nature. Embryologists detailed the existence of the three germ-cell layers in gastrulation as well as the various tissues and primitive cells involved in organogenesis.

Experimentation as a method of investigating animal growth and development began during the nineteenth century. Lower animal species, such as the sea urchin and frog, were frequently investigated; their developmental patterns are simpler than those of mammals, their development occurs outside the maternal body, and they can be found in abundant numbers. Many of these experiments used separation or surgical techniques to isolate or regraft specific tissues or cells of interest. An attempt was made to determine how one tissue type would interact with and influence the development of another tissue type. Thus, the ideas of induction, in which some tissues affect other tissues, came into being. During this time, the descriptive and comparative observations resulting from these experimental manipulations were the major contributions of investigators.

The embryologists of the early twentieth century paid little attention to genetics. They believed that the major influences on development and growth were embryological mechanisms, although genes were thought to provide some nonessential peripheral functions. Chemical analyses of embryos attempted to establish the chemical basis for the cell-to-cell interactions that were seen during development and differentiation. During the middle portion of the twentieth century, geneticists began to investigate the role of the gene in cell function. The function of genes in the cellular manufacturing of specific proteins led to the hypothesis that each kind of cellular protein was the product of one gene. During this time, bacteria and fruit flies (*Drosophila*) were primary organisms of study because of their relatively simple genetic makeups.

In the latter part of the twentieth century, molecular biology techniques were applied to the study of development. Using techniques for transferring and replicating specific genes, researchers have greatly clarified the central importance of genes in development. Scientists came to believe that all the major developmental and differentiation influences that control cell growth are regulated through specific genes that are turned off or on.

DEVELOPMENTAL BIOLOGY

The combination of molecular biology techniques with embryological investigations has led to a new field of study—developmental biology. New methods have been developed and used. Radioactive tracer technology has allowed the investigator to label particular genes or gene products and trace their movements and influences on cell growth through several generations. Recombinant deoxyribonucleic acid (DNA) technology has allowed the isolation and replication of significant genes that are important in development. Immunochemistry uses specific proteins (antibodies) to bind to differentiating cell products and quantify them. Cell-cell hybridization allows the introduction of specific genes into the nuclei of cells in alternate differentiation pathways.

Developmental biology, with its multidisciplinary approach, is solving many of the fundamental questions of development. As scientists become better able to understand the role of genetics and cell-to-cell interactions, they gain insight into the mechanisms that control cell growth and development. Consequently, the potential to control undesirable growth or to enhance underdeveloped growth is within reach.

The problem of cell aging is also under investigation. Questions about why mature cells stop dividing and growing and what the causes of aging are constitute important areas of developmental research. While various theories have been presented, the fundamental key to cellular aging remains to be discovered. One of the most challenging areas of continuing research is the determination of how developmental patterns guide evolutionary changes. Developmental principles may provide the answer to why evolution has given rise to animal diversity. In addition, developmental biology may give scientists the information needed to predict and determine future evolutionary trends. The individual animal is a growing organism that begins as a zygote and passes through the stages of embryonic development, birth, youth, adulthood, aging, and death. Preservation of the species depends on adult individuals' producing gametes that result in the formation of a future generation of zygotes and individuals. Remarkably, each zygote contains the necessary genetic instructions to regulate the orderly processes of growth and development. Thus, animal life continues from generation to generation.

FURTHER READING

Alberts, Bruce, Dennis Bray, Julian Lewis, Martin Raff, Keith Roberts, and James D. Watson. *Molecular Biology of the Cell.* 3d ed. New York: Garland, 1994. This encyclopedic college text is greatly enhanced by its superior diagrams and illustrations. While all aspects of cell biology are covered in great detail, of special interest are the chapters on cellular mechanisms of development and differentiated cells and the maintenance of tissues. Highly recommended for the serious student who wishes to understand developmental processes on a cellular level.

Balinsky, B. I. *An Introduction to Embryology.* Philadelphia: Saunders College Publishing, 1981. While somewhat dated, this very well-written text covers the essentials of classical animal embryology while integrating with them the cellular and molecular mechanisms that regulate them. Of specific interest is the chapter on growth, which integrates this topic into the study of developmental biology.

Baserga, Renato. *The Biology of Cell Reproduction.* Cambridge, Mass.: Harvard University Press, 1985. This illustrated book can be read by the layperson who is interested in understanding more fully the role of cell division and reproduction in growth. The basics of the cell cycle are covered in detail, as are many of the influences that regulate it. In the latter portion of the book, the author describes the genetic mechanisms involved in growth and development.

Gilbert, Scott F, and Michael J. F. Barresi. *Developmental Biology.* Sinauer Associates, 2018. Print. This is a college-level text written for the upper-level student who has a serious interest in development. The author's approach to development uses a historical experimental analysis of the progress of investigators in the research field. About half of the text deals with specific cell or genetic mechanisms that regulate growth and development. An excellent resource for advanced and accurate information regarding the field of development.

_______, ed. *A Conceptual History of Modern Embryology: A Conceptual History of Modern Embryology.* Springer

Verlag, 2012. A collection of essays concentrating on historical aspects of research on embryonic induction and the relationship between experimental embryology and genetics.

Gilbert, Scott F., and A. M. Raunio. *Embryology: Contructing the Organism.* Sunderland, Mass.: Sinauer Associates, 1997. A college-level textbook on comparative embryology. The first two chapters introduce general terms and concepts, while the following twenty chapters each concentrate on a different animal group.

Hartwell, Leland H., and Ted A. Weinert. "Checkpoints: Controls That Ensure the Order of Cell Cycle Events." *Science* 246 (November 3, 1989): 629–634. An excellent review article describing the biochemical controls of cell division and cell proliferation. Information is a bit technical and is written at the level of the more advanced reader who has some background in biology.

Postlethwait, John H, and Janet L. Hopson. *Modern Biology.*, 2012. Print. The chapter on animal development gives an overview of classic embryology, including the role of growth. The chapter on developmental mechanisms and differentiation takes the reader through the major cellular and genetic regulatory mechanisms involved in development. This extremely well written introductory college text is easily read; it is well illustrated with color photographs and self-explanatory diagrams.

ANIMAL HUSBANDRY PRACTICES

—Elizabeth D. Schafer, PhD

FIELDS OF STUDY

Agronomy; Agriculture; Environmental Studies.

SUMMARY

Humans raise livestock to meet individual and community nutritional and commercial demands for animal products.

DEFINITION

Humans raise livestock to meet individual and community nutritional and commercial demands for animal products. Livestock farmers range from individuals, who may raise a few animals to supply their families with protein sources or earn income in local markets, to commercial agriculturists who invest in industrial methods to manage numerous livestock simultaneously. Animal husbandry practices reflect diverse human cultures and types of livestock, from nomads in developing countries who tend herds of indigenous livestock that migrate to water and forage resources to farmers in developed countries who often rely on technology to mass-produce genetically standardized animals at centralized locations.

SIGNIFICANCE FOR CLIMATE CHANGE

Climate issues associated with animals worldwide caused the Food and Agriculture Organization (FAO) of the United Nations to assess livestock's environmental impact. In the early twenty-first century, several billion livestock, including bovines, goats, sheep, swine, and poultry, inhabited the Earth, and 30 percent of the planet's land was used for grazing and to grow livestock feed crops. An estimated 1.3 billion people participated in animal husbandry-related work. Annual meat production totaled 229 million metric tons and milk production was 580 million metric tons in 2001, with experts projecting that those amounts would double in fifty years to meet increasing food demands.

By spring, 2008, U.N. representatives determined that livestock produced 18 percent of the greenhouse gases (GHGs) associated with global warming, noting that those emissions exceeded the amounts produced collectively by land and air transportation vehicles. Media referred to the "carbon hoofprint" when discussing livestock's contribution to global warming. Animal husbandry was found to be responsible for 82 million metric tons of carbon dioxide (CO_2)

Farmer and Livestock. Image by Alex Proimos from Sydney, Australia (Farmer and the Livestock Uploaded by russavia) CC-BY-2.0.

emissions annually. In addition, each individual beef cow emitted approximately 80 kilograms of methane yearly and each dairy cow emitted 150 kilograms. The European Union (EU) suggested that farmers cull their herds to limit emissions. By 2008, New Zealand became the world's first nation to consider taxing farmers whose livestock emitted more GHGs than legislated allowances. The Pastoral Greenhouse Gas Research Consortium investigated scientific ways to manage livestock emissions.

Animal husbandry practices include clearing land for grazing livestock and planting grain crops to sustain them. Deforestation to establish ranches results in greater amounts of CO_2 reaching the atmosphere. Approximately 70 percent of Amazon rain forests have been cleared to maintain livestock, which often harm fields by stripping them of vegetation, causing exposed soil to become eroded or compacted. Subterranean storage of precipitation is disrupted, sometimes initiating desertification. Clearing land for livestock occasionally involves draining wetlands.

The loss of trees and vegetation due to farmers transforming forests into fields reduces albedo (the proportion of incident radiation reflected by Earth's surface): Bared ground absorbs rather than reflects sunlight and heat, intensifying global warming. Although Earth's Arctic ice is distant from most animal husbandry procedures (except for reindeer farming in northern latitudes), the conversion of ice to seawater due to decreasing albedo affects livestock globally. In some areas, extreme heat associated with lower albedo frequently dries up water resources used for animal husbandry. Elsewhere, rising water levels due to increased ice melting displaces livestock.

Excess CO_2 detrimentally affects indigenous plant growth, particularly that of native forages stifled by GHGs. Woody shrubs exotic to prairies thrive when exposed to CO_2. Weeds, some toxic to livestock, spread and often deprive soil of moisture. Climate change can motivate livestock to seek new food and water sources, and these migrating livestock transport seeds from plants and parasites to new locations when they travel, carrying them in fur, hooves, and intestinal tracts, further altering ecosystems.

Veterinary professionals report cases of livestock diseases previously unknown in specific geographical areas, attributing their spread to changed climates. In 2006, bluetongue, an insect-transmitted viral disease usually confined to southern France, was detected in several thousand northern European livestock, necessitating quarantines. Warm temperatures enable insect populations, including tsetse flies and mosquitoes, to extend into locations where usually cooler climates would have inhibited them. When frosts are infrequent, lungworm larvae survive in grasses, causing respiratory disease in livestock.

An FAO report has estimated that one livestock breed becomes extinct each month during the early twenty-first century. Industrial animal husbandry practices limit production to selected western breeds, which replace many native livestock in developing countries. By 2002, only one-fourth of sows in Vietnam were indigenous breeds. The FAO promoted preserving genetic material from indigenous livestock breeds that possessed such resilient traits as heat tolerance and drought hardiness, which might be crucial to future animal husbandry as global warming alters climatic conditions.

In November, 2006, a UN climate conference at Nairobi, Kenya, discussed how global warming threatened nomadic livestock herders in Africa. Early twenty-first century droughts caused an estimated 500,000 nomads to cease raising livestock, as deserts began overtaking fields, disrupting the normal four seasons that herders need to raise livestock. Violence resulted, as herders battled for water and pasture resources limited by climate changes.

FURTHER READING

Morgan, Jack A., et al. "Carbon Dioxide Enrichment Alters Plant Community Structure and Accelerates Shrub Growth in the Shortgrass Steppe." *Proceedings of the National Academy of Sciences* 104.37 (2007): 14,724–4,729. Print.

Seo, Sungno Niggol, and Robert Mendelsohn. "Climate Change Impacts on Animal Husbandry in Africa: A Ricardian Analysis." World Bank Policy Research Working Paper 4261. Washington, DC: World Bank, 2007. Print.

Steinfeld, Henning, et al. *Livestock's Long Shadow: Environmental Issues and Options.* Rome: Food and Agriculture Organization of the United Nations, 2006. Print.

Watson, Paul. "New Zealand Aims for Greener Pastures: Officials Ruminate on How to Curb Methane from the Nation's Livestock, a Culprit in Global Warming." *Los Angeles Times,* 8 June 2008: A-6. Print.

APIOLOGY

—M. Marian Mustoe, PhD

FIELDS OF STUDY

Apiculture; Melittology; Entomology; Agriculture; Agricultural Geography; Biology; Zoology; Biochemistry; Chemistry; Physical Science; Statistics; Botany; Insect And Pest Management; Food Production; Animal Science; Ecology.

PRINCIPAL TERMS

Apis mellifera: genus and species name for the honeybee.

bee behavior: actions a bee takes in its instinctual inclination to survive.

beehive: container that can house a colony of bees.

brood: bees not yet developed out of the egg, larval, or pupal state.

colony collapse disorder: collapse of a colony after a seemingly healthy colony of worker bees falls into a sudden, fatal decline.

comb: wax material produced by bees and formed into hexagonal cells to store brood, nectar, honey, and pollen.

drone: male bee.

honey: product of the hive produced by the bees through a process of the fermentation of nectar.

parthenogenesis: state of being born of virgin female from an unfertilized egg.

pollination: process of plant fertilization in which bees carry pollen from plant to plant.

queen bee: fertile female bee.

swarm: process whereby part of a colony of honeybees leaves the hive in the company of a queen to propagate a new colony.

worker bee: sterile female bee.

SUMMARY

Apiology is the scientific study of the honeybee. It is a subdiscipline of melittology, which is the study of all bees and is a branch of entomology. Apiologists study the evolution of the honeybee and answer questions surrounding its biology, particularly its social behavior and the ecological role the insect plays in its habitat. Other areas of study include the reproduction cycle of the honeybee, its proficiency to gather nectar, and its production of honey. A critical topic for the apiologist is diseases of the honeybee. Given the migratory nature of commercial beekeeping, the spread of bee diseases and their treatment have become a major challenge for beekeepers.

DEFINITION & BASIC PRINCIPLES

The terms "apiology" and "apiculture" (another word for apiology) are derived from the Latin word for bee, *apis.* The science to which these terms refer focuses specifically on the honeybee. Although beekeepers are required to manage and maintain their colonies using up-to-date methods and, in effect, are practicing apiculture, in contrast, it is the outcome of the research of apiologists and those scientifically investigating the well-being of the honeybee that sets the standard for good practice by beekeepers. Managers of commercial bee operations that might contain thousands of colonies are highly dependent on good science to help them stay productive as well as competitive.

BACKGROUND & HISTORY

Apiary research and development emerged from a basic understanding of the behavior of bees. The outcome of this research has had considerable practical impact.

The standard removable-frame beehive—a design that was patented in 1852—was developed by observing bee behavior. Patent holder Lorenzo Langstroth, a beekeeper and Congregationalist preacher living in Ohio, noticed that bees maintained a prescribed space before they sealed over the frames of the hive. His response to this bee space was to develop a structure to function as the hive body that could be completely taken apart for harvesting honey and inspecting the bees. The outcome of this observation radically changed the way bees were housed and maintained.

The science of apiology has greatly contributed to understanding the genetic character and reproductive behaviors of bees. By the 1800's, it was generally accepted that the drone mated with the queen. The queen had the potential of laying eggs that became worker bees or drones. The Austrian monk Gregor Mendel, known for his genetic experiments with pea plants, also explored heredity by keeping and breeding bees. However, his bees were housed in the traditional stationary bee house, and he had no control over which drones bred with the queen.

In the early 1900's, apiology researchers had perfected the technique of instrumental insemination of the queen bee. This method made it possible to control the breeding fprocess and has allowed for dramatic advances in the crossing of bees. The practical outcome of this science for the beekeeper is a well-tempered and productive bee.

Apiology also has provided insight into the diseases and enemies of the honeybee. Extensive research is conducted on mites and their control within the colony. Out of the study of bacterial problems such as foulbrood (a disease that attacks honeybee larvae), new practices in apiary management have emerged that can control these devastating honeybee enemies. Likewise, apiology provides an understanding of the symbiotic relationships that exist between bees and other insects as well as other animals.

HOW IT WORKS

Beekeeping. The process of keeping bees requires an understanding of the biology of bees and the instinctual behaviors they exhibit within their colony. The colony's ability for honey production, fecundity (potential for reproduction), and the tendency to swarm, are also factors that must be managed by the beekeeper to maintain the viability of the colony. The beekeeper must become aware of the environmental conditions that are conducive to pollination and nectar gathering. This requires an understanding of the blooming cycles of various flora within a region as well as weather conditions best suited for flying times by the bees. The outcome of the proper application of this knowledge will be expressed in the form of stronger and more productive colonies.

Bee Diseases. Diseases within the hive take a variety of forms. Some of these are found exclusively in the brood, thereby weakening the colony by lowering its overall population. In contrast, some of these diseases affect the adult bee, destroying the working population. The impact of these maladies can be substantial. Some diseases are the result of bacteria, such as American and European foulbrood; a virus, such as sacbrood; or a parasite such as *Nosema apis.* Mites such as *Varroa jacobsoni* are considered to be the number one killer of honeybees globally. These huge ectoparasites live on the outside of the bodies of the bees and the brood. They survive by attaching themselves to the insect and drawing out its hemolymph, a fluid similar to blood. During the early 1900's, the acarine mite (*Acarapis woodii*), also known as the tracheal mite, was responsible for killing all the black honeybees in England. These mites live in the trachea of the bee, cutting off its airway and choking it. Understanding the biology of this mite is the result of extensive research in apiology. Another pest that can wreak havoc on a hive is the wax moth (*Galleria mellonella*), which destroys the comb and the brood. Because of the nature of commercial migratory beekeeping, these diseases and pests can spread rapidly as colonies from one apiary come into contact with colonies from other apiaries.

Research into the diseases impacting honeybees is crucial to the survival and management of the honeybee. Apiologists identify these diseases by noting symptoms, tracking their spread geographically, and finding methods for their control. Integrated pest management systems using both cultural mechanical controls (such as hive modifications) and genetic controls (such as breeding bees that are resistant to a particular kind of disease) have all emerged from apiologists' research.

Bee Genetics. The domestication of the honeybee has been an ongoing process in societies as far back as those in ancient Egypt. In the twenty-first century, the crossbreeding of honeybees is controlled through scientific methods such as the instrumental insemination of the queen. This requires surgical-grade laboratory instruments, which hold and prepare the queen and the selection of drone bees that contribute semen for the insemination. This technique of breeding ensures that the traits desired in these bees will be secured. Techniques such as these have allowed for the quick development of new hybrid strains of the honeybee. Benedictine monk Karl Kehrle, known as Brother Adam, of Buckfast Abbey in England, is best known for his contribution to honeybee breeding. By selecting queens with particular traits and behaviors from around the world, he created what is now known as the Buckfast bee. The Buckfast bee was bred for good temper, honey production, and resistance to the acarine mite. Apiology also has genetically produced queens that can combat varroa mites. For example, varroa-sensitive hygiene (VSH) queens have been developed with the ability to detect the presence of varroa within a colony. Upon detection, they begin to clean the brood chambers of larvae that contain the mite.

APPLICATIONS & PRODUCTS

Honey. Honey production is the primary commercial output of the honeybee colony. About 200 million pounds of honey are produced yearly within the United States. Consumption in the United States is twice that amount. Thus, honey is imported from sources around the world. The floral sources for nectar, the substance from which bees produce honey, is variable and determines the type and color of honey produced. Perhaps as many as three hundred varieties of honey can be produced in the United States alone, all having their own unique taste and color. Lighter-grade honeys and honey still in the comb are consumed directly and are usually sold at relatively high prices. Darker-grade honey may be used to sweeten baked goods. Popular floral sources for honey include clover, apple blossom, orange blossom, and huckleberry. Honey is sold in dehydrated form, with fruit flavors added to it, in comb form with capped and uncapped cells filled with honey, and as either organic or nonorganic regular or creamed honey. Aside from its use for humans as the world's oldest sweetener, honey is used in the production of paints, adhesives, cosmetics, and medicinal products.

Pollination. A major source of value for the beekeeper and the grower, and hence the public at large, is the immense contribution of pollination that bees make to agriculture. Honeybees are used to pollinate the blossoms of many vegetables and fruits. The distribution of bees to locations timed to coincide with the seasonal arrival of blossoms is known as migratory beekeeping and constitutes a large and important industry. Some research suggests that nearly one-third of a human's diet is composed of products pollinated by insects, and about 80 percent of this pollination is accomplished by honeybees.

Medicine from the Hive. Substances coming from the beehive as well as the sting of the bee itself have been considered medicinal products. "Apitherapy" is a collective term referring to the application of bee venom and other products from the hive for medical purposes. Although modern apitherapy takes a scientific approach, it has a rich history in folk medicine. It began as a healing art practiced by the early Greeks, Egyptians, and Chinese. In the 1800's, the idea of bee-venom therapy was introduced to the United States. Bee venom, taken either directly from being stung or by injection with a syringe, is used in the treatment of arthritis, rheumatism, tendinitis, and multiple sclerosis; it has also been used to treat the pain from gout, shingles, and other maladies. It is believed that bee-venom therapy can increase circulation of the blood. Apitherapy is also used to decrease inflammation of tissues and to stimulate immune responses.

Royal jelly is the substance made by worker bees to feed potential queen bees within the hive. This substance can also be used to treat open wounds and as an energy tonic. The extraction of royal jelly is a labor-intensive process. Propolis is a bee-produced substance reddish to brown in color with the consistency of sticky clay. The bees utilize propolis to bridge over areas within the hive that exceed bee space so as to seal in the hive. This product has antifungal and antibacterial properties, so it can be used medicinally.

Bees collect plant pollen as a nutritional source for the colony. For some people, pollen is an allergen. But some studies suggest that human consumption of raw pollen from particular locales, as well as the raw honey that contains these pollen spores, can build immunity to pollen allergies from the same locales.

Beeswax. The worker bees of the hive secrete wax through glands on the sides of their bodies. In the colony, this wax is used to make comb, which is the base for building up hexagonally shaped cells. After the advent of the removable-comb by Langstroth, frames within the hive contained a foundation made from beeswax with a hexagonal imprint. From this paper-thin wax sheet, the bees produce the honeycomb that is used in rearing brood and storing food for the colony.

Beeswax finds its way into many common substances and products. Beeswax is well known in its use in candle production. In art, melted beeswax is used as a base for batik designs on fabric and in producing encaustic paints. Additionally, beeswax can be carved into figures that are used as models in the production of metal casting.

Many cosmetics, as well as some foods, use beeswax. Beeswax is also used as a seal in tree grafting.

IMPACT ON INDUSTRY

Migratory Beekeeping Industry. The process of commercial pollination by bees translates into millions of bees doing labor that would otherwise have to be done by hand or by spraying pollen with an airplane. Both of these processes are costly. Bees—by virtue of both their efficiency, in visiting each flower in an orchard or field, and the fact that they do not need to be paid for their time—translate into millions of dollars saved in the production of food for human and animal consumption. Pollination by honeybees is limited only when growers misjudge blossom times, do not set sufficient numbers of colonies in the field, or confront weather that does not cooperate with the bees' flying time.

A symbiotic and economic relationship exists with honeybee pollinators and the plants they pollinate. Bees pollinating plants derive food and nectar for their colonies to survive for the winter. This can translate into an economic savings to the beekeeper. If plants or trees to be pollinated are profuse with pollen and nectar, the bees can store this food and feed themselves over the winter, requiring limited supplemental feeding from the beekeeper. Additionally, many migratory beekeepers are in business for pollination only and not honey production. This savings is critical to the migratory beekeeper's economic survival.

Perhaps the most attractive aspect of this industry is that it is among the few agricultural practices that have, for centuries, maintained the ecological principles of sustainability and have caused limited environmental degradation by avoiding mechanical means of pollination.

Research. Both public and private organizations spearhead the research efforts in apiology. At the national level, the U.S. Department of Agriculture maintains research stations in Tucson, Arizona; Baton Rouge, Louisiana; Logan, Utah; Beltsville, Maryland; and Weslaco, Texas. A number of universities have conducted ongoing research into areas such as colony collapse syndrome, Africanized bees or killer bees (an aggressive strain that began infiltrating the southern United States during the early 1990's), bee diseases, bee pests, and bee genetics, as well as apiary economics. At the state and local levels, multiple beekeeping associations in all fifty states, as well as the District of Columbia and Puerto Rico, assist in disseminating information and in some cases collecting and providing data to scientists conducting research.

The International Bee Research Association in the United Kingdom was established in 1949 and publishes a series of research journals, books, and pamphlets for the world beekeeping community. It also sponsors international conferences dealing with beekeeping issues. In the United States, publishing outlets for scientific research include the *American Bee Journal*, established in 1861 and published by Dadant and Sons, and Bee Culture, now online but originally established in 1873 and published by A. I. Root and Sons.

Honeybee Genome Project. Funded by the National Human Genome Research Institute and the U.S. Department of Agriculture, the effort to sequence the genes of *Apis mellifera* began in 2003 and was completed three years later. Interest in tracing the genes of the honeybee was based on its importance in agricultural production and its social structure. However, the project may also provide insight into human health conditions, such as immunity to disease and allergies.

Colony Collapse Disorder. Sometimes referred to as CCD, colony collapse disorder has made international news and describes the demise of millions of honeybee colonies. Given the economic benefit that bees provide in agricultural production, research into this problem is crucial. Although researchers think that a virus might be the cause of this bee malady, scientists are quick to point out that since the mid-twentieth century the varroa mite has been the most damaging pest to the bee industry worldwide. Many researchers suggest that varroa or a combination of this mite and some other factor might be the cause of CCD. Fighting the varroa-mite outbreak has required both integrated pest management systems and treatment with pesticides.

SOCIAL CONTEXT & FUTURE PROSPECTS

Although mason bees, such as the orchard bee (*Osmia lignaria*), have garnered some attention as an alternative pollinator, as a honey producer and pollinator the honeybee still reigns supreme. The impact of honeybee pests such as varroa has already exposed the public to the importance of the honeybee in the production of food and to the potential of rising food costs if the demise of the honeybee continues. Further research into the interaction of bees with pesticides and herbicides as well as other environmental agents will continue to be critical research topics. Bee management systems will continue to be updated to accommodate environmental changes. Such changes are already being observed in some areas where reestablishing new colonies on a yearly basis is common practice. This requires the raising of more bees as well as new queens. Techniques such as instrumental insemination, which speeds the process of genetic modifications outside simple breeding programs, will also continue to develop as demand for bees with resistance to pests and diseases increases. Out of economic necessity, research will continue to develop around the honeybee. Its integration into the economic fabric of agriculture is far too great to ignore.

Further Reading

Hilmi, Martin, Nicola Bradbear, and Danilo J. Mejía-Lorío. *Beekeeping and Sustainable Livelihoods.* Rome: Rural Infrastructure and Agro-Industries Division, Food and Agriculture Organization of the United Nations FAO, 2012. Print

Connor, Lawrence John, ed. *Asian Apiculture: Proceedings of the First International Conference on the Asian Honey Bees and Bee Mites.* Cheshire, Conn.: Wicwas Press, 1993. Overview of the impact of Asian mites such as varroa.

_______. *Queen Rearing Essentials.* Kalamazoo, Mich.: Wicwas Press, 2009. A discussion of the process of grafting and selecting queen bees to improve production.

Cook, Albert John. *Manual of the Apiary.* 3d ed. Chicago: T. G. Newman & Son, 1878. This text is an interesting view on apiary science from the past, much of which still has applications.

Graham, Joe M., ed. *The Hive and the Honey Bee: A New Book on Beekeeping Which Continues the Tradition of "Langstroth on the Hive and the Honeybee."* Rev. ed. Hamilton, Ill.: Dadant, 1992. Offers history and background on modern foundations of apiculture.

Root, A. I., and E. R. Root. *The ABC and XYZ of Bee Culture: A Cyclopedia of Everything Pertaining to the Care of the Honey-Bee: Bees, Hives, Honey, Implements, Honey Plants, Etc.* 41st ed. Medina, Ohio: A. I. Root, 2007. As the rest of the subtitle of this venerable source states, this volume contains "facts gleaned from the experience of thousands of beekeepers and afterward verified in our apiary." A key resource for apiologists.

Seeley, Thomas D. *Honeybee Democracy.* Princeton, N.J.: Princeton University Press, 2010. An entomologist surveys the collective behavior of honeybees.

Spivak, Marla, and Gary S. Reuter. *Honey Bee Diseases and Pests.* St. Paul: University of Minnesota, 2010. This book is an overview of bee diseases and treatment methods.

Web Sites

Bee Culture http://www.beeculture.com

International Bee Research Association http://www.ibra.org.uk

National Human Genome Research Institute Honey Bee Genome Sequencing http://www.genome.gov/11008252

U.S. Department of Agriculture http://www.usda.gov

Archaebacteria

—Bruce L. Stinchcomb

FIELDS OF STUDY

Biochemistry; Environmental Chemistry; Geochemistry; Photochemistry; Environmental Sciences.

PRINCIPAL TERMS

eubacteria: minute, prokaryotic organisms that inhabit a range of habitats considerably greater in diversity than those occupied by other organisms; exclusive of archaebacteria, they constitute the majority of monerans

eukaryotic cell: the cell type present in all animals, plants, fungi, and protists; each cell has a distinct nucleus and mitochondria, chloroplasts, and other subcellular structures absent in prokaryotic cells

microbialite: a biogenic sedimentary structure that is found fossilized in sedimentary rock strata of various ages and that is attributed to the life activities of monerans

monerans: generally, single-celled organisms that often grow in colonies and that are made of prokaryotic cells

prokaryotic cell: the cell type found in the kingdom monera, characterized by a number of criteria, including the absence of a cell nucleus, mitochondria, and chloroplasts

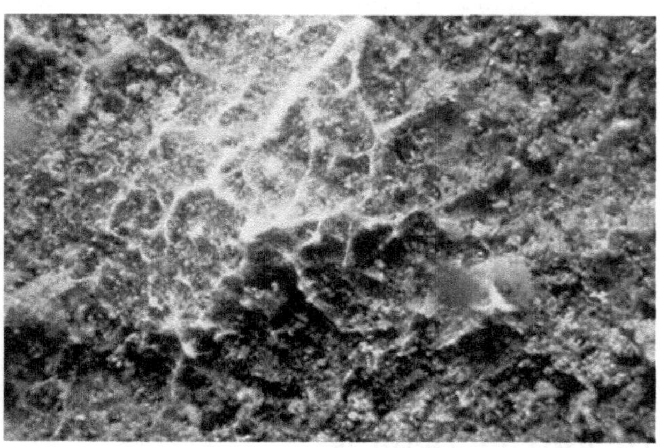

Bacterial mats form on the sea floor, where oxygen concentration is low, and may provide insight into the earliest life-forms on Earth. Image by National Oceanic and Atmospheric Administration.

ribosome: a large multienzyme complex made up of protein and ribonucleic acid (RNA) molecules that carry and process information stored in deoxyribonucleic acid (DNA); this information is also carried by RNA to synthesize proteins

stromatolite: a biogenic sedimentary layered structure produced by sediment trapping, binding, or precipitation as a result of the photosynthesis of microorganisms, principally cyanobacteria (blue-green algae)

SUMMARY

Archaebacteria are primitive, one-celled life-forms without a distinct nucleus, different from bacteria in their genetic components. They have been found to be genetically unique and are probably one of the earth's earliest life-forms.

EARLY LIFE-FORMS

The nature of the earth's earliest life-forms has always been an intriguing question for both the earth sciences (paleontology) and biology. The fossil record shows that one-celled organisms are very ancient, their oldest-known fossils being almost 3.5 billion years old. The long fossil record of prokaryotes consists of both preserved fossil cells and distinctive layered mineral structures called stromatolites and microbialites; these structures were produced from the cell metabolisms of colonies of prokaryotes. From at least 3.5 billion years ago to around 1 billion years ago, microscopic prokaryotes were the earth's only organisms. They included, as they do today, a diversity of forms commensurate with their long evolutionary history. The appearance of the eukaryotic cell (more than 1 billion years ago) ushered in the age of multicellular organisms (metazoans and metaphytes) some 700 million years ago, and these have become the dominant life-forms on the earth. What the much earlier prokaryotic organisms were like and which type of one-celled organisms produced the microbialites is unclear. Usually these oldest of fossils are attributed to the life activities of photosynthetic organisms, particularly the cyanobacteria, referred to in many works as the blue-green algae. A number of other types of one-celled life-forms could have been responsible for some of them, particularly the photosynthetic bacteria and possibly the archaebacteria.

Molecular biologists, utilizing ribonucleic acid (RNA) nucleotide sequencing and other biochemical methods, believe that the nature of early life-forms can be discovered. RNA nucleotide sequences of amino acids can be regarded as a sort of chemical "historical document" that is capable of being "read." The closer two RNA nucleotide sequences are to each other, the smaller is the evolutionary distance between them and the more recently in geologic time they separated from each other. The further one nucleotide RNA sequence is from another, the greater is the evolutionary distance that separates the two organisms. Utilization of nucleotide sequencing can thus produce an evolutionary tree, or phylogeny, for even the most primitive of organisms, and therefore it becomes possible to determine which organisms out of the great variety of primitive life-forms currently living were some of the first to appear. Through information obtained by such sequencing, archaebacteria have been recognized by molecular biologists as some of the most primitive and biochemically unique of organisms. Archaebacteria RNA sequences turn out to be distinctly different from those of other bacteria, even though the various organisms that comprise archaebacteria look like and were previously placed with the bacteria.

Archaebacteria is a polyphyletic group that once included species from two domains, Archaea and Bacteria. There are now considered to be three domains: Archaea, Bacteria, and Eucaryota. Although there are similarities in cell structure and function among Bacteria and Archaea, and some trees group the Archaea and Bacteria, the relationship is currently unclear. In both their tolerance of extreme ecological conditions and their metabolism, the archaebacteria differ from all other monerans, a condition that has led biologists to consider these organisms as particularly well suited to the adverse conditions of the early earth. The very earliest eras of geologic time may well have been the age of archaebacteria.

MOLECULAR CHARACTERISTICS

On a fundamental molecular level, archaebacteria are different in their biochemistry from the other prokaryotes. Nucleotide RNA sequences and other biochemical differences that exist between the various types of eubacteria (bacteria exclusive of the archaebacteria) are minor when compared with the differences between eubacteria and the archaebacteria. Nongenetic differences include such features as cell walls, those of all eubacteria being composed of a complex polymer called peptidoglycan, which is a sugar derivative. In contrast, cell walls of the various types of archaebacteria are composed of a variety of other materials, none of which is peptidoglycan. The lipids (fats) in archaebacteria cells are also fundamentally distinct from the lipids in the cells of both eubacteria and eukaryotes. Ribosomal RNA is what ultimately distinguishes the archaebacteria, for it is markedly different in its sequences of bases from any eubacteria. In higher eukaryotic organisms, where the fossil record is good, greater RNA ribosomal differences exist between those organisms that are separated by long periods of geologic time than between those that are separated by shorter periods of time. Ribosomal RNA differences and the biochemical differences that exist between the eubacteria and the archaebacteria suggest that an evolutionary distance of great magnitude separates them.

The eukaryotic cell has long been observed to be a sort of combination between prokaryotic-type cells, functioning within the cell as chloroplasts and mitochondria, and another cell type that "ingested" the prokaryotes and incorporated them to become a more complex entity in symbiotic collaboration. Archaebacteria have some genetic characteristics that suggest a link with the eukaryotes, which has led some scientists to propose a predecessor to them both: The "other" cell type that linked up with prokaryotes underwent substantial further evolution and eventually became the eukaryotic cell. The pre-eukaryotic other cell type is known as a urkaryote. All three of these cell types—the prokaryote, the urkaryote, and the eukaryote—are hypothesized to have arisen from a common cell ancestor, the progenote. The progenote may have been biochemically simpler than any of the three fundamental life-forms that arose from it, an event that might have taken place during the first 1 billion years of earth history.

THERMOACIDOPHILES

Archaebacteria are represented by three classes: the thermoacidophiles, the extreme halophiles, and the

A Precambrian stromatolite formation from the Siyeh Formation, Glacier National Park, Montana. Image by U.S. Geological Survey.

methanogens. The thermoacidophiles occupy hot, acid environments, often rich both in metallic ions and in sulfur compounds, such as hot springs and fumaroles. These organisms are viable under the acidic hot conditions intolerable to other life-forms, with temperatures as high as 75 degrees Celsius. The extreme halophiles, or halobacteria, live only in extremely salty environments. The methanogens are anaerobes that metabolize organic material to form methane; they were the first of the archaebacteria to be discovered.

Igneous activity of various sorts appears from the geologic record to have been much more intensive and widespread on the early earth (between 2 and 4 billion years ago) than during more recent geologic times or at present. A terrestrial geologic record for the first 1 billion years of the earth's history is unknown, as the widespread igneous and tectonic activity during this time seems to have destroyed the evidence; an actual record begins at nearly 4 billion years ago. During the next 1.5 billion years (the Archean eon), igneous phenomena and massive tectonism were still dominant. Hot-spring and fumarolic activity would have been more commonplace during these early times than during later geologic time, and these environments favor the thermoacidophiles. Although the fossil and sedimentational record of the Archean eon does not negate the possibility that archaebacteria were some of the most widespread and dominant life-forms of that time, determination that a particular organic—or presumed organic—structure of the early earth was produced by archaebacteria or by some other moneran is quite difficult and may well be impossible. A number of puzzling structures, seemingly of biogenic origin, have been reported from Archean strata. Some of these stromatolite-like or microbialite-like structures are associated with what appear to be hot-spring deposits. These structures may well represent minerals deposited as a consequence of life activities of thermoacidophiles associated with geothermal activity. Like so many stromatolite-like and microbialite-like structures, unequivocal proof as to their biogenic origin is difficult to obtain. Structures similar to them, however, are produced today in hot springs, in the hottest waters of which live communities of thermoacidophile archaebacteria.

Sometimes hot-spring deposits and structures contain carbon-rich sediments or graphite. In addition, some hydrothermal veins of various ages have associated carbonaceous or graphitic material. It is possible that such material might have originated from thermophyllic archaebacteria. Stratified metallic element deposits are known from Archean strata, some of which have been thought to be of biogenic (possibly archaebacterial) origin. Some of these deposits yield microbialites exhibiting distinctive dome, finger, or layered structures containing metallic oxides or carbonates. Today, such structures are produced by the cyanobacteria and by the photosynthetic bacteria, but these younger stromatolites lack components like the oxidized metals. Other Archean stromatolites or microbialites and stromatolite-like structures associated with geothermally active environments have a distinctive "signature" different from later forms, and their origin by thermoacidophile archaebacteria cannot be ruled out.

EXTREME HALOPHILES

The second group of archaebacteria, the extreme halophiles, requires an intensely saline environment. Shallow, marginal marine areas, evaporite basins, and salt flats are the niches in which these organisms generally flourish. Physiologically, the extreme halophiles are photosynthetic; however, the photosynthetic pigment is not chlorophyll, but rather a light-sensitive red pigment, bacterial rhodopsin. The cell walls of the extreme halophiles differ from those of other bacteria in the presence of compounds that prevent destruction of the walls in the high salt concentration conditions under which they live. The chemical similarities of ribosomes and lipids of both the extreme halophiles and the methanogens suggest a common origin.

Again, the fossil record of these organisms is difficult to interpret; some biologists suggest that the halophiles were more prevalent early in the earth's history than they are today. Fossil rod-shaped bacterial cells have been found as far back as the mid-Archean (3.2 billion years ago); however, as the gross morphology of archaebacteria differs little from that of eubacteria, the evidence remains inconclusive.

Peculiar and distinct microbialites of Archean age that are associated with radial sprays of gypsum crystals were described from western Ontario in the 1910's by Charles Doolittle Walcott. Walcott, a pioneer North American paleontologist who concentrated on the early (Precambrian and Cambrian) fossil record, made many finds of peculiar structures resembling fossils in Precambrian strata, many of which remain a mystery. Walcott thought that the radiating gypsum crystals were the rays, or spicules, of a type of spongelike organism he called atikokania. Associated with Walcott's atikokania are distinctive microbialites that contain "lenses" of gypsum that almost certainly originated in a very saline environment. These microbialites could possibly represent the product of physiological activity of the extreme halophiles when, during the process of photosynthesis, they locally removed carbonic acid from the saline water. The white lenses that characterize these distinctive microbialites are gypsum fillings between the black calcium carbonate bands, possibly precipitated by photosynthesis of halophilic archaebacteria.

METHANOGENS

Methanogens produce their metabolic energy either from the breaking down of organic compounds incorporated into sediments or from the reduction of carbon dioxide in the presence of elemental hydrogen, with the consequent release of methane. Were it not for the methanogens, organic carbon would eventually become incorporated into the sediments of the earth's crust, where it would accumulate and could not be recycled back into the biosphere; the methanogens facilitate this recycling of carbon. Methanogens, like the other archaebacteria, have biochemical features distinct from all other bacteria, suggesting that they evolved separately from them. Like the other archaebacteria, methanogens differ from other prokaryotes in the sequences of nucleotides that make up the RNA in their ribosomes and protein. Fossil methanogens are more difficult to distinguish within the geologic record than other archaebacteria, however, as they leave no distinctive chemical "footprint," as the others can. The abundance of black, carbon-rich sediments in strata of the Archean eon suggests that the oxygen-free, anaerobic environment in which the methanogens flourish was commonplace during that time. The methanogens' biochemical uniqueness, and thus presumed great geologic age, along with the anaerobic Archean Earth environment, suggests that they may have been a dominant part of the Archean biosphere and not restricted as they are today.

FURTHER READING

Garrett, Roger A., and Hans-Peter Klenk, eds. *Archaea: Evolution, Physilogy, and Molecular Biology*. Malden, Mass. Wiley-Blackwell, 2007. A compilation of scientific reviews and specialist articles discussing the Archaea domain. Discusses the phylogenetic research conducted on these organisms as well as the physiology and specifically metabolic pathways of Archaea. Written for a scientific audience with a biology background.

Gunde-Cimerman, Nina, Aharon Oren, and Ana Plemenita. *Adaptation to Life at High Salt Concentrations in Archaea, Bacteria, and Eukarya*. New York: Springer, 2011. This text provides information on high-salinity habitats and the halophilic organisms found there. The environment overview is followed by sections devoted specifically to Archaea, bacteria, fungi, protozoa, and viruses.

Howland, John L. *The Surprising Archaea: Discovering Another Domain of Life*. New York: Oxford University Press, 2000. Provides an all-encompassing knowledge of Archaea, from their evolution to their ecology. The author captures the excitement of their discovery within the scientific community.

Kandler, Otto, and Wolfram Zillig, eds. *Archaebacteria Eighty-five: Proceedings of the EMBO Workshop on Molecular Genetics of Archaebacteria*. Forestburgh, N.Y.: Lubrecht and Cramer, 1987. A proceedings volume on molecular genetics, biology, and biochemistry of archaebacteria. Two papers are concerned with the geologic and paleontologic record of archaebacteria: "Traces of Archaebacteria in Ancient Sediments," by J. Hahn and Pat Haug, and "Morphological and Chemical Record of the Organic Particles in Precambrian

Sediments," by H. D. Pflug. The latter paper illustrates a wide variety of microstructures from Precambrian sediments and discusses possible pathways by which archaebacteria and other prokaryotes could have been responsible for the concentration of many metallic ore deposits throughout various parts of the Precambrian. Other papers such as "Archaebacterial Phylogeny: Perspectives on the Ur Kingdoms," by C. R. Woese and G. J. Olsen, present biochemical reasons substantiating the uniqueness of archaebacteria from both the eubacteria and the eukaryotes. Other papers probe the biochemical similarities with eukaryote cytoplasm and peculiar substrate requirements.

McMenamin, Mark. *Discovering the First Complex Life: The Garden of the Ediacara.* New York: Columbia University Press, 1998. This entertaining study of the earliest complex life-forms on the planet details the author's work on these organisms. Written for the interested student but understandable by the general reader.

Schopf, J. William, ed. *Major Events in the History of Life.* Boston: Jones and Bartlett, 1992. An excellent overview of the origin of life, the oldest fossils, and the early development of plants and animals. Written by specialists in each field but at a level that is suitable for high school students and undergraduates. Although technical language is used, most of the terms are defined in the glossary.

Woese, Carl R. "Archaebacteria." *Scientific American* 244 (June 1981): 98–122. One of the most comprehensive articles available on the archaebacteria. Distinctive attributes characteristic of the archaebacteria as determined through molecular biology are enumerated. The author was one of the workers originally involved in the discovery of the biochemical uniqueness of archaebacteria.

B

Bees & Other Pollinators

—*Megan E. Watson*

FIELDS OF STUDY

Apiculture; Apiology; Biology; Environmental Studies; Virology;

PRINCIPAL TERMS

monoculture: reliance on a single crop species over a large area, rather than on a variety of crop species

pollination: the transfer of pollen from the stamens of a flower to the pistil of the same or another flower

symbiosis: a mutually beneficial relationship between two different organisms, enabling both to thrive

SUMMARY

Bees and other pollinators include flying insects and other biological agents that facilitate plant reproduction by moving pollen among seed plants. Pollinators such as bees are extremely important to the health of the planet because without them most plant species would disappear. Pollinating species are in decline around the world, a trend that threatens biodiversity and food supplies for many animals, including humans.

POLLINATORS

Bees are the best-known group of pollinators; however, other insects such as butterflies, wasps, beetles, and ants are pollinators, too, along with some species of bats and birds. Plants and their pollinators coevolved, forming symbiotic relationships in which plants provide food for the pollinators, which in turn aid the plants' reproduction. Many other species evolved over time to take advantage of this fundamental relationship between plant and pollinator, including humans and other animals that eat the results of the fertilized plants, such as fruits, nuts, seeds, and berries. Pollinators also contribute to the web of life by ensuring the continued survival of plants through the process of pollination; the loss of pollinators can affect the biodiversity and well-being of a whole ecosystem. By the early twenty-first century, declines in pollinator populations around the world led environmental groups and government agencies to include pollinators in their conservation efforts.

POLLINATOR DECLINE

The decline of pollinating species has many causes, all for the most part related to human activity. One cause is the use (and misuse) of pesticides meant to kill other insects, such as mosquitoes; such pesticides also indiscriminately kill off other insect populations, many of them pollinator species. Another cause of pollinator decline is the large-scale transport of pollinators around the world for agricultural purposes; this practice also results in the large-scale transport of parasites and diseases that affect them, as well as the introduction of invasive species that can compete with or destroy native pollinator populations.

Human expansion into previously wild or relatively untouched spaces contributes to pollinator decline through habitat loss and degradation, as development brings with it pollution, hive destruction, and fragmentation of traditional "nectar corridors." Wild pollinators have fewer places to nest, mate, and roost safely, and their historic ranges have been destroyed or partitioned in such a way that it is far more difficult for them to find the nectar they need to survive. Many pollinators are already on endangered or threatened species lists, including more than one hundred avian species and dozens of mammals.

Squash bees (Apidae) are important pollinators of squashes and cucumbers. Image by USDA ARS.

Climate change has also affected pollinators. A 2015 study found that warming temperatures had caused bumblebees' range to shrink by as much as 190 miles since the 1970s, with the bees staying farther and farther north as the climates in the southern parts of Europe and North America became less hospitable. In March 2017, the once ubiquitous rusty patch bumblebee (*Bombus affinis*) was added to the U.S. Fish and Wildlife Service list of endangered species.

THE IMPORTANCE OF BEES

The thousands of species of bees in the world are widely known for pollination and for their making of honey and beeswax. Bees are the primary pollinators of flowers around the world and are also responsible for the pollination of approximately one-third of the human food supply. Many agricultural crops, ranging from watermelons to cashews, rely on pollination by bees. European honeybees, in particular, are extremely important to agriculture because of humans' ability to manage the species.

Pollination management, also called contract pollination, entails transporting honeybees in their hives to crops that are in need of pollination. Modern monoculture farming has created a situation in which the normal pollinators of some kinds of crops cannot pollinate them, either because of the huge size of the fields or because the pollinators have declined or died off. Honeybees, however, can be put in the fields in such large numbers that they make up for the fact that they are often less efficient pollinators than the native species that previously would have pollinated a particular crop. The economic viability of monoculture agriculture often depends entirely on pollination management, and most professional beekeepers in the twenty-first century focus on contract pollination rather than honey as their primary source of revenue.

The same factors that have caused declines in other pollinator populations since the late twentieth century have affected honeybees as well, causing a widespread honeybee population decline. In addition, across North America and Western Europe since at least 2006, honeybees have disappeared in what is called honeybee colony collapse disorder (CCD), the cause of which remains a mystery. CCD is characterized by the worker bees flying off and never returning to the hive or colony, leaving behind a healthy queen, an unhatched brood, and food supplies.

Given the importance of honeybees to agriculture around the world, scientists have investigated many possible causes for CCD. These possible causes include a parasitic mite, insect diseases, and the use or misuse of insecticides. Some researchers have reported evidence that the cause of CCD may be viral. These possibilities are aggravated by other factors negatively affecting honeybee populations, such as poor diet related to monoculture, which can lead to malnutrition or impaired immune system, and migratory beekeeping, which spreads parasites and diseases among populations. One such disease, known as deformed wing virus after its chief symptom, became a serious threat to honeybee and bumblebee populations in the 2010s; a 2016 study found that the trade in pollinators was directly responsible for spreading the virus around the world. Among other pollinators including some species of birds, bats, beetles, and moths, light pollution is considered a factor in reduced nocturnal pollination efficiency.

Many pollinators are very small, but their importance to terrestrial life is great. Pollinators are a major part of the web of life, and what happens to them can affect every other creature in an ecosystem. Plants, whether they are cacti in the desert of Mexico or blueberries in Scotland, need their pollinators to survive—without pollinators a plant cannot reproduce and its species will go extinct. If this occurs, the pollinators adapted to that plant will no longer have a food supply and will either starve to death or switch to another source of nectar, inevitably intruding on another animal's food supply. As the plants disappear for lack of pollinators, animals such as birds, small mammals, and lizards that eat the fruit, nuts, and seeds produced by pollinated plants will also begin to starve. This effect will ripple through the food web until it reaches larger animals, including human beings.

Further Reading

Carswell, Cally. "Bumblebees Being Crushed by Climate Change." *Science News*. AAAS, 9 July 2015. Web. 1 Oct. 2015.

Droege, Sam, and Laurence Packer. Bees: An Up-Close Look at Pollinators Around the World. Minneapolis, MN: Voyageur Press, 2015. Print.

Fischer, David, and Thomas Moriarty. *Pesticide Risk Assessment for Pollinators*. Chichester: Wiley, 2014. Print.

Grimm, V., et al. "Ecological Modeling for Pesticide Risk Assessment for Honey Bees and Other Pollinators." *Pesticide Risk Assessment for Pollinators* (2014): 149–62. Print.

James, Rosalind R., and Theresa L. Pitts-Singer, eds. *Bee Pollination in Agricultural Ecosystems*. New York: Oxford UP, 2008. Print.

Kareiva, Peter, and Simon A. Levin. *The Importance of Species: Perspectives on Expendability and Triage*. Princeton, NJ: Princeton University Press, 2015. Print.

Kavanagh, James. *Waterford's Discovery Guide Bees and Pollinators: A Folding Pocket Guide to Familiar Species*. Dunedin, FL: Waterford Press, 2016. Print.

Kennedy, Merrit. "Report: More Pollinator Species in Jeopardy, Threatening World Food Supply." *NPR*. NPR, 26 Feb. 2016. Web. 2 Mar. 2016.

Miller-Struttmann, Nicole E., et al. "Functional Mismatch in a Bumble Bee Pollination Mutualism under Climate Change." *Science* 25 Sept. 2015: 1541–544. Print.

Patiny, Sebastian. *Evolution of Plant-Pollinator Relationships*. Cambridge: Cambridge University Press, 2012. Print.

"Pollinators." *US Fish & Wildlife Service*. Dept. of the Interior, 1 Mar. 2016. Web. 2 Mar. 2016. Print.

St. Fleur, Nicholas. "Climate Change Is Shrinking Where Bumblebees Range, Research Finds." *New York Times*. New York Times, 9 July 2015. Web. 1 Oct. 2015.

Wilfert, Lena, et al. "Deformed Wing Virus Is a Recent Global Epidemic in Honeybees Driven by *Varroa* Mites." *Science* 351.6,273 (2016): 594–7. Print.

Biodiversity

—Peter L. deFur

FIELDS OF STUDY

Conservation Biology; Ecology; Environmental Studies;

PRINCIPAL TERMS

aquaculture: the artificial growth of animals or plants that live in the water; the culture of something living in water

biodiversity: the number and kinds of animals and plants and the variations in genetic material they possess

exotic species: organisms that are not naturally found in a place but have been artificially introduced, whether by accident or intentionally

extinct: no longer found anywhere on earth

extirpated: not found in an immediate local area but found elsewhere on earth

habitat: an assemblage of plants, animals, and the physical land, water, minerals, and other elements that support a life-form; usually refers to a single species

invertebrate: an animal without a backbone; for example, worms, clams, crabs, insects, and jellyfish

species: a group or groups of interbreeding natural populations of organisms that are reproductively isolated from other such groups

subspecies: a group or groups of interbreeding organisms that are distinct and separated from similar related groups but not fully reproductively isolated

threatened species: animals or plants so few in number that they may soon be endangered and then extinct

SUMMARY

Biodiversity refers to the total number of animals and other forms of life, including different species and genetic variants.

The study of biodiversity is relatively new. The word was coined by Edward O. Wilson in order to explain the biological living resources and wealth lost when whole areas of tropical rain forest were destroyed by clear-cutting and burning. Many, if not all, of these tropical rain forests were in remote areas, often in developing countries, untouched by modern civilization, and relatively unstudied by scientists. As a result, many of the plants and animals, especially insects, were unknown to scientists. Many of the species lived nowhere else on earth and, once destroyed, became extinct. Scientists realized that these tropical rain forests held a great wealth of species and that their destruction was removing large numbers, perhaps millions, of species, along with the DNA that made each species unique. Wilson referred to this loss of species as a loss of the forests' biological diversity, or biodiversity. The loss is not merely of the species that become extinct but also of the combination of highly specific genetic material in each species.

Because the study of biodiversity involves the study of individual species, their loss, and their habitat, the subject is closely tied to endangered species, ecology, and conservation biology. Endangered species are those plants and animals that are nearly extinct; threatened species are not yet at the point of being endangered but will be if measures are not taken soon. Both plants and animals, including insects, worms, clams, crayfish, and other creatures, may be considered either threatened or endangered. Specific habitats, especially those that are unique and found in only one location on earth, can also be considered threatened. Protecting individual species often means protecting the habitat the species needs for survival.

During the nineteenth century and the first part of the twentieth, some animals were hunted to extinction. Several other species were hunted to near extinction or were extirpated (removed) from large areas, largely by extensive changes to their habitats. As a result of the losses and near losses, some scientists, resource managers, and political leaders instituted programs and activities to prevent extinction and to restore species that were near extinction. Animals (and later plants) that were near extinction were officially designated as endangered and given legal protection to prevent their extinction. Eventually these efforts grew to encompass a much broader field of work by experts in many fields. In 2018, the Center for Biological Diversity estimated that species extinctions occur now at between 1,000 and 10,000 times the normal background rate of 1 to 5 species per year.

One of the more difficult aspects of protecting a single species is the importance of the habitat to the well-being of the species. This fact is one of the underlying principles driving the whole study of biodiversity. Scientists have learned that the habitat is critical for restoration of a species and that protecting the habitat is a powerful means of preventing species loss and protecting biodiversity. As a result, programs and individuals trying to protect or restore biodiversity or individual species have come into conflict with those who use or plan to use large areas of land and whole habitats for purposes such as housing or industrial development, harvesting, and aquaculture.

THREATS TO BIODIVERSITY

Scientists generally believe that three major factors cause the loss of biodiversity: resource overharvesting, habitat loss, and introduction of exotic species. Overharvesting or excessive hunting and catching caused the loss of species such as the great auk, passenger pigeon, and Steller's sea cow. Some species of whales nearly suffered the same fate. However, for the vast majority of animals and plants, overharvesting is not a major cause of endangerment or extinction. Many animals that are commonly harvested have large and dispersed populations, so harvesting stops before it causes populations to decline to the point of endangerment and extinction. This does not mean that overharvesting is not a problem, but that it is seldom the sole cause of population declines and not the most frequent cause.

A number of international controls have been implemented to prevent, limit, or discourage the killing of some of the animals that are threatened by overharvesting in many countries around the world. Many such controls ban the import or shipping of either live animals (tropical birds such as parrots) or parts of them (elephant ivory and rhinoceros horn). The animals endangered by overharvesting—the rhinoceros, elephant, and parrot—tend to be long-lived and restricted in distribution. Their reproductive rates are low, and populations often require a large habitat. The habitats of many of these animals are threatened by human activities, which makes the effect of their being slaughtered for commercial reasons that much more pronounced. In 1997, one of the major international treaties governing commercial transactions in rare and endangered species was modified to permit some African nations to sell stocks of certain animal products that had been previously prohibited. Some experts worried that the change in the laws would again provoke illegal overharvesting.

Arguably the most vile form of overharvesting is poaching. Overharvesting suggests that there is a legal and acceptable permissibility to the capture or killing of certain animals, and that the legally allowed limit of the harvest has been exceeded. Poaching, however, has no legally acceptable limit, and those involved in the practice do so entirely with criminal intent. African elephants are the prime example of this. In most western countries, possession of elephant ivory and articles made from it carries high penalties, and conservation officers in African countries are authorized to shoot poachers on sight. But because elephant ivory commands a high price in other countries that do not have such strictures regarding possession of ivory, there exists a lucrative market for ivory that drives poachers in Africa to seek out the animals and risk their own deaths in order to obtain the ivory for sale on the black market. In 2018, poachers armed with high-powered automatic rifles attacked and killed a group of some ninety elephants near a wildlife sanctuary in Botswana, cut out their tusks with chain saws, and left the bodies where they fell. It is estimated that one-third of all African elephants have been killed through the previous decade.

Many animals are threatened in their native habitats by exotic species introduced into the area. These exotic species often cause biological pressures that the native species cannot handle. They may threaten native species directly by competing for food, living space, critical habitat (such as nesting space), or other necessary environmental resources. Some exotic animals are aggressive and drive away or attack the native ones. In North America, the sea lamprey was introduced into the Great Lakes and has driven populations of many native species to precipitously low levels or even extirpated them, while causing physical damage to their living prey.

Exotic species may also present indirect threats by introducing diseases, parasites, or hitchhiker species that arrive with the exotic. In South Carolina and Texas, exotic shrimp are used in aquaculture, and they can carry a deadly virus that may threaten native shrimp if released into the wild by accident. Zebra mussels were carried into the Great Lakes water system in the ballast water of ocean tankers. With no natural controls, zebra mussels have multiplied to the point that they are not only a nuisance to human activities but also serious competition for native aquatic animals.

HABITAT LOSS & DEGRADATION

The most significant threat to biodiversity, as occurs in tropical deforestation, is habitat loss or degradation, including poisoning. Habitat loss is the elimination of a habitat, often by cutting, plowing, filling, or other construction or harvesting methods. When all the trees in a forest area are felled for wood (a practice known as clear-cutting), the habitat is lost until the trees regrow, an event that may never take place. Degradation is the elimination or lessening of some characteristic of the habitat that is critical for one or more species. Changing the temperature or speed of the water flowing in a stream can render it no longer suitable for a particular fish or mussel. Filling or draining a wetland alters the moisture content of the soil and changes its suitability for plants or animals.

One of the earliest examples of habitat degradation was the poisoning of bald eagles and peregrine falcons in the United States by the pesticide dichlorodiphenyl-trichloroethane (DDT). DDT was widely used throughout the world (and still is in some countries), is highly persistent, and accumulates in animals at ever-increasing levels in a trophic system (a system in which one animal feeds on another that feeds on another, and so on). Falcons and eagles were some of the birds of prey that took in so much DDT that they were poisoned and their eggs became too thin-shelled to survive. Their populations began to decline from reproductive failure. Many species of freshwater mussels are highly sensitive to chemicals and do not survive even low levels in the water. Many of the mussel species have become rare and endangered as a result of this poisoning, and they are not the only species to suffer from environmental poisons.

Altering stream and river flow via dams, diversions, withdrawals, and dredging is another form of habitat degradation that can dramatically change the way animals and plants use the aquatic system. In river systems throughout the Pacific Northwest, salmon return from the ocean to spawn (lay eggs and reproduce) in the river where they grew from egg to fish before migrating to the open ocean. If a dam blocks the river or if the conditions in the river are no longer suitable, the salmon are not able to spawn. In salmon, the fish that spawn in a given river are distinct and genetically identifiable, almost forming a subspecies. Therefore, authorities in the Pacific Northwest have had to determine the relationship between individual spawning groups and the watershed habitat before altering streams and rivers.

The cutting of mature forests in the Pacific Northwest and in Alaska eliminated the habitat for the spotted owl. Great controversy resulted from limits, restrictions, or bans on logging in these forests, which were prized for the high quality lumber that could be made from the trees. These and similar conflicts formed the basis for many studies on scientific, legal, and policy matters concerning the Endangered Species Act in the United States.

Controversies have arisen over habitat protection programs and laws in the United States and other countries because of the restrictions and limitations imposed on development, logging, and other human activities. People on one side argue that the potential benefits from protecting a species are greater than any inconvenience or costs that might ensue, and those on the other side argue that the economic benefit from the activity that threatens the species outweighs any ecological concerns. However, in some cases, companies and governments have worked hard to preserve species believed to be of economic value.

BIODIVERSITY RESEARCH

Biodiversity is studied at the level of populations, of individuals (including molecular biology), and in the interactions between organisms and their environment. Research on biodiversity includes genetic, ecological, and behavioral studies and investigations focusing on diseases and life histories.

Much of the work on any specific species, especially large vertebrates, is conducted in the field, using techniques for observing, counting, tracking, and monitoring the population. Most invertebrates cannot be studied other than by observation and counting. One of the goals of ecological research on biodiversity is to understand the life histories and ecology of individual species. Scientists seek to explain life span, reproductive patterns and strategies, food preferences, environmental requirements, and limits. Other major goals are to describe the habitat requirements and better understand the relationship between the species and the habitat.

The management and legal aspects of various programs are included in biodiversity studies. This research is closely related to the field of conservation biology. Such study entails examining areas such as how programs work; what money is spent, by whom, how, and for what use; whether habitats or species are protected, restored, or lost; and the time required for various activities. These considerations are generally part of program evaluation, a topic that can be more thoroughly researched in the field of management science. The legal aspects of preserving biodiversity and protecting endangered species have been studied by a number of legal and policy-making organizations, including the Environmental Law Institute and Environmental Defense Fund, both of Washington, DC. The National Research Council, the operating arm of the National Academy of Sciences, has published a number of reports on the Endangered Species Act, and these usually address legal, policy, and science issues.

Modern molecular genetic techniques are applied to the study of individual species of plants and animals and their restoration. Tissue, often a blood sample, taken from a single individual is analyzed to determine the genetic relations among members of a population. Scientists may then assess the potential of the remaining individuals to act as the beginning or nucleus for a new population. Modern techniques can determine the genetic makeup of animals and whether the individuals are distantly enough related to form a breeding pair without suffering the adverse consequences of inbreeding.

Artificial breeding and culture are techniques often employed with the few remaining members of a population. These animals may be brought in from the wild and confined, as were the last remaining California condors. The animals are kept in captivity so that artificial breeding and maintenance techniques can be used to raise additional members of the population. Once the population is large enough for release or other factors are favorable, reintroduction into the wild may be attempted.

Zoological parks and botanical gardens have been involved in maintaining the few remaining specimens of some species. With the help of such facilities, animal tissue has been stored in culture, and sperm have been frozen for later use in breeding. Parks and gardens now play an active role in species propagation and husbandry through breeding, nourishment programs, studying and preventing disease, and behavioral training to reintroduce zoo animals into the wild.

INCREASING AWARENESS OF BIODIVERSITY'S IMPORTANCE

Human activities eliminated more than a few animals from the face of the earth before the second half of the twentieth century, when national and international laws were enacted to protect endangered species and the modern environmental movement began. Among the animals lost were the great auk in 1844, and the passenger pigeon in 1914. These animals were hunted to extinction under the mistaken belief that extinction could not happen, and because authorities lacked a way to limit the hunting of the animals. Despite international efforts to protect animals from extinction, in the latter part of the twentieth century, many countries allowed or encouraged the wholesale destruction of tropical rain forests, along with the animals that lived there.

In 1973, amid growing awareness of the need to prevent further extinctions, the Endangered Species Act was passed by the Congress of the United States of America. The act gave the U.S. Fish and Wildlife Service the responsibility for enforcing the law and set procedures for determining which species should be listed as endangered. The act also established the category of threatened species to protect those that were close to becoming endangered. Subsequently, this protection was extended to the habitats on which the animals depend.

Studies by academic and government scientists revealed that most species are threatened by habitat loss or degradation, overharvesting, or exotic species. Most of the problems arise from habitat loss or degradation rather than overharvesting, which had been the cause in several of the cases that prompted passage of the Endangered Species Act. Scientifically, and eventually legally, one of the most difficult tasks is to determine habitat requirements and then ensure that the required habitat is protected. Identifying and protecting habitats became an important part of all programs aimed at protecting biodiversity. Loss or degradation of habitat became one of the criteria that federal agencies used to take legal action or make official decisions.

Many human activities such as logging, building, mining, and farming can destroy or degrade valuable habitats on which endangered species depend. As a result, endangered species protection programs and activities have been in conflict with some business interests seeking to carry out those activities that threaten species or their habitats. According to the World Wildlife Fund's *Living Planet Report 2016*, conservation efforts launched in the 1970s have done little to slow the decline of biodiversity and mitigate human damage to the global ecosystem. The report found that, between 1970 and 2012, vertebrate species population sizes have been more than cut in half, declining by 58 percent. Freshwater species experienced the greatest loss over the measured forty-two-year period, seeing an 81-percent decline in population. Marine and terrestrial species populations both declined by 36 and 38 percent respectively. Species population and biodiversity decline was greatest in South and Central America—a region rich in rainforest ecosystems—where species populations declined by an average of around 83 percent.

Efforts, activities, and programs to protect biodiversity rely on the latest techniques and methods. These include electronic tracking of large animals, molecular analysis of genetic material, and cellular studies of tissues to help determine actual or potential threats to animals. Ecologists investigate life histories, including feeding habits, environmental requirements and limits, and migration patterns.

FURTHER READING

Burton, John, ed. *The Atlas of Endangered Species.* 2nd ed. Macmillan, 1999.

Dhillon, Sarinder K., and Amandeep S. Sidhu. *Data Intensive Computing for Biodiversity.* Berlin: Springer, 2013. Print.

DiSilvestro, Roger L. *The Endangered Kingdom: The Struggle to Save America's Wildlife.* New York: Wiley, 1991. Print.

Leithead, Alastair "Dozens of Elephants Killed Near Botswana Wildlife Sanctuary" 3 September 2018 https://www.bbc.com/news/world-africa-45396394.

Levin, Simon A. *Encyclopedia of Biodiversity.* Oxford: Academic Press, 2013. Print.

Mackay, Richard. *The Atlas of Endangered Species.* London Earthscan from Routledge, 2017. Print.

National, Research Council. *Science and the Endangered Species Act.* Washington: National Academy Press, 2004. Print.

New, T. R. *An Introduction to Invertebrate Conservation.* New York: Oxford UP, 1995. Print.
Ricketts, Taylor H., et al. *Terrestrial Ecoregions of North America: A Conservation Assessment.* Washington: Island, 1999. Print.
Rossberg, Axel G. *Food Webs and Biodiversity: Foundations, Models, Data.* Chichester: Wiley, 2013. Print.
Tudge, Colin. *The Variety of Life: The Meaning of Biodiversity.* New York: Oxford UP, 2000. Print.
Wilson, Edward O., ed. *Biodiversity.* Washington: National Academy Press, 1988. Print.
Wilson, Edward O., ed. *Half-Earth: Our Planet's Fight for Life.* New York: Liveright, 2016.
World Wildlife Federation. *Living Planet Report 2016: Risk and Resilience in a New Era.* WWF International, 2016, www.wnf.nl/custom/LPR_2016_full-report/. Accessed 31 Jan. 2018.

Biofertilizers

—*D. R. Gossett*

FIELDS OF STUDY

Agronomy; Agriculture; Ecology; Ecosystem Management; Environmental Sciences; Environmental Studies; Botany.

KEY TERM

biofertilizers: Fertilizers consisting of either naturally occurring or genetically modified microorganisms

SUMMARY

Biofertilizers provide a means by which biological systems can be utilized to supply plant nutrients such as nitrogen to agricultural crops. The use of biofertilizers could reduce the dependence on chemical fertilizers, which are often detrimental to the environment.

BACKGROUND

Plants require adequate supplies of the thirteen mineral nutrients necessary for normal growth and reproduction. These nutrients, which must be supplied by the soil, include both macronutrients (those nutrients required in large quantities) and micronutrients (those nutrients required in smaller quantities). As plants grow and develop, they remove these essential mineral nutrients from the soil. Since normal crop production usually requires the removal of plants or plant parts, the nutrients are continuously removed from the soil. Therefore, the long-term agricultural utilization of any soil requires periodic fertilization to replace lost nutrients.

Nitrogen is the plant nutrient that is most often depleted in agricultural soils, and most crops respond to the addition of nitrogen fertilizer by increasing their growth and yield; therefore, more nitrogen fertilizer is applied to cropland than any other fertilizer. In the past, nitrogen fertilizers have been limited to either manures, which have low levels of nitrogen, or chemical fertilizers, which usually have high levels of nitrogen. The excess nitrogen in chemical fertilizers often runs off into nearby waterways, causing a variety of environmental problems.

HOW IT WORKS

Biofertilizers offer a potential alternative: They supply sufficient amounts of nitrogen for maximum yields yet have a positive impact on the environment. Biofertilizers generally consist of either naturally occurring or genetically modified microorganisms that improve the physical condition of soil, aid plant growth, or increase crop yield. Such fertilizers provide an environmentally friendly way to increase plant health and yields with reduced input costs, new products and additional revenues for the agricultural biotechnology industry, and cheaper products for consumers.

While biofertilizers could potentially be used to supply a number of different nutrients, most of the interest thus far has focused on enhancing nitrogen fertilization. The relatively small amounts of nitrogen found in soil come from a variety of sources. Some nitrogen is present in all organic matter in soil; as this organic matter is degraded by microorganisms, it can be used by plants. A second source of nitrogen is nitrogen fixation, the chemical or biological process of taking nitrogen from the atmosphere and converting it to a form that can be utilized by plants. Bacteria such as *Rhizobia* can live symbiotically with certain plants, such as legumes, which house nitrogen-fixing bacteria in their roots. The *Rhizobia* and plant root tissue form root nodules that house the nitrogen-fixing bacteria; once inside the nodules, the bacteria use energy supplied by the plant to convert atmospheric nitrogen to ammonia, which nourishes the plant. Natural nitrogen can also be supplied by free-living microorganisms, which can fix nitrogen without forming a symbiotic relationship with plants. The primary objective of biofertilizers is to enhance any one or all of these processes.

One of the major goals in the genetic engineering of biofertilizers is to transfer the ability to form nodules and establish effective symbiosis to nonlegume plants. The formation of nodules in which the *Rhizobia* live requires plant cells to synthesize many new proteins, and many of the genes required for the expression of these proteins are not found in the root cells of plants outside the legume family. Many research programs have been devoted to efforts to transfer such genes to nonlegume plants so that they can interact symbiotically with nitrogen-fixing bacteria. If this is accomplished, *Rhizobia* could be used as a biofertilizer for a variety of plants.

There is also much interest in using the free-living, soil-borne organisms that fix atmospheric nitrogen as biofertilizers. These organisms live in the rhizosphere (the region of soil in immediate contact with plant roots) or thrive on the surface of the soil. Since the exudates from these microorganisms contain nitrogen that can be utilized by plants, increasing their abundance in the soil could reduce dependence on chemical fertilizers. Numerous research efforts have been designed to identify and enhance the abundance of nitrogen-fixing bacteria in the rhizosphere. Soil microorganisms primarily depend on soluble root exudates and decomposed organic matter to supply the energy necessary for fixing nitrogen; hence there is also an interest in enhancing the biodegradation of organic matter in the soil. This research has largely centered on inoculating the soil with cellulose-degrading fungi and nitrogen-fixing bacteria or applying organic matter to the soil, such as straw that has been treated with a combination of the fungi and bacteria.

FURTHER READING

Akinyemi, Okoro M. *Agricultural Production: Organic and Conventional Systems.* Enfield, N.H.: Science Publishers, 2007.

Black, C. A. *Soil-Plant Relationships.* New York: John Wiley & Sons, 1988.

Chrispeels, Maarten J, David E. Sadava, and Maarten J. Chrispeels. *Plants, Genes, and Crop Biotechnology.* Boston: Jones and Bartlett Publisher, 2003. Print.

Lynch, J. M. *Soil Biotechnology: Microbiological Factors in Crop Production.* Malden, Mass.: Blackwell, 1988.

Salisbry, Frank B, Frank B. Salisbry, and Cleon W. Ross. *Palnt Physiology.* Thomson Wadsworth, 2007. Print.

Yadav, A. K., S. Ray Chaudhuri, and M. R. Motsara, eds. *Recent Advances in Biofertilizer Technology.* New Delhi: Society for Promotion and Utilisation of Resources and Technology, 2001.

BIOGEOGRAPHY

—M. Lee

FIELDS OF STUDY

Biology; Ecology; Geography.

PRINCIPAL TERMS

alien: a species of plant or animal that is not native to a particular place, but is found to have successfully colonized it, especially one that has been introduced as a result of human activity

allopatric: of animals or plants, occupying separate, non-overlapping distribution areas; of speciation, occurring as a result of such geographic separation

biota: the totality of the flora, fauna, and microbes that inhabit a particular region, habitat type, or time period

colonization: the long-term establishment of a successfully reproducing population of a particular species of plant or animal, in a habitat where it was not previously found

cosmopolitan: of animals or plants, occupying distribution areas that span the entire or almost the entire globe

disjunct: of animals or plants, having a fragmented distribution area that includes two or more, geographically separate, regions

dispersal: for any given species, the movement out of its distribution area of the minimum number of individuals it will take to colonize a new area

distribution area: the geographic range of a particular group of organisms at any level of classification, such as species, genus, family, or class

ecogeographic rule: a principle describing correlations between the morphology of birds and mammals and the climatic or latitudinal gradients at which they are found

endemic: of animals or plants, being restricted to a particular region of the world

latitudinal gradient: of plant or animal species, an increase in overall numbers found when moving from higher to lower latitudes

speciation: the process by which new species arise or populations of the same species become reproductively isolated from each other, either by geographical separation or some other mechanism

SUMMARY

Biogeography is an interdisciplinary science that applies concepts and methodologies from biology, geography, systems ecology, evolutionary theory, conservation, statistics, and paleontology, among other fields, to study the features of the physical environment that affect the distribution of living organisms across space and time.

DISTRIBUTION AREAS & SPECIES DISTRIBUTIONS

The fundamental question that drives biogeographical study is: How is life distributed over the earth? This is a question that can be answered at many different levels. For example, any given taxon, or group of living organisms at a particular level of classification (such as a species, genus, family, or class), has a geographic range within which it is likely to be found. The plant *Yucca brevifolia,* a species of yucca also known as the Joshua tree, is only found growing in the dry desert soils, open plains, and mesas of the southwestern United States. This is what biogeographers refer to as the tree's distribution area. A species like *Yucca brevifolia* is said to have an endemic distribution area, meaning one that is restricted to a particular region of the world. In contrast, *Drosophila melanogaster* (the common fruit fly), which is found in virtually all types of habitats across almost the entire globe, is said to have a cosmopolitan distribution area.

Edward O. Wilson, a prominent biologist and conservationist, coauthored *The Theory of Island Biogeography.* Image by Jim Harrison (PLoS) CC-BY-2.5.

Many taxons have what are known as disjunct distribution areas; this occurs when very closely related species are found in habitats that are widely separated geographically. For example, many similar species of mosses, small mammals, and butterflies are found both in the mountain ranges of the far northerly Arctic, and in the more southerly mountain ranges of Europe, such as the Alps—but nowhere in between. There are several scientific hypotheses regarding historical reasons for the existence of disjunct biogeographic distribution areas.

Under the vicariance hypothesis, environmental events cause the range of a particular taxon to split into parts. For example, changes in the distribution of land can cause impassable fractures to form in what was once a continuous habitat. Plate tectonic shifts, continental drift, and the formation of mountains all have the potential to break apart a formerly intact habitat region. When this happens, allopatric speciation can occur. This is the evolution of new and distinct species of animals or plants as a result of a single species being separated into more than one non-overlapping distribution area. (Another form of vicariance occurs when individual members of a family that was once spread out over a large, continuous geographic range, go extinct. This, too, can cause fragmentation in the distribution area.) In contrast, some disjunct biogeographic distributions are believed to have been caused by dispersal, or the successful migration of species across an existing geographic barrier. For example, birds are capable of crossing oceans in order to colonize new habitats; other non-motile species, such as plants, may be transported along with them as seeds.

Besides studying how organisms range across the globe, biogeographers are interested in examining how individual organisms within a species are spatially arranged within the area where they are found. This is a concept known as species distribution. Three major types of species distribution are identified: clumped, random, and uniform. In a clumped species distribution, multiple groups of individuals tend to form; this is a common behavioral pattern found, for instance, in herding animals like deer or cattle. For such species, a clumped distribution is beneficial because it offers protection from predators. In a random species distribution, individuals are found in scattered locations with no particular organization or method—such as wildflowers whose seeds have been dispersed by the wind over a field, and which grow wherever they fall. In a uniform species distribution, individuals are spread out at a roughly even distance from each other. Many desert plants, for instance, are found in a uniform distribution because this is the pattern that enables each of them to absorb the most water from the arid environment.

ECOGEOGRAPHIC RULES

When biogeographic principles are used to look at the distribution of living organisms on a macroscopic scale, many striking principles emerge regarding the relative degree of biodiversity that can be found in different regions of the earth. One of the most significant of these "ecogeographic rules" is the latitudinal gradient in species diversity, or species richness. This term describes a striking, though as yet not fully understood pattern in global biodiversity: the tremendous increase in the number of unique species that can be found when traveling from high latitudes (colder regions) to low latitudes (warmer regions). In other words, the equatorial tropics are home to a far greater diversity of life than the poles. This pattern applies across nearly every possible form of life, including birds, mammals, freshwater fish, marine invertebrates, insects, and flowering plants and trees. For example, more than two hundred different species of ants are known in Brazil; Alaska has fewer than ten. The 50,000 square kilometers (approximately 19,305 square miles) of land in the Malaysian peninsular is home to nearly 1,500 different species of trees and shrubs; the same area in the southeastern part of Russia holds fewer than 50 different species.

Paleobiogeographic data—data derived from analyzing the relationships between fossilized animals and plants and the locations upon the earth where those fossils were found—indicates that species richness latitude gradients are not a recent phenomenon. It has been shown, for instance, that during the Triassic (a period spanning about 245 million to 208 million years ago), ammonoids (a now-extinct group of mollusks) were more diverse at lower latitudes than at higher ones, and that the same was true of brachiopods, another group of marine invertebrates, during the Paleozoic ice age (about 570 million to 245 million years ago).

Dozens of competing hypotheses have been put forth regarding the potential mechanism or mechanisms responsible for the existence of species richness latitude gradients. Among them are the time and area hypothesis, the diversification rate hypothesis, and the hypothesis of effective evolutionary time. The time and area hypothesis argues that the tropical regions of the earth are both older and larger than the polar regions, giving them a longer span of evolutionary time in which speciation could have occurred and a greater amount of available habitat—both of which could have resulted in the higher degree of species richness that is observed at lower latitudes. (The time and area hypothesis does not account for certain anomalies, such as why the South East Asian sea—which is not as large as other bodies of water in the Indo-Pacific region—has the greatest degree of species richness.) The diversification rate hypothesis argues that speciation occurs faster and more frequently in tropical regions than in polar regions, either because of lower extinction rates or other factors, such as the increased likelihood of achieving reproductive isolation. The hypothesis of effective evolutionary time takes into consideration the longer amount of time ecosystems are believed to have existed in the tropics, but also gives weight to the effects of temperature on the rate at which evolution can occur. At higher temperatures, such as those associated with lower latitudes, genetic mutation rates, among other factors influencing speciation, are higher. To date, no single hypothesis about the cause of species richness latitude gradients has gained widespread acceptance within the scientific community.

Another important ecogeographic principle related to latitudes is Bergmann's rule. The nineteenth-century biologist Carl Bergmann noted that when the same species (or closely related species) of mammals and other warm-blooded animals were found in different parts of the world, those that lived closer to the equator were comparatively smaller than those that lived at higher latitudes. For example, polar bears are much more massive than related bear species in the tropics. Bergmann's rule is often explained as a result of the direct relationship between body size and heat. Larger animals generate a greater amount of heat as a result of cell metabolic activities, and they also lose less heat to their surroundings, since the ratio of their surface area to overall mass is comparatively low. These adaptations make them well suited to living in colder, high-latitude environments.

BIOGEOGRAPHIC DIVISIONS

Biogeographers use a series of imaginary lines to divide the earth into six major divisions, or realms. A distinct biota, or group of plants, animals, and microbes, inhabits each zone. The most famous of these imaginary divides runs between the islands of Borneo and Sulawesi in the Indonesian archipelago, and is known as Wallace's line, after Alfred Russel Wallace. Wallace was a nineteenth-century British naturalist—a contemporary of Charles Darwin's who also developed a theory of evolution based on natural selection. It was Wallace who first noticed the sharp distinction between birds and animals on the western side of the line—which mostly came from taxons common in Asia—and those on the eastern side of the line—which mostly came from taxons common in Australia.

To the west of the Wallace line, for instance, one would find woodpeckers, pheasants, primates, squirrels, and big cats; to the east, cockatoos, bats, eucalyptus trees, and marsupials. The Wallace line is the most well known of the world's biogeographical divisions not only because of the extreme contrasts between the zones on either side of it, but also because the entire region surrounding it is considered to be a very significant biodiversity hot spot. A biodiversity hot spot is a region characterized by a high density of species richness that is under threat of extinction due to human activity. Many of the species found in the region known as Wallacea have endemic distribution areas, including the caerulean paradise flycatcher (*Eutrichomyias rowleyi*) and the pig-like mammal known as the babirusa (*Babyrousa babyrussa*). Logging, mining, and agricultural development have all had a dramatic negative impact on the Wallacea ecosystem.

Biomes are another way in which biographers organize the earth's living communities. A biome is a large, naturally occurring ecosystem; the particular grouping of plants and animals that inhabit it are determined by its climate (especially temperature and precipitation) and geography. Different scientists classify biomes in different ways, but five of the most common terms used to describe biomes are aquatic (freshwater and marine habitats), desert (arid habitats), forest (densely wooded habitats), grassland (open plain habitats dominated by grasses), and tundra (open plain habitats with a permanently frozen layer of subsoil).

THEORY OF ISLAND BIOGEOGRAPHY

The theory of island biogeography was developed in the late 1960s by the ecologists E. O. Wilson, of Harvard University, and Robert MacArthur, of Princeton University, as a way of organizing and explaining the factors contributing to the relative richness of species biodiversity that occurs in the world's islands. In its most basic form, MacArthur and Wilson's theory states that the number of unique living species on any given island is primarily shaped by two competing forces: the rate at which new species move to the island and establish successful breeding populations there (the rate of immigration), and the rate at which existing populations of successfully breeding species disappear from the island (the rate of extinction).

Various other factors are responsible for determining the rate of immigration and extinction. For example, an island's rate of immigration is closely related to the distance of the island from other ecosystems from which potential colonizers could travel, such as other islands or a mainland. The theory predicts that the more geographically isolated an island is, the lower its rate of colonization will be; this pattern is known as the distance effect. Similarly, an island's rate of extinction is related to the size of the island is, the number of species that already inhabit it, and the availability of resources on it. According to the theory, a larger island will have more habitable area and more resources, like food, available—and will thus be able to support a greater number of species, lowering the extinction rate. As the number of existing species on an island increases, competition between these species for habitat and food will also cause its extinction rate to rise. By comparison, islands that have a variety of habitats (for example, mountainous regions as well as flat forested regions) will be able to support a larger number of different species.

Assuming that these factors remain unchanged, MacArthur and Wilson's theory predicts that for any newly created island, the rate of immigration will start high and drop over time, while the rate of extinction will start at zero and rise over time. Eventually, the number of unique species that exist on a newly created island will become stable, as a balance, or equilibrium, is reached between the rate of immigration and the rate of extinction. (Even after equilibrium has been reached, the specific composition of the island's flora and fauna will be dynamic, rather than static, as different species continue to replace each other.) For example, the 1883 volcanic eruption on the tiny island of Krakatoa in Indonesia caused the total extinction of all its plant and animal species. Observations over the following hundred years showed that the rates of immigration and extinction on what was now effectively a "new" island followed the pattern predicted by MacArthur and Wallace: Between 1883 and 1934, thirty-four new species moved to and colonized the island, five of which went extinct. But between 1934 and 1985, an additional fourteen species began to successfully breed on the island (a lower immigration rate than in the previous period) and an additional eight became extinct (a higher extinction rate than in the previous period).

The theory of island biogeography has since been expanded in scope; it is now used to estimate the probable degree of species richness in any isolated habitat area, such as a river or lake surrounded by desert, or a wetland region surrounded by dry woodlands. It is a particularly useful tool for predicting the effects of habitat fragmentation due to human activity. For example, much of the once continuous deciduous forest of the eastern United States has been cut up into smaller, geographically separated, patches of trees, with roads, farms, and cities taking up the spaces in between. The distance effect predicts that these isolated patches of forest will have a relatively low rate of immigration, since it is difficult for plants and animals to move across the gaps between them. In addition, their small size will reduce the available habitat area and resources they contain, raising the rate of extinction. Overall, the theory of island biogeography predicts that habitat fragmentation will reduce the level of species richness in fragmented areas. Studies of avian populations in forest preserves surrounded by human development support this prediction.

Further Reading

Kruckeberg, Arthur R. *Geology and Plant Life: The Effects of Landforms and Rock Types on Plants.* Seattle, Wash. University of Washington Press, 2004. Despite its technical nature, this is an appealing introduction to the fascinating connections between geology and plant growth and distribution. Kruckeberg writes with both an expert's knowledge and an obvious sense of excitement and curiosity about his subject.

Lieberman, Bruce S. *Paleobiogeography: Using Fossils to Study Global Change, Plate Tectonics, and Evolution.* New York: Kluwer Academic, 2000. Best suited to advanced undergraduates and biogeography scholars, this meticulously organized synthesis of concepts from paleontology, evolutionary theory, plate tectonics, and phylogenetics is an important read for anyone interested in the historical aspects of biogeography.

Lomolino, Mark V., Dov F. Sax, and James H Brown. *Foundations of Biogeography: Classic Papers with Commentary.* Chicago: University of Chicago Press, 2004. An outstandingly well-curated compilation of seminal research papers in biogeography; the whole functions as an ideal introduction to the history of the field for a beginning college student. Seventy-two facsimiles of papers dating back to 1761 are included, along with their references.

Losos, Jonathan B., Robert E. Ricklefs, and Robert H. MacArthur, eds. *The Theory of Island Geography Revisited.* Princeton, N.J. Princeton University Press, 2010. Four decades after the publication of *The Theory of Island Biogeography,* a Harvard University symposium challenged attendees to reflect on contemporary confirmations and challenges to MacArthur and Wilson's original ideas. This lively book, filled with reflections from luminaries in the field, reprints papers presented at the symposium.

Morrone, Juan J. *Evolutionary Biogeography: An Integrative Approach with Case Studies.* New York: Columbia University Press, 2009. Fundamental principles and methodologies in evolutionary biogeography, such as cladistic relationships and the use of molecular clocks, are laid out; each concept is followed by numerous illustrative case studies. A useful reference book that the serious student can either dip into or read cover to cover.

Newton, Ian. *Speciation and Biogeography of Birds.* London: Academic Press, 2003. A student-friendly volume, lovingly illustrated with black-and-white drawings by scientific illustrator Keith Brockie. Its style, though closer to textbook than popular science, is both clear and engaging. Newton's book will also appeal to birdwatchers and other observers of nature.

Shugart, H. H. *How the Earthquake Bird Got Its Name and Other Tales of an Unbalanced Nature.* New Haven, Conn. Yale University Press, 2007. In ten engaging chapters, each structured as a parable about a particular species (from the ivory-billed woodpecker to the European rabbit), Shugart explores the complex relationship between environmental changes and changes in animal populations. Suitable for high-school students and general readers.

BIOLOGY

—Mary E. Carey

FIELDS OF STUDY

Biochemistry; Biology; Botany; Ecology; Environmental Studies; Genetics.

PRINCIPAL TERMS

biodiversity: the total of all living organisms in an environment

deoxyribonucleic acid (DNA): the carrier of all an organism's genetic information

National Institutes of Health: the United States' governmental division that monitors and improves public health

xenotransplantation: the transplantation of organs from one species to another

SUMMARY

Biology is an extensive field subdivided into categories based on the molecule, the cell, the organism,

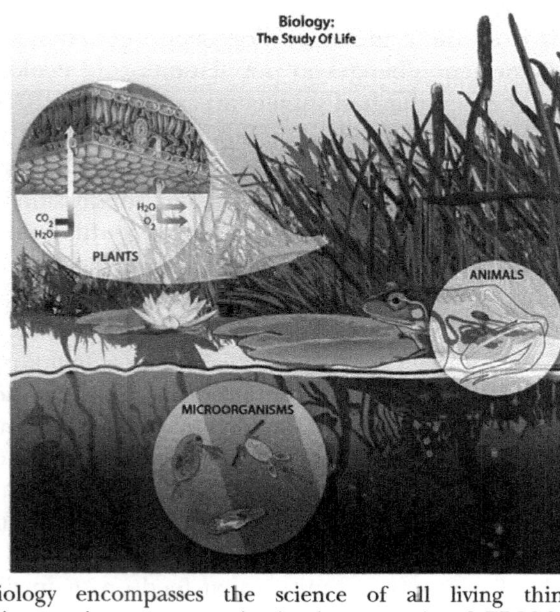

Biology encompasses the science of all living things; microscopic to macroscopic, simple to complex. © EBSCO.

and the population. Molecular biology touches on biophysics and biochemistry. It is the branch of biology that deals with the structure and development of biological systems in terms of the physics and chemistry of their molecules. Cellular biology is closely related to molecular biology through understanding the functions and basic structure of the cell. The cell is the smallest structural unit of an organism that is capable of independent functioning, consisting of one or more nuclei, cytoplasm, and organelles, all surrounded by a semipermeable membrane. Botany is the science or study of plant life. Ecology, also referred to as bionomics, is the study of relationships between organisms and their environments.

BRANCHES OF BIOLOGY

Developmental biology encompasses a number of issues, including gene regulation, genetics, and evolution. These are concepts of importance to vertebrates, invertebrates, and plants. A gene is a hereditary unit that occupies a particular location on a chromosome, determines a particular characteristic of an organism, and can undergo mutation. Genetics is the branch of biology that deals with heredity, especially the hereditary transmission and variation of inherited characteristics. Evolution is the theory that groups of organisms change with the passage of time, mainly as a result of natural selection, so that descendants differ morphologically and physiologically from their ancestors. Population genetics, the study of gene changes in populations, and ecology have been established subject areas since the 1930s. These two fields were combined in the 1960s to form a new discipline called population biology, which became established as a major subdivision of biological studies in the 1970s. Central to this field is evolutionary biology, in which the contributions of Charles Darwin are noted.

Microbiology is the branch of biology that deals with microorganisms. The study of bacteria, including their classification and the prevention of diseases that arise from bacterial infection, is the primary focus of microbiology. This branch is of interest not only among bacteriologists but also among chemists, biochemists, geneticists, pathologists, immunologists, and public health professionals. Parasitology is the study of parasites, organisms that feed on or in different organisms while contributing nothing to the survival of their hosts. Ethology is the scientific study of animal behavior. Animal behavioral studies have developed along two lines. The first of these, animal psychology, is primarily concerned with physiological psychology, and has traditionally concentrated on laboratory techniques such as conditioning. The second, ethology, had its origins in observations of animals under natural conditions, concentrating on courtship, flocking, and other social contacts. One of the important recent developments in the field is the focus on sociobiology, which is concerned with the behavior, ecology, and evolution of social animals such as bees, ants, schooling fish, flocking birds, and humans.

Ethics is the study of the general nature of morals and of specific moral choices. Bioethics addresses such issues as animal experimentation, cloning, euthanasia, gene therapy, genetic engineering, genome projects, protection of human research subjects, organ transplants, and patients' rights.

Biotechnology is the industrial use of living organisms or biological techniques developed through basic research. Biotechnology products include antibiotics, insulin, interferon, and recombinant DNA, and techniques such as waste recycling. The Office of Biotechnology Activities of the National Institutes of Health monitors scientific progress in human genetics research in order to anticipate future developments, including ethical, legal, and social concerns, in basic and clinical research involving recombinant DNA, genetic testing, and xenotransplantation (the use of organs from other species of mammals for transplants). In addition to organs donated from humans, researchers are exploring the use of partially or wholly artificial organs manufactured in the laboratory.

Biodiversity focuses on such issues as conservation, extinction and depletion from overexploitation, habitat pollution, global patterns and values of biodiversity, and endangered species protection.

Well known pioneer biologists include naturalist and explorer Sir Joseph Banks (1743–1820), naturalist and explorer Charles William Beebe (1877–1962), biochemist Gunter Blobel (b. 1936), environmentalist Rachel Carson (1907–64), biochemist Stanley Cohen (b. 1922), biophysicist and codiscoverer of DNA, Francis Crick (1916–2004), naturalist and father of evolutionary theory Charles Darwin (1809–82), evolutionary biologist Richard Dawkins (b. 1941), marine biologist Sylvia Earle (b. 1935), bacteriologist Paul Ehrlich (b. 1932), bacteriologist and discoverer of penicillin Sir Alexander Fleming (1881–1955), microscopist Antonie van Leeuwenhoek (1632–1723), botanist and taxonomist Carolus Linnaeus (1707–23), ethologist Konrad Lorenz (1903–89), zoologist A. S. Loukashkin (1902–88), botanist and geneticist Barbara McClintock (1902–92) , botanist and genetic theorist Gregor Mendel (1822–84), endocrinologist and inventor of the birth control pill Gregory Goodwin Pincus (1903–67), naturalist Alfred Russel Wallace (1823–1913), biophysicist and codiscoverer of DNA James Watson (b. 1928), Nobel Prize–winner Hamilton O. Smith (b. 1931), whose work led to the creation of recombinant DNA technology, discoverer of genetic transduction, Norton David Zimmer (1928–2012), and sociobiologist Edward O. Wilson (b. 1929).

Further Reading

Brooker, Robert J., et al. *Biology*. 3d ed. New York: McGraw-Hill, 2014. Print.

Campbell, Neil A. *Biology*. 9th ed. San Francisco: Cummings, 2011. Print.

Gilbert, Scott F. *Developmental Biology*. 10th ed. Sunderland: Sinauer, 2014. Print.

Lavers, Chris. *Why Elephants Have Big Ears: Understanding Patterns of Life on Earth*. New York: St. Martin's, 2001. Print.

Mayr, Ernst. *The Growth of Biological Thought: Diversity, Evolution and Inheritance*. Cambridge, Mass: The Belknap Press of Harvard Univ. Press, 2003. Print.

Watson, James D. *The Double Helix: A Personal Account of the Discovery of the Structure of DNA*. Reprint. New York: Simon, 1998. Print.

Biostratigraphy

—*Spencer G. Lucas*

FIELDS OF STUDY

Environmental Chemistry; Geology; Archeology; Environmental Sciences; Bioclimatology; Oceanography; Physical Geography; Ecology.

PRINCIPAL TERMS

correlation: the determination of the equivalence of age or stratigraphic position of two strata in separate areas, or, more broadly, determination of the geological contemporaneity of events in the geologic histories of two areas

fossils: remains or traces of animals and plants preserved by natural causes in the earth's crust

index fossil: a fossil that can be used to identify and determine the age of the stratum in which it is found

sedimentary rocks: rocks formed by the accumulation of particles of other rocks or of organic skeletons or of chemical precipitates or some combination of these

stratigraphy: the study of layered rocks, especially of their sequence and correlation

stratum (plural, strata): a single bed or layer of sedimentary rock

SUMMARY

Biostratigraphy is that branch of the study of layered rocks—stratigraphy—that focuses on fossils. Its goals are the identification and organization of strata based on their fossil content. Biostratigraphy thus investigates one of the principal bases of the geologic time scale of earth history.

BIOSTRATIGRAPHIC ZONES

Biostratigraphy is the method of identifying and differentiating layers of sedimentary rock (strata) by their fossil content. Strata with distinctive fossil content are termed biostratigraphic units, or zones. Zones vary greatly in thickness and in lateral extent. A zone may be a single layer that is a few centimeters thick and of very local extent, or it may encompass thousands of meters of rocks extending worldwide. The defining feature of a zone is its fossil content; the fossils of a given zone must differ in some specific way from the fossils of other zones.

Zones are usually recognized after fossils have been collected extensively over the lateral and vertical extent of a rock sequence or at many sequences over a broad region. The positions of the fossils in

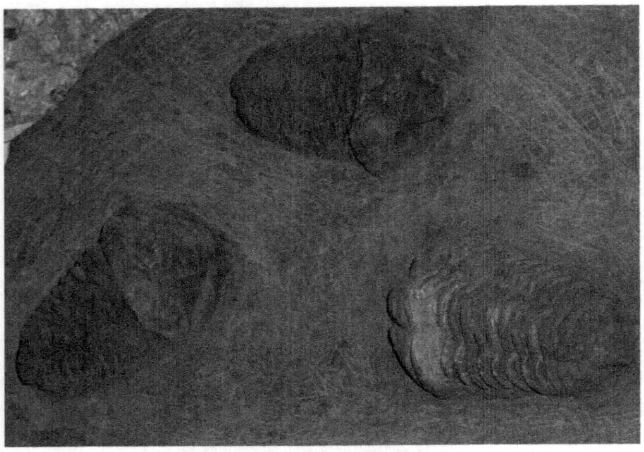

Trilobites are used as index fossils to identify sediments of Cambrian age. Image by BigSus CC-BY-SA-3.0.

the strata are carefully recorded in the field. Fossils that co-occur in a single layer are noted, as are fossils found isolated in the strata. In the laboratory, the biostratigrapher, usually a paleontologist, then tabulates the vertical and lateral ranges of the fossils collected. It is from these ranges that the paleontologist recognizes zones. Different types of zones are recognized depending on the way in which the fossils in the strata prove to be distinctive. Assemblage zones are strata distinguished by an association (assemblage) of fossils. Thus, not one type but many types of fossils are used to define an assemblage zone. All dinosaur fossils, for example, can be thought of as defining an assemblage zone that encompasses earth history from about 220 to 66 million years ago.

Range zones are strata that encompass the vertical distribution, or range, of a particular type of fossil. Thus, one fossil type—not many—is used to define a range zone. In contrast to the example just given, one type of dinosaur, *Tyrannosaurus rex*, lived only between 68 and 66 million years ago. Its fossils thus define a range zone that corresponds temporally to this 2-million-year interval.

Acme zones are rock layers recognized by the abundance, or acme, of a type (or types) of fossil (or fossils) regardless of association or range. Horned dinosaurs (*Triceratops* and its allies) reached an acme between 70 and 66 million years ago. That is, during this period they were most diverse and most numerous. This acme zone thus overlaps the *Tyrannosaurus rex* range zone and represents a small portion of the dinosaur assemblage zone.

Finally, interval zones are recognized as strata between layers where a significant change in fossil content takes place. For example, the mass extinctions that took place 250 and 66 million years ago bound a 184-million-year-long interval zone that is popularly referred to as the "age of reptiles."

DEVELOPMENT OF BIOSTRATIGRAPHY

Biostratigraphy developed independently in England and France just after 1800. British civil engineer William Smith worked in land surveying throughout the country. From his vast field experience, he recognized that a given stratum usually contains distinctive fossils and that the fossils (and the stratum) could often be recognized across a large area. Smith's work culminated in his geological map of England (1815), based on his tracing of rock-fossil layers across much of the country.

Meanwhile, in France, Georges Cuvier and Alexandre Brongniart studied the succession of rocks and fossils around Paris. They, too, discovered a definite relationship between strata and fossils and used it to interpret the geological history of the rocks exposed near Paris. In this history, Cuvier saw successive extinctions of many organisms coinciding with remarkable changes in the strata. To him, these represented vast "revolutions" in geological history, which Cuvier argued were of worldwide significance. It is now known that Cuvier was mistaken, but the discovery that a particular fossil type (or types) was confined to a particular stratum became the basis for biostratigraphy. This allowed geologists to identify strata from their fossil content and to trace these strata across broad regions of the earth's crust.

Almost simultaneous with the development of biostratigraphy was the development of biochronology. Biochronology is the recognition of intervals of geologic time by fossils. It stemmed from the realization that during earth's history, different types of organisms lived during different intervals of time. Thus, the fossils of any organism represent a particular interval of geologic time. (Such fossils are called index fossils because they act as an "index" to a geologic time interval.) Biochronology thus identifies intervals of geologic time based on fossils. These time-distinctive fossils are the fossils by which zones are defined, which is to say that each zone represents, or is equivalent to, some interval of geologic time.

The time value of zones made them more useful in tracing strata and deciphering local geological histories. Biostratigraphy now became one of the central methods of stratigraphic correlation. With the aid of fossils, it became possible to determine the ages of strata and thus demonstrate the synchrony or diachrony of these strata in different areas. Through its use in stratigraphic correlation, biostratigraphy became one of the bases for constructing what is called the relative geological time scale of earth history composed of eons, eras, periods, epochs, and ages. This time scale is the "calendar" by which all geologists temporally order their understanding of the history of the earth.

APPLICATION OF BIOSTRATIGRAPHY

Biostratigraphy is generally used as a method of stratigraphic correlation, the process of determining the equivalence of age or stratigraphic position of layered rocks in different areas. Stratigraphic correlation by biostratigraphy is extremely important in deciphering geological history. It reveals the sequence of geological events in one or more regions. Understanding geological history is of interest for its own sake to scientists and laypersons alike. It is crucial to the discovery of mineral deposits and energy resources within the earth's crust. In addition, it provides insight into the biological events that have taken place on this planet for the last 3.9 billion years.

A good example of the use of biostratigraphy in this last regard comes from the study of dinosaur extinction. When dinosaurs were first discovered in England in 1824, and when the term "dinosaur" was coined by the British anatomist Sir Richard Owen in 1841, no one realized that dinosaurs had lived on Earth for only 150 million years and that their extinction had taken place rather rapidly about 66 million years ago. By 1862, however, enough dinosaur fossils had been collected around the globe that a biostratigraphic pattern was beginning to emerge. In that year, the American geologist James Dwight Dana, in his classic *Manual of Geology*, noted that all dinosaurs disappeared before the end of the Mesozoic era, which is now considered as the interval of earth history between 250 and 66 million years ago. This biostratigraphic generalization was possible because geologists noticed that many Mesozoic rocks (but no older or younger rocks) were full of dinosaur fossils, and thus the Mesozoic came to be termed "the age of reptiles." It might just as well be referred to as the "dinosaur zone," except of course for the first 30 or so million years of the Mesozoic, during which dinosaurs apparently did not exist.

More than a century of research has confirmed Dana's biostratigraphic generalization and considerably refined it. Scientists now generally agree that the last dinosaurs disappeared worldwide approximately 66 million years ago. It is also known that dinosaurs first appeared about 220 million years ago. Thus, scientists are able to recognize a dinosaur zone and erect many types of zones based on the ranges and acmes of specific types of dinosaurs. This biostratigraphy of dinosaurs is the basis for informed discussion of the sequence and timing of events during the evolution of the dinosaurs. For example, scientists are now confident that *Stegosaurus* lived long before *Tyrannosaurus* and that stegosaurs as a group of dinosaurs became extinct long before the end of the Mesozoic.

Although discussion here has relied heavily on dinosaurs for examples of biostratigraphy at work, the fossils of these giant reptiles are not ideal for use in biostratigraphy because it is not easy to identify most dinosaur fossils precisely and because most dinosaurs were not animals with broad geographic ranges. Indeed, the fossils of most use in biostratigraphy, index fossils, are those that are easy to identify precisely and that represent organisms that had wide geographic ranges, enjoyed broad environmental tolerances, and lived only for a brief period of geologic time.

Usually an entire skull or skeleton is needed to identify a dinosaur fossil precisely. The isolated bones most often found are not enough, although they do indicate the fossil is that of a dinosaur. Most dinosaurs (there are some notable exceptions) seem to have lived in one portion of one continent; indeed, fossils of the horned dinosaur *Pentaceratops* (a cousin of *Triceratops*) have been found only in New Mexico. There is strong evidence that some dinosaurs preferred coastlines, whereas others preferred dry areas. Thus, many, if not most, dinosaurs did not live in a wide range of environments. Finally, although many dinosaurs apparently lived for only brief intervals of geologic time, the fossil record of most of these giant reptiles is not extensive enough to pin down their exact interval of existence.

The factors that argue against the use of most dinosaur fossils in biostratigraphy are quite different for microscopic fossils of pollen grains and the shelled protozoans known as foraminiferans. These microscopic fossils fit well the four criteria that identify fossils most useful in biostratigraphy. Indeed, such "microfossils" (studied by micropaleontologists) are some of the mainstays of biostratigraphy.

SIGNIFICANCE

Biostratigraphy—the recognition of strata by their fossil content—is a cornerstone of stratigraphic correlation. To identify bodies of rock, the presence of fossils can be traced over broad areas, and their sequence in distant areas can often be determined. Stratigraphic correlation by biostratigraphy is critical to deciphering geological history; without it, the search for mineral deposits and energy resources would be considerably more difficult. Furthermore, understanding the history of geological disasters—earthquakes, volcanic eruptions, meteorite impacts, and the like—and thereby being able to predict future disasters, relies on knowledge of the sequence and timing of geological events, knowledge often derived from biostratigraphy. Deciphering the history of life on this planet, including the myriad appearances, changes, and extinctions of Earth's biota during the last 3.9 billion years, largely depends on the sequence and timing established by biostratigraphy.

Biostratigraphy has also given rise to biochronology, the recognition of intervals of geologic time based on fossils. As a result, scientists have been able to construct a relative global geologic time scale, and it is within the context of this time scale that all geological and biological events in earth history have been placed.

Further Reading

Ager, Derek V. *The Nature of the Stratigraphical Record.* 3rd ed. Hoboken, N.J. John Wiley & Sons Inc., 1993. A witty and unabashedly curious look at stratigraphy; some of the discussion centers on biostratigraphy. An extensive bibliography, index, and a few well-chosen illustrations illuminate the text.

Barry, W. B. N. *Growth of a Prehistoric Time Scale.* Rev. ed. Palo Alto, Calif.: Blackwell Scientific Publications, 1987. Largely devoted to the history of how the global geologic time scale was formulated, much of this book is a history of biostratigraphy. Well illustrated, with a good bibliography and an index.

Brenner, R. L., and T. R. McHargue. *Integrative Stratigraphy Concepts and Applications.* Englewood Cliffs, N.J.: Prentice-Hall, 1988. Chapter 11 of this college-level textbook provides a detailed look at biostratigraphic concepts, methods, and applications. Well illustrated, with extensive reference lists and an index.

Brookfield, Michael E. *Principles of Stratigraphy.* Hoboken, N.J.: Wiley-Blackwell, 2004. Written for undergraduate students, this text provides an overview of the principles and applications of stratigraphy. Includes stratigraphic techniques and case studies. Organized into three sections, beginning with foundational material, followed by data collection and research topics, and completed with interpretations and analysis.

Dott, Robert H., Jr., and Donald R. Prothero. *Evolution of the Earth.* 8th ed. New York: McGraw-Hill, 2009. This basic textbook on historical geology is aimed at students of geology. However, it is very readable by anyone with a background in science. Presents an up-to-date account of the earth's history from the viewpoint of plate tectonics. Includes a glossary.

Grotzinger, John, and Thomas H. Jordan. *Understanding Earth,* 7th ed. MacMillan, 2014. This comprehensive physical geology text covers the formation and development of the earth. Readable by high school students as well as by general readers. Includes an index and a glossary of terms.

Hedberg, H. D., ed. *International Stratigraphic Guide.* New York: John Wiley & Sons, 1976. The international "rule book" for stratigraphy. It sets procedures and standards to be met when naming stratigraphic units. It also defines many terms used in stratigraphy and has an extensive bibliography. Chapter 6 is devoted to biostratigraphy.

McGowran, Brian. *Biostratigraphy: Microfossils and Geological Time.* New York: Cambridge University Press, 2008. This text expands the application of biostratigraphy from the original use in paleontology to the fields of petroleum exploration and deep-ocean drilling. Addresses the relatively new methodology of studying microfossils and absolute aging.

Ogg, James G., Gabi Ogg, and Felix M. Gradstein. *The Concise Geologic Time Scale.* New York: Cambridge University Press. 2008. This book is a complete overview of the geological time scale, including stratigraphy topics such as chronostratigraphy and magnetic stratigraphy. It is organized by geological time periods. Also contains a reference appendix, the geological time scale table, and indexing.

Prothero, Donald R. *Bringing Fossils to Life.* 2d ed. Boston: McGraw-Hill, 2004. This well-illustrated and entertaining text covers a broad range of paleontological topics, including biostratigraphy. Glossary, bibliography, and index.

Stanley, S. M. *Exploring Earth and Life Through Time.* 2d ed. New York: W. H. Freeman, 1989. An excellent introductory-level college textbook on historical geology. It reviews the history of life and the many fossil forms found in strata in the earth's crust. Chapter 5 includes a discussion of biostratigraphy. Lavishly illustrated, with extensive references, glossaries, appendices on fossil groups, and an index.

Winchester, Simon. *The Map That Changed the World.* London: Penguin, 2001. An entertaining study of William Smith, who developed the idea of biostratigraphy and used it to produce the first geologic map of Great Britain. Written for the interested reader but valuable to students and professionals also.

Bureau of Chemistry & Soils, U.S.

—*William E. Burns*

FIELDS OF STUDY

Chemistry; Soil Science

SUMMARY

The Bureau of Chemistry and Soils (BCS), part of the U.S. Department of Agriculture, was one of many federal departments that dealt with applying science to improve economic productivity. It worked in cooperation with state and local governments to evaluate soils, make soil maps, and minimize the incidence of agricultural dust explosions, in addition to conducting original research on the chemistry of agricultural products.

HISTORY

In 1927, the regulation of food and drugs was spun off from the Bureau of Chemistry and delegated to the newly formed Food, Drug, and Insecticide Administration (later the Food and Drug Administration). The remaining body was merged with the Bureau of Soils, the Fixed Nitrogen Research Laboratory, and the Division of Soil Fertility and Soil Bacteriology of the Bureau of Plant Industry to form the Bureau of Chemistry and Soils. The focus of the new bureau was to be research rather than regulation. Its first head was Henry G. Knight, the dean of agriculture at the University of West Virginia. The former head of the Bureau of Chemistry, Charles A. Browne, whose interests lay more in research than administration and enforcement, accepted the subordinate position of chief of chemistry and technical research in the consolidated bureau. The new bureau employed 445 people in its first year.

Soil science and the making of soil maps for a vast range of communities took up much of the bureau's resources. The bureau participated in the Department of Agriculture's project, launched in 1928, to research soil and water conservation, establishing ten experimental stations by 1933. Another of its primary functions was helping farmers by developing and improving uses for agricultural products outside the traditional areas of food and clothing. The bureau's interest in agricultural products extended to some aspects of manufacturing. For example, it produced a report on the preparation of bookbinding leathers for maximum longevity. The bureau regarded dissemination of its findings to both the scientific community and the public as central to its mission; it maintained an active publication program, and its scientists participated in national scientific organizations.

IMPACT

The Bureau of Chemistry and Soils produced and communicated a large body of agricultural research in the late 1920s and early 1930s. It helped set a precedent for the expansion of federal sponsorship of scientific research in the following decades. In 1938, the Bureau of Chemistry and Soils was consolidated into the Bureau of Agricultural Chemistry and Engineering.

FURTHER READING

Cameron, Frank. *Cottrell, Samaritan of Science.* Garden City, N.Y.: Doubleday, 1952.

Weber, Gustavus Adolphus. *The Bureau of Chemistry and Soils: Its History, Activities and Organization.* Baltimore, Md.: Johns Hopkins Press, 1928.

C

Carbon-Oxygen Cycle

—Gina Hagler

FIELDS OF STUDY

Atmospheric Chemistry; Biochemistry; Environmental Chemistry; Geochemistry; Photochemistry; Environmental Sciences; Bioclimatology; Oceanography; Physical Geography; Ecology.

PRINCIPAL TERMS

biosphere: life on Earth, and the area it inhabits

carbon sequestration: the process of storing carbon in a stable state to negate carbon's effects on climate

combustion: reactions by which oxygen and organic materials become carbon dioxide, water and other oxides

decomposition: process by which organic matter is broken down into its most basic components by microorganisms

greenhouse effect: an atmospheric phenomenon in which certain molecular components of the atmosphere capture infrared radiation from a planet's surface and prevent it from radiating back out into space

homeostasis: the condition in which all components of a multicomponent system are in dynamic balance with each other so that the overall state of the system remains constant

hydrosphere: water on Earth, and its area

pedosphere: the soil

photosynthesis: the process occurring in green plants for the production of glucose from carbon dioxide and water in the presence of chlorophyll and sunlight

respiration: process by which organisms break down glucose and other sugars for energy

THE CARBON-OXYGEN CYCLE

The carbon-oxygen cycle involves the flow of carbon and oxygen through the environment. Carbon is recycled between carbon dioxide and carbon compounds in life-forms. The cycle includes the transfer mechanisms of the carbon compounds and oxygen in the biosphere and in the atmosphere, hydrosphere, and pedosphere. Climate research is focusing on the rates of processing and the homeostasis of the system to determine carbon dioxide release and capture.

The carbon-oxygen cycle is the result of the storing of carbon from atmospheric and dissolved carbon dioxide as organic material through photosynthesis. The carbon is subsequently released through respiration, combustion, or decomposition, all of which combine the carbon in the organic material with atmospheric or dissolved oxygen as carbon dioxide (CO_2). The CO_2 so released into the atmosphere is reabsorbed by photosynthetic organisms such as cyanobacteria, some protists, and green plants, thus completing the cycle and starting the process anew.

The cycle demonstrates how different parts of Earth's environment and climate are interrelated. For example, analysis of the flow of material from rivers into seas allows scientists to consider the impact of biologic materials. Similarly, the release of the gases into the atmosphere and the capturing of carbon by plants allow scientists to see the effect of land use on climate.

By examining environmental factors such as land use or propensity for fire in a given area, one can get an idea of how a given region will impact other areas and can estimate the total amount of gases it might contribute to the atmosphere. Additionally, environmental factors provide hints as to effective methods of managing or sequestering carbon. Forests and trees are typical examples of carbon sequestration.

Based on the carbon-oxygen cycle, it is known that trees take in a certain amount of carbon over time, that the trees will produce living space for some number of animals, and that those animals will produce a certain amount of carbon dioxide throughout their lives and when they die and decompose. Also factored is whether a forest is prone to fires. By looking at such factors, one can get an idea of how much carbon dioxide a forest can remove from the atmosphere and for how long, and if there is anything humans can do to make that forest more efficient.

Most of the oxygen released into Earth's atmosphere is produced by oceanic plankton, whereas forests of all kinds absorb huge amounts of carbon as sites of oxygen turnover.

PROCESSES

Photosynthesis is a set of chemical reactions that occur in green plants and in some bacteria and some protists. Photosynthesis uses energy from the sun to convert carbon dioxide and water into the sugar glucose and oxygen in the presence of chlorophyll.

The process itself is the interaction of light-dependent reactions and the Calvin or dark cycle. The light-dependent reactions separate the hydrogen from the oxygen in water and are thus the oxygen-producing portion of the cycle and prepare reactants needed for the other cycles. The Calvin or dark cycle turns the carbon dioxide and light cycle products into sugar. The light cycle runs directly on captured photons, whereas the Calvin cycle does not depend directly upon sunlight.

Along with rates of release from combustion, which is the conversion of organic material to carbon dioxide and water vapor, it is also necessary to factor in rates of respiration and decomposition. Respiration is how organisms get energy from organic compounds, such as sugars. Plants and animals use aerobic respiration, which evolved after oxygen became prevalent in the atmosphere. Aerobic respiration occurs in special organelles of eukaryotic cells called ribosomes. Here sugars are broken down into carbon dioxide and water in a process known as the Krebs or citric acid cycle.

Decomposition releases various greenhouse gases, as decomposing organisms are often excellent habitats for anaerobic life-forms. Anaerobic respiration releases such gases as carbon dioxide, hydrogen sulfide, and methane. Thus the total impact of the disruption of an environment, such as the burning of a forest, is not only in the combustion of the trees but also in the drop in photosynthesis and decomposition of any detritus.

HISTORY OF THE CYCLE

The understanding of the carbon-oxygen cycle has existed since the eighteenth century, after the discoveries of Antoine Lavoisier (1743–1794), who found that respiration was a process similar to burning, and of Joseph Priestley (1733–1804), who discovered oxygen and who also conceptualized the relationship between blood and air. The full mechanics of the carbon-oxygen cycle and its implications, however, were discovered by scientists in subsequent centuries.

The workings of the chemical reactions of photosynthesis were discovered in the 1940's, and research continues into the role of quantum physics in the light-harvesting stage of photosynthesis. With the growing awareness of climate change, more work is being done to examine the cycle.

Current research often takes samples from the area in question, be they soil samples to estimate bacterial carbon production or air samples to measure carbon levels at a given site. These samples are then analyzed together to get a picture of how each part is affecting the overall carbon-oxygen balance for that area. The carbon data are often mixed with data on other climatological factors, such as albedo, humidity, and temperature, to build a picture of the effects on climate.

ATMOSPHERIC COMPOSITION & THE GREENHOUSE EFFECT

The carbon-oxygen cycle affects climate because it affects Earth's atmosphere. Ignoring water vapor, which is variable, it is known that the standard dry atmosphere is 78.09 percent nitrogen, 20.95 percent oxygen, 0.93 percent argon, and .045 percent carbon dioxide; the rest of the atmosphere comprises various trace gases. Water vapor composes, on average, about 1 percent of the atmosphere.

The carbon-oxygen cycle plays an important role in maintaining homeostasis in the atmosphere. The cycle "cycles" air and carbon to maintain equilibrium.

The gases in the atmosphere and their relative amounts have an important impact upon climate. Some gases absorb infrared rays reflected or emitted from the earth's surface and send them back to Earth and into the lower atmosphere, thus trapping the energy and raising Earth's overall average temperature. This process is known as the greenhouse effect.

Gases that are best at absorbing and emitting in the infrared range are known as greenhouse gases and include water vapor, carbon dioxide, methane, nitrous oxide, and ozone. Water vapor and carbon dioxide make up the largest contribution to the effect, with methane and ozone playing a smaller role though they are much more effective greenhouse gases than are carbon dioxide and water. Many of these gases are produced in industrial reactions such as combustion. Combustion from cars and factories and from other human activities since the start of the Industrial Revolution in the eighteenth century have been major factors in anthropogenic, or human-caused, climate change. It is here that the carbon-oxygen cycle comes into play. There is a maximum rate at which a particular environment can cycle the carbon dioxide into oxygen, meaning that excess carbon dioxide builds up in the atmosphere, enhancing the greenhouse effect.

By studying the plants and animals that are part of the process, scientists can find the effects of increased temperature on those components. The environmental impact, or increased environmental stress, leads to less capacity to handle the increased carbon dioxide. This creates a positive feedback loop that further increases the temperature. Increased global temperature could cause increased cloud cover, which would increase Earth's albedo. Increased temperature also could spawn algal blooms in the oceans, which would absorb carbon dioxide and thus reach a new equilibrium. Such events are indicative of a large amount of environmental damage.

The greenhouse effect is not wholly negative, however. It is actually required for life to exist on this planet. In the absence of greenhouse gases, Earth would have a significantly colder climate than it does now. The average temperature of the planet in the absence of the naturally-occurring greenhouse effect has been estimated to be a balmy -18°C. Additionally, Earth's atmospheric composition has varied throughout its history. Earth's early atmosphere was very different from that of today. It consisted mostly of water vapor, carbon dioxide, hydrogen sulfide, and ammonia. All of the greenhouse gases of the time kept the average Earth temperature fairly high.

Once photosynthetic life evolved about 3.5 billion years ago, the oxygen it produced began to break down the methane and carbon dioxide in the atmosphere, and the level of oxygen began to increase. This resulted in what is called the oxygen catastrophe, in which many of the non-aerobic organisms at the time became extinct, leaving those that adapted to the new situation by developing aerobic respiration. This development had a great impact upon later life, as aerobic respiration became far more efficient.

The oxygen catastrophe also may have caused the Huronian glaciation event, which occurred most likely as a result of greenhouse gas depletion by photosynthetic life-forms. The event was one of the longest glaciation periods in Earth's history, lasting some 300 million years. It and the later Snowball Earth periods that likely caused the evolution of multicellular life and the Cambrian explosion were ended only by the buildup of greenhouse gases from volcanic eruptions and by the weathering of rocks.

Since this time, greenhouse gas levels have oscillated by period; for example, during the Devonian period, carbon dioxide was higher and oxygen was lower than present levels. In the Cretaceous period, both carbon dioxide and oxygen levels were higher. During both periods, the average temperature was higher than it is today.

The Permian-Triassic mass extinction event of 250 million years ago, in which a shutdown of the carbon dioxide cycle may have been responsible for the greatest mass extinction in Earth's history, is a dramatic example of how the carbon-oxygen cycle figures into climate. While the exact causes remain unknown, the mass extinction is thought to have begun with the eruption of the Siberian Traps, a massive set of volcanoes that released massive quantities of carbon dioxide and debris into the atmosphere. This largest eruptions in known history are believed to have caused a small period of global cooling, which stressed many environments. As a result, many trees died before they could process the carbon dioxide that was added to the atmosphere. Evidence seems to show that global warming set in around the peak of the extinction, and it is thought that the oceans became saturated with carbon dioxide, to the point in which there was a degassing event. The seas released their carbon in huge clouds that swept over the land, suffocating all nearby creatures. Massive erosion suggests that the climate became arid and the interior of continents were more or less lifeless. While another event on this scale is unlikely, it provides a strong example of the kind of phenomena that can arise from the disruption of the carbon-oxygen cycle.

GREENHOUSE GAS EMISSIONS

More than 32.5 billion tonnes of carbon dioxide were released into the atmosphere in 2017, the highest annual release globally, with the United States accounting for some 5.14 billion tonnes of the total. Compounding that are ongoing releases of other gases such as methane, which has twenty-five times the effect of carbon dioxide as a greenhouse gas in a one-hundred-year period.

The largest sources of anthropogenic emissions are deforestation and industrial technology, including factories, cars, and cement making (which releases carbon dioxide). However, agriculture also releases greenhouse gases. Cattle, for example, are estimated to release 16 percent of anthropogenic methane, although the effect of millions of cattle is unlikely to be very different from that of the millions of bison that previously roamed the prairies of North America naturally.

Humans have always released carbon dioxide into the environment. Various indigenous North American cultures burned forests annually to encourage habitats for large game animals. Also, much of Europe has been deforested for agricultural space and shipbuilding. However, the emissions from industrial and postindustrial technology are on a wholly greater scale. Equally important is that the carbon being released now has not been part of the carbon-oxygen cycle since it was stored in the form of coal and petroleum more than 250 million years ago. Thus, with the combination of environmental disruption and greenhouse gas emission, recent activities are more akin to a sustained volcanic eruption than to any other event on Earth.

As the system moves further from homeostasis, the compensatory mechanisms, such as increased plant growth, are overloaded or are not given the chance to function, as is the case with heavy deforestation. This forms positive feedback, pushing the environment further from equilibrium. Something as simple as planting additional trees can act to forestall the harmful trend, however. An important point is that the carbon in a living tree has not been expelled; it remains stored as long as the tree lives. Once the tree dies the carbon dioxide is released, either slowly through natural decomposition of the cellulose that makes up the wood, or rapidly through combustion of the wood.

CARBON SEQUESTRATION

Carbon sequestration involves storing carbon in a stable state to negate carbon's harmful effects on the environment. As mentioned, trees serve as reservoirs for carbon. Many efforts focus on manipulating the carbon-oxygen cycle to increase the amount of carbon stored. Means of sequestering carbon range from genetically modifying plants so that they carry out a more efficient form of photosynthesis, to timing forest harvesting and replanting to ensure peak carbon storage, as in the Canadian lumber industry. Other forms of carbon sequestration under consideration include injecting carbon into worked out coal seams; sinking carbon to the bottom of the ocean through algae growth, where it would presumably eventually become petroleum or coal once more in time; producing certain carbon-absorbing compounds; and developing processes that utilize carbon dioxide as a raw material. With proper manipulation of the carbon-oxygen cycle through methods such as carbon sequestration, many of the worst effects of climate change can be mitigated.

Further Reading

Bashkin, Vladimir N. *Modern Biogeochemistry: Environmental Risk Assessment.* Dordrecht: Springer, 2010. Print. An excellent detailed description of the carbon-oxygen cycle, its precise mechanics, means for quantitative analysis, the development of the system, and its relationship to other cycles in the environment. Most suitable for college students and other readers with some background in chemistry and biology.

Kirby, Richard R. *Ocean Drifters: A Secret World Beneath the Waves.* Studio Cactus, 2010. Focuses on plankton, the life-forms responsible for most of the oxygen release on Earth. Includes enlarged photographs. Recommended for general readers.

Kondratyev, K. Y., V. F. Krapivin, and Costas A. Varotsos. *Global Carbon Cycle and Climate Change.* Springer, 2003. Detailed analysis of the relationship of the carbon-oxygen cycle to climate change. Includes the effects of other greenhouse gases and how it relates with Earth's environment. Recommended for those with some background in the subject.

Le´vêque, C. *Ecology from Ecosystem to Biosphere.* Science, 2003. Contextualizes the carbon-oxygen cycle in its ecological role, which allows readers to see how the cycle fits with the larger functioning of the ecosystem and biosphere. Suitable for general readers.

Vallero, Daniel A. *Environmental Biotechnology: A Biosystems Approach.* Academic, 2010. Examines the carbon-oxygen cycle in the context of environmental biotechnology. Systems based, it deals with the larger scale and interactions of the cycle.

Walker, Sharon, and David McMahon. *Biochemistry Demystified.* McGraw-Hill, 2008. An accessible introduction to the processes behind respiration and photosynthesis that underlie the carbon-oxygen cycle. Written for beginners. Also covers biochemistry.

CHAPARRAL

—C. Mervyn Rasmussen

FIELDS OF STUDY

Ecology; Environmental Studies.

PRINCIPAL TERMS

elfin forest: a stunted forest growing at high elevations in warm, moist climates

fire ecology: an ecosystem that depends on periodic fires to clear underbrush; the seeds of many plants in such an ecosystem require fire in order to germinate

SUMMARY

Chaparral is the name of a major ecosystem (or biome) found in areas with moist, cool to cold winters and long, dry hot summers (Mediterranean climate).

Chaparral ecosystems with different names occur in the Mediterranean, South Africa, Chile, Australia, and Mexico. The word "chaparral" is a colloquial adaptation of the original Mexican name, *chaparro.* While the chaparral communities in other parts of the world have the same basic characteristics and very similar adaptations, this article will focus on the chaparral of California and the American Southwest.

Chaparral is an interesting and unique ecosystem. It is an elfin (stunted) forest dependent on a fire ecology, and its adaptations to a harsh and variable climate are remarkable. The chaparral's geology, latitude, altitude, and climate are all related and have played a role in its formation.

In California, the chaparral is located mainly along the central and southern coastal areas, primarily between elevations of 500 and 2,500 feet. It is also found in some areas of the Sierra Nevada foothills, one hundred miles or more inland, and in the lower elevations of some other interior mountains. The geology of most of the areas where this ecosystem occurs is believed to have started with massive upheavals from half a million to ten thousand years ago. The most common substrate was granite. Relentless disintegration resulted in rocky and sandy debris which would allow increasing amounts of plant life to grow. As the organic matter became more abundant, its debris (leaves, twigs, and decaying dead plants) became more and more mixed into the materials of the granite decomposition. This resulted in a pourous, sandy loam.

CHAPARRAL FLORA

Because of the cool, moist winters and dry, hot summers, plants evolved to survive these marked changes. In most of the chaparral, there are quite extreme diurnal temperature changes, with fluctuations of fifty to sixty degrees or more. Compounding these harsh conditions are frequent strong, dry winds, often reaching forty miles per hour.

The plants that have become the residents of the chaparral are mainly shrubby, small-leaved evergreens with leathery, thick stems. Shrubs predominate, but there are also small trees, and wildflowers are in abundance in many areas. All plants are adapted to conserve water. There is little humus in the soil, which is relatively nutrient-poor. The sandy nature of the soil and the variable periods of dry and sudden rain can cause a quick run-off.

The predominant plants are between three and nine feet in height, with some trees being taller. They hug close to the ground to provide shade. The ratio of the surface area of the leaves and stems to their body mass is reduced, and they tend to have thick, heat-resistant surfaces. Some of the plants are capable of turning their leaves so the edges face the sun, which cuts down the warming effect on their surfaces. All these mechanisms greatly reduce evaporative water loss. Most of the bushes and trees also have unusually long tap roots. A three-foot plant might have a tap root that goes ten or more feet below the surface, enabling it to get more water and nutrients.

The most common plant in the chaparral is the greasewood-chamise bush (*Adenostoma fasciculatum*). The chamise is characterized by numerous small, club-shaped leaves with a waxy substance that protects them from drying. When there is a fire, the chamise burns with an intense heat and creates a very black smoke (hence the name "greasewood"). Others that predominate are the christmasberry-toyon bush (*Photinia arbutifolia*), the scrub oak (*Quercus dumosa*), the yucca (*Yucca Whipplei*), and the hoary manzanita (*Arctostaphylos canescens*).

THE ROLE OF FIRE

The role of fire in the maintenance and regeneration of the chaparral is of paramount importance. The hot, dry summer weather, often fanned by winds, makes the chaparral very prone to fires. Because of the high fuel content in the dense plants, with their waxy and oily components, these fires are very intense and can spread rapidly. Fire is necessary to clear excess growth and allow new seeds to germinate. Indeed, several of the key species need fire to release their seeds or they will not germinate. The amount and distribution of the canopy fuel can have a marked effect on regrowth and even spatial variation. Naturally recurring fires are usually good for germination. Unusually intense fires, often from years of fire suppression, may do harm by damaging the plants severely.

In much of the chaparral, human interference has allowed the fires to become more damaging when they occur. This creates a difficult paradox, since many people now reside in the chaparral and chaparral fires can spread very rapidly, especially with strong winds. Conversely, in some areas where fire has been controlled, the chaparral has been retreating. A good example of this is on the southern slopes of Mount Tamalpais, a mountain just north of San Francisco.

CHAPARRAL FAUNA

A wide variety of reptiles, birds, and mammals make the chaparral their home. They have developed adaptations to survive and thrive in this harsh environment. Several species of skinks, lizards, and a variety of snakes, including gopher snakes, the California king snake, and both the red diamond and western rattlesnakes are residents. There are dozens of birds, from several species of hummingbirds to the large birds: the turkey vulture, barn owl, roadrunner, and golden eagle. There are many species of rodents, including kangaroo rats, chipmunks and gophers. The variety of medium to large mammals of common interest is impressive and includes the coyote, gray fox, badger, lynx, bobcat, mountain lion, and mule deer.

Further Reading

Brown, O. E., and R. A. Minnich. "Fire and Changes in the Creosote Bush Scrub of the Western Sonoran Desert, California." *American Midland Naturalist* 116, no. 2 (1986): 411–422. Slightly technical but very informative account of the role of fire in an ecosystem.

Collis, P. H., ed. *Dictionary of Ecology and the Environment.* 3d ed. Chicago: Fitzroy Dearborn, 1998. Excellent reference for readers with no biology background.

Head, W. S. *The California Chaparral: An Elfin Forest.* 1972. Reprint. Happy Camp, Calif.: Naturegraph, 1998. Clearly written, interesting text that does not require a biology background. Many good illustrations and descriptions of flora and fauna.

Jasson-Holt, Sophie. *Unfold the Chaparral.* San Francisco: San Francisco State University Chapbook, 1996. Very well-written and excellent overview of the chaparral.

Odion, Dennis C., and Frank W. Davis. "Fire, Soil Heating, and the Formation of Vegetation Patterns in Chaparral." *Ecological Monographs* 70, no. 1 (2000): 149–169. A detailed scientific account of the role of fire in the chaparral. It would be difficult reading with no background in biology.

CIVILIAN CONSERVATION CORPS (CCC)

—Eugene Larson

FIELDS OF STUDY

Ecology; Environmental Studies; Political Science.

SUMMARY

The Civilian Conservation Corps, a central part of Franklin D. Roosevelt's "New Deal," was conceived as a comprehensive project which would encompass relief for the unemployed, recovery of the nation's economic health, and conservation of American natural resources.

BACKGROUND

In 1934, President Franklin D. Roosevelt noted that the United States was one of the few industrialized countries that had not established a "national policy for the development of our land and water resources." This lack was in the process of rectification when, in March, 1933, shortly after his inauguration, Roosevelt proposed the establishment of the Civilian Conservation Corps (CCC). The legislation provided for the voluntary mobilization of unemployed young men to work on various conservation projects throughout the nation.

As Congress did on most of Roosevelt's proposals during his first one hundred days in office, it acted swiftly, approving the legislation on March 31, 1933. Administered by the Labor Department, the Army, the Forestry Service, and the National Park Service, the CCC had the potential to be an administrative disaster, but disaster did not happen. By July more than 300,000 unemployed young men, aged eighteen to twenty-five and from families on relief, were already working in the CCC's thirteen hundred camps. By 1935, there were more than 500,000 men in the CCC, and before it was dismantled more than 2.5 million young men had joined, working for one dollar a day in twenty-five hundred camps.

IMPACT ON RESOURCE USE

The projects were varied, ranging from restoring battlefields of the American Revolution and Civil War to constructing trails in the High Sierra; from protecting wildlife (including stocking almost one billion fish) and building fire lookout towers to planting two billion trees—200 million as windbreaks in the Dust Bowl. Estimates indicate that of all the forests planted in the history of the United States, both public and private, more than half were planted by the so-called tree people of the CCC. From the east and west, north and south, farm boys worked alongside young men from the cities. The CCC was organized on a military basis, although participation was voluntary, and one could enter and leave when one wished. Most men stayed from several months to about one year.

Although women were excluded and African Americans were subject to a 10 percent quota and were usually segregated, as a conservation organization, the CCC was an instant and lasting success. Many of America's natural resources were preserved during those few years of the 1930's in spite of the predictions by some that many of the projects were beyond the government's powers and that the CCC would be inimical to capitalism or to organized labor because of the CCC's low wages. Some feared that the CCC smacked of communist collectivism or fascist militarism. Not the least of the resources conserved were the young men themselves, whose experience developed their physical bodies as well as their intellectual and emotional capabilities. At the onset of World War II the CCC was terminated, but individual states later established their own conservation corps, such as the California Conservation Corps. John F. Kennedy's Peace Corps was also inspired in part by the CCC.

FURTHER READING

Civilian Conservation Corps. Civilian Conservation Corps Legacy. http://www.ccclegacy.org/

National Archives. Records of the Civilian Conservation Corps. http://www.archives.gov/research/

CLONING OF PLANTS

—*Henry R. Owen*

FIELDS OF STUDY

Biotechnology; Economic Botany and Plant Uses; Genetics

PRINCIPAL TERMS

callus: an unorganized mass of dividing cells such as scar tissue that forms over a healing wound
cell theory: the theory that all living things are composed of discrete units called cells that are produced by other cells in a hierarchy of generations
monoculture: reliance on a single crop species over a large area rather than on a genetically-diverse variety of crop species
somaclonal variation: the unintentional incorporation of genetic abnormalities in plant species through cloning and subsequent normal pollination propagation
totipotency: the principle that each individual cell contains the 'total potential' of reproducing the complete organism in which the cells are found

SUMMARY

Plant cloning is the production of a cell, cell component, or plant that is genetically identical to the unit or individual from which it was derived. The term "clone" is derived from the Greek word *klōn*, meaning a slip or twig. Hence, it is an appropriate choice. Plants have been "cloned" from stem cuttings or whole-plant divisions for many centuries, perhaps dating back as far as the beginnings of agriculture.

HISTORICAL BACKGROUND

In 1838 German scientists Matthias Schleiden and Theodor Schwann presented their *cell theory*, which states, in part, that all life is composed of cells and that all cells arise from preexisting cells. This theory formed the basis for the concept of *totipotency*, which states that since cells must contain all of the genetic information necessary to create an entire, multicellular organism, all of the cells of a multicellular organism retain the potential to re-create, or regenerate, the entire organism. Thus was the basis for plant cell culture research.

The first attempt at culturing isolated plant tissues was by Austrian botanist Gottlieb Haberlandt at the beginning of the twentieth century, but it was unsuccessful. In 1939 Roger Jean Gautheret and his colleagues demonstrated the first successful culture of isolated plant tissues as a continuously dividing callus tissue. The term *callus* is defined as an unorganized mass of dividing cells, such as in a wound response. It was not until 1954, however, that the first whole plant was regenerated, or cloned, from a single adult plant cell by W. H. Muir, Albert Hildebrandt, and Albert Riker. Thereafter, an increased understanding of plant physiology, especially the role of plant hormones in plant growth and development, contributed to rapid advances in plant cell and tissue culture technologies in the 1970s and 1980s. Many plant species have been successfully cloned from single cells, thus demonstrating and affirming the concept of totipotency.

HORTICULTURE

Naturally-occurring clone plant species have been known for centuries, grown from shoots and suckers taken from existing plants. The plants grown from these sources are genetically identical to the parent plant and are subsequently susceptible to the same diseases and pests that are dangerous to the parent plant. As has happened in commercial banana monoculture, in which all banana trees are clones of a single variety, any fungus, virus or infection that can kill one of the plants is able to destroy the entire plantation since none of the plants has a different level of resistance to that infection than does the parent plant.

By far, the greatest impact of cloning plants *in vitro* (Latin for "in glass," meaning in the laboratory or outside the plant) has been on the horticultural industry. In the 1980s plant tissue culture technologies propagated and produced many millions of plants. Since then, many economically important plants have been commonly propagated via tissue culture techniques, including vegetable crops (such as the potato), fruit crops (strawberries and dates), floriculture species (orchids, lilies, roses, Boston ferns), and even woody species (pines and grapes).

The advantages of plant cell, tissue, and organ culture technologies include a more rapid production of plants, taking weeks instead of months or years. Much less space is required (square feet instead of field plots). Plants can be produced year-round, and economic, political, and environmental considerations that hamper the propagation of regional or endangered plant species can be reduced. The disadvantages include the high start-up costs for facilities, the skilled labor required, and the need to maintain sterile conditions.

Two other significant considerations must be considered as a result of plant propagation technologies. As illustrated by the Irish Potato Famine of the 1840s, the cultivation of whole fields of genetically identical plants (*monoculture*) leaves the entire crop vulnerable to pest and disease infestations. The second important consideration when generating entire populations of clones, especially using tissue culture technologies, is the potential for introducing genetic abnormalities, which then are present in the entire population of plants produced, a process termed *somaclonal variation.*

BIOTECHNOLOGY

An absolute requirement for the genetic engineering of plants is the ability to regenerate an entire plant from a single, genetically transformed cell, thus emphasizing the second major impact of plant cell culture technologies. In 1994 the U.S. Food and Drug Administration (FDA) approved the first genetically modified whole food crop, Calgene's Flavr Savr tomato. This plant was produced using what is termed anti-sense technology. One of the tomato's genes involved in fruit ripening was reversed, thus inactivating it and allowing tomatoes produced from it to have significantly delayed ripening, which meant that fruits could be harvested for market while green and rock hard. Although no longer commercially marketed, the Flavr Savr demonstrated the impact of genetic engineering in moving modern agriculture from the Green Revolution into what has been termed the Gene Revolution.

Other examples of agricultural engineering exist, such as Roundup Ready soybeans, engineered to resist the herbicide commonly used on weeds in fields where soybeans are grown, and Bt corn, which contains a bacterial gene conveying increased pest resistance. Since 1987, the U.S. Department of Agriculture (USDA) has required field testing of genetically modified crops to demonstrate that their use will not be disruptive to the natural ecosystem. Thousands of field trials have been completed or are in progress for genetically modified versions of various crop species, including potatoes, cotton, alfalfa, canola, and cucumbers.

These studies have not done enough to alleviate consumer concerns, however. Genetically modified foods (produced from genetically modified organisms, or GMOs) have caused major controversy not only over how they might affect nature—in terms of reduced biodiversity, new pests, and other environmental factors—but over how they could affect human health over the long term as well. There has been concern that GMOs could introduce a new allergen, for example. There has also been also concern that a gene transfer of harmful, antibiotic-resistant bacteria could occur, when animals consume GM plants and then humans consume the animal or the plant itself. Consumers have also shown concern over a lack of labeling; a person may unwittingly eat a GMO product when trying to avoid it.

FURTHER READING

Acquaah, George. *Principles of Plant Genetics and Breeding.* 2nd ed. Hoboken: Wiley, 2012. Print.

Barbosa-Cánovas, Gustavo, et al. *Global Issues in Food Science and Technology.* Burlington: Academic, 2009. Print.

Barnum, Susan R. *Biotechnology: An Introduction.* 2nd ed. Belmont: Wadsworth, 2005. Print.

Chrispeels, Maarten J., and David E. Sadava. *Plants, Genes, and Crop Biotechnology.* 2nd ed. Sudbury: Jones & Bartlett, 2003. Print.

Conger, B. V., ed. *Cloning Agricultural Plants Via in Vitro Techniques.* S.l.: CRC PRESS, 2017. Print.

Dodds, John H, and Lorin W. Roberts. *Experiments in Plant Tissue Culture.* Cambridge, Mass: Cambridge University Press, 2004. Print.

Kyte, Lydiane, and John Kleyn. *Plants from Test Tubes: An Introduction to Micropropagation.* 4rd ed. Portland: Timber, 2013. Print.

Pierce, Benjamin A. *Genetics: A Conceptual Approach.* 4th ed. New York: W. H. Freeman, 2012. Print.

Pierik, R. L. M. *In Vitro Culture of Higher Plants.* Kluwer Academic Publishers, 2000. Print.

Thieman, William J., and Michael A. Palladino. *Introduction to Biotechnology.* 3rd ed. Boston: Pearson, 2013. Print.

Weasel, Lisa H. *Food Fray: Inside the Controversy over Genetically Modified Food.* New York: Amacom-American, 2009. Print.

COEVOLUTION

—Frederick M. Surowiec

FIELDS OF STUDY

Biology; Environmental Studies.

PRINCIPAL TERMS

antagonism: any type of interactive, interdependent relationship between two or more organisms that is destructive to one of the participants

coevolution: the interactive evolution of two or more species that results in a mutualistic or antagonistic relationship

commensalism: a type of coevolved relationship between different species that live intimately with one another without injury to any participant

parasitism: a type of coevolved relationship between different species in which one species exploits the other to its physical detriment

phytophagous: animals, also referred to as herbivorous, that feed on plants

reciprocal relationship: any type of coevolved, highly interdependent relationship between two or more species

selective pressure: evolutionary factors that favor or disfavor the genetic inheritance of various characteristics of a species

symbiosis: a type of coevolved relationship between two species in which both participants benefit; a type of mutualism

SUMMARY

Numerous stable and surprisingly complex relationships between various apparently unrelated organisms, such as plants and animals, have evolved over long periods of time. All these relationships, whether beneficial or harmful to one or more of the participants, represent adaptive interactions that can be used to understand ecological patterns and processes.

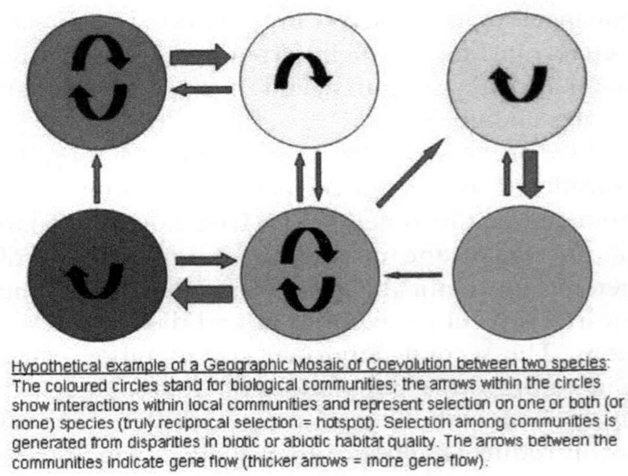

Graphic about the Geographic Mosaic Model of Coevolution. Image by Evoecogen (Own work) CC-BY-SA-3.0.

COEVOLUTION AND INTERDEPENDENCE

Coevolution is an extremely important and widespread phenomenon in the world of living things; it is a biological factor that is global in influence. When two or more different species experience a relationship in which any of the participating species' evolution directly affects the evolution of the other members, coevolution is taking place. This interactive type of evolution is characterized by the fact that the participant life-forms are acting as a strong selective pressure upon one another over a period of time.

The assumption of the interdependence of all organisms is today such a commonplace and fundamental concept that it is surprising that the phenomenon of coevolution has not always enjoyed a more prominent position in evolutionary thinking. Many scientists seem to have considered sets of coevolved organisms as relatively unimportant phenomena, almost on the level of biological "curiosities." The consensus appears to have been that while numerous examples of coevolution existed in both plant and animal kingdoms, on the whole it was of relatively minor importance in comparison with other evolutionary phenomena, such as competition. This opinion has begun to change as researchers increasingly recognize the intrinsic and ubiquitous role that coevolution has played, and continues to play, in the evolution of life at all levels throughout earth history.

GAIA

Organisms do not evolve in a biological vacuum. All organisms exist in, and have evolved within, the framework of one of a great number of delicately balanced and self-tuning biological systems or living communities termed ecosystems. Indeed, the entire planet can be regarded as one huge, incredibly complex ecosystem in which all the lesser ecosystems fit together and work together harmoniously. This planetary ecosystem has been called "Gaia" by some biologists, in reference to the ancient Greek earth goddess. In some respects, Gaia can be conceived of as actually one giant, worldwide organism. All the living communities in this huge ecosystem are products of coevolution. This phenomenon has been in effect over the vast expanses of geologic time and continues today. The only period in history when coevolution was probably not operating was at the dawn of life, billions of years ago, when the very first species of organisms appeared and had not yet established interactive communities. The importance of coevolution as a factor affecting life cannot be overstated.

Some biologists use the term coevolution in a more restricted sense to describe coevolved relationships that have developed between plants and animals, particularly between plants and animals that are herbivores or pollinators. The coevolution between plants and animals is one of the aspects of the field that has traditionally received the most attention, so this aspect of coevolution provides a useful departure point in describing the phenomenon.

COEVOLUTIONARY WARFARE

The coevolution of plants and animals, whether animals are considered strictly in their plant-eating role or also as pollinators, is abundantly represented in every terrestrial ecosystem throughout the world where flora has established itself. Moreover, the overall history of some of the multitude of present and past plant and animal relationships is displayed (although fragmentally) in the fossil record found in the earth's crust. The most elemental relationship between plants and animals is that of plants as food source. This relationship has an extremely long history, beginning with the evolution of microscopic, unicellular plants that were the earth's first autotrophs (organisms that can produce their own food from basic ingredients derived from the environment). In conjunction with the appearance of autotrophs, microscopic, unicellular heterotrophs (organisms such as animals, which must derive food from organic sources such as autotrophs) evolved to exploit the simple plants. This ancient and basic relationship has resulted in uncounted numbers of plant and animal species evolving and coevolving over billions of years of earth's history.

As both plants and animals became multicellular and more complex, more elaborate defense mechanisms evolved among plants, as did more elaborate feeding apparatuses and behavior among animals. This biological "arms race" grew ever more intense as groups of plants and animals eventually adapted to the more rigorous demands of a terrestrial existence, leaving the marine environment behind. New ecosystems developed that culminated in the world's first swamps, jungles, and forests. The plant-animal arms race engendered increasingly more sophisticated strategies of botanical defense and animal offense, and this coevolved interrelationship has continued unabated. This coevolutionary "warfare" between plants and animals has expressed itself partly through the evolution of botanical structures and chemicals that attempt either to discourage or to prevent the attentions of herbivores. These include the development of spines, barbs, thorns, bristles, and hooks on plant leaf, stem, and trunk surfaces. Cacti, holly, and rose bushes illustrate this form of plant strategy.

Another type of deterrence evolved in the form of chemical compounds that can cause a wide spectrum of negative animal response. These compounds range in effect from producing a sensation of mild distaste, such as bitterness, to more extreme effects, such as actual poisoning of herbivore metabolisms. Plants that contain organic compounds such as tannin are examples of the chemical defensive strategy. Tannins produce several negative results in animals, including partially inactivating digestive juices and creating cumulative toxic effects that have been correlated with cancer. Plants containing tannin include trees, such as members of the oak group, and shrubs, such as those that produce the teas used as human beverages. Other plants have developed more lethal poisons that act more rapidly. Plants have also developed other strategies, such as possession of a high silica content (as found in grasses), that act to wear down the teeth of plant eaters. Animals have counteradapted to these plant defensive innovations by evolving a higher degree of resistance to plant toxins or by developing more efficient and tougher teeth with features such as harder enamel surfaces, or the capacity of grinding with batteries of teeth.

COEVOLUTIONARY ALLIANCES

Not all coevolution is characterized by having an adversarial nature; mutually beneficial relationships are also very common. Sometime during the latter part of the Mesozoic era, angiosperms, the flowering plants, evolved and replaced most of the previously dominant land plants, such as the gymnosperms and the ferns. New species of herbivores evolved to exploit these new food sources. At some point, probably during the Cretaceous period of the late Mesozoic, animals became unintentional aids in the angiosperm pollination process. As this coevolution proceeded, the first animal pollinators became more and more indispensable as partners to the plants. Eventually, highly coevolved plants and animals developed relationships of extreme interdependence exemplified by the honeybees and their coevolved flowers. This angiosperm-insect relationship is thought to have arisen in the Mesozoic era by way of beetle predation, possibly on early, magnolia-like angiosperms. The fossil record gives some support to this theory. Whatever the exact route along which plant-animal pollination partnerships coevolved, the end result was a number of plant and animal species that gained mutual benefit from the new type of relationship. Such relationships are in general termed mutualisms.

Eventually some of these plant-animal mutualisms became so intertwined that one or both participants reached a point at which they could not exist without the aid of the other. These obligatory mutualisms ultimately involved other types of animal partners besides insects. Vertebrate partners such as birds, reptiles, and mammals also became involved in mutualisms with plants. Contemporary ecosystems, such as the United States' southwestern desert, include mutualisms between aerial mammals, such as bats, and plants, such as the agave and the saguaro cactus. The bats involved are nectar drinkers and pollen eaters. They have evolved specialized feeding structures such as erectile tongues similar to those found among moths and other insects with similar lifestyles. In turn, the plants involved with the pollinating bats have evolved either reciprocal morphologies or behavior patterns to accommodate their warm-blooded visitors. For example, angiosperms coevolutionarily involved with bats have developed such specializations as bat-attractive scents, flower structures that minimize the chance of injury to bats, and petal openings timed to the nocturnal activity of bats.

SYMBIOSES, COMMENSALISMS & PARASITISMS

Coevolved relationships are not restricted to beneficial or nonbeneficial relationships between plants and animals. They also include an immense number of relationships between animals and other animals, and even between plants and other plants. Among these various types of coevolved situations can be found subcategories such as symbioses, commensalisms, and parasitisms. The first two involve relationships beneficial to varying degrees that feature interactions of increasing physical intimacy between or among two or more species. Parasitism involves an intimate relationship produced through coevolution in which one participant, the host, experiences serious harm or even death through exploitation by the parasite. Predation is probably the most obvious form of coevolution among higher animals such as vertebrates. Modern carnivores such as the canines and felines and their prey are a dramatic example of coevolution at work. Animal hunters over time responded to the improved defenses of their prey by evolving better senses, such as stereoscopic, three-dimensional vision, hearing with expanded range of frequency response, and more effective body structures, such as multifunctional teeth. Such teeth are termed heterodont and represent a great improvement over the simple dental array of the more primitive vertebrates, such as fish and amphibians.

Beginning with the more advanced reptiles appearing in the late stages of the Paleozoic era, teeth began to differentiate into specialized components—incisors, canines, premolars, and molars—that enhanced food acquisition and improved mastication. This, in turn, improved digestion and allowed quicker energy acquisition from food. This evolutionary advantage has reached a zenith of adaptive success among the mammals. Mammalian predators evolved fangs and efficient claws, sometimes retractable, to minimize injury and wear. Along with improved hunting senses and better dentition came increased speed from the evolution of improvements in pelvic and limb arrangements. In response to this process, vertebrate herbivores also became generally swifter or better defended, more alert, attained higher metabolic rates, and were thus better able to elude or defend against predation. Advanced predators placed an intense selective pressure on their prey herbivores, spurring ever more efficient and acutely tuned responses among the herbivore populations. Herbivores evolved either as swift forms, such as deer, or became efficiently defended, walking fortresses, such as porcupines or armadillos. Because of the pervasive effect of coevolution, the overall relationship between predator and prey has been a reciprocal one in which all participants affect one another in an interactive manner.

UNRAVELING THE INTRICACIES OF COEVOLUTION

Field research and laboratory research are pursued concurrently in the effort to unravel the intricacies of the subject of coevolution. Field research involves actual observation in nature of animals and plants, their behavior, and, especially, their interaction with other species. Special attention is given to useful clues that can be employed to establish evolutionary relationships, either presently existing or previously in effect. For example, cooperative behavior between or among several different species of animal or plant is often indicative of an established, coevolutionary relationship. If this behavior is consistent over time and can also be traced or inferred through the agency of the fossil record, more useful data are acquired concerning a possible, evolved, reciprocal relationship. Of particular importance is the confirmation of specialized physical structures that are unique to the members of the observed relationship. Examples are the specialized feeding apparatuses of pollinating animals and the specialized, accommodating flower structures of their angiosperm partners. Such physical structures are strong evidence for the handiwork of the coevolutionary process. Direct human observation is preferable in ascertaining coevolutionary behavior; however, this is not always possible because of the rapidity of the animals involved, their habitat, their extremely small size, their preference for nocturnal activity, or their determined avoidance of humans. Consequently, electronic and mechanical aids are sometimes indispensable. These include remote-controlled still and video cameras, microscopic or telephoto lenses, infrared or ultraviolet lighting units, sonar or radar sensors, trip wires and other mechanical triggering devices, and sound recording equipment with high-gain or long-range microphones.

Laboratory research in the field of coevolution involves investigations heavily reliant on modern, sophisticated laboratory equipment and techniques. High-powered conventional, optical microscopes are employed to determine tissue and cellular structures. Scanning electron microscopes (SEMs) are employed for study of extremely small unicellular animals or plants such as planktonic organisms or extremely small organic structures. In addition to these tools of laboratory specimen observation, there are the analytical equipment and techniques used to determine the genetic codes and blood protein complexes of animals and plants to establish the degree of relatedness or divergence between various species.

MAINTAINING BALANCES

Coevolutionary studies are increasingly important in the biological sciences. One of the aims is to determine the degree of interdependence between various species, whether the relationship is between animals and plants, animals and other animals, or plants and other plants. A key factor to be determined in all these coevolved relationships is that of the nature and degree of balance attained. Although most of the biological world is forever in a state of flux, some categorical, coevolved relationships have been of long duration and can be reasonably assessed as having been in existence for tens of millions of years, such as that of flowering plants and vertebrate and invertebrate pollinators, or even hundreds of millions of years, such as the oceanic, planktonic food chain.

The degree to which these large-scale, coevolved relationships, involving entire planetary ecologies, continue to enjoy their former degree of health and well-being is of the utmost importance to human society. The present depth of understanding of the biological sciences clearly indicates the interrelatedness of all nature. Many angles of study agree that the global life system is experiencing great stress from human intervention: industrialization, urbanization, and overpopulation. It becomes increasingly urgent to know with the utmost precision all facets of the way the global life system operates, and has operated with general stability, over geological expanses of time. Every detail that contributes to this knowledge—every coevolved relationship, no matter how seemingly insignificant—adds to the total effect. This information can be used as an important resource to help maintain the stability of the entire system for ourselves and future generations.

Further Reading

Bakker, Robert T. *The Dinosaur Heresies.* New York: William Morrow, 1986. An interesting book by one of the most fascinating and controversial writers on the subject of evolution. Although the emphasis of the book is on dinosaur evolution, the author's specialty, it also includes a number of good discussions of the coevolved interrelationships between predators and their prey. A wonderful thought-provoking book for all reading audiences. Numerous detailed and beautiful line drawings.

Barth, Friedrich G. *Insects and Flowers: The Biology of a Partnership.* Translated by M. A. Biederman-Thorson. Princeton, N.J.: Princeton University Press, 1991. An introduction to the coevolution of insects and flowers, written by a zoologist. Focuses on the mechanisms of plant-animal relationships and the sensory capabilities of insects. Further Reading and index.

Chaloner, William G., Peter R. Crane, and Else Marie Friis, eds. *The Origins of Angiosperms and Their Biological Consequences.* New York: Cambridge University Press, 1987. A series of essays by professional researchers on the subject of flowering plant evolution with an emphasis on the deduced history of the angiosperms and their coevolved animal partners as recorded in the geologic record. Suitable for the college level.

Clayton, Dale H., and Janice Moore, eds. *Host-Parasite Evolution: General Principles and Avian Models.* New York: Oxford University Press, 1997. A collection of essays that seeks to reinforce the interaction between theory and practice in the field of natural history. Aimed at avian ecologists and evolutionists. Glossary. Appendices cover how to collect and identify microparasites, helminths and protozoa, and arthropods; species names; organizations that identify parasites; and the immune system.

Gould, Stephen Jay. *The Panda's Thumb.* 1980. Reprint. New York: W. W. Norton, 1992. Each chapter of this book is a brilliant reflection on diverse aspects of organic evolution in essay form; the book is suitable for anyone interested in evolution. Chapter 27, "Nature's Odd Couples," focuses on a conspicuous example of a possible coevolution—an inferred mutualistic relationship between the extinct dodo bird and a large tree (*Calvaria major*) that is found on the island of Mauritius. Gould presents the evidence objectively, both for and against the dodo's possible role in facilitating the tree's germination. The chapter is a good example of the kind of investigations and speculations in which evolutionary biologists engage.

Grant, Susan. *Beauty and the Beast: The Coevolution of Plants and Animals.* New York: Charles Scribner's Sons, 1984. This is one of the best books devoted solely to the subject of coevolution written for a general audience. The book is easy and fun to read, and full of information that rests solidly on the most recent scientific research and thinking. Each of the ten chapters is devoted to a major subtopic within the phenomenon of coevolution, such as specific animal and plant symbioses or antagonisms.

Lawton, J. H., Richard Southwood, and D. R. Strong. *Insects on Plants.* Cambridge, Mass.: Harvard University Press, 1984. Gives detailed examples from the scientific literature of coevolved relationships between various insects and plants. An authoritative and dependable source of hard scientific information on the topic, the book tends to the abstract side of theorizing and biological modeling. Readers with a solid foundation in biology or scientific analysis will find this book of greatest use.

Powell, Jerry A. *Biological Interrelationships of Moths and "Yucca Schottii."* Berkeley: University of California Press, 1984. A short but detailed source explaining the pollination relationship between two species of desert moth and their coevolved partner, a particular genus of the yucca plant of the southwestern United States. Extremely detailed scientific documentation is used to describe the known aspects of the interrelationship between the moths and the plant. Appropriate for serious botany, entomology, and evolutionary biology students at the advanced high school or college level or as a general reference.

Rothstein, S. I., and S. R. Robinson, eds. *Parasitic Birds and Their Hosts: Studies in Coevolution.* New York: Oxford University Press, 1998. A collection of essays by an international roster of scholars. Focuses on field studies of cuckoos and cowbirds, and their relationships with their hosts. The two sections of the books address conspecific brood parasitism and prevailing models of host-parasite coevolution.

Thompson, John N. *The Coevolutionary Process.* Chicago: University of Chicago Press, 1994. A cross-disciplinary view of coevolution by a researcher of plant-animal interactions. Proposes that coevolution is a result of processes occurring at multiple geographic scales, or what Thompson calls the "geographic mosaic theory of coevolution." Utilizes field observation and experiment, comparative studies, ecological genetics, and phylogenetic reconstruction.

CONSERVATION

—Jasper L. Harris, updated by Carol A. Rolf, JD

FIELDS OF STUDY

Conservation Biology; Ecology; Environmental Studies.

PRINCIPAL TERMS

footprint: an analogy indicating the 'stamp' of an aspect of society on the overall environmental effect of society, such as a country's 'carbon footprint'

wetlands: swamps, marshes and other low-lying areas that are normal collection areas for water

SUMMARY

The planet Earth may be unable to support future increases in population unless, on a worldwide scale, humans begin to conserve and reduce their rates of consumption and increase efforts to recycle resources for new uses. Moreover, the current global economy is no longer sustainable and is destroying the environment and providing little to support the globally impoverished. Internationally, governments and activists have begun to work together to establish policies that protect the environment and the sustainability of life while concomitantly fostering harmonious economic growth.

Humanity's footprint is being felt around the world. As the global population continues to increase, the natural resources necessary to sustain life continue to decline. Fresh water, fossil fuels, and arable land are just a few of the natural resources that must be properly managed to sustain a global population that may reach 9.1 billion by the year 2050, as predicted by the United Nations.

BACKGROUND

The planet Earth may be unable to support future increases in population unless, on a worldwide scale, humans take immediate action to conserve resources, reduce their rates of consumption, and increase efforts to recycle resources for new uses.

Moreover, the current global economy is no longer sustainable and is destroying the environment and providing little to support the globally impoverished. Internationally, governments and activists have begun to work together to establish policies that protect the environment and the sustainability of life while concomitantly fostering harmonious economic growth.

CONSERVATION

"Conservation" defined generally refers to the use of resources found in the natural environment in such a way that the resources will serve humans effectively and will be available to humans for as long as possible. Therefore, it does not refer to the indefinite "preservation" of resources in their natural state. Quantitatively, effective conservation could be said to involve obtaining the maximum use for the maximum number of people from the minimum amount of a particular resource.

The Earth can be viewed as a life-support system composed of four major subsystems through which energy flows and matter cycles. The subsystems are the atmosphere, biosphere, lithosphere, and hydrosphere, referring to gases, life systems, rock and mineral materials, and water, respectively. As energy flows and matter cycles within and among these subsystems, they interact as component parts to compose the Earth's ecosystem (an ecosystem may be defined as a community of plants, animals, and other organisms interacting in an environment). Humans alter the natural cycling of energy and flow of matter in the Earth's ecosystem. We extract things from natural systems, convert them into what we perceive as more useful products, and then return them to the natural environment in different forms and physical states. In order to achieve more desirable energy conversions, we also use energy from the environment. Natural resources are all of the things that humans take from the environment to help satisfy their needs and wants.

RESOURCES & RESERVES

All the matter and energy on Earth make up its "stock." Natural resources are subsets of this stock that help humans meet their energy and material needs. Those natural resources that are available in a usable form and at an affordable price under prevailing technology and socioeconomic conditions make up "reserves." For example, uranium was not a part of human energy reserves until the technology to capture and control the flow of nuclear energy was developed. Agrofuels produced from plant resources and biofuels developed from recently lifeless plant and animal materials are other examples of the marriage between natural resources and technology to meet energy demands for such uses as powering vehicles and heating buildings.

Natural resources may be classified as renewable, nonrenewable, and perpetual. Renewable resources are those that can be reproduced at a rate equal to or greater than the rate of consumption. Renewable resources are replenished through naturally-occurring physical, and biogeochemical cycles. Examples of renewable resources are forests and soils. They are conserved when they are used and reused at a rate, and in such a way, that does not destroy their sustainability. This does not mean that they cannot be depleted; it means that the rate of consumption does not exceed the rate at which they are replenished over an extended period.

Nonrenewable resources, on the other hand, are those for which the rate of consumption exceeds the rate of renewal. Being finite in quantity, they are exhaustible, cannot be replenished, and exist in fixed amounts. Nonrenewable resources, such as minerals and fossil fuels, are conserved by more thorough exploitation of their deposits and more efficient use. Some can be recycled or reused. Recycling involves collecting and reprocessing a resource, while reuse involves using a resource again in the same form. The reprocessing of used aluminum cans into new cans is an example of recycling. Washing beverage bottles before using them again is an example of reuse. Other nonrenewable resources, such as coal and oil, can be used only once and are gone forever after they have been used.

Perpetual resources, such as water, wind, tides, and solar energy, continue to flow throughout the Earth's ecosystem whether humans use them or not. Therefore, they are sometimes called flow resources. Even when their quality is altered they generally continue to flow within the Earth's ecosystem, making them inexhaustible. However, human activity is affecting the flow of some of these resources, such as water, and the recent harnessing of wind resources to produce energy is leading some scientists to suggest that altering the flow of wind with multiple, large turbines may have unpredictable effects that could change climate globally.

CONSERVATION VS. PRESERVATION

Preservation vs. conservation of natural resources means using things found in the natural environment wisely. In a more quantitative sense, it means sustainability of the natural resources by obtaining the maximum use for the maximum number of people without compromising future needs. It does not imply that resources should be entirely preserved for use later; rather, it means that they should be employed in a way that serves humans as well and as long as possible. Although preservation is closely associated with some aspects of conservation, the two approaches are different, as "preservation" means the complete protection of natural resources from human disturbance. It is true that to conserve some resources is to preserve them: We conserve natural resources such as ecosystems, for example, by restricting their use and protecting them from being altered, because their value is diminished if they are not retained in their original state. However, we conserve most resources when we use them in a certain way, not when we leave them idle.

HISTORICAL PERSPECTIVE

Not until technology developed significantly, and the world's human population reached a certain size, did human exploitation of the environment begin to have significant effects. Until that time, conservation of resources was simply not an issue. For most of humankind's existence, people lived a simple hunter-gatherer existence, obtaining just enough food to survive. Most people lived in small groups—fifty or fewer people—that had little effect on resources or the environment. They made simple tools and weapons. Many groups were nomadic, migrating with the seasons and following game animals. The shift from a hunting-and-gathering society to a sedentary one began about ten thousand years ago. People began breeding animals and cultivating wild plants, thereby having a greater impact on the environment. Slash-and-burn cultivation involved cutting down trees and other vegetation, leaving the cut vegetation on the ground to dry, and then burning it to enrich the soil. Farmers were "subsistence farmers," producing only enough to feed their families.

With the invention of the metal plow about seven thousand years ago, agriculture could be practiced on a larger scale. Animals were used to pull the plows, increasing crop productivity and making the cultivation of new soils possible. Forests were cut and grasslands were plowed—soil erosion and degradation of wildlife habitats inevitably began to follow on a small scale. Occasional food surpluses were produced for sale or storage. Surpluses allowed the development of urban cultures by releasing people from the farm. By the nineteenth century, urbanization and the Industrial Revolution were having profound impacts on the environment and the rate of resource consumption.

HISTORY OF THE CONSERVATION MOVEMENT

In 1864, George Perkins Marsh published *Man and Nature: Or, Physical Geography as Modified by Human Actions.* Marsh's book, which claimed that humanity could no longer afford to continue wastefully exploiting natural resources, is thought by some to mark the beginning of the conservation movement. However, American Indians must also be given credit as one of the first peoples to practice sustainable natural resource use. In 1878, John Wesley Powell completed *A Report on the Lands of the Arid Region of the United States.* Powell's study of the geomorphology and arid landscape transformations in the Colorado River basin was grounded in scientific methodology and called for the creation of a federal agency to survey and map all U.S. lands. In 1879, the United States Geological Survey was created for this purpose. At the beginning of the twentieth century there was growing concern that resource mismanagement could have tragic future consequences. These concerns were based on scientific findings associated with the exploitation and depletion of timber. In 1907, the Inland Waterways Commission, headed by U.S. Forest Service chief Gifford Pinchot, reported that the use and control of water would have an impact on other resources, including timber, soil, wildlife, and minerals. Pinchot's views on resource management greatly influenced forest and water management policies in the United States.

During the Great Depression of the 1930's, the Franklin Delano Roosevelt administration instituted a number of programs in the United States that addressed natural resource problems and helped create employment. In the wake of the severe drought and wind erosion in the Dust Bowl (1930's), the Public Works Administration initiated the Prairie States Forestry Project. Its goal was to establish a shelter belt of trees and shrubs from the Texas panhandle to the Canadian border in North Dakota. This project was designed to reduce wind erosion on rangeland and cropland. Other efforts included the creation of the Tennessee Valley Authority (TVA) and the Civilian Conservation Corps (CCC). The TVA was an innovative water resource management program that involved comprehensive regional planning. Though confined to the Tennessee River and its tributaries, it provided a model for total resource management. The aim of the CCC was to provide employment while repairing some of the damage that had resulted from past exploitation of natural resources and neglect of the environment. Workers constructed bridges, roads, and fire lanes for the development of recreational facilities; conducted tree-planting programs; instituted soil- and water-erosion control projects; made lake and stream improvements; and participated in flood control projects.

Many of these early conservation practices in the United States spread to other countries and, over time, several international conservation organizations were formed, including the United Nations Environment Programme, the International Union for the Conservation of Nature, and the World Wildlife Fund. After World War II, nations focused on resource-related problems, many times creating agencies to assess the impact that the war had on forest and natural resources. Moreover, the use of atomic bombs in the war—and the widespread nuclear testing that occurred in the 1950's—exposed ecosystems to significant levels of radiation. This situation marked the beginning of the modern conservation movement.

THE MODERN CONSERVATION MOVEMENT

Although the United States is credited as the front-runner of the early conservation movement through its linking of ecology with conservation and resource management practices on public lands, during the modern conservation movement, especially in the 1980's, the United States focused on economic growth and deregulation, sacrificing conservation. Nevertheless, the efforts of the Worldwatch Institute and older organizations such as the Sierra Club, the Audubon Society, and Friends of the Earth kept the general citizenry aware of environmental and resource-related issues and their consequences.

In 1962, Rachel Carson, in her book *Silent Spring*, cautioned the public against the indiscriminate use of chemical pesticides. She argued that persistent substances released into the environment move throughout the food chain, concentrating over time, while pests may develop a resistance to the poisons. By the 1960's, pollution from industrial and vehicular sources was beginning to be recognized as a global issue, as industrialized nations increased their spoliation and depletion of natural resources. In the United States, President John F. Kennedy introduced a number of natural resource initiatives aimed at preserving wilderness areas, developing marine resources, reserving shorelines for public use, expanding outdoor recreation, formulating plans for developing water resources and developing actions against water pollution, and encouraging the development of substitutes for resources in short supply. Also, he organized the Youth Conservation Corps to provide a workforce to implement the program. By the 1970's, during the administration of President Richard M. Nixon, the United States had begun to adopt environmental laws such as the Clean Air Act and created the Environmental Protection Agency as its enforcement agency. The first Earth Day to celebrate sustainable use of natural resources was in 1970.

The international community came together to discuss the environment for the first time in 1972. The United Nations held the Conference on the Human Environment in Stockholm, Sweden. This conference, which came to be known as the Stockholm Convention, resulted in the creation of the United Nations Environment Programme. The 1987 publication of the *Brundtland Report* (also known as the *Our Common Future*) by the United Nations World Commission on Environment and Development was one of the first documents to take on the issue of sustainable global development in modern times in a manner similar to that espoused by Pinchot in earlier times. The goal of the *Brundtland Report* was to foster global economic development that is conservation-oriented and economically balanced.

In 1980, another international organization became active in conservation. The International Union for Conservation of Nature (IUCN) published World Conservation Strategy (IUCN). The purpose of this publication, and one of the ongoing goals of the IUCN, was to assist developing nations in conservation planning to protect and maintain natural resources: air, water, soil, forests, and animals.

Since 1992, the international community and world leaders have continued to come together regularly to discuss global environmental problems. At the 1992 United Nations Conference on Environment and Development held in Rio de Janeiro, Brazil, much of the discussion concerned the *Brundtland Report*. This conference, which came to be known as the "Rio Earth Summit," focused on dire predictions concerning global warming, climate change, the ozone hole; concerns about the depletion of natural resources, loss of habitats, and biodiversity; and continued concerns with resource pollution and depletion, especially forests and marine resources. Some experts at the Rio Summit suggested that society had to choose between economic development and conserving the environment. Although governments have been unwilling to make an either/or choice, one outcome of the Rio Earth Summit was the ratification by many nations of various international agreements to resolve some of the issues discussed at the Summit. For example, the Summit led to the ratification of the Montreal Protocol (1987), which concerns depletion of ozone by man-made chemicals, and the 1992 Kyoto Protocol, to limit industrial emissions that may be affecting global climate change.

Sustainable development became the main topic of discussion at the U.N. 2002 Earth Summit, held in Johannesburg, South Africa. Many world leaders came together to promulgate international regulations to address such environmental problems as improving air and water quality; improving food access, agricultural productivity, and sanitation in developing countries; and developing strategies and economic incentives to cope with international environmental issues related to war, poverty, and disease.

Although the United States has not been a signatory to some of the latter-day environmental protocols, environmental conservation organizations, including the Sierra Club, and individuals, such as former U.S. vice president Al Gore, alert the public to environmental issues. Gore won the 2007 Nobel Peace Prize together with the U.N. Intergovernmental Panel on Climate Change for their efforts in getting out the conservation message on global climate change. International organizations continued the conservation movement in preparation for the 2012 Earth Summit.

POPULATION & RESOURCE CONSUMPTION

Population growth is a major factor when considering the time it will take to deplete the Earth's nonrenewable resources. A resource is considered economically depleted after 80 percent of its known reserves have been exploited, because at that point the resource becomes too expensive for wide use. As the world's population grows, the rate of resource exploitation grows. More production is necessary to satisfy the needs and wants of larger populations: More materials are needed, more energy is consumed, and more pollution is created.

Ecologists have come to realize that the Earth is a huge ecosystem with a definite carrying capacity. That is, there is a limit to the number of people that can be supported by the Earth. The rate of resource depletion is a function both of the rate of population growth and of the rate of consumption of resources per person. More resources are consumed per individual in wealthy countries than in poor countries. As poor countries strive to develop, greater pressures are placed on the Earth to provide resources and to assimilate wastes. Overpopulation occurs when there are too many people for the available resources or when population growth exceeds economic growth; such conditions ultimately begin to cause damage to the Earth's life-support system.

WATER RESOURCES: THE NEXT GREAT CONFLICT

The availability and purity of water may be the next great natural resource issue facing the world, and control of water resources could lead to serious conflict. Water not only is necessary to sustain life and health but also is needed for food production and various industrial uses. Water resources continue to be polluted; rivers have dried up and have been dammed, thus reducing downriver flows to wetlands and floodplains; underground water supplies from aquifers have been used faster than they are replenished with rainfall; and development is destroying wetlands and other water resources and converting them, in some cases, to deserts. Because the global society has always relied on water as a renewable resource, it has continued using management policies that are no longer viable. In order to maintain human and wildlife populations, habitats, and health, and in order to ensure that there are sufficient water resources to sustain food production, water resource management policies must change.

OTHER NATURAL RESOURCE ISSUES

Other global natural-resource issues concern subterranean extractions: minerals, precious metals, and gems such as diamonds; oil and gas drilling; and coal mining. Some of the methods used for these extractions and the consequences of accidents and spills have raised the global consciousness about their impact on the environment. A search for renewable resources to replace many of these nonrenewable resources is under way within the global community.

Also controversial is the negative environmental impact of outdated rangeland management techniques, uncontrolled timber harvesting, forest destruction, and mismanaged disposal and dumping. Many developed nations suffer from wasteful consumerism and overuse of natural resources, such as those that provide energy. Government structures to foster conservation are lacking in most developing countries, and in many industrialized nations government leaders and the public are unwilling to make genuine efforts to conserve vital and dwindling natural resources, especially if they negatively impact economic growth. Scientists predict that these abuses of the environment and overuse of natural resources can no longer continue at their present rate if we want to preserve the future of humanity.

ECONOMICS & CONSERVATION

Government and resource management plays an important part in the balancing of resource conservation and resource exploitation. Continued growth in the use of a nonrenewable resource can occur only for a number of years before the resource is depleted. As a resource becomes scarce, the price increases, making it less affordable and reducing the rate of consumption. This is a self-regulating process that makes conservation more practical as resources become more scarce in a market-driven system.

Effective conservation programs often require governmental influence, regulation, or incentives. Since most resources are associated with property, governmental agencies that regulate land, businesses, and private citizens all make decisions that affect resource consumption. The general aim of many decision makers is to maximize the return on investments; conservation must therefore be profitable within a reasonable time for people to practice it voluntarily. In a free enterprise system, resource exploitation produces income from the land and provides much of the incentive for land ownership. Thus, resource exploitation is likely to win out over conservation if there are no incentives to conserve.

Conservation practices must be congruent with economics. Conservation programs are not effectively executed when economic necessity or opportunity intervenes. For the most part, conservation is good for the economy over the long term, because it improves the efficiency of production systems. However, because modern economic growth has not been balanced and has not conserved natural resources, more drastic measures may be necessary to control future growth.

One of the more controversial proposals to combat economic growth issues facing the environment are taxes on carbon-related energy sources. Another highly debated policy recommendation is over emission trading, or cap-and-trade. Controlled ownership of resources through vehicles such as trusts to avoid depletion of natural resources—mostly nonrenewable resources—is another proposal being debated. Those involved in national and global political and social debate must become serious in reaching a consensus for resolving the many environmental issues facing the world and recommend sustainable policies that bring together economists and environmentalists in working for the same achievable goals regarding future growth. However, no matter what policies are eventually adopted, one of the main goals of balancing economic growth with conservation must be continued public awareness. The public and organizations are more likely to become supportive partners for sustainable economic growth if they are provided with not only information on economic growth and its effect on the environment, but also feasible, market-driven solutions.

ASSESSING THE FUTURE

Experts attempting to assess the future of natural resources are divided in their opinions. Positions range from optimistic to direly pessimistic to somewhere between the two extremes. Those who believe that technology can and will solve human problems have reason to be optimistic, and, to a great extent, history supports this view. Whenever humans experience shortages, they turn to technology for solutions—either developing more efficient ways of finding, extracting, and using resources or finding substitutes for them.

On the other hand, technology may not be able to continue solving all humankind's problems—at least not in a sufficiently timely manner to avoid a crisis. The primary basis for the pessimistic argument is that increases in resource consumption rates, coupled with an increase in population, may not allow enough time to find technological solutions to resource shortages. Furthermore, the heavy modern dependence on nonrenewable resources is certain to cause resource shortages.

A more moderate view of the resource future suggests that, although there is good reason for concern, humankind has sufficient time to avoid a major crisis if we begin moving toward a sustainable society now. A sustainable society is one that allows humanity to meet its needs today without compromising its environment and future needs. Sustainability almost certainly requires that people in developed countries begin to live a lifestyle that includes more conservation and recycling, a greater dependence on renewable and perpetual resources than on nonrenewable resources, population control, and more self-discipline. This view embraces the ecological approach to resource management and employs the multiple-use concept. For example, forest conservation not only provides timber but also preserves a habitat for plants and animals; it can serve to help manage water resources, prevent flooding and soil erosion, and provide recreational areas. When we reach sustainability, most of humankind's material and energy needs will be provided by renewable and perpetual resources that should last indefinitely if properly managed.

The ecological approach is holistic. Based on the philosophy that all things in the natural environment are interlaced through a complex system of feedback loops, it implies that the whole is functionally greater than the sum of its parts. This approach to resource management requires an understanding and anticipation of the consequences of human actions throughout the ecosystem. ConservationSustainable development

Further Reading

Castillon, David A. *Conservation of Natural Resources: A Resource Management Approach.* 2d ed. Madison, Wis.: Brown Benchmark, 1996.

Chiras, Daniel D, and John P. Reganold. *Natural Resource Conservation: Management for a Sustainable Future.* Pearson Education Limited, 2014. Print.

Degregori, Thomas R. *The Environment, Our Natural Resources, and Modern Technology.* New York: John Wiley & Sons, 2008.

Freyfogle, Eric T. *Why Conservation Is Failing and How It Can Regain Ground.* New Haven, Conn.: Yale University Press, 2006.

Greenland, David. *Guidelines for Modern Resource Management: Soil, Land, Water, Air.* Columbus, Ohio: C. E. Merrill, 1983.

Greiner, Alfred, and Will Semmler. *The Global Environment, Natural Resources, and Economic Growth.* New York: Oxford University Press, 2008.

Harper, Charles L. *Environment and Society: Human Perspectives on Environmental Issues.* 4th ed. Upper Saddle River, N.J.: Pearson/Prentice Hall, 2008.

Jakab, Cheryl. *Natural Resources (Global Issues).* North Mankato, Minn.: Smart Apple Media, 2008.

Knight, Richard L., and Courtney White, eds. *Conservation for a New Generation: Redefining Natural Resources Management.* Washington, D.C.: Island Press, 2009.

Krupp, Fred, and Miriam Horn. *Earth: The Sequel—The Race to Reinvent Energy and Stop Global Warming.* New York: W. W. Norton, 2008.

Loeffe, Christian V., ed. *Conservation and Recycling of Resources: New Research.* New York: Nova Science, 2006.

Miller, G. Tyler, Jr. *Resource Conservation and Management.* Belmont, Calif.: Wadsworth, 1990.

Parson, Ruben L. *Conserving American Resources.* 3d ed. Englewood Cliffs, N.J.: Prentice-Hall, 1972.

Raven, Peter H., Linda R. Berg, and David M. Hassenzahl. *Environment.* 6th ed. Hoboken, N.J.: Wiley, 2008.

Scott, Nicky. *Reduce, Reuse, Recycle: An Easy Household Guide.* White River Junction, Vt.: Chelsea Green, 2007.

CONSERVATION & PRESERVATION

—*George R. Plitnik*

FIELDS OF STUDY

Environmental Studies; Resource Management; Waste Management

PRINCIPAL TERMS

anthropocentric: treating human beings and human concerns as the primary source of value
biocentric: treating all living things as equal in value
natural area: a region where organisms and geological processes are undisturbed by humans, with as few controls as possible

SUMMARY

The history of conservation in the United States may be divided into three successively evolving stages: utilitarian conservation, nature preservation, and modern environmentalism. In the twenty-first century, as a result of global warming, environmentalism has adopted a more inclusive, planetary view.

BACKGROUND

Human abuse of nature is almost as old as recorded history. Plato lamented land degradation due to hills being denuded for lumber. Eighteenth century French and British colonial administrators understood the link between deforestation, soil erosion, and local climate change. Stephen Hales, a British plant physiologist, instigated the practice of reserving 20 percent of all green plants to preserve rainfall on the Caribbean island of Tobago. Pierre Poivre, French governor of Mauritius, appalled by forest and wildlife devastation, ordered one-fourth of the island to be preserved in woodlands. In America, conservation commenced as a pragmatic response to the excesses imbued by the nineteenth century limitless frontier mentality.

UTILITARIAN RESOURCE CONSERVATION

George Perkins Marsh, who had witnessed the damage caused by excessive grazing and deforestation around the Mediterranean, became alarmed by the profligate waste of resources occurring on the American frontier in the mid-nineteenth century. In 1864, he published *Man and Nature*, warning of the unfortunate ecological consequences of this wanton destruction. This book had several lasting impacts, including the establishment of the National Forest Service in 1873 to protect dwindling timber supplies and endangered watersheds. In 1905, President Theodore Roosevelt, influenced by Marsh's book, moved the Forest Service from the Department of the Interior to the Department of Agriculture, and made his chief conservation adviser, Gifford Pinchot, the new head. This decision situated resource management on a straightforward, rational, and scientific basis.

Together with naturalist John Muir, first president of the Sierra Club, Roosevelt and Pinchot passed game protection laws, restructured the national park system, and reconstituted forest and wildlife refuge systems. These policies were primarily pragmatic. They believed that forests should be saved, not for aesthetic reasons or out of concern about wildlife, but to provide homes and jobs for people. Resources should be used for the greatest good for the greatest number for the longest time. Utilitarian conservation is not concerned about saving resources for future generations, but about wisely developing and using the resources for the benefit of humans now living. According to this viewpoint, there is as much waste in neglecting to develop and utilize natural resources as there is in their wanton destruction. This approach is still evident in the multiple-use policies of the Forest Service.

NATURE PRESERVATION

Muir, believing utilitarian conservation to be too anthropocentric, strenuously opposed Pinchot's influence and policies. Muir espoused the more biocentric viewpoint that all living organisms are imbued with intrinsic rights and deserve to live in nature, whether or not they are useful to humans. Every organism, as part of an ecological web, is not only entitled to continuance, but is essential to the integrity and stability

Stilfserjoch in Südtirol, Italy. The valley is a nature preservation area. Image by Frisia Orientalis (Own work) CC-BY-SA-3.0.

of the biotic community. According to this viewpoint, humans are a miniscule component of nature; as such they have no right to value themselves above other species with whom they coexist. Humans are primarily a negative influence on nature.

In order to preserve its pristine wilderness, John Muir fought for and achieved the establishment of Yosemite as a State Park in 1864, later incorporating additional land to become a National Park in 1890. He was also instrumental in having King's Canyon preserved until it also achieved National Park status. When the National Park Service was established in 1916, it was headed by one of Muir's disciples, guaranteeing that his ideals of attempting to preserve pristine wildernesses in their purest, unaltered state would become a guiding principle. This philosophy is often at odds with the Forest Service.

MODERN ENVIRONMENTALISM

Contemporary environmentalists have moved beyond the simple preservation of nature to embrace problems adversely affecting the health and well-being of all species, particularly humans. Air and water pollution began to become problematic during and particularly after World War II as a result of industrial expansion, greater use of toxic chemicals, and increased automotive traffic. One of the first books to awaken public awareness to the deleterious effects of noxious chemicals in the environment was *Silent Spring* by Rachel Carson, published in 1962. This led to an environmental movement as concerns broadened from preserving nature and using resources wisely to controlling and reducing pollution.

Two pioneers of the environmental movement were David Brower and Barry Commoner. Brower, while serving as the executive director of the Sierra Club, introduced many of the techniques now characteristic of modern environmentalism. These include litigation, intervention in regulatory hearings, and using the mass media for publicity campaigns. Commoner, a biologist, used scientific research to reveal connections among science, technology, the ecosphere, and society. Both activism and research remain defining characteristics of the modern environmental movement. By the 1970's, the movement had expanded from wilderness protection and pollution problems to include human population growth, nuclear power, fossil fuel extraction and use, and recycling. With the first Earth Day in 1970 environmentalism created public awareness and concern about health and ecological damage from pollution.

CONTEXT

Because modern humans are interconnected in a myriad of ways, the Earth has become a global village of people sharing a common planetary environment. Attention has shifted from preserving particular landscapes or preventing pollution of a specific watershed to concern about the life-support systems of the entire world. Humans are changing planetary weather systems, increasing the extinction rate of species, and degrading ecosystems; without drastic remediation the ultimate consequences will be catastrophic. Protecting the planetary environment must become an international cause since it will take worldwide cooperation to effect the many changes necessary. Preliminary steps in this direction have been taken with the Montreal Protocol, adopted by most industrial nations in 1987, which phased out the use of ozone-depleting chlorofluorocarbons. The Kyoto Protocol is another international effort attempting to mandate carbon dioxide reduction. The agreement is still weak because the United States and Australia, two of the world's greatest emitters, refuse to sign.

Twenty-first century humans have begun to comprehend that human societies can no longer act in isolation because the Earth is an interconnected whole. Pollution and environmental problems are inextricably linked to poverty, injustice, oppression, and the exploitation of underdeveloped nations by greedy capitalists in industrialized countries. Only by working together to correct these historic wrongs and actively pursuing sustainable lifestyles can the planetary environment be conserved.

FURTHER READING

Brown, Lester R. *Plan B 3.0: Mobilizing to Save Civilization.* 3d ed. New York: W. W. Norton, 2008. Presents a worldwide plan to stabilize climate and control population. An excellent example of global environmentalism.

Commoner, Barry. *Making Peace with the Planet.* New York: Pantheon, 1990. A handbook of modern ecological conservation that treats the failures of modern environmentalism and prescribes actions that must be taken to live in harmony with the Earth.

Humphrey, Mathew. *Preservation Versus the People?* New York: Oxford University Press, 2002. Discusses humanity's interaction with nature and how one's political philosophy determines one's environmental worldview.

CONSERVATION BIOLOGY

—Lenela Glass-Godwin

FIELDS OF STUDY

Botany; Conservation Biology; Ecology; Environmental Studies;

PRINCIPAL TERMS

inbreeding: mating of individuals within a species that are too closely related familialy, allowing recessive genetic traits to become prominent in their offspring

minimum viable population: the lowest number of individuals of a species sufficient to ensure the propagation of the species without inbreeding

species: any animal or plant type having a unique genetic identity determined by DNA

SUMMARY

Biological scientists throughout the world are interested and involved in conservation biology as a means of preserving biodiversity. Evolutionary biologists, botanists, ecologists, and geneticists, as well as those in applied-science management fields such as wildlife management, fisheries, and forestry, all play important roles in the broad context of conservation biology.

BACKGROUND

Conservation biology is a multidisciplinary field that incorporates the knowledge base and skill sets of all of the biological sciences in order to design and implement methods that will attempt to ensure the long-term continuation of species, ecosystems, and ecological processes.

HISTORICAL PERSPECTIVE

Conservation biology as a scientific discipline has its roots in the conservation movement of the early twentieth century. Wildlife scientists such as Aldo Leopold, foresters such as Gifford Pinchot, and progressive politicians such as Theodore Roosevelt wielded tremendous influence in both conservation and preservation efforts. In the 1960's and 1970's, there was phenomenal growth of community ecology data and important ecological theories, such as island biogeography, promoted by Robert H. MacArthur, Edward O. Wilson, and Daniel Simberloff. Social

Paul Ralph Ehrlich is an American biologist, educator and the president of Stanford's Center for Conservation Biology. He is also a prominent ecologist and demographer. Image by Paul R. Ehrlich CC-BY-2.5.

trends of this time period also contributed to the development of this field, as many people embraced the environmental movement, and, accordingly, a number of significant pieces of environmental legislation were passed. As more and more species either neared extinction or became extinct, the movement to "save the planet" became ever more insistent.

Consequently, much of the focus of conservation biology since the 1970's has been on saving endangered and threatened species. The United States Fish and Wildlife Service (USFWS) defines an endangered species as one that is in danger of extinction throughout all or a significant portion of its range. A threatened species is likely to become endangered in the foreseeable future. The USFWS tracks the numbers of endangered and threatened species both within the United States and around the world. In 2009, the number of endangered animal species within the United States was 410, with 163 listed as threatened. Bird, fish, mammal, and clam species topped the endangered animal list. The endangered plant species total was 600; 146 plant species were threatened. Throughout the remainder of the world, 527 animal species were endangered and 44 were threatened; 1 plant species was listed as endangered and 2 were threatened. Obviously, these numbers vary from year to year; nevertheless, the data support the need for worldwide conservation efforts.

Underlining the importance of that need, 2018 data from the USFWS show a total of 717 endangered animal species, with fishes, birds, mammals, clams and insects in the top five spots, and a total of 944 endangered plant species, all within the United States alone. One can only imagine the number of endangered species that must exist worldwide, with extinctions now believed to be occurring at a rate between 1,000 and 10,000 times the 'normal' background rate of one to five species per year.

MINIMUM VIABLE POPULATIONS

In the past in Europe and North America, most extinctions or local extirpations of species occurred because of overhunting. Examples of this in North America include bison, cougar (in the eastern U.S., with the exception of the remnant Florida panther population), red wolf, and passenger pigeon. In Europe, the wolf, wild boar, goshawk, and capercaillie were gone by the nineteenth century.

Although overhunting is still a reason for species decline in some parts of the world, today the primary cause for the large number of endangered and threatened species is habitat fragmentation. This occurs when a large area of habitat is divided into a number of smaller patches comprising a smaller total area. Often these smaller patches are unsuitable for species of the original habitat, and thus dispersal of species into the "new" habitat can be low. Conservation biologists face the challenge of preserving as many individuals of the endangered or threatened species as possible within suitable habitat, thereby enabling genetic diversity within the species to continue. As a result of fragmentation, conservation biologists are often trying to discover a minimum viable population (MVP) number for each imperiled species. Although the number of five hundred individuals has often been used as a base guideline for MVP, many scientists argue that the specific number of individuals needed to ensure genetic diversity varies by species and that no set number applies to all species.

If the MVP becomes too low, serious genetic problems within the population may occur, as the number of individuals is simply too low to prevent inbreeding. Probably the best known example of this is the cheetah, a species in which almost all of the individuals are now related to one another. Biologists have noted deleterious genetic effects such as sperm malformation and high infant mortality within the cheetah population. With species that have greatly reduced population numbers, conservation biologists are faced with the management challenge of reducing inbreeding, encouraging outbreeding if enough individuals remain, and seeking to enable members of the species to migrate, if possible.

RESERVES

Along with determining MVPs for endangered and threatened species, biologists seek to meet the management goals mentioned above by creating or maintaining suitable reserve sites. As with deciding on MVP, determining the necessary size and shape of a reserve can be extremely difficult and is again species dependent. Based on their research, several conservation biologists have suggested that a circular shape for reserves is best in order to minimize dispersal distances and to minimize the amount of edge habitat. Within both temperate and tropical forest areas, negative edge effects include temperature and humidity changes, wind exposure, tree mortality, and penetration of light into formerly shaded areas.

Many researchers recommend single large reserve areas for large herbivores, large carnivores, and birds, rather than a number of small reserves, unless an extensive corridor of suitable habitat can connect the small reserves. The efforts of national conservation organizations are often necessary to establish and maintain large nature reserves, whereas regional and even local groups may be able to keep smaller reserves going for species that do not require large expanses of habitat in which to survive.

Active management of reserves is usually necessary in order to provide continued suitable habitat for endangered and threatened species. Often exotic vegetation must be removed from reserves, or a certain successional stage of vegetation must be maintained. Elimination of exotic animal species, such as wild boars or brown-headed cowbirds, may be needed. Once land has been set aside for the preservation of imperiled species, conservation biologists must be vigilant in its management in order to preserve the biodiversity of the area.

FURTHER READING

Groom, Martha J., Gary K. Meffe, and C. Ronald Carroll. *Principles of Conservation Biology*. Sunderland, Mass.: Sinauer Associates, 2006.

Pullin, Andrew S. *Conservation Biology*. New York: Cambridge University Press, 2002.

Soulé, Michael E., and Gordon H. Orians. *Conservation Biology: Research Priorities for the Next Decade*. Washington, D.C.: Island Press, 2001.

World Resources Institute. Global Forest Watch: Frequently Asked Questions. http://www.globalforestwatch.org/english/about/faqs.htm

CONSERVATION EASEMENTS

—*George Wrisley*

FIELDS OF STUDY

Conservation Biology; Conservation Law; Environmental Studies.

PRINCIPAL TERMS

habitat: the environment in which a species or a community of different species is normally found

watershed: the land area through which precipitation (rain, snowmelt, etc.) drains to a common egress such as a river or estuary; also known as a drainage basin

SUMMARY

Conservation easements encourage various forms of conservation on privately held lands while providing tax benefits to landowners. An easement is a legal entitlement providing its holder certain privileges regarding the use of property owned by another. In the United States, a conservation

Conservation Easement Boundary markers identify Neil Hanson's Emergency Watershed Protection (EWP) Flood Plain Easement located east of Bowesmont, in the Red River basin of ND, Wednesday May 1, 2013 by U.S. Department of Agriculture.

easement is a legal agreement between a landowner and another party—generally a group such as a local or state land trust, a government agency, a nonprofit organization, or an environmental conservation group—for the purpose of limiting the uses of the property subject to the easement. The easement limits uses and development of the land that would disturb the condition of the aspect of the land that is valued. Some typical prohibitions concern the construction of roads and commercial buildings, mining of the land, and harvesting of its trees. Conservation easements usually last in perpetuity and are legally binding independent of who comes to own the land. The land is still usable by its owner, but the owner's use of it is limited by the terms of the easement.

Landowners can receive certain financial benefits for donating their land through conservation easements. Regarding income taxes, the donation of land constitutes a charitable gift and, as such, may be tax-deductible. Regarding estate taxes, lands that do not have easements may be subject to extremely high state and federal inheritance taxes. A conservation easement on the land reduces its value (since its economic uses are limited) and thus has a positive effect on the amount of inheritance taxes heirs must pay. Regarding gift taxes, again because an easement reduces the value of the land, more land may be given to a family member before it becomes subject to federal gift taxes. Regarding property taxes, having land protected by a conservation easement may qualify it for a reduction in property taxes (laws on this point vary from state to state).

Although conservation easements provide financial benefits, their intended purpose is to support conservation. The privately owned lands subject to conservation easements may serve as core habitats for species, or they may buffer and connect publicly owned lands such as nature reserves or national parks. These functions support the ecosystems found on the public lands. Buffer lands help provide cushions between lands that can be developed without restriction and lands that provide core habitats for flora and fauna. Lands that serve a connective function allow animals to move more freely from one publicly owned area to another, whether they are migrating, hunting, or establishing new territory. Further, much of the land that provides habitat for federally listed endangered species is privately owned.

Conservation easements have not been without controversy. One area of conflict concerns the abuse of such easements for tax-shelter purposes. Critics have noted that some owners receive the tax breaks that come with easements but do not maintain their lands properly in accordance with their easement agreements. Additionally, controversies have arisen regarding changes to the terms of easements by successive owners of particular lands. For example, the National Trust for Historic Preservation approved substantial changes to the terms of the Myrtle Grove easement, which covered land on Maryland's eastern shore in the Chesapeake Bay watershed, raising objections from many environmentalists.

FURTHER READING

Gustanski, Julie A., and Roderick H. Squires, eds. *Protecting the Land: Conservation Easements Past, Present, and Future.* Washington, D.C.: Island Press, 2000.

Journal of Forestry. "Focus on . . . Private Forest Landowners: FAQs about Conservation Easements." 100, no. 3 (2002): 4.

Keske, Catherine. *The Emerging Market for Private Land Preservation and Conservation Easements.* Saarbrücken, Germany: VDM Verlag, 2008.

Rissman, Adena R., et al. "Conservation Easements: Biodiversity Protection and Private Use." *Conservation Biology* 21, no. 3 (2007): 709–718.

CONSERVATION MOVEMENT

—*Laurence W. Mazzeno*

FIELDS OF STUDY

Conservation; Biology; Evolution; Ecology; Environmental Studies.

SUMMARY

The early conservation movement focused attention on the perilous state of natural resources as human societies moved toward increasing commercialization and urbanization in the nineteenth and twentieth centuries. Proponents of conservation continue to work to convey the importance of resource sustainability.

The roots of the conservation movement can be traced to Europe during the late eighteenth century. Individuals repulsed by the growing commercialization and dehumanization brought on the by the Industrial Revolution promoted a return to nature. While some areas had been preserved in their natural state for centuries, either as hunting grounds or as refuges for the upper classes, during the nineteenth century governments in Europe and in North America began setting aside green space, even in urban areas, for recreation and relaxation. In some countries, notably Germany, government control over forests and rivers was used not only to conserve natural resources but also to generate revenues from the harvest of renewable resources such as trees and wildlife. Coordinated efforts to preserve wilderness areas, however, emerged first in the United States. The initiatives of individuals and organizations involved in these efforts grew into the modern conservation movement.

BEGINNINGS OF THE MOVEMENT

From its inception, the conservation movement in the United States was highly decentralized, but people who became involved shared an overarching goal: to preserve some aspect of the natural world from the ravages of rampant commercial development and urbanization. Efforts to save some of America's wilderness areas from the encroachment of civilization led to the creation of Yellowstone National Park in 1872, and some states had taken steps to set aside particularly beautiful areas (many of them difficult to reach). On the other hand, the first true conservation bill passed by the U.S. Congress in 1876 was vetoed by President Ulysses Grant.

Henry David Thoreau, in 1861. Image by Geo. F. Parlow.

Although American writers such as Henry David Thoreau had celebrated the value of the natural landscape, the first true conservationist to emerge on the national scene was John Muir. A Scotsman who had settled in California, Muir spent most of his adult life writing about the beauties of the wilderness, particularly the Yosemite Valley, a preserve set aside by the state in 1864. At that time many people looked upon the wilderness as a place of escape from the grime of city life. Powerful commercial interests, however, saw unsettled areas as opportunities for profit. The mining and timber industries coveted the natural resources there, farmers and ranchers sought lands for grazing, and railroads saw possibilities in creating tourist destinations or shorter routes between population centers.

Muir was one of dozens of amateurs who made their livings by other means but devoted their lives to saving the country's wilderness for future generations. Many were responsible for creating new national parks or monuments, saving sites that had been commercialized (such as Niagara Falls), or establishing wildlife sanctuaries to preserve dwindling numbers of birds, fish, and mammals. Muir lobbied to preserve Yosemite and additional wilderness sites. Eventually he found a strong ally in President Theodore Roosevelt, a lifelong sportsman and conservationist who used his political power to establish numerous national monuments to save lands from mining, timbering, and settlement. Muir was a preservationist. His goal was to have lands set aside, protected from all development or use.

A second important figure in the conservation movement, Gifford Pinchot, could best be described as a utilitarian. Pinchot earned a degree in forestry and eventually became head of the U.S. Forest Service, making him one of the first professional conservationists. He argued that natural resources should be managed carefully, but renewable resources should be available for commercial purposes. These two views came to dominate the debate among conservationists for nearly a century as groups struggling to preserve sites intact clashed with those who saw responsible use as a viable alternative to saving lands from outright, wholesale exploitation.

Over the next century the conservation movement would be advanced by both amateurs and professionals. A few crossed from one group to the other. Notable among them was Stephen T. Mather, who made a fortune in business and then turned his talents to creating the National Park Service in 1916, for which he served as the first director. Intent on preserving the national parks for future generations, Mather often found himself pitted against Pinchot and the Forest Service in debates concerning the proper balance between preservation and commercial use of national lands.

Spurred by Muir's leadership, in 1892 a group of Californians formed the Sierra Club, which would become a driving force in promoting conservation throughout the United States. Similar organizations sprang up in the East, many modeled on the first such club, the Appalachian Mountain Club, established in 1876.

Eventually the idea of conservation grew to include wildlife as well as landscape. Ironically, some of the most avid conservationists were sportsmen. They recognized that careful management of game populations would ensure that hunters could enjoy their sport for decades to come. Aware of what wanton slaughter had done to the American bison population during the nineteenth century, they often lobbied for limits on the numbers of animals that individual hunters could kill.

Conservationist groups were not always successful, however. Despite the protests of the Sierra Club and others, the U.S. government agreed to the creation of the Hetch Hetchy Dam in the Yosemite area to provide water for San Francisco, causing the loss of millions of hectares of natural landscape. Lobbying did push several presidents to create wildlife refuges, setting aside lands where game animals lived or that they used as migratory routes. Eventually these sanctuaries became part of the National Wildlife Refuge System.

BETWEEN THE WORLD WARS

By the end of World War I a number of conservation groups were active in the United States. Individual organizations, however, tended to focus on single aspects of conservation. For example, the Audubon Society, founded in 1905, limited its interest to the plight of birds, while the Izaak Walton League, established in 1922, was concerned with the condition of rivers and lakes because its members were involved in sportfishing. In fact, during this period many conservation organizations were underwritten by firearms companies and sports businesses. As a result, the utilitarian view of conservation tended to be favored in political decisions.

Concurrently, preservationists such as the National Park Service's Mather tried to have more lands set aside; on occasion these efforts were supported by financiers such as John D. Rockefeller, Jr., who purchased properties to expand or create parks. Some narrow-minded conservation efforts led to ecological disasters, however. For example, by the 1920's the gray wolf population in Yellowstone National Park was eliminated, ostensibly to make the park safer for tourists (and for the livestock of nearby ranchers). The absence of these predators led to an explosion of the ruminant population, which then overgrazed the available vegetation; eventually thousands of elk and deer starved.

The election of Franklin D. Roosevelt as president in 1932 brought one of the most active proponents of the conservationist movement into the White House. Roosevelt had a genuine love of the outdoors and was a conservationist at heart. He was a utilitarian when it came to conservation, however, and above all a politician. He believed that a government-controlled program of land retirement, soil and forest restoration, flood control, and hydroelectric power generation would serve two aims simultaneously: The nation's natural landscapes could be preserved or put to good use at the same time people could be given meaningful employment.

Combining his zeal for conservation with the need to revitalize the American economy during the Great Depression, Roosevelt created numerous programs to further the improvement of the nation's natural resources. Notable among these were the Civilian Conservation Corps and the Works Progress Administration, agencies that put people to work improving national parks and other wilderness areas, making them accessible to the public, and restoring many of the renewable resources that had been lost over the years to commercial development. Among his principal successes, Roosevelt oversaw the opening of Great Smoky Mountains National Park in 1934.

Not all of the projects Roosevelt promoted under the banner of conservation were beneficial to the natural environment, however. The Tennessee Valley Authority oversaw the erection of numerous dams to provide hydroelectric power to rural communities, in the process destroying millions of hectares of forestlands and valleys. Politically, the most significant change in the conservation movement under Roosevelt was the affiliation of environmental causes with the Democrats rather than the Republicans, who had been leaders in conservation in previous years.

CONSERVATION AFTER WORLD WAR II

After World War II ended, the United States entered a period of economic prosperity that saw the population grow and industry expand to meet worldwide demand for American goods. As a result, individuals and corporations interested in development lobbied for relaxation of conservation rules that had been put in place during the previous six decades. Many argued for a return to commercial operations in government-protected areas. Conservation groups, which by then numbered in the dozens, fought back with mixed success to maintain and expand the protections in place.

Organizations such as the Sierra Club and the Wilderness Society developed activist tactics to bring attention to the long-term dangers of development in wilderness areas. These groups were successful in defeating a plan to create a dam on the Green River at Echo Park, on the Utah-Colorado border, during the 1950's. Most notably, under the leadership of the Wilderness Society's executive secretary Howard Zahniser, conservationists prodded Congress to pass the Wilderness Act in 1964. That law set aside more than 3.6 million hectares (9 million acres) of land, restricting even minimal development or human use.

The publication of Rachel Carson's *Silent Spring* in 1962, detailing the environmental dangers of toxic waste, gave new impetus to the conservationists and sparked the larger environmental movement. As a consequence, during the 1960's conservation groups began to unite with those interested in larger environmental issues. Sierra Club executive director David Brower became a leading spokesman for environmental causes. The most significant success achieved by these groups in lobbying governmental agencies was the passage of the National Environmental Policy Act of 1969, which established the government's role in protecting humans and the natural environment from damage caused by human development or technology. In December, 1970, the Environmental Protection Agency was established to enforce environmental policies throughout the United States.

INTERNATIONAL CONSERVATION EFFORTS

By the middle of the twentieth century, the modern conservation movement had become international. The record of conservation groups in Europe was strong; many had been successful in continuing and strengthening already existing laws to preserve natural resources. Beginning during the 1960's conservationists in Australia led highly organized campaigns to reverse the effects of human contamination on the natural environment, specifically targeting problems with the quality of soils and the eradication of several species of animals. In Africa, Asia, and South America, however, conservationists had mixed success. Often conservation efforts conflicted with the needs of the people for resources to support life. This was especially true in Africa, where native peoples relied on wildlife for sustenance, and in South America, where lands covered by rain forests were needed for farming. Some successes were achieved when native populations were convinced that conservation would provide long-term economic stability.

At the same time, however, commercial interests continued to harvest animals and timber far in excess of the population's needs, ignoring the environmental catastrophes caused by their actions. Often, government-sponsored programs wreaked havoc on natural resources. Nowhere was this more apparent than in China, where the need for energy to fuel the country's growing population and expanding industrial base led to construction of the massive Three Gorges Dam on the Yangtze River, flooding millions of hectares, destroying or seriously affecting land and riverine populations. From 1970 onward, however, China began to take an active approach to wildlife preservation, as did several other Asian nations.

One of the most successful conservation initiatives undertaken outside the United States has been that of the Central American nation of Costa Rica. Beginning in 1970 the Costa Rican government worked to create a highly developed national park system where conservation practices would be strictly enforced. By 2000 nearly 25 percent of Costa Rica's landmass had been set aside for preservation.

Further Reading

Chester, Charles A. *Conservation Across Borders: Biodiversity in an Interdependent World.* Washington, D.C.: Island Press, 2006.

Coggins, Chris. *The Tiger and the Pangolin: Nature, Culture, and Conservation in China.* Honolulu: University of Hawaii Press, 2003.

Evans, Sterling. *The Green Republic: A Conservation History of Costa Rica.* Austin: University of Texas Press, 1999.

Fox, Stephen. *John Muir and His Legacy: The American Conservation Movement.* Boston: Little, Brown, 1981.

Gibson, Clark L. *Politicians and Poachers: The Political Economy and Wildlife Policy in Africa.* New York: Cambridge University Press, 1999.

Maher, Neil M. *Nature's New Deal: The Civilian Conservation Corps and the Roots of the American Environmental Movement.* New York: Oxford University Press, 2008.

Markus, Nicola. *On Our Watch: The Race to Save Australia's Environment.* Melbourne: University of Melbourne Press, 2009.

Minteer, Ben and Robert Manning, eds. *Reconstructing Conservation: Finding Common Ground.* Washington, D.C.: Island Press, 2003.

Phillips, Sarah T. *This Land, This Nation: Conservation, Rural America, and the New Deal.* New York: Cambridge University Press, 2007.

Runte, Alfred. *National Parks: The American Experience.* 4th ed. Lanham, Md.: Taylor Trade, 2010.

Conservation of Natural Resources

—Thomas E. Hemmerly; updated by Karen N. Kähler

FIELDS OF STUDY

Conservation; Biology; Evolution; Ecology; History; Environmental Studies.

SUMMARY

The goals of conservation often conflict with the drive toward the exploitation of natural resources for immediate economic gain and also with the ethic of preservation, which promotes the indefinite preservation of natural resources in their undisturbed state.

For centuries few people recognized that nature's resources are finite. Only since the mid-nineteenth century has the issue of conserving resources been taken seriously. In North America, the eighteenth and nineteenth centuries were largely devoted to conquering the wilderness. As pioneers cleared forests, little or no thought was given to the idea that they could ever be exhausted. Bison and other animals were hunted nearly to extinction, and bounties were placed on wolves and other predators. In the seas, millions of seals were killed, and whales were hunted relentlessly.

Almost too late, it became apparent to many people that soils were being depleted by erosion and that once-plentiful native plant and wildlife populations were in decline. Even the quality of air, surface water, and groundwater deteriorated. An increasing number of people began calling for an end to environmentally destructive practices. An important influence was that of diplomat and naturalist George Perkins Marsh,

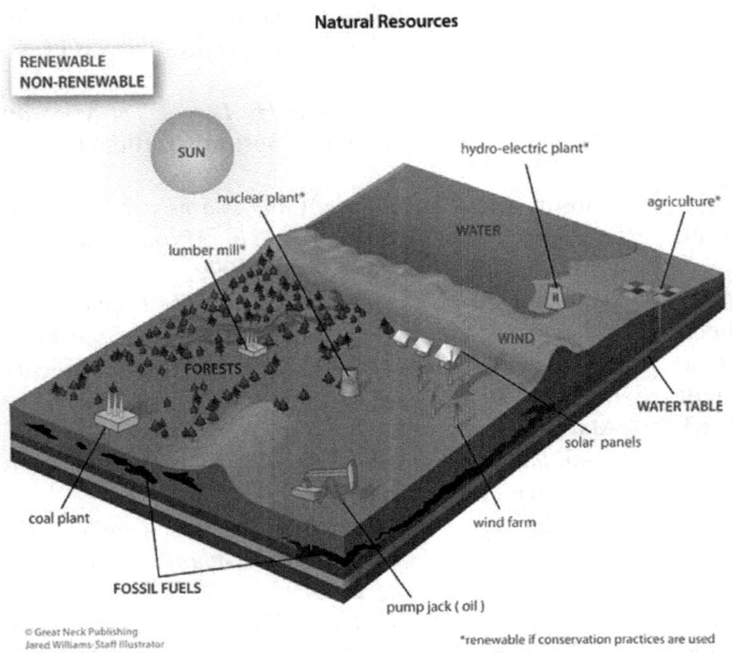

Natural Resources: The earth has many natural resources, some are renewable and others are not. Conservation practices will ensure that resources do not become depleted. © EBSCO.

whose travels had made him keenly aware of the results of centuries of land abuse in Europe. Marsh's book *Man and Nature: Or, Physical Geography as Modified by Human Action* (1864) defined the connections among soil, water, and vegetation. From such influences came the public park movement, including the establishment of Yellowstone National Park.

EMERGENCE OF THE MODERN CONSERVATION MOVEMENT

During the early twentieth century, the living standards of most Americans generally improved, but at the same time it was recognized that consumption of natural resources had to be controlled. Out of this realization emerged a human-centered conservation philosophy known as wise use, which dictated that human beings should use nature in such a way that its resources could continue to be used over long periods of time.

Among the conservationists of the era were U.S. president Theodore Roosevelt and forester Gifford Pinchot. Roosevelt increased the extent of the national forests in the United States and created many wildlife refuges. Pinchot, a friend and associate of Roosevelt, became chief forester of the U.S. Department of Agriculture and greatly influenced the conservation movement. He applied European methods of managing forests that were consistent with the wise-use philosophy. This put him in conflict with naturalist John Muir and others who wished to preserve wilderness areas in their natural state. Two opposing camps developed: On one side were preservationists, who argued that nature deserves to be protected for its own sake; on the other were conservationists, who believed in regulated exploitation.

During the 1920's and 1930's, scientific advances and economic problems influenced conservation views and policies. The ecosystem concept, the principle that nature is composed of local units with interacting living and nonliving components, developed. It was to become the cornerstone of the science of ecology and an important tool of conservation. During this same period a new, scientific approach to wildlife management emerged. Aldo Leopold, regarded as the father of the new science, wrote the influential textbook *Game Management* (1933), although he is better known today for his collection of essays *A Sand County Almanac, and Sketches Here and There*, which was published in 1949, soon after Leopold's death. Paul B. Sears achieved prominence with *Deserts on the March* (1935), which vividly dramatized the problems of the Dust Bowl of the 1930's.

The conservation legacy of President Franklin D. Roosevelt includes his attempts to restore the depressed U.S. economy by creating various work programs, including the Civilian Conservation Corps (CCC). Many of these programs were aimed at soil conservation, reforestation, and flood control.

POSTWAR DEVELOPMENTS

In the two decades following World War II, concerns about the Cold War and economic expansion kept conservation far from the forefront of American thought. As the population rapidly expanded, air and water pollution worsened, grasslands became overgrazed, and agricultural chemicals were used in increasing quantities. The complacency of the times was shattered by the publication of Rachel

Carson's *Silent Spring* (1962), which warned that dichloro-diphenyl-trichloroethane (DDT) and other pesticides were threatening the lives of animals and humans. The book caused widespread public concern, and laws were ultimately passed that outlawed DDT in the United States.

Under President John F. Kennedy's administration, attention was given once more to environmental issues in general and to conservation. Funds were appropriated to improve air quality, and new land was acquired for parks. Stewart Udall, secretary of the interior under Kennedy and his successor, Lyndon B. Johnson, advocated a new positive attitude toward protection of the environment. During the 1960's, a time of great scientific activity and social ferment, ecologists employed advanced technologies in conducting ecosystem analyses. Among the social movements of the decade was a new environmentalism, which culminated in the first Earth Day celebration in 1970.

The 1970's saw the expansion of environmentalism, and the term "environmentalism" came to be understood as almost synonymous with "conservation." The former clear distinction between the terms "conservation" and "preservation" was blurred. Environmental organizations continued to elaborate their views and increase their visibility. More action-oriented alternative groups used sabotage tactics to stop development in wilderness areas. In contrast, the Nature Conservancy became known for its businesslike policy of acquisition, protection, and management of natural areas. In response to the growing awareness of environmental issues among Americans, a new round of federal legislation was passed. Under President Richard Nixon, the Environmental Protection Agency (EPA) was created in 1970. The revised Clean Air Act of 1970 was followed by the Clean Water Act of 1972, and in 1973 Congress enacted the Endangered Species Act.

POLICY & POLITICS

The 1980's were marked by new conservation problems and a general indifference at the federal level in the United States. Among the global problems identified were stratospheric ozone depletion and global warming.Reversals of environmental protections occurred during the administration of President Ronald Reagan after he appointed James Watt as secretary of the interior. Watt attempted to dismantle many of the environmental protection programs that had been put in place during the previous decades.

The 1990's had their own environmental challenges and successes. After twelve years of apathy, conservationists were reinvigorated by the election of Bill Clinton as president and high-profile environmentalist Al Gore as vice president. Despite continued optimism among environmentalists and significant gains in environmental protections, however, an antienvironmental movement also developed. Many politically conservative individuals and organizations began to assert that environmentalism had gone too far. Business leaders and property owners complained of overregulation, and even some religious authorities expressed concerns that nature had been elevated in importance above humans. Nevertheless, the majority of Americans continued to feel that nature deserves to be protected. The late twentieth century saw a growing awareness of the importance of biodiversity and increasing support for the idea that all species should be protected against extinction. From this trend emerged the value-laden science of conservation biology.

Under the presidency of George W. Bush, the early twenty-first century was marked by business- and industry-friendly policies that diminished protections for wildlife and the environment. However, in the final month of his administration, Bush took the historic action of designating three areas of the Pacific Ocean as marine national monuments. This status protects the Marianas Trench, Pacific Remote Islands, and Rose Atoll marine national monuments from destruction or extraction of their resources, waste dumping, and commercial fishing.

In April, 2010, President Barack Obama held a national conservation conference that echoed one convened by President Theodore Roosevelt more than a century earlier. Dubbed America's Great Outdoors Initiative, Obama's conference took into account the developed, urban, and populous nature of the country. The twenty-first century strategy for the nation that came out of the conference proposed partnerships among government, nonprofit organizations, and private landowners for land preservation efforts that would be smaller in scale than national parks and designated wilderness areas and thus less likely to encounter heated political opposition.

FURTHER READING

Adams, Jonathan S. *The Future of the Wild: Radical Conservation for a Crowded World.* Boston: Beacon Press, 2006.

Carroll, Scott P., and Charles W. Fox, eds. *Conservation Biology: Evolution in Action.* New York: Oxford University Press, 2008.

Chiras, Daniel D., and John P. Reganold. *Natural Resource Conservation: Management for a Sustainable Future.* 10th ed. Upper Saddle River, N.J.: Benjamin Cummings/Pearson, 2010.

Cox, George W. *Conservation Biology: Concepts and Applications.* 2d ed. Dubuque, Iowa: Wm. C. Brown, 1997.

Groom, Martha J., Gary K. Meffe, and Carl Ronald Carroll. *Principles of Conservation Biology.* 3d ed. Sunderland, Mass.: Sinauer, 2006.

Hambler, Clive. *Conservation.* New York: Cambridge University Press, 2004.

Kline, Benjamin. *First Along the River: A Brief History of the U.S. Environmental Movement.* 3d ed. Lanham, Md.: Rowman & Littlefield, 2007.

Macdonald, David W., and Katrina Service, eds. *Key Topics in Conservation Biology.* Malden, Mass.: Blackwell, 2008.

Pullin, Andrew S. *Conservation Biology.* New York: Cambridge University Press, 2002.

CONSERVATION OF SAVANNAS

—Anita Baker-Blocker

FIELDS OF STUDY

Conservation; Biology; Evolution; Ecology; Environmental Studies.

SUMMARY

Savannas form in tropical or subtropical wet-dry climates, with moisture from the intertropical convergence zone dominant in wet months. Savannas have an average temperature of at least 16 degrees Celsius (64.4 degrees Fahrenheit) during the coolest months and receive between 50 and 255 centimeters (20 and 100 inches) of rain yearly. Savannas have poor drainage; the vegetation in savanna regions is dominated by drought-resistant grasses and a few trees. Fires, which occur during the dry season and prevent forests from developing, are essential in maintaining these grasslands.

Savannas form at the borders of tropical rain forests. In Africa, the savanna of the Sahel lies to the north of the rain forest, and northern and southern savannas meet in the East African savanna. The Colombian and Venezuelan llanos of the Orinoquía region are north of the Amazon rain forest, and the Argentine pampas and Brazilian *cerrado* are south of the rain forest. Almost the entire country of Uruguay is savanna. Savannas occupy roughly 25 percent of the continent of Australia, running across the northern tropics. South of the Himalaya Mountains, the Terai-Duar savanna in northern India, Nepal, and Bhutan is maintained by the yearly monsoon.

Savannas support large numbers of herbivorous and predator animals, birds, reptiles, and insects. Termites play an important role in savanna ecosystems; their mounds dot savannas worldwide. Termites consume cellulose and lignin from living grasses and trees, as well as vegetative litter. Primates (including human beings) exploit savanna termites as a high-protein source of food.

During the wet season (which may arise from monsoonal rains), large animal herds graze over the grasslands; during the dry season, herds move toward lakes and rivers. Predators follow the herbivores. Migratory birds from Asia, Europe, and North America winter in Southern Hemisphere savannas, joining large numbers of ground-living sedentary birds. Historically, sleeping sickness in Africa, spread by the tsetse fly, and malaria, spread by mosquitoes, kept human populations from exploiting many savannas, but modern insecticides and modern medicines have mitigated these disease factors somewhat, and human occupation of savannas has climbed steadily since the mid-twentieth century.

Human activities are the major threat to savannas worldwide. Illegal trade in endangered species of

Amazing native grasslands in the Missouri Coteau, Saskatchewan. The Canadian Prairies rolling hills of Missouri Coteau are filled with pothole wetlands making it one of the most important areas for waterfowl in North America.

savanna birds, such as South American parrots, and African and Asian animals, including tigers, lions, and primates, is a multibillion-dollar industry worldwide. Some endangered birds and animals are prized as pets; others are killed for their pelts or for use of their body parts in foods or medicines. Efforts to stop the trade in endangered species and their products have had limited results. For example, hunting of the endangered African bush elephant (also known as the Savanna elephant) was prohibited in 1989, but poachers continue to kill the elephants for their ivory tusks. The population of these elephants stood at only an estimated ten thousand in 2010.

It has been estimated that if conversion of the Brazilian *cerrado* savanna to croplands continues at the rates seen in the late twentieth century, the savanna will disappear by 2030. Oil production in the Venezuelan llanos and Brazilian *cerrado* may also alter the savanna biome. In Australia, pressure from the cattle industry to alter the natural savanna to provide for optimal beef production has changed the savanna ecosystem. Problems have arisen there from the introduction of non-native plant species, land clearance, and pastoral grazing, which has changed the natural pattern of fires in the savanna.

FURTHER READING

Bassett, Thomas J., and Donald Crummey, eds. *African Savannas: Global Narratives and Local Knowledge of Environmental Change.* Portsmouth, N.H.: Heinemann, 2003.

Mistry, Jayalaxshmi, and Andrea Berardi, eds. *Savannas and Dry Forests: Linking People with Nature.* Burlington, Vt.: Ashgate, 2006.

Pennington, R. Toby, and James A. Ratter. *Neotropical Savannas and Seasonally Dry Forests: Plant Diversity, Biogeography, and Conservation.* Boca Raton, Fla.: CRC Press, 2006.

Roach, Mary. "Almost Human." *National Geographic,* April, 2008, 126–145.

CONSERVATION POLICY

—Craig W. Allin

FIELDS OF STUDY

Conservation; Political Science; Government; Biology; Evolution; Ecology.

SUMMARY

Trends in conservation policy in the United States have varied in emphasis between allowing for limited exploitation of resources and prohibiting most economic uses of particular resources so that they can be preserved in an undeveloped state. Both approaches, however, have led to the protection and preservation of public lands and animal habitats.

The first use of the term "conservation" in relation to the natural environment was claimed by Progressive intellectuals in the early twentieth century. In his autobiography *Breaking New Ground* (1947), Gifford Pinchot, first chief of the U.S. Forest Service and a close friend of President Theodore Roosevelt, recalled realizing that all the natural resource problems are really one problem: the use of the earth for the permanent good of humans. The idea needed a name; presidential adviser Overton Price suggested "conservation," and the matter was settled.

MILESTONES IN CONSERVATION POLICY

Year	*Event*
1864	Yosemite Valley is ceded to California to create a park.
1872	The Yellowstone National Park Act establishes the world's first national park.
1891	The Forest Reserve Act authorizes the U.S. president to establish national forests.
1894	The Yellowstone Game Protection Act closes parks to hunting and commercial fishing.
1897	The Forest Management Act mandates that national forests be managed to perpetuate water supplies and wood products.
1900	The Lacey Act prohibits interstate shipment of wildlife that has been killed illegally.
1902	The Newlands Act establishes a national reclamation policy.
1903	The first National Wildlife Refuge is created at Pelican Island, Florida.
1905	The U.S. Forest Service is created within the Department of Agriculture to manage national forests.
1906	The Antiquities Act authorizes the creation of national monuments by presidential proclamation.
1910	The Pickett Act authorizes presidential land withdrawals for any public purpose.
1911	The Weeks Act provides for governmental purchase of national forestlands.
1913	The Hetch Hetchy Dam is authorized in Yosemite National Park.
1916	The National Park Service is created in the Interior Department to manage national parks.
1918	The Migratory Bird Treaty Act restricts the hunting of migratory birds.
1933	The Civilian Conservation Corps Act is passed.
1933	The Tennessee Valley Authority is created.
1934	The Taylor Grazing Act regulates grazing on public lands.
1937	The Federal Aid in Wildlife Restoration (Pittman- Robinson) Act provides federal aid to states for wildlife management.
1946	The U.S. Bureau of Land Management is created.
1950	The Federal Aid in Fish Restoration (Dingell-Johnson) Act provides federal aid to states for sport fish management.
1956	The Fish and Wildlife Act creates the U.S. Fish and Wildlife Service in the Interior Department.
1956	A proposal to construct Echo Park Dam in Dinosaur National Monument is defeated.
1960	The Multiple Use-Sustained Yield Act clarifies the purposes of national forests.
1964	The Wilderness Act establishes the National Wilderness Preservation System.
1964	The Land and Water Conservation Fund Act provides a trust fund for parkland acquisition.
1968	The National Wild and Scenic Rivers Act establishes a national river conservation system.
1968	The National Trails System Act establishes a national system of recreational trails.
1970	The National Environmental Policy Act requires environmental impact statements for federal activities that affect the environment.
1970	The Environmental Protection Agency (EPA) is created.
1970	Clean Air Act amendments establish stricter air-quality standards.

1971	The United Nations Educational, Scientific, and Cultural Organization (UNESCO) Biosphere Reserve Program recognizes areas of global environmental significance.
1972	The Clean Water Act establishes stricter water-quality standards.
1972	The United Nations Environmental Conference in Stockholm, Sweden, is attended by 113 nations.
1972	Federal Water Pollution Control Act amendments provide protection for wetlands.
1972	The Federal Environmental Pesticides Control Act requires pesticide registration.
1972	The Marine Mammal Protection Act imposes a moratorium on hunting or harassing of marine mammals.
1973	The Convention on International Trade in Endangered Species of Wild Fauna and Flora (CITES) prohibits international trade in endangered species.
1973	The Endangered Species Act commits the United States to the preservation of biological diversity.
1974	The Safe Drinking Water Act sets federal standards for public water supplies.
1976	The Toxic Substances Control Act authorizes the EPA to ban substances that threaten human health or the environment.
1976	The Federal Land Policy and Management Act directs the Bureau of Land Management to retain public lands and manage them for multiple uses.
1976	The Resource Conservation and Recovery Act directs the EPA to regulate waste production, storage, and transportation.
1976	The National Forest Management Act gives statutory protection to national forests and sets standards for management.
1977	The Surface Mining Control and Reclamation Act establishes environmental standards for strip mining.
1977	Clean Air Act amendments set high standards for air quality in large national parks and wilderness areas.
1980	The Fish and Wildlife Conservation Act provides federal aid for the protection of nongame wildlife.
1980	The Alaska National Interest Lands Conservation Act establishes more than 40.5 million hectares (100 million acres) of national parks and wildlife refuges in Alaska.
1980	The Comprehensive Environmental Response, Compensation, and Liability Act (CERCLA) establishes the Superfund hazardous waste cleanup program.
1982	The Nuclear Waste Policy Act establishes a process for siting a permanent nuclear waste repository.
1985	The U.S. government establishes the Conservation Reserve Program (CRP) to reduce agricultural surpluses by encouraging farmers to reduce the amounts of land they devote to crops, thereby helping to prevent soil erosion and reduce carbon in the atmosphere.
1987	The Montreal Protocol limits the production and consumption of chlorofluorocarbons (CFCs).
1988	The Ocean Dumping Act prohibits the dumping of sewage sludge and industrial waste.
1990	Clean Air Act amendments strengthen the Clean Air Act.
1992	The Earth Summit in Rio de Janeiro, Brazil, is attended by 179 nations.
1997	The Kyoto Protocol on Climate Change encourages global reduction in greenhouse gas emissions.
2001	The Roadless Area Conservation Rule, a federal policy initiative designed to protect national forests from commercial development, is issued.
2005	More than 4,000 hectares (approximately 10,000 acres) of Puerto Rican rainforest are added to the U.S. national forest system.
2008	The Food, Conservation, and Energy Act extends CRP enrollment authority through 2012.
2009	The Omnibus Public Land Management Act adds 850,000 hectares (2.1 million acres) of new wilderness areas in nine U.S. states.

ORIGINS OF CONSERVATION POLICY

The antecedents of modern conservation policies go back centuries. Aboriginal cultures around the world developed taboos governing behavior on the hunt. Venetians established reserves for deer and wild boar in the eighth century. Hunting reserves were common in Europe and Asia, and in colonial America those trees thought best for ships' masts were preserved for that purpose by decree.

Conservation policy, as the term is now understood, is a product of the nineteenth century, when population growth, urbanization, and industrialization created unprecedented opportunities for people to influence the natural world in which they lived, both for good and for ill. By the end of the nineteenth century, more Americans lived in cities than on farms. The nation was connected from coast to coast by telegraph and rail, and economic growth was rapid.

Most natural resource policies of the nineteenth century were designed to facilitate economic development. Best known among these policies was the Homestead Act of 1862, which gave free land to settlers, but similar policies provided free land to railroads and states. Other laws stimulated growth by providing for free use of timber and minerals on public lands.

The economic progress of the nineteenth century came at high cost to the environment, typified by the profligate use of forests and the near extermination of North American buffalo, and some prominent Americans took notice. In 1832 the artist and journalist George Catlin wrote of the probable extinction of the buffalo and American Indians, and he advocated a large national park where both might be preserved. Henry David Thoreau echoed Catlin's concerns in 1858, calling for national preserves. In 1864 George Perkins Marsh published *Man and Nature: Or, Physical Geography as Modified by Human Action*, the earliest important text with an ecological perspective.

As economic exploitation diminished the supply of natural resources, public attitudes began to change, and with them governmental policies. Although many laws continued to encourage economic growth, others reflected the growing desire for conservation. In 1864 the U.S. Congress sought to preserve the Yosemite Valley and Mariposa Big Tree Grove by giving them to the state of California for a public park. Eight years later Congress established the world's first national park at Yellowstone. In 1884 additional legislation prohibited all hunting and commercial fishing within Yellowstone National Park. In 1891 Congress established what would eventually become the national forest system when it authorized the president to set aside forest reserves on public lands.

PRESERVATION VERSUS WISE USE

These early conservation policies stressed resource preservation. Parks and forest reserves were simply set aside; none was effectively managed. The lack of management—especially forest management—displeased the advocates of scientific forestry, who also considered themselves conservationists. Early in its history, thus, the American conservation movement was divided, with some conservationists preaching "preservation" and others "wise use."

These contradictory tendencies were epitomized in the conflict between John Muir, founder of the Sierra Club, and Gifford Pinchot, principal architect and first chief of the U.S. Forest Service. Muir was the intellectual heir of Thoreau and Catlin. A perceptive scientist and popular author, he devoted his life to the exploration, enjoyment, and preservation of natural ecosystems worldwide. Pinchot had studied scientific forestry at its source in Germany. He was a gifted politician, and his passion was not for preservation but for wise use. Muir believed that people could not improve on nature; his conservation was aesthetic and spiritual. Pinchot was committed to maximizing the human benefits from resource use through science; his conservation was economic and utilitarian. Although friends for a time, Muir and Pinchot eventually parted ways, with Muir becoming an advocate for preservation and national parks and Pinchot an advocate for wise use and national forests.

In the United States the legacy of Pinchot is alive and well in the multiple-use management principles of the Forest Service and the Bureau of Land Management and in organizations such as the Society of American Foresters, the International Society of Fish and Wildlife Managers, the National Rifle Association, and the Soil Conservation Society of America. Muir's emphasis on preservation has been institutionalized in the National Park Service and the Fish and Wildlife Service as well as in organizations such as the National Audubon Society, the Nature Conservancy, the Sierra Club, and the Wilderness Society.

PROGRESSIVE ERA

Many historians emphasize three eras of American conservation policy corresponding roughly to the Progressive Era, the New Deal, and the so-called environmental decade of the 1970's. The Progressive Era, epitomized by the presidency of Theodore Roosevelt, was the first golden age of American conservation policy. During this period Congress passed a number of pathbreaking conservation laws. The Lacey Act of 1900 put the power of federal enforcement behind state game laws, criminalizing the interstate transport of wildlife killed or captured in violation of state regulations. Another milestone of wildlife conservation was the ratification of a migratory bird treaty with Canada and passage of a law to enforce the treaty. With the Migratory Bird Treaty Act of 1918, the federal government asserted national authority to manage wildlife for conservation purposes, authority that was upheld by the U.S. Supreme Court in the case of *Missouri v. Holland* (1920).

Two critically important governmental agencies were created during this era: the Forest Service and the Park Service. In 1905 advocates of wise use and scientific forestry were rewarded with a Forest Service in the Department of Agriculture. The new agency's first director was Gifford Pinchot, the nation's foremost advocate of multiple-use forest management based on scientific principles. Under Pinchot's leadership the concepts of multiple use and sustained yield were applied in the rapidly growing national forest system. When Theodore Roosevelt became president, the United States had 18.6 million hectares (46 million acres) of national forest. By the end of his term of office, Roosevelt had increased the total size of the national forest system to 78.5 million hectares (194 million acres).

During the Progressive Era, advocates of preservation were often unsuccessful in their opposition to the policies of wise-use conservationists, but in the end they also had a victory. The most painful loss came in Yosemite National Park, where advocates of wise use joined forces with the city of San Francisco to dam the Hetch Hetchy Valley for a municipal water supply, forever destroying a natural valley some regarded as comparable to Yosemite Valley itself. The public outcry over the damming of Hetch Hetchy contributed to pressure for better park protection, however, and in 1916 preservationists achieved a long-sought goal: creation of the National Park Service to manage the growing system of fourteen national parks, including Yellowstone (1872), Yosemite and Sequoia (1890), Mount Rainier (1899), Crater Lake (1902), Wind Cave (1903), Mesa Verde (1906), Glacier (1910), Rocky Mountain (1915), and Hawaii and Lassen Volcanic (1916).

Two new forms of conservation reserves made their debut during this era: national wildlife refuges and national monuments. Roosevelt regarded wildlife sanctuaries as critical to the survival of game species. In 1903 he acted on his belief, creating the nation's first national wildlife refuge on Pelican Island in Florida. He had no specific legal authority to create a national wildlife refuge, but his usurpation was accepted at the time and later approved in principle. In 1910 the Pickett Act authorized the president to set aside land for any public purpose. National monuments began on a firmer foundation, but here too Roosevelt pushed conservation to the limit. The Antiquities Act of 1906 authorized the president to establish national monuments. As the name suggests, the law anticipated relatively small reservations to protect archaeological sites, but the monuments Roosevelt designated included 34,400 hectares (85,000 acres) at Petrified Forest, 120,600 hectares (298,000 acres) at Mount Olympus, and 326,200 hectares (806,000 acres) at the Grand Canyon. All of these later became national parks.

NEW DEAL ERA

The New Deal was, for the most part, a response to disaster. The primary disaster was the Great Depression, but the decade of the 1930's also saw the Dust Bowl, a minor climatic change that produced disastrous results on the Great Plains. New Deal conservation policies were responsive to the economic and ecological crises of the era, and they stressed wise use through scientific management rather than preservation. The Tennessee Valley Authority was created in 1933 to stimulate employment and economic growth in Appalachia through scientific management of the area's natural resources. The Taylor Grazing Act of 1934 was designed to end overgrazing of western public lands by imposing a system of permits based on principles of scientific management. The Civilian Conservation Corps and the Soil Conservation Service were both established during this era, and each contributed to repairing environmental damage. Greater concern for the management rather than the disposal of western public lands was also reflected in the creation in 1946 of the Bureau of Land Management to replace the General Land Office.

ENVIRONMENTAL DECADE

The so-called environmental decade lasted almost twenty years. It began with the inauguration of President John F. Kennedy, persisted through the presidential administrations of Lyndon B. Johnson, Richard Nixon, Gerald Ford, and Jimmy Carter, and ended with the inauguration of President Ronald Reagan. The conservation policies of this era were responsive to post-World War II economic growth that seemed to ensure economic prosperity while threatening quality of life. Stewart Udall, U.S. secretary of the interior, warned of a *Quiet Crisis* (1963), Rachel Carson of a *Silent Spring* (1962), Barry Commoner of *The Closing Circle* (1971), and Paul R. Ehrlich of *The Population Bomb* (1968). Conservation policy matured into environmental policy during this era. Conservation was still about husbanding natural resources, but to the historic concerns of conservation—such as forests, wilderness, and wildlife—were added concerns regarding clean water and clean air, energy supplies, and the problems posed by hazardous and toxic wastes. During this era Congress passed most of the major laws that continue to shape conservation policy at the beginning of the twenty-first century.

Preservation policy was strengthened. In 1964 Congress passed the Wilderness Act and the Land and Water Conservation Fund Act. The former established the National Wilderness Preservation System, which has grown from more than 3.6 million hectares (9 million acres) to more than 40.5 million hectares (100 million acres). The latter facilitated acquisition of land for parks and open space. Four years later Congress established a national system of trails as well as a national system to protect wild and scenic rivers from certain kinds of development. Both the Forest Service and the Bureau of Land Management were given new statutory direction emphasizing planning and preservation, sometimes at the expense of economic development. At the end of the era Congress passed the Alaska National Interest Lands Conservation Act (1980), making conservation withdrawals of more than 40.5 million hectares of public lands and doubling the size of the national park and wildlife refuge systems nationwide.

Wise-use conservation was also well served as Congress radically increased federal regulation of resource use. President Johnson established a presidential commission on natural beauty and addressed world population and resource scarcity in his 1965 state of the union speech. Environmental management was nationalized through a series of far-reaching statutes addressing air pollution, water pollution, marine resources, noise pollution, biological diversity, toxic chemicals, and hazardous waste. New burdens were placed on government and private citizens. The National Environmental Policy Act of 1969 required all governmental agencies to study the probable environmental effects of their actions before moving forward. A large number of environmental enforcement programs were reorganized in 1970 into the newly created Environmental Protection Agency (EPA). The EPA is an independent agency, but presidents have routinely regarded its director as having cabinet status.

CONSERVATION POLICY FOR A NEW MILLENNIUM

At the policy level, the era following the environmental decade was one of consolidation rather than new initiatives. The Reagan administration was hostile to environmental policy and attempted to tilt public policy toward less environmental regulation. President Reagan was able to prevent the adoption of major new conservation policies, but his administrative efforts—led by Interior Secretary James Watt—to roll back environmental laws were successfully resisted by Congress. A major new air-pollution statute was passed during the presidential administration of George H. W. Bush, but this was exceptional for the era. Bill Clinton's presidential administration gave greater attention to conservation policy, but in the years following the 1994 elections there was little cooperation between the president and Congress on environmental issues.

Americans have come to expect government to practice conservation and protect environmental quality. Doing so has become increasingly difficult. Beyond the policy gridlock of the 1980s, 1990's, and 2000s, the issues themselves became more difficult. The most pressing issues—such as stratospheric ozone depletion, climate change, and biological diversity—were beset by scientific uncertainty. Such concerns are also global in scope and thus beyond the ability of any single nation to address independently. The future of conservation policy appears to be in the international arena, where extant institutions lack the authority to govern. Treaties addressing the use of ozone-destroying chlorinated fluorocarbons, the preservation of biological diversity, and the limitation of heat-trapping greenhouse gas emissions demonstrate that nations are giving increasing attention to conservation issues, but international achievements remain modest.

Despite decades of government conservation efforts, a 2015 *Proceedings of the National Academy of Sciences* analysis indicated that the Wilderness Act and other existing U.S. policies have favored scenic vistas, particularly in the West, and endangered mammals and birds at the expense of vulnerable reptiles, amphibians, and trees. Thus, the researchers found a mismatch between what has been protected and the species and geographical areas that most need it in order to preserve biodiversity.

In the late 2010s, the administration of President Donald Trump increasingly turned away from conservation in order to exploit natural resources. Soon after his inauguration in 2017, Trump announced the U.S. withdrawal from the landmark international Paris Agreement on emissions reduction and, that December, shrank the Grand Staircase–Escalante and Bears Ears National Monuments, constituting the largest such reduction in land protection to date. The following year, the administration rewrote Interior Department guidance to facilitate public uses such as mining and oil and gas extraction, over and against habitat protection and mitigation of human activity. The future of U.S. conservation policy, both at home and abroad, was thus called into question.

FURTHER READING

Allin, Craig W. *The Politics of Wilderness Preservation.* 1982. Reprint. Fairbanks: University of Alaska Press, 2008.

Chiras, Daniel D., and John P. Reganold. *Natural Resource Conservation: Management for a Sustainable Future.* 10th ed. Upper Saddle River, N.J.: Benjamin Cummings/Pearson, 2010.

Davis, David Howard. *American Environmental Politics.* Belmont, Calif.: Wadsworth, 1998.

Dowie, Mark. *Conservation Refugees: The Hundred-Year Conflict Between Global Conservation and Native Peoples.* Cambridge, Mass.: MIT Press, 2009.

Eilperin, Juliet. "Energy and Environment Interior Rescinds Climate, Conservation Policies Because They're 'Inconsistent' with Trump's Energy Goals." *The Washington Post,* 5 Jan. 2018, www.washingtonpost.com/news/energy-environment/wp/2018/01/05/interior-rescinds-climate-conservation-policies-because-theyre-inconsistent-with-trumps-energy-goals. Accessed 4 Apr. 2018.

French, Hilary. *Vanishing Borders: Protecting the Planet in the Age of Globalization.* New York: W. W. Norton, 2000.

Hayes, Samuel P. *Conservation and the Gospel of Efficiency: The Progressive Conservation Movement, 1890–1920.* 1959. Reprint. Pittsburgh: University of Pittsburgh Press, 1999.

Jenkins, Clinton N., et al. "US Protected Lands Mismatch Biodiversity Priorities." *Proceedings of the National Academy of Sciences,* vol. 112, no. 16, 21 Apr. 2015, pp. 5081–86, doi: doi.org/10.1073/pnas.1418034112. Accessed 4 Apr. 2018.

Nash, Roderick. *Wilderness and the American Mind.* 4th ed. New Haven, Conn.: Yale University Press, 2001.

Rosenbaum, Walter A. *Environmental Politics and Policy.* 7th ed. Washington, D.C.: CQ Press, 2008.

CONVERGENT & DIVERGENT EVOLUTION

—Frederick M. Surowiec

FIELDS OF STUDY

Conservation; Biology; Evolution; Ecology; Environmental Studies.

PRINCIPAL TERMS

adaptive radiation: the successful invasion by a species into a number of previously unavailable ecological niches

analogue: an individual structure shared by two or more species that is of only superficial similarity; thus, it is not indicative of a common ancestor

clade: a type of grouping of living or extinct species along lines of shared, unique structures, or homologues, indicative of a common ancestor; helpful in establishing evolutionary relationships

convergence: the evolution of a similar morphology by unrelated or only distantly related species caused by both having adapted to similar lifestyles in similar environments

divergence: the evolution of increasing morphological differences between an ancestral species and offshoot species caused by differing adaptive pressures

environmental constraints (pressures): the physical demands placed upon any species by its surroundings that ultimately determine the success or failure of its adaptations and consequently its success as a species

homologue: an individual structure shared by two or more different species that is indicative of a common ancestor

SUMMARY

The phenomena of convergent and divergent evolution have produced a number of good examples of natural selection at work. Thus, they provide results which can be studied in detail of interactions between environment and species which solidly validate aspects of basic evolutionary theory. Examples abound in the earth's fossil record as well as in a number of living species.

Biological species have been defined as populations of organisms that are capable of successfully interbreeding (producing fertile offspring) only with other members of the same species. Members of any species possess unique sets of biological characteristics, termed "characters." These characters are physical expressions of a genetic code unique to members of that species. The code represents an extremely complex and thorough set of instructions for equipping an individual organism with the body and the behavioral knowledge it requires for success in the particular environment to which its species has adapted.

Thus, because of natural selection acting upon many past generations of that species, living members are fine-tuned to a specific ecological niche, or econiche, of the greater ecosystem of which the species is a member. When conditions within the ecosystem change (a general climatic change, for example) or when other scenarios occur, such as when a smaller subpopulation of the species migrates into new, ecologically different territory or becomes isolated in some way, selective pressure is brought to bear upon members of the group or subgroup. Random mutation is a mechanism by which selective pressure is thought to be brought about. Such mutations are changes in the genetic code that occur spontaneously in some individuals within the species in an ongoing manner. Most random mutations are insignificant phenomena with regard to the species as a whole, because most have either a neutral or negative survival value: Either they do not help the individual possessing them to survive or they are counteradaptive to an extreme degree and prove fatal. Consequently, mutations in general are not usually transmitted beyond the generation in which they occur or beyond the affected member or members. In certain scenarios, however, mutations that have a positive survival value can spread throughout the population. This is believed to be especially true when a smaller, isolated subgroup of the population is dealing with a changed or new environmental situation. Such processes are thought, for example, to have been instrumental in the evolution of groups of closely related but now morphologically distinct species found in isolated, mid-ocean island groups. These adaptive radiations of species that are monophyletic and thus share a relatively recent common ancestor are good examples of the process of evolutionary divergence at work.

THE CASE OF THE GALÁPAGOS FINCHES

In one of the best-known cases, studies have traced the presumed paths of divergence among a set of island bird species. This particular radiation produced a number of new species possessing novel adaptive morphologies evolved to exploit new econiches. This is the classic example of Darwin's finches. Darwin's finches are a group of closely related birds, numbering about fourteen species, found on the various islands of the Galápagos Archipelago, which straddles the equator. The islands are remote from any large body of land that would typically harbor similar birds: South America lies about 960 kilometers to the east across an unbroken stretch of the Pacific Ocean. In 1835, Charles Darwin, author of the highly influential work on organic evolution, *On the Origin of Species* (1859), visited the islands while employed as a naturalist on a British scientific voyage. His studies of the flora and fauna of these islands provided him with many observations that directly influenced his later writings.

Darwin's studies of the Galápagos finches convinced him, and generations of subsequent scientists, that the finches are a clear example of divergent evolution in operation. The scenario he deduced is that probably only one ancestral species arrived from South America by ocean currents or winds, established itself, and began to exploit the numerous, as yet unoccupied, econiches that the volcanic islands provided. In the relative ecological vacuum that the original finch species found among the islands, adaptive radiation occurred, resulting in the present, diverse species. The species of Darwin's finches found today on the islands exhibit a great variety of beak types, many of which are atypical of finch-type birds in general but rather are typical of birds found among totally different avian family classifications. Typical finches are noted for beaks adapted for the crushing of seeds—the diet of the usual members of the finch family, such as the familiar North American cardinal. Among the dozen or so Galápagos finches can be found a wide assortment of beaks adapted for obtaining or processing a much greater variety of diets. Darwin's finches include species with beaks and behaviors adapted for diets of insects, seeds, cacti, and other vegetal matter. The adaptive radiation in the case of the Galápagos finches was relatively easy to work out because of the obvious environmental factors involved (the islands' remoteness and general barrenness) and the unusual variety of adaptations that the finches had made. Establishing the details of evolutionary divergence in other living ecosystems can be confusing because the numbers and types of econiches and interacting species are often far more numerous and diverse—for example, the lush and intricate ecosystem of a large, tropical rain forest such as the Amazon.

TRACING THE FOSSIL RECORD

Myriad examples of evolutionary divergence exist between both living plant and animal species throughout existing ecosystems in the modern world. The fossil record, however, also can be studied, to examine the phenomenon between extinct animal groups. This record of past life-forms preserved in the crustal rocks of the earth provides numerous examples of diverging species as organisms adapted to changing general conditions or spread into novel environments. An example is the many species of ceratopsian dinosaurs found in the latter part of the Cretaceous period of the Mesozoic era of the earth's history. Although the earlier ancestral forms appear to be bipedal and possess no significant armor, the later radiation of ceratopsians is well known by way of such impressive animals as *Triceratops*, a typical ceratopsian: a heavy, quadrapedal herbivore with a large, horned and beaked skull with a defensive, bony frill. Many variations on the basic late ceratopsian body architecture evolved through divergence. Varieties included such forms as *Pentaceratops*, *Torosaurus*, *Styracosaurus*, *Chasmosaurus*, and *Centrosaurus*, among many others. In all these later animals, the basic morphology regarding body, tail, and limbs remained the same. All forms also retained the typical massive, beaked head. What diverged were such morphological features as number and length of facial horns and length and degree of ornamentation of the frill. One of the important things that study of such fossil forms shows is that such past examples of evolutionary divergence eloquently underscore the continuity of the evolutionary process through time down to the present living world. This continuity further reinforces the validity of basic evolutionary theory in general.

EVOLUTIONARY CONVERGENCE

A related phenomenon concerning adaptive evolution is the phenomenon of evolutionary convergence. This process can be described briefly as the evolution of similar body structures in two or more species that are only quite distantly related; they therefore come to resemble each other, sometimes to a startling degree of at least outward sameness. These sets of similar-looking but polyphyletic species frequently even display similar behavioral characteristics. All similarities found in convergence cases are believed to be attributable to the fact that the various species involved have adapted to a similar econiche within a similar ecosystem. Because in nature, form follows function and the morphology of an animal or plant is the product of environmental pressures that continuously favor the better-adapted organism, it is easy to understand how convergence can take place. Like divergence, the evidence for convergence can be traced not only among the many participants in contemporary biospace but also over the course of vast stretches of past time.

One of the classic examples of convergence is a threefold example that, conveniently, not only includes representatives from three different classes of vertebrates but also spans many millions of years of time and includes an extinct group. This is the textbook example that compares and contrasts the morphology of sharks, a type of cartilaginous fish; ichthyosaurs, an extinct type of marine reptile; and dolphins, marine mammals like whales. All three groups possess numerous member species, both fossil and alive (except for ichthyosaurs), which resemble each other in body plan and lifestyle. All three groups include species which lead (or led) an open ocean, fish-eating existence. Consequently, the forms of their bodies came to follow the functions dictated by their environment—sometimes termed their environmental constraints. All three groups' general body plan began to approach a hydrodynamic ideal for a water-living animal: a streamlined fusiform, or spindle shape, efficient in passing through an aqueous medium. Besides this feature, pelagic, or open ocean-living, sharks, ichthyosaurs, and dolphins all evolved a dorsal fin to act as a vertical stabilizer for water travel. In addition, each group evolved a propulsive tail and a pectoral fin necessary for the demands of constant swimming and steering in water. Even more remarkable in this comprehensive example, dolphins' and ichthyosaurs' ancestors were both originally land-dwelling vertebrates that returned to the marine environment. This case presents an inclusive and persuasive argument for the reality of the phenomenon of convergent evolution.

As with the use of both living and extinct examples in the discussion of divergent evolution, the existence of fossil as well as contemporary species that display convergent morphology is convincing evidence for the process of adaptive evolution. Again, a continuity across vast stretches of time exists that connects evolutionary phenomena in a continuum.

OBSERVATION, COMPARISON & CLASSIFICATION

Research in the field of adaptive evolution, especially the phenomena of divergence and convergence of species, began centuries ago with the simple process of recognizing relationships in the surrounding environment between living plants and animals. The search for a unifying order to tie the complex web of animal and plant life together in some meaningful manner was for a long time a part of natural science. The modern theory of organic evolution fulfills this goal admirably in many respects. The methods used to illuminate the intricacies of evolution still encompass the type of keen, analytical observation of phenomena and reflection on their causes and effects that characterized Darwin's studies on the voyage of HMS *Beagle*. Observation, collection of specimens for comparison, classification of specimens according to a meaningful scheme, and, finally, an attempt to sort out the processes involved in a way that agrees with the dictates of strict logic are hallmarks of the scientific method at work.

Researchers investigating evolutionary divergence and convergence have powerful aids in the form of increasingly sophisticated technology. The main focus of their work is the correct interpretation of the path that various lineages took over time to arrive at known, living forms or extinct forms. In the case of living forms, technology originally developed in the field of medicine has been pressed into service to help establish relationships. For example, detailed analyses of various body tissues and fluids have been employed. Blood types have been traced with varying degrees of success, as have various proteins. Powerful optical microscopes are employed to analyze various tissue types and their structures. Since the invention of scanning electron microscopes, these more powerful instruments have further aided in probing the compositions and textures of animal and plant tissues to determine affinities among various species. In addition to these methods, very sophisticated laboratory techniques are now used to unravel and analyze deoxyribonucleic acid (DNA) strands and to try to determine the actual genetic encoding possessed by a particular organism. All these methods help establish more clearly the picture of biological relationships in regard to ancestries.

This physiological approach is obviously of limited utility with regard to fossil species. Except for such instances as the various ice age animals that were frozen in such environments as the tundra, extinct life-forms cannot be analyzed by medical means, as the original tissue has been transformed or destroyed by geological processes. In the case of most fossil forms, hard body parts such as bones and teeth (for vertebrates) and exoskeletons (in the case of invertebrates) must be analyzed in a more structural way to determine possible evolutionary relationships.

The clarification of the paths that various animal and plant lineages took during the process of their evolution further confirms the validity of basic organic evolutionary theory such as natural selection and adaptation. Study of divergent and convergent species is part of the ongoing study of living organisms that make up the functional ecosystems of which humankind is also a part. Learning more about these ecosystems and the parts that all the member species play within them is extremely important in the light of the contemporary world picture of pollution, overpopulation, and industrialization. The increased insight into how ecosystems operate from the species interaction approach is one of the positive by-products that studies of divergent and convergent evolution among species provide.

FURTHER READING

Bowler, Peter J. *Life's Splendid Drama: Evolutionary Biology and the Reconstruction of Life's Ancestry, 1860–1940.* Chicago: University of Chicago Press, 1996. A critical survey of the history of evolutionary thought, including the topic of parallel and convergent evolution.

Carroll, Robert L. *Vertebrate Paleontology and Evolution.* New York: W. H. Freeman, 1988. A thorough treatment of the complex subject of vertebrate evolution. Each vertebrate order is covered by a chapter profusely illustrated with excellent anatomical drawings, which discusses at length the order's biological characteristics, adaptations, and occurrence in the fossil record. There are also chapters that clearly explain evolutionary concepts in an understandable way. Convergence and divergence are treated within the greater context of biological evolutionary phenomena.

Cvancara, Alan M. *Sleuthing Fossils: The Art of Investigating Past Life.* New York: John Wiley & Sons, 1990. Intended as a portable reference book aimed primarily at amateur fossil collectors and nonprofessional enthusiasts, this "how to" book also is designed as an authoritative introduction to the subject of the total fossil record of the earth and its implications for the subject of organic evolution. Good black-and-white photographs enhance the text, along with some graphs and drawings. Little science background is assumed, except for a very basic high school level.

Orians, Gordon H., and Otto T. Solbrig, eds. *Convergent Evolution in Warm Deserts.* Stroudsburg, Pa.: Dowden, Hutchinson & Ross, 1977. A series of summaries of a large-scale, comparative study undertaken by more than twenty different scientists from both South and North America. The study aimed at determining the degree of actual convergence attained by the flora and fauna in two geographically remote deserts: the Sonoran in the American Southwest and the Monte in Argentina. The deserts analyzed were chosen because of their similar topographies, soils, and climatic conditions. Photographs, drawings, diagrams, and charts help make this book an exceptionally useful reference on the subject.

Paul, Gregory S. *Predatory Dinosaurs of the World.* New York: Simon & Schuster, 1988. A meticulously thorough book that covers almost every conceivable aspect of theropod, or flesh-eating, dinosaurs. Evolutionary convergence and divergence among the theropods is adroitly explained.

Ridley, Mark. *Evolution.* 2d ed. Boston: Blackwell Scientific, 1996. A college-level textbook in evolutionary biology. Explains the theory of evolution in four sections: evolutionary genetics, adaptation and natural selection, evolution and diversity, and paleobiology and macroevolution. Includes a Windows/Mac-compatible CD-ROM with extensive supplementary material, including interviews with evolutionary biologists, animations explaining difficult concepts, interactive experiments and timeline, and a mini encyclopedia of terms with video clips, animation, photographs, and diagrams.

Savage, R. J. G. *Mammal Evolution: An Illustrated Guide.* New York: Facts on File, 1986. Presents mammalian evolution in a colorful, highly illustrative manner, with frequent use of lifelike reconstructions of extinct forms going about their hypothesized daily lives. Skeletal and dental diagrams are also frequently used to explain the evolutionary adaptations and innovations of various groups. The authoritative text succeeds in clarifying the various concepts and points of evolutionary theory covered, among which are convergence and divergence. A clearly written reference work for scientifically inclined high school and college students.

Strickberger, Monroe W. *Evolution.* 3d ed. Sudbury, Mass.: Jones and Bartlett, 2000. Introductory textbook for college students. Traces current ideas about how, why, and where the world and its organisms changed throughout history. Broadly covers all evolutionary topics in the fields of anthropology, biochemistry, comparative anatomy, molecular biology, population, ecology, and population genetics.

D

Deserts & Dunes

—*M. Lee*

FIELDS OF STUDY

Conservation; Soil Science; Evolution; Ecology; Environmental Studies.

PRINCIPAL TERMS

aridisol: a type of soil that contains little or no organic matter; typically alkaline and composed mostly of salts such as chlorides, sulfates, and carbonates

deflation: the removal of particles of clay, dust, sand, and rock from dry soil by strong winds

desert pavement: a surface covered by a thin layer of small, round, stones that are closely packed together; the residual deposits left behind after the process of deflation

desert varnish: a smooth, dark, hard, shiny coating that forms on exposed rock in desert environments; usually the film is composed of manganese or iron oxides

eolian: of a deposit, arising from or carried by the action of the wind

erg: any vast, flat area of desert covered in sand; applied especially to the sand dunes of the Sahara Desert

fluvial: of a deposit, arising from, carried by, or occurring in a river

playa: a flat, dry lake-bed usually characterized by clay or salt deposits, where rainwater accumulates temporarily but eventually evaporates or moves underground; also called a pan or a dry lake

rain shadow: a dry region that forms on the side of a mountainous area protected from the prevailing winds and, therefore, the moisture they carry

saltation: a bouncing, jumping motion in which small, hard particles such as sand move over an uneven surface as a result of the flow of air or water

slip face: the sloping side of a dune that faces away from the prevailing wind; also called lee slope

wash: a narrow channel or streambed that is dry through most of the year but that can become partly or totally submerged during flash floods that occur during sudden heavy rains; also called wadi

SUMMARY

Deserts are arid or semiarid regions, defined as areas that receive an extremely small quantity of precipitation each year. Deserts and semideserts cover approximately one-third of the earth's total surface area and are found on every continent. Deserts and semideserts are characterized by unique geological and ecological features that are not present in more humid environments. These features include dunes, which are landforms of loose, shifting mounds or ridges made of sand or other sedimentation.

DESERT CHARACTERISTICS

Although the word "desert" immediately conjures an unmistakable image of a hot, sandy expanse, starkly empty of life, a great deal of variation makes up desert environments. While many are hot, some are freezing cold. Some border the ocean and others have annual rainy seasons. Three-fourths of desert surfaces are covered not by sand but rather by some other material, such as clay, rock, ice, or aridisol, a type of soil that contains little or no organic matter. (Aridisols are typically alkaline and composed mostly of salts such as chlorides, sulfates, and carbonates.)

Multiple definitions of deserts and systems for classifying them exist. One of the most widely used desert classification systems was developed by American

Erg Chebbi, Morocco, Image by Rosino.

scholar and field geographer Peveril Meigs, in the 1950's. Meigs sorted the earth's dry regions based on the amount of precipitation each receives in a year. Regions with twelve or more months in a row with no rainfall are known as extremely arid; regions that receive less than 250 millimeters (10 inches), of rainfall per year are known as arid; regions that receive between 250 and 500 millimeters (10–20 inches) of rainfall per year are known as semiarid. Only extremely arid and arid regions qualify as true deserts; semiarid regions, which are often unforested grasslands that border deserts, are sometimes known as steppes. Another way of defining a desert draws on a system developed by Russian-born German climatologist Wladimir Köppen, who proposed that a desert be identified as a region that loses more water each year through evaporation than it gains through precipitation.

Many deserts, despite being arid most of the year, do have some surface water. Ponds, lakes, and streams may form temporarily during brief heavy rainstorms, causing flash floods. Some deserts, however, are also crossed by permanent bodies of water, such as Egypt's Nile River or the Rio Grande in the Chihuahuan Desert; both of these rivers have their sources in high mountainous areas and traverse the desert on their way to the coast. In other deserts, subsurface water exists in the form of underground aquifers, or reservoirs of water, that sometimes rise to the surface and create oases, or isolated regions of fertile land.

MAJOR DESERTS OF THE WORLD

The majority of the world's desert regions are located within two bands, or belts, each of which is within 25 degrees of the equator. In these belts, dry trade winds cause cloud cover to disperse, letting sunlight quickly heat the earth's surface. The Sahara Desert in North Africa is an example of a trade wind desert. The Sahara, which stretches from the Atlantic Ocean to the Red Sea, is the largest hot desert in the world. The central Sahara region, which occupies an area of about 9.1 million square kilometers (sq km), or about 3.5 million square miles, is the driest; its regions to the north and south receive more precipitation and are more vegetated. The central Sahara is recorded as receiving less than 25 millimeters (1 inch) of rainfall per year, although precipitation patterns are extremely erratic there. There may be some years when this region receives no rainfall at all, but it is also possible that brief, intense thunderstorms may occur.

Much of the central Sahara is covered in ergs, or vast, flat, sandy areas. However, the region also is marked by a variety of surface landforms, including large dunes (such as Chech dune and Raoui dune); flat, high, stone plateaus; wide plains covered in gravel; dry riverbeds or wadis; low-lying salt flats; and the occasional oasis. A few mountain ranges, such as the Ahaggar (also known as the Hoggar) and the Tibesti mountains, also occur in this area. The Sahara is an extremely windy place, where temperatures of higher than 50 degrees Celsius (122 degrees Fahrenheit) have been recorded; winter and night-time temperatures, though, may dip to below freezing.

The equatorial belt region also is home to a type of desert known as a rain shadow desert. These are dry areas that form between two mountain ranges that each block moist air currents from being blown toward the valley they enclose. The Gobi Desert is a rain shadow desert kept arid by the vast mountain ranges of the Himalayas; it spans northern China and southern Mongolia (*gobi* is the Mongolian word for "desert") and is bounded by, among other mountains, the Altai and Hangayn (or Khangai) to the north and the Yin and Qilian to the south.

Much of the Gobi, which occupies an area of about 1.3 million square kilometers (about 500,000 square miles), is not covered in sand, but dry sedimentary rock such as chalk and gypsum. These rock plains were formed 65 million or more years ago, and the Gobi has been the source of dinosaur and other fossils and of the remains of Stone Age societies. The climate in this region is marked by bitterly cold winters and warm summers. The total yearly rainfall in the Gobi ranges from about 200 millimeters (about 8 inches) in the northeast, which receives summer monsoons, to less than 50 millimeters (about 2 inches) in the extremely arid west.

Not all of the world's deserts, however, have formed near the equator. In cold environments like at the earth's poles, the air can no longer hold much moisture and there is little or no precipitation; any water that exists on the surface is not found in rivers or lakes, but is locked in the form of ice. The inland plateau of the Antarctic receives little snow, and it is considered the world's largest desert at more than 14.2 million square kilometers (about 5.5 million square miles). Although the rest of the continent is marked by such varied surface features as mountains, glaciers, and coral reefs, the Antarctic desert is relatively featureless. The snow is packed into flat, firm plains marked only by small dune-like formations known as sastrugi (or zastrugi).

DUNE FORMATION PROCESSES

For a dune to form, a handful of key environmental factors must first be met. First, an abundant supply of dry sediment, usually sand, must exist. Second, there must be a way to transport that material, such as winds or moving water. These winds or water must be moving quickly enough to move the sediment, but not so powerfully as to destroy the mounds that are formed. Finally, there must be a place for the accumulated sediment to be deposited. The sand that makes up a dune may come from the eolian processes that erode mineral particles from the surface of rocks, or may be composed of biological materials like the calcium-carbonate-rich crumbled shells or skeletons of marine organisms.

In a dune formed by eolian, or wind-driven, processes, sand particles may be transported in several different ways. Extremely tiny or light particles may be held in place vertically while they move long distances horizontally; their weight is fully supported by upward currents of air, while turbulent eddies of wind push them forward. This process is known as suspension. In saltation, particles move downwind in a continuous series of hops, skips, and jumps. Each time a particle jumps, it moves a short distance only, but as it moves it hits other particles and triggers them to hop forward as well. In the process of creep, saltating particles that hit large, heavy grains may push those grains slowly downwind.

All of these processes cause sand particles to build up, though they do not necessarily create dunes. They may form sand sheets, or flat sandy plots of land, or ripples, which are low crests and troughs on the surface of a sandy surface (with the biggest particles at the crests). Ripples can grow into dunes under the continued action of the wind, which keeps moving particles up to the crest of the pile by saltation and creep, until the slip face becomes steep enough to collapse in a tiny avalanche. This collapsing sand becomes stabilized at a particular angle, known as the angle of repose. As this cycle of build up and collapse repeats, the dune inches forward in the direction of the wind.

The process by which underwater or riverbed dunes are formed through fluvial, or water-driven, processes, is similar. However, fluvial processes can carry larger particles in suspension than can be carried by eolian processes. In addition, saltation is less common underwater than it is in air, and suspension is more common. Finally, some particles are carried by rivers as dissolved sediments, which later recrystallize.

TYPES OF DUNES

A minimum of five primary types of dunes—crescentic, linear, dome, parabolic, and star—have been identified by geologists, although many additional subtypes and variations exist. The basic category a dune falls into depends on its geometric shape, which is determined by factors such as wind direction and strength, sediment type and quantity, and the presence or absence of vegetation.

Crescentic dunes, as their name implies, roughly resemble a crescent moon in shape. Narrow curved tips form at the edges of the dune, pointing downwind; the dune's concave slip face (a slip face is a slope on the opposite side of the prevailing wind) is between these tips. Crescentic dunes are also known as barchans and are the most common dune shape on Earth. Typically, they form in areas where the land is flat and, the wind blows consistently from a single direction. These fast-moving dunes have been known to travel as quickly as 100 meters (about 328 feet) per year across a desert.

Sand ridges that are either long and straight or sinuous—curving back and forth—are called linear, or longitudinal, dunes. Particularly, sinuous dunes are sometimes known as seifs, from the Arabic word for "sword." (Ancient Eastern swords had curving blades.) Linear dunes often form when there are winds that blow from two directions; each wind forms ridges that are parallel to the direction in which it blows. Rare dome dunes are roughly oval or circular in shape and do not have a clear slip face; parabolic dunes are shaped like the letter *U* and often form in deserts that are near a coast. The long arms of such dunes point upwind and are stabilized by the presence of vegetation growing in the sand, while the convex curve of a parabolic dune is pushed forward by the wind. Finally, star dunes are composed of a tall central mound from which several long arms—a minimum of three—radiate outward. Each arm has its own slip face and is formed by a wind that blows in a different direction. Star dunes are common in certain parts of the Sahara Desert.

Any of these types of dunes may be further described as either simple, compound, or complex. Simple dunes have the minimum number of slip faces required to fall into a particular category; they usually form when the prevailing wind or winds remain relatively stable in strength and direction. Compound dunes, which tend to be large, comprise two or more smaller dunes with the same geometric shape, forming one atop the next. Complex dunes are similar to compound dunes, but they are formed of multiple smaller dunes with different geometric shapes. For complex dunes to form, the strength and direction of the wind or winds must change after the formation of the first dune.

DESERTIFICATION

Desertification is defined as the degradation of formerly fertile land—that is, the decrease in the ability of what was once productive land to support the continued growth of vegetation. Desertification can be attributed to a variety of complex environmental factors, including the loss of nutrients and moisture in the soil, increased salinization of soil, and an increase in wind and water erosion. In the course of Earth's history, desert regions have formed, enlarged, and shrunk in size because of natural processes that had nothing to do with human activities. Desertification, however, is accelerating because of anthropogenic causes.

An estimated 70 percent of the world's dryland regions are experiencing either moderate or severe desertification. This phenomenon is caused by droughts and changes in temperature, cloud cover, and wind patterns from global warming, overgrazing, deforestation, irrigation, and other effects of unsustainable human land use. For example, livestock grazing on already semiarid land can remove vegetation that would otherwise prevent the erosion of soil by wind or water, and repeated cultivation of crops on the same land may deplete the supply of nutrients contained in the soil or make it harder for it to absorb water. Proximity to an existing desert is not necessary for desertification to occur.

Further Reading

Abraham, Athol D., and Anthony J. Parsons. *Geomorphology of Desert Environments.* New York: Springer, 2009. An updated and comprehensive reference guide to desert geomorphology research, written by experts in the field. Organized by theme, including "weathering," "hillslopes," and "piedmonts," the book contains both color and black-and-white photographs and figures.

Aleshire, Peter. *Deserts.* New York: Chelsea House, 2008. Attentive to points of interest for the nonacademic. Desert legends, mysteries, and anecdotes are woven in, though they do not overtake the text. Organized by geographic region and broken into small readable sections. Suitable for high school students.

Doran, Peter T., W. Berry Lyons, and Diane M. McKnight, eds. *Life in Antarctic Deserts and Other Cold, Dry Environments.* New York: Oxford University Press, 2010. A fascinating look at how geological research in the freezing, arid regions of Antarctica provides scientists with lessons they can apply in their attempts to find life on other planets in the galaxy. Suitable for advanced undergraduates and graduate students.

Laity, Julie. *Deserts and Desert Environments.* Hoboken, N.J.: Wiley-Blackwell, 2008. This comprehensive, well-organized, well-illustrated, and technically accessible introduction to drylands and their formation, features, and flora and fauna makes an excellent first resource for undergraduates new to the study of deserts, sedimentation, and geomorphology.

Mather, Anne. "Arid Environments." In *Environmental Sedimentology,* edited by Chris Perry and Kevin Taylor. Malden, Mass.: Blackwell, 2007. This volume contains a map and figure-rich chapter on desert environments that explores the role of sediment transfer in the formation and transformation of dryland features, particularly as they are affected by environmental law and resource management controls.

Maun, Anwar M. *The Biology of Coastal Sand Dunes.* New York: Oxford University Press, 2009. Best suited to undergraduates with a working knowledge of ecology and conservation biology, this technical but not opaque textbook emphasizes resource management issues and outlines practical tools students can use to conduct their own field research projects.

Pye, Kenneth, and Haim Tsoar. *Aeolian Sand and Sand Dunes.* New York: Springer, 2008. Almost encyclopedic in scope, this review of research literature in eolian processes and sediment properties may intimidate the beginning student and is best used as a reference volume for those already familiar with sedimentology and related geological concepts.

Verstraete, Michel M., Robert J. Scholes, and Mark Stafford Smith. "Climate Change and Desertification: Looking at an Old Problem Through New Lenses." *Frontiers in Ecology and the Environment* 7, no. 8 (October 2009): 421–428. The authors suggest a new approach to combating desertification based on a holistic understanding of its causes, both intrinsic—those that arise from changes in the physical environment, climate, and ecosystems—and extrinsic—those relating to the behavior of local social systems and the global community.

DROUGHT

—*Robert M. Hordon*

FIELDS OF STUDY

Conservation; Climatology; Biology; Evolution; Ecology; Environmental Studies. Climatology; Meteorology; Environmental Sciences; Environmental Studies; Climate Modeling; Physical Geography; Hydrometeorology; Hydroclimatology; Atmospheric Science; Climate Zones; Climate Classification; Hydrology; Earth System Modeling.

PRINCIPAL TERMS

adiabatic: a change of temperature within the atmosphere that is caused by compression or expansion without transfer of heat into or out of the system

desertification: the relatively slow, natural conversion of fertile land into arid land or desert

evapotranspiration: the combined water loss to the atmosphere from both evaporation and plant transpiration

Palmer Drought Index: a widely adopted quantitative measure of drought severity that was developed by W. C. Palmer in 1965

potential evapotranspiration: the water needed for growing plants, accounting for water loss by evaporation and transpiration

precipitation: any form of liquid water or ice that falls from the atmosphere to the ground

Sahel: the semiarid southern fringe of the Sahara in West Africa that extends from Mauritania on the Atlantic coast to Chad in the interior

soil moisture: water that is held in the soil and that is therefore available to plant roots

subsidence: in meteorology, the slow descent of air that becomes increasingly dry in the process, usually due to an area of high pressure

SUMMARY

Drought is an unusually long period of below-normal precipitation. It is a relative rather than an absolute condition, but the end result is a shortage of water for plant growth, affecting the people who live in that region and beyond. Drought is particularly disastrous for farmers and the practice of agriculture.

IMPACT OF DROUGHT

Droughts have had enormous impacts on human societies since ancient times. The most obvious effect is crop and livestock failure, which have caused famine and death through thousands of years of human history. Drought has resulted in the demise of some ancient civilizations and, in some instances, the forced mass migration of large numbers of people. Water is so critical to all forms of life that a pronounced shortage could, and has, decimated whole populations.

The effects of drought are profound. The dry conditions in the Great Plains of North America in the early 1930's in conjunction with extensive and improper farming activities resulted in the creation of the Dust Bowl, which at one point covered more than 200,000 square kilometers, or an area about the size of Nebraska. During the early 1960's, a severe drought affected the Mid-Atlantic states. Parts of New Jersey experienced sixty consecutive months of below-normal precipitation, so depleting local water supplies that plans were actively considered to bring rail cars of water into Newark and other cities in the northern part of the state, as the reservoirs that usually supplied the region were practically dry. The Sahel region south of the Sahara in West Africa had a severe drought beginning in the late 1960's and continuing into the early 1970's, creating an enormous negative impact on the local population, livestock, and vegetation. Hundreds of thousands of people starved, thousands of animals died, and many tribes were forced to migrate south to areas of more reliable precipitation. Large areas of eastern Africa in the Sudans and Ethiopia continue to experience an on-going state of drought.

DROUGHT CHARACTERISTICS

Almost all droughts occur when slow-moving air masses that are characterized by subsiding air movements dominate an area. Often, the air comes from continental interiors where the amount of moisture that is available for evaporation into the atmosphere is very limited. When these conditions occur, the potential for precipitation is low for a number of reasons. First, the humidity in the air is already low, as the continental air mass is distant from maritime (moist) influences. Second, air that subsides undergoes adiabatic heating at the rate of 10°C per 1,000 meters.

The term "adiabatic" refers to a change of temperature within a gas (such as the atmosphere) that occurs as a result of compression (descending air) or expansion (rising air), without any input or extraction of heat from external sources. For example, assume that air at a temperature of 0°C is passing over the Sierra Nevada in eastern California at an elevation of 3,500 meters. As the air descends and reaches Reno, Nevada, at an elevation of 1,200 meters, the higher atmospheric pressure found at lower elevations results in compression and heating at the dry adiabatic rate of 10°C per 1,000 meters, yielding a temperature in the Reno area of 23°C. Thus, adiabatic heating from subsiding air masses results in a decline in relative humidity and an increase in moisture-holding capacity. In addition, the movement of air under these conditions is usually unfavorable for vertical uplift and the beginning of the condensation process. The final factor that reduces precipitation potential is the decrease in cloudiness and corresponding increase in sunshine, which in turn leads to an increase in evapotranspiration demands that favor soil moisture loss.

Another characteristic associated with droughts is that once established, they appear to persist and even expand into nearby regions, resulting in desertification in extreme cases. This tendency is apparently related to positive feedback mechanisms. For example, the drying out of the soil influences air circulation and the amount of moisture that is then available for precipitation farther downwind. At the same time, the atmospheric interactions that lead to unusual wind systems associated with droughts can induce surface-temperature variations that, in turn, lead to further development of the unusual circulation pattern. Thus, the process builds on itself, causing the drought to both last longer and intensify. The situation persists until a major change occurs in the circulation pattern in the atmosphere.

Many climatologists concur with the concept that precipitation is not the only factor associated with drought. Other factors that demand consideration include moisture supply, the amount of water in storage, and the demand generated by evapotranspiration. Although the scientific literature is replete with information about the intensity, length, and environmental impacts of drought events, the role of individual climatological factors that can increase or decrease the severity of a drought is not fully understood.

DROUGHT IDENTIFICATION

Research in drought identification has been changing over the years. Drought was once considered solely in terms of precipitation deficit. Although that lack of precipitation is still a key atmospheric component of drought, sophisticated techniques are now used to assess the deviation from normal levels of the total environmental moisture status. These techniques have enabled investigators to better understand the severity and length of drought events, as well as the extent of the affected area.

Drought has been defined in a number of ways. Some authorities consider it to be merely a period of below-normal precipitation, while others relate it to the likelihood of forest fires. Drought is also said to occur when the yield from a specific agricultural crop or pasture is significantly less than expected. It has also been defined as a period when soil moisture or groundwater decreases to a critical level.

Drought was identified early in the twentieth century by the U.S. Weather Bureau as any period of twenty-one or more days when precipitation was 30 percent or more below normal. Subsequent examination of drought events that were identified by this method revealed that soil moisture reserves were often elevated during these events to the extent that there was sufficient water to support vegetation. It was also determined that the amount of precipitation preceding the drought event was ample or even heavy. Thus, it became apparent that precipitation should not be used as the sole measure to identify drought. Subsequent research has shown that the moisture status of an area is affected by additional factors.

Further developments in drought identification during the middle decades of the twentieth century began to focus on the moisture demands that are associated with evapotranspiration in an area. Evaporation is primarily the process by which liquid water becomes water vapor at the surface and enters the atmosphere. To a lesser extent, this also includes the conversion of "solid water," as ice and snow, either directly by sublimation or through first melting into an intermediate liquid state. Transpiration refers to the loss of moisture by plants to the atmosphere. Although evaporation and transpiration can be studied and measured separately, it is convenient to consider them in applied climatological studies as the single process of evapotranspiration.

There are two ways to define evapotranspiration. The first is actual evapotranspiration, which is the actual or real rate of water-vapor return to the atmosphere from the earth and vegetation; this process could also be called "water use." The second is potential evapotranspiration, which is the theoretical rate of water loss to the atmosphere if one assumes continuous plant cover and an unlimited supply of water. This process could also be called "water need," as it indicates the amount of soil water needed if plant growth is to be maximized. Procedures have been developed that enable one to calculate the potential evapotranspiration for any area from monthly mean temperature and precipitation values.

Some drought-identification studies have focused on agricultural drought, looking at the adequacy of soil moisture in the root zone for plant growth. This procedure involved the evaluation of precipitation, evapotranspiration, available soil moisture, and the water needs of plants. The goal of this research was to determine drought probability based on the number of days when soil moisture storage is reduced to zero.

Evapotranspiration was also used by the Forest Service of the U.S. Department of Agriculture when it developed a drought index to be used by fire-control managers. The purpose of the index was to provide a measure of flammability that could create forest fires. This index has limited applicability to nonforestry users, as it is not effective for showing drought as an indication of total environmental stress.

PALMER DROUGHT INDEX

One of the most widely adopted drought-identification techniques was developed by W. C. Palmer in 1965. The method, which became known as the Palmer Drought Index, defines drought as the period of time, usually measured in months or years, when the actual moisture supply at a given location is consistently less than the climatically anticipated or appropriate supply of moisture. The calculation of this index requires the determination of evapotranspiration, soil moisture loss, soil moisture recharge, surface runoff, and precipitation. The Palmer

Drought Index values range from approximately +4.0 for an extremely wet moisture status class to –4.0 for extreme drought. Normal conditions have a value close to 0. Positive values indicate varying stages of abundant moisture, whereas negative values indicate varying stages of drought.

Although the Palmer Drought Index is recognized as an acceptable procedure for incorporating the role of potential evapotranspiration and soil moisture in magnifying or alleviating drought status, there have been some criticisms of its use. For example, the method produces a dimensionless parameter of drought status that cannot be directly compared with other environmental moisture variables, such as precipitation, which are measured in units (centimeters, millimeters) that are immediately recognizable. In addition, the index is not especially sensitive to short drought periods, which can affect agricultural productivity.

In order to address these shortcomings, other researchers use water-budget analysis to identify deviations in environmental moisture status. The procedure is similar to the Palmer method inasmuch as it incorporates the environmental parameters of precipitation, potential evapotranspiration, and soil moisture. However, the moisture status departure values are expressed in the same units as precipitation and are therefore dimensional. Drought classification using this index method ranges from approximately 25 millimeters for an above-normal moisture status class to -100 millimeters for extreme drought. The index would be close to 0 for normal conditions.

SIGNIFICANCE

Drought is invariably associated with some form of water shortage, yet many regions of the world have regularly occurring periods of dryness. Three different forms of dryness have a temporal dimension described as perennial, seasonal, and intermittent. Perennially dry areas include the major deserts of the world, such as the Sahara, Arabian, and Kalahari. Precipitation in these areas is not only very low but also very erratic. Seasonal dryness is associated with regions where the bulk of the annual precipitation comes during a few months of the year, leaving the rest of the year without rain or other precipitation. Intermittent dryness is associated with those instances where the overall precipitation is reduced in humid regions or where the rainy season in seasonally dry areas does not occur or is shortened.

The absence of precipitation when it is normally expected creates variable problems. For example, the absence of precipitation for one week in an area where daily precipitation is the norm would be considered a drought. In contrast, it would take two or more years without any rain in parts of Libya in North Africa for a drought to occur. In those areas that have one rainy season, a 50 percent reduction in precipitation would be considered a drought. In regions that have two rainy seasons, the failure of one could lead to drought conditions. Thus, the word "drought" is a relative term, as it has different meanings in different climatic regions.

User demands also influence drought definition. Distinctions are often made among climatological, agricultural, hydrologic, and socioeconomic drought. Climatological, or meteorological, drought occurs at irregular periods of time, usually lasting months or years, when the water supply in a region falls far below the levels that are typical for that particular climatic regime. The degree of dryness and the length of the dry period are used as the definition of drought. For example, drought in the United States has been defined as occurring when there is less than 2.5 millimeters of rain in a forty-eight hour period. In Great Britain, drought has been defined as occurring when there are fifteen consecutive days with less than 0.25 millimeter of rain for each day. In Bali, Indonesia, drought has been considered as occurring if there is no rain for six consecutive days.

Agricultural drought occurs when soil moisture becomes so low that plant growth is affected. Drought must be related to the water needs of the crops or animals in a particular place, since agricultural systems vary substantially. The degree of agricultural drought also depends on whether shallow-rooted or deep-rooted plants are affected. In addition, crops are more susceptible to the effects of drought at different stages of their development. For example, inadequate moisture in the subsoil in an early growth stage of a particular plant will have minimal impact on crop yield as long as there is adequate water available in the topsoil. However, if subsoil moisture deficits continue, then the yield loss could become substantial.

Hydrologic drought definitions are concerned with the effects of dry spells on surface flow and groundwater levels. The climatological factors associated with the drought are of lesser concern. Thus, a hydrological drought for a particular watershed is said to occur when the runoff falls below some arbitrary value. Hydrological droughts are often out of phase with climatological and agricultural droughts and are also basin-specific; that is, they pertain to the particular watersheds that they affect.

Socioeconomic drought includes features of climatological, agricultural, and hydrological drought and is generally associated with the supply and demand of some type of economic good. For example, the interaction between farming (demand) and naturally occurring events (supply) can result in inadequate water for both plant and animal needs. Human activities, such as poor land-use practices, can also create a drought or make an existing drought worse. The Dust Bowl in the Great Plains and the Sahelian drought in West Africa provide ready examples of the symbiotic relationship between drought and human activities.

In a sense, droughts differ from other major geophysical events such as volcanic eruptions, floods, and earthquakes because they are actually nonevents. They result from the absence of events (precipitation) that should normally occur. Droughts also differ from other geophysical events in that they often have no readily recognizable beginning and take some time to develop. In many instances, droughts are only recognized when plants start to wilt, wells and streams run dry, and reservoir shorelines recede.

There is wide variation in the duration and extent of droughts. The length of a drought cannot be predicted, as the irregular patterns of atmospheric circulation are not fully known and remain unpredictable. A drought ends when the area receives sufficient precipitation and water levels rise in the wells and streams. Because the severity and areal extent of a drought cannot be predicted, all that is really known is that they are an integral part of the overall natural system and that they will continue to occur.

FURTHER READING

Ahrens, C. Donald. *Meteorology Today: An Introduction to Weather, Climate and the Environment.* 8th ed. Thomson Brooks-Cole, 2007. One of the most widely used and authoritative introductory textbooks for the study of meteorology and climatology. Explains complex concepts in a clear, precise manner. Provides numerous images and diagrams.

Bryson, Reid A., and Thomas J. Murray. *Climates of Hunger.* University of Wisconsin Press, 1977. An interesting account of the profound effect of climate on human societies going back to ancient times. Discusses major climate changes and droughts for various regions in the world in separate chapters. Treatment is nonmathematical and suitable for senior-level high school students and above.

Climate, Drought, and Desertification. World Meteorological Organization, 1997. Deals with the climatic factors, such as drought, that lead to desertification worldwide. Includes color illustrations.

Dixon, Lloyd S., Nancy Y. Moore, and Ellen M. Pint. *Drought Management Policies and Economic Effects in Urban Areas of California, 1987–92.* RAND, 1996. Examines the impacts of the 1987–1992 drought in California on urban and agricultural users. Assesses the effects of the drought on a broad range of residential, commercial, industrial, and agricultural water users.

Fisher, R. J. *If Rain Doesn't Come: An Anthropological Study of Drought and Human Ecology in Western Rajasthan.* Manohar, 1997. Focuses on the human ecology associated with India and other countries prone to drought. Examines the factors involved with drought and its durations, as well as drought relief tactics and evaluation of such programs. Illustrations, maps, and references.

Frederiksen, Harald D. *Drought Planning and Water Resources Implications in Water Resources Management.* World Bank, 1992. Contains two papers on drought planning and water-use efficiency and effectiveness. Deals with policy and program issues in water-resources management from the perspective of the World Bank.

Hidore, John, John E. Oliver, Mary Snow, and Richard Snow. *Climatology.* 3d ed. Merrill, 1984. Discusses all aspects of climatology. Contains numerous black-and-white illustrations and maps. Suitable for college-level students. Although quantitative measures are included, no particular mathematical background is necessary.

Knight, Gregory, Ivan Raev, and Marieta Staneva, eds. *Drought in Bulgaria.* Ashgate Publishing Limited, 2004. Presents the occurrence of drought in Bulgaria as a case study to provide information on prospective future global conditions. Discusses the geological, economic, and ecological effects of drought in Bulgaria from 1982 to 1994. Concludes with policy and conservation recommendations for the future.

Mainguet, Monique. *Aridity: Drought and Human Development.* Springer-Verlag, 2010. Discusses global and local aridity, drought, and associated changes in vegetation and hydrology. The second half of the text focuses on anthropogenic impact. Includes references and multiple indexes.

Lutgens, Frederick K., Edward J. Tarbuck, and Dennis Tasa. *The Atmosphere: An Introduction to Meteorology.* 11th ed. Prentice Hall, 2009. An excellent introduction and description of the atmosphere, meteorology, and weather patterns. Suitable for readers new to the study of these subjects. Color illustrations and maps.

Wilhite, Donald A., ed. *Drought and Water Crises: Science, Technology, and Management Issues.* CRC Press, 2005. Covers water management practices currently used in areas around the world. Discusses monitoring, drought planning, and water conservation policies. Provides case studies that discuss many issues of and solutions to drought. Concludes with a chapter discussing the future of water conservation.

Wilhite, Donald A., and William E. Easterling, with Deborah A. Wood, eds. *Planning for Drought: Toward a Reduction of Societal Vulnerability.* Westview Press, 1987. An extensive collection of thirty-seven short chapters on drought covering a wide range of topics from the climatological to the institutional. Provides a good background to the many issues pertaining to drought, its social impacts, governmental response, and adaptation and adjustment. Suitable for college-level students.

Workman, James G. *Heart of Dryness: How the Last Bushmen Can Help Us Endure the Coming Age of Permanent Drought.* Walker Publishing Company, 2009. Introduces the reader to the occurrence of drought and its impact on humans, animals, vegetation, and land. Describes in detail the issues faced and created by humans in desiccated regions through the eyes of the people facing it presently.

E

Ecological Niches

—Jessica O. Ellison

FIELDS OF STUDY

Ecology; Environmental Studies; Agronomy.

PRINCIPAL TERMS

community: all the populations of plant and animal species living and interacting in a given habitat or area at a given time

environment: all the external conditions that affect an organism or other specified system during its lifetime

food pyramid: diagram representing organisms of a particular type that can be supported at each trophic level from a given input of solar energy in food chains and food webs

habitat: place or type of place where an organism or community of organisms naturally thrives

organism: any form of life

trophic level: a level in a food chain or food web at which all organisms consume the same general types of food

SUMMARY

An ecological niche is the physical space in which an animal lives and all the interactions with the other living organisms and components of its environment. Ecological niches are important in the preservation of organisms that are vulnerable to certain activities of human beings.

The idea of the niche probably had its first roots in ecology in 1910. At that time, Roswell Johnson wrote that different species utilize different niches in the environment. He theorized that individuals of a particular species are only in certain places because of food supply and environmental factors that limit their distribution in an area. Later, in 1924, Joseph Grinnel developed his concept of niche that centered on an organism's distribution having limits set on it by climatic and physical barriers. At the same time, Charles Elton was defining his own idea of niche. His description of niche involved the way an organism makes its living—in particular, how it gathers food.

TROPHIC LEVELS

For many years, ecologists focused on Elton's definition and referred to niche in terms of an organism's place in the food pyramid. The food pyramid is a simplified scheme in which organisms interact with one another while obtaining food. The food pyramid is represented as a triangle, often with four horizontal divisions, each division being a different trophic level.

The base of the food pyramid is the first trophic level and contains the primary producers: photosynthetic plants. At the second trophic level are the primary consumers. These are the herbivores, such as deer and rabbits, which feed directly on the primary producers. Secondary consumers are found at the third trophic level. This third trophic level contains carnivores, such as the mountain lion. The members of the uppermost trophic level are the scavengers and decomposers, including hyenas, buzzards, fungi, and bacteria. The organisms in this trophic level break down all the nutrients (such as carbon and nitrogen) in the bodies of plants and animals and return them to the soil to be absorbed and used by plants.

It should be noted that no ecosystem actually has a simple and well-defined food pyramid. Many organisms interact with more organisms than those at the adjacent trophic levels. For example, a coyote could be considered to belong to the third trophic level with the carnivores, but the coyote also occasionally feeds on fruits and other primary producers. Basically, all living things are dependent on the first trophic level, because it alone has the capability to convert solar energy to energy found in, for example, glucose and starch. The food pyramid takes the geometric form of a triangle to show the flow of energy through a system.

Photosynthetic plants lose 10 percent of the energy they absorb from the sun as they convert solar energy into glucose and starch. In turn, the herbivores can convert and use only 90 percent of the energy they obtain by eating plants. Hence, less energy is found at each higher trophic level. Because of this reduction in energy, fewer organisms can be supported by each higher trophic level. Consequently, the sections of the pyramid get smaller at each higher trophic level, representing the decreasing levels of energy and number of members.

Lichenes on a rock

INTERRELATIONSHIPS AMONG ORGANISMS

Through the years, two concepts of niche have evolved in ecology. The first is the place niche, the physical space in which an organism lives. The second is the ecological niche, and it encompasses the particular location occupied by an organism and its functional role in the community.

The functional role of a species is not limited to its placement along a food pyramid; it also includes the interactions of a species with other organisms while obtaining food. For example, the methods used to tolerate the physical factors of its environment, such as climate, water, nutrients, soils, parasites, and the like, are all part of its functional role. In other words, the ecological niche of an organism is its natural history: all the interactions and interrelationships of the species with other organisms and the environment.

The study of the interrelationships among organisms has been the focus of ecological studies since the 1960's. Before this time, researchers had focused on the food pyramid and its effect on population changes of merely a single species. One example, the classic population study of the lynx and the snowshoe hare of Canada, originally focused on the interactions of the species in the food pyramid. It was discovered that the lynx had a ten-year population cycle closely following the population cycle of its prey, the snowshoe hare. The lynx population appeared to rise, causing a decline in the population of the snowshoe hare. In the investigations that followed, however, studies diverted the focus from the food pyramid to other elements of the niche of the two species. For example, the reproductive nature of the hare provided a contradiction to the simple predator-prey explanation. The hare has a faster rate of reproduction than the lynx. It seemed impossible that the significantly lower population of lynx could effectively place sufficient predator pressure on the hare to cause its drastic decline in numbers. Therefore, it appeared that the population dynamics of the hare and lynx was regulated by more than simply a predator-prey relationship.

Later studies of the lynx and hare suggested that the peaks and dives in the two populations may also be a factor of parasites of the hare that are carried by the lynx. A rise in the lynx population increases the carriers of parasites of the hare. Therefore, it is thought that, although the hare has a much greater reproduction rate than the lynx, the population of hares will still decline because of the combination of predation by the lynx and the increased frequency of parasites of the hare. This study involved looking at more than one dimension of the ecological niche of a species and broke away from concentrating on only the interactions between organisms in the food pyramid.

NICHE OVERLAP

The goal of understanding how species interact with one another can also be better accomplished by defining the degree of niche overlap, the degree of the sharing of resources between two species. When two species use one or more of the same elements of an ecological niche, they exhibit interspecific competition. It was once believed that interspecific competition would always lead to survival of only the better competitor of the two species. That was the original concept of the principle of the competition exclusion law of ecology: No two species can utilize the same ecological niche. It was conjectured that the weaker competitor would either migrate, begin using another resource not used by the stronger competitor, or become extinct. It is now believed that the end result of two species sharing elements of ecological niches may not always be exclusion.

Ecologists theorize that similar species do, in fact, coexist, despite the sharing of elements of their ecological niches, because of character displacement, which leads to a decrease in niche overlap. Character displacement involves a change in the morphological, behavioral, or physiological state of a species without geographical isolation. Character displacement occurs as a result of natural selection arising from competition between one or more ecologically similar species. Examples might be changes in mouth sizes so that they begin to feed on different sizes of the same food type, thereby decreasing competition.

SPECIES SPECIALIZATION

The more specialized a species, the more rigid it will be in terms of its ecological niche. A species that is general in terms of its ecological niche needs will be better able to find and use an alternative for the common element of the niche. Since a highly specialized species cannot substitute whatever is being used, it cannot compete as well as the other species. Therefore, a specialized species is more likely to become extinct.

For example, a panda is a very specialized feeder, eating mainly bamboo. If a pest is introduced into the environment that destroys bamboo, the panda will probably starve, being unable to switch to another food source. On the other hand, the coyote is a generalized feeder. It has a broad variety of food types that make up its diet. If humans initiate a pest-control program, killing the population of rabbits, the coyote will not fall victim to starvation, because it can switch to feeding predominantly on rodents, insects, fruits, and domesticated animals (including cats, dogs, and chickens). Hence, species with specialized ecological niche demands (specialists) are in greater danger of extinction than those with generalized needs (generalists). Although this fundamental difference in survival can be seen between specialists and generalists, it must be noted again that exclusion is not an inevitable result of competition. Many cases of ecologically similar species coexist.

When individuals of the same species compete for the same elements of the ecological niche, it is referred to as intraspecific competition. Intraspecific competition has the opposite effect of interspecific competition: niche generalizations. In increasing populations, the first inhabitants will have access to optimal resources. The opportunity for optimal resources decreases as the population increases; hence, intraspecific competition increases. Deviant individuals using marginal resources may slowly begin to use less optimal resources that are in less demand. That can lead to an increase in the diversity of ecological niches used by the species as a whole. In other words, the species may become more generalized and exploit wider varieties of niche elements.

Representing a situation on the opposite end of the spectrum from that of two organisms competing for the same dimension of an ecological niche is the vacant niche theory. This ecological principle states that when an organism is removed from its ecological niche, space, or any other dimension of the niche, another organism of the same or similar species will reinvade.

FIELD RESEARCH

Theoretical studies of ecological niches are abstract, since humans are limited to three-dimensional diagrams, and there are more than three dimensions to an ecological niche. This multidimensionality is referred to as the *n*-dimensional niche. This abstract *n*-dimensional niche can be studied mathematically and statistically, but ecology is mainly a field science. Therefore, the focus of techniques is on those used for field research of the ecological niche.

Research that attempts to describe all the elements of the *n*-dimensional ecological niche would require extensive observations. Yet, ecological niches are difficult to measure not only because of the plethora of data that would have to be collected but also because of the element of change in nature. The internal and external environment of an organism is always dynamic. Nothing in life is static, even if equilibrium has been established.

These constant fluctuations create daily and seasonal changes in space and ecological niches. Therefore, because of the constant fluctuations, any merely descriptive field observations would not be reliable depictions of an organism's ecological niche. Ecologists must also resort to quantitative data of measurable features of an organism's ecological niche. For example, the temperature, pH, light intensity, algae makeup, predators, and activity level of the organism are measurable features of an ecological niche in a pond community. The difficulty is in the collection of each of the necessary measurements making up an ecological niche. The ecologist would have to limit the data to a manageable number of specific dimensions of the niche based on conjecture and basic intuition. Such limitations often lead to incomplete and disconnected measurements that can at best only partially describe a few of the dimensions of the ecological niche.

Ecologists realize that complete observations and measurements of all the dimensions of an organism's ecological niche are unattainable. The focus in understanding how a species interacts with its community centers on determining the degree of niche overlap between any two species—in other words, the level of competition for niche space and resources. Studies of this niche overlap are typically limited to dimensions that can be quantitatively measured. Yet, there is still the problem of deciding which of the dimensions are involved in the competition between the two species. Again, the ecologist must usually rely on inherent knowledge about the two species in question. Often, researchers investigating niche competition measure no more than four ecological niche dimensions to determine the niche overlap in an attempt to understand how two individuals competing for the same space, resources, or other ecological niche features can coexist.

Field methods for observations and quantitative measurements of elements of ecological niches, niche overlap, and niche competition are probably endless. To name a few, describing an organism's niche may involve fecal samples to determine its diet, fecal samples of possible predators to identify its primary predator, animal and plant species checklists of its space niche along with soil components, climatic trends, and the like. Niche competition and overlap often can be studied first in the laboratory under controlled situations. One method might involve recording the population dynamics of the species as different elements in the ecological niche are manipulated to determine which is the better competitor and what is the resource that is most responsible for limiting the population size.

NICHE & COMMUNITY

The shift in meaning and study from merely space and trophic level placement in the food pyramid to ecological niche of *n* dimensions has been beneficial for the field of ecology. This focus on community ecology is obviously much more productive for the goal of ecology, the understanding of how all living organisms interact with one another, and with the nonliving elements in the environment.

Perhaps more important is the attempt to describe niches in terms of community ecology, which can be essential for some of humankind's confrontations with nature. For example, it becomes more and more apparent that synthetic chemicals are often too costly and too hazardous to continue using for control of crop pests and carriers of diseases. The goal is to control pests effectively with biological controls. Biological controls can involve the introduction of natural predators of the undesirable pest or the introduction of a virus or bacteria that eliminates the pest and is harmless to humans and wildlife.

The success of a biological control is directly proportional to the knowledge of the pest's *n*-dimensional ecological niche and the other organisms with which it comes in contact. A classic example of the havoc that can result from manipulations of nature without adequate ecological information is when Hawaii attempted to use biological controls to eradicate a population of snakes, which humans had accidentally introduced. The biological control used was the snake's natural predator, the mongoose. One very important dimension of the ecological niche of both species was ignored. One species was active only at night, while the other was active only during the day. Needless to say, this particular venture with a biological control was not a success.

Another relevant function of community-oriented studies of ecological niches involves endangered species. In addition to having aesthetic and potential medicinal values, an endangered organism may be a keystone species, a species on which the entire community depends. A keystone species is so integral to keeping a community healthy and functioning that if the species is obliterated, the community no longer operates properly and is not productive.

Habitat destruction has become the commonest cause of drastic population declines of endangered species. To enhance the habitat of the endangered species, it is undeniably beneficial to know what attracts a species to its particular preferred habitat. This knowledge involves the details of many of the dimensions of its ecological niche integral to its population distribution. Another common cause of endangering the survival of a species is when an introduced organism or exotic species competes for the same resources and displaces the native species. Solving such competition between native and introduced species would first involve determining niche overlap.

It is often stated that an ounce of prevention is worth a pound of cure. Thus, the researching and understanding of all the dimensions of ecological niches are integral components of preventing environmental manipulations by humankind that might lead to species extinction. Many science authorities have agreed that future research in ecology and related fields should focus on solving three main problems: species endangerment, soil erosion, and solid waste management.

This focus on research in ecology often means that studies of pristine communities, those undisturbed, will be the most helpful for future restoration projects. Although quantitative and qualitative descriptions of pristine areas seem to be unscientific at the time they are made, because there is no control or experimental group, they are often the most helpful for later investigations. For example, after a species has shown a drastic decline in its population, the information from the observations of the once-pristine area may help to uncover what niche dimension was altered, causing the significant population decrease.

FURTHER READING

Bronmark, Christopher, and Lars-Anders Hansson. *The Biology of Lakes and Ponds.* New York: Oxford University Press, 2018. A textbook for college-level students. Chapter 4 discusses ecological niches and the relationships among organisms.

Ehrlich, Paul R. *The Machinery of Nature.* New York: Simon & Schuster, 1986. Chapter 6, "Who Lives Together, and How," cites many examples of interactions between species in terms of community ecology. Includes several photographs of animals in their habitats and a list for further reading.

Odling-Smee, F. John, Kevin N. Laland, and Marcus W. Feldman. "Niche Construction." *American Naturalist* 147, no. 4 (April, 1996): 641–649. Argues that niche construction regularly modifies both biotic and abiotic sources of natural selection.

Rayner, Alan D. M. *Degrees of Freedom: Living in Dynamic Boundaries.* River Edge, N.J.: World Scientific Publications, 1997. Rayner presents his ideas of "eco-Darwinism" to a general audience, arguing that environmental context is more important than DNA in determining the behaviors of communities from microbes to ant colonies, plant roots, and humans.

Relyea, Rick, and Robert E. Ricklefs. *Ecology: The Economy of Nature.* New York: Macmillan Education, 2018. Print.

Shugart, Herman H. *Terrestrial Ecosystems in Changing Environments.* New York: Cambridge University Press, 1998. A textbook for upper-level undergraduates and graduate students in ecosystem modeling. The book begins with a review of concepts including ecological niches, and then uses the concepts in examples of modeling.

Smith, Robert L. *Ecology and Field Biology.* 6th ed. San Francisco: Benjamin/Cummings, 2007. A text easily readable by college and senior-level high school students. It has many informative illustrations, and the extensive appendices make the book a useful source. Includes an excellent index and glossary.

Stone, Richard. "Taking a New Look at Life Through a Functional Lens." *Science* 269, no. 5222 (July, 1995): 316–318. A reassessment of the concept of keystone species.

Ecology

—*Lawrence E. Hurd*

FIELDS OF STUDY

Conservation; Biology; Evolution; Ecology; Environmental Studies.

PRINCIPAL TERMS

adaptation: a genetic (intrinsic) feature of an organism which, through natural selection, enhances its fitness in a given environment

community: all the populations that exist in a given habitat

ecosystem: the biological community in a given habitat, combined with all the physical properties of the environment in that habitat

environment: all the forces and things external to an organism that directly affect it

fitness: the contribution of an organism to future generations; the perpetuation of its genes through reproduction

habitat: the place where an organism lives; for example, a pond or forest

natural selection: a change in the genetic makeup of a population as a result of different survival and reproduction rates (fitness) among its members

niche: the role of an organism in its environment; the sum of all factors that define its existence (temperature, energy requirements, and so on)

population: all the individuals in a habitat which are of the same species

resource: a requirement for life, such as space for living, food (for animals), or light (for plants), not including conditions such as temperature or salinity

SUMMARY

Ecology is the study of how organisms relate to their natural environments. The two principal concerns of ecologists are the distribution and abundance of organisms: Why are animals, plants, and other organisms found where they are, and why are some common and others rare? These questions have their roots in the theory of evolution. In fact, it is difficult (and not often worthwhile) to separate modern ecological matters from the concerns of evolutionary biologists. Ecology can be divided according to several levels of organization: the individual organism, the population, the community, and the ecosystem.

ENVIRONMENT & NATURAL SELECTION

An ecologist views organisms as consequences of past natural selection brought about by their environments. That is, each organism represents an array of adaptations that can provide insight into the environmental pressures that resulted in its present form. Adaptations of organisms are also revealed by other features, such as the range of temperature an organism can tolerate, the amount of moisture it requires, or the variety of food it can exploit. Food and space for living are considered resources; factors such as temperature, light, and moisture are conditions which determine the rate of resource utilization. When ecologists have discovered the full range of resources and conditions necessary for an organism's existence, they have discovered its niche.

Many species, such as many insects and plants, have a large reproductive output. This compensates for high mortality imposed by natural selection. Other species, such as large mammals and birds, have fewer offspring. Many of these animals care for their young, thus increasing the chances that their offspring will survive to reproduce. These are two different strategies for success, based upon the principle that organisms have a finite energy budget. Energy acquired from food (animals) or sunlight (plants) must be partitioned among growth, maintenance, and reproduction. The greater the energy allocated to the care of offspring, for example, the fewer the offspring that can be produced.

The concept of an energy budget is a key to understanding evolutionary strategies of organisms, as well as the energetics of ecosystems. The amount of energy fixed and stored by an organism is called net production; this is the energy used for growth and reproduction. Net production is the difference between

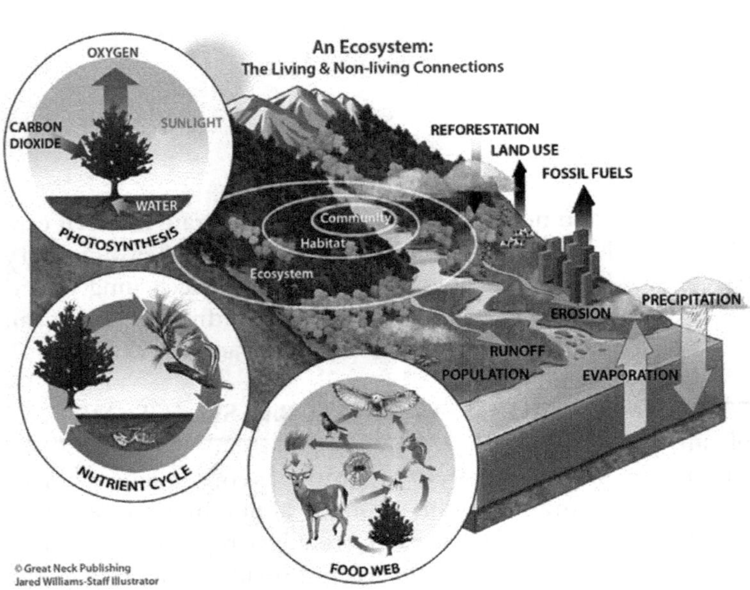

gross production (the amount of energy assimilated) and respiration (metabolic maintenance cost). The greater the respiration, the less energy will be left over for growth and reproduction. Endothermic animals, which physiologically regulate their body heat (mammals and birds), have a very high respiration rate relative to ectotherms (reptiles, amphibians, fish, and invertebrates), which cannot. Among endotherms, smaller animals have higher respiration rates than larger ones, because the ratio of body surface area (the area over which heat is exchanged with the environment) to volume (the size of the "furnace") decreases with increasing body size.

DEMOGRAPHY & POPULATION REGULATION

Although single organisms can be studied with regard to adaptations, in nature most organisms exist in populations rather than as individuals. Some organisms reproduce asexually (that is, by forming clones), so that a single individual may spawn an entire population of genetically identical individuals. Populations of sexually reproducing organisms, however, have the property of genetic variability, since not all individuals are identical. That is, members of a population have slightly different niches and will therefore not all be equally capable of living in a given environment. This is the property upon which Charles Darwin's theory of natural selection depends: Because not all individuals are identical, some will have greater fitness than others. Those with superior fitness will reproduce in greater numbers and therefore will contribute more genes to successive generations. In nature, many species consist of populations occupying more than a single habitat. This constitutes a buffer against extinction: If one habitat is destroyed, the species will not become extinct, because it exists in other habitats.

Two dynamic features of populations are growth and regulation. Growth is simply the difference between birth and death rates, which can be positive (growing), negative (declining), or zero (in equilibrium). Every species has a genetic capacity for exponential (continuously accelerating) increase, which will express itself to varying degrees depending on environmental conditions: A population in its ideal environment will express this capacity more nearly than one in a less favorable environment. The rate of growth of a population is affected by its age structure—the proportion of individuals of different ages. For example, a population which is growing rapidly will have a higher proportion of juvenile individuals than one which is growing more slowly.

Populations may be regulated (so that they have equal birth and death rates) by a number of factors, all of which are sensitive to changes in population size. A population may be regulated by competition among its members for the resource that is in shortest supply (limiting). The largest population that can be sustained by the available resources is called the carrying capacity of the environment. A population of rodents, for example, might be limited by its food supply such that as the population grows and food runs out, the reproductive rate declines. Thus, the effect of food on population growth depends upon the population size relative to the limiting resource. Similarly, parasites that cause disease spread faster in large, dense populations than in smaller, more diffuse ones. Predators can also regulate populations of their prey by responding to changes in prey availability. Climate and catastrophic events such as storms may severely affect populations, but their effect is not dependent upon density and is thus not considered regulatory.

INTERACTIONS BETWEEN SPECIES

Competition occurs between, as well as within, species. Two species are said 'to be in competition with each other if and only if they share a resource that is in short supply. If, however, they merely share a resource that is plentiful, then they are not really competing for it. Competition is thought to be a major force in determining how many species can coexist in natural communities. There are a number of alternative hypotheses, however, which involve such factors as evolutionary time, productivity (the energy base for a community), heterogeneity of the habitat, and physical harshness of the environment.

Predator-prey interactions are those in which the predator benefits from killing and consuming its prey. These differ from most parasite-host interactions in that parasites usually do not kill their hosts (a form of suicide for creatures that live inside other creatures). Similarly, most plant-eating animals (herbivores) do not kill the plants on which they feed. Many ecologists classify herbivores as parasites for this reason. There are exceptions, such as birds and rodents that eat seeds, and these can be classified as legitimate predator-prey interactions. Predators can influence the number of species in a community by affecting competition among their prey: If populations of competing species are lowered by predators so that they are below their carrying capacities, then there may be enough resources to support colonization by new species.

In many cases, the interaction between two species is mutually beneficial. Mutualism is often thought to arise as a result of closely linked evolutionary histories (coevolution) of different species. Termites harbor protozoans in their guts that produce an enzyme which can break down cellulose in wood. The protozoans thus are provided with a habitat, and termites are able to derive nourishment from wood. Some acacia trees in the tropics have hollow thorns which provide a habitat for ants. In return, the ants defend the trees from other insects which would otherwise damage or defoliate them.

Communities of organisms are composed of many populations that may interact with one another in a variety of ways: predation, competition, mutualism, parasitism, and so on. The composition of communities changes over time through the process of succession. In terrestrial communities, bare rock may be weathered and broken down by bacteria and other organisms until it becomes soil. Plants can then invade and colonize this newly formed soil, which in turn provides food and habitat for animals. The developing community goes through a series of stages, the nature of which depends on local climatic conditions, until it reaches a kind of equilibrium. In many cases this equilibrium stage, called climax, is a mature forest. Aquatic succession essentially is a process of becoming a terrestrial community. The basin of a lake, for example, will gradually be filled with silt from terrestrial runoff and accumulated dead organic material from populations of organisms within the lake itself.

ECOSYSTEMS

Ecosystems consist of several trophic levels, or levels at which energy is acquired: primary producers, consumers, and decomposers. Primary producers are green plants that capture solar energy and transform it, through the process of photosynthesis, into chemical energy. Organisms that eat plants (herbivores) or animals (carnivores) to obtain their energy are collectively called consumers. Decomposers are those consumers, such as bacteria and fungi, that obtain energy by breaking down dead bodies of plants and animals. These trophic levels are linked together into a structure called a food web, in which energy is transferred from primary producers to consumers and decomposers, until finally all is lost as heat. Each transfer of energy entails a loss (as heat) of at least 90 percent, which means that the total amount of energy available to carnivores in an ecosystem is substantially less than that available to herbivores.

As with individual organisms, ecosystems and their trophic levels have energy budgets. The net production of one trophic level is available to the next-higher trophic level as biomass (mass of biological material). Plants have higher net productivity (rates of production) than animals because their metabolic maintenance cost is lower relative to gross productivity; herbivores often have higher net productivity than predators for the same reason. For the community as a whole, net productivity is highest during early successional stages, since biomass is being added more rapidly than later on, when the community is closer to climax equilibrium.

In contrast to the unidirectional flow of energy, materials are conserved and recycled from dead organisms by decomposers to support productivity at higher trophic levels. Carbon, water, and mineral nutrients required for plant growth are cycled through various organisms within an ecosystem. Materials and energy are also exchanged among ecosystems: There is no such thing in nature as a "closed" ecosystem that is entirely self-contained.

DESCRIPTIVE, EXPERIMENTAL & MATHEMATICAL ECOLOGY

The science of ecology is necessarily more broadly based than most biological disciplines; consequently, there is more than one approach to it. Ecological studies fall into three categories: descriptive, experimental, and mathematical.

Descriptive ecology is concerned with describing natural history, usually in qualitative terms. The study of adaptations, for example, is descriptive in that one can measure the present "value" of an adaptive feature, but one can only conjecture as to the history of natural selection that was responsible for it. On the other hand, some patterns are discernible in nature for which hypotheses can be constructed and tested by statistical inference. For example, the spatial distribution (dispersion) of birds on an island may be random, indicating no biological interaction among them. If the birds are more evenly spaced (uniform dispersion) than predicted assuming randomness, however, then it might be inferred that the birds are competing for space; they are exhibiting territorial exclusion of one another. Such "natural experiments," as they are called, depend heavily upon the careful design of statistical tests.

Experimental ecology is no different from any other experimental discipline; hypotheses are constructed from observations of nature, controlled experiments are designed to test them, and conclusions are drawn from the results of the experiments. The basic laboratory for an ecologist is the field. Experiments in the field are difficult because it is hard to isolate and manipulate variable factors one at a time, which is a requisite for any good experiment in science. A common experiment that is performed to test for resource limitation in an organism is enhancement of that resource. If food, for example, is thought to be in short supply (implying competition), one section of the habitat is provided more food than is already present; another section is left alone as a control. If survivorship, growth, or reproductive output is higher in the enhanced portion of the habitat than in the control area, the researcher may infer that the organisms therein were food-limited. Alternatively, an ecologist might have decreased the density of organisms in one portion of the habitat, which might seem equivalent to increasing food supply for the remaining organisms, except that it represents a change in population density as well. Therefore, this second design will not allow the researcher to differentiate between the possibly separate effects of food level and simple population density on organisms in the habitat.

Mathematical ecology relies heavily upon computers to generate models of nature. A model is simply a formalized, quantitative set of hypotheses constructed from sets of assumptions of how things happen in nature. A model of population growth might contain assumptions about the age structure of a population, its genetic capacity for increase, and the average rate of resource utilization by its members. By changing these assumptions, scientists can cause the model population to behave in different ways over time. The utility of such modeling is limited to the accuracy of the assumptions employed.

Modern ecology is concerned with integrating these different approaches, all of which have in common the goal of predicting the way nature will behave in the future, based upon how it behaves in the present. Description of natural history leads to hypotheses that can be tested experimentally, which in turn may allow the construction of realistic mathematical (quantitative) models of how nature works.

ALL-ENCOMPASSING NATURE

People historically have viewed nature as an adversary. The "conquest of nature" has traditionally meant human encroachment on natural ecosystems, usually without benefit of predictive knowledge. Such environmental problems as pollution, species extinction, and overpopulation can be viewed as experiments performed on a grand scale without appropriate controls. The problem with such experiments is that the outcomes might be irreversible. A major lesson of ecology is that humans are not separate from nature; we are constrained by the same principles as are other organisms on the earth. One object of ecology, then, is to learn these principles so that they can be applied to our portion of the earth's ecosystem.

Populations that are not regulated by predators, disease, or food limitation grow exponentially. The human population, on a global scale, grows this way. All the wars and famines in history have scarcely made a dent in this growth pattern. Humankind has yet to identify its carrying capacity on a global scale, although regional famines certainly have provided insights into what happens when local carrying capacity is exceeded. The human carrying capacity needs to be defined in realistic ecological terms, and such constraints as energy, food, and space must be incorporated into the calculations. For example, knowledge of energy flow teaches that there is more energy at the bottom of a food web (producers) than at successively higher trophic levels (consumers), which means that more people could be supported as herbivores than as carnivores.

The study of disease transmission, epidemiology, relies heavily on ecological principles. Population density, rates of migration among epidemic centers, physiological tolerance of the host, and rates of evolution of disease-causing parasites are all the subjects of ecological study.

An obvious application of ecological principles is conservation. Before habitats for endangered species can be set aside, for example, their ecological requirements, such as migratory routes, breeding, and feeding habits, must be known. This also applies to the introduction (intentional or accidental) of exotic species into habitats. History is filled with examples of introduced species that caused the extinction of native species. Application of ecological knowledge in a timely fashion, therefore, might prevent species from becoming endangered in the first place.

One of the greatest challenges we face is the loss of habitats worldwide. This is especially true of the tropics, which contain most of the earth's species of plants and animals. Species in the tropics have narrow niches, which means that they are more restricted in range and less tolerant of change than are many temperate species. Therefore, destruction of tropical habitats, such as rain forests, leads to rapid species extinction. These species are the potential sources of many pharmaceutically valuable drugs; further, they are a genetic record of millions of years of evolutionary history. Tropical rain forests also are prime sources of oxygen and act as a buffer against carbon dioxide accumulation in the atmosphere. Ecological knowledge of global carbon cycles permits the prediction that destruction of rain forests will have a profound impact on the quality of the air.

FURTHER READING

Begon, Michael, John L. Harper, and Colin R. Townsend. *Ecology: Individuals, Populations, and Communities.* 3rd ed. Cambridge: Blackwell, 1996. Print.

Carson, Rachel. *Silent Spring.* 1962. 50th anniversary ed. London: Penguin, 2012. Print.

Elton, Charles. *Animal Ecology.* New York: Macmillan, 1927. Print.

Hutchinson, G. Evelyn. *The Ecological Theater and the Evolutionary Play.* New Haven: Yale UP, 1969. Print.

Krebs, Charles J. *The Message of Ecology.* New York: Harper, 1987. Print.

Molles, Manuel C., Jr. *Ecology: Concepts and Applications.* 7th ed. New York: McGraw, 2016. Print.

Morin, Peter Jay. *Community Ecology.* 2nd ed. Chichester: Wiley, 2012. Print.

Pianka, Eric R. *Evolutionary Ecology.* 6th ed. San Francisco: Benjamin/Cummings, 2000. Print.

Ricklefs, Robert E. *Ecology.* 4th ed. New York: Chiron, 1999. Print.

ECOSYSTEMS

—Robert Lovely

FIELDS OF STUDY

Biology; Evolution; Ecology; Environmental Studies.

PRINCIPAL TERMS

biomass: the weight of organic matter, often expressed in terms of grams per square meter per year

consumers: animals, fungi, and bacteria that get energy by feeding on organic matter

food chain: an abstract chain representing the links between organisms, each of which eats and is eaten by another

food web: a network of interconnecting food chains representing the food relationships in a community

photosynthesis: the process by which green plants and algae use sunlight as energy to convert carbon dioxide and water into energy-rich compounds such as glucose

primary production: the energy assimilated by green plants and stored as organic tissue

producers: green plants and chemosynthetic organisms that can produce food from inorganic materials

respiration: the release of chemical energy to do work in plants and animals; a reversal of the photosynthetic process

trophic level: a feeding level on the pyramid of numbers, consisting of all the kinds of animals that feed at comparable levels on food chains

SUMMARY

The ecosystem is the fundamental concept in ecology. It is viewed by ecologists as the basic unit of nature, consisting of the complex of interacting organisms inhabiting a region with all the nonliving physical factors that make up their environment. Ecologists study structural and functional relationships of ecosystem components to be able to predict how the system will respond to natural change and human disturbance.

The ecosystem is essentially an abstract organizing unit superimposed on the landscape to help ecologists study the form and function of the natural world. An ecosystem consists of one or more communities of interacting organisms and their physical environment. Ecosystems have no distinct boundaries; thus the size of any particular ecosystem should be inferred from the context of the discussion. Individual lakes, streams, or strands of trees can be described as distinct ecosystems, as can the entire North American Great Lakes region. Size and boundaries are arbitrary because no ecosystem stands in complete isolation from those that surround it. A lake ecosystem, for example, is greatly affected by the streams that flow into it and by the soils and vegetation through which these streams flow. Energy, organisms, and materials routinely migrate across whatever perimeters the ecologist may define. Thus, investigators are allowed considerable latitude in establishing the scale of the ecosystem they are studying. Whatever the scale, though, the importance of the ecosystem concept is that it forces ecologists to treat organisms not as isolated individuals or species but in the context of the structural and functional conditions of their environment.

Rachel Carson, author of *Silent Spring*. Official photo as FWS employee. c. 1940.

DEVELOPMENT OF THE ECOSYSTEM CONCEPT

Antecedents to the ecosystem concept may be traced back to one of America's first ecologists, Stephen Alfred Forbes, an eminent Illinois naturalist who studied food relationships among birds, fishes, and insects. During the course of his investigations, Forbes recognized in 1880 that full knowledge of organisms and their response to disturbances would come only from more concentrated research on their interactions with other organisms and with their inorganic (nonliving) physical surroundings. In 1887, Forbes suggested that a lake could be viewed as a discrete system for study: a microcosm. A lake could serve as a scale model of nature that would help biologists understand more general functional relationships among organisms and their environment. Forbes explained how the food supply of a single species, the largemouth bass, was dependent either directly or indirectly upon nearly all the fauna and much of the flora of the lake. Therefore, whenever even one species was subjected to disturbance from outside the microcosm, the effects would probably be felt throughout the community.

In 1927, British ecologist Charles Elton incorporated ideas introduced by Forbes and other fishery biologists into the twin concepts of the food chain and the food web. Elton defined a food chain as a series of linkages connecting basic plants, or food producers, to herbivores and their various carnivorous predators, or consumers. Elton used the term "food cycle" instead of "food web," but his diagrams reveal that his notion of a food cycle—that it is simply a network of interconnecting food chains—is consistent with the modern term.

Elton's diagrams, which traced various pathways of nitrogen through the community, paved the way for understanding the importance of the cycling of inorganic nutrients such as carbon, nitrogen, and phosphorus through ecosystems, a process that is known as biogeochemical cycling. Very simply, Elton illustrated how bacteria could make nitrogen available to algae at the base of the food chain. The nitrogen then could be incorporated into a succession of ever larger consumers until it reached the top of the chain. When the top predators died, decomposer organisms would return the nitrogen to forms that could eventually be taken up again by plants and algae at the base of the food chain, thus completing the cycle.

Elton's other key contribution to the ecosystem concept was his articulation of the pyramid of numbers, the idea that small animals in any given community are far more common than large animals. Organisms at the base of a food chain are numerous, and those at the top are relatively scarce. Each level of the pyramid supplies food for the level immediately above it—a level consisting of various species of predators that generally are larger in size and fewer in number. That level, in turn, serves as prey for a level of larger, more powerful predators, fewer still in number. A graph of this concept results in a pyramidal shape of discrete levels, which today are called trophic (feeding) levels.

ECOSYSTEMS IN TWENTIETH CENTURY THOUGHT

Although the basic concept of an ecosystem had been recognized by Forbes as early as 1880, it was not until 1935 that British ecologist Arthur G. Tansley coined the term. Though he acknowledged that ecologists were primarily interested in organisms, Tansley declared that organisms could not be separated from their physical environment, as organism and environment formed one complete system. As Forbes had pointed out half a century earlier, organisms were inseparably linked to their nonliving environments. Consequently, ecosystems came to be viewed as consisting of two fundamental parts: the biotic, or living components, and the abiotic, or nonliving components.

No one articulated this better than Raymond Lindeman, a limnologist (freshwater biologist) from Minnesota who in 1941 skillfully integrated the ideas of earlier ecological scientists when he published an elegant ecosystem study of Cedar Bog Lake. Lindeman's classic work set the stage for decades of research that centered on the ecosystem as the primary organizing unit of study in ecology. Drawing on the work of his mentor, G. Evelyn Hutchinson, and Charles Elton, Arthur Tansley, and other scientists, Lindeman explained how ecological pyramids were a necessary result of energy transfers from one trophic level to the next.

By analyzing ecosystems in this manner, Lindeman was able to answer a fundamental ecological question that had been posed fourteen years earlier by Elton: Why were the largest and most powerful animals, such as polar bears, sharks, and tigers, so rare? Elton had thought the relative scarcity of top predators was due to their lower rates of reproduction. Lindeman corrected this misconception by explaining that higher trophic levels held fewer animals not because of their reproductive rates but because of a loss of chemical energy with each step up the pyramid. It could be looked upon as a necessary condition of the second law of thermodynamics: Energy transfers yield a loss or degradation of energy. The predators of one food level could never completely extract all the energy from the level below. Some energy would always be lost to the environment through respiration, some energy would not be assimilated by the predators, and some energy simply would be lost to decomposer chains when potential prey died of nonpredatory causes. This meant that each successive trophic level had substantially less chemical energy available to it than was transferred from the one below and, therefore, could not support as many animals.

Ecologists soon expanded the principle of Elton's pyramid of numbers to model other ecosystem processes. They found, for example, that the flow of chemical energy through an ecosystem could be characterized as an energy pyramid; the biomass (the weight of organic material, as in plant or animal tissue) in a community could be plotted in a pyramid of biomass. Collectively, such pyramidal models became known as ecological pyramids.

ENERGY PRODUCTION & TRANSMISSION

Lindeman subdivided the biotic components of ecosystems into producers, consumers, and decomposers. Producers (also known as autotrophs) produce their own food from compounds in their environment. Green plants are the main producers in terrestrial ecosystems; algae are the most common producers in aquatic ecosystems. Both plants and algae are producers that use sunlight as energy to make food from carbon dioxide and water in the process of photosynthesis. During photosynthesis, plants, algae, and certain bacteria capture the sun's energy in chlorophyll molecules. (Chlorophyll is a pigment that gives plants their green color.) This energy, in turn, is used to synthesize energy-rich compounds such as glucose, which can be used to power activities such as growth, maintenance, and reproduction or can be stored as biomass for later use. These energy-rich compounds can also be passed on in the form of biomass from one organism to another, as when animals (primary consumers) graze on plants or when decomposers break down detritus (dead organic matter).

The energy collected by green plants is called primary production because it forms the first level at the base of the ecological pyramid. Total photosynthesis is represented as gross primary production. This is the amount of the sun's energy actually assimilated by autotrophs. The rate of this production of organic tissue by photosynthesis is called primary productivity. Plants, however, need to utilize some of the energy they produce for their own growth, maintenance, and reproduction. This energy becomes available for such activities through respiration, which essentially is a chemical reversal of the process of photosynthesis. As a result, not all the energy assimilated by autotrophs is available to the consumers in the next trophic level of the pyramid. Consequently, respiration costs generally are subtracted from gross primary production to determine the net primary production, the chemical energy actually available to primary consumers.

MEASURING ECOSYSTEM PRODUCTIVITY

The carrying capacity for all the species supported by an ecosystem ultimately depends upon the system's net primary productivity. By knowing the productivity, ecologists can, for example, estimate the number of herbivores that an ecosystem can support. Consequently ecosystem ecologists have developed a variety of methods to measure the net primary productivity of different systems. Productivity is generally expressed in kilocalories per square meter per year when quantifying energy, and in grams per square meter per year when quantifying biomass.

Production in aquatic ecosystems may be measured by using the light and dark bottle method. In this technique two bottles containing samples of water and the natural phytoplankton population are suspended for twenty-four hours at a given depth in a body of water. One bottle is dark, permitting respiration but no photosynthesis by the phytoplankton. The other is clear and therefore permits both photosynthesis and respiration. The light bottle provides a measure of net production (photosynthesis minus respiration) if the quantity of oxygen is measured before and after the twenty-four-hour period. (The amount of oxygen produced by photosynthesis is proportional to the amount of organic matter fixed.) Measuring the amount of oxygen in the dark bottle before and after the run provides an estimate of respiration, since no photosynthesis can occur in the dark. Combining net production from the light bottle with total respiration from the dark bottle yields an estimate of gross primary production.

Other studies have concentrated on quantifying the rate of movement of energy and materials through ecosystems. Investigations begun in the 1940's and 1950's by the Atomic Energy Commission to track radioactive fallout were eventually diverted into studies of ecosystems that demonstrated how radionuclides moved through natural environments by means of food-chain transfers. This research confirmed the interlocking nature of all organisms linked by the food relationship and eventually yielded rates at which both organic and inorganic materials could be cycled through ecosystems. As a result of such studies, the Radiation Ecology Section at Oak Ridge National Laboratory in Tennessee became established as a principal center for systems ecology.

Ecologists sometimes extend the temporal boundaries of their studies by utilizing the methods of paleoecology (the use of fossils to study the nature of ecosystems in the past). Research of this type generally centers on the analysis of lake sediments, whose layers often hold centuries of ecosystem history embodied in the character and abundance of pollen grains, diatoms, fragments of zooplankton, and other organic microfossils.

More general trends in the methods of studying ecosystems include a continuing emphasis on quantitative methods, often using increasingly sophisticated computer modeling techniques to simulate ecosystem functions. Equally significant is a trend toward a "big science" approach, modeled on the Manhattan Project, which employed teams of investigators working on different problems related to nuclear fission in different parts of the country. The well-known international biological program, the Hubbard Brook Project in New Hampshire, and continuing projects on long-term ecological research all serve as examples of ecosystem studies that involve teams of researchers from a wide range of disciplines.

RESPONDING TO DISTURBANCE

One of the practical benefits of studying ecosystems derives from naturalist Stephen Forbes's suggestion, which he made in 1880, that the knowledge from biological research be used to predict the response of organisms to disturbance. When disturbance is caused by natural events such as droughts, floods, or fires, ecologists can use their knowledge of the structure and function of ecosystems to help resource managers plan for subsequent recolonization and succession of species.

The broad perspective of the ecosystem approach becomes particularly useful in examining the effects of certain toxic compounds because of the complexity of their interaction within the environment. The synergistic effects that sometimes occur with toxic substances can produce pronounced impacts on ecosystems already stressed by other disturbances. For example, after the atmosphere deposits mercury on the surface of a lake, the pollutant eventually settles in the sediments where bacteria make it available to organisms at the base of the food chain. The contaminant then bioaccumulates as it is passed on to organisms such as fish and fish-eating birds at higher trophic levels. Synergistic effects occur in lakes already affected by acid deposition; researchers have found that acidity somehow stimulates microbes to increase the bioavailability of the mercury. Thus, aquatic ecosystems that have become acidified through atmospheric processes may stress their flora and fauna even further by enhancing the availability of mercury from atmospheric fallout. The complexity of such interactions demands research at the ecosystem level, and ecosystem studies are prerequisite for prudent public policy actions on environmental contaminants.

Further Reading

Begon, M., J. L. Harper, and C. R. Townsend. *Ecology: Individuals, Populations, and Communities.* Oxford: Blackwell, 1996. Print.

Bouma, Jetske A., and Pieter van Beukering. *Ecosystem Services: From Concept to Practice.* Cambridge: Cambridge UP, 2015. Print.

Clark, Tim W., A. Peyton Curlee, Steven C. Minta, and Peter M. Kareiva, eds. *Carnivores in Ecosystems: The Yellowstone Experience.* New Haven: Yale UP, 1999. Print.

Colinvaux, Paul. *Why Big Fierce Animals Are Rare: An Ecologist's Perspective.* Princeton: Princeton UP, 1978. Print.

Golly, Frank Benjamin. *A History of the Ecosystem Concept in Ecology: More than the Sum of the Parts.* New Haven: Yale UP, 1993. Print.

Hagen, Joel B. *An Entangled Bank: The Origins of Ecosystem Ecology.* New Brunswick: Rutgers UP, 1992. Print.

Hunter, Malcolm L., ed. *Maintaining Biodiversity in Forest Ecosystems.* New York: Cambridge UP, 1999. Print.

Mustafa, Saleem, and Rossita Shapawi. *Aquaculture Ecosystems: Adaptability and Sustainability.* Hoboken: Wiley, 2015. Print.

Odum, Eugene P. *Ecology and Our Endangered Life-Support Systems.* Sunderland: Sinauer, 1989. Print.

Real, Leslie A., and James H. Brown. *Foundations of Ecology: Classic Papers with Commentaries.* Chicago: U of Chicago P, 1991. Print.

Smith, Robert Leo. *Ecology and Field Biology.* 6th ed. San Francisco: Benjamin/Cummings, 2007. Print.

Sullivan, Timothy J., et al. *Air Pollution and Freshwater Ecosystems: Sampling, Analysis, and Quality Assurance.* Boca Raton: CRC, 2015. Print.

ECOTOURISM

—Jeffrey Wielgus

FIELDS OF STUDY

Ecology; Environmental Studies.

SUMMARY

Nature-based tourism has both positive and negative effects on the world's ecosystems, drawing travelers to experience natural wonders, promote environmental activism, and stimulate the economy of local communities, but exposes habitats to the risks of high traffic and disruption.

Many of the world's ecosystems still contain areas that remain mostly in their natural state. The term *ecotourism* refers to traveling to such areas to enjoy the recreational and educational experiences they provide. People have found pleasure in nature throughout history, but the expansion and intensification of human activities have degraded many of the world's ecosystems. This human footprint has reduced the possibility to benefit from visiting well-preserved ecosystems, forcing tourists to travel longer distances to see nature in its pristine, unspoiled state.

Modern transportation systems enable travelers from around the world to visit distant locations to enjoy the remaining areas of wilderness. Countries that protect their ecosystems can benefit from the inflow of ecotourists, but an increase in visitors also can put pressure on natural areas and on the human communities that depend closely on the resources that they provide.

ROOTS & MOVEMENTS

In Asia, reverence for natural places is a major tenet of religions such as Buddhism, Hinduism, Shintoism, and Taoism. Pilgrimages to key historical sites in the development of these philosophies can be considered a form of ecotourism. Native cultures of Africa, the Americas, and Australia place natural areas in high regard; certain places considered sacred attract the faithful on pilgrimages.

In Europe and its former colonies in the Western Hemisphere, some ecosystems are seen as places of special value; many can be traced to the Romantic Movement. This philosophical movement encompassed parts of the eighteenth and nineteenth centuries; it considered nature to be a source of well-being unrivaled by any object or structure created by humans. The naturalists of the Romantic Movement insisted on protecting ecosystems (although this term was introduced much later) from human encroachment, and their ideas helped give rise to the modern conservation movement.

An ecotourism guide looks out from the bow of his boat as he brings a tourist back from a trip to El Salvador's first marine protected area in 2009. Image by USAID/Angela Rucker.

One key figure in the latter development of this ecologically oriented movement was John Muir (1838–1914), an author and activist who helped influence the creation of the national parks system in the United States. Muir's writings enthralled Americans, who wanted to see and experience nature's beauty for themselves. Their travels were made possible by the development of the cross-country railroad system, and later by the automobile.

By the 1980s, the term *ecotourism* had come into the vernacular as a way to refer to nature-based tourism. Sectors of the travel industry, both domestic and international, had previously begun specializing in nature travel since the middle of the twentieth century. Wildlife-watching trips to African national parks, for example, have been offered in Europe by specialized travel agencies since the end of World War II, and marine ecotourism has become a popular travel experience ever since scuba-diving equipment began to be mass-produced in the early 1970s.

Further, some countries began to embrace ecotourism as a major source of external revenues. Costa Rica has become an ecotourism leader for travelers who are interested in exploring mountain areas, tropical rainforests, coral reefs, and other ecosystems in this politically stable Central American country. Such travel is concentrated mostly in the country's protected areas, which include privately owned reserves. International tourism, driven mainly by ecotourism, has become one of the country's leading sources of foreign currency. Costa Rica is easily accessible from the United States and Canada, which are among the largest sources of tourists worldwide.

Beginning in the 1990s, the Costa Rican government began offering incentives for companies investing in environmentally responsible ecotourism there. These include a certification program for hotels with a low environmental footprint, and a certification program and rating system for coastal communities based on marine water quality, access to potable drinking water, beach cleanliness, wastewater treatment, environmental education, safety, and management. Environmentally conscious tourists can select their destination based on coastal communities with high ratings.

ECONOMIC BENEFITS

Several African countries also have adopted ecotourism as a path to socioeconomic development. In Kenya, hunting-based tourism, which was popular in the early twentieth century, largely transitioned to wildlife-viewing tours in the newly created national parks, starting in the 1950s. After the country's independence from Britain in 1961, the government began to develop programs to attract international investment to develop tourism facilities that cater mainly to wildlife-viewing and beach ecotourists. Ecotourism offered a way to supplement the country's dependence on the export of agricultural products, whose prices fluctuated greatly, as a source of foreign currency.

Some local communities in Kenya have partnered with both governmental and nongovernmental organizations (NGOs) to offer ecotourism services such as lodging and wildlife-viewing tours. South Africa, too, offers specialized services for ecotourists; wildlife viewers there can choose among a wide range of accommodations, from tents (typically available in the national parks) to luxury hotels that are available in some of the private natural reserves.

A growing number of developing countries have been able to diversify their economic bases and benefit from a generally stable inflow of foreign currency brought by ecotourists. Tropical rainforests, savannas, and coral reefs are ecosystems that are particularly sought out, and they are coincidentally present in a large number of developing countries. The activity has led to new industries and local jobs; governments and private landowners share the incentive to protect the ecosystems, sustain tourism, and protect and enhance their local economy.

In developing and developed countries, ecotourism provides financial resources that help maintain the systems of protected areas. In the United States, visitor fees in the parks and preserves managed by the federal government and individual states account for approximately one-fifth and one-third, respectively, of the park systems' budgets. Funds from visitor fees can be used to create and maintain tools for interpretation, such as information boards and multimedia visitor centers, which help educate visitors on the area's ecosystems and explain which activities should be avoided to prevent damages to the resources. Such enhancements tend to increase repeat visits.

NEGATIVE IMPACTS & MITIGATION

Ecotourism can help educate visitors and residents on the importance of protecting ecosystems, and it can provide incentives for the creation of protected areas—but it also can have a negative effect on natural resources. An increase in the number of ecotourists is generally accompanied by an increase in the number of hotel rooms to accommodate them, which results in higher water consumption and wastewater disposal. An increase in visitors to protected areas also leads to more solid and liquid wastes in these ecologically sensitive areas. The physical impact of tourism and the interference of ecotourists with wildlife also can become problematic.

Corals located in shallow waters are frequently trampled by visitors to marine protected areas, and corals can die from diseases that affect their injured areas. In many protected areas, the feeding and breeding behavior of birds has been affected by tourists. In national parks of the United States, the natural feeding habits of bears are affected by the presence of humans, and bears that search for food left by visitors are occasionally injured or killed by vehicles.

The impact of these activities can even intensify the impact of climate change, by hastening the deterioration and vulnerability of already-sensitive areas. As coral reefs are adversely impacted by warming waters, human activity in these areas can hasten their decline. Visitors in search of bears, other wildlife, trees, flowers, and more can be drawn to higher elevations, deeper into forests and protected areas, bringing their influence to formerly remote areas, increasing the disruption of their habitats.

Ecotourism can have negative social consequences as well. Local communities can feel that their traditional way of life is affected by the influx of visitors to protected areas in their vicinity. In Kenya, some indigenous Maasai communities have developed resentment toward ecotourism, which they see as favoring only a few people and causing social divisions between those who benefit and those who feel affected by the intrusion of tourism. In the Ecuadorian Amazon, ecotourism has caused conflicts within some of the Secoya and Siona indigenous communities. The distribution of benefits from tourism is seen as unfair by some, and the adoption of customs and technologies introduced by tourists also is viewed negatively by members of these communities.

There are a number of initiatives at the national and international levels whose goal is to make ecotourism a sustainable activity from both the ecological and social perspectives. NGOs in many developing countries have been established specifically to help design strategies for sustainable ecotourism, and universities worldwide offer courses related to ecotourism management. University research tracks on ecotourism cover areas such as ecologic, economic, socioeconomic, and cultural effects and benefits; visitor preferences; and visitor willingness to pay for visits to protected areas. Several international organizations, including the International Ecotourism Society (IES) and the World Tourism Organization (WTO), promote sustainable ecotourism and disseminate information on ecotourism research.

The WTO and the United Nations Environment Programme organized the World Ecotourism Summit in 2002 in Quebec City, Canada. The summit brought together representatives from academic institutions, governments, NGOs, tourism companies, and indigenous communities to share their experiences in ecotourism and to identify priority areas for the development of sustainable ecotourism. The IES organizes the annual Ecotourism and Sustainable Tourism Conference, the goal of which is "providing practical solutions to advance sustainability goals for the tourism industry."

FURTHER READING

Fennell, D. A. *Ecotourism.* 3rd ed. Abingdon, UK: Routledge, 2008.

Higham, J., ed. *Critical Issues in Ecotourism: Understanding a Complex Tourism Phenomenon.* Oxford, UK: Butterworth-Heinemann, 2007.

Honey, M. *Ecotourism and Sustainable Development: Who Owns Paradise?* 2nd ed. Washington, DC: Island Press, 2008.

Wild, R. and C. McLeod, eds. *Sacred Natural Sites: Guidelines for Protected Area Managers.* Gland, Switzerland: International Union for Conservation of Nature and Natural Resources (IUCN), 2008.

ENERGY CONSERVATION

—Ronju Ahammad

FIELDS OF STUDY

Economics; Conservation; Business; Mathematics.

SUMMARY

While the distribution and utilization patterns of energy resource consumption has been justified, it is unsustainable for future generations. Energy

Recycle logo. Image by Offiikart (Own work) CC0.

consumption will double in industrialized countries and triple in developing countries over the next 30 years.

Widespread availability of energy at an affordable cost is one of the most important factors in economic progress and improving the quality of life, but can result in wasteful energy use that in turn leads to environmental degradation. Energy conservation provides a range of resources that can simultaneously minimize energy consumption while keeping it widely available and affordable. More recently, energy conservation has been harnessed to reduce emissions of greenhouse gases and other pollutants, and individuals such as Amory Lovins have widely (and successfully) promoted conservation as an alternative to the continued construction of new power plants.

Although thought of as a recent development, energy conservation has a long history. In ancient Greece and Rome, scarce and expensive fuel resources led architects and engineers to design both efficient buildings and heating systems. These advancements were largely lost after the fall of the Roman Empire, but were rediscovered during the European Renaissance, when improved chimneys and stoves were reintroduced. As early as 1618, German writer Franz Kessler wrote *Holzsparkunst* (*Saving Wood*), which promoted efficient heating systems in order to save Germany from deforestation. While the Germans learned this lesson, the French (and others) did not, and the introduction of efficient stoves into Paris in the eighteenth century simply made heat available to a large population, which led to rapid deforestation of the French countryside. Benjamin Franklin saw this same pattern in America and invented a more efficient stove to avoid a similar deforestation in America (which happened anyway).

The dilemma of conservation is that by reducing the energy required to do a specific task, the cost will also be reduced and people will use more of it. One of the best examples of this is the introduction of vehicle fuel economy standards in the United States, which made cars more efficient, but as a result, Americans increased the numbers of miles they drove by 151 percent over the next 25 years, five times more than the growth in population.

One of the more effective and widespread conservations employed in recent years are the promulgation of efficiency standards for a wide range of appliances as well as buildings. Refrigerators, air conditioners, furnaces, and light bulbs have become much more efficient, and building codes in many areas now mandate that new and renovated buildings must be very efficient.

Energy conservation has been aided by the introduction of smart meter and energy control systems, which allow even residential consumers to monitor and control their energy consumption. As is the case with most energy conservation measures, they are more likely to be adopted by consumers with the financial resources to do so, although governments and Nongovernmental organizations (NGOs) have become increasingly active in reducing energy use in low-income areas of developed nations and throughout developing countries. The introduction of more efficient appliances, low-power computers, and other technology has also made it possible to introduce them into areas depending on limited renewable resources such as solar photovoltaics.

Increasing and rampant energy consumption is polluting the global atmospheric environment through the emission of excessive carbon compounds, while it is gradually limiting access to energy for a substantial amount of the world's population. Energy conservation is thus needed to reduce consumption of limited energy resources and to encourage the use of energy-efficient products in our daily life in such a way that would reduce environmental tradeoffs and ensure sustainable use.

AVAILABILITY OF ENERGY RESOURCES

Energy resources are found in renewable and nonrenewable forms. Nonrenewable energy resources are most commonly consumed and include biomass and fossil fuels such as oil, natural gas, and coal.

Nonrenewable energy resources are extensively converted into other forms to generate electricity. Although nonrenewable energy resources have multiple uses in daily life and their consumption is expected to increase, their future quantities are limited. Notably, fossil-fuel-based energy consumption has already given rise to environmental protection concerns, calls for energy conservation, and searches for alternative energy resources.

The largest amount of fossil-fuel-based energy is consumed in the developed parts of the world, but only 30 percent of global population lives there. It has been well documented that since the industrial era of the late 1800s, greenhouse gas (GHG) emissions have increased exponentially due to fossil fuel combustion in developed nations. Of particular concern is the influx of carbon dioxide (CO_2) into the atmosphere, which has risen to 450 ppm from the normal level of 250 ppm in the preindustrial period. Due to excessive CO_2 emission in the atmosphere, global warming has already increased and the resulting weather variability has heavily impacted social and economic conditions in many poor countries. With regard to climate change, energy consumption issues have been the focal point driving the quest for alternative energy sources.

DEVELOPING CONSERVATION MEASURES

Energy conservation and the concomitant development of alternative energy sources can contribute to both social and economic development. Energy conservation can most importantly reduce the need for finite fossil-fuel-based energy products and thus reduce GHG emissions into the atmosphere. Reduced energy consumption is likely to result in lower per capita demand for energy as well as lowered industrial production costs. The resulting energy savings will lead in turn to lowered costs associated with the investment, supply, and distribution of energy to meet the increased demands driven by population growth.

Just as humans have driven energy consumption, humans have the capacity to drive energy conservation. Energy is a significant source of power in the accomplishment of our daily work. In general terms, energy is the ability to do work, for instance to move objects, light our homes and cities, and power our factory machines. Energy conservation should thus start from our daily lives and surroundings.

Individuals can reduce their daily energy consumption and adopt low-carbon lifestyles through changing their behavior and product usage. Households have a significant role in energy efficiency because of their diverse demands for energy use and household activities with respect to their diverse social and economic situations. Households use both direct (natural gas, electricity, biomass, and other fossil fuels) and indirect (production, transportation, and disposal of goods and services) energy.

Energy conservation efforts most commonly include the following steps (which are not exhaustive): walking, riding a bicycle, or riding mass transit instead of driving automobiles, which emit 60 percent of the air pollution in cities; using fluorescent lightbulbs, which can reduce energy consumption 10 times compared to incandescent light bulbs; air-drying clothes instead of using a clothes dryer; and recycling newspapers, aluminum cans, and plastic bottles to reduce future energy costs for manufacturing the same products.

METHODS OF CONSERVATION

Energy conservation also involves the type of appliances we use in our households and offices. Different appliances have varied energy consumption rates and consumers should act proactively to reduce their energy consumption. Examples include raising or lowering the water or air heater or air-conditioning thermostat, switching off the lights when leaving a room, not using the standby mode of electronic appliances like computers, sealing air leaks around windows and doors to reduce excessive heating and cooling, planting trees to block sunlight near windows to reduce associated cooling costs, cleaning or changing air filters in air heating and cooling systems to improve efficiency, and the use of energy-efficient appliances.

Many factors influence energy conservation at the individual and household levels as well as within energy-intensive business units. One key issue is a lack of awareness regarding the selection and use of energy-efficient products and services. Awareness campaigns, such as the distribution of information booklets and broadcast programs and the marketing of energy-efficient products may increase people's conscious decisions to adopt products that facilitate energy conservation. People's awareness of their energy use and their self-motivation in terms of environmental concerns can determine their consumption patterns. People select energy products with respect to their economy, comfort, and context. Increased energy saving awareness can motivate people toward reduced consumption and the preference of alternative energy sources. In this regard, energy policies must encourage the adoption of efficient products.

The core idea of energy conservation is to use energy more efficiently. Energy efficiency is determined by lifestyle, but also by access to more efficient products. Energy-efficient products include those that consume less energy, are not purely produced from fossil fuels, are recyclable, and are driven by alternative energy sources. For example, electricity is usually generated from finite, nonrenewable fossil fuels, but there is a huge potential for the production of electricity from renewable solar, wind, and hydro sources.

Alternative energy sources such as solar energy are efficient means for energy conservation and the reduction of energy consumption and are directly accessible resources in the world around us. While fossil-fuel-based electricity generation is not equally available in developing and least-developed countries, there is great potential for the adoption of solar-based home electricity systems. To some extent, promotion of solar panel–based electrification systems would improve the social and economic conditions of rural populations in those countries that seldom have access to fossil-fuel-based energy resources while reducing over-reliance on biomass fuel or the outside importation and exploration of fossil fuel resources.

Energy conservation is a global issue but conservation roles and responsibilities vary among nations. Developed nations need to play a vital role in conserving energy, as these nations consume the highest percentage of energy resources due to greater access and to technological and financial investment in energy exploration. Residents in developed nations use a plethora of appliances in their daily life, which are powered by electricity generated from a variety of mostly nonrenewable energy resources. There is also an upsurge in demand for oil for fuel combustion to support the growing urban traffic of those nations. Citizens of highly energy-dependant countries need to adopt more energy-efficient products to reduce their consumption levels. At the same time, energy conservation efforts must be accompanied by the development of alternative renewable energy technology, such as solar power, wind power, and hydropower. Some developed nations have already adopted alternative energy sources—such as nuclear energy in the United States and Sweden, and wind power in Denmark and the Netherlands—to generate electricity and boost industrial economic growth.

Demand for energy is projected to at least remain constant to maintain growth in developed countries, and demand will rise in developing and least-developed countries to drive economic growth. The role and demand of energy-deficit countries regarding energy conservation must not be downplayed. Raising economic growth in these areas will require sufficient access to energy. Renewable alternative energy sources have already showed promise. The access of rural people to energy resources is vital to accelerate the national development of poor countries, which may be viable if alternative energy can be provided in remote areas. Many initiatives are already underway to promote alternative energy sources in Asian and African countries. For example, a few private agencies are providing solar panels to rural residents with no access to national electric grids in Bangladesh, while a group of people formed a cooperative to use solar panels for their businesses in Namibia.

The adoption of alternative energy sources and new technologies, however, are not yet cost-effective or easily accessible for most remote impoverished populations, and little government funding is in place to support such efforts. Human life will not be sustainable, however, without energy conservation efforts focused on environmental protection and global socioeconomic realities.

FURTHER READING

Bynyard, Peter. "Pursuing the Energy Efficiency-Conservation Path." *Energy Policy*. (1988).

"Communications on Energy: Household Energy Conservation." *Energy Policy* (1984).

Koerth-Baker, Maggie. *Before the Lights Go Out: Conquering the Energy Crisis Before It Conquers Us.* Hoboken, NJ: John Wiley & Sons, 2012.

Lovins, Amory. "Energy Strategy: The Road Not Taken?" *Foreign Affairs* (October 1976).

Maher, Neil M. *Nature's New Deal: The Civilian Conservation Corps and the Roots of the American Environmental Movement.* New York: Oxford University Press, 2009.

Sheldrick, Bill, and Sally Macgill. "Local Energy Conservation Initiatives in the UK: Their Nature and Achievements." *Energy Policy* (1988).

ENVIRONMENTAL IMPACTS OF RAISING CATTLE

—*Jamie Michael Kass*

FIELDS OF STUDY

Animal Husbandry; Conservation; Ecology; Environmental Studies.

SUMMARY

The environmental impacts of the raising of cattle, especially intensive farming, include the production of greenhouse gases that contribute to climate change, land degradation owing to increased erosion and overgrazing, and pollution of air and water caused by concentrated waste. These problems most often arise from the nature of the farming practices of cattle ranchers, and they can be alleviated somewhat by changes in management techniques.

The collective species *Bos primigenius* includes three subspecies: *Bos primigenius taurus*, the European cattle; *Bos primigenius indicus*, the zebu; and the extinct *Bos primigenius primigenius*, the ancestor to modern domestic cattle and also called the auroch. The term "cattle" is usually used to refer to domesticated *B. primigenius* that are bred for multiple uses, spanning the production of food to clothing to fuel, and also to serve as work animals. Cattle are chief sources of food (from meat and milk), labor, clothing (from leather), and fuel (from dung) in many cultures worldwide.

According to a 2014 report from the United Nations Intergovernmental Panel on Climate Change (IPCC) , the world's livestock, most of which are cattle, are among the major causes of serious global environmental problems, including land degradation, air and water pollution, and loss of biodiversity. Further, livestock are responsible for 14–18 percent of total global greenhouse gas emissions (methane), which have been linked to climate change. The potential for those who raise livestock to contribute to solving these environmental problems is thus very large, and experts argue that major improvements can be achieved at reasonable costs.

The grazing of livestock occupies almost 30 percent of the earth's terrestrial surface, and the agricultural production of feed for livestock takes up about one-third of all arable land. In addition to affecting land that has already been converted to agriculture, the raising of cattle is responsible for the direct acceleration of deforestation in some parts of the world where ranchers burn forests to expand grazing land. This is a pressing problem in Latin America, where nearly 70 percent of previously forested land is used as pasture and much of the rest is used for growing livestock feed crops such as soybeans and corn. Tropical soils are especially susceptible to degradation by overgrazing, compaction, and erosion because of high precipitation and low nutrient loads. Ranching is thus usually not a sustainable practice in the Tropics, but it continues because it promises quick economic returns.

Concentrated animal feeding operations (CAFOs), in which livestock are fed in small confined areas for maximum profit, produce great amounts of highly concentrated waste that ultimately ends up in the water and as gases in the air. FAO has estimated that livestock manure is the largest sectoral source of water pollutants. In addition to manure, chemicals from pesticides and fertilizers used for feed crops and antibiotics administered to livestock also end up in groundwater and surface water, where they contribute to high nutrient loading and algal blooms, medicinal pollution that affects aquatic biology, and high sediment loads that reduce water quality. In addition, antibiotics used in intensive cattle operations have been linked with emerging antibiotic resistance.

GREENHOUSE GAS EMISSIONS

Cattle produce methane through an anaerobic process in the gut called methanogenesis; the methane is released through belching and flatulence. Methane is an extremely effective greenhouse gas, having a warming effect 23 to 50 times greater than carbon dioxide; this causes concern as there are approximately 1.3 billion cattle worldwide, and the number only grows. According to the 2014 IPCC report, the raising of cattle generates more greenhouse gases than do all forms of transportation. With increasing prosperity worldwide, people consume more meat and dairy products every year, especially in developing countries. The global production of both meat and dairy is expected to double by 2050.

A number of possible ways of reducing methane production in cattle have been proposed or are under study. These include the administration of bovine medicines similar to the antacid Alka-Seltzer, the use of new varieties of feed grasses, and targeted breeding that selects for less gassy cattle. Any successes in this area will certainly be helpful, but scientists note that a global reduction in numbers of domestic cattle must also accompany these techniques if cattle's large-scale production of methane is to be effectively reduced.

ENVIRONMENTAL SOLUTIONS

Humans certainly need not stop raising cattle entirely in order to ensure a healthy future for the environment, but some practices must change. Sustainable cattle ranching requires restoration of overgrazed and damaged land through soil conservation, the planting of trees, and protection of areas sensitive to erosion. Changing the ways cattle are fed to better reflect a natural diet can go a long way toward curbing greenhouse gas emissions. In addition, it has been argued that moving away from CAFOs to less intense feed operations can have positive effects on the cows, their environment, and the quality of all products that come from the cows. Increased use of processed manure as fertilizer can reduce waste pollution and reclaims a resource that should be valued and used instead of dumped into the water supply.

Many cattle ranchers and dairy farmers have become dedicated to reducing the negative environmental impacts of raising cattle because for them the advantages of sustainable management systems outweigh the pressures for higher profit. Persons living in subsistence communities often have little choice in how they raise cattle, however—they clear land and overgraze because they will starve otherwise. It has been proposed that developed nations should provide incentives to poorer countries to ensure that practicing deforestation to create pastureland is not the only option the people have for income.

Another solution to the negative environmental impacts of cattle raising that is sometimes proposed, primarily by animal rights and environmental conservation groups, is vegetarianism. If humans were to limit their beef consumption, this would certainly reduce demand and therefore reduce the number of cattle worldwide that are overgrazing, belching methane, and compacting soil. As critics of this approach have noted, however, expecting large numbers of people to change their diets radically is unrealistic. This is especially true in the United States, where the culture of beef eating is quite strong. For this reason, many environmentalists, scientists, and others have increasingly suggested that people could help reduce the negative environmental impacts of cattle raising simply by reducing the amount of beef in their diets.

FURTHER READING

Clutton-Brock, Juliet. *A Natural History of Domesticated Mammals.* 2d ed. New York: Cambridge University Press, 1999.

Hayes, Denis, and Gail Boyer Hayes. *Cowed: The Hidden Impact of 93 Million Cows on America's Health, Economy, Politics, Culture, and Environment.* New York: W.W. Norton, 2015.

Liu, George. *Cattle: Domestication, Diseases and the Environment.* New York: Nova, 2013.

Mentz, André. *Ecological Cattle Farming: A Roadmap to Sustainable Cattle Farming.* South Africa, Northern Cape: Pula Books, 2016.

Miller, Naomi Frances, Katherine M. Moore, and Kathleen Ryan. *Sustainable Lifeways: Cultural Persistence in an Ever-Changing Environment.* Philadelphia: University of Pennsylvania Press, 2011.

Niman, Nicolette Hahn. *Defending Beef: The Case for Sustainable Meat Production.* White River Junction, Vermont: Chelsea Green, 2014.

Phillips, C. J. C. *Principles of Cattle Production.* Wallingford, UK: CABI, 2010.

Soliva, Carla Riccarda, Junichi Takahashi, and Michael Kreuzer, eds. *Greenhouse Gases and Animal Agriculture.* Boston: Elsevier, 2006.

Steinfeld, Henning, et al. *Livestock's Long Shadow: Environmental Issues and Options.* Rome: United Nations Food and Agriculture Organization, 2006.

EROSION & EROSION CONTROL

—*Mark S. Coyne*

FIELDS OF STUDY

Soil Science; Conservation; Ecology; Environmental Studies.

SUMMARY

Erosion involves the loss of topsoil through the actions of wind and water, and erosion control includes the efforts undertaken to mitigate such loss. Controlling erosion is vital because soil loss from agricultural land is a major contributor to nonpoint source pollution and desertification, and represents one of the most serious threats to world food security.

In the United States alone, some two billion tons of soil erode from cropland on an annual basis. About 60 percent, or 1.2 billion tons, is lost through water erosion, and the remainder is lost through wind erosion. This is equivalent to losing 0.3 meter (1 foot) of topsoil from 810,000 hectares (2 million acres) of cropland each year. Although soil is a renewable resource, soil formation occurs at rates of just a few inches per hundred years, which is much too slow to keep up with erosive forces. The loss of soil fertility through erosion is incalculable, as are the secondary effects of pollution of surrounding waters and increase of sedimentation in rivers and streams.

Erosion removes the topsoil, the most productive soil zone for crop production and the plant nutrients it contains. Erosion thins the soil profile, which decreases a plant's rooting zone in shallow soils, and can disturb the topography of cropland sufficiently to impede the operation of farm equipment. Through erosion, nitrates, phosphates, herbicides, pesticides, and other agricultural chemicals are carried into surrounding waters, where they contribute to eutrophication and oxygen-depleted "dead zones" offshore. Erosion also causes the deposit of increased sedimentation in lakes, reservoirs, and streams, which eventually require dredging.

There are several types of wind and water erosion. The common steps in water erosion are detachment, transport, and deposition. Soil particles become detached from soil aggregates, and the particles are carried, or transported, away; in the process, the particles scour new soil particles from aggregates. Finally, the soil particles are deposited when the water flow slows. In splash erosion, raindrops impacting the soil can detach soil particles and hurl them considerable distances. In sheet erosion, a thin layer of soil is removed by tiny streams of water moving down gentle slopes. This is one of the most insidious forms of erosion because the effects of soil loss are imperceptible in the short term. Rill erosion is much more obvious because small channels form on a slope. These small channels can be filled in by tillage. In contrast, ephemeral gullies are larger rills that cannot be filled by tillage. Gully erosion is the most dramatic type of water erosion. It leaves channels so deep that even equipment operation is prevented. Gully erosion typically begins at the bottoms of slopes, where the water flow is fastest, and works its way with time to the top of a slope as more erosion occurs.

Wind erosion generally accounts for less soil loss than does water erosion, but in states such as Arizona, Colorado, Nevada, New Mexico, and Wyoming, it is actually the dominant type of erosion. Wind speeds 0.3 meter (1 foot) above the soil that exceed 16 to 21 kilometers (10 to 13 miles) per hour can detach soil particles. These particles, typically fine to medium-size sand grains fewer than 0.5 millimeters (0.02 inches) in diameter, begin rolling and then bouncing along the soil, progressively detaching more and more soil particles by impact. The process, called saltation, is responsible for 50 to 70 percent of all wind erosion. Larger soil particles are too big to become suspended and continue to roll along the soil. Their movement is called surface creep.

The most obvious display of wind erosion is called suspension, when very fine silt and clay particles detached by saltation are knocked into the air and carried for enormous distances. The Dust Bowl of the 1930's was caused by suspended silt and clay in the Great Plains of the United States. It is also possible to see the effects of wind erosion on the downward sides of fences and similar obstacles. Wind passing over these obstacles deposits the soil particles it carries.

Other effects of wind erosion are tattering of leaves, filling of road and drainage ditches, wearing of paint, and increasing incidence of respiratory ailments.

CONTROL MEASURES

The four most important factors affecting erosion are soil texture and structure, roughness of the soil surface, slope steepness and length, and soil cover. Several passive and active methods of erosion control involve these four factors. Wind erosion, for example, is controlled through the creation of windbreaks, rows of trees or shrubs that shorten a field and reduce the wind velocity by about 50 percent. Tilling the land perpendicular to the wind direction is also a beneficial practice, as is keeping the soil covered by plant residue as much as possible.

Water erosion is controlled through a number of practices. In the United States, farmers can get help from the federal government-sponsored Conservation Reserve Program to protect highly erosive, steeply sloped land. Tilling land along the contours of slopes aids in preventing erosion, as does the shortening of long slopes by terracing, which also reduces the slope steepness. Permanent grass waterways can be planted in areas of cropland that are prone to water flow. Likewise, grass filter strips can be planted between cropland and adjacent waterways to impede the velocity of surface runoff and cause suspended soil particles to sediment and infiltrate before they can become contaminants.

Conservation tillage practices such as minimal tillage and no-till farming, or zero tillage, have been widely adopted by farmers as a simple means of erosion control. As the names imply, these are tillage practices in which as little disruption of the soil as possible occurs and in which any crop residue remaining after harvest is left on the soil surface to protect the soil from the impacts of rain and wind. The surface residue also effectively impedes water flow, which results in less suspension of soil particles. Because the soil is not disturbed, practices such as no-till farming also promote rapid water infiltration, which reduces surface runoff. Zero tillage is rapidly becoming the predominant practice in southeastern states such as Kentucky and Tennessee, where rainfall levels are high and erodible soils occur.

FURTHER READING

Barnes, Dianna. *Coastal and Beach Erosion: Processes, Adaptation Strategies and Environmental Impacts.* New York: Nova, 2015.

Blanco, Humberto, and Rattan Lal. *Principles of Soil Conservation and Management.* New York: Springer, 2008.

Dent, David. *Soil As World Heritage.* Dordrecht: Springer Netherlands, 2014.

Faulkner, Edward. *Plowman's Folly.* Norman: University of Oklahoma Press, 1943.

Field, Harry L., and John B. Solie. "Erosion and Erosion Control." In *Introduction to Agricultural Engineering Technology: A Problem Solving Approach.* 3d ed. New York: Springer, 2007.

Fournier, Arthur J. *Soil Erosion: Causes, Processes and Effects.* New York: Nova Science, 2011.

Hand, Carol. *Depleted & Contaminated Soil and Your Food Supply.* New York: Rosen Pub, 2013.

Osman, Khan Towhid. *Soil Degradation, Conservation and Remediation.* New York: Springer, 2016.

Plaster, Edward. *Soil Science and Management.* 5th ed. Clifton Park, N.Y.: Delmar Cengage Learning, 2008.

Schwab, Glen, et al. *Soil and Water Conservation Engineering.* 5th ed. Clifton Park, N.Y.: Delmar Cengage Learning, 2005.

EVOLUTION: ANIMAL LIFE

—Peter B. Heller

FIELDS OF STUDY

Agriculture; Agricultural Geography; Biology; Zoology; Biochemistry; Chemistry; Physical Science; Botany; Animal Science; Ecology.

PRINCIPAL TERMS

gene: the basic unit of heredity

gene flow: the movement of genes from one population to another

genetic drift: change in gene frequencies in a population owing to chance

interbreeding: the mating of closely related individuals, which tends to increase the appearance of recessive genes

migration: the movement of individuals, resulting in gene flow, changing the proportions of genotypes in a population

mutation: alteration in the physical structure of the DNA, resulting in a genetic change that can be inherited

natural selection: the process of differential reproduction in which some phenotypes are better suited to life in the existing environment and thus are more likely to survive

speciation: the formation of new species as a result of geographic, physiological, anatomical, or behavioral factors

species: the basic category of biological classification representing a group of potentially or actually interbreeding natural populations which are reproductively isolated from other such groups

taxon (pl. taxa): group of related organisms at one of several levels such as the family Canidae, the genus *Canis*, or the species *Canis lupus*

SUMMARY

Evolution is the change in the gene pool of a population over time by such processes as natural selection, genetic drift, mutation, and migration.

Current consensus holds that the so-called big bang—the high-temperature, high-density event that marked the beginning of the universe—occurred some fifteen billion years ago, with the sun and Earth formed about four and a half billion years ago. Four billion years ago, the relatively newly created sun shone with only 70 percent of its current strength. The atmosphere had no free oxygen. No bacteria, no viruses, no plants, and no animals were in existence. Subsequently, as chemical processes are assumed to have created oxygen and an organic "soup," microbial life in the form of the simplest cells without a nucleus, prokaryotes, developed out of this primordial ooze. These bacteria were the only living organisms for about two billion years. After that time, about 1.5 billion years ago, more complex cells with nuclei, eukaryotes, appeared. Thus, in all, for some 5.5 billion years, bacteria were the only existing animal organisms.

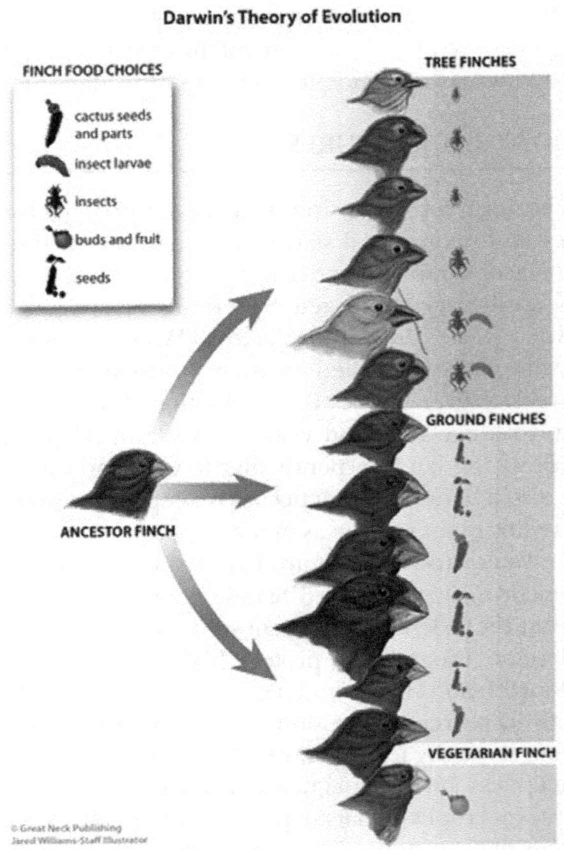

Darwin's Theory of Evolution: Darwin believed that finches' beaks evolved enabling them to survive in different ecosystems on the Galapagos Islands. © EBSCO.

THE BEGINNINGS OF ANIMAL LIFE

Eventually, the two kingdoms, the botanical and the zoological, started to diverge and millions of animals came into being. They were multicelled and were equipped with specialized body parts. They were distinguished from plants primarily but not exclusively by their methods of feeding, locomotion, and reproduction. Most of the phyla seem to have appeared during the Precambrian period, an immense span of geological time that ended about 590 million years ago. Few fossils have remained from these prehistoric times, but the most explosive period for the development of life was the Cambrian, some 590 to 505 million years ago. In a relatively short span of ten million years, all the animal phyla currently known came into being, perhaps encouraged by an increase in oxygen in the seas, where animal life began.

Eventually, animals developed a nervous system enabling them to control their movements more appropriately, as well as sense organs to help them find suitable food. At the margins, however, the dichotomy between the botanical and the animal world has remained ambiguous, since there are many microorganisms that defy clearcut classification. At times, these difficult cases are known as Protista, or Protists.

In all forms of life, including the animal kingdom, no phylum has been produced by a single evolutionary event. Nor have different animal orders appeared as a result of sudden evolutionary changes. Rather, all have come about, whether in gradual or punctuated manner, by the cumulative effect of small steps in different directions. This, at least, is the theory posited by Charles Darwin's explanation of evolution.

The time line of the animal kingdom is closely connected with this evolutionary chronology. Darwin's theory of the origin of species through natural selection, nearly universally accepted in the scientific community but at times opposed on Biblical creationist grounds and in some circles of the lay community, has it that animals, like plants, have changed since the beginning of life on earth and are still evolving today. In this view, there was no sudden creation of all species. Instead, over long periods of time, new species have evolved from isolated populations of existing species. These came to occupy new niches separate from the niches of the original species. Thus, all current species are changed descendants of others that existed previously. If there are fewer apparent links between phyla than between families further down the classification ladder, the reason is that phyla have had a longer history and so have experienced more opportunities for the elimination of intermediate forms. From an evolutionary viewpoint, the difference between species down an evolutionary line are even more recent than those between families, and so on. On average, it takes about 500,000 generations for one species to evolve into another. For species to have survived in their environments—with simultaneous changes in ecology, climate, and flora—many animal forms are now more complex and efficient than their ancestors used to be.

GENES & EVOLUTION

Life began in the seas, so for animals to live in freshwater, let alone on dry land or in the air, many obstacles had to be overcome. The impetus to conquer these inhospitable realms came from competition, according to Darwin's theory of the survival of the fittest. Those fauna that managed to surmount the problems were the ones that underwent waves of adaptive measures and evolved into new kinds of animals.

It was only in the twentieth century that it was discovered that the characteristics of a species are passed on from parents to offspring by genes. Genes provide cells for particular features, such as webbed feet. With such characteristics inherited by offspring from parents, there is a resemblance among generations. However, at times a parent may produce quite a different offspring because of genetic change. The young, in turn, may replicate this difference in their own descendants in a process known as mutation, which may occur spontaneously for unknown reasons, but also may occur due to known causes, such as exposure to radiation. At times, mutations may be useful in allowing the species to adapt better to its environment; for instance, darker moths have a better chance of survival in a forest than lighter ones because the latter are more visible to their predators than the former. Other mutations may be harmful, such as larger size that slows down a species, making its flight from danger more difficult.

Whatever the case, within a time frame of 1 to 10 million years, animals may remain the same, may evolve, or may become completely extinct, as did the dinosaurs about 60 million years ago, after being dominant for some 350 million years.

ADAPTATION TO ENVIRONMENT

Natural selection leading to a new species may be accelerated when members of the original species move to a new environment, whether voluntarily or driven by the elements. Separated populations may develop different traits as they adapt to their new condition. Eventually, they will become sufficiently different to be unable to produce offspring with members of their original population. This process has repeated itself many times, over millions of years, and accounts for the large diversity in the animal world, not to mention the additional diversity consequent on artificial breeding by humans, widely observed among domesticated animals such as horses, cattle, and dogs.

Animals occupied new environments as species living in water moved to the land or later to the air. Thus, the step from fish to amphibian was essentially one from living in water for the whole life to living on land for the adult stage of the life cycle. The step from amphibian to reptile was one of increasingly proficient adaptation to land life at all stages of the life cycle.

Birds and mammals evolved in different directions from the reptiles, the first in adaptation to an arboreal and finally a flying life and the second as a further advance in the maintenance of an even and high body temperature—homeothermy—by combining an insulating external layer such as hair with a variety of physiological thermostats.

Events in geology, climate, and flora also determined the geographic distribution of species. Thus, marsupials are currently found almost entirely in Australasia and South America. The tiger exists only in India and Southeast Asia. The lion is restricted mainly to Africa. This pattern reflects the way in which these groups have evolved in relation to the physical world.

New animal groups evolve into many different forms, especially when they become dominant. For instance, when mammals came to occupy the dominant position, some became meat-eaters while others became vegetarians; some became smaller while others became larger; some became runners while others ended up as burrowers or flyers; still others returned to the water. This trend allowed the descendants of the original type to exploit a much greater range of environments and resources. Essentially, those species whose sense organs or brain morphology and functions improved the most ended up being dominant—primates in general and humans in particular.

TIME FRAME

The exact time of the origins of animals during earth's evolutionary history is not known because the early species were soft-bodied, at first single-celled and later multicellular life forms, that did not fossilize well. Fossils are the best material evidence of archaic times. Fossils do not appear earlier than 650 to 500 million years ago, not only because the animal life of the time was inappropriate for fossilization, but also because continued crustal shifts in the ensuing eons disturbed the very early rock formations. Accordingly, fossil evidence is unavailable for the entire early history of animals, which must consequently remain speculative. Current taxonomic interrelationships suggest the early history, and taxonomic diagrams may be regarded as presumptive evolutionary diagrams as well. However, a ball of carbon discovered in a cavity etched in a rock some 3.86 billion years old suggests that some life on earth was already possible at that time.

Knowledge is also limited by the fact that, even though over a million different species of animals have been identified, it is suspected that a similar number remain to be discovered or became extinct before such identification could be made. In the United States alone, some forty species of birds, about thirty-five species of mammals, and twenty-five other species have become extinct in the last two hundred years alone—less than a blip on earth's time scale—as a result of human activities such as the destruction of animal habitats through urbanization, the clearing of land for agricultural purposes, pollution, the introduction of new species from other parts of the world which turned out to be predatory to domestic specimens, hunting, and especially human population growth. It is widely predicted that climatic change triggered by greenhouse gases will continue, even enhance, this process, thereby endangering more animal species. Whatever the future, however, evolutionary biologists estimate that some 99 percent of all species that have ever lived on earth are now extinct.

Despite these and other caveats, here is a very approximate timetable of the evolution of animal life:

Timetable of the Evolution of Animal Life

Life Form	*Date of Emergence*
Simplest single-celled Protozoa	3.5 billion years ago
Invertebrates evolving from Protozoa	670 to 640 million years ago
First vertebrates evolving from invertebrates	500 million years ago
First mammals	200 million years ago
Hominids (modern man) from apelike hominoids	200,000 to 25,000 years ago

Translated in terms of a single year:
January 1: Big Bang
March 22: Bacteria, the first living animals
November 9: Invertebrates
November 22: Vertebrates
December 16: Mammals
December 28: Primates, the highest order of mammals
December 31, a few minutes before midnight: Modern man, the dominant primate

EVOLUTION OF EXISTING & EXTINCT HUMAN SPECIES & AUSTRALOPITHECINES

The root of the hominid evolutionary tree is still imperfectly known. The earliest australopithecine species, *Australopithecus anamensis*, is believed to be over four million years old, by which time that branch had diverged from African apelike ancestors. This species was followed by the *Australopithecus afarensis* nearly 3.5 million years ago. Much later came *Homo habilis*, called "skillful man" since they could presumably produce primitive tools, some two million years ago. They were followed by *Homo erectus*, "upright man," about one million years ago. Finally, *Homo sapiens*, "knowing man," emerged about 200,000 years ago. In the meantime, the australopithecine branch, after evolving through a number of intermediate species such as *A. africanus*, *A. aethiopicus*, and *A. robustus*, died out about one million years ago. To date, the earliest unearthed human-like primate fossil, that of Lucy, a three-foot-tall female discovered in Ethiopia, is about 3.2 million years old. The earliest fossil of an upright walker was discovered in the same area of Ethiopia in 2015 and is approximately 3 million years old. The oldest known fossils ever discovered, however, are those of bacteria and are in fact microscopic fossils of cells and bacteria "living" together in what was then an oxygen-free world over 3.4 billion years ago.

Modern humans are believed to have radiated out of Africa into Asia and Europe. Subsequently, cultural evolution became more prominent than biological evolution, but as modern humans evolved over the last four million years to their current condition, they developed manipulative skills, bipedalism, a change from specialized to omnivorous feeding habits, and especially, a threefold increase in cranial capacity from *H. afarensis* to *H. neanderthalensis*, together with behavior appropriate to the control of the environment.

Although humans are not the only animals capable of conceptual thought, they have refined and extended that ability until it has become their hallmark. Thus, thanks to the symbolic language of *Homo sapiens*, modern humans make possible the accumulation of experience from one generation to the next. Such cultural evolution is possessed by few, if any other animal species. It is for this reason that humans, more than other animals, have found ways to mold and change their environment according to need rather than in response to environmental demands. Because of this ability and humans' control of technology, the species has more say about their biological future than any other.

FURTHER READING

Carroll, Sean B., Jennifer K. Grenier, and Scott D. Weatherbee. *From DNA to Diversity: Molecular Genetics and the Evolution of Animal Design*. Malden: Blackwell Science, 2001. Print.

Conway-Morris, Simon. *The Crucible of Creation: The Burgess Shale and the Rise of Animals*. Los Angeles: Getty Center, 1999. Print.

Ghosh, Pallab. "'First Human' Discovered in Ethiopia." *BBD*. BBC, 4 Mar. 2015. Web. 31 Dec. 2015.

Lavers, Chris. *Why Elephants Have Big Ears: Understanding Patterns of Life on Earth*. New York: St. Martin's, 2001. Print.

Long, John A. *The Rise of Fishes: Five Hundred Million Years of Evolution*. Baltimore: Johns Hopkins UP, 1996. Print.

"Oldest Fossils on Earth Discovered." *Science Daily*. Science Daily, 22 Aug. 2011. Web. 31 Dec. 2015.

EXPANSIVE SOILS

—*Nathan H. Meleen*

FIELDS OF STUDY

Soil Science; Engineering; Ecology.

PRINCIPAL TERMS

differential movement: the unequal movement of various parts of a building or pavement in response to swelling or shrinkage of the underlying soil

evapotranspiration: the movement of water from the soil to the atmosphere in response to heat, combining transpiration in plants and evaporation

hydration: a process whereby, when soils become wet, water is sucked into the spaces between the particles, causing them to grow several times their original size

loading: an engineering term used to describe the weight placed on the underlying soil or rock by a structure or traffic

moisture content: the weight of water in the soil divided by the dry weight of the soil, expressed as a percentage

montmorillonites: a group of clay minerals characterized by swelling in water; the primary agent in expansive soils

shrinkage: an effect opposite to hydration, caused by evapotranspiration

soils and foundation engineering: the branch of civil engineering specializing in foundation and soil subgrade design and construction

soil stabilization: engineering measures designed to minimize the opportunity and/or ability of expansive soils to shrink and swell

SUMMARY

Expansive soils—soils that expand and contract with the gain and loss of water—cause billions of dollars in damage to houses, other lightweight structures, and pavements, exceeding the costs incurred by earthquakes and flooding.

Expansive soils—soils that expand and contract with the gain and loss of water—cause billions of dollars in damage to houses, other lightweight structures, and pavements, exceeding the costs incurred by earthquakes and flooding.

SOIL SWELLING & SHRINKAGE

Certain types of soil expand and contract with the gain and loss of water, primarily through seasonal changes. One meter of expansive soil has sufficient power to lift a 35-ton truck 5 centimeters. Many cracks in walls and concrete surfaces are assumed to be caused by settlement, when in fact they are produced by the lifting force of expansion.

Soils vary enormously in their ability to expand, ranging from zero in very sandy soils to highly expansive in montmorillonite clays. By contrast, kaolinite clays do not expand. Some key indicators of expansive soils are a bricklike hardness when dry, including great resistance to crushing, and stickiness and weakness when wet; also, they have a glazed surface when cut with a knife. Soil cracks and popcornlike surface appearance are other key signs. Where soils from volcanic ash produce primarily montmorillonite clays, bentonite is formed. This extremely expansive soil prevents even vegetation from growing and is easily visible from a road, with its barren, popcornlike surface and painted-desert coloration.

The ability of montmorillonite clays to expand comes from their physical and chemical structure. Clay minerals have a molecular structure like sheets of paper. Some bond very strongly from layer to layer and have little swelling potential. Expansive clay minerals, however, have weak bonds between the layers. These layers are negatively charged and repel one another, but are bonded by positively charged ions (cations) such as sodium and calcium dissolved in the soil water. When soils rich in these minerals become wet, water is drawn into the spaces between the clay particles, a process called hydration, causing them to grow up to seven times their original size. The opposite effect, shrinkage, is caused by evapotranspiration.

In climates with great variation in seasonal moisture, especially marginally humid and semiarid lands, large changes occur in the moisture content of the soil, with subsequent swelling and shrinkage. In the midlatitudes and subtropics, winter and spring are the wet seasons, as vegetation is dormant and rain and melting snow are abundant. Summer and fall are the dry seasons, soilwise; evapotranspiration demand is high, and much water is transferred from the soil to the atmosphere. Human activity can also produce localized effects through excessive watering, leaking pipes, poor drainage around buildings, and use of trees and bushes for landscaping. It is these volume changes in the swelling soils beneath structures, caused by the gain and loss of water, that make expansive soils so damaging.

DIFFERENTIAL MOVEMENT & VARIABLE LOADING

Expansive soils do their damage primarily through differential movement and variable loading. If the entire structure is lifted and sinks uniformly, there is little problem. Unfortunately, swelling and shrinkage are seldom uniform. Many buildings are located on sloping land, and, as a result, groundwater is more likely to be encountered on the uphill side during wet seasons. Consequently, more expansive force is exerted on that side, cracking walls and slabs. The construction season can also be a factor on sloping land, as the portion of the building pad cut into the hillside, and the portion built up with fill, tends to respond differently to changes in moisture. Even on level land, deep foundations may be stable while grade beams and slabs resting upon the surface rise and fall with the seasons.

While much attention is given to damaged buildings, damage to roads may actually have a greater impact on the public. The life of a road is measured by the total weight it carries over a period of time, divided by the strength of the roadbed. Many older roads are built directly upon the natural grade, including highly expansive soils. Because these soils are weak when wet, the heavier loading at concentrated points under truck tires causes the pavement to break down more rapidly. Once a cracking pattern is visible, a chuckhole is but a few wetting cycles away, as water can now easily reach and further weaken the soil below. Major roadbed problems may also occur near the boundaries between highly expansive and less expansive soils as a result of differential swelling. Water seepage at concentrated points beneath the roadbed also undermines the strength of the roadbed and contributes to differential movement.

Damage to structures from expansive soils can be categorized as reduced performance, economic loss, and massive failure. Massive failure is rare, but the threat of it is usually enough to demand removal of the danger. High on this list would be leaning chimneys and massively cracked concrete channels. Also, raised portions of patios, sidewalks, and driveways are hazardous to pedestrians. Reduced performance occurs when doors and windows cannot be closed because of misalignment, and unsightly cracks appear on walls and concrete slabs. Economic losses occur from such unsightly cracks when potential buyers are discouraged from purchasing property or property owners have to repair them. Also, higher heating and cooling bills result from energy losses through cracks, especially around windows and doors. For taxpayers, a little-recognized cost is incurred through more frequent road repairs.

ON-SITE GEOTECHNICAL INVESTIGATION

Fortunately, the problems caused by expansive soils are reasonably well understood by soils and foundation engineers. Typical procedures include an on-site geotechnical investigation, laboratory testing of samples obtained during the on-site investigation, and design recommendations specific to the anticipated usage of a site. The importance of sound engineering here cannot be overemphasized. The cost of such work is small compared to the cost of the project and is far cheaper than potential remedial work. Inasmuch as the soil ends up underneath the building, roadway, or other structure, it is obviously inaccessible when problems arise; it is far easier and cheaper, therefore, to compact the soil properly and install piers and drains before rather than after.

An experienced soils engineer will generally know what to expect within a given part of a city or region. Soils are notoriously diverse, however, and the only sure way to secure the accurate information needed for a sound, cost-effective design is an on-site investigation. Key information is provided by borehole samples, extracted by a drilling rig similar to that used for wells. The selection and number of borehole locations are determined from architectural layouts and the size of the project. One hole may be sufficient for a house, whereas numerous holes may be ordered for a high-rise building. Three kinds of geotechnical information are obtained directly from these holes: a description of the type and thickness of soil and rock; resistance to penetration (influenced by moisture content), which gives a crude measure of soil strength; and, if the hole is deep enough, depth to the water table.

With distressed structures, one of the most important judgments a soils engineer can make is whether remedial work is needed or will be cost-effective. Cracks often appear in new buildings and pavement; though frustrating, they may be of little consequence. By contrast, serious problems may be avoided if timely repairs are made before incipient failures, such as leaning chimneys, reach the critical point.

TESTS FOR EXPANSIVE POTENTIAL

The most common measure of expansive potential is the Atterberg limits test, run in the laboratory. Though crude, it has proved its value over many years as a predictor of expansive and nonexpansive soil characteristics. The test essentially measures the difference between the resistance of a soil sample to flow as it becomes wetter and wetter, called the liquid limit, and disintegration under rolling as it becomes drier and drier, called the plastic limit. The numbers used are moisture content percentages, obtained by computing the weight of water as a percentage of the dry weight of the soil. The difference between these two percentages is called the plasticity index, referred to as the PI. Variances among plastic limit numbers are generally small, but liquid limit numbers vary enormously. The stickier the clay, the more moisture will be required to make it flow; thus the higher the liquid limit. One characteristic of expansive soils worth remembering is their stickiness when wet. Consequently, the higher the liquid limit, the larger the PI and the greater the swell potential.

The Atterberg limits test is relatively easy and inexpensive to run. It is the simplest way to test large numbers of samples, and such tests are conducted by the thousands in soils laboratories. Where precise and detailed information is needed from key samples, however, a consolidation-swell test is run. This test is much more complex, time-consuming, and expensive, but the information gained is vital for the design of more expensive structures. The test mimics the behavior of soils under the loading applied by a building and their response to swell pressure as water is absorbed. Additional tests, such as confined and unconfined compression, are also commonly run on the more expensive projects. These test results, combined with experience and input from colleagues, form the basis for the recommended design. Good designs anticipate and avoid problems, minimize construction costs, and recommend specifications for quality-assurance testing during construction.

DESIGN ALTERNATIVES

Design alternatives include strengthening the structure, stabilizing the soil, using some combination of the first two, or anticipating movements to isolate their impact. If a slab can be strengthened so that it moves as a unit, the effects of differential movement will be mitigated. In larger buildings, reinforced grade beams are commonly suspended on piers, with space deliberately left under the grade beam to allow room for soil expansion. Major attention is given to stabilizing the soil. The goal is to compact the soil in its expanded condition, balancing the need for strength with the need to allow some room for additional expansion. A key problem is to keep the soil from drying out before it is sealed, usually by a concrete slab or asphalt. Drains are commonly installed around the perimeter and covered by concrete, asphalt, or plastic sheeting. Three types of drains are used: peripheral drains around the perimeter; interceptor drains to control subsurface water from uphill sources; and sump drains, which require a pump and are usually located under a basement floor slab.

A combination of increased strength and soil stabilization is often used. Much attention is given to the water content and density of the soil beneath the structure. When feasible, select fill (stable soil having low expansion and good binding qualities) is imported for the top layer. Lime or fly ash may be mixed with the soil to increase both its strength and its stability. This treatment is especially effective under pavement (parking areas and roadbeds). Special equipment has been invented to mix these additives into the soil efficiently. Laboratory tests are commonly run to measure the effectiveness of treatment in terms of PI reduction and strength; a common rule for lime treatment is to reduce the PI to 10 or less. Lime is the easiest to handle and its effectiveness is easiest to test, though fly ash, a by-product of coal burning in electrical power plants, is growing in use because of its low cost. The problem with fly ash is that its use requires meticulous application, beyond the abilities of many inexperienced operators. It often contains heavy metals that may contaminate water supplies. Engineering supervision of the mixing and compaction—with strict attention to moisture and time specifications—is critical to the successful use of fly ash.

Where construction cost is a key concern, it is sometimes feasible to design the structure so that slabs and interior walls can move freely without affecting the foundation or other parts of the building. Trim work can mask such movement so that aesthetic concerns are not a factor. Once structures are completed, the correction of problems is far more difficult. Corrections may range from simple drains and moisture barriers around perimeters to tearing out the pavement and redoing the soil subgrade. An accurate diagnosis is even more essential to success in this case. Consequently, skilled engineers study the problem in detail, and heroic efforts are sometimes made to obtain soil samples. The engineer's judgment of the seriousness of the problem and the remedial action that follows are especially vital. A skilled engineer can tell much from cracks and other distress patterns.

GROWING IMPORTANCE OF FOUNDATION ENGINEERING

Only since the 1970's has the severe threat represented by swelling soils been recognized by the construction industry. Several changes are responsible for this growth. First, buildings with slab-on-grade floors (concrete resting directly upon soil) have replaced basements and crawl spaces in many areas. Insofar as seasonal moisture changes are restricted to the top 1 meter and essentially disappear at a depth of 3 meters, this change created conditions far more susceptible to expansion and contraction. Second, heavier trucks and increased usage have put greater stress on roadbeds and parking areas. Once the pavement is broken, it is much easier for water to reach the soil below, and failure soon follows. Third, population growth and rising costs of land have encouraged the development of nonfarmland, which increases the odds of encountering expansive soils. Still, this extended use of land is fine, as long as the dangers are understood and taken into account in designing and planning projects.

When confronted with any hazard, one must consider whether the problem is serious enough to require expert help. Foundation engineering is a little-recognized but invaluable part of any construction program involving expansive soils. Knowledge of soil conditions prior to new construction is well worth the relatively small cost. Well-informed individuals provide an invaluable service to themselves, employers, and the organizations they serve. As part of the National Cooperative Soil Survey, the U.S. Department of Agriculture and the Soil Conservation Service provide booklets on regional surveys. In map and table format, accompanied by explanatory text, these publications present detailed information about agricultural and engineering soil conditions.

Further Reading

Al-Rawas, Amer Ali, and Mattheus F. A. Goosen, eds. *Expansive Soils: Recent Advances in Characterization and Treatment.* New York: Taylor & Francis, 2006. Discusses classification of expansive soils. Provides information on swelling and consolidation, and the effects these processes have on buildings. Discusses how site characteristics and stabilizing agents can alter expansion processes.

Buzzi, O., S. Fityus, and D. Sheng, eds. *Unsaturated Soils.* New York: CRC Press, 2009. Discusses the dynamics of unsaturated and expansive soils. Discusses experimental studies and theoretical studies in separate volumes. Includes articles drawn from the Conference on Unsaturated Soils held in 2009. Designed to aid academics and graduate students in soil mechanics research.

Fenner, Janis L., Debora J. Hamberg, and John D. Nelson. *Building on Expansive Soils.* Fort Collins: Colorado State University, 1983. Explains the swelling mechanism. Includes a map of the United States showing the occurrence and distribution of potentially expansive materials, and gives identification techniques, design alternatives, and remedial measures. Especially recommended for builders.

Harpstead, Milo I., Thomas J. Sauer, and William F. Bennett. *Soil Science Simplified.* 4th ed. Ames: Iowa State University Press, 2001. Excellent primer for the nonspecialist. Well illustrated and suitable for high school Earth science curricula. Thorough coverage of soil topics relevant to agriculture, engineering, and geology.

Holtz, Wesley G., and Stephen S. Hart. *Home Construction on Shrinking and Swelling Soils.* Denver: Colorado Geological Survey, 1978. A short primer on the problems and possible solutions related to buildings situated on expansive soils. Includes good illustrations of typical problems, drain design, and recommended interior construction techniques.

Jochim, Candace L. *Home Landscaping and Maintenance on Swelling Soil.* Special Publication 14. Denver: Colorado Geological Survey, 1981. Prepared for the homeowner, this beautifully illustrated publication has large type and clear, simple illustrations, including a map of swell potential in Colorado, especially the Denver area. Also provides information about groundwater and how to maintain proper drainage around a house.

Katti, R. K. *Behaviour of Saturated Expansive Soil and Control Methods.* Rev. ed. Oxford: Taylor & Francis, 2002. Published under the auspices of the Central Board of Irrigation and Power in India. Focuses on soil stabilization, soil dynamics, and black cotton soil in that country. Includes an extensive bibliography and an index.

Koerner, Robert M. *Construction and Geotechnical Methods in Foundation Engineering.* New York: McGraw-Hill, 1984. Part of a series in construction engineering and geotechnical engineering, and attempts to address this mixed audience. Heavy mathematical treatment.

Legget, Robert F. *Cities and Geology.* New York: McGraw-Hill, 1973. Designed as a textbook for courses in urban or environmental geology. Highly relevant to those interested in urban planning. Uses case histories worldwide to demonstrate the effects of geology on urban growth and problem solution. An important chapter on geologic hazards is included.

Nelson, John D, and Debora J. Miller. *Expansive Soils: Problems and Practice in Foundation and Pavement Engineering.* New York: John Wiley and Sons, 1997. Nelson deals primarily with the interaction between soil and human-made structures such as building foundations and roadways.

Newson, Malcolm. *Land, Water, and Development: Sustainable and Adaptive Management of Rivers.* 3d ed. London: Routledge, 2009. Presents land-water interactions. Discusses recent research, study tools and methods, and technical issues, such as soil erosion and damming. Suited for undergraduate students and professionals. Covers concepts in managing land and water resources in the developed world.

Olson, Gerald W. *Soils and the Environment: A Guide to Soil Surveys and Their Application.* New York: Chapman & Hall, 1981. Designed for a worldwide audience and general readership. Explains how soil surveys are prepared, what the goals are, and how to understand the most important applications.

Rajapakse, Ruwan. *Geotechnical Engineering Calculations and Rules of Thumb.* Burlington, Mass.: Butterworth-Heinemann, 2008. Written for the professional geological engineer. Provides explanations of calculations and general practices of engineers building shallow and pile foundations, retaining walls, and analyzing geological features and compositions.

Sobolevsky, Dmitry Yu. *Strength of Dilating Soil and Load-Holding Capacity of Deep Foundations: Introduction to Theory and Practical Application.* Rotterdam, Netherlands: A. A. Balkema, 1995. A detailed survey of soil mechanics, soil testing, swelling soils, and foundations suitable for students who already have some background in the field.

F

Factory Farming

—Sally Driscoll and David C. Morley

FIELDS OF STUDY

Agribusiness; Agronomy; Mathematics.

PRINCIPAL TERMS

battery cages: Cages used to confine poultry that can be stacked in rows and columns. Most battery cages measure between forty-eight and sixty-eight square inches and are too small for birds to spread their wings.

certified humane: A label indicating that in addition to meeting the United States Department of Agriculture's (USDA) certified organic standards, meat or eggs offered for sale also come from free-range animals that have been provided with fresh air, sunlight, exercise, and a healthy diet.

certified organic: A USDA label indicating that meat or eggs offered for sale have not been irradiated and are derived from animals that have not been given antibiotics, hormones, or feed sprayed with pesticides. This designation does not necessarily mean that the animal was grass fed or free range.

concentrated animal feeding operation (CAFO): A type of factory farm in which thousands of animals are confined in small pens or cages until they are "harvested." The U.S. Environmental Protection Agency (EPA) inspects CAFOs with more than 1,000 head of beef cattle, 2,500 swine, or 750 dairy cattle.

factory farming: The industrial, large-scale production of livestock, poultry, hogs, fish, and crops; also called "intensive farming." Many factory farms perform all operations from breeding to processing.

free range: A method of farming in which livestock are not confined to cages but are permitted to roam freely. If provided with outdoor access, they gain fresh air, sunlight, and often natural food. The USDA certifies whether meat is "free range," although verification standards are minimal: to be certified, cattle and hogs must eat grass and live outdoors.

gestation crates: Cages in which sows are confined during their four-month-long gestation period. When birth is imminent, the sows are moved to "farrowing crates," where they are confined along with their piglets for three weeks. The sows are then returned to the gestation crates and artificially impregnated to begin the cycle again. Gestation crates are too small for pigs to turn around or stretch.

lagoons: Large pits dug in the ground on factory farms to accommodate effluent (manure and urine) produced by hogs, cows, and other livestock.

sustainable agriculture: Farming that does not have a detrimental effect on the environment.

SUMMARY

Large-scale industrial farms supply approximately 80 percent of the poultry, beef, and pork consumed in the United States. Aquaculture supplies about 40 percent of fish. To keep up with demand, roughly ten billion animals are slaughtered annually. The ability to mass-produce large quantities of food ensures convenience and affordable prices for most Americans.

The term "factory farming" has come into popular usage because many of the small, independent farms of the past have evolved into huge enterprises complete with assembly lines and thousands of employees. Many fall under the category of "concentrated animal feeding operations" (CAFOs), which can endanger public health, the environment, and the welfare of animals.

Public health concerns involve the advent of new diseases, including avian influenza, *Escherichia coli* (*E. coli*) O157:H7, and bovine spongiform encephalopathy (BSE)—better known as "mad cow disease"—all of which spread easily among animals confined in close quarters. Many of these diseases are antibiotic resistant, which health officials blame on the overuse of antibiotics by livestock and poultry farmers who administer the drugs to prevent and cure a multitude of diseases—or, in many cases, simply to increase food production.

In addition to antibiotics, factory-farmed cows are injected with the controversial recombinant bovine growth hormone (rBGH or rBST), approved by the United States Food and Drug Administration (FDA) but banned in the European Union, Canada, and several other countries due to human health concerns.

Rotary milking parlor. Image by Gunnar Richter Namenlos.net (Own work) CC-BY-SA-3.0.

Hormones, antibiotics, and pesticides from processed animal feed are excreted in animal manure. The manure is then stored in lagoons that often leak or spill out into water systems, where they pose additional health risks to humans, fish, and wildlife.

As manure decomposes, it releases methane, hydrogen sulfide, and carbon dioxide into the atmosphere, contributing to global warming and air pollution. Research has shown that residents who live near hog farms suffer a greater incidence of respiratory problems than the general population.

Animal-welfare groups, including People for the Ethical Treatment of Animals (PETA) and the Humane Society of the United States (HSUS), oppose CAFOs for their use of battery cages, gestation crates, and confining stalls that prevent exercise and other healthy activities.

Additional concerns about factory farms include the health and welfare of employees, many of whom are illegal immigrants and thus unlikely to speak out against substandard working conditions for fear of reprisal, and the economic burden the large corporations have placed on the small farmer.

HISTORY

The roots of contemporary factory farming methods can be traced to the Industrial Revolution. Farm machinery, including the threshing machine and the combine, allowed for the mass production of the grains that now constitute most of a farm animal's diet. The advent of the railroad and refrigerated cars opened new markets for milk, beef, eggs, and other farm products. Electricity allowed for mechanized feeding, watering, ventilation systems, and assembly lines for egg collection and meat processing. From their crude beginnings in the late nineteenth century, milking machines evolved into massive stainless-steel facilities that can milk hundreds of cows per hour.

Twentieth-century advancements included the discovery of vitamins, now commonly added to processed feed. Fertilizers and pesticides helped increase the production of grains. Artificial insemination resulted in a year-round supply of beef and pork. Incubators allowed for breeding of poultry. Antibiotics helped reduce outbreaks of disease, a threat that increased along with the prevalence of CAFOs.

The post–World War II baby boom and increased affluence in the United States created more demand for meat. Searching for ways to keep up with demand while yielding higher profits, farmers began squeezing in more animals, often on shrinking farm acreage. Some had to choose between expansion and bankruptcy. Mergers became common. Out of this climate grew Smithfield Foods, Christensen Farms, and Tyson Foods, to name a few of the largest players in the agribusiness industry.

Incidents in which manure lagoons contaminated local water supplies became frequent. In 1972, the Clean Water Act authorized the National Pollutant Discharge Elimination System (NPDES), a program that issues permits to factory farms. The act also established regulations for preventing manure-contaminated wastewater from flowing into streams, rivers, lakes, and other bodies of water. Four years later, the Environmental Protection Agency (EPA) began annual inspections of CAFOs. Since then, dozens of lawsuits have implicated factory farms in polluting water systems.

Health and animal welfare concerns increased during the 1990s. In response, the United States Department of Agriculture (USDA) initiated the "certified organic" labeling program. Farmers and others in the industry also implemented the "certified humane raised and handled" program, which identifies meat and dairy products that both meet the USDA's organic standards and come from animals that were provided with fresh air, sunshine, exercise, and a healthy, natural diet.

In 2003, the newly organized American Grassfed Association began to work with the USDA to implement a labeling program for meat coming from cows and pigs raised outdoors in a pasture. Supporters of grass-based diets cite higher amounts of protein, fiber, vitamins A and E, and omega-3 fatty acids than in feed consisting mostly of corn and soybeans. A grass diet is also thought to help the body eliminate harmful bacteria, such as *E. coli.*

Aquaculture, or "fish farming," of salmon, catfish, shrimp, and other fish and shellfish took hold during the 1980s and 1990s in response to an increased demand for fish at a time when overfishing began to threaten many commercial species. Environmentalists cite marine pollution as a by-product of fish farms that are built in the ocean or along coastlines. Human health concerns are related to the prevalence of disease among confined fish and the use of antibiotics and other additives, including food coloring, which is used to make salmon bred in captivity look more appetizing. There is also a danger to wild fish populations when farmed fish escape their pens.

FACTORY FARMING TODAY

As animal welfare groups educate the public, attitudes toward factory farming are changing. In 2006, Arizona voted for a ban on confining breeding sows in gestation crates, following a similar vote held in Florida four years earlier. The European Union has already implemented various regulations to reduce the confinement of sows, cows, and poultry.

Demand for grass-fed beef, free-range chicken and eggs, and organic meat has resulted in the growth of such farms. Stores that stock these foods, including Whole Foods and Trader Joe's, are also expanding rapidly.

In December 2006, Pilgrim's Pride Corporation, then considered to be the second-largest supplier of poultry in the United States, announced a merger with the nation's third-largest supplier, Gold Kist, making Pilgrim's the largest chicken producer in the United States. In 2002, Pilgrim's Pride was implicated in a listeria outbreak in which eight people died and fifty were sickened. In 2009, Brazilian-based meat processor JBS, the largest in the world, acquired a majority stake in Pilgrim's Pride.

Under the administration of President Barack Obama the EPA has taken steps to better enforce clean water laws. In addition, U.S. secretary of agriculture Tom Vilsack and then–attorney general Eric Holder promised in 2010 to pursue the implementation of new transparency rules in an effort to ensure fair competition in agriculture, starting with a series of five workshops on the topic. Yet transparency in agriculture remains a hotly contested topic, prompting numerous attempts at the state level to introduce legislation that would criminalize

Chickens in industrial coop. Image by איתמר ק., ITamar K. (Own work).

whistleblowing activities aimed at exposing animal abuses within the agricultural industry. Between 2011 and 2015, at least four states—Iowa, Utah, Idaho, and North Carolina—passed "ag-gag" laws, which institute harsh penalties for undercover investigative activities at agriculture facilities, and dozens more were introduced but failed to pass. In August 2015, Idaho's law was deemed unconstitutional by the U.S. District Court for the District of Idaho.

FURTHER READING

PERIODICALS

Casa, Kathryn. "The Changing Face of Farming in America." *National Catholic Reporter* 12 Feb. 1999: 13+. *Points of View Reference Center.* Web. 6 Nov. 2015.

Colleran, Brian. "Think before You Eat: The Widespread Effects of Factory-Farmed Meat." *E: The Environmental Magazine* July–Aug. 2008: 30–31. *Points of View Reference Center.* Web. 6 Nov. 2015.

Hollenbeck, James E. "Concentrated Animal Feeding Operations (CAFOs) as Potential Incubators for Influenza Outbreaks." *Trakia Journal of Sciences* 11.2 (2013): 205–9. *Academic Search Complete.* Web. 13 Nov. 2015.

Kirby, David. "Obama on Factory Farming: Something to Crow About." *Washington Post.* Washington Post, 26 Mar. 2010. Web. 6 Nov. 2015.

Kolbe, Emily A. "'Won't You Be My Neighbor?': Living with Concentrated Animal Feeding Operations." *Iowa Law Review* 99.1 (2013): 415–43. *Academic Search Complete.* Web. 6 Nov. 2015.

McCarthy, Colman. "Infiltrators Expose the Cruelty of Meat." *National Catholic Reporter* 27 Apr. 2012: 17. *Points of View Reference Center.* Web. 6 Nov. 2015.

Nierenberg, Danielle. "The Commercialization of Farming: Producing Meat for a Hungry World." *USA Today Magazine* Jan. 2004: 22–24. *Academic Search Complete.* Web. 6 Nov. 2015.

Thomas, Mary Ann. "Factory Farming Backlash Spurs More Choice—and a Little Chaos." *Christian Science Monitor* 14 Sept. 2013: n.pag. *Academic Search Complete.* Web. 6 Nov. 2015.

Vogt, Willie. "Regulations Affect Farms in Variety of Ways." *Southwest Farm Press* 19 Sept. 2013: 14. *Academic Search Complete.* Web. 6 Nov. 2015.

WEBSITES & DIGITAL FILES

Abrams, Lindsay. "Factory Farming Is Killing the Planet: Why the Meat Industry's Future Needs to Look More Like Its Past." *Salon.com.* Salon Media Group, 25 July 2015. Web. 6 Nov. 2015.

"Ag Gag: Safeguarding Industry Secrets by Punishing the Messenger." *Food Integrity Campaign.* Food Integrity Campaign, 2015. Web. 6 Nov. 2015.

"Cage-Free vs. Battery-Cage Eggs." *The Humane Society of the United States.* HSUS, 7 Apr. 2015. Web. 6 Nov. 2015.

Guilford, Gwynn. "The New Frontier for Big Agriculture Is Aquaculture." *Quartz.* Atlantic Monthly Group, 17 Aug. 2015. Web. 6 Nov. 2015.

"Industrial Livestock Production." *Sustainable Table.* GRACE Communications Foundation, 2015. Web. 6 Nov. 2015.

Union of Concerned Scientists. *CAFOs Uncovered: The Untold Costs of Confined Animal Feeding Operations.* By Doug Gurian-Sherman. Cambridge: UCS, 2008. *Union of Concerned Scientists.* Web. 6 Nov. 2015.

FARMING INDUSTRY

—Jeffrey Larson

FIELDS OF STUDY

Agribusiness; Agronomy; Agriculture, Agronomy; Food Science; Mathematics.

SUMMARY

The farming industry plants, grows, harvests, transports, and markets crops to be used for food, animal feed, biofuel, textiles, and other purposes. Some of the major industrial crops include grain, oilseeds, tobacco, dry beans, potatoes, vegetables, melons, fruits, and nuts. In the United States, the five most significant crops are heavily subsidized by the federal government: corn, rice, soy, wheat, and cotton. The industry encompasses many additional areas of labor beyond simply working the fields, such as agricultural science, engineering, business, and economics.

HISTORY OF THE INDUSTRY

The U.S. farming industry dates back to European agricultural practices that were brought to the New World during colonial times and combined with American Indian farming techniques. The southern colonies planted tobacco and cotton and used slave labor to grow them. This slave-based agriculture was central to the nation's economy through the founding of the United States and lasted until the American Civil War brought an end to slavery in the 1860s.

The mid-to-late 1800s saw the introduction of more industrial agricultural practices that increased productivity and helped drive U.S. economic growth. The advancement of the farming industry was facilitated in 1887 by the establishment of state land-grant universities. These educational institutions taught and studied agriculture and established a federally funded system of agricultural experiment stations and networks in each state. In addition, as the Industrial Revolution progressed, the farming industry was rapidly mechanized during the late nineteenth and early twentieth centuries. Farmers employing such machines as tractors and combine harvesters could complete tasks at a speed and on a scale previously considered impossible. Mechanized agricultural production thus replaced manual labor wherever possible.

In the early 1930s, the farming industry was struggling in the face of economic and environmental hardships. The Dust Bowl ruined many farms, driving farmers off their land, during a time when the Great Depression had already caused a wave of farm foreclosures. In response, the U.S. Congress passed the Agricultural Adjustment Act of 1933, instituting an emergency program to aid farmers by protecting their equity and increasing the prices paid for farm products. Programs were also instituted to teach farm operators techniques to help protect soil resources.

Renewed demand for the industry was generated by World War II, and farm prices improved. By the mid-1940s, U.S. agriculture entered a sustained period of productivity gains. The development of railroad and highway networks and the increasing use of container shipping also were essential to the growth of mechanized agriculture. They expanded the size of the market any given farm could reach by making long-distance shipping of produce affordable. This in turn made it possible for farmers to increase their profits by producing more food than their local regions could purchase, providing a motive to increase productivity beyond the scope of local demand.

Even with long-distance shipping, however, in the later twentieth century, agricultural productivity exceeded demand nationally, creating a surplus of agricultural commodities. The government sought to reduce this surplus, instituting policies designed simultaneously to increase demand and reduce supply. Surplus food, for example, was purchased by the government for use in school lunch programs to feed impoverished children.

Hardy Meyers chicken operation near Petal, Mississippi.

In the late twentieth and early twenty-first century, agriculture has become an increasingly global industry. Exports continue to account for approximately 30 percent of total gross farm receipts in the United States, but American farm operators face new challenges, including the emergence of new foreign competitors, trade tensions over new technologies, the impact of climate change, and food safety issues.

Increased use of machines and the introduction of government price supports during the past century allowed farm operators to increase the size of their farms and gain efficiencies. Such advances in productivity through mechanization, plant developments, and new chemical fertilizers and pesticides led to fewer, larger, more specialized farms and a massive migration out of the farming industry.

Also in recent decades, the farming industry has seen consumers becoming increasingly concerned about food safety, nutrition, quality, variety, and prices. Environmental interests have established relationships to help bring environmentally friendly production methods to agriculture. Government expenditures on food and nutrition programs and natural resource conservation have increased and become more integrated within the industry. The demand for organic food products also continues to expand rapidly.

THE INDUSTRY TODAY

The farming industry today is driven by demand from federal agricultural policy programs, food consumption trends, and the crop export market. The profitability of individual farm operators or companies depends on their ability to maximize crop yield, minimize input costs, and minimize risk. Larger farms and companies have the advantage of highly automated technologies and have access to the latest in seed and crop technologies. However, more data indicate that small operations can compete effectively by producing specialty products.

U.S. farms that produce and market crops are becoming larger and more concentrated. There has been an increase in the number of farm operators contracting with large business operations called agribusinesses. These businesses purchase crops from many farms to achieve vast economies of scale. They allow contractors to reduce their risk levels, because the contractors are guaranteed a buyer and a set price for their crops. Other operators sell their products themselves, either to local distributors and suppliers or on commodities exchanges, where prices often are determined by worldwide supply and demand.

The farming industry increasingly encompasses and partners with highly technological businesses that are integral in the production and sales of farm products. These include companies developing farm machinery, fertilizer, pesticides, and herbicides. Some farmers, however, engage in pest management without the use of pesticides. For example, they may use natural predators, such as insects that eat other insects rather than crops, or introduce bacteria that only harm insects or substances that interfere with pests' reproductive cycles. Such techniques are often referred to as "organic" and are seen as aspects of the larger umbrella practice of organic farming.

In addition to organic farming, the industry employs a host of new and renewed methods of raising crops, including precision farming and bioengineering. Precision farming utilizes technology and data to make efficient decisions about raising crops. For example, precision farmers make detailed maps of their land and use electronic devices to monitor and manage crop conditions. Monitoring devices measure the amounts of water, weeds, and nutrients in the soil and overlay those data on maps created from satellite and aerial photographs. The Global Positioning System (GPS) is used to pinpoint the exact locations of land and nutrients. This information is used to determine where additional fertilizer, herbicides, and water may be required. By reducing unnecessary applications of chemicals, operators reduce their production costs. GPS technology also helps operators navigate equipment accurately through their fields. Sensors on this equipment monitor yields, providing operators with even more information to determine where extra nutrients may be needed or which crops are ready for harvest.

Bioengineers modify the genetic code of plants to provide them with desirable characteristics or remove undesirable characteristics. Some crops are engineered to resist certain pests, tolerate drought, or contain additional nutrients. A growing number of farm operations, however, avoid the use of synthetic pesticides and herbicides and bioengineered crops altogether. These organic farms take advantage of the higher prices they can charge for certified organic grains, fruits, and vegetables. Since 1990, sales of organic products have grown 20 percent per year, and they are expected to continue climbing. The organic market has allowed some small farms to remain profitable. Organic farmers often need to know more than conventional farmers about the science of agriculture: They must know how plants grow and what helps them thrive. Most organic operators (and some conventional operators) adopt techniques that raise soil quality to naturally resist diseases, weeds, and insects. They might increase soil quality by planting high-nutrient crops in the off-season, for example, or by using conservation tillage practices, such as leaving remains of a crop to decompose in the soil after harvesting.

Advances in the farm industry have created many other disciplines related to agriculture. Agricultural scientists for example, advise farm operators regarding the optimal techniques to control weeds, apply pesticides, conserve water, or prevent soil erosion. Agricultural sales involves selling farm-related products, which requires knowledge of agriculture. There are many avenues one can take if one is interested in the farm industry in the twenty-first century. Opportunities exist, from owning or managing a large, capital-intensive farm to providing the technology to discover a specialty product that is in high demand. Whatever sector of the twenty-first century farming industry one embarks upon, one will need to possess a strong business aptitude and the ability to make money in an industry that is expensive to enter and that involves substantial risk.

INDUSTRY OUTLOOK

The outlook for the farming industry shows it to be somewhat stable. This outlook stands in contrast to many years of decline in small farms and in farm-related employment. The number of agricultural workers is expected to decline by 3 percent by 2022, and agricultural manager jobs are projected to decline by about 19 percent. Improvement in technology have lead to significant productivity increases that enables the agricultural industry to produce more crops with fewer workers.

Market pressures on small and midsize family farms will continue to drive consolidation in the farm industry; however, a growing number of small-scale farms have found success with organic food production and horticulture, which are among the fastest growing segments of the agricultural industry. The larger, more prosperous farms are likely to grow in size in order to achieve greater economic returns on their investments. Larger farms may also continue to receive a greater portion of farm subsidies to help offset losses beyond the operators' control, such as drought, disease, and depressed markets. Despite this ongoing shift to larger farms, the majority of farms in the United States are family farms according to the Food and Agriculture Organization, meaning that the principal operator and spouse provide most of the farms labor; family farms that meet the FAO definition accounted for 86.1 percent of U.S. farms and 47.4 percent of U.S. farm production in 2014. An additional 11.5 percent of U.S. farms, accounting for 38.4 percent of farm production, are family owned and operated by rely extensively on hired labor. Any direct ownership of large farms by publicly held corporations is likely to remain small. Through advances in technology, productivity will continue to improve and also result in less farm labor being required to produce crops. Low prices for many agricultural goods have not kept up with the increasing costs of farming. If these conditions continue, it will be difficult for many small farmers to survive. The difficulty of making a living exclusively from farming may continue to require farmers to seek other means of employment in addition to their farms.

There are factors on the horizon that are influencing the changes taking place in agriculture. More domestic consumers are purchasing agricultural products from farmers markets, community supported agriculture (CSA), and other local food distribution systems. Consumers are becoming increasingly health conscious, and the demand for organic produce continues to grow. Such produce is grown mostly on small to midsize farms that can use nonchemical methods of pest control effectively. This type ofproduction within the industry allows farms of small acreage to remaineconomically viable. Federal, state, and local government programs are increasingly available to provide assistance targeted at small to midsize farms.

Exports for agricultural products have been rising, reflecting international demand, reaching of $140.5 billion in 2017, the third highest total on record. In the first ten months of 2018, exports were 2 percent ahead of the previous year at nearly $122 billion, although it remains to be seen what effect current and proposed trade tariffs could have on the industry. U.S. agricultural exports to China, Canada, and Mexico account for about 42 percent of total exports, with Japan and the European Union accounting for a combined 18 percent of U.S. agricultural exports. Rising incomes in countries such as China, India, Indonesia, and Brazil increased the demand for diversified diets in those countries, leading them to seek products beyond their traditional foods. An increased global consumption of biofuels and related research and technology for the development of energy self-sufficiency added further to the U.S. agriculture industry. This type of pressure on supply and price is strongly shown by the case of corn. This farm commodity dramatically affects the broader structure of global food markets, and at one point 25 percent of the corn crop was found to be used for ethanol production. In the future, whether corn will be used that intensely for fuel is critically dependent on the price of oil and also on the politics of biofuels. Cereals and oilseeds can also be used in multiple ways, and the competition for these commodities is expected to rise among businesses.

Climate change also threatens the productivity of the entire global food system. Some food supply and world economics experts foresee a looming food crisis that will first affect those populations that are already food-insecure and that may then spread to other populations. This could place a high demand on the farming industry in countries where high-efficiency agriculture continues to exist and thrive.

EMPLOYMENT ADVANTAGES

Those who work in the farming industry directly or indirectly enjoy the variety of the work. Many are exposed to rural settings and the satisfaction of being self-employed and working the land. Many small-farm operators work alongside family members, strengthening their families' cohesiveness. Food policy and sustainable agriculture are rapidly gaining traction as fields of great concern to many people, and many young people are passionately committed to finding ways to make sustainable farming pay. It remains unclear how many will succeed, but resources to help them are increasing. These include programs to develop cohesive local food production, processing, and distribution systems called foodsheds, as well as resources to enable small and midsize farms to compete with large farms. If these programs succeed, small-scale farming is poised to be an extremely exciting and fulfilling, albeit difficult, career choice.

ANNUAL EARNINGS

The farming industry is ever changing and depends on several factors, including market pressures, alternative-fuel consumption, and politics. In September 2018, the the USDA's Economic Research Service forecast a national net income for agriculture of $65.7 billion in 2018, down 13% from 2017. The 2018 net farm income forecast is substantially below the 10-year average of $87 billion (in nominal dollars). It would be the lowest since 2016 in nominal and inflation-adjusted 2017 dollars. This is primarily the result of the outlook for continued weak prices for most major crops. U.S. farm income experienced a golden period during 2011 through 2014 due to strong commodity prices and robust agricultural exports.

FURTHER READING

Crowell, Susan. "2007 Census of Agriculture: Agriculture's 'Middle' Slipping Away." *Farm and Dairy.* Farm and Dairy, 9 Feb. 2009. Web. 10 Sept. 2012.

Hoag, Dana L. *Applied Risk Management in Agriculture.* Boca Raton: CRC, 2010. Print.

Hoppe, Robert A., and Penni Korb. "Large and Small Farms: Trends and Characteristics." *Structural and Financial Characteristics of U.S. Farms.* Washington, DC: USDA Economic Research Service, 2004. PDF file.

Jurena, Remy. *Agriculture in the U.S. Free Trade Agreements: Trade with Current and Prospective Partners, Impact and Issues.* New York: Nova Science, 2008. Print.

Leval, Kim, et al. *The Impact and Benefits of USDA Research and Grant Programs to Enhance Midsize Farm Profitability and Rural Community Success.* Lyons: Center for Rural Affairs, 2006. Print.

MacDonald, James. "Family Farming in the United States." *Economic Research Service.* US Dept. of Agriculture, 4 Mar. 2014. Web. 23 Oct. 2015.

Mapes, Kathleen. *Sweet Tyranny: Migrant Labor, Industrial Agriculture, and Imperial Politics.* Urbana: U of Illinois P, 2009. Print.

Ricketts, Cliff, and Kristina Ricketts. *Agribusiness: Fundamentals and Applications.* 2nd ed. Clifton Park: Delmar Cengage Learning, 2009. Print.

Schnepf, Randy. *U.S. Farm Income Outlook for 2015.* Washington, DC: Congressional Research Service, 2015. PDF file.

United States Bureau of Labor Statistics. "Agricultural Workers." *Occupational Outlook Handbook.* US Dept. of Labor, 8 Jan. 2014. Web. 23 Oct. 2015.

United States Department of Agriculture. Economic Research Service. "Special Feature: The Shift to Large Farms." *Structure and Finances of U.S. Farms: Family Farm Report,.* Washington, DC: Author, 2007. PDF file.

United States Department of Agriculture. World Agricultural Outlook Board and U.S. Interagency Agricultural Projections Committee. *USDA Agricultural Projections to 2019.* Washington, DC: Author, 2010. Print.

United States Department of Agriculture. "World Agricultural Supply and Demand Estimates." *USDA.* US Dept. of Agriculture, 9 Oct. 2015. Web. 23 Oct. 2015.

Farms & Farming

—Lisa Jo Rudy

FIELDS OF STUDY

Agribusiness; Agriculture; Botany; Agronomy; Animal Science; Soil Science; Ecology.

SUMMARY

Farms are areas of land or water that are used to grow plants and raise animals for food and other uses. Some farms grow or raise just one type of plant or animal, while others grow many different species. Farms can be very large, covering hundreds of acres. Many, however, are only large enough to feed a family. Farming has been a human activity for over 12,000 years. In the last 100 years, however, new technologies have changed the way people farm in many parts of the world.

HISTORY OF FARMS & FARMING

Until about 12,000 years ago, human beings were hunters and gatherers. Then, at about the same time in locations around the world, a revolution occurred. Instead of hunting and gathering food from the wild, human beings began planting and harvesting crops. There are theories about why this change happened in each location, but no one knows why it occurred in so many places at the same time.

This change, called the Neolithic Revolution, had a huge impact on human life. People no longer needed to move with the seasons, gathering or hunting food as it was available. Instead, they could stay in one place and store food for winter. Permanent communities, towns and cities, were built. While some people farmed, others were freed to do different types of activities. Communities divided up necessary tasks. Individuals developed specialties in trades like building and making textiles.

Michael Eudy weeds one of his herb beds by hand. Image by Jacardona (Own work) CC-BY-SA-3.0.

The first crops were probably grains such as wheat and barley, as well as rice and figs. Vegetables such as squash were first grown about 10,000 years ago. Goats, sheep, pigs, and cows were among the first animals to be farmed. They are still among the most common farm animals.

FARMING BEFORE THE INDUSTRIAL REVOLUTION

Farms and farming were a part of almost every civilization from the time of the Neolithic Revolution. The choice of crops and animals to farm depended upon location and cultural norms. Farmers began to plant, harvest, and trade a wider range of fruits, vegetables, and grains. In some areas, crops were planted as feed for farm animals. Alfalfa, barley, wheat, and oats were sometimes grown for animals rather than for human consumption.

In addition to growing plants for food, people began growing plants that could be used for other purposes. Cotton, for example, was used for cloth, while flax was grown to make linen. Hemp was grown to make strong, durable ropes, and tobacco was grown for smoking.

FARMING SINCE THE 1800S

Prior to the 1800s, farmers had little information about soil, water, fertilizer, or pest control. As the science of farming advanced, however, farmers gained access to chemicals and natural techniques for improving harvests and managing insects. They also began to deliberately cross-breed plant species. New, hardier grains such as triticale (a combination of wheat and rye) were developed.

In the twentieth and twenty-first century, farming was increasingly industrialized. More farms were dedicated to nonfood crops such as corn for ethanol (a fuel). Genetic technologies made it possible to design crops to be hardier and more nutritious.

A promotional print from about 1873 for Grange members showing scenes of farming and farm life. Used in article on the Granger movement. Image by Library of Congress.

Further Reading

"Crop." *National Geographic: Education.* National Geographic Society, 2014. http://education.nationalgeographic.com/

Finlayson, Bill. *Changing Natures: Hunter-gatherers, First Famers and the Modern World.* Bristol, UK: Bristol Classical Press, 2010.

"The Development of Agriculture." *Genographic Project.* National Geographic Society, 2014. https://genographic.nationalgeographic.com/development-of-agriculture/

FISHING INDUSTRY, FISHERIES & FISH FARMING

— Ming Y. Zheng

FIELDS OF STUDY

Ecology; Environmental Studies; Aquaculture; Farming.

DEFINITION

Fish have been a protein source throughout history. Early fishing primarily involved individuals capturing fish near their communities for consumption or trade. Ships gave fishers access to ocean-based fisheries. Commercial fishing became industrialized by the late nineteenth century, as technological innovations helped locate, catch, and process fish. In addition to fish living in natural freshwater or saltwater fisheries, fish cultivated in fish farms' ponds or tanks represented approximately one-fourth of the fish eaten in the world. Countries benefited economically with domestic trade and by exporting valuable fish. In the early twenty-first century, fisheries generated billions of dollars globally with approximately 42 million people employed to catch fish and several hundred millions more working in related industries. Fisheries reinforced food security when climate changes caused shortages of other agricultural products

SIGNIFICANCE FOR CLIMATE CHANGE

For centuries, fishers realized that weather affected fish populations, but they lacked the scholarly resources to investigate their observations. During the nineteenth century, fishery researchers began applying scientific methodology to study diverse factors affecting fish health, reproduction, and habitats. They contemplated reasons for decreased fish populations besides overfishing. U.S. Environmental Protection Agency representatives voiced concerns that climate deviations affected the quantity and quality of fisheries to the U.S. Congress in 1988. The American Fisheries Society promoted research examining how climate change might affect fisheries. Scientists consulted ships' logs and records documenting fish-catch statistics and meteorological patterns to evaluate hypotheses about the climate's possible role in fish population losses. Researchers used computer simulations to consider future climatic factors that could potentially harm fish.

Fish are exceptionally vulnerable to habitat changes. The World Wildlife Federation emphasizes that temperature increases of 2° Celsius or more are dangerous for fish. Healthy water temperatures differ according to species. Scientists link increased water temperatures to global warming, because water absorbs heat trapped by greenhouse gases (GHGs). The Intergovernmental Panel on Climate Change (IPCC) states that abnormally high ocean temperatures have occurred as deep as 3,000 meters below sea level. Water-temperature increases can interfere with fishes' oxygen supply and the physiological processes associated with maturation, digestion, and spawning. High temperatures also weaken fishes' resistance to toxic substances and pathogens that invade their habitats. Scientists evaluate fish otoliths (ear bones), which indicate growth, to study how temperatures at varying depths influence development.

Researchers have correlated temperature deviations in oceans with fish displacement due to reduced growth of food sources, such as plankton. Melting

Top Ten Inland Fishing Nations

Nation	*Fishery Production, 1998 (thousands of metric tons)*	*Percentage of Global Production**
China	2,280	28.5
India	650	8.1
Bangladesh	538	6.7
Indonesia	315	3.9
Tanzania	300	3.7
Russian Federation	271	3.4
Egypt	253	3.2
Uganda	220	2.8
Thailand	191	2.4
Brazil	180	2.3

Source: Food and Agriculture Organization.
*The top ten countries accounted for 65 percent of global inland fisheries production.

glaciers caused by global warming raise water levels and dilute the oceans, decreasing salinity. This surplus water disrupts currents crucial for transporting food and removing pollutants, including from such human sources as fishing vessels and fish-factory emissions. Fish migrate to waters with more compatible temperatures and ample food supplies, but they often fail to thrive if they are unable to adapt. Since the late 1980's, Massachusetts fishermen reported an 80 percent decline in cod catches. Fisheries managers aware of areas where wild fish have relocated can adapt management practices to minimize additional climate change damage and to replenish fisheries.

Temperature increases of almost 3°Celsius in some locations have caused freshwater in streams, rivers, and lakes to evaporate, reducing habitat size and stressing fisheries, particularly trout and salmon species in the western United States. In Montana, grayling fish in the Big Hole River decreased from ninety-six fish per kilometer of river in the 1990's to as few as eight in the early twenty-first century because of temperature changes. Some hardier species, such as smallmouth bass, sought warmer habitats, competing for resources and displacing indigenous fish. Scientists warned that loss of diverse fish genetic material diminishes fisheries and some species will become extinct. Fishery experts estimated reductions of approximately 90 percent of bull trout and 40 percent of salmon populations by 2050 if extreme heat and arid climate conditions persist.

Climate changes alter fish ecosystems in lakes, resulting in fishery populations being reduced by as much as 30 percent, as has been reported in Africa's Lake Tanganyika. In addition to temperature and precipitation deviations, climate changes slow wind velocities needed to stir nutrients from deeper lake water to the surface. Changing climates might force fish farmers to relocate stock from ponds to protected sites where hot temperatures do not threaten fish health.

Oceans become more acidic when water absorbs carbon dioxide (CO_2) in the atmosphere, bleaching coral reefs, which many fish need for habitats and nurseries. A January 16, 2009, *Science* article reported researchers' discovery that calcium carbonate in fish fecal material can control some ocean acidity. CO_2 circulating in fish blood stimulates production of calcium carbonate. Ironically, as GHG emissions increase, the amount of fish-produced calcium carbonate might too, helping counter climate change's impact on fisheries.

Further Reading

Marine Resources Service, Fishery Resources Division, FAO Fisheries Department. *Review of the State of World Marine Fishery Resources.* Rome, Italy: Food and Agriculture Organization of the United Nations, 2005. Illustrated sections feature diverse regional fisheries. Climate chapter discusses scientific investigations that question whether climate changes have caused population fluctuations in fisheries.

Nelleman, Christian, Stefan Hain, and Jackie Alder, eds. *In Dead Water: Merging of Climate Change with Pollution, Over-Harvest, and Infestations in the World's Fishing Grounds.* Arendal, Norway: United Nations Environment Programme, GRID-Arendal, 2008. Comprehensive examination of global warming issues relevant to fisheries. Photographs, figures, and maps displaying past and projected climate changes supplement the text.

Robbins, Jim. "As Fight for Water Heats Up, Prized Fish Suffer." *The New York Times,* April 1, 2008, p. F4. Scientists explain how temperature changes and diversion of water for agricultural uses combined to harm fish habitats in western U.S. rivers. Includes suggestions for restoring fish populations.

Sharp, Gary D. *Future Climate Change and Regional Fisheries: A Collaborative Analysis.* Rome, Italy: Food and Agriculture Organization of the United Nations, 2003. Fisheries expert differentiates between global change and warming and how they have continually shaped ecosystems supporting fish throughout the Earth's history. Figures, glossary, bibliography, list of recommended Web sites.

Wilson, R. W., et al. "Contribution of Fish to the Marine Inorganic Carbon Cycle." *Science* 323, no. 5912 (January 16, 2009): 359–362. Highlights the contribution of fish excretions to the carbonate content of seawater; argues that, because the magnesium content of these excretions allows them to be dissolved more readily than other carbonates, they may have a disproportionate effect on ocean alkalinity.

Food Chains & Food Webs

—*Ming Y. Zheng*

FIELDS OF STUDY

Agriculture; Agronomy; Food Science; Ecology; Environmental Studies.

PRINCIPAL TERMS

biological magnification: the increasing accumulation of a toxic substance in progressively higher feeding levels

consumer: an organism that eats other organisms

decomposers: microbes such as fungi and bacteria that digest food outside their bodies by secreting digestive enzymes into the environment

detritus feeders: an array of small and often unnoticed animals and protists that live off the refuse of other living beings

energy pyramid: a graphical representation of the energy contained in succeeding trophic levels, with maximum energy at the base (producers) and steadily diminishing amounts at higher levels

nutrient cycle: a description of the pathways of a specific nutrient (such as carbon, nitrogen, or water) through the living and nonliving portions of an ecosystem

producers: organisms that produce food for themselves as well as for nearly all other forms of life, including plants, plantlike protists, and cyanobacteria

trophic level: the categories of organisms in a community, and the position of an organism in a food chain, defined by the organism's source of energy

SUMMARY

In the biosphere—the portion of earth that sustains life—organisms including plants, animals, and microbes interact and interconnect with one another through organized and complex networks. Understanding this network, also referred to as a food chain or food web, is vital to the study of many subjects in biology.

All activities of life are powered, whether directly or indirectly, by a single source of energy: sunlight. The energy enters the biosphere through primary producers, which trap and convert solar energy into chemical energy, first in form of sugars and ultimately in other complex organic molecules. Energy in its chemical form is then passed along from one type of organism to another through a complex feeding relationship. Two fundamental laws underlie this function: Energy moves through ecosystems in a one-way street, while nutrients cycle and recycle. Each time the energy is used, some of it is lost as heat. Energy needs constant replenishment from an outside source, the sun. In contrast to energy, nutrients constantly cycle and recycle in a circular flow. Nutrients enter the system from soil or water or atmosphere through primary producers (plants), and pass along to herbivores, then to carnivores. The wastes and dead bodies or body parts degraded by detritus feeders return nutrients to the ecosystem.

FEEDING RELATIONSHIPS WITHIN A COMMUNITY

Feeding relationships in a community are often defined or described through food chains and food webs. A linear feeding relationship spanning all trophic levels is called a food chain, whereas many interconnecting food chains in a community make up the food web. Obviously, different ecosystems have drastically different food chains. To illustrate who feeds on whom in a community, it is better first to examine some basic laws and structures that govern a community in general. Energy and nutrients are two common elements that sustain all communities and ecosystems. Energy enters communities through the process of photosynthesis, by which plants and other photosynthetic organisms trap a small portion of sunlight and convert it into sugars. Photosynthetic organisms, from the mighty sequoia to the zucchini and tomato plants in a garden to single-celled diatoms in the ocean, are called autotrophs or producers, because they produce food for themselves. They also produce food for nearly all other organisms, called heterotrophs or consumers.

The amount of life a community can support is determined by how much energy the producers within it can capture. Within the community, energy flows from producers, occupying the first trophic level, through several levels of consumers. The consumers that feed directly and exclusively on producers are herbivores, ranging from caterpillars to buffaloes to wheat aphids. These herbivores, also called primary consumers, form the second trophic level. Carnivores, such as the spider, eagle, fox, and birds that eat caterpillars, are meat eaters, feeding primarily on primary consumers. Carnivores are secondary consumers that form the third or higher trophic level. Some consumers, such as the black bear that eats both blueberries and salmon, occupy more than one trophic level.

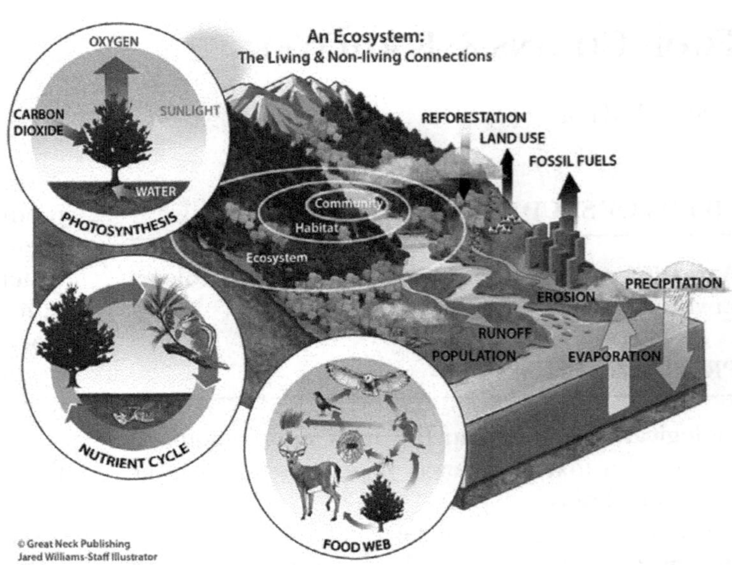

Ecosystems are complex systems involving all of the interactions among living organisms and the environment. © EBSCO.

A food web, however, represents many interconnecting food chains in a community, describing the actual, complex feeding relationships within a given community. A food web also reflects the feeding nature of organisms that occupy more than one trophic level. Animals such as raccoons, bears, rats, and a variety of birds are omnivores, eating at different consumer levels at different times. In addition to producers and consumers, a functional ecosystem also consists of detritus feeders and decomposers that release nutrients for reuse. The extremely diverse network of detritus feeders is made up of earthworms, mites, protists, centepedes, nematodes, worms, some crustaceans and insects, and even a few vertebrates such as vultures. Except for vultures, these organisms thrive in the garden, and compost by extracting energy stored within dead organic matter, in turn releasing it in a further decomposed state. The excretory products released serve as food for other detritus feeders and decomposers, which are primarily fungi or bacteria. Fungi and bacteria digest food outside their bodies by releasing digestive enzymes into the environment. They then absorb the nutrients they need, and leave the remaining nutrients for recycling. Without detritus feeders and decomposers, nutrients would soon be locked into organic matters, and the ecosystem will cease to be functional.

ENERGY TRANSFER & THE NUTRIENT CYCLE

One important principle that governs the flow of energy through the ecosystem is that the energy transfer from one trophic level to the next is never efficient. The net transfer of energy between two trophic levels is roughly 10 percent. During the transfer, 90 percent of the energy is lost as heat or in other forms. Of one thousand calories stored in the producer (plants), the caterpillars that consume all plant tissues will obtain one hundred calories, a bird that eats caterpillars will extract ten calories, and when a hawk catches the bird, that portion of the energy is reduced to a mere one calorie. This inefficient energy transfer between trophic levels is called the 10 percent law, or an energy pyramid. This 10 percent law has profound impacts within an ecosystem. Plants have the most energy available to them; the most abundant animals will be those directly feeding on plants, and carnivores will always be relatively rare, especially those of large size that occupy a higher trophic level.

However, when toxic substances pass through trophic levels, the exact opposite is true. While energy diminishes in the process of flowing from lower to higher trophic levels, toxic substances progressively increase in concentration along the food chain. This phenomenon, called biological magnification, was discovered through the study of the use of the pesticide Dichloro-diphenyl-trichloroethane (DDT). Tests of water samples following the use of this pesticide showed a trace amount of DDT. Tissue analyses of predatory birds in the same aquatic ecosystem, however, revealed a DDT concentration a million times greater than that in the water. Fish caught from the same waters also contained much higher DDT levels than the water, but substantially lower levels than that of the birds that consumed those fish. DDT has since been confirmed as the cause for population declines of several predatory birds, especially fisheaters such as bald eagles, brown pelicans, and cormorants. Understanding biological magnification is crucial to the prevention of widespread loss of wildlife.

Unlike energy, no mechanism or source exists to allow a constant replenishment of nutrients. The same pool of nutrients has been supporting life from the beginning. Nutrients are elements and small molecules that form all the building blocks of life. Macronutrients are those acquired by organisms in large quantities, including water, carbon, hydrogen, oxygen, nitrogen, phosphorus, sulfur, and calcium. Micronutrients, including zinc, molybdenum, iron, selenium, and iodine, are acquired in trace quantities. Nutrients cycle from producer, to consumer, to detritus feeder and decomposer, and eventually back to producer. The major reservoirs of nutrients are in the nonliving environment, such as soil, rock, water, and atmosphere. In an undisrupted ecosystem, nutrients have cycled and recycled in a sustainable manner for thousands of years. However, human intervention, either through industry or agriculture, has created enormous problems for sustainable nutrient cycling. One example is the nitrogen cycle. Although plants may obtain nitrogen from soil, the primary source of nitrogen is the atmosphere. Nitrogen gas (mainly N_2) makes up over 70 percent of the air. N_2 may be extracted by some plants or converted by lightning into useable forms that drop as rainfall. Through industrial production and use of nitrogen fertilizers in agriculture, humans have overcharged the nitrogen cycle, causing acid rains and surface water and groundwater contamination, which pose a serious threat to natural ecosystems.

FOOD CHAINS & FOOD WEBS IN A TERRESTRIAL ECOSYSTEM

Since different ecosystems have drastically different food chains and food webs, they can be better illustrated using examples of different communities. The first example is a land community, to provide an overview of terrestrial food chains and food webs. Plants such as maple trees and squash are the producers that occupy the first trophic level. Aphids, caterpillars, grasshoppers, and other animals that forage directly plants or plant tissues are primary consumers that occupy the second trophic level. Birds, spiders, and other insects that feed on primary consumers are secondary consumers occupying the third trophic level. Large birds, such as eagles, owls, and hawks, that eat secondary consumers are tertiary consumers making up the fourth trophic level. This food chain can go on to even higher trophic levels.

However, natural communities rarely contain well-defined groups of primary, secondary, and tertiary consumers in a linear pattern. In reality, a food web showing many interconnecting food chains in a community describes much more accurately the actual feeding relationships within a given land community. This is in part due to the omnivorous, or "eating all" nature of some animals. These animals include but are not limited to bears, rats, and raccoons. They act as primary, secondary, and even tertiary consumers at different times. Many carnivores will eat either herbivores or other carnivores, thus acting as secondary or tertiary consumers, respectively. An owl, for instance, is a secondary consumer when it eats a mouse, which feeds on plants, but a tertiary consumer when it eats a shrew, which feeds on insects. Once a shrew eats carnivorous insects, it is by itself a tertiary consumer, making the owl that feeds on the shrew a quaternary consumer that occupies the fifth trophic level. Since organisms are interlocked in such complex yet organized networks, disruption at any particular point of the food web (damage to one group of organisms) might have far-reaching effects on a whole community or ecosystem.

FOOD CHAINS & FOOD WEBS IN A MARINE ECOSYSTEM

Coral reefs will be used as an example for marine ecosystem. Reefs are created by concerted efforts of producers—algae—and consumers—corals. In warm tropical waters, with just the right combination of bottom depth, wave action, and nutrients, specialized algae and corals build reefs from their own calcium carbonate skeletons. The reef-building corals grow best at depths of less than forty meters, where light can penetrate and allow their algal partners to photosynthesize. Algae and corals are involved in a mutualistic relationship, where algae benefit from the high nitrogen, phosphorus, and carbon dioxide levels in the coral tissues. In return, algae provide food for the coral and help produce calcium carbonate, which forms the coral skeleton.

Coral reefs provide an anchoring place for many other algae, a home for bottom-dwelling animals, and shelter and food for the most diverse collection of invertebrates and fish in the oceans. In essence, algae is the producer that occupies the first trophic level. Corals that feed on algae are primary consumers sitting at the second trophic level. Many fish (such as blue tang) that feed on corals are secondary consumers occupying the third trophic level. Larger fish, such as sharks that eat small fishes, are tertiary consumers at the fourth trophic level. A vast array of zooplanktons, invertebrates such as sponges, the poisonous blue-ringed octopus, and so on, also live in coral reef ecosystems to make extremely complex marine food webs. For example, the Great Barrier Reef in Australia is home to more than two hundred species of coral, and a single reef may harbor three thousand species of fish, invertebrates, and algae.

Similar to terrestrial ecosystems, aquatic ecosystems are also prone to human disturbance. Of all aquatic or marine ecosystems, coral reefs are probably most sensitive to certain types of disturbance, especially silt caused by soil eroding from nearby land. As silt clouds the water, light is diminished and photosynthesis reduced, hampering the growth of the corals. Furthermore, as mud accumulates, reefs may eventually become buried and the entire magnificent community of diverse organisms destroyed. Another hazard is sewage and runoff from agriculture. The dramatic rise in fertilizer in near-shore water causes eutrophication, by which excessive growth of algae blocks sunlight from the corals, deprives corals of nutrients, and suffocates corals and other organisms. A third threat to coral reefs, overfishing, is also strictly a result of human interference. It is estimated that in over eighty countries, an array of species, including mollusks, turtles, fish, crustaceans, and even corals, are being harvested much faster than they can replace themselves. Collectively, these human activities had destroyed over 30 percent of coral reefs worldwide by year 2000. Assuming no effective measure is taken to preserve or restore coral reef ecosystems, another 50 percent of reefs will disappear by year 2030. The message is clear: Once humans disturb an ecosystem, through damaging one or more species in an intricately networked food web, balance and sustainability within the whole system is affected. The price for such disruption is high and far-reaching.

Further Reading

Baskin, Y. "Ecologists Dare to Ask: How Much Does Diversity Matter?" *Science* 264 (April 8, 1994): 202–203. Outlines and discusses the scientific evidence for the importance of biodiversity to ecosystem function.

Bell, R. H. V. "A Grazing Ecosystem in the Serengeti." *Scientific American* 255 (July, 1971): 86–93. The great migrations of a variety of herbivores of the African savanna are illustrated and described.

Goreau, T. F., et al. "Corals and Coral Reefs." *Scientific American* 241 (August, 1979): 124–136. Good discussions on the ecology and formation of the great reefs built by mutualistic relation between algae and coral.

Holloway, M. "Sustaining the Amazon." *Scientific American* 269 (July, 1993): 90–99. Describes the threats to the rain forests and possible ways to preserve tropical forests and their richness in biodiversity.

Montez, Michele, and Lorraine Bodger. *The New 50 Simple Things Kids Can Do to Save the Earth.* Kansas City: Andrews McMeel Pub, 2009. Print. Provides compelling, practical advice on how individuals can make a difference in being good stewards to global resources and the environment.

Quamman, David. *The Song of the Dodo: Island Biogeography in an Age of Extinction.* New York: Simon & Schuster, 1997. This beautifully written book takes the reader on a globe-spanning tour of wild places and ideas through interwoven personal observations, scientific theory, and history.

FORESTS & THE MITIGATION OF ANTHROPOGENIC CLIMATE CHANGE

—*Nancy M. Gordon*

FIELDS OF STUDY

Ecology; Ecosystem Management; Climatology; Bioclimatology; Climate Modeling; Environmental Sciences; Environmental Studies; Hydroclimatology; Physical Geography; Earth System Modeling; Tree Ring Analysis.

PRINCIPAL TERMS

afforestation: creating forests on lands that was not previously forested

carbon sink: plant growth, mineral formation and synthetic activities that act to remove carbon dioxide and other carbon sources from the environment

dormancy: the portion of the year during which no growth occurs

growing season: the portion of the year during which photosynthesis occurs

reforestation: planting trees to replace forests that have been eliminated either naturally or by human action

SUMMARY

Eight thousand years before the present, before the rise of human civilization, one-half of the Earth's surface was covered by forest. In the intervening years, 20 percent of previously forested land has lost its forest cover, resulting in the loss of a major carbon sink. One of the easiest ways for humans to prevent further carbonization of the atmosphere is to prevent further deforestation and to maintain existing forests intact.

DISTRIBUTION OF FORESTS

The forests of the world fall into three primary zones: tropical, temperate, and boreal. These zones are defined by their climate and precipitation. Trees require a minimum of 25 centimeters of rainfall per year to grow; land that, by virtue of its location, does not attract the minimum rainfall annually will not support trees. Such land will be either grassland, shrubland, or desert. Trees also require a frost-free part of the year for a growing season. Generally, the longer the growing season, the greater the annual growth.

The tropical rain forests on either side of the equator support tree growth that continues virtually all year long. About half of the world's forests are tropical forests, and they contain the greatest biological diversity of species. Because trees through photosynthesis convert atmospheric carbon to contained carbon, the half of the world's forests in tropical areas are of vital importance.

The temperate forest does not grow all year long. The falling temperatures reach a point at which growth is no longer supported, and the trees go into dormancy for part of the year. The temperate zone has a less varied collection of tree species than the tropics, but it contains a mix of coniferous and broad-leaved species, the latter of which generally lose their leaves during the dormant period.

The forest growing farthest to the north is the boreal forest, composed very largely of coniferous species. It grows more slowly than the other forests, because its growing season is the shortest, and it exists in dormancy for a major part of each year. While the true polar regions may have enough precipitation to support trees, their growing season is too short for trees to grow enough to support the needles that provide respiration.

TWENTIETH CENTURY CHANGES

The conversion of formerly forested land to other land uses was greater in the twentieth century than in any previous century. This conversion was driven in large part by the growth in human populations in all parts of the world, but especially during the latter half of the century, primarily in the tropical regions. Between 1950 and 2000, the world's population more than doubled, from 2.5 billion to 6 billion, and the developing world, largely located in the same area where tropical forests are found, contained three-fourths of that larger population. Since forests provide the largest carbon sink, the loss of many forests, especially the tropical forests, is mainly responsible for the rising proportion of carbon in the atmosphere.

The chief driver of deforestation during the twentieth century has been the demand for additional agricultural land to support the growing population. The conversion of forestland to agricultural land adds to atmospheric carbon in several ways. In many cases, the trees that are cut down to release the land for agriculture are burned, and the carbon stored in them is released into the atmosphere. Second, the carbon contained in forest soils over time is also released into the atmosphere when the trees are no longer there to prevent its release. Third, the trees that had been on the land are no longer there to capture future carbon as they grow.

Beside the need for agricultural land to grow crops to feed the growing world population, that population has experienced a change in dietary demand, particularly to include meat products. Some of the land freed up by deforestation is converted not to cropland but to pasture land. Specialized crops such as sugar or soybeans have a market price that poor populations seek to realize by converting forestland to agricultural land. In many areas, the production of specialized crops with major markets in the developed world, such as rubber, has also promoted deforestation.

LOCATIONS OF DEFORESTATION

The loss of forestland to agriculture has occurred in most parts of the world adapted to tropical forests. The Amazonian forest in Brazil and the forests in central America have been subject to important depletion since the mid-twentieth century. The deforestation has also been substantial in Southeast Asia, notably on many of Indonesia's islands, as well as in Malaysia and Thailand. The forest in Africa has been less affected.

The temperate forest that was heavily deforested in the nineteenth century has started to bounce back as urbanization and the mechanization of agriculture have reduced the demand for agricultural land. Thus where the population is heavily urbanized, former agricultural land is being gradually reforested, as in the United States and Europe.

Much of the world's sawtimber comes from the coniferous trees in the temperate forest and from the coniferous trees that cover the boreal forest. The Russian forest constitutes one-fifth of the total forestland of the world, but in recent years much cutting to supply sawtimber to the developed world has depleted some of that forest. Combined with softwood coming from Canada's boreal forests, these two sources have supplied a major portion of the dimensional lumber used by the developed world for the construction of houses. Lumber production peaked in the United States in 1906, at 46 billion board feet, but since then lumber production supplied by U.S. forests has gone down. However, wood remains the third most productive commodity in world trade, behind petroleum and natural gas. A substantial proportion becomes fuelwood.

CONTEXT

At the Rio Conference in 1992, 158 countries agreed to try to prevent further deforestation. It has, however, proved difficult to accomplish partly because market forces are working against it, and partly because definition has proved elusive. The Kyoto Protocol of 1999 identified "reforestation" and "afforestation" as processes that could mitigate deforestation, but compensation to those who carry out such measures has not found widespread acceptance. No way has yet been found to value existing forests such that their preservation could be financially rewarded.

FURTHER READING

Humphreys, David. *Logjam: Deforestation and the Crisis of Global Government.* Earthscan, 2006. Highly critical of the failure of international organizations to take on the task of preventing deforestation.

Malmesheimer, R. W., et al. "Preventing GHG Emissions Through Avoided Land-Use Change." *Journal of Forestry* 106, no. 3 (April/May, 2008). This special issue is devoted entirely to the Society of American Foresters' task force report on forest management solutions for mitigating climate change.

Stern, Nicholas. *The Economics of Climate Change: The Stern Review.* New York: Cambridge University Press, 2007. This famous report tackles the charges of those who claim that the economic costs of attempting to prevent climate change are too great.

Williams, Michael. *Deforesting the Earth: From Pre-history to Global Crisis.* Chicago: University of Chicago Press, 2003. This massive account by an author already well known for his careful evaluation of forest history is definitive.

G

Genetic Engineering

—Dawn A. Laney, MS

FIELDS OF STUDY

Genetics; Botany; Agriculture; Biology; Chemistry.

SUMMARY

Genetic engineering, also known as genetic modification, is an interdisciplinary scientific technique using molecular techniques to directly alter the basic genetic blueprint (DNA) of bacteria, plants, animals, humans, and other living organisms to achieve or enhance a specific trait or useful characteristic. Genetic engineering, though often controversial, is used in diverse areas, including medicine and agriculture, to diagnose and treat diseases, produce industrial products, neutralize pollutants, create higher-yielding crops, and perform scientific research. The genetic engineering process uses the tools of molecular genetics to explore and change living systems on a fundamental level and has revolutionized scientists' ability to understand, modify, and enhance the natural world.

DEFINITION & BASIC PRINCIPLES

Genetic engineering is the direct and purposeful alteration of an organism's DNA—the basic genetic blueprints of a bacterium, plant, animal, human, or other living organism—to add or enhance a specific characteristic or trait. Although genetic engineering is most often discussed in the controversial arenas of crop production or theoretical human genetic manipulation, genetic engineering is used in diverse areas such as medicine, industry, and agriculture to treat disease, diagnose problems, produce industrial products, convert industrial waste, create hardier crops, and perform better scientific research.

The focus of genetic engineering is the gene. Genes are the basic units of inheritance that contain information and instructions for the creation, maintenance, and reproduction of living organisms. Genes are composed of DNA, a highly organized molecule located in almost every cell of an organism's body. In genetic engineering, scientists add very specific pieces of useful genetic material to another organism's genes to change an organism's natural characteristics.

Genetic engineering was made possible by the development of new molecular genetic procedures, often called recombinant DNA technology, that can identify the particular DNA sequence of a gene or an entire genome, allow scientists to find the genetic material that codes for useful or desired features, and then insert the new material into the correct place in another organism's genetic code.

Organisms that have had new genetic material inserted into their code are referred to as genetically modified organisms (GMOs) or genetically engineered organisms (GEOs). Examples of GMOs range from corn that has been engineered to produce an innate insecticide to cows that produce milk containing human insulin.

BACKGROUND & HISTORY

Before modern genetic engineering was possible, farmers had long selected for desired traits by breeding plants and animals with the desirable traits, a process known as selective breeding. Brewers and bakers also changed grains and flour into preferred products such as beer and bread through the use of small organisms called yeast and the process of fermentation.

By the early twentieth century, plant scientists had begun to use the work done by Gregor Mendel in the nineteenth century on the inheritance patterns of specific plant features to more formally introduce improvements in a plant species in a process called classic selection. However, the features of the basic unit of inheritance were not known until James D. Watson and Francis Crick identified the structure of DNA in 1953 with the double-helix model.

The nature of DNA and the technology to manipulate and modify the genetics of an organism was not available until twenty years later, when the first successful recombinant organism was created by Herbert Boyer and Stanley Cohen. Boyer and his laboratory had isolated an enzyme that could precisely cut segments of DNA in an organism, and Cohen found a way to introduce antibiotic-carrying plasmids into bacteria and a way to isolate and clone the genes in the plasmids. They combined their knowledge to create a way to clone genetically engineered molecules in foreign cells. Their discoveries led to the creation of a quick and easy way to make chemicals such as human growth hormone and synthetic insulin.

After Boyer and Cohen, many other scientists worked with recombinant DNA techniques to improve the procedures and develop a variety of genetically modified organisms designed to meet specific scientific, agricultural, industrial, and medical needs. Over time, these techniques and applications in genetic engineering spawned the multibillion-dollar biotechnology industry.

As the biotechnology industry grew and genetically modified organisms became more widespread, it became important to define which organisms were genetically modified organisms and which were products of classic selection. It also became necessary to determine if living organisms produced through genetic engineering could be patented by the companies and universities designing them. In 1980, the U.S. Supreme Court ruled in the case *Diamond v. Chakrabarty* that genetically altered life-forms can be patented.

In 1982, the U.S. Food and Drug Administration (FDA) approved the first consumer product developed through modern genetic engineering: a biosynthetic human insulin, sold under the trademark Humulin. The bacterially produced insulin created by Genentech and marketed by Eli Lilly revolutionized the treatment of diabetes, as it produced fewer immune reactions and its supply no longer depended on the availability of animals.

In 1996, Genzyme Transgenics (which in 2002 became GTC Biotherapeutics) created a transgenic goat that produced milk containing a cancer-fighting protein. It soon created additional transgenic animals that could produce specific human proteins to treat human disease. The ability to produce human hormones, enzymes, and other therapeutic products has decreased the risk of disease transmittal from donors to recipients of human products, increased supply, decreased immune reactions, and decreased the variability between medication batches that had been seen in the past.

In the 1990s, scientists sought to develop genetically modified plants and crops, including genetically engineered foods. By 1992, the first plant designed for human consumption (the Flavr Savr tomato) was approved for commercial production by the U.S. Department of Agriculture. In 1994, the European Union approved genetically modified tobacco in France. After these genetically modified crops gained approval, genetically engineered plants and other organisms became more widespread in the United States and reached supermarket shelves. It has been estimated that more than 75 percent of food products on store shelves may contain at least a small quantity of genetically engineered crops.

The Human Genome Project (1990–2003), a collaborative international scientific research initiative spearheaded by the National Institutes of Health, advanced genetic engineering by its publication of human DNA sequencing data. These data allowed scientists to learn more about the physical and functional aspects of genes and DNA. By its completion, the Human Genome Project had fully sequenced the human genome and provided a basic genetic road map for scientists to find human DNA segments of interest.

During the 1900s, scientists also used genetic engineering technology to develop numerous varieties of investigational organisms with very specific characteristics for use in research. Some genetically modified organisms were used to learn more about the natural progression of particular diseases. Others were created to test experimental therapies before moving to humans. These genetically modified organisms have helped scientists learn more about genetic disease, cancer, aging, and other chronic diseases.

In academic and industry laboratories, modern genetic engineering continues to solve problems related to health, disease, industry, and agriculture. Additional applications in humans and human disease have been assisted by government-funded initiatives such as the Human Genome Project. Although controversial at times, genetic engineering is a modern tool to be used in addressing a wide range of problems.

HOW IT WORKS

Although the types of organisms modified vary substantially in genome size and structure, all genetic engineering involves several general steps: identifying the desired feature or end application, isolating the gene segment that codes for the feature, inserting the gene segment into a vector, and adding it to the target organism, a process called transformation.

Identification. To create genetically modified organisms that will meet a specific need or solve a particular issue, the best way to engineer a solution must be determined. For example, if a large oil spill required cleanup, the first step would be to determine the type of organism and the desired features that would be most effective at removing the spilled oil. Issues to consider in solving a problem through genetic engineering include the desired size and type of organism to be modified, the availability of desired characteristics or features with a known DNA segment, and the possible positive and negative environmental impact resulting from a modified organism.

Isolation of the Proper Gene Segment. To modify an organism through insertion of a gene or DNA segment, the specific DNA segment in the donor organism must be known and be able to be effectively removed from its host genome. In some cases, the DNA segment coding for the desired characteristic is known and available because of previous scientific work. In other cases, this step can be very labor intensive and require long-term research.

After the DNA segment is identified, a particular recombinant technique, often a restriction enzyme, is used to cut the desired gene or DNA segment out of the donor organism and move it into a vector. Depending on the genetic engineering requirements, other techniques such as polymerase chain reaction or agarose gel electrophoresis can be used to isolate a gene or gene segment.

Insertion. Insertion is the genetic engineering step during which the desired gene or DNA segment is integrated into the vector. In this step, restriction enzymes are used to cut the vector open in a particular place so that the desired gene or DNA segment can attach itself to the vector. Then a special enzyme glue called ligase is used to attach the DNA segment to the vector. The most commonly used vectors in genetic engineering are circular form of DNA called bacterial plasmids; however, the type of vector used for a particular application depends on the size of the gene or segment being moved and the organism being modified.

Transformation. The next step of genetic engineering is called transformation. During this step, the desired gene or DNA segment is introduced and successfully added to the organism being modified. A variety of methods can be used to send the vector containing the desired DNA segment into the organism being modified in such a way that the new genetic information is added to the organism's standard genes. These methods include microinjection, use of a gene gun, electroporation, or use of viruses. In each of these methods, the vector carrying the desired genetic segment is forced into the new cells of the organism being modified. Completion of the transformation step relies on testing that determines whether the inserted DNA segment is producing the desired effect or trait in the newly modified organism.

CRISPR. A specific method of genetic engineering, known as CRISPR (clustered regularly interspaced short palindromic repeats) or CRISPR-Cas9, was developed in the first decades of the twenty-first century and was soon viewed as a revolutionary technique. The process is based on segments of DNA, also called CRISPR, from microorganisms including certain bacteria that are able to manipulate the DNA of invading viruses. The enzyme Cas9 and a segment of guide RNA (gRNA) are used as scissors on the molecular level to cut DNA at precise points, allowing fine control of gene editing. The process allowed for much faster and successful genetic engineering than previous methods, opening up the possibility of a much wider application of genetic modification.

APPLICATIONS & PRODUCTS

Genetic engineering has far-reaching applications in food production, industry, medicine, and research.

Crops. One of the most widespread but also most controversial uses of genetic engineering is in the creation of genetically modified crops and food. The goal of genetic modification varies from crop to crop.

Soybeans have been modified with a DNA segment conveying resistance to herbicides sprayed over fields to kill weeds growing amid the soybeans. The Flavr Savr tomato was engineered to decrease ripening time and increase shelf life. Varieties of rice and corn have been engineered using DNA segments for other plant genomes to have increased levels of vitamins.

From the first commercially grown genetically engineered product for human consumption (the Flavr Savr tomato), adoption of genetically engineered crops in the United States has increased quickly. According to data from the U.S. Department of Agriculture's Economic Research Service, fields planted with genetically engineered cotton (herbicide-tolerant and insect-resistant cotton) reached 96 percent of the total cotton acreage in 2014. That same year, herbicide-tolerant soybeans and biotech corn accounted for 94 percent and 93 percent, respectively, of their crop populations.

Supporters of genetically modified crops feel that the plants can increase food production to meet the world's needs using lower amounts of pesticides and increasing farmer profits. Opponents of genetically engineered crops are concerned about perceived safety issues regarding food produced from these crops, ecological issues around increased use of herbicides, contamination between genetically modified and naturally grown crops because of cross-pollination of fields, and economic difficulties dealing with patents on genetically modified crops.

Livestock. Although farmers have bred particular varieties of livestock such as cows, goats, chickens, and sheep for thousands of years to maximize desirable qualities, genetic engineering allows a more rapid introduction of specific qualities that may or not occur naturally in the animals. The benefits of genetically engineered livestock are numerous and affect the producers, environment, and consumers. Producers benefit by having disease-resistant, increasingly productive, or fast-growing animals. For example, the gene responsible for regulating milk production in cows can be modified to increase milk production. Also, if animals are engineered to have milder waste, the environment will benefit. The FDA has reviewed genetically modified pigs that are better able to digest and process phosphorus in ways that release up to 70 percent less phosphorus in their waste. Consumers benefit from more nutritious, vitamin-enriched meat, as in the case of pigs that are engineered to produce omega-3 fatty acids through the expression of a roundworm gene.

Many of the concerns that apply to crops also apply to livestock. In 2015, the FDA stated that a type of salmon called AquAdvantage, genetically modified to grow twice as fast as conventional Atlantic salmon, was safe for human consumption, the first time a genetically engineered meat was approved by the organization. The FDA has placed several genetically modified animals under review in a category called food-drug. The FDA considers a modified DNA segment to be like a drug and is regulating transgenic organisms in the same way it oversees animals that receive growth hormones or antibiotics.

Diagnosis. Genetic engineering has allowed the development of faster, cheaper, and more accurate diagnostic tests for certain diseases to be used both in the laboratory and in the body. The tests based on genetic engineering are used to identify infectious diseases, hormonal changes, pregnancies, cancer, and other diseases and conditions. For example, a series of faster and more accurate tests for the presence of the human immunodeficiency virus (HIV) have been developed based on genetically modified HIV antigens. Other tests can diagnose diseases by detecting particular substances in specific locations in the body. These exams rely on genetically modified antibodies with markers that can be injected into the body.

Medications. The use of genetically modified organisms to produce human hormones, enzymes, vaccines, and medications has revolutionized the pharmaceutical industry. Since 1982, when Genentech's biosynthetic human insulin was introduced, the ability to manufacture new products in a controlled environment instead of collecting similar substances from the limited supply of human and animal sources has led to more readily available, effective, and reliable medications. Products include human growth hormone to treat children with insufficient growth, plasminogen activator to dissolve blood clots, and erythropoietin to treat low blood iron (anemia). In 1994, genetic engineering also led to innovative treatments for rare genetic disorders such as Gaucher disease with the production of specific human enzymes in genetically modified Chinese hamster ovary cells. The point of this enzyme replacement therapy is to replace the enzyme that the affected individuals are missing through intravenous injections of genetically engineered human enzymes. In multiple situations, the use of genetically engineered organisms to create medications has saved lives and decreased the burden of disease in ways that could not be imagined before. Future applications of genetic engineering in medicine are likely to focus on the creation of better medications for life-threatening indications.

DISEASE CURES. Genetic engineering made it possible to develop gene therapy. Genetic diseases are inherited conditions that occur because of one or more genetic changes or mutations that prevent the correct functioning of a particular gene. Most genetic diseases do not have a treatment or cure. However, with genetic engineering techniques, scientists hope that they will be able to transform an affected individual's mutated gene into a working gene by replacing it with a functional copy of the gene. Gene therapy has shown some success in helping individuals with severe combined immunodeficiency (SCID), hemophilia type B, and several other genetic diseases, and even has been applied to cancer. However, this type of treatment is still under investigation to determine if it can safely and permanently cure genetic conditions, and other concerns also remain, including ethoeconomic issues over the potential high cost and limited availability of effective treatments.

RESEARCH. Genetic engineering and genome sequencing have been used to improve investigative techniques through the ability to manipulate organisms on a basic, genetic level. In genetic research, genetic engineering techniques have allowed scientists to create mice and other organisms affected by a specific gene change for detailed study of a specific genetic disease. For example, a genetically modified mouse that lacks the gene to produce amyloids has been used to study Alzheimer's disease. On a broader scale, genetic engineering allows detailed analysis of an organism's structure, function, and development. Through the insertion of a marker in or near a gene coding for a product of interest, scientists can track the location of that gene's product over time.

INDUSTRIAL APPLICATIONS. Genetically engineered organisms are used in several manufacturing arenas in production, processing, and waste removal. Most industrial applications of biotechnology are based on naturally occurring processes using modified bacteria, yeast, and other small organisms to digest, transform, and synthesize natural materials from one form into another. More specifically, genetically modified microorganisms have been used to produce industrial chemicals such as ethylene oxide (for making plastics), ethylene glycol (antifreeze), and alcohol. Bacteria have also been engineered to remove toxic wastes from the environment, for example, the varieties of genetically engineered bacteria that consume oil by chemically transforming its compounds into usable basic molecules. Future directions in industry include production of textile fibers, fuels, plastics, and other industrial chemicals out of industrial wastes or raw materials.

SOCIAL CONTEXT & FUTURE PROSPECTS

Genetic engineering has already altered the course of agriculture, industry, and medicine with its life-changing applications. Crops have been modified so that they are more nutritious and naturally produce pesticides. Life-saving medications made of human hormones integrated into bacteria are widely available in consistent and purified forms. Bacteria that convert toxic chemicals into harmless basic elements have been developed. Great strides have been made in using gene therapy to cure genetic diseases. Much of the future of genetic engineering will be marked by further refinement of these applications and processes, as well as by unforeseeable breakthroughs made possible by future technological development. However, these significant scientific strides also come with important ethical questions and safety concerns.

Environmentalists are concerned about the impact of genetically modified crops on ecosystems, in particular whether the genes introduced into genetically modified crops will be transferred to conventional crops through cross-pollination. Some researchers express concern that the true impact of GMOs is untested, and that widespread availability of such organisms should be delayed until more data is collected.

Despite general scientific consensus that GMOs as currently produced pose no health risks to humans, some individuals and organizations strongly protest otherwise. While many such opponents cross into the realm of conspiracy theories, some legitimate organizations continue to pose questions. Advocacy groups such as Greenpeace and the World Wildlife Fund are concerned about the safety of genetically modified food and feel that the available data do not prove that there are no risks to human health from consumption. Despite statements from the Royal Society of Medicine and the U.S. National Academy of Sciences in support of the safety of such foods, these groups have called for additional and more rigorous testing before genetically engineered foods are marketed.

Other countries have significant concerns about the safety of genetically modified foods. The European Union regulates genetically modified food imported from other nations, including the United States. Venezuela has banned the growing of genetically modified crops, and India has issued a moratorium on the cultivation of genetically modified foods pending an investigation into safety concerns. In December 2014, however, following an investigation into potential threats to health and environment presented by genetically engineered crops, India's Genetic Engineering Appraisal Committee approved twelve genetically modified crops for experimental trials to gather more information on their safety and utility. Other counties such as Japan and Zambia have also registered concerns over the safety of genetically modified foods. In November 2014, a Russian NGO announced what it claimed to be the most comprehensive and largest study yet examining the health effects of genetically modified foods. The three year study will involve feeding rats a diet of genetically modified maize and an extremely common herbicide to gauge what effect, if any, the products have on the rats' health.

There is significantly less controversy over the use of genetically modified organisms in industrial production and medicines. However, the use of genetic engineering techniques for human gene therapy and related applications has touched off a firestorm of ethical debate. Controversy continues to flare up on issues including the ethoeconomic issues of genetic testing, the potential for designer babies, and the possibility of body enhancement through genetics, as well as over the general ethical and religious implications of altering the human genome at any level. These debates are complicated by a lack of cohesive regulation regarding human genetic engineering.

FURTHER READING

Avise, John C. *The Hope, Hype, and Reality of Genetic Engineering: Remarkable Stories from Agriculture, Industry, Medicine, and the Environment.* New York: Oxford UP, 2004. Print.

Cathomen, Toni, Matthew Hirsch, and Matthew H. Porteus. *Genome Editing: The Next Step in Gene Therapy.* Springer, 2016.

Hodge, Russ. *Genetic Engineering: Manipulating the Mechanisms of Life.* New York: Facts On File, 2009. Print.

Jabr, Ferris. "Building Tastier Fruits & Veggies." *Scientific American* 311.1 (2014): 56–61. Print.

Khan, Muhammad Sarwar, Iqrar A. Khan, and Debmalya Barh. *Applied Molecular Biotechnology: The Next Generation of Genetic Engineering.* CRC Press, 2016.

Nicholl, Desmond S. T. *An Introduction to Genetic Engineering.* New York: Cambridge UP, 2008. Print.

Powell, Russell. "The Evolutionary Biological Implications of Human Genetic Engineering." *Journ. of Medicine & Philosophy* 37.3 (2012): 204–25. Print.

Shanks, Pete. *Human Genetic Engineering: A Guide for Activists, Skeptics, and the Very Perplexed.* New York: Nation, 2005. Print.

Yount, Lisa. *Biotechnology and Genetic Engineering.* 3d ed. New York: Facts On File, 2008. Print.

GENETICALLY MODIFIED ORGANISMS

—Christine Watts, PhD

FIELDS OF STUDY

Genetics; Botany; Agriculture; Biology; Chemistry.

SUMMARY

Genetically modified organisms are produced through biotechnology and genetic engineering and basically involve genetic modifications in which genetic material is added or removed to alter the genetic structure of the organism. Many organisms have undergone genetic modification, including bacteria and viruses, plants and animals, and even human beings. The majority of genetically modified organisms are created for therapeutic reasons, such as medicine and food for human consumption. Such organisms have the potential to affect all members of human society and their surrounding environment and have therefore become one of the most controversial ethical and ecological issues of the twenty-first century.

DEFINITION & BASIC PRINCIPLES

Humans have selectively bred and crossbred plants and animals for desired traits since almost the dawn of agriculture, but advances in genetic technology have given people novel ways in which to manipulate plants and animals. These advances are motivated by the desires to develop new medical treatments for genetic diseases and disorders and to increase food production to satisfy the world's growing population. Most advances involve recombinant DNA technology, in which an organism's genes are altered by removing a specific gene from the cell of one organism and inserting it into the cell of another. This splicing together of gene fragments from different species produces a new organism that would not be produced through natural reproduction processes or would not be feasible because of the impossibility of interspecies breeding. This new organism is defined as a genetically modified organism (GMO).

The advancement of genetic technology and the introduction of GMOs into the human food chain has prompted controversy over the ethics of manipulating nature and the potential for GMOs in worldwide agricultural production and medicine. Although many experts state that GMOs are safe for human consumption and offer myriad benefits to humankind, others claim that the production and consumption of GMOs is unethical and untested, which means that GMOs involve unknown consequences, which potentially could be dangerous.

BACKGROUND & HISTORY

The process of natural selection, first described by Charles Darwin in 1859 in his seminal *On the Origin of Species by Means of Natural Selection*, states that species evolve over time. Individual organisms that possess the most desired and useful characteristics survive, reproduce, and give birth to offspring; these offspring, in turn, possess the same positive characteristics. For many hundreds of years, people have manipulated the process of natural selection though traditional agricultural selection and crossbreeding to create or eliminate specific characteristics in plant and animal species, producing a wide variety of cereal crops, livestock animals, and pets.

Human interference has altered many plants and animals through crossbreeding or selection, but the desirable traits initially appeared through naturally occurring genetic variation. Because the desired traits were already in existence, human interference in the breeding process was often viewed as relatively benign and within natural bounds. Although humans have manipulated the breeding of plants and animals based on phenotypic characteristics for a long time, the ability to directly manipulate the genotype developed much later. Specifically, to feed a growing and hungry world population and develop medical treatments, medical and agricultural scientists have researched and advanced genetic modification technology.

Although genetic engineering is a phenomenon of the late twentieth century, the building blocks for such technology began with the first isolation of DNA in 1869 and the subsequent awareness of its relevance to heredity in 1928. The first accurate double-helix model of DNA was developed in 1953 by James D. Watson and Francis Crick, and the first gene sequence and recombinant DNA was created in 1972 by researchers from Stanford University. The latter discovery truly heralded the beginning of the biotechnological industry and the development of GMOs.

Genetic engineering research continued during the 1970s, and the first publicly and commercially available GMO, a form of human insulin produced by bacteria, was developed in the United States in 1982. However, for the most part, the majority of commercial GMOs sold and used in the twenty-first century are found in agriculture and food production. GMO research scientists believe that genetic engineering is the only method that will guarantee global food production, particularly as predications regarding global climates have indicated that traditional agriculture practices will fail to meet demand.

HOW IT WORKS

Fundamentally, the development and manufacture of GMOs is the replacement of natural selection processes with artificial genetic manipulation. At its most basic, GMO technology relies on a sound understanding of DNA and involves the subtraction of specific genetic material or substitution of material from one species with that from another. Genetic engineering, a complex endeavor, deals with the most fundamental building blocks of an organism. Within a cell are tiny strandlike structures called chromosomes, which contain a nucleic acid called DNA. This molecule contains all the genetic material required for inheritance and thus is the basis for genetic manipulation technology.

Initially, the term genetic manipulation referred to a vast array of techniques for the modification of organisms through reproduction and gene inheritance. Later, however, the definition became more restricted and refers specifically to recombinant DNA technology, a form of genetic engineering in which the genome of a cell or organism is artificially modified. The fundamental concept of this technology is that genetic material from different species is combined to create a new species or organism. That is, molecules of DNA from more than one source are united together inside a cell, which is then inserted into a new organism or host, where it is able to reproduce. Because the genome is passed on to offspring, the modification is considered to be self-perpetuating. An organism's biological activities and physical characteristics are controlled by its genome, so modification of the genome can significantly influence the organism's biological functions and traits. The objective behind such technology is to advance the fields of medicine and agriculture to develop more effective medical treatments and to improve crop yield and disease resistance.

APPLICATIONS & PRODUCTS

The possible applications and products of genetic engineering are vast, perhaps limited only by the imagination. For the most part, however, the major function of genetic engineering, and hence the development of GMOs, is related to their potential in agricultural, medical, and environmental applications.

Agricultural Applications. The continuing rapid expansion of the human population is necessitating an increase in the supply of food. Providing adequate food for a hungry world has become a significant issue for science. The need for food has been instrumental in promoting and advancing genetic modification techniques to produce new and improved organisms, particularly those that, for example, have higher yields or are drought and disease resistant. Through recombinant gene technology, it has become possible to create plant species that are capable of surviving in extreme temperatures and with low rainfall, that can convert atmospheric nitrogen into a useable form (thereby eliminating the need for nitrogen fertilizer), and that have the ability to produce their own resistance to pests and pathogens (thereby eliminating the need for chemical pesticides). Versions of soybeans, canola, corn, potatoes, sugar beets, and cotton that have been genetically modified to increase herbicide tolerance and resist insects are all available for purchase.

With some specific exceptions, research and the development of genetically engineered animals has proven to be less straightforward than the genetic modification of plants and certainly more ethically problematic. In addition, although the public definitely shows some resistance to the idea of introducing genetically modified plants into the human food chain, most people express much greater resistance to the idea of directly consuming genetically modified animals. Therefore, research on genetically modified animals for use in agriculture has stayed in a relatively early stage of development. However, there has been some research into and experimentation with the genetic manipulation of animals to increase production and meat yield; such experimentation is most promising in fish species rather than in hoofed farm animals.

Medical Applications. Although agricultural applications are very important, the potential medical applications of GMOs are perhaps even more significant. The world's first commercial applications of genetic engineering were, in fact, medically oriented and included synthetic human insulin, approved for public sale in the United States by the Food and Drug Administration (FDA) in 1982, and a human hepatitis-B vaccine, approved by the FDA in 1987. Before the 1980s, synthetic human insulin (produced from animals) was available only in relatively limited quantities. Since the 1980s, research into medical applications for GMOs has rapidly advanced. Of particular benefit is the ability of genetic engineering to produce GMOs on a previously unavailable scale.

Perhaps the most significant potential application of genetic engineering and GMOs is in the treatment and possible cure of genetic diseases. Human society is plagued with both serious diseases and mild disorders, more than three thousand of which are genetic in origin and therefore difficult to cure using conventional medicine. Although this technology is still in its infancy, gene therapy is perhaps limitless in it potential to assist people with genetic disease.

ENVIRONMENTAL APPLICATIONS. Increasing human populations are important not only in relation to agricultural food production but also in terms of their impact on the environment. Genetic engineering and GMOs could potentially solve some of the world's most serious ecological problems. Research has produced genetically modified viruses that can be used to create ecologically friendly lithium batteries, modified bacteria that can produce biodegradable plastic, and genetically manipulated bacteria that have been encoded for use in bioremediation. Genetic modification technology may even be of use in the fight for survival of some of the world's most vulnerable and endangered species.

SOCIAL CONTEXT & FUTURE PROSPECTS

Genetically modified organisms potentially could be very advantageous to people. Specifically, scientists claim that GMOs will be vital to the future of food production and therapeutic medicine. Given the possibility of climate change due to global warming, crops that can produce higher yields, resist pests and pathogens, and better tolerate drought are very attractive. The use of GMOs in the treatment of genetic disorders makes them potentially life-saving. Supporters of GMO technology argue that genetic engineering has become an economic and environmental necessity in regard to agriculture, environmental bioremediation, and medicine.

Despite their obvious benefits, GMOs also hold many potential dangers. In addition, GMOs are not well received by the public. Surveys in some countries have revealed that the majority of people are actually against the creation and production of genetically modified foods, animals in particular. This opinion is shared by many environmental organizations, which claim that the undeniable benefits of GMOs are far outweighed by their possible effects on ecosystems, native flora and fauna, and human health. Of particular concern is that many of the potential risks of GMOs are as yet unknown. Opponents of GMO technology have stated that imposing GMOs onto the public without long-term rigorous testing is irresponsible and that more research is required.

Critics of GMOs have also argued that the Food and Drug Administration (FDA) should play a greater role in testing the safety of these products and regulating their sale. However, in 2015, the FDA instituted a new rule stating that foods containing genetically modified ingredients from sources approved by the FDA do not necessarily need to have a label identifying these genetically modified ingredients, leaving that choice up to the producer. Additionally, the organization released a statement stipulating that it preferred the use of terms such as "not bioengineered" over "non-GMO" on any labeling. Many people expressed disagreement with these decisions.That same year, the FDA reiterated that it did not have any substantive evidence proving that GMOs differ in any significant way from their natural counterparts or that they presented any greater safety risk. Around the same time, the FDA also officially approved the first genetically modified animal for sale in the U.S. food market, salmon. After analyzing data submitted by the manufacturer of the fish, which was modified to grow at faster speeds, the FDA found that the salmon was safe to eat by both humans and animals and that its genes remained stable throughout generations.

FURTHER READING

Bertheau, Yves. *Genetically Modified and Non-Genetically Modified Food Supply Chains: Coexistence and Traceability.* Wiley, 2012.

Howe, Christopher. *Gene Cloning and Manipulation.* 2nd ed., Cambridge UP, 2007.

Nelson, Gerald C., ed. *Genetically Modified Organisms in Agriculture: Economics and Politics.* Academic, 2001.

Newton, David E. *GMO Food: A Reference Handbook.* ABC-CLIO, 2014.

Nicholl, Desmond S. T. *An Introduction to Genetic Engineering.* 3rd ed., Cambridge UP, 2008.

Primrose, Sandy B., et al. *Principles of Gene Manipulation.* 6th ed., Blackwell, 2003.

Strom, Stephanie. "F.D.A. Takes Issue with the Term 'Non-G.M.O.'" *The New York Times,* 20 Nov. 2015, www.nytimes.com/2015/11/21/business/fda-takes-issue-with-the-term-non-gmo.html. Accessed 28 Oct. 2016.

Tutelyan, Victor. *Genetically Modified Food Sources: Safety Assessment and Control.* Elsevier, 2013.

Watson, James D., et al. *Recombinant DNA: Genes and Genomes—A Short Course.* 3rd ed., Freeman, 2007.

Young, Tomme R. *Genetically Modified Organisms and Biosafety: A Background Paper for Decision-Makers and Others to Assist in Consideration of GMO Issues.* International Union for Conservation of Nature, 2004.

GEOARCHAEOLOGY

—*J. Lipman-Boon*

FIELDS OF STUDY

Atmospheric Chemistry; Biochemistry; Environmental Chemistry; Geochemistry; Photochemistry; Environmental Sciences; Bioclimatology; Archaeology; Physical Geography; Ecology.

PRINCIPAL TERMS

absolute date: a date that gives an actual age, though it may be approximate, of an artifact

Curie point, or Curie temperature: the temperature at which materials containing iron oxides lose their magnetic pattern and align with the earth's magnetic field

relative date: a date that places an artifact as older or younger than another object, without specifically giving an age for it

remote sensing: any of a wide variety of techniques, such as aerial photography, used for collecting data about the earth's surface from a distance

stratigraphy: the deposition of artifacts in layers or strata

SUMMARY

Archaeology is rapidly becoming a markedly more scientifically based field, as part of a trend that started with the "New Archaeology" of the 1970's. Archaeological geology covers the wide range of geological sciences that are applied to archaeology during excavation and postseason, or postexcavation, sorting, classifying, and analyzing.

DATING METHODS

Archaeological geology is the application of geological methods and techniques to archaeology. The two disciplines have become so closely intertwined at times that some have spoken for a new term to describe their partnership: "geoarchaeology." A term aimed more specifically at the contributions of the physical and chemical sciences associated with archaeology (such as potassium-argon dating) is "archaeometry." Without such scientific methods, archaeology becomes guesswork at best; dating of finds, for example, should be derived from empirical data, or information that can be proved through experiment and observation, rather than deduced from theory without corroboration. Archaeology is, basically, deductions made about an artifact from the context in which it is found; geology helps to define and date that context, thereby providing the empirical information from which speculation about the artifact can be derived.

The principle of superposition was probably one of the first geological methods that archaeology utilized. This law states that a layer superimposed on another layer should be younger, having been laid down after the lower, or older, layer. This study of stratigraphy is a keystone to archaeological dating but constitutes relative rather than absolute dating. One of the better-known methods of absolute

A geoarchaeologist analyzing a column sample.

dating is carbon-14 dating. Carbon-12 and carbon-14, an isotope of carbon-12, are elements that exist in all living organisms. Once the organism dies, whether it is plant or animal, the input of carbon-14 from the environment stops, and the remaining carbon-14 begins to decay. The amount of carbon-14 left relative to the amount of carbon-12 is used to calculate the amount of time that the organism has been dead, using the known half-life of carbon-14 (5,730 years).

Archaeomagnetism (the term is the archaeological equivalent of geology's "paleomagnetism") is another dating method, but one that uses changes in the earth's magnetic field as recorded by archaeological artifacts such as kilns. The earth experiences continual changes in the intensity and polarity of its magnetic field. If a clay artifact has been heated to its Curie point, or temperature, its magnetic particles will align in the direction of the polarity of the earth's magnetic field in which they cool. Geological identification of the polarity pinpoints the time of firing. Thermoluminescence dating is also a method used on clay. A piece of pottery is heated to 500 degrees Celsius, and the ensuing emission of light is measured as an indication of the length of time since its firing, as the energy of the thermoluminescence increases as it is stored up over time. Another method for dating inorganic objects—in this case, volcanic rock—is potassium-argon dating. When volcanic rock is newly solidified, it contains no argon, but it does contain potassium-40. As the potassium-40 decays, it becomes argon-40. Because the half-life of potassium-40 is known (1 billion years), the amount of argon-40 gives the absolute age of the rock.

REMOTE SENSING & PETROLOGY

Geological techniques are also used for locating and recording archaeological sites. Remote sensing is a technique by which data are collected on a site in a "remote" rather than a hands-on way. Photographic images are a major part of remote sensing. Aerial photography, for example, involves photographing the landscape from airplanes or satellites. Images and patterns that record electromagnetic radiation provide another source of remote-sensing data, as does soil resistivity surveying. Soil resistivity surveying is a method used to map buried features by finding electrical conductivity differences between the features and the soil around them. The ease or resistance with which the current penetrates the soil is the basic principle. To test the resistivity, four electrodes are inserted into the ground. Two generate the current, and two measure the drops in voltage; an equation taking into account the distance between the electrodes, the amperage, and the drop in voltage then gives the total resistance to conductivity.

Petrology, or the study of different aspects of rocks, is a standard feature of archaeological geology. Most techniques focus on the study of thin sections or powdered samples of rock—for example, through the use of scanning electron microscopes and X-ray diffraction, respectively. In addition, archaeologists utilize geological studies of cryoturbation (freeze-thaw cycles in soil), argilliturbation (shrinking and swelling cycles in clays), aeroturbation (disturbances by gas, wind, and air), aquaturbation (disturbances from the movement of water), and seismiturbation (disturbances by earthquakes). Study of these different types of environmental disturbance helps identify the different site-disturbing processes at work.

APPLICATION OF GEOARCHAEOLOGICAL METHODS

Many aspects of the geological sciences can be applied to archaeological sites. It is the nature of the site that determines the appropriate method to use. For example, a historical archaeological site, or one that is dated to within the parameters of recorded history, would not probably be a site at which radiocarbon dating would be useful: The artifacts would not be old enough for a dating method that yields figures in thousands of years. Conversely, a prehistoric site such as the possibly Iron Age Caer Cadwgan, a probable hillfort in Wales, would benefit primarily from radiocarbon dating: The site is probably about three thousand years old, and charcoal and bone—excellent types of samples for carbon-14 testing—are the main elements found in excavation at this site. This site has also yielded small glass beads, items that would perhaps be datable by thermoluminescence.

Historical sites in general, however, are not the prime candidates for archaeological geology that prehistoric sites are. Prehistoric archaeology, in contrast, depends completely on geological analysis for some conclusions because it predates any written records;

there are no fortuitously preserved documents to fall back on for verification. Archaeological geology can accomplish much, but it is most useful for four principal processes during an excavation: locating a site, recording and analyzing the features of a site, and dating a site.

The archaeological use of soil resistivity for locating sites or individual features was first applied to locate prehistoric stone monuments just after World War II. The differences in ground conductivity are used to locate anomalies, which can range from buried ditches to stone walls. This method was first developed in geology to locate ore deposits, faults, and sinks (sunken land where water can collect). Remote sensing can be a useful method for several different goals. It can be used to locate sites, monitor changes in the archaeological record, reveal the distribution of archaeological sites, or map sites. Aerial photography can be used to locate sites, and it can then be used to map a site and record its features for planning an excavation. On some sites in North Africa, the camera will be sent up in a balloon to take the photographs, as the site may be too small or the budget of the excavation too limited for airplane-carried camera work.

Archaeology is a destructive process; once a site is excavated or even surface collected, it cannot be restored. Remote sensing can, sometimes, take the place of excavation in what is termed nondestructive archaeology. It can also be a useful substitute for excavation when there is not time for a full-scale dig to be mounted, as during times when a formation may be temporarily visible (for example, winter snows revealing significant gradations in the land) but conditions may not be right for excavation. Remote sensing can also preserve at least an image of a site that must be destroyed, as during construction of a road. With this use, it is an invaluable tool in rescue archaeology.

APPLICATION OF DATING METHODS

Petrology is a useful tool in general for archaeology. Petrology can be used on many different types of sites, from prehistoric to historic. It can be used to give a provenance, or origin, for building and sculptural stones as old as Stonehenge or as young as the classical Greek marbles. Archaeomagnetism can be used not only to date artifacts but also to distinguish sources for substances. Obsidian sources, for example, have been found to each have different magnetization strengths. It is possible, then, that the source of ore for coins, which may contain trace amounts of iron, may also be able to be discovered. Archaeomagnetic dating of lava has been used to date the end of the Minoan civilization (about 1500 BCE), which was destroyed by the volcanic eruptions at Santorini (or Thera) in Greece. It is used mainly for early human sites, or for sites up to 10,000 years old.

Thermoluminescence can also be used to date lava as well as burnt flint (implements heated by accident or on purpose to improve certain qualities), burnt stones (heated on a fire and then placed in a food container as a "pot-boiler"), glass (volcanic glass especially), sediments (buried soils), and ceramics. This method is popular because of its absolute dates for a wide range of ages: 50 to 500,000 years old. Potassium-argon dating is also used for inorganic samples but only ones that predate humankind. Samples can only be as young as 100,000 years old. Another limitation, is that because it is used to date lava, in order to be of any use the site must be connected to a particular volcanic eruption, and sites such as Santorini and Pompeii are not common (and too young anyway). The usefulness of potassium-argon dating to archaeology is its ability to fix dates for reversals of the earth's magnetic field, which in turn are used to date archaeological sites through archaeomagnetism.

Using petrology to establish the provenance of a rock artifact is of great importance in prehistoric archaeology. The evaluation of rock types found can help to identify the mining or quarrying skills of a culture, its determination of the usefulness of various kinds of rocks, and even some of the places to which people may have traveled. If a certain type of rock is not native to the area in which it is found yet occurs in large quantities, it may be inferred that perhaps significant trade or travel was taking place. The value that a culture may have placed on a particular rock may also be determined, judging by its use (whether ceremonial or practical, for example). Use-wear, as opposed to earth-moving processes that have changed the shape of the rock, is another object of study. Petrological methods such as X-ray diffraction can be used to identify the origin of rocks on a site. Studying the rocks from a number of different sites can be a way of tracing the trade routes of a culture.

Further Reading

Brothwell, Don, and Eric Higgs, eds. *Science in Archaeology*. With a foreword by Grahame Clark. 2d ed. New York: Praeger, 1970. Despite the date of the book, it is helpful because it has chapters on many of the scientific methods that are in use, including thermoluminescence, potassium-argon, and radiocarbon (carbon-14) dating. This volume is useful for seeing the wide range of scientific methods that are applied in archaeology and the beginning of New Archaeology. Suitable for undergraduate college-level students.

Butzer, Karl W. *Archaeology As Human Ecology: Method and Theory for a Contextual Approach*. Cambridge University Press, 2006. Print. Butzer sets out the principles and objectives of geoarchaeology as a subdiscipline of archaeology. His emphasis is on the environmental sciences in general as necessary for the best empirical data in archaeology.

Garrison, Erv. *Techniques in Archaeological Geology*. New York: Springer, 2010. This text provides an overview of archaeology and palaeoanthropology studying the Quaternary epoch. It discusses techniques of archaeology and geology of the stratum.

Goldberg, Paul, and Richard I. MacPhail. *Practical and Theoretical Geoarchaeology*. New York: Wiley-Blackwell, 2006. Discusses various geological features encountered by archaeologists. Includes case studies of geoarchaeology research and laboratory techniques. Contains a bibliography appendices and indexing.

Kelley, Jane H., and Marsha P. Hanen. *Archaeology and the Methodology of Science*. Albuquerque: University of New Mexico Press, 1988. Kelley and Hanen have prepared this volume as a way to reconcile the philosophies of science and archaeology. They include case histories that target the methodological problems of interpreting even sound empirical data. Useful for college-level students seeking to discover how the sciences fit into archaeology.

Kempe, D. R. C., and Anthony P. Harvey, eds. *The Petrology of Archaeological Artifacts*. New York: Clarendon Press, 1983. A very useful book for different techniques of archaeological geology and their various applications, including specific examples. As the title indicates, petrology is the main focus. This book is technical and therefore best suited to advanced college-level readers.

Lasca, N. P., and J. Donahue, eds. *Archaeological Geology of North America*. Boulder, Colo.: Geological Society of America, 1990. This multiauthored volume covers the paleoarchaeology of North America. A useful introductory chapter reviews the entire subject. Further Reading and index.

Parkes, P. A. *Current Scientific Techniques in Archaeology*. London, Routledge, 2016. This book includes detailed information on archaeometry. Technical but well written. Suitable for college-level readers.

Rapp, George, Jr., and Christopher L. Hill. *Geoarchaeology*. 2d ed. New Haven, Conn.: Yale University Press, 2006. Accessible to readers with a limited background in earth sciences. The text covers the history of geoarchaeology, methodology and analysis, and completes the introduction to geoarchaeology with descriptions of field research topics.

Ryan, William, and Walter Pitman. *Noah's Flood: The New Scientific Discoveries About the Event That Changed History*. New York: Simon & Schuster, 1998. The author uses an interesting combination of geologic, archaeologic, and linguistic information to conclude that the Biblical Flood myth derives from the catastrophic flooding of the Black Sea. Written for the general reader but with plenty of scientific information.

Schiffer, Michael, ed. *Advances in Archaeological Method and Theory*. 11 vols. San Diego, Calif.: Academic Press, 1978–1987. This series is a comprehensive review of changes in scientific archaeological methods and new applications of those methods. Each chapter is by a different contributor, and many include specific examples of studied sites.

__________. *Formation Processes of the Archaeological Record*. Salt Lake City: University of Utah Press, 2002. This book includes chapters on the earth processes that can disturb and change the archaeological record, such as earthquakes and freeze-thaw cycles. Contains photographs and is suitable for the general reader.

West, Frederick H., ed. *American Begnnings*. Chicago: University of Chicago Press, 1996. This multiauthored volume deals with the geological and archaeological evidence for the movement of humans across the Bering Strait land bridge. Well illustrated. Extensive bibliography.

GEOMICROBIOLOGY

—Michael P. Auerbach

FIELDS OF STUDY

Biochemistry; Environmental Chemistry; Microbiology; Photochemistry; Environmental Sciences; Bioclimatology; Ecology.

PRINCIPAL TERMS

anoxic: a physical state in which oxygen is low

bioremediation: a geomicrobiological process whereby bacteria, microbes, and other microorganisms consume pollutants and deposit water, nutrients, and other by-products

elemental cycling: the process by which elements move in various forms through the planet's subterranean, ground-level, and above-ground environments

microbe: a microscopic organism capable of causing fermentation or disease

microbial mat: a rubbery sheet comprising layers of different species of bacteria and microbes

stromatolite: a fossilized microbial mat

weathering: a geological process whereby igneous, metamorphic, or sedimentary rock is broken down into smaller particles

SUMMARY

Geomicrobiology is the study of the influence of microbes (bacteria and similar microscopic organisms that cause fermentation and disease) on geological processes. As the field is relatively new, scientists are becoming increasingly aware of the effects that fungi, yeasts, and algae have on rocks, the soil, and the ocean floor. Geomicrobiology has implications for a wide range of scientific foci, including rock weathering, geochemical recycling, and the breakdown of certain minerals. The growth of this scientific field is aided by improved technologies such as electron microscopy, three-dimensional computer modeling software, and state-of-the-art laboratories.

BASIC PRINCIPLES

Geomicrobiology involves the study of the effects of microbes on Earth's geological processes. Microbes play an important role in changes to the chemical composition of soil. They also contribute to the deformation and breakdown of igneous rocks (stones formed when molten magma cools) and metamorphic rocks (stones whose composition changes under different environmental conditions). Furthermore, bacteria and other forms of microbial organisms can provide clues about the different chemical compounds that existed in certain prehistoric environments.

Geomicrobiologists focus on bacteria that grow on sediments, usually in underwater environments. When these microscopic organisms coalesce, they form microbial mats, which are rubbery sheets comprising layers of different species of bacteria. The mats become an ecosystem, as microbes at the higher layers emit waste that is consumed by those on the lower levels. This transfer causes the formation of different types of minerals and can contribute to chemical and mineralogical changes in the sediment in which they are buried. In time, microbial mats are buried in sediment, forming rocks known as stromatolites. These fossils serve as some of the only evidence of the role microbes played in the earth's earliest stages of development.

BACKGROUND & HISTORY

Geomicrobiology owes its origins largely to the scientific contributions of two seventeenth-century scientists: Englishman Robert Hooke and Dutchman Antoni van Leeuwenhoek. Hooke built one of the first compound microscopes, a device that could increase magnification to a much greater degree than earlier devices, which he used to study insects, bird feathers, and other small organic subjects and specimens. Hooke's work with the microscope was recorded in his seminal book *Micrographia*.

Inspired by the work, Leeuwenhoek built an even more powerful microscope, which enabled him to study even smaller organisms. As the microscopes he built during his career evolved, Leeuwenhoek began to move beyond multicellular organisms and studied algae and even the film on his teeth. The strange images he observed were submitted to the Royal Society of London, and Leeuwenhoek entered into history as one of the first scientists to observe living bacteria.

In the centuries that followed, microbiology's applications were generally limited to the fields of medicine and biology. However, in the late nineteenth century, Russian geographer Vasily Dokuchaev became convinced that the earth's soil contained microscopic organisms in the same way that the world above ground did. Dokuchaev initiated the first study of the genetic composition of the earth, identifying and studying microorganisms in the soil and their influence on the environment. In light of his groundbreaking work, Dokuchaev has been dubbed the founder of modern genetic soil science, a field from which geomicrobiology has evolved.

MICROBES & MICROBIAL MATS

Microbes, bacteria, and similar microscopic organisms play a number of roles in the earth's geological processes. One such role is in the immobilization and, ultimately, the chemical alteration of minerals. Microbes and various species of bacteria grow on the sediment at the floor of oceans, lakes, rivers, and other bodies of water. In time, these growths become layered microbial mats, with each lower layer feeding on the waste material of the higher level.

The mats have an effect on the minerals and rocks beneath them, in some cases removing nutrients necessary for the formation and maintenance of certain sedimentary rocks. For example, the presence of aggressive species of microorganisms in a groundwater environment with a limited supply of phosphorus can lead to the destruction of the sedimentary rock feldspar. Similarly, sediment that has been colonized by microbes at the bottom of a deep ocean basin, where oxygen is scarce, has been found to immobilize certain metal ions, causing a change in the chemical composition of the basin.

The chemical by-products of microbes and bacteria can lead to the breakdown of nearby rocks. For example, the waste products of the bacterium *Bacillus mucilaginosus* are acidic; as the presence of this bacterium increases, so too does the acidity of the immediate vicinity. The acids cause the weathering of silicate minerals (weathering itself being a process by which rock is broken down into smaller particles), which leads to a change in the chemical and mineralogical composition of the immediate environment.

THE CONTRIBUTIONS OF MICROORGANISMS

The weathering effects of microorganisms on minerals have important implications, some positive and some negative, for the structure of Earth's crust and Earth's climate. In some cases, microbes consume nutrients and deposit materials that benefit the development of certain crystals and minerals. Microbes frequently absorb pollutants and return water and other valuable elements to the soil in a process called bioremediation. However, minerals are not very stable; if microbes absorb too much of the pollutants, they also may withhold carbon dioxide from the minerals and make it difficult for those minerals to undergo weathering. In time, it is possible that a disruption in mineral weathering caused by microbial consumption of high amounts of pollutants can lead to an increase in Earth's overall temperature.

Microbes play an important role in the cycling of certain elements. Elemental cycling, the process by which elements move in various forms through the planet's subterranean, ground-level, and above-ground environments, is critical to life on Earth, as it brings such vital elements as carbon, nitrogen, and oxygen (and water, a compound) to each level. As scientists use improved technologies, such as mass spectrometers and chromatography, they are increasingly seeing a more widespread involvement of microbes in this geochemical process.

For example, scientists launched a study of the presence of iron hydroxide in a creek in California. Existing evidence had indicated that the cause of the presence of the iron-based compound was primarily inorganic (especially given that the studied stream had flowed from groundwater that was lacking sufficient amounts of oxygen). However, researchers noticed that a microbial mat formed at a point where the anoxic groundwater met with oxygen-rich water (anoxic meaning a state in which oxygen is severely low). This new evidence led researchers to conclude that the cycling of iron in this area was likely caused by microorganisms in the system.

MICROORGANISMS & PREHISTORY

The evolving field of geomicrobiology is also uncovering evidence of the role that microbes played in Earth's prehistoric eras. Scientists have increasingly been studying such periods to better understand changes in geological processes. In one study, scientists focused on an area off the southeast coast of India. Using electron microscopes and spectrometers, the researchers revealed the presence of fossilized bacteria from the early Holocene era (about 10,000 years ago). These bacteria were closely linked to fossilized microscopic plant fragments. The scientists concluded that the bacteria had likely consumed toxic compounds, such as arsenic, from nearby pyrite crystals. The bacteria enabled the crystals to fertilize the soil and retain toxic substances that could stymie plant life. Such information is highly useful for persons looking to understand how toxic substances are absorbed and stored in the soil.

COMPUTER MODELING

Because the processes associated with geomicrobiology commonly transpire over longer periods of time, it is necessary to compile long-term data in such a way that the process in question is illustrated clearly. To that end, the use of two-and three-dimensional computer models proves highly effective. Some such models may focus on a long-term concept, such as fossilization or element cycling. Others focus on a specific region, such as the Antarctic ice shelf. An important benefit of the use of computer models is that, once completed, such programs may serve collectively as a skeletal framework for any region or trend, which in turn enables scientists to compare different processes and variables.

An example of this utility of computer modeling can be found in studies of the proposal to store nuclear waste at Yucca Mountain in Nevada. To gauge the effects that microbes would have on this area in light of the waste materials that would be introduced, scientists created a computer code, Microbial Impacts to the Near-Field Environment Geochemistry (MING), which allows researchers to study different species of microbes along with specific temperatures, mineral deposits, and time periods of as long as one million years. By allowing for such a diversity of variables, MING is helping researchers to map and study potential geomicrobiological changes at the site over time.

EVOLVING MICROSCOPY

The microscope has long been a staple of biology, and geomicrobiology continues to advance because of improvements made to microscopy. One of the most invaluable microscopes is the electron microscope, which can view objects at a magnification hundreds of thousands of times greater than a standard microscope. Electron microscopes can now show geomicrobiological study subjects in three dimensions and with greater clarity than previous incarnations; this allows scientists to study such trends as microbial colonization and element cycling, among other phenomena. Some electron microscopes can examine the internal structure of microbes. This technology uses x-rays to look inside these microorganisms with extraordinary clarity.

In addition to using microscopy, geomicrobiologists frequently use chromatographers and spectrometers. This technology is designed to capture special affinities of microbes, such as gaseous or liquid byproducts in the case of chromatographs, and the release of energy in the case of spectrometry. For example, spectrometers have been used to study isotope excursions. Similarly, chromatographs study the release of energy in the metabolic processes of microbes during elemental cycling. These technologies add a new dimension to geomicrobiology.

RELEVANT ORGANIZATIONS & NETWORKS

Geomicrobiology presents a wide range of implications for the scientific community and for others concerned with global climate change especially. Many different interested organizations take geomicrobiology as a starting point for their involvement in scientific research into climate change.

Several government agencies in the United States have departments dedicated to geomicrobiology. The U.S. Geological Survey (USGS), for example, runs a special project focused on geomicrobiology and the broader field of geobiology. This project includes a study of the genetic composition of microbial mats and the role microorganisms play in their respective environments. Also, the National Science Foundation provides funding for geomicrobiology research.

Universities and colleges play key roles in studying geomicrobiology. Princeton University's Department of Geosciences, for example, has a geomicrobiology group dedicated to the study and public dissemination of information in this field. Manchester University in England and Aarhus University in Denmark also have strong programs in this interdisciplinary arena. It is at many of these universities that the leading theories and research on active and historical volcanic sites, for example, are developed and investigated.

Because information can be shared so quickly through the Internet and other global networks, geomicrobiology research continues to gather speed. Princeton's geomicrobiology group and others like it collaborate on projects and share data with their colleagues worldwide. Other networks, such as the Geological Society of America, have divisions and programs in geobiology and geomicrobiology.

IMPLICATIONS & FUTURE PROSPECTS

The field of geomicrobiology is relatively new and ever evolving. Twenty-first century technology has further enabled scientists to track the relationship between microorganisms and their geological environment, meaning that scientists can now better understand the complex processes that have shaped the earth throughout its history. Computer systems capable of compiling large volumes of data and using this data to develop models are helping researchers analyze trends that occur over thousands, if not millions, of years. These technologies are also helping geomicrobiologists study such pressing issues as pollution and global warming.

Further Reading

Archana, G. "Geomicrobiology: An Indian Perspective." *Current Science* 100, no. 4 (2011): 451–452. In this article, the author provides a review of geomicrobiology topics discussed at a 2011 conference in India. Among the topics covered are microbe sediment interactions, the nutrient cycle, and the development of geomicrobiology.

Brown, David J., and Martin Lee. "From Microscopic Minerals to Global Climate Change?" *Geology Today* 23, no. 5 (2007): 172–177. This article discusses the effect of microbes and bacteria on the degradation of minerals in their ecosystems. The authors argue that because of the relative instability of certain minerals, an overabundance of microbes can contribute to global warming conditions.

Canfield, Don, Erik Kristensen, and Bo Thandrup. *Aquatic Geomicrobiology*. Atlanta: Academic Press/ Elsevier, 2005. The authors discuss how microbes play the role of catalysts in the geochemical reactions within their environments. This volume focuses on aquatic environments, wherein microbes are involved with elemental cycling and other geochemical and hydrochemical processes.

Ehrlic, Henry Lutz, and Dianne K. Newman. *Geomicrobiology*. 5th ed. Boca Raton, Fla.: CRC Press, 2008. This book is considered by many to be one of the top texts in the growing field of geomicrobiology. The authors discuss the many effects microbes and other microorganisms have on such geological processes as fossilization, mineral degradation, and elemental cycling.

Konhauser, Kurt. *Introduction to Geomicrobiology*. Hoboken, N.J.: Wiley-Blackwell, 2006. The author provides an overview of the different types and species of microorganisms and how these microscopic life forms interact with minerals, soil, and other geological environments.

O'Brien, Rachel, and Jennifer A. Roberts. "Teaching the Principles of Geomicrobiology and the Process of Experimental Research to Undergraduate and First-Year Graduate Students." *Journal of College Science Teaching* 38, no. 2 (2008): 30–38. This article discusses the various approaches university professors may use to teach the relatively new and multidisciplinary field of geomicrobiology. The authors discuss the value of scientific collaboration and the coursework future geomicrobiologists should take to become fully versed in this field.

GEOMORPHOLOGY & HUMAN ACTIVITY

—*Mariana Rhoades*

FIELDS OF STUDY

Geography; Botany; Agronomy; Agriculture; Biochemistry; Ecology; Entomology; Genetics; Landscape Architecture; Soil Science.

PRINCIPAL TERMS

erosion: wearing away of soil and rock by weathering, landslides, and the action of streams, glaciers, waves, wind, and underground water

landform: one of the many features that, taken together, make up the surface of the earth

landscape: broad term for the land setting of an area

topography: general configuration of a land surface, including its relief and the position of its human-made features

weathering: destructive processes by which rocks are changed though exposure to the atmosphere at or near the earth's surface, with little or no transport of the loosened or altered material

SUMMARY

Geomorphology is a science that classifies landforms and describes and analyzes the origin and evolution of surface features. It is interdependent with other branches of geology—such as sedimentology and hydrology—that study the processes that act on marine and planetary features. Geomorphology is of immense global importance because of the ever-increasing activity of humans as geomorphic agents. As the human population increases in numbers and complexity, it increasingly affects and is affected by geomorphic processes.

LANDFORMS

Geomorphology is the scientific study of topographic surface features and naturally occurring earth processes. The landform is the basic unit of systematic analysis of geomorphology. Landforms can be considered constructional (coral reefs and volcanic eruptions, for example). However, because of the intensity of atmospheric weathering and erosion, most landforms are erosional. During landscape analysis, a geomorphologist must not only analyze current surface features but also reconstruct rock units and landforms that existed before the present landscape.

Time is a major factor in geomorphology, in that landforms are altered by geologic processes over time. A theoretical analysis of landform evolution uses a time range of millions of years. Practical or applied geomorphology uses a time scale of hours to years because of the need to predict changes on a human scale, such as how long an excavated area will remain stable during hot, humid conditions. The use of radioactive isotopes establishes the absolute age of rocks as well as the exposure age of rock surfaces, thus making estimation of absolute rates of land surface change possible. The geomorphic application of isotopic dating to land surface changes can establish when an ice sheet retreated from a region.

Central to the geomorphic study of any region is its structure or specific physical properties. Studies of landform structure reveal the type of rock present, its mineral composition, grain size or crystalline structure, and rock strength. Also, studies of landform structure indicate the depositional sequence of the rock material and any naturally occurring stresses that the rock unit has been subjected to, resulting in faults and fractures.

A full range of processes contributes to produce a landscape. The action of wind, water, and ice erodes rocks and transports the eroded material to deposition sites. Gravity pulls down large materials that rise above ground level. Rates and intensity of geomorphic processes vary. Surface weathering rates may be extremely slow, such as a few centimeters per thousand years, or quite rapid, as measured in an avalanche that may travel at 50 meters per second or more. Intensity of geomorphic processes is governed by climate, vegetation present, and elevation of the surface feature.

Mountaintop removal coal mining is an example of geomorphology caused by human activity. Image by Flashdark (Own work).

Solar radiation provides the energy for geomorphic processes at a greater rate than all other energy sources combined. Other energy sources are the gravitational and inertial forces associated with the mass and motion of Earth, the moon, the sun, and other planetary objects. Additional energy comes from the outward flow of heat, caused by radioactivity, from the earth's interior.

The two major geomorphic processes are constructional processes—such as mountain building and volcanic eruptions—and erosional processes—such as weathering, erosion, and the action of wind, rivers, and glaciers. Other geomorphic processes are sediment movement and deposition, biologic processes as they pertain to vegetative cover, near-shore and ocean systems, and human-driven processes. There are few spheres of human activity that do not create landforms directly or indirectly.

GLOBAL WARMING

Predicted temperature increases (2 to 6 degrees Celsius per century) and global warming as a result of the buildup of greenhouse gases in the earth's atmosphere are not new phenomena. Although it became a significant issue in environmental management debates only during the last two decades of the twentieth century, this process has operated, with fluctuations, over the past 4.6 billion years of the earth's history.

Global warming studies focus on the atmospheric, hydrospheric, and biospheric consequences of climate change. Geomorphological consequences of climate change are significant because of potential changes to landscapes and human occupation of these surface areas. Potential geomorphological changes include climatic alterations affecting vegetation cover and agriculture, which could alter soil erosion, surface water runoff, river siltation, and flooding patterns; changes in frequency, magnitude, and geographical extent of tropical storms, which could affect land erosion, especially in coastal zones; changed patterns of rainfall, with excessive rainfall resulting in increased flooding or low amounts of rainfall affecting human water supplies; elevated temperatures at high altitudes and latitudes, which may alter ice and snow distribution and the extent of permafrost; and raised sea level caused by the reduction of glaciers and ice sheets, impacting coastlines through beach narrowing, cliff erosion, and delta formation.

These geomorphic changes may seem catastrophic, but it is helpful to review them in the light of environmental changes in the geologically recent past. Since the end of glaciation, about 10,000 years before the present, sea level has risen approximately 121 meters, with periods of rapid sea level rise of 14.6 millimeters per year between 13,000 and 7,500 years ago and 24 millimeters per year between 12,500 and 11,500 years ago. Although these rapid sea-level changes do not compare with current predicted values, they do demonstrate that surface features such as river deltas and coral reefs can continue to exist. Both landforms are capable of quick evolution, river deltas because of sediment supply and reefs by organic structural growth. Both landforms have survived post-Pleistocene rises in sea level and, if not perturbed by further human activity, should continue to do so. The systems that may be most threatened by the projected sea-level rise are land-based human systems such as agriculture.

Accurate geomorphic predictions and changes to landscapes are troublesome because exact weather patterns and distributions of extreme climate events are difficult to assess. Knowledge of average climatic changes does not automatically assign specific locations of weather systems and events. While it is true that landform changes are altered by climatic conditions, local geology, hydrology, topography, and land use also play major roles in landform evolution. The climatic, oceanic, and biospheric repercussions of global warming provide the basic ingredients for geomorphic change, but human activity will determine the scale, extent, and rate of the resultant changes.

Human apprehensions about adverse changes to the biosphere and human socioeconomic systems have created the current concerns about global warming. Concern is heightened by the fact that the predicted trends of climate change have never been experienced by humans. The predicted climate changes, if they fully occur, are similar to climatic conditions prior to the last glaciation. Finally, humans now recognize that they are the main influence in climate change and are capable of modifying the rate of this predicted change.

COASTAL ZONE

Major coastal processes are those associated with moving water in near-shore environments. Three forces act upon water to create waves: astronomical, meteorological, and tectonic. Astronomical forces are the driving force for tides, which in turn influence the width of the shoreline and growth patterns of the near-shore flora and fauna. Waves produced by meteorological forces operate to modify the coastline. Storm-generated waves and sea swells, along with ocean currents, operate to erode the coastline, transport eroded materials, and deposit debris in coastal areas. Tectonic forces such as earthquakes and volcanic eruptions produce dangerous seismic sea waves known as tsunamis. These three large-scale forces are not affected by human activity, but landforms in the coastal zone have long been altered and used by humans.

Early changes to the coasts by humans were indirect and unintentional because of small populations. Direct and intentional use of the coastal zone began with farming on flat delta land. Lands in these areas were reclaimed or extended by canalization and diking. Reclamation, as a coastal modifier, continues to the present. In the Netherlands, more than 3,800 square kilometers of land has been gained by reclamation in the last eight hundred years. A 1984 study of reclaimed land in Singapore revealed that the country is now 10 percent larger than when it was founded.

Another direct, intentional modification of the coastal zone is the use of shoreline stabilization structures such as seawalls, breakwaters, and jetties. A seawall or breakwater structure is placed parallel to the shore to protect an eroding coastline. Sand-bearing coastal waves are slowed down by seawalls and drop their load of sand up-current from the structure. Down-current from the barrier, the water picks up more sediment to replace the lost load, and the down-current beach is eroded. Jetties built perpendicular to the shoreline also operate to erode and redistribute sand, creating an unnatural scallop shape along the shoreline. Coastal structures placed by humans severely alter the shape and stability of the coastline. Beach erosion is not halted, nor is the shoreline stabilized.

A review of the state of the marine environment for the United Nations Environment Programme in 1990 reported that the open ocean, because of its large diluting capacity, is still relatively clean, despite measurable levels of artificial radioactive material and synthetic organic compounds. Oil slicks from tanker collisions, explosions, navigational or mechanical errors, and offshore oil installations continue to put a human face on ocean pollution. Likewise, coastal areas near large populations are clearly linked to human activity and exhibit detectable increases in phosphate concentrations from sewage and agricultural discharges. The amount of nitrates (a component of fertilizers) found in coastal waters is increasing, as are areas of eutrophication, unusual plankton blooms, and excessive seaweed growth.

RIVER SYSTEMS

Fluvial systems carry water, sediments, and dissolved materials and minerals downstream. Streams, driven by gravity, cut channels and scour the stream bottom and sides with their sediment load. Also, streams deposit sediments in artificial reservoirs and natural freshwater and marine basins. Both stream erosion and deposition create landforms, with stream valleys being one of the most frequent and widespread landforms in North America. Although the portion of water involved in streams constitutes only a small percentage of the total water in the hydrologic cycle, human interference with stream processes perturbs the global water system. The hydrologic cycle is the constant circulation of water from the sea through the atmosphere by evaporation and its return to the land, streams, lakes, glaciers, and the subsurface by precipitation. Water eventually returns to the atmosphere by way of transpiration and evaporation from plants, landforms, and ocean basins.

Human occupation of cities has a profound effect on the hydrological cycle because humans interrupt and rearrange areas where water is naturally stored, such as lakes and groundwater-bearing aquifers. In *Urban Hydrology* (1984), M. J. Hall discusses linkages between hydrology and urbanization. First, the replacement of vegetated soil with impermeable surfaces (asphalt) reduces water storage in the soil horizon, the slow percolation of water into aquifers, and the transpiration mechanism in the hydrologic cycle, which allows moisture to return to the atmosphere. Second, with large amounts of precipitation on solid surface areas, the velocity of water flow is increased, moving the water to stormwater systems rather than natural stream channels, where evaporative processes can occur. Third, construction activity clears the land surface, disturbs the soil layer, reshapes natural slopes with potentially unstable slopes, and leaves a limited vegetation cover or builds additional impermeable surfaces. If overloaded with large quantities of construction-derived debris, streams that routinely carry a small amount of sediment may experience a variety of changes in their physical and biological characteristics. These changes include deposition of sandbars or dunes in the channel. Coarser sediments carried in the water may increase the erosion of channel banks. Bottom-dwelling flora and fauna may be blanketed with sediments, reducing the viability of fish species that feed on such stream organisms.

Dams and reservoirs began to play an important role in water use in the late nineteenth century. The immediate impact of reservoirs is the intentional alteration of flow downstream, usually to increase low flows for year-round water use or to curb floods. Other impacts include the loss of land caused by the filling of the reservoir, increased evaporation from the reservoir surface, groundwater seepage into reservoir walls, channel and bank scouring, and sediment deposition. Inland reservoirs have resulted in major changes to coastlines because of sediment and nutrient starvation. Damming of major streams aggravates beach and delta erosion. Before the Aswan High Dam was constructed in Egypt, the Nile River transported 140 million tons of silt per year to the Mediterranean Sea. Increased erosion of the delta began once Aswan was filled, and the loss of nutrients caused a reduction in fish catches in the Mediterranean Sea.

SOIL

Soil formation is the direct result of physical, biological, and chemical weathering processes acting on rock units and rock-cored landforms. Soil is the active surface layer that mantles most rock masses and supports the growth of rooted plants. It is not just a simple, loose layer with a stock of plants and plant nutrients on the land surface, but rather a specific stratum that regulates biological and geological interactions. Because soil exists between the geological, biological, and atmospheric realms, it is subject to a complicated set of direct and indirect links with surface processes, both natural and anthropogenic. Human activities that can degrade soils include improper cultivation, deliberate deforestation, over-compaction by heavy machinery or trampling of grazing domesticated animal herds, and submergence of land because of drainage basin changes.

Globally, only 13 percent of the land surface (1.5 billion hectares) is naturally appropriate for growing crops. An additional 1.7 billion hectares is available for agriculture if proper land management practices are instituted and followed. The remaining vegetated areas are too cold, hot, dry, or low in nutrients or soil cover to produce crops. Soils can be physically removed by wind and water erosion when land is improperly cultivated. The removal of vegetation by clearing the fields for planting increases erosion from water, and fertile soils are washed from the land. With the protective vegetation and topsoils removed, wind erosion can increase soil loss. Tilling land surfaces parallel to the prevailing wind direction exposes soil to wind erosion. Tilling land across contours and valley slopes also degrades soil layers.

Chemical pollution of groundwater can occur as a result of the treatment of land surfaces used for agriculture. Both fertilizers, used to replace soil nutrients, and pesticides, used for raising crop yields, can enter subsurface groundwater zones after irrigation waters percolate through the soil horizons. Groundwater is especially susceptible to this type of degradation because of the slow movement of subsurface waters and the prolonged retention times within the aquifer. Salinization of soils can occur when poor irrigation techniques are used. Over-irrigation, poor drainage of irrigated soils, and irrigating with water of high salt content are major causes of this problem. Other issues—such as dam placement and excessive evaporation from rivers in hot, dry climates—complicate soil salinization.

Trees act as a buffer to the erosive processes caused by rain and storms. When trees are removed, much of the rainfall runs directly off the land, taking with it large amounts of soil and sediment that choke stream channels or decrease water quality. Rain forests that are cut and developed for agriculture are particularly susceptible to soil degradation because of the clay-rich layers within tropical soil horizons. Once rain-forest trees are removed, massive amounts of rainfall wash the soil areas, removing the small amounts of nutrients and the fine-grained clay materials. Once this stripping of the soil and nutrients occurs, the area is unsuitable for agriculture or reforestation programs. Another form of deforestation consists of stripping the land for the gathering of fuel wood. Of all the degraded land in the world, 7 percent is the result of firewood gathering and burning, which opens the land to soil erosion by running water and wind erosion.

Further Reading

Cvancara, Alan M. *Field Manual for the Amateur Geologist.* New York: John Wiley, 1995. Cvancara's introduction to core geology topics is suitable for high school students. Includes many useful illustrations.

Goudie, A. S., and H. A. Viles, eds. *The Earth Transformed: An Introduction to Human Impacts on the Environment.* Oxford: Blackwell, 2003. A well-illustrated, wide-ranging survey of human activities and their environmental consequences. Includes numerous global case studies.

Gregory, Kenneth J., and Andrew S. Goudie, eds. *The SAGE Handbook of Geomorphology.* Thousand Oaks, Calif.: SAGE Publications, 2011. A compilation of essays discussing geomorphology, techniques, processes and environmental changes. Provides in-depth descriptions and examples along with fundamentals of geomorphology. Includes many in text references, tables, and indexing.

Hooke, R. L. "On The History of Humans as Geomorphic Agents." *Geology* 28 (2000): 843–846. The article discusses the manipulation of land by humans throughout history. It expands on the technology and population size increasing human impact. The article is relatively short and concepts are clearly explained; although this article comes from a technical and academic journal, it is accessible to the layperson.

Landa, Edward R., and Christian Feller, eds. *Soil and Culture.* New York: Springer, 2010. This book provides a full history of the relationship between humans and earth. Articles are organized into parts discussing soil's function in art, philosophy, culture, and health, among others. This compilation of essays provides interesting information from an anthropological perspective.

Mackenzie, Fred T. *Our Changing Planet.* 4th ed. Upper Saddle River, N.J.: Prentice Hall, 2010. Mackenzie covers geologic and human-induced change, in language suitable for high school students. Illustrations and index.

Nisbet, E. G. *Leaving Eden: To Protect and Manage the Earth.* New York: Cambridge University Press, 1991. Nisbet discusses environmental problems and their causes, as well as methods to protect the environment and restore areas that have been damaged.

Simmons, I. G. *Environmental History: A Concise Introduction.* Oxford: Blackwell, 1993. A high-school-level introduction to the interactions between nature and humans.

Stutz, Matthew L., and Orrin H. Pilkey. "The Relative Influence of Humans on Barrier Islands: Humans versus Geomorphology." *Reviews in Engineering Geology* 16 (2005): 137–147. Discusses human effects on island dynamics, and the ecosystem's resistance and resilience due to geomorphology. This article clearly explains complex concepts in ecology and geology.

Geomorphology of Dry Climate Areas

—*John F. Shroder Jr.*

FIELDS OF STUDY

Geography; Botany; Agronomy; Agriculture; Biochemistry; Ecology; Entomology; Genetics; Landscape Architecture; Soil Science.

PRINCIPAL TERMS

borax: a sodium borate mineral that is an ore of boron and occurs as surface crust or large crystals in the muds of alkaline lakes; borax is used in glass, ceramics, cleansing agents, water softeners, and other industrial applications

desert (rock) varnish: a thin, dark, hard, shiny or glazed iridescent (red, brown, black) film, coating, stain, or polish on rocks that is composed largely of iron and manganese oxides and silica formed by weathering of dust films and by microbial action

differential weathering: physical and chemical weathering that occurs at irregular or different rates, caused by variations in composition and resistance of a rock or by differences in intensity of weathering, and usually resulting in an uneven surface where more resistant material stands higher or protrudes above less resistant parts

pluvial period: an episode of time during which rains were abundant, especially during the last ice age, from a few million to about ten thousand years ago

saltation: a mode of sediment transport in which the particles are moved progressively forward in a series of short intermittent leaps, jumps, hops, or bounces from a bottom surface

SUMMARY

In dry climates, the great temperature contrasts, strong winds, rare but torrential rains, and limited but powerful water floods produce a diverse geomorphology as well as natural difficulties of existence for those who live in such areas.

DISTRIBUTION OF DRY CLIMATE AREAS

Dry climate geomorphology is the study of landforms in arid and semiarid regions. Arid lands of various kinds, usually termed deserts, amount to at least 25 percent of the total land area of the world outside the polar regions and up to 43 percent by some definitions. They are generally considered to be those areas with less than 25 centimeters of precipitation per year or where evaporation amount is twice that of precipitation. Semiarid regions, or steppes, are not quite as extensive as deserts but are still significant. They can be defined as having precipitation between about 25 and 50 centimeters per year, which means that evaporation is about equal to precipitation. The semiarid areas are generally peripheral to the desert areas.

The most extensive arid lands of the world are associated with the two circumglobal belts of dry subtropical air that subsides to the north and south from the rising equatorial air masses, heating and drying in the process. Examples include the Sahara and Kalahari deserts of Africa. A second type of dry area is found in continental interiors far from sources of

Wind erosion carries topsoil from farmland during the Dust Bowl, circa 1930's. Image by U.S. Department of Agriculture.

moisture. The Gobi and Takla Makan deserts of central Asia fall into this category.

A different and more local kind of desert is found on the lee side of mountain ranges. The mountains create a barrier to the flow of moist air, producing a rainshadow effect. As the air rises against the windward slope of a range, it cools and condenses into precipitation. This chain of events removes much of the moisture from the air mass. Then, as the air mass passes over the range and down the other side, it warms in descent and becomes even drier as it moves along. The deserts behind the coastal ranges of California are of this type.

A fourth category of deserts occurs along coastlines. They occur locally along west coasts, where upwelling cold seawater cools passing marine air, thereby decreasing its ability to hold moisture. As the air encounters warm land, its limited moisture condenses and gives rise only to coastal fogs. Deserts of this type occur in Chile, Peru, and southwest Africa.

TEMPERATURE & RAINFALL

The limited cloud cover in arid areas allows strong thermal contrasts from day to night through intense solar heating during the day and reradiation of heat back out into space at night. The resulting high temperature gradients result in the strong winds and intensive movement of sediment so characteristic of dry climate geomorphology. No major geologic process, however, is restricted entirely to desert regions. Rather, the same processes operate with different intensities in moist and in arid landscapes. Thus, in a desert, the surface sediments, soils, and landforms show some distinctive differences from those elsewhere.

Although arid areas do not generally have rivers originating in their boundaries, it does rain occasionally, and often the storms are torrential. Most runoff that originates in desert areas never reaches the sea, as the water soon disappears through evaporation, or it soaks directly into the earth. Rivers such as the Colorado or the Nile begin in high, humid mountains far from the lower deserts through which they flow. Such rivers carry so much water that they keep flowing to the sea, despite great losses where they cross a desert.

FLASH FLOODS

In arid areas where vegetation and water are limited, soils are thin, and bedrock is therefore commonly exposed at the surface. Weathered fragments of rock tend to break off along natural fractures to leave the steep rugged cliffs that are so common in such regions. The plentiful rock fragments can be easily incorporated into flash floods and moved to lower elevations.

Studies show that typical torrential rainstorms in deserts are likely to be accompanied by flash floods that suddenly and swiftly transport large quantities of rock and soil debris. The debris is deposited as sediment that forms alluvial fans at the bases of mountain slopes and alluvial plains on the floors of wide valleys and basins. The debris-laden streams rapidly lose water into the porous and permeable sediment of the desert basins and end as dry washes or gullies.

ALLUVIAL FANS & PEDIMENTS

Alluvial fans are one of the most characteristic landforms in dry regions. They are caused by the change from the steep, narrow river channels of upland areas to the lower gradients and the unconfined slopes of desert basins, where the water velocity is checked and the sediment load spread. The fan itself commonly shows a characteristic concave-upward slope profile. Alluvial fans are typically dominated on the upper slopes by coarse fragments and on the lower, gentler slopes by finer sands, silts, and clays. Debris flows—sudden rapid waves of wet, bouldery mud—sometimes rush out of mountains after torrential rains and add to the sedimentation on the fans.

Erosion and sedimentation patterns are responsible for pediments, another of the characteristic landforms of arid zones. These are broad, gently sloping areas of bedrock, spread as aprons around the bases of the mountains. Measured cross sections of a typical pediment and its associated mountain show fairly steep bedrock mountain slopes that abruptly change to the gentler bedrock slopes of the pediment. Pediments are eroded by running water that also builds up a thin apron of sediment below as the water continues to erode back the slopes above.

A pediment meets the upper mountain slope, not at a curve but at a distinct angle. This angularity suggests that mountain slopes in the desert do not become gentler with time, as they would in a humid region where chemical weathering, soil formation, and down slope creeping movement of soil are dominant. Instead, the pediments seem to adopt an angle determined by the resistance of the bedrock and maintain that angle as they gradually retreat. In this way, retreat of the mountain slope should extend a pediment forward at its upslope edge. The growth of the pediment at the expense of the mountain may continue until the entire mountain has been consumed.

PLAYAS & BADLANDS

In an arid region, water is rarely plentiful enough to flow for long into a basin to maintain a permanent lake on its floor. Instead, the ephemeral streams that flow down from the highlands after an exceptionally large rain can discharge enough water into the basin to produce a temporary shallow lake. These lakes, known as playas, may last a few days or weeks before the water evaporates or soaks into the ground. Many dry playas are white, gray, or tan because of precipitated salts and clays on the surface. The Spanish origin of the term *playa* (meaning "beach") is a reflection of the observation by early explorers of southwestern North America that many old lakes had dried up and left only beaches and dry lake bottoms behind. In fact, these basins were once filled with water during the high-rainfall, or pluvial, periods during the last ice age, when the climate was much wetter and cooler. Now dry, playa basins are scattered throughout the arid areas of the world.

Badlands in dry zones are extremely closely dissected landscapes on weak, impervious rocks or sediments largely devoid of vegetation. Badland topography is a wilderness of steep, smooth slopes, knife-edged or sharply rounded crests, and steep, narrow valleys. Such areas of bare ground erode through rain splash and slope wash in occasional intense rains.

EOLIAN PROCESSES

Although water is the predominant eroding agent in dry areas, the wind is effective in moving large quantities of sand and dust. Contrary to popular belief, however, most deserts are not covered with sand dunes. Nevertheless, landforms resulting from wind erosion or deposition predominate. Both wind activity and the resulting landforms are commonly referred to as eolian (after Aeolus, the Greek god of wind). Eolian processes in the dry regions of the world are responsible for major erosional and depositional landforms as well as significant sedimentary deposits.

Wind erosion acts in two ways: by deflation, the lifting and removal of loose sand and dust particles from the earth's surface, and by abrasion, the sandblasting action of windblown sand. Deflation causes depressions called blowouts in areas where loose sediment occurs that is of small enough grain size to be moved by wind. These deflation basins can become very large, even below sea level, as long as the underground water table is not reached and as long as large-sized gravels do not accumulate as lag deposits to protect the surface with desert pavements.

YARDANGS & ERGS

Large landforms produced by wind abrasion are not common, but distinctive linear ridges called yardangs (from a Turkistani word meaning "ridge") occur in some desert regions. Typical yardangs have the form of an inverted boat hull and commonly occur in groups, all aligned with the prevailing wind. This shape and orientation are streamlined by the wind to offer minimum resistance to the moving air. Yardangs develop best in soft sediment that is easily eroded but that is cohesive enough to retain steep slopes.

The most extensive areas of wind-transported sand are the ergs, or great sand seas, that occur in the major deserts of the world. Ergs may cover more than 0.5 million square kilometers—twice the size of the state of Nevada. It has been calculated that 99.8 percent of all windblown sand is in the ergs of the world. The largest are in Africa, Asia, and Australia. One third of Saudi Arabia—about 1 million square kilometers—is covered with ergs in the vast "empty quarter" of the Rub' al Khali.

SIMPLE DUNES

Individual dunes in an erg form and move as many different mounds or ridges of sand. Generally, dunes form where an obstacle distorts the flow of air. On encountering an obstruction, wind sweeps over and around it but leaves a pocket of slower-moving air directly behind the obstacle. As sand is blown up or around the obstruction and into the protected wind shadow, its velocity is reduced, and deposition occurs. Once a dune is formed, it acts as a barrier itself, further disrupting the flow of the wind and causing sustained deposition downwind. Continued erosion on the windward side and redeposition on the leeward side can produce rates of dune migration of as great as 25 meters per year. As a dune grows larger, the sand grains saltate up the low-angle windward slope to the top, then slip down into the wind shadow. The slip face keeps a more or less constant angle of repose and creates the high-angle cross-bedding so typical of dunes. Dunes can range in size from a few meters high to as much as 250 meters high.

In an attempt to make some order out of immense diversity, geologists have lumped dunes into several major types, though there are gradations among them as well as many irregular shapes that are hard to fit into any scheme. The basic classification of dunes is threefold: simple, compound, and complex draa dunes.

Simple dunes consist of arrangements of different dune forms relating to one or more wind directions and different amounts of available sand. Simple dunes may be convex or concave, and they may be transverse or longitudinal to the wind.

COMPOUND & COMPLEX DRAA DUNES

Compound dunes are combinations of simpler forms without a change of scale or size of the individual component dunes. The solitary crescent-shaped dune, the barchan, which moves over a flat surface of pebbles or bedrock, is the product of a limited sand supply and winds of moderate velocity that blow in a constant direction. Typically, they are small isolated dunes from 1 to 50 meters high. The tips of a barchan point downwind, and sand grains are swept around them as well as up and over the crest. The steep slip face occurs inside the crescent. Transverse dunes typically develop where there is an abundant supply of sand and a constant wind direction. These dunes develop a wavelike form, with sinuous ridges and troughs that extend perpendicular to the prevailing wind.

Longitudinal, or seif (an Arabic word meaning "sword"), dunes are long, parallel ridges of sand; that is, they elongate in a direction parallel to the resultant of several slightly different wind directions. These dunes develop where strong prevailing winds converge and blow over an area having a limited supply of sand. The grains are shepherded into a long ridge by the blowing of the wind, first in one direction across and along the dune and then from the opposite side of the dune. Many longitudinal dunes are less than 4 meters high, but they can extend downwind for several kilometers. In large deserts, they can be more than 100 meters high and 120 kilometers long; they can be spaced 0.5 to 3 kilometers apart.

Parabolic, or blowout, dunes develop out of preexisting, partly vegetated and thereby stabilized dunes. In these dune forms, first a deflation depression develops, and the sand is piled on the downwind edge of the oval depression. As the shallow deflation hollow enlarges, the sand piles up to form a crescent-shaped ridge. In map view, a parabolic dune is similar to a barchan, but the tips of the parabolic dune point upwind around the deflation hollow from which they originate. Where greatly elongated, parabolic dunes have been called hairpin dunes, again with the extended tips pointing upwind.

Complex draa dunes have dimensions of hundreds of meters, and spacings of kilometers; because of their great bulk, they appear to be relatively static. They are thought to be manifestations of atmospheric turbulence on a distinctly larger scale than that manifested by simple dunes, involving interaction between the regional winds and massive, relatively unchanging sand forms. Barchanoid, transverse, longitudinal, and peaked forms all occur, but in masses that can be more than one hundred times larger than the simple forms. Draa dune fields are characteristic of thick sand accumulations in extremely arid zonal deserts, notably the Sahara and the deserts in Saudi Arabia.

DUST STORMS & LOESS

Dust storms are a major process in deserts. They can transport thousands of tons of fine sediment high in the atmosphere for hundreds of kilometers. These processes have long been known as sandstorms, but, in fact, far more dust is carried aloft than is sand close to the ground. Great dust storms can reach altitudes of more than 2,500 meters above the desert floor and advance at speeds of up to 200 meters per second. Perhaps 500 million tons of windblown dust is carried out of deserts each year, about the same amount carried annually by the Mississippi River. Large quantities blow out into the oceans from Australia or Africa, and some even move across the Atlantic Ocean from the Sahara to the east coast of South America.

Windblown dust is deposited as loess (a German word meaning "loose") and is defined as wind deposited silt and clay; loess accumulates slowly and ultimately blankets large areas, commonly masking preexisting landforms. Such deposits may cover as much as one-tenth of the world's land surface and are particularly widespread in semi-arid regions along the edges of the world's great deserts. Large deposits can reach 300 meters in thickness. Loess is a quite distinctive sedimentary deposit. Most loess is massive and lacks layering, apparently because grains of different sizes settled progressively from the air and were deposited at random. Where exposed, loess commonly stands in steep cliffs because the molecular attraction between the very fine grains is enough to make the particles quite cohesive. One of the primary differences between loess deposits and sand-dune deposits is that sand continues to be mobile even while it is on the ground, whereas loess dust is cohesive and stable once deposited. Loess also tends to settle out on semiarid grasslands or occasionally on woodland, which are themselves not active areas of further wind erosion. Once deposited, therefore, loess tends to remain in place and can continue to accumulate to great thicknesses over time; loess makes very fertile soil.

STUDY OF PROCESSES & PHENOMENA

Running water and wind are the primary processes of dry climates, but significant weathering, gravity-driven slope failure, mass wasting, and even cold climate processes occur. With this wide range of landform-controlling factors, study of the relevant processes involves a nearly complete range of geological and geographical techniques. Aside from plentiful climate measurements of different types, sediment movement is monitored, weathering characteristics described and measured, and landform types and distributions mapped.

In order to develop an understanding of the historical dimension of landform development in dry climates, considerable effort is expended by scientists to develop long-term chronologies. This goal is achieved through measurement of landform morphologies that show progressive erosion and deposition development through time, elucidation of the nature of varied sedimentary layers through stratigraphic analysis, and radiometric and relative-age dating techniques to fix the chronologies.

SIGNIFICANCE FOR HUMAN POPULATIONS

Dry climate geomorphology is very important because so much of the world's population lives in such environments and must contend with the extremes of climate and geomorphic process there. Agriculture is particularly difficult in such arid situations. In the United States, much of the dry agricultural grassland and shrubland of the west is characterized by a special suite of arid landforms. Some of the dry desert basins of the Southwest are important sources of exotic salts, and their different terrains have special significance to the military and space agencies. For example, playa salt deposits, formed by repeated filling and evaporation of the lakes, can be tens of meters thick and a valuable source of industrial chemicals. Borax, the active ingredient in some scouring powders, is but one example. Dry playas are also extraordinarily flat surfaces and have been used for a long time as huge landing fields for rockets and spacecraft such as a shuttle. In addition, the largely cloud-free and windy dry lands are important sites for solar-power and wind-power generation.

Further Reading

Abrahams, A. D., and A. J. Parsons, eds. *Geomorphology of Desert Environments*. 2d ed. London: Chapman and Hall, 2009. The essays in this volume deal with a broad range of topics relating to dry climates. Intended for the specialist but readable by those with some science background.

Bloom, Arthur. *Geomorphology: A Systematic Analysis of Late Cenozoic Landforms*. 3rd ed. Long Grove, Ill.: Waveland Press, 2004. This college-level text covers the basics of geomorphology. Includes a chapter on arid and savanna landscapes. Index and bibliography.

Chorley, Richard J., S. A. Schumm, and D. A. Sugden. *Geomorphology*. New York: Methuen, 1984. One of the best but more difficult general texts available. Nevertheless, the discussions on arid land phenomena are current, the graphics are clear, and the references include later sources. The discerning student of general geomorphology could start with this volume as a guide to the major issues of dry climate geomorphology and then move on to other more detailed, but less rigorous, works.

Gregory, Kenneth J., and Andrew S. Goudie, eds. *The SAGE Handbook of Geomorphology*. Thousand Oaks,

Calif.: SAGE Publications, 2011. A compilation of essays discussing geomorphology, techniques, processes and environmental changes. Provides in-depth descriptions and examples along with fundamentals of geomorphology. Includes many in text references, tables, and indexing.

Lancaster, N. *Geomorphology of Desert Dunes.* London: Routledge, 1995. Lancaster provides an in-depth treatment of desert topography. Technical in some areas. Index and bibliography.

Mabbutt, J. A. *Desert Landforms.* Cambridge, Mass.: MIT Press, 1977. This volume is one of the premier works on deserts and desert landforms, with examples from around the globe. The emphasis is on landforms and the processes by which they are produced. The style is clear and the many illustrations are generally well done.

Nicholson, Sharon E. *Dryland Climatology.* New York: Cambridge University Press, 2011. This text covers the geomorphology, hydrology, ecology, and climatology of dry land. Discusses adaptation, microhabitats and desertification. Written for graduate students and professional researchers and scientists.

Ritter, Dale F., R. Craig Kochel, and Jerry R. Miller. *Process Geomorphology.* 4th ed. Long Grove, Ill.: Waveland Press, 2006. The author presents useful information about all the processes that act on the earth's surface, including those that act in dry climate areas. The chapter on wind processes and landforms is clear and well written, but not many additional references are cited.

Rodrigues, Silvio Carlos, and Gelze Serrat de Souza Campos Rodrigues. "Geomorphology and Biogeography of Tropical Deserts." In *Tropical Biology and Conservation Management: Encyclopedia of Life Support Systems.* Oxford: UNESCO, Eolss Publishers, 2009. Discusses processes and landforms in desert areas. Identifies the geographical distribution, characteristics, and factors controlling tropical deserts. Complete with bibliography. The encyclopedias are written for researchers, college students and faculty, and scientific professionals.

GEOMORPHOLOGY OF WET CLIMATE AREAS

—*Philip A. Smith*

FIELDS OF STUDY

Geography; Botany; Agronomy; Agriculture; Biochemistry; Ecology; Entomology; Soil Science.

PRINCIPAL TERMS

chemical weathering: the chemical alteration of rocks and minerals into new forms that are chemically stable at the earth's surface

erosion: the removal of weathered rock and mineral fragments and grains from an area by the action of wind, ice, gravity, or running water

humid-midlatitude: land area with average temperature of the coldest month less than 18 degrees Celsius but at least eight months with average monthly temperatures greater than 10 degrees Celsius; this area has no dry season

landform: a grouping of genetically related landscapes based on similarity of appearance

mechanical weathering: the physical breakdown of rocks into progressively smaller particles

regolith: the layer of mechanically and chemically weathered rock and mineral matter above fresh, unweathered bedrock

savanna (tropical wet-dry): land area with monthly temperatures averaging greater than 18 degrees Celsius and with one to six months of average monthly precipitation of less than 6 centimeters

tropical rain forest: land area with monthly temperatures averaging greater than 18 degrees Celsius and monthly precipitation averaging more than 6 centimeters

SUMMARY

Wet climate geomorphology is the study of landforms and landforming processes in the humid regions of the earth. These landforms correspond roughly to tropical rain forest, savanna, and humid-midlatitude climatic regions. Each of these landforms is distinct because of differences in temperature and precipitation.

The river here is deeply eroding soft glacial deposits under the forest. Image by John Allan CC-BY-SA-2.0.

CLIMATIC GEOMORPHOLOGY

Geomorphology is a branch of geology and physical geography concerned with the study of the earth's landforms and the processes that create them. There are several ways of looking at those processes and, as a result, several branches of geomorphology. One is climatic geomorphology, which is the study of the role of climate in shaping the land. Wet climate geomorphology is a component of climatic geomorphology that exclusively addresses land areas having relatively high annual precipitation. This is a natural grouping, since water is instrumental in the wearing down and sculpting of the land.

WEATHERING & EROSION

Before looking specifically at the landforms and processes of the humid regions, it is first necessary to examine the major natural processes operating on the land, no matter what the climate. These are weathering, transport, and deposition of rock and mineral matter. The first two processes are sometimes collectively referred to as erosion, although a more accurate definition of erosion would be the removal of rock and mineral matter from an area. Weathering may be either mechanical, the physical breakdown of rocks into smaller and smaller particles, or chemical, the chemical alteration of rocks and minerals into new, more stable forms that may become part of a soil. The most important type of mechanical weathering is freeze-thaw, in which freezing water expands within rocks and fractures them. Mechanical weathering, as well as transport of weathered debris, is suppressed in hot, wet climates, where there is little or no freezing and a thick vegetation cover protects the soil from being removed and carried away (eroded). In temperate climates such as Europe and the eastern United States, mechanical weathering is somewhat more important and produces huge amounts of loose rock and mineral particles that can be eroded.

Chemical weathering is most active in tropical wet regions, where both temperature and precipitation are high. It occurs only in the presence of abundant percolating subsurface water and breaks down many common rock-forming minerals into three components: clays, quartz, and dissolved particles (ions). The clay and quartz remain behind, unless eroded, and form a residue from which soils develop, while the ions dissolved in the subsurface water may recombine to form new minerals such as iron oxide. They may also be used by plants or be transported downward in percolating water and eventually deposited into streams and rivers. Some common minerals, and therefore the rocks that contain them, are resistant to chemical weathering and do not erode easily.

Transport of weathered rock and mineral matter occurs primarily by running water, either as sheet flow (when an unchanneled sheet of water flows down a slope) or as channeled flow (in gullies, streams, and rivers). Where the protective cover of vegetation is thick, sheet flow is ineffective as a transporting agent. Where it is thin, sheet flow can erode the underlying soil as fast or faster than it can form. Where very steep slopes exist, landsliding becomes an important transporting agent. Deposition of weathered and transported debris occurs when the velocity of the running water decreases—at breaks in slope or in the floodplains of rivers, for example.

MORPHOCLIMATIC ZONES

Wet climate landforms can be subdivided into three types based on temperature and the seasonal amount and distribution of precipitation. These landforms, or "morphoclimatic zones," are tropical rain forests, where both temperature and precipitation remain high all year long; savannas, or tropical wet-dry regions, where both temperature and precipitation are high but precipitation is restricted to a rainy season; and humid-midlatitude regions, where precipitation is moderate but uniformly distributed throughout the year and temperatures undergo large seasonal fluctuations.

TROPICAL RAIN FORESTS

The tropical rain forests of the world are confined to a belt extending 10 to 15 degrees of latitude north and south of the equator. Examples are the Amazon River Basin in South America, much of central Africa, and the Indonesian Islands. In these areas, high temperatures and rainfall produce a thick vegetation cover even on steep slopes. As a result, sheet flow is minor, and most rainwater infiltrates the ground, where it promotes chemical weathering. In these areas, mechanical weathering is inhibited, but chemical weathering is intense and produces a deep mantle of weathered rock and mineral grains above fresh bedrock. This mantle of unconsolidated debris is called regolith and may extend more than 100 meters below the land surface. The downward percolation of subsurface water is so intense and constant that the dissolved ions that it carries are completely removed, leaving a residue of clay, quartz, and iron oxides. Soils in tropical wet areas therefore tend to be infertile below the top meter or so.

In mountainous parts of the tropical rain forest, where fast-moving streams have high energy, streams cut rapidly down through the easily eroded regolith and produce deep, narrow valleys with steep, V-shaped slopes separated by narrow ridges. The regolith on these slopes sometimes becomes saturated with water, resulting in landslides. This type of topography is typical of areas such as Papua New Guinea and parts of the Amazon basin.

In contrast to the rugged topography of mountainous regions, lowlands in the tropical rain forest are quite flat. They contain large rivers such as the Amazon in Brazil and the Congo in central Africa. These rivers meander on wide floodplains tens of kilometers wide and thousands of kilometers long. Many rivers have a peculiar stepped appearance, with occasional rapids or even waterfalls interrupting long stretches of sluggish flow. Tropical rain forests contain some of the world's most spectacular waterfalls, such as Angel Falls in Venezuela (900 meters high) and Victoria Falls in Zambia. The waterfalls are produced by resistant rock layers, such as quartzite, that do not chemically weather and that the rivers cannot wear down given the lack of coarse abrasive particles in transport. Another unique landform feature of the tropical rain forest is the occasional presence of large, isolated, haystack-shaped hills devoid of vegetation. These are called monolithic domes, and they present a curious contrast to the surrounding heavily vegetated lowlands. A number of these domes are found around Rio de Janeiro in Brazil. Geomorphologists believe that monolithic domes exist because they are more resistant to weathering than the surrounding rocks. Soils cannot form on them because of steep slopes, landsliding, and lack of the water-retention capacity necessary to promote chemical weathering.

SAVANNA REGIONS

Savanna regions are much larger in area than the tropical rain forests and extend north and south as a wide fringe bordering the tropical rain forests. Savanna regions include parts of southern and northern Brazil, southern Central America, central and southern Africa, and parts of India and Southeast Asia—areas containing more than half the world's population. Savannas have a dry season lasting from three to six months. During the dry season, vegetation dies or becomes dormant and chemical weathering decreases significantly as a result of the lack of percolating water. With a decreased vegetative cover, mechanical weathering and erosion of weathered debris by sheetwash increase significantly, resulting in removal of the regolith faster than it can form.

Savanna landscapes therefore tend to have a stepped appearance, with vast flat or gently sloping plateaus (pediplains) with thin regolith or none, interrupted by abrupt slopes that lead up to the next higher plateau. This style of erosion is called backwasting and is typical of savanna and semiarid regions. Former high plateaus may be reduced by erosion to isolated remnants called inselbergs. The stepped appearance is augmented by the formation of hard, impermeable layers in the soil called crusts or duricrusts. They form progressively during dry seasons, when percolating subsurface water evaporates and chemically precipitates oxides of iron or, less commonly, aluminum. Once formed, these crusts persist and may grow to several meters in thickness. They protect the underlying, less resistant regolith from erosion unless they are breached, in which case erosion proceeds rapidly by undercutting them, sometimes all the way down to bedrock. In this way, the crusts help preserve flat plateau surfaces at various levels, although the surfaces are progressively reduced in area through time.

When stream flow resumes in the wet season, streams are unable to cut down through the crusts and therefore have very shallow, hard-to-recognize channels. Much rainwater runs over the plateau surfaces as sheetwash, carrying with it fine particles of regolith. When flow ceases, areas of shallow ponded water serve as gathering places for animals such as elephants, lions, and wildebeests. The well-known Serengeti Plain in Tanzania, Africa, is an excellent example of this kind of landscape.

HUMID-MIDLATITUDE REGIONS

Landforms in humid-midlatitude regions such as Europe and the eastern United States differ in many respects from the tropical rain forest and savanna regions. Here, the cool winters promote mechanical weathering via freeze-thaw, chemical weathering is moderate because of cooler temperatures, and precipitation is moderate but evenly distributed throughout the year. Streams therefore run all year long and carry higher loads of coarse sediment, such as sand and gravel, that chemical weathering has not reduced to smaller sizes. These mechanically weathered sediments act as abrasive tools and wear down even the harder rocks. Stream courses therefore do not have the stepped appearance seen in the tropical rain forest. They cut valleys whose slopes recede via landslides and, more important, soil creep, the very slow downslope movement of the moist soil.

Soil creep helps to produce the rounded hills and gentle, curved slopes typical of humid-midlatitude areas such as the Appalachian Highlands of the eastern United States or the gently rolling landscapes of parts of Western Europe. Slopes grade uniformly into lowland floodplains, where coarse sediment is stored. Isolated remnants such as monolithic domes or inselbergs are quite rare in humid-midlatitude regions. Since chemical weathering is only moderately active, soils tend to be thin but rich in plant nutrients because the downward percolation of water is not intense enough to cause complete removal of dissolved ions necessary for plant growth. Humid-midlatitude regions are therefore ideal areas for agriculture.

TOPOGRAPHIC MAPS

The primary interest of the geomorphologist is to determine the nature of the important causative processes acting to form the landscape and to construct predictive models of landscape change based on a knowledge of those processes and their rates. A variety of methods are used for this purpose, and specific techniques as well as analytical tools are usually dictated by the problem to be solved. Most geomorphological analyses involve measurement of some landscape feature, such as slope steepness, and observation of the natural processes acting to form it, such as the amount of precipitation, underlying rock type, or type of vegetation cover. The effect of each process must be measured and its relative importance established. Almost all such analyses involve statistical treatment.

Topographic maps, aerial photographs, and geologic maps are essential for any geomorphological study. Topographic maps are two-dimensional representations of the topography of an area. They allow geomorphologists to study landscape features without actually being at a site, and they lend themselves to statistical treatment. For example, the relation between slope steepness and the frequency of landslides can be determined quantitatively because both features have a topographic expression and can be recognized on topographic maps. Such a study may conclude that slopes greater than 40 degrees have 20 percent more chance of landslides than slopes less than 40 degrees. This example, though rather simple, nevertheless has great practical importance to engineers working in wet climate areas.

AERIAL PHOTOGRAPHS

A series of aerial photographs taken at a certain elevation—for example, at 3,000 meters—can be partially overlapped and, when viewed with stereoscopic glasses, can give an exaggerated three-dimensional view of a land area. Such photographs present a visual image of the topography with more detail than can be provided by topographic maps.

Photographs taken at different times can be compared in order to determine rapid landscape changes and their rates of change. This method is often used in studies of river-channel migrations across a floodplain. The information can then be used by engineers and land planners to determine future areas of urban expansion.

GEOLOGIC MAPS & OTHER STUDY TOOLS

Geologic maps are two-dimensional representations of the types and orientations of rocks or surficial materials in an area. They allow geomorphologists to study the relationship between the underlying rock or soil and the topography as expressed on topographic maps or aerial photographs. For example, in the humid-midlatitude Valley and Ridge area of Pennsylvania, elongate ridges stand up to 1,000 meters above intervening valleys. The ridges are composed of resistant, hard sandstone, whereas the valleys are underlain by shale, a weaker and more easily eroded rock.

Other investigative tools have a more specific purpose. These include sieves to measure grain sizes of debris, petrographic microscopes to determine mineral types, and X-ray diffractometers to determine types of clay minerals. In addition, various techniques are used by geomorphologists to determine the age of landscape-forming events. The techniques used are quite sophisticated and include carbon-14 dating, fission track dating, amino acid racemization, and dendrochronology (the use of tree rings to determine the dates and climate at which they formed). Finally, computers are essential tools in geomorphological studies. They are used extensively as aids in statistical analyses and for purposes of modeling geomorphic processes and predicting landscape changes.

APPLICATIONS OF GEOMORPHOLOGICAL RESEARCH

The geomorphological study of wet climate landforms has tremendous practical importance through its ability to predict the effects of human-induced changes in the landscape. For example, pressures for development in nations occupying the tropical rain forest have resulted in massive forest-clearing operations that threaten the very existence of these forests. One effect has been a tremendous increase in erosion of the soil and underlying regolith as a result of the high precipitation and stripping of the protective vegetation cover by sheetwash. Parts of the Amazon basin in Brazil now have a badlands type of topography. Crops will not grow in the remaining bare, infertile regolith, because seeds and fertilizer simply wash away with the regolith. The eroded debris rapidly clogs formerly clear-flowing streams and endangers numerous aquatic plant and animal species. Tropical geomorphologists are now involved in projects to reclaim much of this land and to prevent future landscape degradation.

Many applications of geomorphological research in wet climate areas have to do with the dynamics of river channels and river erosion. Engineering works such as reservoir construction and river channelization have had some unanticipated side effects, notably the severe erosion of land downstream from impoundments or increased flood heights downstream from channelized rivers. Geomorphologists have been able to model these river responses to change and have enabled engineers to construct safer and more effective flood control and water supply works.

A new challenge for geomorphologists is the reclamation of disturbed lands such as old landfills and open pit mines. Proper reclamation is based on an understanding of the landforming processes and how they operate in different climatic regions. With this understanding, geomorphologists have had much success in converting potentially hazardous areas to environmentally safe and economically valuable uses such as golf courses, parks, and playgrounds.

SCIENTIFIC VALUE OF CLIMATIC GEOMORPHOLOGY

In addition to its practical value, climatic geomorphology has great scientific value. Because climate exerts strong controls on the landforming processes of weathering, transport, and deposition, climatic geomorphologists can recognize large areas of the earth where past climates were different from the present one. For example, the currently humid-midlatitude region of central Germany retains remnants of flat plateau surfaces formed under a savanna climate. Many, if not most, land areas of the earth show such polygenetic landscapes (landscapes having more than one origin). This information is used to establish a history of climate change through the last 70 million years.

A rapidly developing area of geomorphology is the study of planetary landforms. Pictures of the Martian surface clearly show valleys similar to those cut by rivers on Earth, yet there is almost no surface water on Mars today. Does this mean that Mars has somehow lost its water, or could some other agent of erosion such as wind have produced the valleys? Current geomorphological research is trying to answer this question, as well as others regarding the formation of planetary landforms.

FURTHER READING

Bloom, Arthur. *Geomorphology: A Systematic Analysis of Late Cenozoic Landforms.* 3rd ed. Long Grove, Ill.: Waveland Press, 2004. This college-level text covers the basics of geomorphology. Includes a section on wet climates, with particular emphasis on weathering. Index and bibliography.

Gregory, Kenneth J., and Andrew S. Goudie, eds. *The SAGE Handbook of Geomorphology.* Thousand Oaks, Calif.: SAGE Publications, 2011. A compilation of essays discussing geomorphology, techniques, processes and environmental changes. Provides in depth descriptions and examples along with fundamentals of geomorphology. Includes many in-depth text references, tables, and indexing.

Latrubesse, E. M., J. C. Stevaux, and R. Sinha. "Tropical Rivers." *Geomorphology* 70 (2005): 187–206. Compares the geology and geomorphology of various river basins around the globe. Discusses drainage from mountain belts to lowland plains in wet and wet-dry climates. The authors also mention anthropogenic river management issues.

Marsh, William M. *Earthscape: A Physical Geography.* New York: John Wiley & Sons, 1987. An introductory text that can serve as a good starting place for readers who are interested in climatic geomorphology but who have no prior training. Of special relevance are Chapters 8 and 9 (climate), 10, 11, and 12 (vegetation), 13 and 14 (soils), 16 (rock structures), and 18, 19, and 20 (weathering and erosion). Suitable for upper-level high school or college-level readers.

Restrepo, Juan D., Sergio A. Lopez, and Juan C. Restrepo. "The Effects of Geomorphic Controls on Sediment Yield in the Andean Rivers of Colombia." *Latin American Journal of Sedimentology and Basin Analysis* 16 (2009): 79–92. A technical and detailed study on the sedimentation within the Andean basin. Compares Colombian runoff of the Pacific system and the Caribbean system. Hydroligic, climatic, and morphometric data were collected to build a model for predicting sediment yield. A background in geology or hydrology is recommended.

Skinner, B. J., et al. *The Dynamic Earth: An Introduction to Physical Geology.* 5th ed. New York: John Wiley & Sons, 2006. A revised physical geology text with an expanded section on landforming processes. It can provide the reader with the necessary geological background to pursue more advanced knowledge in climatic geomorphology. Suitable for upper-level high school or college-level readers.

Thomas, M. F. *Geomorphology in the Tropics: A Study of Weathering and Denudation in Low Latitudes.* New York: John Wiley and Sons, 1994. This detailed and technical text is intended for specialists. Extensive bibliography.

GLACIAL LANDFORMS

—Diann S. Kiesel

FIELDS OF STUDY

Geography; Climatology; Soil Science.

PRINCIPAL TERMS

arête: an extremely narrow mountain ridge created between adjacent U-shaped valleys

An arête with an elevation of 13,694 feet near Gilpin Peak in the San Juan Mountains, Ouray County, Colorado. Image by U.S. Geological Survey.

cirque: a bowl-shaped depression near mountaintops where glaciers originate

drumlin: a streamlined hill formed under actively moving ice

esker: a sinuous ridge of stratified drift formed in a tunnel under the ice

kame: a conical hill of stratified drift

kettle: a depression created by the melting of a chunk of ice

moraine: landscapes of till, varying from fairly flat terrain to gently rolling hills to long ridges

stratified drift: material deposited by glacial meltwaters; the water separates the material according to size, creating layers

till: a mixture of unsorted, unconsolidated materials deposited by glacial ice

SUMMARY

Glacial landforms are common in many parts of Canada, the northern United States, northern Europe, Asia, and many mountain ranges of the world. The material composing these landforms, as well as the shapes of some of the landforms, is essential to many human activities.

GLACIAL MOVEMENT

Glacial landforms are distinguished primarily on the basis of shape and composition, both of which relate directly to the mode of origin of the glacial features. Some landforms originate by erosion or deposition directly by the ice, while other features are created by erosion or deposition of the meltwaters from the ice. It is important to distinguish whether the ice or its meltwaters produced the landform. Another factor in recognizing glacial landforms is the realization that many glacial features are commonly associated with a particular kind of glacier. Two types of glaciers are commonly recognized: continental glaciers and valley glaciers. Although some landform features are common to both types of glaciers, others are relatively unique to each type.

An understanding of glacial ice movement is also necessary in order to understand how glacial landforms develop. Whether the ice is moving because of pressures from great thicknesses of continental ice or is responding to gravity down a valley slope, the ice always moves forward. Even when the glacier is melting back, and it looks like it is getting shorter or smaller, internally the ice is still moving forward.

Glaciers can erode the material over which they move, and if they scrape down to solid bedrock, several glacial features can result. If the glacier is carrying small, hard particles along its base, these particles may scratch the bedrock, creating many roughly parallel striations. If large rocks are being transported at the base of the ice, they can gouge out varisized grooves, which can range from a few centimeters to several meters deep. Conversely, if the ice has ground its load into flour-size particles, called rock flour, the flour can actually polish the bedrock. The orientation of striations and grooves can be used to determine ice-flow direction.

EROSIONAL LANDFORMS

Of all the glacial landforms, perhaps the most spectacular are the erosional products of mountain glaciers. Mountains that have experienced glaciation exhibit some of the world's most breathtaking scenery. Glaciers tend to sharpen peaks and ridges and to steepen valley walls, producing avenues for numerous spectacular waterfalls. As snow accumulates near the tops of mountains, the forming ice tends to carve out a basin-shaped depression on the side of the mountain. This basin is called a cirque and is the home base of the glacier; it is in this basin that more snow will accumulate, thickening the ice, until the ice begins to move out of the cirque and down the valley. If three or more glaciers form around the same mountain peak, the crowding of their cirques around the peak will erode it into a sharper than usual pyramidal shape. This type of mountain peak is called a horn. Many of the peaks in the Alps have the word "horn" incorporated into their names; the Swiss Matterhorn is probably the most famous glacial horn. If two glaciers are moving down a mountain in adjacent valleys, the ridge between them may be eroded into a narrower, sharper ridge called an arête.

Mountain glaciers tend to widen the base and steepen the sides of the valleys through which they move, creating obviously U-shaped valleys. Large glaciers have more power to erode more deeply, resulting in deeper U-shaped valleys. Small tributary glaciers produce valleys that are not carved as deeply; therefore, when the glaciers retreat from such an area, the floors of the tributary valleys are left high above the floor of the main valley. Such valleys are appropriately called hanging valleys. If streams are present, they become waterfalls that cascade over the edges of the hanging valleys. U-shaped valleys usually contain strings of lakes that are impounded by irregularities of the valley floor. These lakes are quite picturesque in that the rock flour they contain from glacial meltwaters gives the water a distinctive turquoise color.

Perhaps the most spectacular of the glacial valleys are fjords, most common in Norway, British Columbia, Alaska, and Chile. A fjord is a coastal glacial valley that was carved out when sea level was lower during times of glaciations. Upon retreat of the ice and subsequent rise in sea level, seawater completely inundated the valley. The water in many fjords is often more than 1,000 meters deep. In some areas where fjords extend inland for considerable distances and intersect with other fjords, it may be difficult to distinguish a fjord from a large lake in a U-shaped valley. The fjord, being inundated with marine water, will contain seaweed and experience tidal changes; the lakes will be freshwater and relatively static.

A fjord at the Narrows in Fords Terror, Petersburg district, southeastern Alaska. Image by U.S. Geological Survey.

GLACIAL DEPOSITS

Glacial deposits are widespread and create interesting landform features. Glaciers deposit till, an unconsolidated mixture of clay, sand, silt, pebbles, cobbles, and boulders. Depending on the area over which the glacier moved, the relative abundance of these components will vary; some till can be very sandy, while other till can be clayey. Abrasion during ice transport causes the particles in till to be usually angular in shape. Because it is the glacial ice itself that deposits this material, the till is unsorted. This lack of sorting results because the materials in till are let down to the surface as the ice melts in the same relative positions they had in the ice. Till can be formed into various types of morainal landforms, depending on how the glacial ice moved. Till can be deposited as ground moraine as the ice retreats. It is laid down under the ice as the ice moves along, resulting in a landscape called a till plain, which usually takes on the appearance of gently rolling, hummocky hills.

When a glacier front becomes stationary, material will continue to be moved forward through the internally forward-moving ice and be deposited along the front of the stationary glacier. It will build up into a ridge called an end moraine. If the glacier then advances, the end moraine will be destroyed by the moving ice. If the glacier retreats, the end moraine will become part of the landscape. The end moraine that is created at the maximum forward position of the ice is called the terminal moraine. Any end moraines that form as the ice periodically retreats are called recessional moraines and usually form concentric bands of ridges in the landscape. If a glacier advances and then retreats without spending any significant length of time in a stationary position, end moraines cannot form; the till will be formed into ground moraine.

Two types of moraines are unique to mountain glaciers. Because these glaciers are restricted to valleys, accumulations of debris tend to gather along the valley walls at the sides of the glaciers. These are called lateral moraines and are left behind as long ridges along the valley walls as the ice melts away. A medial moraine forms when two glaciers join and their lateral moraines are consequently trapped between the two ice rivers, which are now flowing along as one glacier. Several medial moraines can form as more and more glaciers from tributary valleys join, giving the glacier a striped appearance.

MELTWATER

Glacial meltwater is an important agent in creating several landform features. The power of glacial meltwater can be extraordinary, often carving deep gorges. The Channeled Scablands of Washington attest the effects of glacial meltwaters. When glaciers melt, they release not only water but also the huge amounts of debris they carry. The material deposited by glacial meltwater is called stratified drift and differs from till in that it is sorted and layered. The heavier particles drop out first, followed by progressively smaller particles as the water travels farther from the ice, gradually slowing down. Fine clays and silts are usually transported the farthest, sometimes completely out of the area, giving the stratified drift its distinctive sand and gravel characteristics. As the glaciers eventually release less and less water, smaller and smaller particles are deposited over the coarser material deposited previously. Unlike the ice-deposited till, transport in water tends to round the particles of stratified drift.

Outwash plains are huge, flat areas of stratified drift located adjacent to where glaciers existed. If outwash is confined to a river valley, it is called a valley train. Large outwash areas are usually devoid of vegetation cover at first and are subjected to strong winds. Often silt-sized particles in the outwash are transported by wind and deposited in thick layers downwind from the deposit. Windblown silt is called loess and is an important ingredient in soils. Soils developed on loess-capped till plains are particularly good for agriculture.

Meltwater sands and gravels can form interesting landscape features if conditions of the melting glacier are just right. If the area near the snout of the glacier becomes too thin, the ice will cease to move internally and become stagnant. When that happens, the stagnant ice may be eroded by its own meltwater as well as that coming off the active ice. Meltwater streams may form on the ice and drill tunnels in and under the ice. The movement of the streams determines which type of landform feature will evolve from the deposition of the sands and gravels transported by the stream. For example, if streams in or on the ice cascade down fairly vertical shafts (called moulins) in the ice or tumble down a fairly steep face of the ice edge, the sand and gravel will spill into a conical pile called a kame, which can be more than 60 meters high. Usually several kames are formed in the same area, dotting the landscape as isolated hills.

Another example of meltwater stream deposition is eskers, which also form in zones of stagnant ice where sand and gravel are deposited within stream tunnels that are in and under the stagnant ice. When the ice melts away, long, sinuous ridges are left in the landscape. New England has some impressive eskers that are more than 40 meters high in places and wind for miles through the countryside.

Other features commonly associated with outwash and stagnant ice areas are kettles, which form when chunks of ice break off from the glacier and become buried in outwash or in till as the glacier melts. Gradually the ice chunks melt, leaving depressions in the ground, sometimes in the shape of pioneers' kettles, which is how the name originated. Kettles that intersect the groundwater table or collect precipitation contain water and are called kettle lakes, some of which can cover several acres. An outwash area that contains numerous kettles is called pitted outwash.

OTHER LANDFORMS

Actively moving ice sometimes has the ability to form streamlined shapes through either erosion or deposition. A roche moutonnée is an eroded bedrock form with a gentle up-ice side and a steep, rough down-ice side. Hence, ice-flow direction can be determined by the orientation of roches moutonnées. These features vary in size from 1 meter to large hills in the landscape.

Drumlins are streamlined hills that look like half-buried whales or inverted spoons composed of a variety of materials. The cores can be composed of bedrock or, more commonly, stratified drift and are covered with a veneer of till. The formation process of drumlins is unclear, but agreement on some aspects does exist. They appear to have been formed by actively moving ice near, but not at, the edges of the ice. Drumlins always appear in groups, called drumlin fields, composed of thousands of drumlins. The most famous drumlin fields are in New York, Wisconsin, and New England.

The Great Lakes exist because of the action of continental ice sheets in the Midwest. Tongues of ice gouged out old river valleys, creating the lake basins. Meltwater and precipitation filled the basins. The history of the glacial Great Lakes from the time of the ice sheets to the present day is extremely complicated. As the ice retreated and made minor advances in various parts of the upper Midwest, different sizes and shapes of lakes emerged in each basin. The lakes had different outlets at different times, creating different water levels. Most of the old lake levels can be traced today because old beach and dune deposits exist at various elevations along the present-day Great Lakes shorelines. The old clay-rich lake beds that cover large parts of Ohio, Indiana, and Illinois have some of the richest soils in the United States.

STUDY OF GLACIAL LANDFORMS

Glacial landforms are studied using a variety of methods. The general forms of the features are analyzed using topographic maps and various types of aerial photographs. Such tools not only illustrate the forms of each glacial feature but also show its relationship to other nearby features. Patterns in the landscape can be observed quite easily, and ice-direction movement may be interpreted. Topographic maps are very detailed, showing all natural and human-made features in a small area. These maps also include contour lines that represent elevations above sea level; with some practice, hills and valleys can be identified. All landforms that are large enough to be

A kettle, or glacial pothole, near Tuolumne Meadows in Yosemite National Park. Image by U.S. Geological Survey.

incorporated by the given contour interval can be shown on the map. All kinds of glacial features can be identified using topographic maps because their shapes are distinctively outlined by the contours.

Aerial photographs can be used in the same way as topographic maps, except that with photographs, the actual feature in the landscape can be seen. There are several types of aerial photographs that can be used for analysis, depending on which features are being studied. Standard black-and-white, color, and a variety of other wavelength bands (from satellite imagery) each emphasize distinct characteristics of the landscape.

Recall that a diagnostic characteristic for many glacial landforms is their composition. The composition of any particular feature is determined by digging into it. Sometimes quarries or roadcuts have already exposed the interior of the structure, making the job of the geologist easier. At other locations, core samples are obtained by drilling. Sometimes well records from existing water wells may be utilized to determine which kind of material has been deposited in an area. Another clue to the composition of a feature may be obtained by the type of soil that has developed on the material. County soil surveys are useful for this kind of analysis.

On a more detailed scale, geologists may want to understand the mechanics of ice flow when determining the creation of the landform, the direction of ice flow, the source area of the material, or the period in which the ice existed in a particular area. These pieces of information can be obtained by analyzing samples taken from the landscape features. Till fabric, pebble lithology and roundness, particle size, radiocarbon dating of organic material, and the amount and type of weathering are all commonly used observations.

Many areas rely on glaciers as a major water resource for either general consumption or power generation. Mountain glaciers are carefully monitored in order to determine the amount of water that can be expected from them each year. The monitoring is done by either digging into the ice at various points and measuring ice layers and rates of melting or, more commonly, by taking precipitation and meltwater stream gauging measurements throughout the entire glacial basin.

FURTHER READING

Adams, George F., and Jerome Wyckoff. *Landforms.* New York: Golden Press, 1971. This Science Golden Guide covers all aspects of landforms. The section on glaciers and glaciated land is brief but contains excellent diagrams and photographs to illustrate the formation of glacial landscapes. The section on useful publications not only lists general references but also indicates where to obtain regional, state, and local information about geologic features.

Bennett, Matthew M., and Neil F. Glassner, eds. *Glacial Geology: Ice Sheets and Landforms.* 2d ed. Hoboken, N.J.: Wiley-Blackwell, 2009. This text contains detailed information on the characteristics and dynamics of glaciers. Discusses the occurrence of glaciations around the world, glacial mechanisms, hydrology, erosion, and sedimentation. Contains chapter-specific suggested readings.

Bloom, Arthur G. *Geomorphology: A Systematic Analysis of Late Cenozoic Landforms.* 3rd ed. Long Grove, Ill.: Waveland Press, 2004. This college-level text covers the basics of geomorphology. Includes three chapters on glaciers and glaciology. Index and bibliography.

Brunsden, Denys, and John C. Doornkamp, eds. *The Unquiet Landscape.* Bloomington: Indiana University Press, 1974. The introduction to this book contains a good general introduction to geology, maps, and general methods of geological study. A brief bibliography and an index are included. The two chapters explaining glacial erosion and deposition and related landforms are illustrated with excellent photographs. The chapters in the book are interrelated; for example, information about glacial effects on landscape is also explained in the chapters on lakes and changing sea level.

Colby, William E., ed. *John Muir's Studies in the Sierra.* San Francisco: Sierra Club Books, 1960. John Muir was the first to recognize the dramatic impact of glaciation in the Sierra Nevada. His observations and methods of reasoning showed geologists the folly of their previously incorrect interpretations of the development of the Sierra landscape. This book contains Muir's observations and explorations written in his easy-to-understand style. It includes his own drawings as well as several photographs. Because glacial geology is related to other geologic effects, Muir considers all aspects of the Sierra geology.

Dolgoff, Anatole. *Physical Geology.* Boston: Houghton Mifflin Harcourt, 1999. This is a comprehensive guide to the study of the earth. Extremely well illustrated and includes a glossary and an index. Although this is an introductory text for college students, it is written in a style that makes it understandable to the interested layperson. Contains a section on glaciers, glaciology, and glacial deposits.

Forrester, Glenn C. *Niagara Falls and the Glacier.* Hicksville, N.Y.: Exposition Press, 1976. This small book explores all the geology of Niagara Falls, especially the influence of the continental glacier and the development of the Great Lakes. It is well illustrated with maps, photographs, and diagrams and includes a walking tour. Although there is no index, there is a brief bibliography.

Guyton, Bill. *Glaciers of California: Modern Glaciers, Ice Age Glaciers, the Origin of Yosemite Valley, and a Glacier Tour in the Sierra Nevada.* Berkeley: University of California Press, 2000. Discusses the glacial history of California and glacial landforms. This book includes a glossary, additional reading, and indexing. Also cites references for each chapter.

Hambrey, Michael, and Jurg Alean. *Glaciers.* 2d ed. New York: Cambridge University Press, 2004. Written by the director of the Centre of Glaciology, this text covers characteristics and dynamics of glaciers as well as their effect on humans. Focuses mainly on European glaciations, but still provides a strong overview of glaciers.

John, Brian S. *The Ice Age: Past and Present.* London: Collins, 1977. This book is very thorough in its examination of glacial and meltwater features. Many photographs, maps, and diagrams are used, and both a good index and a bibliography are included.

Matsch, Charles L. *North America and the Great Ice Age.* New York: McGraw-Hill, 1976. Numerous photographs, maps, and diagrams illustrate the chapter on glacial erosional and depositional features. There is also a chapter on lakes and sea-level changes. Brief selected readings and an index are included.

Menzies, M., ed. *Modern Glacial Environments: Processes, Dynamics, and Sediments.* Oxford, England: Butterworth-Heinemann, 1996. This multiauthored volume deals with all aspects of the modern study of glaciers. Intended for the specialist. Index and extensive bibliography.

Moore, W. G. *Glaciers.* London: Hutchinson Educational, 1972. This is a small book of excellent black-and-white photographs with commentaries on glaciers and glacial landscapes from around the world. Although there is no bibliography or index, there is a list of terms as well as questions accompanying each photo.

Price, R. J. *Glacial and Fluvioglacial Landforms.* New York: Hafner Press, 1973. Most of this book treats glacial and meltwater landforms in a more technical, yet nonmathematical, manner, with excellent diagrams, photographs, and maps. The index is good, and the bibliography is very thorough.

Schultz, Gwen. *Glaciers and the Ice Age.* New York: Holt, Rinehart and Winston, 1963. The chapters on glacial landforms are intriguingly named "The Great Detective Story" and "The Artistry of Ice." Both describe erosional and depositional features and discuss how they can be interpreted in the landscape. Illustrated with photographs, maps, and diagrams. Although both the bibliography and the index are brief, there is a glossary.

Global Strategy for Plant Conservation (GSPC)

—*Cynthia A. Bily*

FIELDS OF STUDY

Ecology; Plant Science; Environmental Studies; Agronomy.

SUMMARY

The Global Strategy for Plant Conservation (GSPC) aims to protect plant species from extinction. Estimates indicate that there are as many as 300,000 plant species in the world and that more than 9,000 of them are facing extinction. GSPC provides a framework for international and regional cooperation to protect plant diversity.

BACKGROUND

At the end of the twentieth century, scientists estimated that as much as 15 percent of the world's plant species were at risk of extinction. In 1999, at a meeting of the International Botanical Congress held in St. Louis, Missouri, an urgent call was made for an international effort to preserve plant diversity. In 2000, a smaller group of botanists from conservation organizations met in Grand Canary, Canary Islands, and drew up the Gran Canaria Declaration on climate Change and Plant Conservation. In April, 2002, this declaration, in turn, was presented to and expanded by the 180 parties of the United Nations Convention on Biological Diversity, who unanimously called for a Global Strategy for Plant Conservation (GSPC). To help countries understand and address the specific targets of the GSPC, several international and American plant conservation organizations joined to form the Global Partnership for Plant Conservation in 2003. As of 2009, the United States had signed but not ratified the Convention on Biological Diversity.

PROVISIONS

The strategy presents six broad tasks: conducting research and establishing databases to produce a clear record of existing plant diversity; conserving plant diversity, particularly those plants that are directly important to human survival; controlling the use and exchange of plant diversity to sustain diversity and to

Medicinal Plants Conservation Zone, Wayanad. Image by Vinayaraj (Own work) CC-BY-SA-3.0.

provide fair distribution of benefits; educating the public about the importance of plant diversity; training an expanded corps of conservation officers; and establishing networks and organizations to expand the capacity for conserving plant diversity. To accomplish these tasks, the strategy identified sixteen specific international targets to be reached by 2010. These targets included compiling a list of all of the known plant species, assuring that no endangered plant species were harmed through international trade, and ensuring the protection of 50 percent of the most important plant diversity areas. Each nation created its own internal targets, in collaboration with other nations.

IMPACT ON RESOURCE USE

A 2008 progress report to the Conference of the Parties to the Convention on Biological Diversity reported substantial progress on eight of the sixteen specific targets and was generally optimistic about the chances for meeting several of the targets by 2010, thanks to enhanced national, regional, and international structures and strategies. Several countries, including Ireland, the United Kingdom, and South Africa, have drawn up aggressive plans to protect biodiversity, and in 2007, China announced a massive "National Strategy for Plant Conservation," hoping to save five thousand threatened species from extinction. By 2009, 189 countries had endorsed the GSPC.

FURTHER READING

Botanic Gardens Conservation International. The Global Partnership for Plant Conservation. http://www.plants2010.org/

United Nations Environment Programme (UNEP). Global Strategy for Plant Conservation. http://www.cbd.int/gspc/

GRAZING & GRASSLANDS

—James L. Robinson, PhD; updated by Karen N. Kähler

FIELDS OF STUDY

Conservation Biology; Evolution; Ecology; Animal Husbandry; Environmental Studies.

SUMMARY

Grazing is the consumption of any plant species by any animal species; grasslands are ecosystems where grasses and other nonwoody vegetation predominate. While grazing is of mutual benefit to plants and animals, overgrazing is ultimately detrimental to both the plant and animal populations, as well as to grassland ecosystems. Maintaining a balance between grazing animals and the plants on which they feed prevents deleterious consequences.

Grasslands are a major feature of Earth's ecology, as they comprise 37 percent of land area. Grasslands characterized by the presence of low plants, mostly grasses, and are distinguished from woodlands, tundra, and deserts. Grasslands experience sparse to moderate rainfall and are found in both temperate and tropical zones. Grassland plants coevolved over millions of years with the grazing animals that depended on them. Wild ancestors of cattle and horses, as well as antelope and deer, were found in Eurasian grasslands.

On the North American prairie, bison and antelope prospered. Wildebeest, gazelle, zebra, and buffalo dominated African savannas, whereas the kangaroo was the preponderant grazer in Australia. Grasslands occupied vast areas of the world more than ten thousand years ago, before the development of agriculture and industrialization, and the subsequent explosive growth of the human population.

Grazing is a symbiotic relationship whereby animals gain their nourishment from plants, which in turn benefit from the activity. Grazing is also an important contributor to human food security, as it provides most of the nutrition for ruminants that in turn, supply humans with meat and milk. Grazing removes the vegetative matter required for grasses to grow, facilitates seed dispersal, and disrupts mature plants, permitting young plants to take hold. Urine and feces from grazing animals recycle nutrients to the plants. The grassland ecosystem also attracts other animals, including invertebrates, birds, rodents, and predators. The grasses, grazing animals, and grassland carnivores, such as wolves or cat species, constitute a food chain.

Grasses are generally well suited to periods of low rainfall because of their extensive root systems and can go dormant during periods of drought. Humans have been an increasing presence in grassland areas, where more than 90 percent of modern crop production occurs and much urbanization and industrialization have taken place. The remaining grasslands, unsuitable for crops because of inadequate rainfall or difficult terrain, are used for grazing by domesticated or wild herbivores. In addition, many woodland areas around the world have been cleared and converted to grasslands where animals can graze.

IMPACTS OF OVERGRAZING

Continued heavy grazing of a given area leads to deleterious environmental consequences. Even repeated removal of leaf tips will not adversely affect the regeneration of grasses, provided that the basal zone of the plant remains intact. Whereas animals can generally safely eat the upper half of the grass shoot, if they ingest the lower half, which sustains the roots and fuels regrowth, they will eventually kill the plants. Overgrazing leads to denuding of the land, invasion by less nutritious plant species, erosion caused by decreased absorption of rainwater by soil, and starvation of animal species. Because the loss of plant cover changes the reflectance of the land, climate changes can follow that make it virtually impossible for plants to return, with desertification an ultimate consequence.

The number of animals is not the only factor in overgrazing; the timing of the grazing can also be detrimental. Grasses require time to regenerate, and continuous grazing will inevitably kill them. Consumption too early in the spring can stunt their development. Semiarid regions are particularly prone to overgrazing because of low and often unpredictable rainfall; regrettably, these are the areas of the world to which much livestock grazing has been relegated, because the moister grassland areas have been converted to cropland.

Overgrazing has contributed to environmental devastation worldwide. Excessive grazing by cattle, sheep, goats, and camels is partly responsible for the deserts of the Middle East. Uncontrolled livestock grazing during the late nineteenth century and early twentieth century negatively affected many areas in the American West, where sagebrush and juniper trees invaded the grasslands. Livestock overgrazing has similarly devastated areas of Africa and Asia. Feral horses in the American West and the Australian outback continue to damage those environments.

Overgrazing by wildlife can also be deleterious. The Kaibab Plateau deer disaster in Arizona is one such example, where removal of natural predators and livestock that competed with the deer for food led first to a deer population explosion, then to overgrazing by the deer, followed by starvation and large die-offs within the deer population. Protection of elk and bison in Yellowstone National Park has similarly led to high populations, excessive grazing, and changes to the environment. Only the provision of winter feed has prevented the die-offs that would otherwise naturally ensue. Ironically, winter feeding has perpetuated the problem by maintaining these populations at levels higher than grazing can sustain. Feeding has also encouraged the animals to congregate in unusually large numbers, which has contributed to the spread of disease. In 2010 a coalition of conservation groups lost a lawsuit to stop the supplemental feeding of elk and bison on the National Elk Refuge in Wyoming.

GRASSLANDS MANAGEMENT

Grassland areas need not deteriorate if they are properly managed, whether for livestock, wild animals, or both. The land's carrying capacity, or the number of healthy animals that can be grazed indefinitely in a given area, must not be exceeded. Because of year-to-year changes in weather conditions and hence food availability, determining carrying capacity is not simple; worst-case estimates typically have been used as guidelines to minimize the risk of exceeding carrying capacity. The goal should be a grassland rendered and kept healthy by optimizing, not maximizing, the number of animals. For private land, optimizing livestock numbers is in the long-term self-interest of the landowner. For publicly held land, managed in common or with unclear or disputed ownership, restricting animals to the optimum level is particularly difficult to achieve. Personal short-term benefit often leads to long-term disaster, in a phenomenon known as the tragedy of the commons.

Managing grasslands involves controlling the numbers of animals and enhancing the habitat. Cattle and sheep can be physically restricted through the use of herding and fencing, although requiring such restrictions can be difficult to achieve through political means. Much more problematic is controlling wildlife when natural predators have been eliminated and hunting is severely restricted. As for habitat improvement, the prudent use of chemical, fire, mechanical, and biological approaches can increase carrying capacity for domesticated and wild herbivores. Removing woody vegetation by burning or mechanical means can increase grass cover, fertilizing can stimulate grass growth, and reseeding with desirable species (plants native to the particular region) can enhance the habitat. Effective grassland management also requires matching animals with the grasses on which they graze.

An approach to grazing known as holistic management may have the potential not only to stave off ecosystem damage of grasslands but also to reverse desertification. This approach operates on the essential principle that, because herbivores and perennial grasses evolved together, the grasses will thrive only in combination with herbivores grazing and roaming naturally. Contrary to common wisdom regarding best management practices for grazing, holistic management involves grazing livestock in ultradense, constantly moving herds that mimic big-game grazing patterns. The livestock till the soil with their hooves and fertilize it with their excrement. By grazing the grasses, they allow sunlight to reach the grasses' growth buds; by contrast, when grazing is so restricted that the vegetation is able to die upright, the growth buds are shielded from the sun and the entire plant dies the following year. The common management practice of allowing grazed land an extended period to rest and recover, then, may not promote a resurgence of vegetative cover; rather, this practice may cause the land to remain barren and dry.

In 1992 holistic management pioneer Allan Savory began a program in Zimbabwe, Africa, in which livestock herds were increased by 400 percent on 2,630 hectares (6,500 acres) of land that had been barren for hundreds of years. By 2010, after years of holistic planned grazing, this area had become healthy grassland with open water. Other holistic management practitioners around the globe have enjoyed similar successes.

GRAZING IN THE UNITED STATES

There are roughly 312 million hectares (770 million acres) of rangelands (grasslands, forests, wetlands, and other ecosystems that are suitable for grazing) in the United States, more than half of which are privately owned. The federal government manages 43 percent, and the remainder is under state and local government control.

Laws pertaining directly to grazing in the United States include the Taylor Grazing Act of 1934, the Federal Land Policy and Management Act of 1976, and the Public Rangelands Improvement Act of 1978. The Taylor Grazing Act introduced measures to control the unregulated grazing practices of homesteaders that had led to overgrazing, enhanced erosion, damage to streams and springs, and the land's reduced productivity; however, rancher needs still tended to take precedence over range condition. Four decades later, heightened environmental awareness led to passage of the Federal Land Policy and Management Act, which established a multiple-use mandate for land management agencies to serve present and future generations in their practices. Not long after came passage of the Public Rangelands Improvement Act, which sought to improve the condition of public rangelands so that they might meet their potential for grazing and other uses. U.S. laws pertaining to environmental quality and endangered species also have impacts on rangeland management.

The Bureau of Land Management (BLM) manages livestock grazing on 64 million hectares (157 million acres) of the 99 million hectares (245 million acres) of public lands that it administers. The BLM administers roughly eighteen hundred permits and leases, which are held mostly by cattle and sheep ranchers. The U.S. Forest Service, which administers the 77 million hectares (191 million acres) of national forest system lands, manages some 39 million hectares (96 million acres) of rangelands. The Forest Service became the nation's first grazing control agency in the early 1900's. In 1934 the Department of the Interior's Division of Grazing Control (soon renamed the Division of Grazing) joined it; this division became the Grazing Service in 1939, which merged with the General Land Office in 1946 to form the BLM.

Both the Forest Service and the BLM implement a regulatory system of permits, rental fees, herd size limits, and grazing seasons. They must maintain a balance among several often-conflicting objectives: providing forage for grazing and browsing animals, ensuring the land's long-term health and productivity, protecting watersheds, managing wildlife habitat, administering permitted mineral and energy resource exploration and extraction, offering recreational opportunities, and preserving the land's distinctive character and aesthetic appeal. In order to meet the array of resource needs, rangeland management agencies inventory, classify, and monitor rangeland conditions. Where rangeland health needs improvement, they implement measures to restore ecosystem functions. Public land decision makers must take into account a variety of factors that affect rangelands, including severe and extensive wildfires, invasive plant species, rural residential development driven by population increases, and global climate change.

FURTHER READING

Chiras, Daniel D., and John P. Reganold. "Rangeland Management." In *Natural Resource Conservation: Management for a Sustainable Future*. 10th ed. Upper Saddle River, NJ: Prentice Hall, 2010.

Du Toit, Johan, Richard Kock, and James Deutsch. *Wild Rangelands: Conserving Wildlife While Maintaining Livestock in Semi-Arid Ecosystems*. Chicester: Wiley, 2012.

Gibson, David J. *Grasses and Grassland Ecology*. New York: Oxford University Press, 2009.

Gordon, Iain J., and Herbert H. T. Prins, eds. *The Ecology of Browsing and Grazing*. Berlin: Springer, 2010.

Lemaire, G., et al., eds. *Grassland Ecophysiology and Grazing Ecology*. Wallingford, UK: CABI Publishing, 2007.

Lye, Lin Heng. *Sustainability Matters: Environmental Management in Asia*. Singapore: World Scientific, 2010.

Manske, Llewellyn L. *Perpetually Sustainable Grazingland Ecosystems*. Dickinson, North Dakota: North Dakota State University, 2013.

Manske, Llewellyn, and Sheri A, Schneider. *Biologically Effective Management of Grazinglands*. Dickinson, ND: North Dakota State University, 2014.

Vallentine, John F. *Grazing Management*. Amsterdam/New York: Elsevier Science, 2016.

Woodward, Susan L. *Grassland Biomes*. Westport, CT: Greenwood Press, 2008.

GREENHOUSE EFFECT

—Rachel Leah Blumenthal

FIELDS OF STUDY

Atmospheric Science; Physical Chemistry; Heat Transfer; Photochemistry; Thermodynamics; Chemical Kinetics; Climatology.

PRINCIPAL TERMS

anthropogenic: influenced by humans; the anthropogenic greenhouse effect, also called the enhanced greenhouse effect, refers to the phenomenon as amplified by human activities

biochar: another name for charcoal, used specifically for usages of charcoal that involve positive effects such as soil improvement or carbon sequestration

carbon sink: a reservoir (either natural or human made) for storing or sequestering carbon

fossil fuels: carbon-rich, energy-storing compounds that, when burned, release greenhouse gases into the atmosphere

global warming potential: a measure of the ability of a greenhouse gas to contribute to global warming, based on its heat-trapping efficiency and its lifetime in the atmosphere

greenhouse gas: a gas in Earth's atmosphere that can absorb and emit infrared radiation

near-infrared radiation: the shortest-wavelength segment of infrared radiation, which has a longer wavelength than visible light; some solar radiation is in the near-infrared frequency range

negative emission: a process that removes carbon dioxide from the atmosphere; technologies include biochar and bioenergy with carbon capture and storage

ozone: a triatomic oxygen molecule and greenhouse gas; absorbs harmful ultraviolet rays from the sun before they reach Earth's surface

thermal infrared radiation: the longest-wavelength segment of infrared radiation, which has a longer wavelength than visible light; the earth emits radiation in the thermal infrared frequency range

SUMMARY

Gases in Earth's atmosphere trap infrared radiation emitted by the Earth, warming the surface in a process called the greenhouse effect. The greenhouse effect occurs naturally but is increasing in significance because of human activities, leading to probable detrimental effects involving climate change.

INTRODUCTION

The sun's radiation that reaches Earth's atmosphere is made up of ultraviolet, visible light, and near-infrared frequencies, all relatively short wavelengths. Nearly one-half of this radiation reaches Earth's surface; some of it is reflected but much is absorbed and then reemitted by Earth at longer wavelengths, in the thermal infrared range.

In a perfect system, a thermal equilibrium would be reached, but Earth's surface temperature is higher than expected because of a phenomenon known as the greenhouse effect. Certain gases in Earth's atmosphere absorb the energy emitted by the earth and then release that energy in all directions; some of the energy returns toward Earth's surface, and the cycle continues.

The nomenclature of the greenhouse effect is misleading. Whereas real greenhouses absorb sunlight to trap heat, the physical walls of the greenhouse serve to prevent convection, keeping the warm air inside the structure. The environmental greenhouse effect relies on a different process: the absorption and emission of thermal energy. Earth is not the only planetary body to experience the greenhouse effect; Mars, Venus, and Saturn's largest moon, Titan, also show evidence of the same effect, with the effect ob Venus being extreme.

In 1824, French mathematician and physicist Jean Baptiste Joseph Fourier (1768–1830) realized that according to Earth's size and its proximity to the sun, Earth should be colder if its only heat were coming from the sun. Fourier published several articles exploring other potential sources of Earth's extra heat, and one theory involved Earth's atmosphere acting as an insulator. Generally this marks the first time any person described what came to be called the greenhouse effect. In 1863, British physicist John Tyndall (1820–1893) revealed experimental evidence proving the existence of the greenhouse effect, particularly regarding the strong infrared-radiation absorption capabilities of water vapor but also the less but still significant absorptions of other atmospheric gases, such as carbon dioxide, ozone, and nitrous oxide.

By the turn of the twentieth century, Swedish physicist and chemist Svante Arrhenius (1859–1927) was building on Fourier's and Tyndall's work, focusing specifically on the absorption values for carbon dioxide. Arrhenius predicted that the growing emissions from combustion of fossil fuels would ultimately bring about climate change, but he saw this phenomenon in a positive light: He thought it meant the end of ice ages.

If not for the greenhouse effect, Earth's surface temperature would be about 18°C below the freezing point of water, and life as known would not exist. Whereas earlier researchers such as Arrhenius expected that human influence would push the greenhouse effect beyond natural expectations, but with positive results, many researchers now believe that humans are affecting the system too much, fearing that climate change is heading in a direction detrimental to human life.

OVERVIEW OF GREENHOUSE GASES

The greenhouse effect depends on the presence of certain gases in the atmosphere that allow the sun's short-wavelength radiation to pass by while absorbing and reemitting long-wavelength infrared radiation emitted from Earth's surface. These gases are called greenhouse gases.

Water vapor is the greenhouse gas that contributes most, up to 70 percent, to the overall greenhouse effect. On a large scale, water vapor's atmospheric abundance is not significantly affected by humans; it is mainly affected by the climate. While Earth's atmosphere might be nearly 20 percent water vapor in hot and humid regions, in cool and dry regions it might comprise well below 1 percent water vapor. Because its abundance is directly related to heat, water vapor actually contributes positively to the greenhouse effect. As the effect causes the temperature to rise in a particular region, more water vapor exists in the air, which leads to an increase in the greenhouse effect.

Carbon dioxide accounts for about one-fourth of the greenhouse effect, but it is the level of carbon dioxide that is most influenced by human activities. This means that carbon dioxide's increasing atmospheric abundance plays a major role in the future of climate change.

Another greenhouse gas, methane, would actually have a much stronger impact on the greenhouse effect than carbon dioxide if it were present in the same abundance. However, much less methane exists in the atmosphere. Methane is directly responsible for only about 10 percent of the greenhouse effect. It also has an indirect effect, though, because it contributes to the formation of ozone, another greenhouse gas.

Ozone is not plentiful anywhere in the atmosphere, but the largest percentage of it can be found in an area dubbed the ozone layer, which is located 20 to 30 kilometers (12 to 19 miles) above Earth, in the lower stratosphere. The ozone layer absorbs almost all of the sun's ultraviolet radiation, which would be harmful were it to reach life-forms, including humans, on Earth's surface. It is important to note that while some media reports group ozone depletion with the issue of greenhouse gases and climate change, the phenomena are quite distinct.

The other major greenhouse gases are nitrous oxide, hydrofluorocarbons, perfluorocarbons, and sulfur hexafluoride. Most of these gases can stay in the atmosphere for a long time. Carbon dioxide, for example, generally stays in the atmosphere in the range of three to nine decades. Water vapor, however, lasts only about nine days.

Clouds also are important contributors to the greenhouse effect, although they are not technically a gas. They can absorb and reflect infrared radiation, but they also reflect some solar radiation before it even reaches Earth's surface.

Some of the most abundant atmospheric gases do not significantly contribute to the greenhouse effect. Nitrogen and oxygen, for example, are the two most abundant gases in Earth's atmosphere (78 and 21 percent abundance, respectively), but their diatomic molecular structure is not affected by infrared radiation.

GREENHOUSE GAS SOURCES & SINKS

Most greenhouse gases have both natural and human-made sources, especially because the Industrial Revolution (1750–1850) brought about the rise of modern machinery and industrial technology. Similarly, greenhouse gases can be removed from the atmosphere by natural or human-made storage reservoirs known as sinks (or by other processes).

As discussed, the atmospheric abundance of water vapor, unlike other greenhouse gases, is not particularly influenced by humans (except on a small, local scale in agriculture). Water vapor is a product of the water cycle, and it is more abundant in warmer climates.

Carbon dioxide, however, is a product of a variety of human-made sources, particularly the burning of fossil fuels, which are carbon-rich compounds that store energy. Around one-third of fossil fuels being burned are liquid, such as petroleum (and the gasoline derived from it), and about another one-third are solid fuels, such as coal. Since the Industrial

Revolution, the burning of both petroleum and coal has been on the rise, the former for transportation and the latter for electricity. The remaining one-third consists of gaseous fuels (such as natural gas), cement production, and several other sources. Deforestation also is a major source of carbon dioxide emissions; as trees are removed on a large scale and not replaced, less carbon dioxide is pulled from the atmosphere for photosynthesis.

Modern agriculture is another major source of greenhouse gases, particularly methane, which is produced by manure management techniques and by enteric fermentation (perhaps better known as cattle flatulence). Additionally, fertilizers are a major source of nitrous oxide.

The natural removal of greenhouse gases from the atmosphere occurs by physical and chemical changes and exchanges. Physical removal can involve either a state change (water vapor condensing or evaporating, for example) or an exchange with a different part of Earth (such as gases being physically mixed into the ocean). Similarly, chemical removal can involve reactions within the atmosphere itself or reactions with something else (removal by plants for photosynthesis, for example).

Increasingly, as humans become more aware of the potentially detrimental effects of civilization, particularly industry, on the greenhouse effect, researchers are finding technological ways to remove greenhouse gases from the atmosphere or to prevent them from reaching the atmosphere in the first place. One such technology, called carbon capture and storage, predictably collects carbon from fossil-fuel power plants and stores it safely, generally underground within large geologic formations. Another technique involves the use of charcoal to sequester carbon in the soil for hundreds or even thousands of years. When charcoal is used for this purpose, it is referred to as biochar.

THE GREENHOUSE EFFECT & CLIMATE CHANGE

Humankind's impact on the greenhouse effect is referred to by researchers as the enhanced or anthropogenic greenhouse effect. As discussed, the main human contribution is rising carbon dioxide levels, particularly from the burning of fossil fuels and from deforestation.

Arrhenius and other researchers in his time predicted that the enhanced greenhouse effect would protect humans from future ice ages, but the numbers give modern researchers some cause for alarm as they try to determine to what extent human actions may be permanently changing Earth's climate for the worse.

In a general sense, the phrase *climate change* refers to significant and lasting weather changes through long periods of time from natural or human-made reasons. *Global warming* technically pertains specifically to increases in Earth's average surface temperature. In modern vernacular, though, many nonscientists use *climate change* and *global warming* nearly synonymously.

While some disagreement exists regarding the interpretation of data, most major scientific bodies and industrialized nations recognize that global temperatures have been steadily increasing beyond normal fluctuations since the late nineteenth century and especially since the late twentieth century, and most scientists point at the anthropogenic greenhouse effect as the main reason for this rise in temperatures. As greenhouse gas sources have drastically increased since the Industrial Revolution, the sinks have not been able to keep up. Research by the National Oceanic and Atmospheric Administration shows that carbon dioxide levels have risen 36 percent since the start of the Industrial Revolution and that methane levels have risen 148 percent.

When analyzing the effects of different greenhouse gases on climate change, it is useful to understand the global warming potential (GWP) measure, which takes into account both the efficiency of a greenhouse gas (how much heat it can trap) and its atmospheric lifetime. Carbon dioxide's GWP is standardized to a value of 1, and all other greenhouse gases are calculated relative to carbon, relative to a certain time frame.

If climate change continues in the current pattern, it is expected that sea levels will rise worldwide, resulting in significant changes in precipitation patterns and, ultimately, to desert expansions (desertification) and other habitat changes. Glacial melting and heat waves are also expected, and these changes will likely cause the extinction of various species. These are only a few consequences that demonstrate the growing need for continued research of greenhouse gases and climate change.

FURTHER READING

Archer, David. *Global Warming: Understanding the Forecast.* Blackwell, 2007. This text is an accessible yet comprehensive introduction to climate change, covering topics such as the greenhouse effect, blackbody radiation, and the carbon cycle. It will be a comfortable but informative read for those who are new to the topic.

Black, Brian, and Gary J. Weisel. *Global Warming.* Greenwood Press, 2010. Part of the Historical Guides to Controversial Issues in America series, this guide uses historical context and the concepts of geology, Earth science, and climate science to provide an overview of global warming.

Dow, Kirstin, and Thomas E. Downing. *The Atlas of Climate Change: Mapping the World's Greatest Challenge.* University of California Press, 2006. Introduces climate change with an overview of terms and concepts before moving into causes, effects, and possible solutions. Particular focus on the greenhouse effect and climate systems.

Gore, Albert, and Melcher Media. *An Inconvenient Truth: The Planetary Emergency of Global Warming and What We Can Do About It.* Rodale, 2006. Now an acclaimed documentary, this book by former U.S. vice president Al Gore uses a mixture of scientific research and storytelling to illuminate the dangers of global warming.

Rackley, Stephen A. *Carbon Capture and Storage.* Butterworth-Heinemann/Elsevier, 2010. An overview of technology available for carbon capture and storage, which is meant to reduce greenhouse gas emissions from industrial processes.

Stille, Darlene R. *The Greenhouse Effect: Warming the Planet.* Compass Point Books, 2006. A basic overview of the greenhouse effect, covering topics such as the difference between weather and climate, the runaway greenhouse effect, and effects of global warming.

GROUNDWATER

—*Gina Hagler*

FIELDS OF STUDY

Conservation; Evolution; Ecology; Environmental Studies.

PRINCIPAL TERMS

aquifer: the water-bearing stratum of permeable rock, sediment, or soil

artesian aquifer: an aquifer that is bounded above and below by impermeable beds and that experiences hydraulic pressure; a type of confined aquifer

artesian pressure: a confining internal pressure of groundwater in an artesian aquifer; it is significantly greater than atmospheric pressure and causes groundwater to rise above its natural level in the aquifer, often above the surface of the ground, without the use of a pump

bedrock: the solid rock underlying soil or any other unconsolidated superficial cover

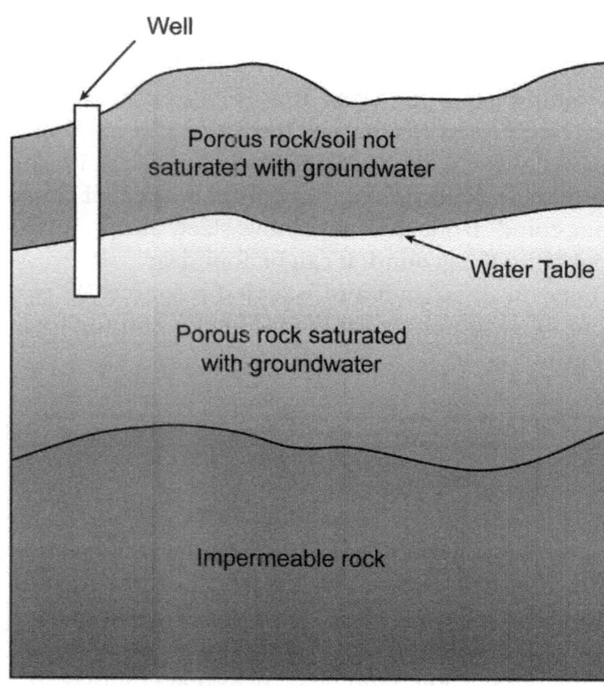

Groundwater is found beneath the solid surface. Notice that the water table roughly mirrors the slope of the land's surface. A well penetrates the water table. Image by Geoff Ruth (File:High School Earth Science 1-13.pdf (page 493)) CC-BY-SA-3.0.

groundwater: the water that exists in the pore spaces and fractures in rock and sediment (aquifers) beneath the earth's surface

hydrologic cycle: the complete cycle through which water passes from the oceans, through the atmosphere, to the land, and back to the ocean; a water cycle

infiltration: the penetration of water through the ground surface into subsurface soil, or the penetration of water from the soil into sewer or other pipes through defective joints, connections, or utility hole walls

subsurface water: the water in the root zone of the soil and groundwater flowing or stored in the rock mantle of the earth

surface water: the water flowing in stream channels

transpiration: the process by which water vapor is lost to the atmosphere by living plants

water cycle: a hydrologic cycle

water table: the upper surface of the zone of saturation in permeable rocks not confined by impermeable rocks; it rises and falls with the amount of groundwater

SUMMARY

Groundwater is the water that is located in aquifers beneath the earth's surface. It exists in the cracks and spaces of the soil, sand, and rock that make up the aquifers. Groundwater is a critical source of water that comes from the rain, snow, sleet, and hail that soaks into the ground. It can be found virtually everywhere on the planet and is a vital resource: 98 percent of the water available to humans is in the form of groundwater.

INTRODUCTION

Groundwater is the water that exists in aquifers beneath the surface layer of the earth. It is the major source of drinking water and water for irrigation, and it exists virtually everywhere on the planet. Groundwater is not a free-flowing underground river, but rather water that exists in the fractures, cracks, and spaces in soil, sand, and rocks that comprise the aquifers.

Gravity causes the water that comes from rain and melted snow, sleet, and ice to seep into the ground. As it penetrates the soil, the water continues to move down through the soil and rock until it reaches what is known as the saturated zone. This zone is bordered on the bottom by an impermeable layer, which does not permit the water to continue its downward movement, and bordered at the top by what is called the water table. Within these borders, the saturation zone is made up of soil, sand, and rock with naturally occurring spaces between them. These spaces are filled by water.

The height of the water table varies depending on how much groundwater exists in the saturated zone; in contrast, the bottom of the saturated zone remains stable against the impermeable layer. The water table moves because the saturated zone fills from the bottom up, with the top surface layer moving accordingly. It is difficult to determine the level of the water table while standing on the ground because it can vary from as little as a foot beneath the surface to several hundred feet beneath the surface. One can, however, see where the water table lies by looking at areas of surface water, such as a lake. The top surface of the lake will be at the level of the water table. This also is true along beaches, where the level of the ocean is the same as the level of the water table. Heavy rains can cause the level of the water table to rise, while heavy pumping of groundwater supplies or drought conditions can cause the level of the water table to fall.

AQUIFERS

An aquifer is a mass of saturated rock that does not permit the easy movement of water. These rocks can be permeable and porous. They are made up of a variety of rock types that include sandstone, conglomerate, fractured limestone, and unconsolidated materials such as sand, gravel, and silt. Fractured volcanic rocks, particularly the rubble zones between volcanic flows, make excellent aquifers because they are usually porous and permeable. Rocks such as granite and schist are poor aquifers because of their low porosity, and unless they are highly fractured, they cannot create ample space for water. The more permeable the rock, the more quickly water can move through the layers. The more porous the rock, the more space there is for the water to accumulate.

Aquifers occur at various depths. Those closest to the surface are more likely to be used for drinking water and for irrigation. They also are more likely to be filled by local rainfall. Although they "hold" the water in the saturated zone, they are not underground lakes. The water in aquifers moves, but it does not move in an underground river. Rather, the water in an aquifer moves through the spaces between the rock and sediment in the aquifer in both a downward direction, until it reaches an impermeable layer, and in a "sideways" direction, until it finds a way to the surface through a lake or other outlet. Depending on the porosity of the rocks and soil, the movement of the water may be slow or fast. Porosity and permeability are factors in the ability of the aquifer to act as a filter.

Aquifers act as natural filters for the water as it seeps through the layers. The layers of soil and rock trap the sediment and other particles, including bacteria, providing natural purification in the process. The smaller the space the water moves through, the smaller the bacteria that will be filtered. As a result, an aquifer can be filled with clean water as part of a natural process.

An aquifer also can be filled with contaminated water if the contaminants reach the water table. This can happen at landfills or in septic tanks, in areas where high concentrations of an element form naturally, or in areas where chemicals have been dumped. Contamination of aquifers can take place anywhere that an activity creates a pathway that increases the rate at which rain or snowmelt reaches the water table. This occurs because the pathway bypasses the natural purification that takes place when water sources seep into the ground and make their way into the saturated zone naturally, saturating the ground artificially and without any chance for the water to be properly filtered.

Because of gravity, water moves through an aquifer as it makes its way from a higher elevation to a lower elevation. Water also moves from areas of higher pressure to areas of lower pressure. These two forces are the driving force (the hydraulic head) behind groundwater movement. Water in an aquifer can move through four different types of rock: unconsolidated rock (made up of loose sand, silt, and gravel), porous sedimentary rock, porous volcanic rock, and fractured rock. Water is able to move easily between the spaces, which is why gravel and sand are common aquifers. Carbonate rocks are brittle and tend to fracture; the fractures allow some water movement. Volcanic rocks also have fractures that allow water movement.

AQUIFER TYPES

The three types of aquifer are confined, unconfined (which includes the artesian aquifer), and perched aquifers.

The confined aquifer has a water supply sandwiched between two impermeable layers. Because of this physical arrangement, the water cannot easily pass back and forth between the layers once it has made its way into that space. It will stay in the space until it moves to a surface body of water or until the water is pumped from the ground by means of a human-made well. The pressure in a confined aquifer can be quite strong.

An artesian aquifer is a confined aquifer accompanied by pressure. The water in an artesian well experiences upward force, so that when an artesian well is dug, its water is pushed to the surface during the drilling process. Often the pressure will force the water above the level of the water table, to the surface and even higher, where it will then flow back into the ground, all without a pump.

An unconfined aquifer (a type of artesian aquifer) is a water supply with an impermeable layer below it and a permeable layer above it. The water table in such an aquifer is free to fluctuate up and down with the recharge and discharge of groundwater. The water from the surface is directly available through this top level of the saturated zone. Because these aquifers have readily permeable layers above them and are close to the water table, they are particularly vulnerable to contamination. They are also subject to seasonal variations in rain and snowmelt.

A perched aquifer has an impermeable layer below it but exists above the water table. It is not unusual to find a perched aquifer in an area of glacial sediment or sitting atop a layer of clay. A perched aquifer is essentially a form of unconfined aquifer but is distinctly smaller in size. A perched aquifer usually can be found at elevations higher than extensive regional aquifers.

GROUNDWATER DISCHARGE

Water can be discharged from an aquifer in three ways: by movement of the water through the aquifer and into a body of surface water, by the drilling of a well, and by a natural spring. A well is a hole that is drilled or dug deep enough to intersect with the water table. A hole dug in an unconfined aquifer will be deep enough to reach beneath the water table, and the water level in the well will match that of the water table. The water is then hauled or pumped to the surface. Also, the well must be dug deep enough to allow for seasonal variations in the water table. If too many wells are dug or too much water is taken from existing wells during a time when there is little or no snowmelt or rainfall, the wells can deplete the aquifer and the well will run dry.

In a confined aquifer, the pressure from beneath the ground will result in an artesian well in which no pump will be required. The hydraulic pressure will cause the water to move up the sides of the well without the need for a pump. Usually, the water will rise above the level of the ground. Springs are naturally occurring sources of groundwater discharge. They occur where the water table intersects the surface and water flows out of the ground and where impermeable rock meets an aquifer. In the case of a confined aquifer, the spring is known as an artesian spring.

Groundwater in an aquifer is not static, so it is of increasing importance to understand what the use of water in one area has to do with the availability of water in another. Groundwater sources affect other water sources, so management of a single aquifer may not be adequate in regions with large amounts of groundwater discharge.

GROUNDWATER RECHARGE

Among other uses, aquifers supply drinking water and water to irrigate crops. Water pumped or otherwise taken from an aquifer leads to less water in that aquifer. The aquifer is recharged, or refilled, naturally as part of a hydrologic cycle in which water moves downward from the surface to the saturated zone. The water comes almost exclusively from rain and snowmelt.

In areas with little or no human activity, in which vast amounts of land remain undeveloped or unpaved, water can find its way into the ground and into the aquifers beneath. In areas with much human activity, such as development and logging, a reduced amount of water makes its way into the saturated zone; rainfall and snowmelt lead to runoff rather than seepage. Three other factors that influence the amount of groundwater and rate of recharge are evaporation of water from surface water such as lakes, transpiration from existing vegetation, and infiltration, which is determined by the amount of unpaved surface and the permeability of the surface soil.

Inadequate recharge of the groundwater leads to a depletion of underground resources. Once this happens, a quick, natural recharge of the aquifer is unlikely, requiring instead the costly artificial recharge of the saturated zone. Understanding the relationship between the discharge and recharge of an aquifer is essential to the proper management of that aquifer.

Many models are available to estimate groundwater recharge. These models take into account the amount of paved surfaces and development and an area's logging activity. The models also factor in the permeability of the subsurface material and other variables, such as the density of the materials. From the usages side of the equation, groundwater recharge models are used to decide the number of wells or projects needed to tap into the groundwater supply. Considered on the source side are the amount of annual rainfall and snowmelt, any unusual weather patterns, and indications that the typical patterns will be disrupted. In total, these models provide good data for groundwater management.

ISSUES

The heavy reliance of humans on groundwater for drinking, irrigation, and other purposes has caused many problems for the natural hydrological cycle. In the United States alone approximately half of the population uses groundwater as the chief source of drinking water and more than 50 billion gallons are used for agriculture every day. This heavy use, along with decreased recharge, has led to many instances of groundwater depletion, in which the water table is significantly lowered. Depletion has many negative effects, including the disruption of wells that are no longer deep enough to reach the water table. If a well goes dry it must be deepened or replaced—often an expensive project—and new pumps may be required that are more costly and need more energy to draw water. Meanwhile the property above the aquifer may experience land subsidence, in which the removal of water causes the underlying soil and rock to compress and collapse. This can cause major damage to buildings and other structures over the affected area.

Groundwater depletion also impacts the water level in nearby bodies of surface water, such as lakes, streams, and wetlands. Declining water levels can disrupt ecosystems, reducing the habitat for plants dependent on reaching the water level and therefore that of various animals as well. Many formerly lush riparian ecosystems in dry areas of the United States and elsewhere have been reduced by excessive groundwater pumping. Another potential consequence of groundwater depletion is a drop in water quality. Often this comes in the form of saltwater intrusion, a process in which saltwater rushes in to replace pumped-out fresh groundwater, contaminating the water supply for humans and other organisms. Saltwater intrusion is common in coastal areas due to the salty water found beneath oceans, but some very deep inland groundwater is also saline and can intrude and contaminate aquifers far from the sea.

Even in areas that have yet to experience groundwater depletion human activity presents threats to water quality, especially in the form of groundwater pollution. Groundwater can be contaminated in many ways, including seepage of surface pollutants that also threaten surface water. Substances such as oil and gas, salt used on roads, fertilizers and pesticides, toxic waste from mines or manufacturers, and waste from landfills or septic tanks can all enter the hydrologic cycle and make their way into groundwater. Improper disposal of household chemicals and pharmaceutical and personal care products (PPCPs) is another growing source of groundwater pollution, and studies continue to investigate the potential health effects of small amounts of such substances ingested through drinking water. Contaminated groundwater poses a serious environmental and public health problem; drinking polluted water can cause serious disease ranging from dysentery to cancer.

CONCLUSIONS

Groundwater is a vital natural resource and one that is in great demand. As groundwater in one aquifer is used to meet the needs of a community, that same aquifer may have a role in the function of other nearby aquifers and surface-water sources, thereby stretching its capabilities. As aquifers come under increased demand because of development and industrialization, this demand heightens the need for aquifer management to promote sustainable use and avoid groundwater pollution. The depletion of aquifers around the world due to overdraft for human water use has become a prominent issue, connected to broader trends in environmental management and water rights. Falling water tables force the drilling of deeper wells, and may cause challenges such as land subsidence and saltwater intrusion along with potential shortages of drinking water. The management models used today are sophisticated systems that account for soil conditions, weather patterns, population and industrial density, and changes in weather patterns caused by global climate change.

FURTHER READING

Anderson, Mary P., ed. *Groundwater.* Oxford: Intl. Assn. of Hydrological Sciences, 2008. Print.

Bear, Jacob. *Dynamics of Fluids in Porous Media.* Mineola: Dover, 1988. Print.

Bear, Jacob. *Hydraulics of Groundwater.* Mineola: Dover, 2007. Print.

"Groundwater Depletion." *USGS Water Science School.* US Geological Survey, 9 Dec. 2015. Web. 2 Feb. 2016.

Groundwater Foundation. Groundwater Foundation, 2016. Web. 2 Feb. 2016.

Healy, Richard, and Bridget Scanlon. *Estimating Groundwater Recharge.* New York: Cambridge UP, 2013. Print.

Mays, Larry W. *Ground and Surface Water Hydrology.* Hoboken: Wiley, 2012. Print.

Moore, John E. *Field Hydrogeology: A Guide for Site Investigations and Report Preparation.* Boca Raton: CRC, 2012. Print.

Odell, Lee H. *Treatment Technologies for Groundwater.* Denver: American Water Works Assn., 2010. Print.

Strassberg, Gil, Norman L. Jones, and David R Maidment. *Arc Hydro Groundwater: GIS for Hydrogeology.* Redlands: ESRI, 2011. Print.

Woods, Andrew W. *Flow in Porous Rocks: Energy and Environmental Applications.* Cambridge: Cambridge UP, 2015. Print.

Growth of Organic Farming

—*Corinne Elliott*

FIELDS OF STUDY

Conservation; Agriculture; Agronomy; Food Science; Evolution; Ecology; Environmental Studies.

KEY FIGURES

Jerome Irving Rodale (1898–1971), founder of Rodale Press and a leader in organic farming education and research

Robert Rodale (1930–1990), son of Jerome Rodale and a leader in organic farming publishing and research

William A. Albrecht (1888–1974), soil scientist whose research focused on the relationship between plant and animal health and soil characteristics

SUMMARY

Organic farming became commercially feasible in the United States during the 1970's as increasing numbers of Americans became concerned about the health hazards of pesticides and the damage to the environment caused by traditional farming methods.

A farmer sows seeds on his organic farm in 1972. Image by NARA.

SUMMARY OF EVENT

Early organic farmers were often ridiculed and accused of being radically opposed to technology and progress. In fact, they sought to implement as fully as possible the understanding of biological systems that science provided and to fight the misuse of technology. Those who favored organic agriculture believed that people in a democratic society should be free to farm as they desire yet should not be free to harm future generations, use up nonrenewable resources, or damage the environment.

In England and France in 1972, more than ninety thousand acres were farmed organically. In the United States, Jerome Irving Rodale's book *Organic Gardening* was distributed to 750,000 readers. Under the leadership of Jerome's son Robert Rodale, the publishing company that Jerome founded, Rodale Press, continued to publish materials on gardening, the environment, and human health and nutrition. Groups that focused on helping connect organic growers with markets for their crops—such as Natural Food Associates, founded in Texas in 1953—grew in importance in the 1970's. So-called ecological agriculture was also supported by soil scientists such as William A. Albrecht, who in 1971 began to publish a monthly newspaper for farmers titled *Acres, U.S.A.*

Scientific advances, as well as the pressure to feed an increasing world population, produced a modern agriculture dependent on external inputs for increased production. Before the twentieth century, researchers had discovered that row cropping, thorough tillage, weeding, and application of chemical fertilizers increased crop production and could perhaps assist in efforts to feed the world. Statistics gathered by the U.S. Department of Agriculture (USDA) illustrate the dramatic increases in crop yields between the late 1930's and the early 1970's: Corn yields increased 240 percent; wheat yields, 139 percent; cotton yields, 96 percent; potato yields, 214 percent; and strawberry yields, 342 percent.

These increases in production were accomplished because modern agriculture tended to substitute capital—in the form of machinery and chemicals—for labor and to increase the size of the farm unit. The environmental costs of these production practices included increases in pollution from chemical fertilizers and pesticides in air, water, and foods; decreases in food quality; depletion of soil and energy resources; and concentration on the production of single crops (monoculture) in specific geographic regions. Moreover, farmers suffered from the lower food prices that resulted from their increased productivity. In economic terms, the demand for food was comparatively inelastic—it did not respond to changes in prices.

Society paid the price for the loss of agricultural labor. As people who could no longer find work in rural areas moved to the cities, welfare and health care costs increased. Despite these developments, the USDA warned in 1971 that a change in conventional agriculture would force 50 million Americans into hunger or starvation. Consumers benefited from low food prices, but farmers using conventional techniques found it difficult to sustain profits while at the same time protecting the environment.

In 1964, farms with more than $40,000 in sales were roughly twice as likely to use pesticides as were farms with sales of less than $10,000. The former tended to have better access to capital, to receive volume discounts, and to use different crop mixes. As the regional markets for small producers gradually disappeared, diversification became very costly. Scientists warned of the increased risk of crop failure from the intensive practice of monoculture. The overplanting of genetically related corn resulted in the corn blight of 1970, when the harvest decreased by 10 percent even though the acreage planted was up by 5 percent; some states suffered losses as high as 50 percent. In 1972, two types of peas and nine strains of peanuts made up 95 percent of the production in the United States. An increasingly dangerous situation for producers and consumers was created as the gene pool shrank and pests developed resistance to pesticides.

Farmers who had chosen alternative methods of agriculture argued for the maintenance of genetic diversity in both plant and animal stocks and for the protection of the environment. Federal agriculture policy, however, was biased toward conventional agriculture. A 1975 study by the National Academy of Sciences determined that the USDA had neglected basic scientific research, focusing instead on large-scale farming and productivity, although only 16 percent of U.S. farms had 1974 sales in excess of $40,000, while 52 percent had sales of less than $10,000. Little research had been done that could help farms with low incomes; indeed, many of the studies conducted seemed directed against their survival. From 1950 to 1975, the number of U.S. farms decreased by 50 percent.

Conventional farming practices can create severe environmental problems. In 1935, for example, such practices resulted in the loss of three billion tons of soil from cultivated lands in the United States. During the next three decades, the Soil Conservation Service

supervised the expenditure of $11 billion to address the problem, yet four billion tons of soil were lost in 1964. In Great Britain in 1970, soils were declared incapable of maintaining crop production at their current levels as a result of a dangerous depletion of organic matter.

The pesticides used in conventional agriculture have often been successful in initially protecting crops but have generated unexpected effects on ecosystems, including, in some cases, a return of pests in even larger numbers. Pesticide residues in human food have been linked to allergies, cancers, and many other health problems. The manufacture and shipment of pesticides, moreover, have often resulted in disastrous accidents and subsequent contamination of the environment. By contrast, a major premise of organic farming is the ability of healthy plants growing under ideal conditions to protect themselves from pests through natural processes.

In the late 1960's, public pressure across the United States led to restrictions on the use of organochlorine pesticides in agriculture. The chemicals that replaced them were less persistent in the environment and food chain but potentially more toxic to the farmer. The substitutes of the 1970's also tended to be more toxic to bees and other beneficial predators and to exert harmful physiological effects on crops. In 1970, the U.S. Congress passed the Occupational Safety and Health Act, which required hazard-free work environments for all types of workers, including farmworkers.

Public pressure and the potential dangers of pesticides forced farmers to make choices in favor of the environment and worker health. Farmers were faced with public concerns about pollution and consumer health, the increased cost of conventional farming and dependence on industry, overproduction of low-quality food, low prices, and scarcity of affordable labor. Many observers blamed corporate sales forces and university experts for having led farmers into high-cost production, high debt, and low profit. As conventional farming became more expensive while the prices paid per unit rose slowly, increasing numbers of American farmers learned the value of organic techniques.

Organic farming involves minimizing the use of costly presticides and other treatments by using practices long known to agriculture. These include crop rotation, mechanical methods of weed control, and maximized application of scientific knowledge to all aspects of farming. Organic farming promotes the production of highly nutritious food crops by cooperating with nature, understanding biological cycles, maximizing soil fertility, using a minimum of artificial fertilizers and organic matters, maintaining livestock ethically, avoiding environmental pollution, and protecting the genetic diversity of plants and animals. Moreover, many organic farmers believe that farming is part of a broader social, economic, and ecological agenda.

SIGNIFICANCE

In the late 1960's and 1970's, the desire for organically produced food increased dramatically in the United States. Many believed that the effects of pesticide residues were cumulative and could not be fully assessed for many years; certainly, there was increasing evidence that wildlife was being dramatically affected by pesticides. As consumers turned to organically produced foods, the prices of such foods decreased, and, in turn, demand increased further.

Many Americans initially perceived organic farming as a hobby for part-time or gentlemen farmers. Organic agricultural practices were not widely adopted at first primarily because of economic factors. During the 1960's, the incomes of farmers were 20 percent lower than the incomes of other American workers, and although farm income rose in the early 1970's, many farmers tried to maximize short-term profits by adopting new technologies.

In the 1970's, higher prices for chemical fertilizers and increasing federal restrictions on some pesticides increased the costs of conventional farming. Nevertheless, those who chose to farm organically did so knowing that they were taking a risk. Many farmers who changed to organic techniques did so to improve the health of their families, often after negative experiences with pesticides. Some farmers changed their practices to improve the health of their livestock. Still others viewed farming as a spiritual relationship with the land; their motivations included a concern for protecting the environment.

Organic farmers were gradually able to make larger profits. A survey of farm prices of organic food found little difference between organic and conventional vegetable and fruit prices in California, but nationally the major field crops cost 10 percent more when raised organically. The largest impacts of large-scale changeovers were felt with crops such as corn, soybeans, and wheat; many organic growers of these crops claimed to be able to produce competitive yields after the initial conversion period. Organic farming also became linked with integrated pest management, an approach in which low levels of pest infestations—which would have been uneconomical to treat with pesticides—were allowed to exist, encouraging the buildup of favorable predators and parasites.

For developing countries, rapidly increasing populations, combined with deterioration of soil as a result of erosion and desertification, made the likelihood of success with organic farming more problematic. International organizations such as the World Bank and the United Nations, however, eventually declared that the use of conventional agriculture presents threats to human health and to the environment and does not lead to sustainable and profitable agricultural production. Most international agricultural experts thus continued to believe not only that organic farming was the answer for feeding the world but also that it was the only economic solution for sustainable agriculture.

Further Reading

Duram, Leslie A. *Good Growing: Why Organic Farming Works.* Lincoln: University of Nebraska Press, 2005. Presents arguments for the benefits of organic farming and discusses the growth of this form of agriculture. Includes a section that focuses on the experiences of five organic farming families.

Francis, Charles A. *Enough Food: Achieving Food Security Through Regenerative Agriculture.* Emmaus, Pa.: Rodale Institute, 1985. Presents a somewhat biased summary of the need for organic farming and its potential to alleviate starvation in the world.

Koepf, Herbert H. *The Biodynamic Farm: Agriculture in the Service of the Earth and Humanity.* Hudson, N.Y.: Anthroposophic Press, 2006. Provides a historical summary of the principles of organic agriculture and argues for the need for farming based on the concept that the whole farm is a single organism.

Lampkin, Nicolas. *Organic Farming.* Rev. ed. Alexandria Bay, N.Y.: Diamond Farm Enterprises, 1994. Explains the principles and practices of organic farming and discusses the limited scientific data available on the subject. Includes illustrations and index.

Oelhaf, Robert C. *Organic Agriculture: Economic and Ecological Comparisons with Conventional Methods.* New York: John Wiley & Sons, 1978. Presents factual comparisons of conventional and organic agriculture. Highly recommended, particularly for its economic analysis.

H

Habitats & Biomes

—*Jim Fowler*

FIELDS OF STUDY

Conservation Biology; Biology; Evolution; Ecology; Environmental Studies.

PRINCIPAL TERMS

abiotic: the physical part of an ecosystem or biome, consisting of climate, soil, water, oxygen and carbon dioxide availability, and other physical components

biome: one of the widespread types of ecosystem on the earth, such as the Arctic tundra or the desert

biosphere: the sum of all the occupiable habitats for life on earth

biotic: the living part of an ecosystem or biome, consisting of all organisms

community: a population of plants and animals that live together and make up the biotic part of an ecosystem

ecosystem: a relatively self-sufficient group of communities and their abiotic environment

environment: the habitat created by the interaction of the abiotic and biotic parts of an ecosystem

habitat: the specific part of the environment occupied by the individuals of a species

population: a group of all the individuals of one species

species: a group of similar organisms that are capable of interbreeding and producing fertile offspring

SUMMARY

The biosphere is the sum total of all habitats on earth that can be occupied by living organisms. Descriptive and experimental studies of habitat components allow scientists to predict how various organisms will respond to changes in their environment, whether caused by humans or nature.

Life in the form of individual organisms composed of one or more living cells is found in a vast array of different places on earth, each with its own distinctive types of organisms. Life on earth has been classified by scientists into units called species, whose individuals appear similar, have the same role in the environment, and breed only among themselves. The space in which each species lives is called its habitat.

HABITATS, COMMUNITIES, ECOSYSTEMS & BIOMES

The term "habitat" can refer to specific places with varying degrees of accuracy. For example, rainbow trout can be found in North America from Canada to Mexico, but more specifically they are found in freshwater streams and lakes with an average temperature below 70 degrees Fahrenheit and a large oxygen supply. The former example describes the macrohabitat of the rainbow trout, which is a broad, easily recognized area. The latter example describes its microhabitat—the specific part of its macrohabitat in which it is found. Similarly, the macrohabitat of one species can refer to small or large areas of its habitat. The macro habitat of rainbow trout may refer to the habitat of a local population, the entire range of the species, or (most often) an area intermediate to those extremes. While "habitat," therefore, refers to the place an organism lives, it is not a precise term unless a well-defined microhabitat is intended. The total population of each species has one or more local populations, which are all the individuals in a specific geographic area that share a common gene pool; that is, they commonly interbreed. For example, rainbow trout of two adjacent states will not normally interbreed unless they are part of local populations that are very close to one another. The entire geographic distribution of a species, its range, may be composed of many local populations.

On a larger organizational scale, there is more than only one local population of one species in any habitat. Indeed, it is natural and necessary for many species to live together in an area, each with its own micro- and macrohabitat. The habitat of each local population of each species overlaps the habitat of many others. This collective association of populations in one general area is termed a community, which may consist of thousands of species of animals, plants, fungi, bacteria, and other one-celled organisms.

Groups of communities that are relatively self-sufficient in terms of both recycling nutrients and the flow of energy among them are called ecosystems. An example of an ecosystem could be a broad region of forest community interspersed with meadows and stream communities that share a common geographic area. Some ecosystems are widely distributed across the surface of the earth and are easily recognizable as similar ecosystems known as biomes—deserts, for example. Biomes are usually named for the dominant plant types, which have very similar shapes and macrohabitats. Thus, similar types of organisms inhabit them, though not necessarily ones of the same species. These biomes are easily mapped on the continental scale and represent a broad approach to the distribution of organisms on the face of the earth. One of the more consistent biomes is the northern coniferous forest, which stretches across Canada and northern Eurasia in a latitudinal belt. Here are found needle-leafed evergreen conifer trees adapted to dry, cold, windy conditions in which the soil is frozen during the long winter.

NORTH AMERICAN BIOMES

The biomes of North America, from north to south, are the polar ice cap, the Arctic tundra, the northern coniferous forest; then, at similar middle latitudes, eastern deciduous forest, prairie grassland, or desert; and last, subtropical rain forest near the equator. Complicating factors that determine the actual distribution of the biomes are altitude, annual rainfall, topography, and major weather patterns. These latter factors, which influence the survival of the living, or biotic, parts of the biome, are called abiotic factors. These are the physical components of the environment for a community of organisms.

The polar ice cap is a hostile place with little evidence of life on the surface except for polar bears and sea mammals that depend on marine animals for food. A distinctive characteristic of the Arctic tundra, just south of the polar ice cap, is its flat topography and permafrost, or permanently frozen soil. Only the top meter or so thaws during the brief Arctic summer to support low-growing mosses, grasses, and the dominant lichens known as reindeer moss. Well-known animals found there are the caribou, musk-ox, lemming, snowy owl, and Arctic fox.

The northern coniferous forest is dominated by tall conifer trees. Familiar animals include the snowshoe hare, lynx, and porcupine. This biome stretches east to west across Canada and south into the Great Lakes region of the United States. It is also found at the higher elevations of the Rocky Mountains and the western coastal mountain ranges. Its upper elevation limit is the "treeline," above which only low-growing grasses and herbaceous plants grow in an alpine tundra community similar to the Arctic tundra. In mountain ranges, the change in biomes with altitude mimics the biome changes with increasing latitude, with tundra being the highest or northernmost.

Approximately the eastern half of the United States was once covered with the eastern deciduous forest biome, named for the dominant broad-leaved trees that shed their leaves in the fall. This biome receives more than seventy-five centimeters of rainfall each year and has a rich diversity of bird species, such as the familiar warblers, chickadees, nuthatches, and woodpeckers. Familiar mammals include the white-tailed deer, cottontail rabbit, and wild turkey. The Great Plains, between the Mississippi River and the Rocky Mountains, receives twenty-five to seventy-five centimeters of rain annually to support an open grassland bi ome often called the prairie. The many grass species that dominate this biome once supported vast herds of bison and, in the western parts, pronghorn antelope. Seasonal drought and periodic fires are common features of grasslands.

The land between the Rockies and the western coastal mountain ranges is a cold type of desert biome; three types of hotter deserts are found from western Texas west to California and south into Mexico. Deserts receive fewer than twenty-five centimeters of rainfall annually. The hot deserts are dominated by many cactus species and short, thorny shrubs and trees, whereas sagebrush, grass, and small conifer trees dominate the cold desert. These deserts have many lizard and snake species, including poisonous rattlesnakes and the Gila monster. The animals often have nocturnal habits to avoid the hot, dry daytime.

Southern Mexico and the Yucatán Peninsula are covered by evergreen, broad-leaved trees in the tropical rain-forest biome, which receives more than two hundred centimeters of rain per year. Many tree-dwelling animals, such as howler monkeys and tree frogs, spend most of their lives in the tree canopy, seldom reaching the ground.

Aquatic biomes can be broadly categorized into freshwater, marine, and estuarine biomes. Freshwater lakes, reservoirs, and other still-water environments are called lentic, in contrast to lotic, or running-water, environments. Lentic communities are often dominated by planktonic organisms, small, drifting (often transparent) microscopic algae, and the small animals which feed on them. These, in turn, support larger invertebrates and fish. Lotic environments depend more on algae that are attached to the bottoms of streams, but they support equally diverse animal communities. The marine biome is separated into coastal and pelagic, or open-water, environments, which have plant and animal communities somewhat similar to lotic and lentic freshwater environments, respectively. The estuarine biome is a mixing zone where rivers empty into the ocean. These areas have a diverse assemblage of freshwater and marine organisms.

THE BIOSPHERE

All the biomes together, both terrestrial and aquatic, constitute the biosphere, which by definition is all the places on earth where life is found. Organisms that live in a biome must interact with one another and must successfully overcome and exploit their abiotic environment. The severity and moderation of the abiotic environments determine whether life can exist in that microhabitat. Such things as minimum and maximum daily and annual temperature, humidity, solar radiation, rainfall, and wind speed directly affect which types of organisms are able to survive. Amazingly, few places on earth are so hostile that no life exists there. An example would be the boiling geyser pools at Yellowstone National Park, but even there, as the water temperature cools at the edges to about 75 degrees Celsius, bacterial colonies begin to appear. There is abundant life in the top meter or two of soil, with plant roots penetrating to twenty-two meters or more in extreme cases. Similarly, the mud and sand bottoms of lakes and oceans contain a rich diversity of life. Birds, bats, and insects exploit the airspace above land and sea up to a height of about 1,200 meters, with bacterial and fungal spores being found much higher. Thus, the biosphere generally extends about 10 to 15 meters below the surface of the earth and about 1,200 meters above it. Beyond that, conditions are too hostile. A common analogy is that if the earth were a basketball, the biosphere would constitute only the thin outer layer.

STUDYING HABITATS & BIOMES

Abiotic habitat requirements for a local population or even for an entire species can be determined in the laboratory by testing its range of tolerance for each factor. For example, temperature can be regulated in a laboratory experiment to determine the minimum and maximum survival temperatures as well as an optimal range. The same can be done with humidity, light, shelter, and substrate type: "Substrate preference" refers to the solid or liquid matter in which an organism grows and/or moves—for example, soil or rock. The combination of all ranges of tolerance for abiotic factors should describe a population's actual or potential microhabitat within a community. Furthermore, laboratory experiments can theoretically indicate how much environmental change each population can tolerate before it begins to migrate or die.

Methods to study the interaction of populations with one another or even the interaction of individuals within one local population are much more complicated and are difficult or impossible to bring into a laboratory setting. These studies most often require collecting field data on distribution, abundance, food habits or nutrient requirements, reproduction and death rates, and behavior in order to describe the relationships between individuals and populations within a community. Later stages of these field investigations could involve experimental manipulations in which scientists purposely change one factor, then observe the population or community response. Often, natural events such as a fire, drought, or flood can provide a disturbance in lieu of a manipulation caused by man.

There are obvious limits to how much scientists should tinker with the biosphere merely to see how it works. Populations and even communities in a local area can be manipulated and observed, but it is not practical or advisable to manipulate whole ecosystems or biomes. To a limited extent, scientists can document apparent changes caused by civilization, pollution, and long-term climatic changes. This information, along with population- and community-level data, can be used to construct a mathematical model of a population or community. The model can then be used to predict the changes that would happen if a certain event were to occur. These predictions merely represent the "best guesses" of scientists, based on the knowledge available. Population ecologists often construct reasonably accurate population models that can predict population fluctuations based on changes in food supply, abiotic factors, or habitat. As models begin to encompass communities, ecosystems, and biomes, however, their knowledge bases and predictive powers decline rapidly. Perhaps the most complicating factor in building and testing these large-scale models is that natural changes seldom occur one at a time. Thus, scientists must attempt to build cumulative-effect models that are capable of incorporating multiple changes into a predicted outcome.

The biosphere, then, can be studied at different levels of organization, from the individual level through populations, communities, ecosystems, and biomes to the all-encompassing biosphere. Each level has unique relationships that require different methods of inquiry; in fact, these levels describe many of the subdisciplines within the science of ecology.

BIOSPHERE & BIODIVERSITY

Understanding the organization of the natural world of which man is a part is essential to the continued success of humankind. By understanding the abiotic and biotic relationships within and between each ecological level, from microhabitat to biosphere, scientists can partially explain why so many species of organisms have evolved over the last 3.5 billion years. This study of biodiver sity may eventually be a key to maintaining a stable biosphere, in which there would be no drastic changes in climate or community relationships. For example, compare the diversity of micro habitats and species in a natural grassland biome with the established monoculture practices of agriculture, with the latter's emphasis on one species. In a wheat field, there are fewer micro habitats available, but those few are available in abundance. This leads to an increase in the population size of "pest" species that compete with man for an abundant food resource, wheat. Understanding the microhabitat requirements of pest species can lead to the reduction of crop losses.

The goal of the study of habitats, biomes, and the biosphere is the construction of predictive models. Once scientists have a general understanding of natural ecosystem processes, mathematical models may be able to predict future changes in the environment caused by the activities of civilization or natural climatic changes. For example, they may indicate whether increased human population size or increased large-scale agriculture in or near desert biomes will lead to the spread of desertlike conditions. On a biosphere scale, they may show whether the increase in the carbon dioxide content of the atmosphere caused by man's activities will lead to global warming. Both these effects are being predicted by many scientists. Prior to the relatively recent growth in the science of ecology and the general interest in it, man had little concern about the effects of the exploitation of natural resources such as forests, of synthetic chemical pollutants such as Dichlorodiphenyl-trichloroethane (DDT), or even of the rapid growth of the human population size. With an increasing knowledge and understanding of habitat requirements, natural community interrelationships, and cycling of nutrients and pollutants within the biosphere, however, scientists have greater predictive power concerning the effects of economic development and human population growth. Ecologists are studying the effects of changing rain forests into agricultural land, the extinction of species, the loss of the ozone layer, and many other phenomena that have the potential to change the abiotic environment and therefore affect the stability of biotic communities. These results will be incorporated into future predictive models, giving them increased accuracy.

The challenge—both to scientists and to human civilization as a whole—is to use an understanding of the biosphere to maintain a level of economic growth that is ecologically sustainable. The study of communities and ecosystems may discover ways that civilization can better adapt to its current environment rather than attempt to mold the environment to fit its own preconceived, established ideas.

FURTHER READING

Allaby, Michael. *Biomes of the World.* 9 vols. Danbury, Conn.: Grolier International, 1999. Each volume offers systematic coverage of a biome of group of related biomes, including physical description, characteristic animal life, and environmental issues and threats. Illustrations, maps, photos, further reading lists, lists of Web sites, glossary, and index.

Bradbury, Ian K. *The Biosphere.* New York: Belhaven Press, 1991. Emphasizes the living organisms within the global ecosystem, including basic cell chemistry, genetics, and evolution. Outlines the way in which ecosystems work. Does not require a background in biology to understand.

Cox, George W. *Conservation Biology: Concepts and Applications.* 2d ed. Dubuque, Iowa: Wm. C. Brown, 1997. An introductory college-level textbook. Discusses terrestrial and aquatic ecosystems, including each system's characteristics, plant and animal life, and environmental problems.

Hanks, Sharon La Bonde. *Ecology and the Biosphere: Principles and Problems.* Delray Beach, Fla.: St. Lucie Press, 1996. A college-level textbook aimed at non-science majors. Covers what science is and what students need to know about it; what the biosphere is, how it works, and what its current problems are; and what students can do about the problems.

Luoma, Jon R. *The Hidden Forest: The Biography of an Ecosystem.* Corvallis, OR: Oregon State University Press, 2006. Print. Focuses on the Andrews Experimental Forest, a 16,000 acre primal forest in central Oregon, which has been the subject of an intensive thirty-year study. A case study in the minutia of ecosystem research.

Miller, G T, and Scott Spoolman. *Living in the Environment.* Cengage Learning, 2018. Print. A college textbook that serves as an introduction to environmental science. Good coverage of the biosphere, ecosystem, and biome levels of ecology as well as an extensive annotated bibliography for each chapter.

Scientific American 261 (September, 1989). A special issue on managing the earth, with articles on the changing atmosphere, changing climate, threats to biodiversity, the growing human population, strategies for agriculture, and an ecologically sustainable economy. Emphasis is on the application of ecological knowledge.

Sutton, Ann, and Myron Sutton. *Eastern Forests.* New York: Alfred A. Knopf, 1986. Describes all the forest community types from the eastern deciduous and northern coniferous forest biomes. Unique in that it contains natural history information on insects, spiders, mushrooms, trees, wildflowers, reptiles, amphibians, birds, and mammals. Prefaced by a general description of each community type, written for the layperson.

Wetzel, Robert G. *Limnology: Lake and River Ecosystems.* San Diego: Academic Press, an imprint of Elsevier, 2015. Print. A classic and comprehensive text; no better general reference is currently available for freshwater ecology. Includes extensive information on swamps and marshes.

HERBIVORES

—Sanford S. Singer

FIELDS OF STUDY

Zoology; Conservation Biology; Evolution; Ecology; Environmental Studies.

PRINCIPAL TERMS

carnivore: any animal that eats only the flesh of other animals

gestation: the term of pregnancy

metamorphosis: insect development into adults, passing through two or more dissimilar growth forms

ruminant: a herbivore that chews and swallows plants, which enter its stomach for partial digestion, are regurgitated, chewed again, and reenter the stomach for more digestion

SUMMARY

Herbivores are animals whose diets consist entirely of plants. They have two ecological functions. First, they eat plants and keep them from overgrowing. Second, they are food for carnivores, which subsist almost entirely upon their flesh, and omnivores, which eat both plants and animals. Herbivores live on land or in oceans, lakes, and rivers. They can be insects, other arthropods, fish, birds, or mammals.

WILD HERBIVORES

Insects are the largest animal class, with approximately one million species. Fossils show their emergence 400 million years ago. Insects occur worldwide, from pole to pole, on land and in fresh or salt water. They are the best developed invertebrates, except for some mollusks. They mature by metamorphosis, passing through at least two dissimilar stages before adulthood. Metamorphosis can take up to twenty years or may be complete a week after an egg is laid.

Many insects are herbivores. Some feed on many different plants; others depend on one plant variety or a specific plant portion, such as leaves or stems. Relationships between insects and the plants they eat are frequently necessary for plant growth and reproduction. Among the insect herbivores are grasshoppers and social insects such as bees.

Artiodactyls are hoofed mammals, including cattle, pigs, goats, giraffes, deer, antelope, and hippopotamuses. Most are native to Africa, but many also live in the Americas, Europe, and Asia. Artiodactyls walk on two toes. Their ancestors had five, but evolution removed the first toe and the second and fifth toes are vestigial. Each support toe—the third and fourth—ends in a hoof. The hippopotamus, unique among artiodactyls, stands on four toes of equal size and width.

Artiodactyls are herbivores, lacking upper incisor and canine teeth, but pads in upper jaws help the lower teeth grind food. Many are ruminants, such as antelope, cattle, deer, goats, and giraffes. They chew and swallow vegetation, which enters the stomach for partial digestion, is regurgitated, chewed again, and reenters the stomach for more digestion. This maximizes nutrient intake from food.

Deer are hoofed ruminants whose males have solid, bony, branching antlers that are shed and regrown yearly. The deer family, approximately 40 species, occurs in Asia, Europe, the Americas, and North Africa. Deer live in woods, prairies, swamps, mountains, and tundra. Their size ranges from the seven-foot-tall moose to the one-foot-tall pudu. Deer first appear in the fossil rec ord ten million years ago. Deer eat the twigs, leaves, bark, and buds of bushes and saplings, and grasses. Females have one or two offspring after ten-month pregnancies. Common species are the white-tailed and mule deer in the United States; wapiti in the United States, Canada, Europe, and Asia; moose in North America and Europe; and reindeer in Russia, Finland, and Alaska.

Antelope, a group of approximately 150 ruminant species, have permanent, hollow horns in both sexes. Most are African, although some are European or Asian. They eat grass, twigs, buds, leaves, and bark. There are no true antelope in the United States, where their closest relatives are pronghorns and Rocky Mountain goats (goat- antelope with both goat and antelope anatomic features). The smallest antelope, the dik-dik, is rabbit-sized. Elands, the largest antelope, are ox-sized. Unlike deer horns, antelope horns are unbranched. Most antelope run rather than fighting, and are all swift. Antelope live on plains, marshes, deserts, and forests. Females birth one or two offspring per pregnancy. Impala and gazelles, such as the springbok, are found in Africa. In Asia, Siberian saigas and goat antelope (takin) inhabit mountain ranges. Chamois goat antelope live in Europe's Alps.

Giraffes and hippos are unusual artiodactyls. Giraffes inhabit dry, tree-scattered land south of the Sahara. Their unusual features are their very long legs and necks. Males are over sixteen feet tall, including the neck. Both sexes have short, skin-covered horns. Long necks, flexible tongues, and upper lips pull leaves—their main food—from trees. Giraffes have brown blotches on buff coats and blend with tree shadows. They live for up to twenty years. They have keen senses of smell, hearing, and sight, and can run thirty-five miles per hour. Due to their two-ton weights, they live on hard ground. Giraffes rarely graze, and go for months without drinking, getting most of their water from the leaves they eat, because it is difficult for them to reach the ground or the surface of a river with their mouths. Females have one offspring after a fifteen-month gestation.

The unusual feature of hippos is that they walk on all four toes of each foot. Perhaps this is because they weigh three to four tons. Hippos are short-legged, with large heads, small eyes, small ears, and nostrils that close underwater. Huge hippo mouths hold long, sharp incisors and canines in both jaws. Hippos once lived throughout Africa. Now they are rarer, due to poaching for ivory. A hippo can be fifteen feet long and five feet high at the shoulder. Semiaquatic, hippos spend most daylight hours nearly submerged, eating aquatic plants. At night they eat land plants. Females bear one offspring at a time.

AQUATIC HERBIVORES

Fish are aquatic vertebrates, having gills, scales and fins. They include rays, lampreys, sharks, lungfish, and bony fishes. The earliest vertebrates, 500 million years ago, were fishes. They comprise over 50 percent of all vertebrates and have several propulsive fins: dorsal fins along the central back; caudal fins at tail ends; and paired pectoral and pelvic fins on sides and belly. Fish inhabit lakes, oceans, and rivers, even in Arctic and Antarctic areas. Most marine fish are tropical. The greatest diversity of freshwater species is found in African and rain-forest streams.

Fish vary in length from half an inch to fifty feet, and some weigh seventy-five tons. Many, including giant whale sharks, are herbivores. Fish respiration uses gills, through which blood circulates. When water is taken in and expelled, oxygen enters the blood via the gills and carbon dioxide leaves. Fishes reproduce by laying eggs that are fertilized outside the body, or by internal fertilization and development with the birth of well-developed young.

DOMESTICATED ARTIODACTYLS

Bovids are domesticated artiodactyls. Most have horns and hooves. Bovid horns are spiraled, straight, tall, or L-shaped from the sides of the head. All have hooves to help them grip the ground. Most are ruminants. Their breeding habits are similar. Males fight over females and the strongest wins. Gestation, four to eleven months, yields two to three young. The young nurse for several months and then join the herd. Young males leave female herds to live with other bachelors.

Cattle are domesticated bovids, raised for meat, milk, and leather. Modern cattle come from European, African, and Asian imports. Breeding modern cattle began in Europe in the mid-1800's. Today, three hundred breeds exist. Dairy cattle, such as Holsteins, make copious milk. Beef cattle, such as Angus, were bred to yield meat. As of 1990, about 1.3 billion cattle were found worldwide.

Sheep are also artiodactyls. Wild sheep occur in some places, such as the North American bighorn and Mediterranean mouflon. Sheep were domesticated eleven thousand years ago from mouflon. Today, domesticated sheep have a world population of approximately 1.3 billion and inhabit most countries, being more widely distributed than any other domesticated animal. These ruminants have paired, hollow, permanent horns. Male horns are massive spirals; those of females are smaller. Adult body length is five feet and weights are 250 to 450 pounds. Females birth two or three young after five-month gestations. Sheep can live for up to twenty years.

Sheep provide wool, meat, and milk. About eight hundred domesticated breeds exist, in environments from deserts to the tropics. Those bred for wool, half the world's sheep, live in semiarid areas, are medium sized, and produce fine wool. Most are in Australia, New Zealand, and South America. Mutton-type sheep, 15 percent of the world sheep population, produce meat. Fat-tailed sheep, 25 percent of the sheep population, produce milk. In 1990, the five leading sheep countries were Australia, China, New Zealand, India, and Turkey. The United States raised less than 1 percent of the world total.

Goats are ruminants, closely related to sheep, but have shorter tails, different horn shape, and bearded males. They eat grass, branches, and leaves and breed from October to December. A five-month gestation yields two offspring. Numerous goat breeds are domesticated worldwide for meat and milk, and as pets and burden carriers. Domesticated Angora goats yield silky mohair. Goat milk is as nutritious as cow milk and used in cheese-making.

It is clear that wild herbivores are ecologically important to food chains. This is because they eat plants, preventing their overgrowth, and they are eaten by carnivores and omnivores. Domesticated herbivores—cattle, sheep and goats, used for human sustenance—account for three to four billion living creatures. Future production of better strains of domesticated herbivores via recombinant deoxyribonucleic acid (DNA) research may cut the numbers of such animals killed to meet human needs. Appropriate species conservation should maintain the present balance of nature and sustain the number of wild herbivore species living on earth.

Further Reading

Gerlach, Duane, Sally Atwater, and Judith Schnell. *Deer.* Mechanicsburg, Pa.: Stackpole Books, 1994. This book discusses deer habits, habitats, species, and much more, and has a useful bibliography.

Gullan, Penny J, Peter S. Cranston, and Karina H. McInnes. *The Insects: An Outline of Entomology.* Chichester: Wiley-Blackwell, 2014. The book is a reference on insects, including herbivores.

Olsen, Sandra L. *Horses Through Time.* Niwot, Colo: Roberts Rinehart, 2004. Print. This interesting book covers horse history, evolution, and more.

Rath, Sara. *The Complete Cow.* Stillwater, Minn.: Voyageur Press, 1990. This book covers cattle, their history, and their breeds.

Shoshani, Jeheskel, and Frank Knight. *Elephants: Majestic Creatures of the Wild.* Rev. ed. New York: Checkmark Books, 2000. This book covers many topics on elephants, including natural history.

Herds

—Kari L. Lavalli

FIELDS OF STUDY

Agriculture; Animal Husbandry; Conservation Biology; Evolution; Ecology; Environmental Studies.

PRINCIPAL TERMS

carnivores: animals that eat the flesh of other animals

dilution effects: the reduction in per capita probability of death from a predator due to the presence of other group members

herbivores: animals that eat plants and show specializations of teeth and digestive tracts to do so

phalanges: the free toes of the foot; some can be modified to bear claws, hoofs, or nails

subungulates: nonhoofed mammals that support their weight on more than the terminal phalanges; some, such as elephants and hyraxes, have pads under their metatarsals, and others, such as the sirenians, have forelimbs modified into flippers

ungulates: hoofed mammals that support their weight only on the hoof-clad terminal phalanges and have teeth specialized for clipping vegetation

SUMMARY

Herds are associations of animals, usually the same species, that keep, feed, or move together. Generally, the use of the word "herd" is reserved for large, terrestrial animals.

Herding in animals serves several purposes, most commonly exploitation of food resources and reduction of the probability of predation via dilution

For livestock, like these healthy cows, vaccines play an important part in animal health since antibiotic therapy is prohibited in organic farming. U.S. Department of Agriculture.

effects. When food is spaced irregularly into patches and cannot be easily defended, individuals of a particular species will form a herd, simply as a result of coming together to feed on the same resource in the same location. Sizable grazing mammals are well known for forming some of the largest and most dense herds seen. The most familiar of these include African ungulates (wildebeests, zebra, gazelle, wild horses, rhinos, hippopotami) and subungulates (elephants, sirenians, the extinct mastadons); bison, buffalo, caribou, elk, moose, and deer in North America; kangaroo in Australia; and deer, elk, moose, antelope, wild horses, sheep, goats, pigs, boars, and peccaries in Eurasia. In addition to large mammalian herbivores, small grazers that exploit vegetation in herds include marine and freshwater snails, tortoises and turtles, geese, and hyrax.

Those grazers that feed indiscriminately tend to form herds that vary in density with season and density of foliage. Furthermore, these herds tend to vary their densities and vegetation utilization patterns with the type of vegetation exploited and its productivity potential. In seasons or areas where rainfall is common, herds tend to congregate together and exploit small patches of grass, which they clip into grazing lawns. Maintaining grazing lawns increases the productivity of the plants via increased nitrogen content and increased digestibility in freshwater, marine, and terrestrial environments. When rainfall is scarce, the herds disperse over vast areas, which allows some grasses to grow into tall meadows. The benefits to individuals in a herd from grazing a larger area and creating a grazed lawn are greater than those to a lone individual that would only be able to graze a small patch of grass and that would, presumably, be at greater risk of predation than one individual surrounded by many others. Furthermore, if the lone individual were subject to a greater risk of predation, there would be less likelihood of it being able to return to the same patch over and over again to keep it as a grazed lawn. Thus, by grouping, an individual benefits twice: once by the gain in nutrition from a grazed lawn and again from the dilution effect, where the risk of predation is diluted by the number of members forming the group.

HERDING TO AVOID PREDATION

The large herbivores that most commonly form herds are clearly visible to predators. Predation risk reduction via the dilution effect or the selfish herd effect is the other main benefit of herding. In the dilution effect, the probability of a particular individual being killed or injured is reduced by the presence of other group members that might be attacked first. In other words, there is safety in numbers. Individuals in a group may also benefit by putting other animals between themselves and the predator (the selfish herd effect). Grouping thus provides the opportunity to decrease the area of danger around each individual. If individuals within the group are acting in a selfish herd manner, the groups formed tend to be tightly clumped, as all individuals attempt to put other individuals on all sides around them and move into a central location. Finally, formation of a herd can make it more difficult for a predator to select one specific individual from the crowd for attack, or once an attack is initiated and herd members scatter during flight, can make it difficult for the predator to decide who it should chase.

Herd members also benefit by a reduction in the time necessary to scan for predators while foraging. An individual foraging has to split its time between consuming as much food as possible and avoiding becoming another animal's meal. Thus, a solitary individual has to be much more vigilant than does a member of a group who can rely on many other eyes and thus reduce its own scan rate. Because of this advantage and the advantage in the reduction of the rate of per capita predation, even herds that simply congregate around a resource benefit greatly.

MIGRATORY & SOCIAL HERDS

Many large herbivores migrate in response to food availability in different seasons. Red deer, caribou, wildebeest, mountain goats, northern fur seals, and humpback whales are examples of animals that all migrate in response to seasonal changes in rainfall or food abundance. Some of these migrations are over incredible distances: Wildebeests travel about six hundred miles; northern fur seals and humpbacks can travel three to four thousand miles. During these migrations, small herds unite with bigger herds to form even larger herds. While terrestrial herbivores migrate to follow food, some marine mammals, such as gray whales and humpbacks, migrate south to calf and breed (but do not feed, living off food reserves instead) and migrate north to abundant feeding grounds.

Some herding species display social organization beyond that expected by mere association. This social organization tends to break a herd into smaller units (matrilineal groups, harems, or small territories controlled by one or several males) that are clustered within the entire herd. These species include the horses, zebras, pronghorn sheep, walruses, sea lions, seals, and elephants. Horse and zebra herds are composed of a number of small groups of females and their foals. These individual groups are overseen by a single stallion; young males leave these groups to form bachelor herds. Group members distinguish each other via a "corporate smell." Stallions generally control a group of females and will fight with challenging stallions in elaborate rearing displays. If a stallion is challenged and loses after inseminating the females in his group, the challenger will mount the females and rape them to induce them to abort. They will then come into estrus quickly and he will remate them. Stallions will groom females to cement their relationship with them; likewise, mares will groom their foals for the same purpose. Male impalas and gazelle maintain harems during the breeding season; male wildebeest do the same, but if the herd is large, the defense of the harem may be accomplished by several males, rather than one. Male elk are divided into four main categories: Primary bulls are the first to establish harems, but as they become exhausted from challenges and herding and mating females, all the while not eating, the harems are taken over by secondary bulls. Once the secondary bulls become exhausted, tertiary bulls take over. The fourth category, opportunistic bulls, only mate with females by chance.

Pinniped herds (sea lions, seals, and walruses) form breeding herds, where males establish territories. In some groups, females are herded together in harems to remain in the male's territory; in other groups, females are free to move from territory to territory. Males vigorously defend their territories against intruding males.

Sirenians, the dugong and manatee, have proved difficult to study, but dugongs often form large herds for unknown reasons. During mating, a series of males follows a receptive female to form a cluster of up to twenty animals. Males then initiate fighting to determine who will mount the female. Following mating, the main social unit is the female and her calf. A similar situation exists in the manatee, with females and their calves being the main social unit and females in estrus becoming the focus of a mating herd of males.

Herding serves important purposes for animals: utilization of a common resource (usually food) and predator reduction via a safety-in-numbers principle. Herds can also be formed during migrations or for breeding purposes. While many herds are simply loose associations of animals, some are more highly organized into a number of social units such as harems or matrilineal groups.

Further Reading

Drickamer, L. C., S. H. Vessey, and D. Meikle. *Animal Behavior: Mechanisms, Ecology, and Evolution.* 4th ed. Dubuque, Iowa: Wm. C. Brown, 1996. Describes the mechanisms, ecology, and evolution that support a comprehensive study of behavior.

Fryxell, J. M. "Forage Quality and Aggregation by Large Herbivores." *American Naturalist* 138, no. 2 (August, 1991): 478–498. This article examines three major hypotheses for why herbivores form herds and demonstrates that the mechanism operating depends largely on the type of herbivore and the circumstances it faces within its specific environment.

Hamilton, W. D. "Geometry for the Selfish Herd." *Journal of Theoretical Biology* 31 (1971): 294–311. This landmark paper was the first to explore the idea that grouping may be the result of selfishly behaving individuals that are trying to avoid becoming the predator's meal.

McNaughton, S. J. "Grazing Lawns: Animals in Herds, Plant Form, and Coevolution." *American Naturalist* 124, no. 6 (December, 1984): 863–886. This seminal article examines the effect that grazers have upon a relatively productive grassland in the Serengeti plain and explains why it may be beneficial to feed in a herd configuration.

Reynolds, J. E., and D. K. Odell. *Manatees and Dugongs.* New York: Facts on File, 1991. This book is one of the few that concentrates on sirenians and provides an overview of their distribution and biology.

Wallace, J. *The Rise and Fall of the Dinosaur.* New York: Gallery Books, 1987. This book provides an overview of the hypotheses surrounding the evolution of dinosaurs, their extinction, and their biology, as drawn from fossil evidence.

HORIZON (SOIL)

—*Lindsay Rohland*

FIELDS OF STUDY

Plant Science; Crop Science; Botany; Agronomy; Agriculture; Biochemistry; Ecology; Entomology; Genetics; Landscape Architecture; Soils Engineering.

SUMMARY

In geological terms, a *horizon* is a distinct layer of soil. Horizons run roughly parallel to the surface of the land. Each horizon has different visible, chemical, or physical characteristics from the layers located immediately above and below it. A column, or cross-section, of soil featuring all of its horizons is called a soil profile. A soil profile stretches vertically from the soil surface to the bedrock below. Scientists classify the five main horizons using the following letters: O, A, E, B, and C. Sometimes they use the letter R to represent minerals, air, water, organic matter, and clay. Geologists base these categories on particle size, with sand having the largest particle size and clay having the smallest. Most soils contain a mixture of these three main soil categories. The variable percentages of each are responsible for the texture of a particular soil sample.

Soil serves a number of important functions in the environment, affecting everything from the water organisms drink to the air they breathe. For example, it provides a natural filtration system. As water moves through soil, the soil absorbs impurities. Thus, water that trickles through soil on its way to underground aquifers or to lakes, rivers, and streams on Earth's surface has been filtered. Soil affects the air by absorbing and releasing gases such as carbon dioxide and water vapor. Soil provides a home to countless organisms, from microscopic bacteria to burrowing gophers. Grasses, flowers, trees, and other plants are rooted in soil and draw nutrients from it to help them grow. Additionally, soil serves as the foundation upon which humans build homes and other structures. Soil's integral roles in natural processes and human endeavors are as varied as the layers, or horizons, in a soil profile.

Soil samples illustrating horizons (subsoil on right). Image by Jonty68 at English Wikipedia.

INTRODUCTION

A soil profile includes a cross-section of all soil horizons from Earth's surface to the bedrock below. Each soil horizon has characteristics that distinguish it from the layer directly above or below it. Some common characteristics of each horizon are outlined below.

O horizons are surface horizons composed largely of organic material—for example, the leaves, twigs, and needles of trees—that has recently fallen to Earth's surface and that has not been saturated with water for a significant amount of time. Sometimes O horizons may be subdivided into two layers designated O_i and O_a. Organic materials in the O_i layer often have arrived so recently that they have just barely started to decay. In contrast, materials in the O_a layer have already partially decomposed. O horizons are quite common in forested areas with many trees.

A horizons are also surface horizons, sometimes covered by an O horizon and sometimes situated right at Earth's surface. A horizons contain both mineral matter, such as sand, silt, and clay, and organic matter. Because both O horizons and A horizons contain organic matter, it can sometimes be difficult to differentiate between the two. A horizons often make up the surface layer of soil in areas with few trees, such as grasslands.

E horizons are subsurface horizons. E horizons are mineral horizons in which a significant amount

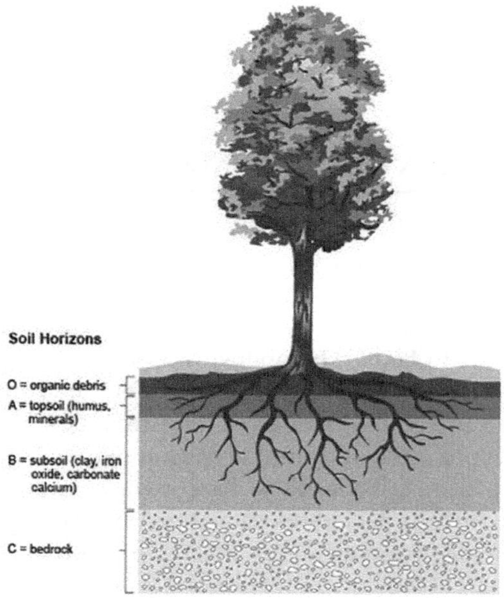

Soil Horizons. Cross-section of the soil horizons including the organic debris layer, topsoil layer, subsoil, and bedrock. © EBSCO.

of eluviation, or leaching, occurs. In eluviation, water transports minerals as it trickles from higher layers of soil to lower layers. These leached minerals then accumulate in the next layer, called the B horizon. As a result of eluviation, particles of soil in E horizons are generally a light color.

B horizons are subsurface horizons where minerals leached from higher layers are redeposited. The accumulation of leached materials in B horizons is called illuviation. B horizons are characterized by dark bands of mineral-rich material and typically contain very little organic material.

C horizons are subsurface horizons located just above bedrock, or the R horizon. C horizons are mineral layers. The material in C horizons undergoes the least amount of weathering and shares many similarities with the parent rock.

R horizons contain bedrock. Technically, R horizons are composed of rock, not soil. Examples include granite, basalt, and quartzite. However, material from the R horizon may contribute to the formation of some soils.

The order of horizons varies considerably in soil profiles. Some profiles may contain all the horizons: O, A, E, B, C, and R. Others may contain only A, B, and C. Still others may contain different variations of one particular horizon, leading to a configuration such as A, B, E, and B.

Beyond the five main horizons—O, A, E, B, and C—a few other soil classifications exist. H horizons, for example, share many characteristics with O horizons. They are predominantly organic material and are typically located atop mineral soils. Unlike O horizons, however, H horizons have been saturated with water for a significant amount of time. Occasionally, soil profiles include additional layers of ice, water, or limnic materials, represented by the letters I, W, and L, respectively. I layers contain a minimum 75 percent ice. Such layers may appear and disappear, depending on temperature. In some cases, however, ice may drive a wedge between two other distinct layers, thereby forming an I layer. Because some organic soils float on water and because some soils are permanently covered by water, a W may be used to designate a water layer. Limnic materials, or sediments containing a mixture of organic and inorganic material that are deposited in a body of water, may be designated as L layers.

FURTHER READING

"Appendix 1: Soil Horizon Designations." *FAO Corporate Document Repository*, www.fao.org/docrep/W8594E/w8594e0g.htm. Accessed 8 Nov. 2016.

Ricklefs, Robert E., and Gary L. Miller. *Ecology*. 4th ed., W. H. Freeman and Company, 2000.

"Soil Genesis and Development, Lesson 4 – Soil Profile Development." *Plant & Soil Sciences eLibrary*, passel.unl.edu/pages/informationmodule.php?idinformationmodule=1130447025&topicorder=4. Accessed 8 Nov. 2016.

"Soil Horizons." *K–12 Soil Science Teacher Resources*, soils4teachers.org/soil-horizons. Accessed 8 Nov. 2016.

"Soil Profile." *Soil Nutrient Management for Maui County, College of Tropical Agriculture and Human Resources (CTAHR)*, www.ctahr.hawaii.edu/mauisoil/a_profile.aspx. Accessed 8 Nov. 2016.

"Soils Glossary Appendix." *Soil Science Society of America*, www.soils.org/publications/glossary/appendix/. Accessed 8 Nov. 2016.

"Soils Overview." *Soil Science Society of America*, www.soils.org/files/about-soils/soils-overview.pdf. Accessed 8 Nov. 2016.

"Weathering." *National Geographic Society*, nationalgeographic.org/encyclopedia/weathering/. Accessed 8 Nov. 2016.

Horticulture

—*Carol A. Rolf, JD*

FIELDS OF STUDY

Plant Science; Crop Science; Botany; Agronomy; Agriculture; Biochemistry; Ecology; Entomology; Genetics; Landscape Architecture; Soils Engineering.

PRINCIPAL TERMS

Botany: Branch of biology that involves the scientific study of plants.

Genetic Engineering: Technique used to modify the genes of a plant so that it has a desired trait.

Hydroponics: Method of growing plants in water, without soil, by adding nutrients to the water.

Micropropagation: Cloning of multiple plants from the tissues of another plant.

Plant Biotechnology: Using biological systems to genetically engineer hardier plants and enhance their productivity.

Plant Structures: Parts of plants used most often in propagation, including runners, tubers, corms, bulbs, and rhizomes.

Propagation: Sexual production of plants from seeds or asexual production through cuttings, grafts, and tissue cultures.

Xeriscaping: Method of growing gardens in arid conditions by using drought-resistant plants and water conservation techniques.

SUMMARY

Horticulture is both a science and an art and includes the propagation and cultivation of plants often associated with gardening. Trees, bushes, grasses, and fungi are typical naturally occurring plant materials. Cultivated plants include not only trees, shrubs, grasses, and fungi but also vegetables, fruits, nuts, herbs, spices, ornamentals, and flowers. Although horticulture has existed since ancient times, modern scientific advances in the field have resulted in many practical uses of plant materials, including the development of pharmaceuticals. Genetics plays a significant role in modern-day horticultural practices, especially in areas such as food production.

DEFINITION & BASIC PRINCIPLES

Horticulture is the art of aesthetically arranging plant materials and the science of breeding, propagating, and growing plants. Horticulture requires a knowledge of landscape design, plant science, and individual plants. The ability to identify plants and their growing needs is necessary for determining how to use plant materials in landscape designs and also in plant breeding and propagation.

Plants are classified as herbaceous or woody. Woody plants include deciduous and evergreen trees, shrubs, and smaller plant materials. Deciduous plants become dormant at the end of the growing season, and plants classified as evergreens will retain their foliage—needles and broad leaves—throughout the year.

Horticulture is not the same as agriculture, which involves growing crops on a much larger scale than the typical horticultural venture and may also include animal production. Horticulture can be pursued either as a hobby or as a commercial pursuit. Those involved in horticulture need to be versed in many disciplines. Design, planning, graphics, and construction engineering are important for the art of horticulture, and a background in math and biochemistry are needed for horticultural science.

BACKGROUND & HISTORY

Horticulture is associated with ancient botany, which began as early as the fourth century BCE. Plants were selected and domesticated, and early propagation practices such as grafting can be seen in artwork found in ancient Egyptian tombs. Classical philosophers, such as Aristotle, defined plant types and developed some of their early uses for food, drugs, and fuel. Although an interest in botanical sciences waned during the Middle Ages, monasteries were known for their gardening prowess. During the eighteenth and nineteenth centuries, gardening became important as an art form, especially in Europe and Asia, and horticulture as a science began to expand.

One of the first modern horticulturalists in the United States was Liberty Hyde Bailey, a graduate of Michigan Agricultural College. As is true of many horticulturalists, Bailey wrote many texts on botany and horticulture. Bailey also worked with President Theodore Roosevelt's administration to improve rural life through horticulture. Early landscape architects, such as Frederick Law Olmstead, were also instrumental in advancing the field of horticulture.

HOW IT WORKS

A horticulturalist must be cognizant of soil, light, temperature, and water conditions to propagate and cultivate plants. Horticulture begins with the preparation of soil through tilling and fertilization, unless the plant is grown without soil as in hydroponics. Light and temperature can be controlled by beginning plant cultivation inside greenhouses and transplanting plants to the outdoors when light and temperature are optimal for growth. Installation of water collectors and irrigation systems, which have been part of horticulture since ancient times, is also necessary. Xeriscaping with modified drought-resistant plant materials, however, has reduced the reliance on water in some locations. Control of pests and diseases is also important, although many pest- and disease-resistant plant materials exist because of modern technology. Some horticulturalists no longer rely on chemical fertilizers and pesticides, and organic gardening is becoming profitable because of the public demand for naturally grown foods. Horticulture may also involve modern harvesting and handling practices that use more mechanical methods as opposed to manual labor.

Propagation. Propagation techniques vary from microscopic techniques such as gene splicing and producing tissue cultures to plant breeding and employing natural and cultivated plant structures and cuttings to generate new plants. The most common form of plant propagation is with seeds. Vegetables and flowers are typically propagated in this manner.

Clonal propagation can be done naturally or achieved through scientific methods involving plant structures rather than seeds. Some examples of natural clonal propagation through plant structures include strawberry runners, potato tubers, crocus corms, daffodils bulbs, and Calla lily rhizomes. Scientific propagation can be as simple as placing cuttings of shoots, roots, and leaves into water or moist soil. Shoots, such as willows, will regenerate roots. Roots, such as sweet potatoes, will grow new shoots, and leaves, such as African violets, will grow both new shoots and roots.

More complex clonal propagation methods include grafting techniques that physically join the parts from two plants together (often used in the propagation of fruit trees) and micropropagation, which requires the horticulturalist to grow plant cells and tissues artificially. Micropropagation is used in the production of modified plants that may have desirable qualities such as pest and disease resistance. It enables the production of a greater number of hardier plants in a shorter amount of time than could be produced from seeds or other propagation methods in the same amount of time.

Cultivation. Cultivation involves the direct planting of seeds and plant structures or the transplanting of plants that were started elsewhere. Cultivation of plant materials requires pruning and weeding to enhance future growth and improve on a plant's appearance. Ornamental cultivation techniques might include training plants so that they are reoriented to better conditions, such as light. Dwarfing is an ornamental cultivation practice that is often used in Asian cultures. Ornamental horticulture is important for those involved in floriculture and landscape architecture.

APPLICATIONS & PRODUCTS

Plants are everywhere and are necessary to sustain life. The practice of horticulture, therefore, has multiple applications that are important to society. In addition, modern technology has resulted in a greater diversity in plant-based products.

Food Production. Plant biotechnology includes genetic engineering, which allows the exchange of genes between plant species, usually ones that are closely related. More than 70 percent of processed foods contain some ingredients that have been genetically engineered, and biotech crops, such as corn and soybeans, are on the rise. New horticultural technologies and research are especially important in global regions, where horticultural production is restricted because of marginal climate conditions, poor soils, lack of water for irrigation, and reduced land resources for cultivation. DNA plays a major role in plant breeding of both quantitative and qualitative traits. Developing nations must especially depend on horticultural research to breed plants with pest- and drought-resistant characteristics to expand productivity on less acreage.

The global food market is changing. The ability to transport food products longer distances and the improved shelf life of such food are the result of horticultural advances that have opened up new markets for foods that may be exported by small producers in developing nations. Moreover, the methods of producing food are changing, and horticulturalists are finding ways to meet society's desire for more natural and organic food products while sustaining and protecting the environment.

Pharmaceuticals. Herbal remedies and medicinal plants have been used by healing practitioners since ancient times. Although scientists have questioned the curative powers of some herbs, horticulture continues to play an important role in the breeding of plants used as alternative medicines and in the development of plant-based mainstream drugs. Horticulturalists are also involved in synthesizing plant compounds into new medicines. Rain-forest plants have become especially important in this research because of the many naturally occurring chemical defenses against pests and diseases that these plants exhibit. Horticulturalists are on the forefront in conducting research concerning the curative properties of naturally occurring chemical compounds found in plants. These chemicals, known as pytochemicals, may provide new drugs for fighting cancer.

Flavors and Fragrances. The cultivation of plants such as flowers and fruits is necessary to produce many diverse products. Among these are cosmetics, perfumes, and food flavorings. Plant breeding and genetic engineering have enhanced the quality of plant-based flavorings, thus making some foods more palatable.

Farms and Nurseries. Horticultural practices play an important role in any enterprise that involves the growing of plants for commercial, public, or private uses. Tree farms range from farms specializing in the production of Christmas trees to those providing trees for forest conservation, which protects the environment by providing flood control and carbon sequestration. Other farms may produce vegetables, flowers, or turf, and golf course, park, and garden designers depend on farms and nurseries to produce plant materials to implement their plans.

Energy Production. Horticulture plays a role in the production and breeding of plants that can be converted into biofuels. Although the biofuel industry is in its infancy, substantial global research is taking place. For example, Clemson University is involved in research to turn discarded peaches into hydrogen gas, and nations such as India are producing biodiesel from nonedible plant oil seeds. Sugarcane and maize have also been used to produce ethanol. Because diverse plants can be used to produce biofuels, developing nations may be able to create biofuels cheaply by using naturally growing plant materials and thus create new industries to sustain population growth. However, horticulturalists must ensure that sufficient plants are available for both food production and these emerging biofuel industries.

Education. Development of healthy eating habits has become an important goal in many developed nations, such as the United States, where obesity is on the increase. Educators are employing gardening and horticulture in their lesson plans to teach children about nutrition. In less-developed countries, where poor nutrition is caused by a lack of horticultural crops, international organizations such as the United Nations provide education not only on advanced production and cultivation practices including irrigation and pest control but also on food safety, conservation, and marketing.

Therapy. Horticulture and gardening are therapeutic for those with learning, mental health, and physical disabilities. The American Horticultural Therapy Association is one of several organizations involved in this growing practice. Horticultural therapists incorporate gardens into wellness programs, and this type of therapy has become popular in programs to treat alcohol and substance abusers.

IMPACT ON INDUSTRY

The Global Horticulture Initiative, known as GlobalHort, is an international consortium of institutions, governmental agencies, and nongovernmental organizations involved in horticultural research, training, and information sharing. Located in both Belgium and Tanzania, GlobalHort focuses on improving international health and nutrition through the production and promotion of various horticultural crops. Partners include the Food and Agriculture Organization of the United Nations, the World Health Organization, and the World Bank.

Government Agencies. In the United States, one of the primary agencies involved in horticultural research and education is the U.S. Department of Agriculture (USDA). State and local governments also have active horticultural research and public education programs, including those associated with county and university agricultural extension services. The USDA is in partnership with many researchers at land-grant universities and sponsors the National Agricultural Library, which collaborates with public, private, and nonprofit agencies to provide education, often concerning horticultural aspects of food production.

University Research. Much of horticultural research is conducted at universities. In the United States, many of these institutions are state universities that started as land-grant colleges. University research is concentrated on new plant science, especially related to propagation and cultivation. Many land-grant universities have botanical gardens that provide ample opportunities for field research with new plant materials and research trials to enhance plant characteristics. Often university research is conducted at experimental stations; research ranges from agroforestry experimentation at the University of Missouri to urban horticulture research at the University of California Cooperative Extension.

Institutions and Organizations. Great Britain's Royal Horticultural Society, which has focused on gardening for about 150 years, continues to provide horticultural advice and education, especially through its published journal. Horticultural research and education has been provided internationally since 1864 by the International Society for Horticultural Science, made up of scientists representing about 150 nations. The American Society for Horticultural Science provides a similar service in the United States. Within the Society of Chemical Industry, an international private organization with offices in the United States, United Kingdom, and Australia, is a horticultural group concerned with transferring horticultural science into business practices, especially those associated with agriculture and food. In the United States, the American Horticultural Society is an important organization for providing private gardeners with horticultural information. Online networking is possible through the nonprofit organization, Society for Advancement of Horticulture, which connects researchers, growers, and commercial enterprises associated with horticulture.

Growers and Producers. Commercial and private growers and producers associate with one another to learn the latest technologies in horticulture and to share information. Growers and producers cultivate a variety of plant materials not limited to commercial produce, such as vegetables, fruits, nuts, and decorative trees, shrubs, grasses, and flowers. Some horticultural specialties include growing grapes and berries for wine, aquatic gardening, and producing unusual species such as carnivorous plants, cacti and succulents, and bonsai and dwarf varieties. Flower growers often specialize in one species such as roses, orchids, or lilies.

Related Industry and Business Sectors. Growers and producers must depend on industry to supply them with new horticultural products and the latest scientific advances, including seeds and plant structures, machinery, greenhouse and irrigation systems, and fertilizers and pesticides. Many of the newer companies are associated with the growth in organic and natural plant production. Once commercial products—everything from vegetables to flowers—are produced, the growers and producers must depend on the business sector to process plant products safely and to market them.

SOCIAL CONTEXT & FUTURE PROSPECTS

As the worldwide population continues to increase, horticultural advances through biotechnology can benefit society by ensuring an adequate food supply. Genetically engineered plants are more pest and disease resistant, and modified plants that withstand cold temperatures and are more tolerant of drought and salt allow for longer growing seasons and more numerous habitats. Plants are being engineered to have superior nutrient properties that could reduce the quantity of food necessary to nourish the world's growing population. Furthermore, many of the plants created through biotechnology and genetic engineering are playing a greater role in the production of beneficial medicines and drugs; some experts predict that society may one day have access to edible vaccines.

However, these scientific advances are raising concerns. Scientists are debating whether genetically engineered foods could produce gene mutations in organisms that consume them, create new food allergens, or cause environmental damage because of the antibiotic resistance of genetically engineered foods. Cross-pollination has also become an issue, as it cannot be prevented or stopped. Some fear the creation of super weeds that cannot be controlled after naturally occurring weeds cross-pollinate with genetically modified pesticide-resistant plants.

FURTHER READING

Adams, Charles R., Katherine M. Bamford, and Michael P. Early. *Principles of Horticulture.* 5th ed. Burlington, Mass.: Butterworth-Heinemann, 2008. Introductory text dealing with all aspects of horticulture care and propagation, pest control, and soils, water, and nutrition.

DiSabato-Aust, Tracy. *The Well-Tended Perennial Garden: Planting and Pruning Techniques.* 2d ed. Portland, Oreg.: Timber Press, 2006. Useful guide for planning a perennial garden with a section for notes and personal documentation.

Garner, Jerry. *Careers in Horticulture and Botany.* 2d ed. New York: McGraw-Hill, 2006. Lists many job opportunities in the field of horticulture and describes the necessary education and training.

Pauly, Philip J. *Fruits and Plains: The Horticultural Transformation of America.* Cambridge, Mass.: Harvard University Press, 2008. Discusses how horticultural practice in the United States helped shape the nation.

Reiley, Edward H., and Carroll L. Shry, Jr. *Introductory Horticulture.* 7th ed. Florence, Ky.: Delmar Cengage Learning, 2006. Easy-to-read textbook that includes information on the latest technologies and on finding a job in horticulture.

Toogood, Alan. *American Horticultural Society Plant Propagation: The Fully Illustrated Plant-by-Plant Manual of Practical Techniques.* New York: DK Adult, 1999. Good reference for propagating more than 1,500 plants from existing specimens with illustrations and explanations of plant biology.

Web Sites

American Horticultural Society http://www.ahs.org

American Horticultural Therapy Association http://www.ahta.org

American Society for Horticultural Science http://www.ashs.org

Global Horticulture Initiative http://www.globalhort.org

International Society for Horticultural Science http://www.ishs.org

U.S. Department of Agriculture Horticultural Crops Research Unit http://www.ars.usda.gov/main/site_main.htm?modecode=53–58–10–00

HYBRIDIZATION (BOTANY)

—Manjit S. Kang

FIELDS OF STUDY

Plant Science; Crop Science; Botany; Agronomy; Agriculture; Biochemistry; Ecology; Entomology; Genetics.

SUMMARY

Hybridization is the process of crossing two genetically different individuals to create new genotypes. For example, a cross between a parent 1, with the genetic makeup (genotype) *BB*, and parent 2, with *bb*, produces progeny with the genetic makeup *Bb*, which is a hybrid (the first filial generation or F_1). Hybridization was the basis of Gregor Mendel's historic experiments with garden peas. Inheritance studies require crossing plants with contrasting or complementary traits.

BACKGROUND

Hybridization of plants occurs in nature through various mechanisms. Some plants (such as the oil palm) are insect-pollinated, and others (such as maize, or corn) are wind-pollinated. Such plants are referred to as *cross-pollinated plants.* Natural hybridization has played a significant role in producing new genetic combinations and is the norm in cross-pollinated plants. It is a common way of generating genetic variability.

In plants with perfect flowers (*autogamous*, having flowers with both stamens and pistils), cross-pollination rarely occurs. Such plants (such as wheat and rice) are called *self-pollinated plants*. Flowers bearing only pistils or stamens are said to be imperfect flowers. Plants that have separate pistillate and staminate flowers on the same plant (such as maize) are called *monoecious*. Plants that have male and female flowers on separate plants (such as asparagus) are called *dioecious*.

Through artificial means (controlled pollination), hybridization of both cross-pollinated and self-pollinated plants can be accomplished. Artificial hybridization is an important aspect of improving both cross-pollinated and self-pollinated plants. The breeder must know the time of development of reproductive structures of the species, treatments to promote and synchronize flowering, and pollinating techniques.

APPLICATIONS TO AGRICULTURE

The concept of *hybrid vigor*, or *heterosis*, resulted from hybridization. Heterosis (or heterozygosis) occurs when the hybrid outperforms its parents for a certain trait. Around 1761 Joseph Gottlieb Kölreuter was the first to report on hybrid vigor in interspecific crosses of various species of *Nicotiana*. He concluded that cross-fertilization was generally beneficial and self-fertilization was not. In 1799 T. A. Knight conjectured that because of widespread existence of cross-pollination in nature, it must be the norm. Charles Darwin reported the results of his experiments with maize. He indicated that in twenty-four crosses, there was an increase in plant height, which was attributed to hybridization, and that decrease in plant height was associated with self-pollination (or selfing). He also noted that crossing of inbred plants could reverse the deleterious effects of selfing or *inbreeding*. In 1862 Darwin wrote, "Nature tells us, in the most emphatic manner, that she abhors perpetual self-fertilization." In the late 1800's William J. Beal evaluated hybrids between maize varieties. He observed that some hybrids yielded 50 percent more than the mean of their parents. S. W. Johnson provided an explanation for hybrid vigor in 1891. G. W. McClure reported in 1892 that hybrids between maize varieties were superior to the mean of the two parents.

EXPLOITATION OF HETEROSIS

The phenomenon of heterosis has been exploited in crop plants, such as maize, sorghum, sunflower, onion, and tomato. Maize (corn) was the first crop in the United States in which hybrids were produced from *inbred lines*. It was George Shull who, following the rediscovery of Mendel's laws of inheritance in 1900, conducted the first experiments on inbreeding and crossing, or hybridizing, of inbred lines. Shull suggested that inbreeding within a maize variety resulted in pure (*homozygous*) lines and that hybrid vigor resulted from crossing of pure lines because *heterozygosity* was created at many allelic sites. Hybrid maize was introduced in the United States in the late 1920's and early 1930's, after which U.S. maize production increased dramatically from the use of hybrids.

Heterosis now drives a multibillion-dollar business in agriculture. Yield improvement made in various crops in which heterosis was detected has been tremendous. In 1932 in the United States, 44.8 million hectares (111 million acres) were required to produce 51 million metric tons of maize grain, with a mean yield of 1.66 metric tons per hectare. In 1994 it took only 32 million hectares (79 million acres) to produce 280 million metric tons of grain, with a mean yield of 8.69 metric tons per hectare. In the United States in 1996, twenty-one vegetable crops occupied 1,576,494 hectares (3.9 million acres), with a mean of 63 percent of the crop in hybrids. Heterosis saved an estimated 220,337 hectares (544,459 acres) of agricultural land per year, feeding 18 percent more people without an increase in land use. From 1986 to 1995, the best rice hybrids showed a 17 percent yield advantage over the best inbred-rice varieties at the International Rice Research Institute.

Despite the impact that heterosis has had on crop production, its molecular genetic basis is still not clear. It is hoped that with the progress being made in the genetic sequencing of various plant species, a better understanding of heterosis will emerge. Plant breeding entails hybridization within a species as well as hybridization between species or even genera, called *wide crosses.* The latter are important for generating genetic variability or for incorporating a desirable gene not available within a species. There are barriers, however, for accomplishing *interspecific* and *intergeneric* crosses. Plants of the same species cross easily and produce fertile progeny. Wide crosses are difficult to make and generally produce sterile progeny because of chromosome-pairing difficulties during meiosis.

Triticale is the only human-made cereal crop, which is a cross between the genus *Triticum* (wheat) and the genus *Secale* (rye). The first fertile triticale was produced in 1891. Some of the interspecific and intergeneric barriers should be overcome via the newer techniques of gene transfer. It is expected that genes from wild relatives of cultivated plants will continue to be sought to correct defects in otherwise high-yielding varieties.

FURTHER READING

Basra, A. S., ed. *Heterosis and Hybrid Seed Production in Agronomic Crops.* Binghamton, N.Y.: Food Products Press, 1999. This book discusses current research in some of the most important crops of the world.

Coors, J. G., and S. Pandey. *Genetics and Exploitation of Heterosis in Crops.* Madison, Wis.: American Society of Agronomy and Crop Science Society of America, and Soil Science Society of America, 1999. Provides an account of the various issues related to hybrid vigor, or heterosis.

Fehr, Walter R. *Hybridization of Crop Plants.* Madison, Wisc: American Society of agronomy & Crop science Society of America, 1982. Print. Brings together the experience of plant breeders and scientists in a form that can be used by the layperson.

HYDROLOGIC CYCLE

—Samuel F. Huffman

FIELDS OF STUDY

Climatology; Meteorology; Environmental Sciences; Environmental Studies; Climate Modeling; Physical Geography.

PRINCIPAL TERMS

base flow: that part of a stream's discharge derived from groundwater and interflow seeping into the stream, representing the normal amount of water in that system

capillary force: a phenomenon in which water moves through tiny pores in rock, soil, and other materials, driven by intermolecular attraction between the water and the porous materials

evaporation: the process by which substances, especially water, change from a liquid into a vapor; when a substance changes directly from solid to gas without an intermediate liquid stage, the process is called sublimation

infiltration: the movement of water into and through the soil

interception: the process by which precipitation is captured on the surfaces of vegetation before it reaches the land surface

overland flow: the flow of water over the land surface caused by direct precipitation

precipitation: atmospheric water in the form of rain, hail, mist, sleet, or snow that falls to the earth's surface

runoff: the total amount of water flowing into a stream, including overland flow, return flow, interflow, and base flow

soil moisture: the water contained in the unsaturated zone above the water table

transpiration: the process by which plants give off water vapor through their leaves

SUMMARY

Water circulates on Earth through a system called the hydrologic cycle. This water cycle functions through vegetation, in the atmosphere, below the ground, and on land, lakes, rivers, and oceans. The sun and the force of gravity provide energy to drive the cycle from ground and surface water to atmospheric moisture that returns to the land and oceans as precipitation.

EVAPORATION, CONDENSATION & PRECIPITATION

The unending circulation of water on Earth is called the hydrologic cycle. This system is driven by the heat energy received from the sun. Gravity pulls water that falls on the surface back to the oceans to be recycled once again. The total amount of water on Earth is an estimated 1.36 billion cubic kilometers. Most of this vast amount of water, some 97.2 percent, is found in the oceans. The Greenland and Antarctic ice caps and glaciers contain 2.15 percent of Earth's water. The remaining 0.65 percent is divided among rivers (0.0001 percent), freshwater and saline lakes (0.017 percent), groundwater (0.61 percent), soil moisture (0.005 percent), the atmosphere (0.001 percent), and the biosphere and groundwater below 4,000 meters (0.0169 percent). While the percentage of water appears small for each of these water reservoirs, the total volume of water contained in each is immense.

A description of the hydrologic cycle must begin with the oceans, as most of Earth's water is located there. Each year, about 320,000 cubic kilometers of water evaporates from the world's oceans. Evaporation is the process whereby a liquid changes to a gas. Adding energy in the form of heat to the water causes the water molecules to become increasingly active and to move more rapidly, weakening the chemical and physical forces that bind them together. As the temperature of the water increases, water molecules tend to move from the ocean's surface into the overlying air. Factors that influence the rate of evaporation from free water surfaces are solar radiation, temperature, humidity, and wind velocity. It is estimated that an additional 60,000 cubic kilometers of water enters the atmosphere from rivers, streams, and lakes or through transpiration by plants every year. Thus, a total of about 380,000 cubic kilometers of water is evapotranspired on Earth every year.

The amount of water vapor that can be present in the air depends largely on the temperature. At higher temperatures, more vapor can be present. As the vapor-laden air is lifted and cooled at higher altitudes, the vapor condenses to form droplets of water. Condensation is aided by small dust and salt particles or nuclei in the atmosphere. As droplets collide and coalesce, clouds begin to form and precipitation can begin. Wind may transport moisture-laden air long distances, and most precipitation events are the result of three causal factors: frontal precipitation, or the lifting of an air mass over a moving weather front; convectional precipitation related to the uneven heating of Earth's surface, causing warm air currents to rise and cool; and orographic precipitation, resulting from a moving air mass being forced to move upward over a mountain range, cooling the air as it rises. Each year, about 284,000 cubic kilometers of precipitation falls on the world's oceans. This water has completed its cycle and is ready to begin a new cycle. Approximately 96,000 cubic kilometers of precipitation falls upon the land surface each year. This precipitation follows a number of different pathways in the hydrologic cycle. It is estimated that some 60,000 cubic kilometers evaporates from the surface of lakes or streams or transpires directly back into the atmosphere. The remainder—about 36,000 cubic kilometers—is intercepted by human structures or vegetation, is infiltrated into the soil or bedrock, or becomes surface runoff.

INTERCEPTION, RUNOFF & INFILTRATION

Although the amount of water intercepted by and evaporated from human structures—the surfaces of buildings and other artificial surfaces—may approach 100 percent, much urban water is collected in storm sewers or drains that lead to a surface drainage system or that spread water over the land surface to infiltrate the subsoil. Interception loss from vegetation is dependent upon interception capacity (the ability of the vegetation to collect and retain falling precipitation), wind speed (the higher the wind speed, the greater the rate of evaporation), and rainfall duration (the interception loss will decrease with the duration of rainfall, as the vegetative canopy will become saturated with water after a period of time). Broad leaf forests may intercept 15 to 25 percent of annual precipitation, and a bluegrass lawn may intercept 15 to 20 percent of precipitation during a growing season.

When the duration and intensity of the rainfall are greater than the soil's ability to absorb it, the excess water begins to run off, a process termed overland flow. Overland flow will begin only if the precipitation rate exceeds the infiltration capacity of the soil. Infiltration is the process whereby water sinks between soil particles directly into the soil surface or into fractures of rocks. It is dependent upon the characteristics of the soil or rock type and upon the nature of the vegetative cover. Sandy soils have infiltration rates of 3.6 to 3.8 centimeters per hour, and clay rock soils average 2.0 to 2.3 centimeters per hour. Nonporous rock would have an infiltration rate of zero, and all precipitation would become runoff. The presence of vegetation impedes surface runoff and increases the potential for infiltration to occur.

Water infiltrating into the soil or bedrock encounters two forces: capillary force and gravitational force. A capillary force is the tendency of water in the subsurface to adhere to the surface of soil or sediment particles. This tendency may actually draw the water upward against the downward pull of gravity. Capillary forces are responsible for the soil moisture found a few inches below the land surface.

COMPLETION OF THE CYCLE

Growing plants are continuously extracting soil moisture and passing it into the atmosphere through a process called transpiration. Soil moisture is drawn into the plant rootlet because of osmotic pressure. The water moves through the plant to the leaves, where it is passed into the atmosphere through specialized leaf openings called "stomata." The plant uses less than 1 percent of the soil moisture in its metabolism, with the remainder being used for the transport of nutrients and metabolites between the roots and leaves, and for temperature regulation. Thus, transpiration is responsible for most water vapor loss from the land in the hydrologic cycle. An oak tree, for example, may transpire 151,200 liters per year.

The water that continues to move downward under the force of gravity through the pores, cracks, and fissures of rocks or sediments will eventually enter a zone of water saturation. This source of underground water is called an aquifer—a rock or soil layer that is sufficiently porous and permeable to hold and transport water. The top of an aquifer, or saturated zone, is the water table. This water is slowly moving toward a point where it is normally discharged into a lake, spring, or stream. Groundwater that feeds and maintains the flow of a stream is called base flow. Base flow is typical of so-called spring-fed streams, and enables such streams to continue to flow during droughts and through cold winter months. Groundwater may flow directly into the oceans along coastlines.

When the infiltration capacity of the soil surface is exceeded, overland flow can begin as broad, thin sheets of water no more than a few millimeters thick called sheet flow. After flowing a short distance, the sheets break up into threads of current that flow in tiny channels called rills. The rills can coalesce into gullies and, finally, into streams and rivers. While evaporation losses do occur from the stream surface, much of the stream's water is returned to the oceans in this way, completing the hydrologic cycle.

Scientists are interested in how long it takes water to move through the hydrologic cycle. The term "residence time" refers to how long a molecule of water would remain in the various components of the hydrologic cycle. The average length of time that a water molecule would stay in the atmosphere is about one week. Two weeks is the average residence time for a water molecule in a river and ten years in a lake. It would take four thousand years for all the water molecules in the oceans to be recycled just once. Groundwater may require anywhere from a few weeks to thousands of years to move through the cycle. This time period may appear extremely long to humans, yet it suggests that over the course of geologic time, every water molecule on Earth has been recycled millions of times.

STUDY OF THE HYDROLOGIC CYCLE

Scientists have developed a vast array of mathematical equations and instruments to collect data to quantify the complexities of the hydrologic cycle. The geographic, secular, and seasonal variations in temperature, precipitation, evapotranspiration, solar radiation, vegetative cover, and soil and bedrock type, among other factors, must be evaluated to understand the local, regional, or global hydrologic cycle.

Precipitation, an extremely variable phenomenon, must be accurately measured to determine its input into the hydrologic cycle. The United States has more than thirteen thousand precipitation stations equipped with rain gauges placed strategically to compensate for wind and splash losses. Techniques have been developed to determine the average depth of precipitation falling on a given area or drainage basin. The effective uniform depth method utilizes a rain-gauge network of uniform density to determine the arithmetic mean for rainfall in the area. The isohyetal and polygonal methods are used to determine the arithmetic mean for an area or basin with a nonuniform distribution of rain gauges. The amount of water in a snowpack is estimated by snow surveys. The depth and water content of the snowpack are measured and the extent of the snow cover mapped using satellite photography.

The amount of precipitation lost by interception can be measured and evaluated. Interception is determined by the type of vegetation, the amount of evaporation that occurs during the storm, and the length of the storm. Most often, interception is determined by measuring the amount above the vegetative canopy and the amount received at the surface. The difference is deemed to be the amount lost to interception.

The volume of water flowing by a given point at a given time in an open stream channel, measured in cubic meters per second (CMS), is called discharge. Discharge is determined by measuring the velocity of water in the stream channel with a current meter. The Price meter and the pygmy-Price meter meet the specifications of the U.S. Geological Survey. The cross-sectional area of the stream channel is determined at a specific point and multiplied by the stream velocity to determine discharge. Automated stream-gauging stations are located on most streams to supply data for various hydrologic investigations.

The U.S. National Weather Service maintains about five hundred stations using Class A land pans to measure free-water evaporation. These pans are 122 centimeters in diameter and 25.4 centimeters deep, and they are made of unpainted galvanized metal. Water depths of 17 to 20 centimeters are maintained. The wind velocity is also determined. Errors may result from splashing by raindrops or from birds. Because the metal pan will also heat and cool more rapidly than will a natural reservoir, a pan coefficient must be employed to compensate for this phenomenon. A lake evaporation nomograph is employed to determine daily lake evaporation. The mean daily temperature, wind velocity in kilometers per day, solar radiation, and mean daily dew point are the variables used to determine daily lake evaporation.

The amount of evapotranspiration can be measured using a lysimeter, which is a large container holding soil and living plants. The lysimeter is set outside, and the initial soil moisture is determined. All precipitation or irrigation is measured accurately. Changes in the soil moisture storage determine the amount of evapotranspiration.

Those techniques are utilized to determine the water budget for different geographic areas. Collectively, these data enable scientists to estimate the total water budget of the hydrologic cycle.

Further Reading

Dodds, Walter K., and Matt R. Wiles. *Freshwater Ecology: Concepts and Environmental Applications of Limnology.* 2d ed. Burlington, Mass.: Academic Press, 2010. Covers the physical and chemical properties of water, the hydrologic cycle, nutrient cycling in water, as well as biological aspects. Written by two of the leading scientists in freshwater ecology. An excellent resource for college students. Each chapter provides a short summary of main topics.

Fetter, C. W. *Contaminant Hydrogeology.* Grove, Ill.: Waveland Press, 2008. Emphasizes the conditions that lead to the contamination of groundwater, the problems this contamination can cause, and the remediation techniques used to correct these problems. Includes a detailed discussion of the influence that geologic conditions have on groundwater occurrence and contamination. Also addresses regional groundwater movement within aquifers. Contains helpful illustrations of the principles discussed. Includes a glossary of important terms.

Guymon, Gary L. *Unsaturated Zone Hydrology.* Upper Saddle River, N.J.: Prentice-Hall, 1994. Delivers a comprehensive exploration of the movement of water through and within the unsaturated zone, using a quantitative approach to combine research and knowledge from several different disciplines. Suitable for students of engineering and environmental science, as well as for practicing engineers and scientists.

Hamblin, Kenneth W., and Eric H. Christiansen. *Earth's Dynamic Systems.* 10th ed. Upper Saddle River, N.J.: Prentice Hall, 2003. Offers an integrated view of Earth's interior not common in books of this type. Includes excellent illustrations, diagrams, and charts, as well as a glossary and laboratory guide. Suitable for high school readers.

Hunt, Constance Elizabeth. *Thirsty Planet: Strategies for Sustainable Water Management.* New York: St. Martin's Press, 2004. Discusses the hydrologic cycle, components of the cycle, human-induced changes to the cycle, and the repercussions of the growing water crisis. Also covers floods, climate change, and freshwater ecosystem restoration. Provides a broad overview of hydrology topics of study.

Ingebritsen, S. E., Ward E. Sanford, and C. E. Neuzil. *Groundwater in Geologic Processes.* 2d ed. New York: Cambridge University Press, 2006. Extensively revised, well received, and widely used. Focuses on the relationship of groundwater and geologic processes within the field of hydrology. Intended for specialist study in hydrology and hydrogeology.

Lerman, A. "Geochemical Cycles." In *The Oxford Companion to the Earth.* Edited by Paul Hancock and Brian J. Skinner. New York: Oxford University Press, 2000. Provides an excellent overview of geochemical cycles, focusing on the water cycle, sodium cycle, and carbon cycle. Accessible to lay readers.

Lutgens, Frederick K., Edward J. Tarbuck, and Dennis Tasa. *Essentials of Geology.* 11th ed. Upper Saddle River, N.J.: Prentice Hall, 2011. Provides an excellent discussion of the hydrologic cycle and associated processes. Designed for the college student.

Mays, Larry W. *Ground and Surface Water Hydrology.* Hoboken, N.J.: John Wiley & Sons, 2011. Stresses the fundamental principles of hydrologic processes relevant to groundwater and surface water, making extensive use of online resources and discussing newer topics in hydrology such as remote sensing and GIS. Covers all principal climate types. Written at the college level.

Moore, John E. *Field Hydrogeology.* 2d ed. Boca Raton, Fla.: CRC Press, 2012. Provides practical information on field investigation, aquifer testing, and groundwater quality testing. Discusses planning and report writing in detail. Provides case studies from around the world. Best suited for students learning the planning and investigation of field hydrology.

Tarbuck, Edward J., and Frederick K. Lutgens. *Earth: An Introduction to Physical Geology.* 10th ed. Upper Saddle River, N.J.: Prentice Hall, 2010. Offers an excellent discussion of the hydrologic cycle and Earth's water balance. Illustrations, index, and bibliographical references.

Thompson, Stephen A. *Hydrology for Water Management.* Rotterdam: A. A. Balkema, 1999. Provides a thorough account of hydrology, groundwater flow, and stream flow, focusing on the management of water supplies and efforts to keep them free of pollutants and available to the largest number of people possible. Illustrations, maps, index, and bibliography.

I

Industrial Fermentation

—*Sergei A. Markov, PhD*

FIELDS OF STUDY

Culinary Science; Chemistry; Bacteriology; Microbiology.

PRINCIPAL TERMS

Industrial Fermentation: the use of living organisms (mainly microorganisms), typically on a large scale, to produce commercial products or to carry out important chemical transformations.

BACKGROUND

The goal of industrial fermentation is to improve biochemical or physiological processes that microbes are capable of performing while yielding the highest quality and quantity of a particular product. The development of fermentation processes requires knowledge from disciplines such as microbiology, biochemistry, genetics, chemistry, chemical and bioprocess engineering, mathematics, and computer science. The major microorganisms used in industrial fermentation are fungi (such as yeast) and bacteria. Fermentation is performed in large fermenters or other bioreactors often of several thousand liters in volume. Industrial fermentation is a part of many industries, including microbiology, food, pharmaceutical, biotechnology, and chemical.

SUMMARY

Industrial fermentation is an interdisciplinary science that applies principles associated with biology and engineering. The biological aspect focuses on microbiology and biochemistry. The engineering aspect applies fluid dynamics and materials engineering. Industrial fermentation is associated primarily with the commercial exploitation of microorganisms on a large scale. The microbes used may be natural species, mutants, or microorganisms that have been genetically engineered. Many products of considerable economic value are derived from industrial fermentation processes. Common products such as antibiotics, cheese, pickles, wine, beer, biofuels, vitamins, amino acids, solvents, and biological insecticides and pesticides are produced via industrial fermentation.

HISTORY

Traditional fermentations such as those for making bread, cheese, yogurt, vinegar, beer, and wine had been used by people for thousands of years before its microbial nature was understood. Brewing beer was one of the first applications of fermentation in ancient Egypt as long as 10,000 years ago. The exact origins of dairy products are unknown—it may have been as early as 8000 BCE. It was probably nomadic Turkish tribes in Central Asia who invented cheese and yogurt making. Traditionally, dairy fermentation was a means of milk preservation. The scientific understanding of fermentation began only in the nineteenth century after French scientist Louis Pasteur published the results of his studies on the microbial nature of wine making.

The first industrial fermentation bioprocesses based on knowledge of microbes appeared in the early twentieth century. Russian biochemist Chaim Weizmann is considered to be the father of industrial fermentation. Weizmann used the bacterium *Clostridium acetobutylicum* for the production of acetone from starch in 1916. Acetone was used to make explosives during World War I.

Significant growth of this field began in the middle of the twentieth century, when the fermentation process for the large-scale production of antibiotic penicillin was developed. The goal of industrial-scale production of penicillin during World War II led to development of fermenters by engineers that were working together with biologists from the pharmaceutical company Pfizer. The fungus *Penicillium* grows and produces an antibiotic much more effectively under controlled conditions inside the fermenter. Continuous progress in industrial fermentation technology in the twentieth century has followed the development of genetic engineering. Genetic engineering allows gene transfer between species and creates possibilities to generate new products from genetically modified microorganisms that are grown in fermenters.

The twenty-first century has been characterized by the introduction of biofuels, which are made by industrial fermentation processes. Once again, past and future developments in fermentation technology require contributions from a wide range of disciplines, including microbiology, genetics, biochemistry, chemistry, engineering, mathematics, and computer science.

HOW IT WORKS

Industrial fermentation is based on microbial metabolism. Microbes produce different kinds of substances that they used for growth and maintenance of their cells. These substances can be useful for humans. The goal of industrial fermentation technology is to enhance the microbial production of useful substances.

Process of Fermentation. In biology, fermentation is a process of harvesting energy of organic molecules in oxygen-free conditions. Sugars are a prime example of what can be fermented, although, there are many other organic molecules that can be used. Different fermentations are known and are categorized by the substrate metabolized or the type of the product.

In industry, any large microbiological process is called fermentation. Thus, the term fermentation has a different meaning than in biology. Most industrial fermentations require oxygen.

Industrial Fermentation Organisms. Different organisms, such as bacteria, fungi, and plant and animal cells, are used in industrial fermentation processes. An industrial fermentation organism must produce the product of interest in high yield, grow rapidly on inexpensive culture media available in bulk quantities, be open to genetic manipulation, and be nonpathogenic (does not cause any diseases).

Fermentation Media. To make a desired product by fermentation, microorganisms need nutrients (substrates). Nutrients for microbial growth are known as media. Most fermentation requires liquid media or broth. General media components include carbon, nitrogen, oxygen, and hydrogen in the form of organic or inorganic compounds. Other minor or trace elements must also be supplied, for example, iron, phosphorus, or sulfur.

Fermentation Systems. Industrial fermentation takes place in fermenters, which are also called bioreactors. Fermenters are closed vessels (to avoid microbial contamination) that reach vast volumes, as many as several hundred thousand liters. Designed by engineers, the main purpose of a fermenter is to provide controllable conditions for growth of microbial cells or other cells. Parameters such as pH, temperature, nutrients, fluid flow, and other variables are controlled. There are two kinds of fermenters, those for anaerobic processes (oxygen-free) and aerobic processes. Aerobic fermentation is the most common in industry. Anaerobic fermenters can be as simple as stainless-steel tanks or barrels. Aerobic fermenters are more complicated. The most critical part in these systems is aeration. In a large-scale fermenter transfer of oxygen is very important. Oxygen transfer and dispersion are provided by stirring with impellers or oxygen (air) sparging.

Fermentation Control and Monitoring. Industrial fermentation control is very important to ensure that organisms behave properly. In most cases computers are used for controlling and monitoring the fermentation process. Computers control temperature, pH, cell density, oxygen concentration, level of nutrients, and product concentration.

APPLICATIONS & PRODUCTS

There are a wide range of industrial fermentation products and applications.

Food, Beverages, Food Additives, and Supplements. Industrial fermentation plays a major role in the production of food. Food products traditionally made by fermentation include dairy products (cheeses, sour cream, yogurt, and kefir); food additives and supplements (flavors, proteins, vitamins, and carotenoids); alcoholic beverages (beer, wines, and distilled spirits); plant products (bread, coffee, soy sauce, tofu, sauerkraut); and fermented meat and fish (pepperoni and salami).

Industrial Fermentation. The primary and largest industry revolves around food products. Milk from cows, sheep, goats, and horses have traditionally been used for the production of fermented dairy products. These products include cheese, sour cream, kefir, and yogurt. More recently so-called probiotics appeared and have been marketed as health-food drinks. Dairy products are produced via fermentation using lactic bacteria such as *Lactobacillus acidophilus* and *Bifidobacterium.* Fungi are also involved in making some cheeses. Fermentation produces lactic acid and other flavors and aroma compounds that make dairy products taste good.

Many products of industrial fermentation are added into food as flavors, vitamins, colors, preservatives, and antioxidants. These products are more desirable than food additives produced chemically. Many of the vitamins are made by microbial fermentations including thiamine (vitamin B_1), riboflavin (vitamin B_2), cobalamin (vitamin B_{12}), and vitamin C (ascorbic acid). Vitamin C is not only a vitamin but is also an important antioxidant that helps to prevent heart diseases. Carotenoids are another effective antioxidant. They are also used as a natural food color for butter and ice cream. Carotenoids are red, orange, and yellow pigments produced by bacteria, algae, and plants.

Food preservatives are yet another product of industrial fermentation. Organic acids, particularly lactic and citric acids, are extensively used as food preservatives. Some of these preservatives (such as citric acid) are used as flavoring agents. A mixture of two bacterial species (*Lactobacillus* and *Streptococcus*) is usually used for industrial production of lactic acid. The mold *Aspergillus niger* is used for citric acid manufacturing. Another common preservative is the protein nisin. Nisin is produced via fermentation by the bacterium *Lactococcus lactis.* It is employed in the dairy industry especially for production of processed cheese.

Antibiotics and Other Health Care Products. Antibiotics are chemicals that are produced by fungi and bacteria that kill or inhibit the growth of other microbes. They are the second most significant product of industrial fermentation. Most antibiotics are generated by molds or bacteria called actinomycetes. More than 4,000 antibiotics have been isolated from microorganisms, but only about 50 are produced regularly. Among them, beta-lactams, such as penicillins and tetracyclines, are most common. Penicillin is produced by the mold *Penicillium chrysogenum* via corn fermentation in bioreactors of up to 200,000 liters.

The other major health care products produced with the help of industrial fermentation are bacterial vaccines, therapeutic proteins, steroids, and gene therapy vectors. There are two categories of bacterial vaccines: living and inactivated vaccines. Living vaccines consist of weakened, also known as attenuated, bacteria. Examples of living vaccines include those for diseases such as anthrax, which is caused by *Bacillus anthracis,* and typhoid fever, which is caused by *Salmonella typhi.* Inactivated vaccines are composed of bacterial cells or their parts that have been inactivated by heat or formaldehyde. Examples of these vaccines are those for meningitis, whooping cough, and cholera. Vaccine production takes place in fermenters no bigger than 1,000 liters in volume. It requires highly controlled operations to avoid the release of bacteria into the environment. All exhaust gases pass through sterilization processes.

Therapeutic proteins include growth hormone, insulin, wound-healing factors, and interferon. Previously, such compounds were made from animal tissues and were very expensive to manufacture. Genetic engineering now allows their production by fermentation from bacteria. Human growth hormone is synthesized in the human brain and controls growth. Too little growth hormone can cause some cases of dwarfism. The American company Genentech started production of human growth hormone from genetically modified *Escherichia coli* by fermentation in 1985. Insulin is an animal and human hormone that is involved in the regulation of blood sugar. The body's inability to make sufficient insulin causes diabetes. Insulin extracted from pigs had been used to treat diabetes, but it has been replaced by insulin produced by industrial fermentation from genetically modified bacteria.

Chemicals. Numerous chemicals, such as amino acids, polymers, organic acids (citric, acetic, and lactic), and bioinsecticides are produced by industrial fermentation. Amino acids are used as a food and animal feed, as well as in the pharmaceutical, cosmetic, and chemical industries. Bacteria such as *Micrococcus luteus* and *Corynebacterium glutamicum* are used for industrial fermentation to produce chemicals. Bacterial toxins are effective against different insects. Since the 1960's, preparations of the bacteria *Bacillus thuringiensis* have been produced by fermentation as a biological insecticide.

ENZYMES. Enzymes are used in many industries as catalysts. Microorganisms are the favored source for industrial enzymes. Seventy percent of these enzymes are made from *Bacillus* bacteria via fermentation. Most commercial microbial enzymes are hydrolases, which break down different organic molecules such as proteins and lipids. The enzyme glucose isomerase is important in the production of fructose syrups from corn and is widely used in the food industry.

BIOMASS PRODUCTION. During biomass production by fermentation, the cells produced are the products. Biomass is used for four purposes: as a source of protein for human food or animal feed, in industry as fermentation starter cultures, in agriculture as a pesticide or fertilizer, and as a fuel source.

One major product of this application of industrial fermentation is baker's yeast biomass. Baker's yeast is required for making bread, bakery products, beer, wine, ethanol, microbial media, vitamins, animal feed, and biochemicals for research. Yeast is produced in large aerated fermenters of up to 200,000 liters. Molasses is used as a nutrient source for the cells. Yeast is recovered from fermentation liquid by centrifugation and then dried. It can then be sold as compressed yeast cakes or dry yeast.

Many bacteria have been considered as potential sources of protein to fulfill the food needs in some countries of the world. As of 2011, only a few species are cultivated around the world as a source of food and feed. Among them, cyanobacteria are the most popular. The protein level of *Cyanobacterium Spirulina* can be as high as those found in meat, nuts, and soybeans, from 50 to 70 percent. This cyanobacterium has been used as a human food for millennia in Asia, parts of Africa, and in Mexico.

Apart from yeast and bacteria, people are also using the biomass of algae. Algae are a source of animal feed, plant fertilizer, chemicals, and biodiesel. Because light is necessary to grow algae, the biomass is produced in open ponds or in transparent tubular glass or plastic bioreactors, called photobioreactors.

BIOFUELS. Industrial fermentation is used in the production of biofuels, mainly ethanol and biogas. These two biofuels are produced by the action of microorganisms in bioreactors. Fermentation can also be used for generation of biodiesel, butanol, and biohydrogen. Biofuels are considered, by many, as a future substitute for fossil fuels. Pollution from fossil fuels affects public health and causes global climate change due to the release of carbon dioxide (CO_2). Using biofuels as an energy source generates less pollutants and little or no CO_2.

Production of ethanol is a process based on fungal or bacterial fermentation of a variety of materials. In the United States, most of the ethanol is produced by yeast (fungal) fermentation of sugar from cornstarch. Sugar is extracted using enzymes, and then yeast cells convert the sugar into ethanol and CO_2. Ethanol is separated from the fermentation broth by distillation. Brazil, the second largest ethanol producer after the United States, uses sugarcane fermentation to generate ethanol. The Brazilian production of ethanol from sugarcane is more efficient than the American corn-based ethanol.

Biogas is produced during the anaerobic (nonoxygen) fermentation of organic matter by communities of microorganisms (bacteria and *Archaea*). There are different types of biogas. One type contains a mixture of methane (50 to 75 percent) and CO_2. Another type is composed primarily of nitrogen, hydrogen, and carbon monoxide (CO) with trace amounts of methane. Methane is generated by microorganisms called *Archaea* and is an integral part of their metabolism. For practical use, biogas is generated from wastewater, animal waste, and "gas wells" in landfills.

SOCIAL CONTEXT & FUTURE PROSPECTS

Industrial fermentation plays a major role in providing food, chemicals, and fuels. End users are consumers, farmers, medical doctors, and industrialists. Industrial fermentation is changing the course of history. People have made food by fermentation for centuries. In the twentieth century, the development of antibiotics and their production by industrial fermentation had the most significant impact on the practice of medicine than any other development.

The growth of the industrial fermentation field is continuing rapidly. Since the beginning of the twenty-first century, industrial fermentation underwent an unprecedented growth and expansion due to biofuel introduction. This record growth is particularly visible in the U.S. ethanol industry. In 1980, the U.S. ethanol industry produced 175 million gallons of ethanol by fermentation, and in 2009, 10.6 billion gallons.

The role of industrial fermentation in human society is likely to expand in the future because of increasing requirements for resources.

Further Reading

Bailey, James E, and David F. Ollis. *Biochemical Engineering Fundamentals.* New York: McGraw-Hill, 2006. Print. Classic textbook on biochemical engineering.

Bourgaize, David, Thomas R. Jewell, and Rodolfo G. Buiser. *Biotechnology: Demystifying the Concepts.* San Francisco: Benjamin/Cummings, 2000. Classic text on biotechnology.

Doran, Pauline M. *Bioprocess Engineering Principles.* Amsterdam: Elsevier/Academic Press, 2013. Print. Describes various bioreactors and fermenters used for industrial fermentation.

Glazer, Alexander N., and Hiroshi Nikaido. *Microbial Biotechnology: Fundamentals of Applied Microbiology.* 2d ed. New York: Cambridge University Press, 2007. In-depth analysis of application of microorganisms in industrial fermentation.

Lydersen, Bjorn K., Nancy A. D'Elia, and Kim L. Nelson, eds. *Bioprocess Engineering: Systems, Equipment and Facilities.* New York: John Wiley & Sons, 1994. Describes equipment and facilities for industrial fermentation.

Wright, Richard T., and Dorothy F. Boorse. *Environmental Science: Toward a Sustainable Future.* 11th ed. Boston: Benjamin/Cummings, 2011. This textbook describes several bioprocesses used in waste treatment and pollution control.

Intensive Farming

—M. Marian Mustoe, PhD

FIELDS OF STUDY

Plant Science; Crop Science; Botany; Agronomy; Agriculture; Biochemistry; Ecology; Entomology; Soils Engineering.

SUMMARY

The world's growing population requires ever-increasing amounts of food, and intensive farming provides high levels of crop production on the land available. The greatest challenge to such farming is its long-term sustainability; intensive farming's dependence on fossil-fuel-burning heavy equipment and its reliance on chemicals to eliminate pests and replace nutrients in soil are damaging to the environment.

During the nineteenth century, the economy of the United States was largely dependent on the agricultural production of thousands of small farms. As the country industrialized, and particularly after World War II, small farmers began to consolidate their lands, with the result that there were fewer farm operators and much larger farms. Mechanization, in the form of tractors, seed application implements, and other heavy equipment, as well as technological advancements in chemical pesticides, herbicides, and fertilizers contributed to supporting the infrastructure of these large-scale farming operations and lowered the farmers' labor costs.

As world population continues to rise, demand for food also rises, and food crops require land for production. With the spread of urbanization and suburbanization, productive cropland has become increasingly scarce. Intensive farming, also known as industrial farming, maximizes crop production on the land available by relying heavily on the mechanization of many processes, by treating even marginal soils with chemical fertilizers to increase yield, and by using chemical pesticides and herbicides. Such large-scale commercial agriculture is very capital-intensive in nature, and most intensive farming operations are thus found in the world's economically developed regions.

Industrial farming operations make extensive use of technology to run systems and monitor resources, both to keep labor costs low and to increase productivity. For example, many such vast farms have switched from labor-intensive manually operated irrigation systems to computer-controlled center-pivot irrigation systems. Some farms use geographic information system (GIS) technology to facilitate the monitoring and control of their land resources.

BENEFITS OF LARGE-SCALE FARMING

From the 1940's into the 1980's, the so-called Green Revolution in agriculture did much to transform regions that had been dependent on more traditional forms of farming into areas that could produce greater amounts of food for their populations through more intensive agriculture. The impact of the Green Revolution was especially strong in some developing nations, such as India, where the shift to larger farms and greater mechanization was highly successful.

Increased food production is the largest benefit of intensive commercial farming. In the United States, the level of crop production on farms doubled between the nineteenth and twenty-first century. It has been estimated that in the 1960's one U.S. farmworker produced enough fiber and food for about thirty people; by 2010, the number of people whose food needs could be met by a single U.S. farmworker had increased to more than one hundred.

ENVIRONMENTAL IMPACTS

The practices associated with industrial farming have a number of negative impacts on the environment. Running heavy machinery, producing fertilizers and pesticides, and processing, storing, marketing, selling, and transporting massive amounts of crops require a great deal of energy, most of which comes from natural gas and oil. For example, grain grown in the midwestern United States might be shipped to Michigan for milling and processing into cereal; the finished product, a breakfast cereal, thus might travel thousands of miles before it reaches a consumer's table. This intense dependence on petroleum for agricultural production has long-term negative impacts for the environment, as fossil-fuel supplies are finite and the burning of fossil fuels contributes to carbon dioxide emissions, which have been linked to global warming.

The pesticides used in intensive farming have come under scrutiny for health reasons, with many critics pointing out the danger of pesticide

exposure to farmworkers and questioning how residues that may remain on crops could affect consumers of the resulting food product. In addition, pesticides and fertilizers used across vast fields are carried into water sources by irrigation and rainfall runoff. Furthermore, the soil of industrial farms is degraded over time and requires large amounts of fertilizers to rebuild nutrients; yet another impact on the soil is the erosion that can be caused by the use of heavy machinery and intensive irrigation.

The water requirements of huge industrial farms are another factor with impacts on the environment. Many dams have been built specifically to direct water resources to enable crop irrigation, and other sources have also been affected. For example, it has been projected that the deep wells being used as water sources for industrial farms in the American

Intensive farming. Water pipes and irrigation equipment being readied in the yard of Langley Lawn Farm in preparation for the summer growing season. Image by John M CC-BY-SA-2.0.

Midwest could eventually run low given the demands of these farms and the continual encroachment of cities and suburbs into formerly rural areas.

In recognition of the negative impacts that traditional industrial farming practices can have on the environment, some farm operators have instituted conservation methods such as minimal tillage, which reduces erosion, and integrated pest management, which reduces pesticide use. Additionally, some farms have invested in fuel-efficient heavy equipment and in technologies, such as GIS, that can increase efficiency of land management while keeping crop production high.

Another way in which industrial farms have tried to improve yields while reducing the use of pesticides is by planting genetically modified (GM) seeds. Such genetically modified organisms are often engineered for greater disease and pest resistance and for tolerance to herbicides. This method is not without controversy, however. Opponents fear that modified genes could be transmitted by pollination to nearby conventional crops and weaken the genetic diversity of the particular plant species. Others are concerned that not enough research has been conducted on the health effects of consuming GM foods.

FURTHER READING

Avery, Alex, and Dennis Avery. "High-Yield Conservation: More Food and Environmental Quality through Intensive Agriculture." In *Agricultural Policy and the Environment*, edited by Roger E. Meiners and Bruce Yandle. Lanham, Md.: Rowman & Littlefield, 2003.

Barrows, Geoffrey, Steven Sexton, and David Zilberman. "Agricultural Biotechnology: The Promise and Prospects of Genetically Modified Crops." *Journal of Economic Perspectives* 28, no. 1 (2014): 99–120.

Conkin, Paul K. *A Revolution Down on the Farm: The Transformation of American Agriculture Since 1929.* Lexington, Ky.: University Press of Kentucky, 2008.

Filson, Glen C., ed. *Intensive Agriculture and Sustainability: A Farming Systems Analysis.* Vancouver: University of British Columbia Press, 2004.

Gliessman, Stephen R., and Martha Rosemeyer, eds. *The Conversion to Sustainable Agriculture: Principles, Processes, and Practices.* Boca Raton, Fla.: CRC Press, 2010.

Laidlaw, Stuart. *Secret Ingredients: The Brave New World of Industrial Farming.* Toronto: McClelland & Stewart, 2004.

INTERNATIONAL UNION FOR CONSERVATION OF NATURE (IUCN)

—*Lisa A. Wroble and Karen N. Kähler*

FIELDS OF STUDY

Ecology; Environmental Studies; Agronomy.

SUMMARY

The International Union for Conservation of Nature brings together governments, nongovernmental organizations, scientists, communities, and commercial interests in its efforts to protect species and ecosystems and promote sustainable use of natural resources. It has been instrumental in drafting and gaining signatories for a number of important international treaties designed to protect the environment.

The International Union for Conservation of Nature (IUCN) is the world's oldest global environmental organization and its largest professional global conservation network. Scientists working within IUCN gather data, identify species and wildlife areas in need of protection, and determine ways to sustain the earth's resources for present and future generations.

This democratic membership union, headquartered in Gland, Switzerland, serves as a neutral forum where governments, scientists, nongovernmental organizations, commercial interests, and communities come together to find solutions for conservation and development concerns. Its membership spans 140 countries and includes more than two hundred governmental organizations and more than eight hundred nongovernmental organizations. IUCN employs the voluntary expertise of almost eleven thousand scientists and specialists to provide policy advice on conservation issues. A leading authority on sustainability and the environment, the organization develops and supports conservation science and runs thousands of field projects and activities around the world.

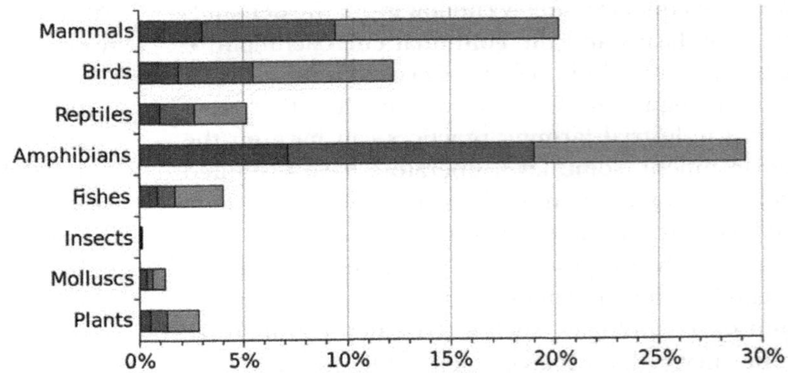

Percentage of species listed on the IUCN Red List, by group; Red: critical; Orange: endangered; Gold: vulnerable. Vectorization by B. Jankuloski.

The idea for an international conservation organization emerged as early as 1910. Swiss naturalist Paul Sarasin advocated for one and succeeded in establishing it, but it lost momentum during World War I. Sarasin was unable to reestablish a viable organization before his death in 1929, and none of the other persons who had been inspired by his vision were able to make significant headway over the next two decades. The establishment of the United Nations after World War II, however, meant new hope for Sarasin's dream. In 1948, an international conference on nature conservation was held at Fontainebleu, France, and a majority of the delegates in attendance—representatives from twenty-three governments, 126 national institutions, and eight international organizations—signed a formal act that established the International Union for the Protection of Nature (IUPN) on October 5.

Originally headquartered in Brussels, Belgium, IUPN pledged to protect endangered species of wildlife on a worldwide scale. It created an international network of conservationists who used the union and its periodic conferences as means for sharing information. In 1950 IUPN launched the Survival Service (later the Species Survival Commission), a group of scientists who worked on a volunteer basis to document the plights of the world's endangered species. IUPN began to interact directly with government representatives from many nations to promote conservation of their endangered species.

In 1956 IUPN changed its named to the International Union for Conservation of Nature and Natural Resources, or IUCN. Between 1990 and 2008, the name World Conservation Union was also used in conjunction with the IUCN name.

THE ORGANIZATION MATURES

In 1961, IUCN moved its headquarters from Belgium to Switzerland. That same year it welcomed the fledgling World Wildlife Fund (WWF) to share its new headquarters. WWF was created to raise funds for IUCN and other worldwide conservation efforts, to conduct public relations campaigns, and to gain public support for conservation efforts. Ornithologist Peter Scott, then vice president of IUCN, became WWF's first chairperson. IUCN was among the earliest recipients of WWF grants, which funded projects such as creating a footpath in a reserved forest of Madagascar, transporting eight endangered white rhinos from South Africa to what is now Zimbabwe for breeding, and saving the rare Arabian oryx from extinction. The two organizations shared accommodations until the burgeoning WWF moved its offices in 1979. (WWF changed its name to the World Wide Fund for Nature in 1986 but retained use of the acronym by which it has been known since its founding.)

In 1962, the United Nations General Assembly adopted the first World List of National Parks and Equivalent Reserves, a list more than three hundred pages long compiled by IUCN's newly formed Commission on National Parks and Protected Areas (later its World Commission on Protected Areas). The commission subsequently worked to develop detailed criteria for what constitutes a protected area. The UN List of Protected Areas, a listing of the world's national parks, scientific reserves, and

The chinchilla is determined by the IUCN to be a critically endangered species. Image by ChrisStubbs (Own work) CC-BY-SA-3.0.

natural monuments, later became a regular publication of IUCN.

In 1963 IUCN began its Red List system for identifying the world's threatened species and assessing conservation efforts. The Red List has since become recognized as the authoritative global listing of plants and animals facing possible extinction.

IUCN and the United Nations Educational, Scientific, and Cultural Organization (UNESCO) codrafted the text of the Convention Concerning the Protection of the World Cultural and Natural Heritage in 1972. This convention established the World Heritage List, which names valued natural and cultural sites to be protected. IUCN continues to serve as a technical advisory body to UNESCO's World Heritage Committee.

In 1975 the Convention on International Trade in Endangered Species (CITES) went into force. This international agreement, created to ensure that trade in specimens of wild animals and plants does not jeopardize the survival of species, emerged from a 1963 IUCN member resolution. IUCN teamed with WWF in 1976 to set up a monitoring network called the Trade Records Analysis of Flora and Fauna in Commerce (TRAFFIC). TRAFFIC investigators found that the smuggling and poaching of animal and plant species was occurring throughout the world, even in countries that had signed treaties opposing such activities. Increased awareness of the problem encouraged local governments to work with TRAFFIC to stop illegal commerce in exotic animals and plants.

In 1980 IUCN, WWF, and the United Nations Environment Programme issued a joint publication titled *World Conservation Strategy: Living Resource Conservation for Sustainable Development*, which emphasized the need for a holistic approach to conservation and the sustainable use of natural resources. Fifty countries soon created their own national conservation strategies based on the recommendations presented in *World Conservation Strategy*. The three organizations copublished a follow-up strategy document, *Caring for the Earth: A Strategy for Sustainable Living*, in 1993, in which they recommended 132 actions that individuals at all social and political levels can take to protect the environment and improve quality of life.

Other major IUCN contributions include preparation of the World Charter for Nature (1982), the Convention on Biological Diversity (1992), and the draft Covenant on Environment and Development (1995; updated 2000, 2004, and 2010). IUCN also sponsors the World Conservation Congress, which convenes every four years.

Further Reading

Bräutigam, Amie, and Martin Jenkins. *The Red Book: The Extinction Crisis Face to Face.* Gland, Switzerland: IUCN, 2001. Print.

Holdgate, Martin W. *From Care to Action: Making a Sustainable World.* Washington, DC: Taylor, 1996. Print.

International Union for Conservation of Nature. *IUCN.* IUCN, 13 Dec. 2015. Web. 31 Dec. 2015.

International Union for Conservation of Nature, United Nations Environment Programme, and World Wide Fund for Nature. *Caring for the Earth: A Strategy for Sustainable Living.* Gland: IUCN, UNEP, WWF, 1993. Print.

International Union for Conservation of Nature, United Nations Environment Programme, and World Wide Fund for Nature. *World Conservation Strategy: Living Resource Conservation for Sustainable Development.* Gland: IUCN, UNEP, WWF, 1980.

Morphet, Sally. "NGOs and the Environment." *The Conscience of the World: The Influence of Non-governmental Organizations in the U.N. System.* Ed. Pete Willetts. Washington, DC: Brookings Inst. P, 1996. Print.

Van Dyke, Fred. *Conservation Biology: Foundations, Concepts, Applications.* 2d ed. New York: Springer, 2008. Print.

Vié, Jean-Christophe, Craig Hilton-Taylor, and Simon N. Stuart, eds. *Wildlife in a Changing World: An Analysis of the 2008 IUCN Red List of Threatened Species.* Gland: IUCN, 2009. Print.

Irrigated Farming

—Grove Koger

FIELDS OF STUDY

Agriculture; Agronomy; Ecology; Ecosystem Management; Climatology; Bioclimatology; Climate Modeling; Environmental Sciences; Environmental Studies; Hydroclimatology.

SUMMARY

Irrigated farming led to the profitable cultivation of acres that otherwise would not have been suitable for crop growing, the establishment of numerous subsidiary enterprises, and significant shifts in population.

Although most regions of the eastern United States receive sufficient rainfall for farming, lands west of the one-hundredth meridian, which runs from the Dakotas southward to Texas, are generally arid and need irrigation to be farmed. In areas where rainfall permits growth but is not abundant, irrigation can increase crop yields dramatically.

Early steps to encourage farming in the West met with limited success. During the 1870's and 1880's, a number of private irrigation companies were formed, but generally they were unable to deliver water to distant tracts and most failed within a few years. The Desert Land Act of 1877 allowed would-be farmers to buy land at a low price, provided part of the land came under irrigation within three years, a requirement often impossible to meet. The Carey Act of 1894 offered land to the states for irrigation projects, although few took advantage of it.

Far greater success came with the formation of the Bureau of Reclamation, created by the Newlands Reclamation Act of 1902. Under the bureau's guidelines, the federal government financed and oversaw the damming of rivers, the creation of reservoirs, and the digging of canals, allowing water users to repay the costs gradually. Such projects, which initially employed large numbers of laborers, resulted in the cultivation of vast tracts of land and increased settlement in the western states. The spread of irrigated farming also encouraged the manufacture and marketing of farm equipment, fertilizers, and pesticides.

By the middle of the twentieth century, advances in technology made it possible to pump water from rivers and aquifers and deliver it to elevations beyond the reach of earlier systems. Irrigation systems featuring automated sprinkling also became widespread. Toward the end of the century, irrigated farming became increasingly common in the southeastern United States. The number of large, corporate farms increased significantly, while small, family farms and farm jobs decreased. People also began to express concern about the long-term negative consequences of irrigation: salinization of soils, damage to riverine ecology by dams, and concentration of pollutants caused by irrigation runoff. In 2003, barely a century after the creation of the Bureau of Reclamation, nearly 53 million acres of the United States were under irrigation.

Further Reading

Pisani, Donald J. *Water and American Government: The Reclamation Bureau, National Water Policy, and the West, 1902–1935.* Berkeley: University of California Press, 2002.

Reisner, Marc, and Lawrie Mott. *Cadillac Desert: The American West and Its Disappearing Water.* New York, N.Y: Penguin Books, 2017. Print.

Rowley, William D. *The Bureau of Reclamation: Origins and Growth to 1945, Volume 1: Reclamation—Managing Water in the West.* Denver, Colo.: Bureau of Reclamation, U.S. Department of the Interior, 2006.

Isolating Mechanisms in Evolution

—John S. Mecham

FIELDS OF STUDY

Anthropology; Biology; Evolution; History.

PRINCIPAL TERMS

allozyme: one of two or more forms of an enzyme determined by different alleles of the same gene; usually analyzed by gel electrophoresis

chloroplast deoxyribonucleic acid: a circular DNA molecule found in chloroplasts; chloroplasts are cytoplasmic organelles of green plants and some protists that carry out photosynthesis

fission: the division of an organism into two or more essentially identical organisms; an asexual process

hermaphrodite: an individual with both male and female organs; functions as both male and female

hybrid: a term that in the broad sense can apply to any offspring produced by parents that differ in one or more inheritable characteristics; used to denote offspring produced by a cross between different species

mitochondrial deoxyribonucleic acid (DNA): a circular molecule of DNA found in the mitochondria; mitochondria are cytoplasmic organelles that function in oxidative respiration

nuclear ribosomal deoxyribonucleic acid (DNA): nuclear DNA that codes for the ribosomal DNAs; ribosomes are small cytoplasmic particles that function in protein synthesis

SUMMARY

Isolating mechanisms act to prevent interbreeding and the exchange of genes between species. The establishment of isolating mechanisms between populations is a critical step in the formation of new species.

Isolating mechanisms (reproductive isolating mechanisms) prevent interbreeding between species. The term, which was first used by Theodosius Dobzhansky in 1937 in his landmark book *Genetics and the Origin of Species*, refers to mechanisms that are genetically influenced and intrinsic. Geographic isolation can prevent interbreeding between populations, but it is an extrinsic factor and therefore does not qualify as an isolating mechanism. Isolating mechanisms function only between sexually reproducing species. They have no applicability to forms that reproduce only by asexual means, as by mitotic fission, stoloniferous or vegetative reproduction, or egg development without fertilization (parthenogenesis in animals). Obligatory self-fertilization in hermaphrodites (rare in animals) is a distortion of the sexual process that produces essentially the same results as asexual reproduction. Many lower animals and protists regularly employ both asexual and sexual means of reproduction, and the significance of isolating mechanisms in such forms is essentially the same as in normal sexual species.

PREMATING MECHANISMS

Reproductive isolating mechanisms are usually classified into two main groups. Premating (prezygotic) mechanisms operate prior to mating, or the release of gametes, and, therefore, do not result in a wastage of the reproductive potential of the individual. Postmating (postzygotic) mechanisms come into play after mating, or the release of gametes, and could result in a loss of the genetic contribution of the individual to the next generation. This distinction is also important in the theoretical sense in that natural selection should favor genes that promote premating isolation; those that do not presumably would be lost more often through mismatings (assuming that hybrids are not produced, or are sterile or inferior), and this could lead to a reinforcement of premating isolation.

Ethological (behavioral) isolation is the most important category of premating isolation in animals. The selection of a mate and the mating process depends upon the response of both partners to various sensory cues, any one of which may be species-specific. Although one kind of sensory stimulus may be emphasized, different cues may come into play at different stages of the pairing process. Visual signals provided by color, pattern, or method of display are often of particular importance in diurnal animals such as birds, many lizards, certain spiders, and fish. Sounds, as in male mating calls, are often important in nocturnal breeders such as crickets or frogs but are also important in birds. Mate discrimination based on chemical signals or odors (pheromones) is of fundamental importance in many different kinds of animals, especially those where visual cues or sound are not emphasized; chemical cues also are often important in aquatic animals with external fertilization. Tactile stimuli (touch) often play an important role in courtship once contact is established between the sexes. Even electrical signals appear to be utilized in some electrogenic fish.

Ecological (habitat) isolation often plays an important role. Different forms may be adapted to different habitats in the same general area and may meet only infrequently at the time of reproduction. One species of deer mouse, for example, may frequent woods, while another is found in old fields; one fish species spawns in riffles, while another spawns in still pools. This type of isolation, although frequent and widespread, is often incomplete as the different forms may come together in transitional habitats. The importance of ecological isolation, however, is attested by the fact that instances in which hybrid swarms are produced between forms that normally remain distinct have often been found to be the result of disruption of the environment, usually by humans. Mechanical isolation is a less-important type of premating isolation, but it can function in some combinations. Two related animal species, for example, may be mismatched because of differences in size, proportions, or structure of genitalia.

Finally, temporal differences often contribute to premating isolation. The commonest type of temporal isolation is seasonal isolation: Species may reproduce at different times of the year. A species of toad in the eastern United States, for example, breeds in the early spring, while a related species breeds in the late spring, with only a short period of overlap. Differences can also involve the time of day, whereby one species may mate at night and another during the day. Such differences, as in the case of ecological isolation, are often incomplete but may be an important component of premating isolation.

POSTMATING MECHANISMS

If premating mechanisms fail, postmating mechanisms can come into play. If gametes are released, there still may be a failure of fertilization (intersterility). Spermatozoa may fail to penetrate the egg, or even with penetration there may be no fusion of the egg and sperm nucleus. Fertilization failure is almost universal between remotely related species (as from different families or above) and occasionally occurs even between closely related forms.

If fertilization does take place, other postmating mechanisms may operate. The hybrid may be inviable (F1 or zygotic inviability). Embryonic development may be abnormal, and the embryo may die at some stage, or the offspring may be defective. In other cases, development may be essentially normal, but the hybrid may be ill-adapted to survive in any available habitat or cannot compete for a mate (hybrid adaptive inferiority). Even if hybrids are produced, they may be partially to totally sterile (hybrid sterility). Hybrids between closely related forms are more likely to be fertile than those between more distantly related species, but the correlation is an inexact one. The causes for hybrid sterility are complex and can involve genetic factors, differences in gene arrangements on the chromosomes that disrupt normal chromosomal pairing and segregation at meiosis, and incompatibilities between cytoplasmic factors and the chromosomes. If the hybrids are fertile and interbreed or backcross to one of the parental forms, a more subtle phenomenon known as hybrid breakdown sometimes occurs. It takes the form of reduced fertility or reduced viability in the offspring. The basis for hybrid breakdown is poorly understood but may result from an imbalance of gene complexes contributed by the two species.

It should be emphasized that in most cases of reproductive isolation that have been carefully studied, more than one kind of isolating mechanism has been found to be present. Even though one type is clearly of paramount importance, it is usually supplemented by others, and should it fail, others may come into play. In this sense, reproductive isolation can be viewed as a fail-safe system. A striking difference in the overall pattern of reproductive isolation between animals and plants, however, is the much greater importance of premating isolation in animals and the emphasis on postmating mechanisms in plants. Ethological isolation, taken together with other premating mechanisms, is highly effective in animals, and postmating factors usually function only as a last resort.

FIELD STUDIES & EXPERIMENTAL STUDIES

Field studies have often been employed in the investigation of some types of premating isolating mechanisms. Differences in such things as breeding times, factors associated with onset of breeding activity, and differences in habitat distribution or selection of a breeding site are all subject to direct field observation. Comparative studies of courtship behavior in the field or laboratory often provide clues as to the types of sensory signals that may be important in the separation of related species.

Mating discrimination experiments carried on in the laboratory have often been employed to provide more precise information on the role played by different odors, colors, or patterns, courtship rituals, or sounds in mate selection. Certain pheromones, for example, which act as sexual attractants, have been shown to be highly species-specific in some insects. The presence or absence of certain colors or their presentation has been shown experimentally to be important in mate discrimination in vertebrates as diverse as fish, lizards, and birds. Call discrimination experiments, in which a receptive female is given a choice between recorded calls of males of her own and another species, have demonstrated the critical importance of mating call differences in reproductive isolation in frogs and toads. Synthetically generated calls have sometimes been used to pinpoint the precise call component responsible for the difference in response.

Studies on postmating isolating mechanisms have most often involved laboratory crosses in which the degree of intersterility, hybrid sterility, or hybrid inviability can be analyzed under controlled conditions. In instances in which artificial crosses are not feasible, natural hybrids sometimes occur and can be tested. The identification of natural backcross products can attest incomplete postmating, as well as premating isolation. Instances of extensive natural hybridization are of special interest and have often been subjected to particularly close scrutiny. Such cases often throw light on factors that can lead to a breakdown of reproductive isolation. Also, as natural hybridization more often occurs between marginally differentiated forms in earlier stages of speciation, new insights into the process of species formation can sometimes be obtained. Finally, such studies may yield information on the evolutionary role of hybridization, including introgressive hybridization, the leakage of genes from one species into another. Morphological analysis has long been used in such cases, and chromosomal studies are sometimes appropriate. In recent years, allozyme analysis by gel electrophoresis has become a routine tool in estimates of gene exchange, and molecular analysis of nuclear deoxyribonucleic acid (DNA), or mitochondrial DNA, have been useful. As mitochondria are normally passed on only maternally, their DNA can also be used to identify cases in which females of only one of the two species has been involved in the breakdown of reproductive isolation.

Investigations of the role of natural selection in the development and reinforcement of reproductive isolation have employed two different approaches. One has involved the measurement of geographic variation in the degree of difference in some signal character (call, color, or pattern, for example) thought to function in premating isolation between two species that have overlapping ranges. If the difference is consistently greater within the zone of overlap (reproductive character displacement), an argument can be made for the operation of reinforcement. Another approach has involved laboratory simulations, usually with the fruit fly *Drosophila*, in which some type of selective pressure is exerted against offspring produced by crosses between different stocks, and measurement is made of the frequency of mismatings through successive generations. The results of such studies to this time are contradictory, and the role of selection with regard to development of reproductive isolation requires further study.

ENHANCING REPRODUCTIVE EFFICIENCY

The efficiency of reproduction in most animals is enhanced immeasurably by premating isolating mechanisms. Clearly, in animals a random testing of potential mates without regard to type is totally unacceptable for most species in terms of reproductive capacity and time and energy resources. Premating isolation in this sense is a major factor in promoting species diversity in animal communities.

Both premating and postmating isolating mechanisms are also critical to the maintenance of species diversity in that they act to protect the genetic integrity of each form: A species cannot maintain its identity without barriers that prevent the free exchange of genes with other species. Furthermore, a species functions as the primary unit of adaptation. Every species in a community has its own unique combination of adaptive features that enable it to exploit the resources of its environment and to coexist with other species with a minimum of competition. The diversity of different species that can coexist in the same area depends upon the unique "niche" that each occupies; adaptive features that determine that niche are based on the unique genetic constitution of each species, and this genetic constitution is protected through reproductive isolation.

The development of reproductive isolating mechanisms is also critical to the formation of new species (speciation), and ultimately to the development of new organic diversity. The most widely accepted, objective, and theoretically operational concept for a sexual species is the biological species concept. Such a species can be defined as population or group of populations, members of which are potentially capable of interbreeding but which are reproductively isolated from other species. The origin of new species, therefore, depends upon the development of reproductive isolating mechanisms between populations. A major focus of research in evolutionary biology and systematics has been, and continues to be, on the various factors that influence the development of reproductive isolating mechanisms.

FURTHER READING

Baker, Jeffrey J. W., and Garland E. Allen. *The Study of Biology*. 4th ed. Reading, Mass.: Addison-Wesley, 1982. Some of the more complete college-level biology texts provide a good overview of evolution and related topics for the beginner. Coverage in this text is particularly complete, including treatment of isolating mechanisms and speciation. A list of suggested readings should be helpful.

Dobzhansky, Theodosius. *Genetics of the Evolutionary Process*. New York: Columbia University Press, 1970. Dobzhansky's contributions over the years have had a major impact on genetics and evolutionary biology. The present book represents a revision and expansion of his *Genetics and the Origin of Species*, which first appeared in 1937. The book is directed at the advanced reader and presupposes a foundation in genetics. Much of chapter 10, however, which is devoted to an excellent review of reproductive isolation, can be understood by the more general reader. An extensive technical bibliography is provided.

Dobzhansky, Theodosius, Francisco J. Ayala, G. Ledyard Stebbins, and James W. Valentine. *Evolutionary Biology*. Springer Verlag, 2014. Print. An excellent, if somewhat uneven textbook by four distinguished scientists. Although intended for advanced undergraduates and graduate students, the material is presented in such a way as to benefit those with a less complete background. Coverage of molecular techniques is particularly good but somewhat dated. Twelve pages are devoted to isolating mechanisms, and many references are given under literature cited.

Futuyma, Douglas J. *Evolutionary Biology*. New York: W.H. Freeman, 2006. Print. This process-oriented textbook assumes some undergraduate training in biology, especially genetics, but most topics are prefaced with background information that should help the less prepared. Treatment of isolating mechanisms is brief, but the book provides one of the best introductions to current evolution theory presently available. Major references are given at the end of each chapter, and there is an extensive list of references under literature cited.

Mayr, Ernst. *Populations, Species, and Evolution: An Abridgment of "animal Species and Evolution"*.

Belknap Press of Harvard U.P., 1977. Print.This text is essentially an abridgment of Mayr's *Animal Species and Evolution* (1963). Ernst Mayr has been a prime mover and shaker in the development of evolution theory over the past fifty years, and anyone with a serious interest in the subject should be exposed to some of his writings. This book, although directed toward the more advanced student, is lucidly written and can benefit those with a more limited background. Chapter 5 deals with isolating mechanisms, particularly in animals. A technical bibliography is provided.

L

Land-use Planning

FIELDS OF STUDY

Environmental Sciences; Environmental Studies; Physical Geography; Hydrology; Soil Science.

PRINCIPAL TERMS

horizon: a layer of soil material approximately parallel to the surface of the land that differs from adjacent related layers in physical, chemical, and biological properties

leaching: the dissolving out or removal of soluble materials from a soil horizon by percolating water

sediment: rock fragments such as clay, silt, sand, gravel, and cobbles

soil profile: a vertical section of a soil, extending through its horizons into the unweathered parent material

weathering: the mechanical disintegration and chemical decomposition of rocks and sediments

Scientists are finding that soils can be useful tools in the interpretation of past environments of an area and also in the formulation of land-use plans for that region. First, the soil scientists must determine the exact soils that are in the area through a mapping program. During this project, they determine the regional developmental sequence of soils—that is, the poorly developed and the well-developed soils of the different places on the landscape. They assign ages to each soil in the sequence using radiocarbon-dated samples from the soils. As a result, when a particular soil is located in that region and compared with the sequence, an age can be assigned to the land surface upon which the soil developed. Geologists commonly use soils to estimate the ages of deposits, especially on glacial landscapes. The idea of multiple glaciations has been developed through the use of soils. Poorly developed soil on glacial sediments represents a recent glaciation compared with a site that features well-developed soil lying on sediment deposited by an older glacial advance.

The frequency of geological hazard events such as landslides, rockfalls, and flooding can be estimated by the study of buried soils in sediment deposits. An ancient, buried A horizon within a soil profile means that some geological event covered an old land surface with sediment. If the organic matter of that buried A horizon is dated using radiocarbon dating, an approximate time of burial can be determined. If a valley bottom has many buried soils, it probably means that the nearby stream floods quite frequently. These decisions are important for land-use planning, for one does not want to build in an area where hazards occur frequently. For example, nuclear power plants must not be located in areas where there are active faults (ground breaks). Soils are commonly used to determine if faults are active in an area and, if so, the frequency of movement on the fault.

The U.S. Department of Agriculture produces maps of soils across the country. The maps describe the productivity of crop growth and manageability related to land use of each soil. These data help farmers grow crops that are best suited for their soils; they tell the farmers which soils are most susceptible to erosion so that extra precautions can be taken with the fragile soils. The data also help land-use planners keep the most productive soils in agriculture instead of paving them. In addition, the maps help to predict how much fertilizer to put on the fields. Natural soil fertility is related to geological activity and weathering, as the oldest soils are leached of most of their nutrients and are least fertile. Young soils have had little weathering and leaching and are most fertile, especially those on floodplains, young glacial deposits, and volcanic deposits. Soil maps are also used to recognize where shrink-swell clays occur in the soils. The presence of shrink-swell clays in the soil is important for home owners. If present, these clays can crack the foundation of a house. To prevent these damages, special foundations must be constructed and the clays stabilized with the use of lime.

Further Reading

Birkeland, P. W. *Soils and Geomorphology.* 3d ed. New York: Oxford University Press, 1999. One of the best books on the subject written from the perspective of a geologist. Offers many examples from around the world. Covers soil formation and development. Discusses the applications of soil formation to geology in the final chapter. Suitable for advanced high school and university-level readers.

Buol, S W. *Soil Genesis and Classification.* Ames: Wiley-Blackwell, 2011. Print. An update of a classic soil-science text. Features an update on the new soil horizon symbol nomenclature. Includes an in-depth discussion on soil development. Suitable for university-level readers.

Gerard, A. J. *Soils and Landforms.* London: Allen & Unwin, 1981. Covers all factors of soil formation, emphasizing the topographic factor. Excellent section on catenas (distribution of soils on a slope). Best suited to university-level readers.

Jenny, H. *Factors of Soil Formation: A System of Quantitative Pedology.* New York: Dover, 1994. Expands on the five factors of soil formation and development. Aimed at university-level readers. Illustrations, bibliography, and index.

__________. *The Soil Resource: Origin and Behavior.* New York: Springer-Verlag, 1980. Rewrites Jenny's classic book, written in 1941. Expands on the five factors of soil formation and development. Aimed at university-level readers.

Paton, T. R., G. S. Humphreys, and P. B. Mitchell. *Soils: A New Global View.* London: CRC Press, 2018. Print. Aimed at readers at the university level. Covers good examples of soils and soil formation from around the world. Color illustrations, maps, bibliography, and index.

Retallack, G. J *Soils of the Past.* 2d ed. New York: Blackwell Science, 2001. Offers information on paleosols. Discusses soil classification and formation, identifying paleosols and their characteristics, and factors involved in paleosol formation. Cover the soil-fossil records with discussion of ancient landscapes and ecology.

Schaetzl, Randall J., and Sharon Anderson. *Soils: Genesis and Geomorphology.* New York: Cambridge University Press, 2005. Provides a large amount of fundamental content. Discusses soil morphology and classification, the formation of soil, other soil dynamics, paleosol characteristics, and the reconstruction of environments through the study of paleosols.

Simonson, R. W. "Outline of a Generalized Theory of Soil Genesis." *Soil Science Society of America Proceedings* 23 (1959): 152–156. A classic paper in which the author outlines the four types of processes involved in soil formation.

Singer, M. J., and D. N. Munns. *Soils.* 6th ed. Upper Saddle River, N.J.: Prentice Hall, 2005. One of the most readable elementary soil science texts available. Includes a good basic section on soil formation. Appropriate for high school-level readers.

Soil Science Society of America. *Glossary of Soil Science Terms.* Madison, Wis.: Author, 2001. Defines soil science terms; written by soil science professionals.

U. S. Department of Agriculture, Soil Survey Staff. *Keys to Soil Taxonomy.* Honolulu: University Press of the Pacific, 2005. Contains the most used soil classification system in the world. A challenging read but an abundant source of information.

Van Breemen, Nico, and Peter Buurman. *Soil Formation.* 2d ed. Boston: Kluwer Academic Publishers, 2002. Examines all stages and processes of soil formation. Illustrations and diagrams help to clarify difficult concepts. Intended for college students, but useful for the layperson.

Walther, John Victor. *Essentials of Geochemistry.* 2d ed. Burlington, Mass.: Jones & Bartlett Publishers, 2008. Contains chapters on radioisotope and stable isotope dating and radioactive decay. Geared toward geology and geophysics rather than chemistry. Covers thermodynamics, soil formation, and chemical kinetics.

LIMING (SOIL)

—*Richard S. Spira*

Liming is the addition of calcium and magnesium to soil to neutralize acidity and increase the activity of soil bacteria. The amount of lime needed to reach the desired acidity depends on the pH and the buffering capacity of the soil. The pH is a measurement of acidity or alkalinity. Buffering capacity refers to the soil's ability to resist change to its pH. Additionally, liming improves nutrient efficiency by encouraging fertilizer uptake. However, there are important limits to liming, as oversupply may cause harm to plant life. The most common liming materials come from grinding natural limestone. Limestone is composed mostly of calcium carbonate (CaCO3) and by most state laws must contain 6 percent magnesium for purposes of liming. In addition to buffering the soil, the main way to reverse acidification in freshwater is through liming the water body or its surrounding drainage basin.

Carbon dioxide (CO_2) does not absorb the Sun's energy, but it does absorb the heat energy released from the Earth. Some of the released heat energy returns to Earth and some goes out into space. This describes the greenhouse effect of CO_2; that is, CO_2 lets light energy in, but not all of the heat energy can get out. Soil liming can act as either a source or a sink for CO_2. A heat sink absorbs and dissipates the heat energy.

Some studies indicate that rather than all of the carbon in lime becoming CO_2, the lime may sequester up to half of its carbon content, thus becoming an effective carbon sink. If this is so, liming might even be part of the strategy to moderate climate change. Streams draining agricultural watersheds generally show a net CO_2 uptake. However, as nitrate (a common fertilizer ingredient) concentrations increase, lime may switch from a net CO_2 sink to a CO_2 source.

A John Deere tractor with a JD front loader and a trailed Acuspread AS85 multi-purpose spreader lime on a Devon field, England, United Kingdom. Image by Mark Robinson (originally posted to Flickr as Spreading Lime) CC-BY-2.0.

Peat soil contains large amounts of carbon and therefore is a significant emission source of greenhouse gases—especially of methane. When farming acidic peat lands, adjustment of the acidity through liming is essential. However, when adding lime to these soils often rich in carbon and methane, CO_2 production increases. In addition, because of the enhanced decomposition of organic matter through liming, soluble organic carbon increases and may factor into climate change.

Finally, it is important to note that the amount of CO_2 emissions from liming is not completely clear. This is because the amount of carbonate lime applied to soils is unclear, as is the net amount of carbon from liming that is released as CO_2.

LIVESTOCK & ANIMAL HUSBANDRY

—*James L. Robinson, PhD*

FIELDS OF STUDY

Apiculture; Agriculture; Biology; Zoology; Physical Science; Statistics; Food Production; Animal Science; Ecology.

SUMMARY

Animal husbandry refers to the management of domesticated animals such as beef or dairy cattle, sheep, goats, pigs, and chickens: livestock. Such animals constitute a renewable resource providing humans with food, fiber, fuel, power, implements, and other benefits.

BACKGROUND

Effective animal husbandry requires an affinity for the animals being managed, skill in handling them, and knowledge of them and their environment. Respect for animals is important to good management, as is skill in handling to minimize injuries and stress to both animal and handler. Knowledge is needed of their nutrition, reproduction, and behavior as well as the physical, biological, cultural, and economic context in which they are managed. While some inputs (such as aberrant weather and governmental regulations) are beyond the control of the producer, good management will ensure the most efficient productivity from the available inputs.

Meat: Leading Producers, 2006

	Metric Tons
World total	**212,776,000**
Beef and veal	
United States	11,981,000
Brazil	9,020,000
European Union	8,060,000
China	7,492,000
Argentina	3,100,000
India	2,375,000
Australia	2,183,000
Mexico	2,175,000
Russia	1,430,000
Canada	1,391,000
Pork	
China	51,972,000
European Union	21,677,000
United States	9,559,000
Brazil	2,830,000
Canada	1,898,000
Russia	1,805,000
Vietnam	1,713,000
Japan	1,247,000
Philippines	1,215,000
Mexico	1,200,000
Poultry	
United States	16,043,000
China	10,350,000
Brazil	9,355,000
European Union	7,803,000
Mexico	2,592,000
India	2,000,000
Japan	1,227,000
Argentina	1,200,000
Russia	1,180,000
Thailand	1,100,000

Source: U.S. Department of Agriculture, National Agricultural Statistics Service, *Agricultural Statistics*, 2007.

Meat: Leading Producers, 2006. Image by U.S. Department of Agriculture, National Agricultural Statistics Service, *Agrigultureal Statistics*, 2007.

INTENSIVE & EXTENSIVE MANAGEMENT

Intensive and extensive management are the two main options for animal husbandry. Intensive management refers to confinement-type operations that provide animals with shelter, food, and water. It has been called "landless" because it requires very little property. Examples include beef feedlots, concentrate-based dairy farms, and confinement swine or poultry operations. In extensive systems, on the other hand, the animals are provided with an area in which they fend for themselves, finding their own food, water, and shelter. Examples are rangeland beef operations, pasture-based dairying, and free-range poultry farms. In practice, animal husbandry often includes both intensive and extensive management.

In the early twenty-first century, the U.S. beef industry generally involved extensive operations for at least the first year of life and an intensive phase just prior to market; availability and prices of feed grains may determine the extent to which intensive management is practiced. Dairy operations around the world range from intensive to extensive—from no to exclusive pasture, respectively. Seasonal variation of pasture may dictate when it is available and used. Because dairy cows must be milked two or three times a day, dairy operations are never as extensive as some beef operations, where the producer may have contact with the animals no more than once a year.

Intensive animal management generally requires more management expertise, more capital investment, and more energy utilization. Since the animal is totally under control of the producer, all needs of the animal must be provided. The inevitably greater concentration of animals requires closer attention to their housing and health. The larger capital investment is attributable to facilities and equipment. More energy utilization is needed to maintain temperature and ventilation as well as to operate equipment. Intensive management also places greater emphasis on maximizing animal performance. Because more capital and energy are used, effort is made to extend animal performance by genetics, nutrition, and other management tools. Intensive managment also requires more dependence on others for feed. While some intensive livestock producers raise their own feedstuff, many do not. They may depend on crop farmers within the region or half a world away. Contemporary swine operations in Japan and Korea require corn and soybeans from the U.S. Midwest.

Extensive animal management demands more land and more dependence on the animals' abilities than intensive management. The larger land requirement is a primary feature of this system. The greater dependence on the animals' abilities follows from less direct provision by the producer for their needs. Survival and growth may depend on their locating food, water, and shelter as well as avoiding danger. Reproduction may be left to natural service, easy birthing, and good mothering. Extensive management involves more tolerance for decreased animal performance. When weather conditions do not provide sufficient food, the animals will have less than maximal growth and fertility. Neonatal losses attributed to weather, predators, or terrain are tolerated. Indeed, human intervention may not be a realistic option when animals are widely dispersed. An important parameter is the "stocking rate," the number of animals per land area. Too few animals will not fully use the vegetation, as many grasses are most nutritious at an early stage of development and become less nutritious and coarser if not eaten then. Too many animals will overgraze, impairing regrowth of the vegetation. Optimum "stocking rate" corresponds closely to the ecological concept "carrying capacity," the number of animals that an area can sustain over an extended period of time. Extensive systems can demand substantial management expertise. For instance, pasture-based dairying in New Zealand requires considerable knowledge to optimize pasture growth and utilization.

BIOLOGICAL & NONBIOLOGICAL PARAMETERS

Any animal management system must take into account numerous biological parameters pertinent to the animal under management. These include nutritional requirements, biological time lag (time from conception to market), reproduction (gestation length and number of newborn, newborn survival), efficiency of feed conversion, nature of weight gain, genetic selection, and susceptibility to disease. Decisions are made about using natural service or artificial insemination. The extent to which agricultural by-products, crop residues, and/or production enhancers are used depends on their efficacy, availability, and price.

Any animal management system also involves a number of nonbiological parameters. The available climate, water supply, and land are physical attributes that bear upon the husbandry options. Two other facets of the land affecting management are its tenure, whether owned, leased, or occupied, and its use, whether restricted or not. Husbandry is also affected by the availability and skill level of labor. Another factor is the infrastructure—the dependability of transportation providing access to markets, postfarm processing, and communication systems. Profitability, the difference between receipts and cost of inputs, as well as any subsidies, determines whether one can engage in any agricultural activity for long. Personal values, including lifestyle and risk management, also impact involvement in animal agriculture. Finally, historical and societal values, particularly those directly touching on the use of animals and natural resources, influence the extent and nature of animal husbandry.

ISSUES

Three issues of contemporary interest relative to livestock and animal husbandry concern the need for animal agriculture, its sustainability, and its increasing corporate nature. The willingness of people to purchase and consume products of animal origin will always determine the need for animal agriculture. If the price people must pay for such products is too high, demand will decline. As the general affluence of a country increases, the demand for foods of animal origin increases.

The sustainability of contemporary agriculture has been called into question because of its heavy dependence on fossil fuels for energy and its adverse effects on the environment. Properly managed, animals have a role to play in sustainable agriculture. They can help dispose of some agribusiness by-products—crop residues and crops not suitable for human consumption—and generate waste that can be used to fertilize crops.

Animal agriculture is increasingly conducted by corporations rather than by family-owned farms or ranches. Once farming moves away from subsistence farming and generates excess over what the farm family needs, it becomes a business. The pressure for efficiency, as well as for high and consistent product quality, is driving animal agriculture toward increasingly specialized and integrated enterprises. While this tendency appears to be inevitable, serious concerns arise concerning the oligopolies, if not monopolies, that may control the production of animal products and the management of domestic animals, a valued renewable resource.

Further Reading

Campbell, John R., M. Douglas Kenealy, and Karen L. Campbell. *Animal Sciences: The Biology, Care, and Production of Domestic Animals.* 4th ed. Boston: McGraw-Hill, 2003.

Campbell, Karen L., and John R. Campbell. *Companion Animals: Their Biology, Care, Health, and Management.* 2d ed. Upper Saddle River, N.J.: Pearson Prentice Hall, 2009.

Cheeke, Peter R. *Contemporary Issues in Animal Agriculture.* Upper Saddle River: N.J, 2004. Print.

Ensminger, M. Eugene. *The Stockman's Handbook.* 7th ed. Danville, Ill.: Interstate, 1992.

Field, Thomas G., and Robert E. Taylor. *Scientific Farm Animal Production: An Introduction to Animal Science.* 9th ed. Upper Saddle River, N.J.: Prentice Hall, 2008.

Gillespie, James R., and Frank Flanders. *Modern Livestock and Poultry Production.* 8th ed. Clifton Park, N.Y.: Delmar Cengage Learning, 2009.

Shapiro, Leland. *Introduction to Animal Science.* Upper Saddle River, N.J.: Prentice Hall, 2001.

Logging Industry

—Kymberley A. Snarr

FIELDS OF STUDY

Ecology; Ecosystem Management; Climatology; Bioclimatology; Environmental Sciences; Tree Ring Analysis.

SUMMARY

Since the rise of civilization, humans have been logging forests and affecting forest ecosystems. Before the period when Europeans arrived in North America, logging around the world tended to exert a more local effect—although evidence indicates that the downfalls of various earlier dominant civilizations came about in part from the ecological effects of land-use practices, of which forest extraction and depletion was a component.

With the North American logging history as a clean-slate example, it was during the colonial period that the logging industry began to systematically harvest old-growth forests to supply the settlers with cleared land for agriculture, and with lumber for construction and their continued expansion by land and sea. The Industrial Revolution brought mechanization to extraction, and an increased rate of harvest, which began to allow transnational logging companies to supply the ever-increasing global market demands. This expansion has continued in regions around the world, and now has resulted in a reduction of nearly half of the Earth's original forest cover—with most of the known loss occurring since the late 1970s, mainly in the tropical forests.

The logging industry has become more regulated by governments keen to reduce forest ecosystem degradation and loss. Biotechnology has fueled the development of hybrid tree species grown in plantations in tropical zones; these have become a key part of the logging industry's approach to future harvests. Harvest approaches in some cases have been altered to improve sustainability, with third-party certification as part of the sustainable approach. The attention of governments, nongovernmental organizations, and the industry itself has turned to the problem of illegal logging, which continues to be an additional factor in deforestation and forest habitat degradation.

To ensure a robust future in the global economy, the logging industry is likely to continue to use tropical and temperate forest plantations, selective

U.S. Fish and Wildlife Service staffers measuring selectively cut logs on a truck. Even though many countries now enforce logging regulations, about one-third of the world's forests have been lost since 1990. Image by U.S. Fish and Wildlife Service.

logging rotations, and to commoditize waste as renewable fuel; the industry will also negotiate new pathways for sustainability.

EARLY LOGGING

In the earliest of human times, forest extractions were minimal, with a small amount used for housing structures and early forms of work implements. Approximately 5,000 years ago, with the rise of cities and states, increased logging delivered needed wood supplies as civilizations grew. Large ecological effects on local forests were seen due to more concentrated and higher rates of logging. Evidence indicates that several earlier civilizations, such as the Mayans, rose and fell in part due to the destructive effects of localized logging.

Up until 500 years ago, most logged material was used within the region in which it was logged. This situation changed during the American colonial period, when North American settlers began to log the expansive stands of old-growth pine forests in their new homeland. From the 1700s, lumber exports outweighed domestic demand for products including construction materials, white pine for lumber, masts for ship building, and pulp for paper, all mainly to markets in Europe.

The technology of the logging industry during this time period was basic. Axes and the crosscut saw were used to cut trees in the forests. The feller cut down the tree, the bucker cut the log into manageable sizes, and men with broad axes squared the timber at the stump. Oxen and horses pulled the skid with squared timbers to the water routes, and the water routes floated these squared timbers to ships, which transported them to colonial markets.

By the 1800s, sawmills became commonplace, and logs were no longer squared in the bush but were cut to the length dictated by the market. Waste in the logging industry during this period was high, often with two-thirds of each tree being left behind.

At the beginning of the logging industry, loggers were independent operators who worked in difficult conditions. They lived in remote camps, in small shanties constructed from logs, away from families, friends, and basic services. The hours were long, and the work was dangerous, with few safety measures put into place until the mid-twentieth century.

As capital costs increased, commercial operations moved into the hands of larger companies and out of the hands of individuals. By the late eighteenth century, the bigger logging companies were purchasing large tracts of land from the government and harvesting the standing timber. The work was carried out mainly by jobbers, or independent operators, on contract to the company. Some of the pioneers in this industry were Weyerhaeuser, Abitibi, McMillian Bloedel, and Domtar.

INDUSTRIAL REVOLUTION TO TODAY

The Industrial Revolution brought heavier technology and mechanized methods to the logging industry, applied to cutting, extraction, skidding, milling, and transportation. By the early twentieth century, railroads improved flow of product from the bush to sawmills and markets—which increased pressure on the forests. Today, mechanization includes the use of helicopters, tractors, cables, and specialized equipment such as feller-bunchers and forwarders to remove logs from the forest. Traditional approaches to logging, using axes, machetes, or chainsaws; non-mechanized skidding; and water and/or animal power make up a very small percentage of global logging today.

During the late nineteenth and early twentieth centuries, the logging industry moved into full-scale mass production. Lumber and wood-based pulp and paper production was found in Russia, Scandinavia, Europe, and North America. In the 1950s, the established producer countries' products were sold on a wider world market, meeting demands for wood products, especially in developing countries in Asia and Latin America. Traditionally, logging has had global markets, but full globalization of the logging industry was pioneered after World War II by such national economies as Japan, which sought raw materials from around the globe. Today, the logging industry is important to the national economies in every quarter of the world.

With diminishing northern forests to harvest from and increasing regulations, many large logging companies turned their attention to logging transnationally in the 1950s. This initiative began with a move into Asian countries, which had lower labor costs. By the late 1990s, these transnationals and many of the large new Asian logging companies had depleted much of the Asian forests' stock and turned to logging tropical forests in Africa, the Amazon River basin, Burma, and Indonesia, where corruption makes forestry laws nearly unenforceable and a lack of transparency in commercial transactions gives little regard to the environment or local peoples.

The logging industry is a major global industrial sector; gross production accounted for $160 billion worldwide in 1998. The industry is projected to grow to $300 billion by 2020. Today, two-thirds of the world's forests are located in 10 countries: the Russian Federation, Brazil, Canada, the United States, China, Australia, the Democratic Republic of Congo, Indonesia, Angola, and Peru. There are 9.6 billion acres (3.9 billion hectares) of forests, of which 95 percent are natural forests and 5 percent are forest plantations. Forests cover about 30 percent of the world's land area. Nearly 44 percent of forests are located in the tropics, and about one-third in boreal regions, with temperate forests accounting for 13 percent, and subtropical forests amounting to nine percent.

ECOLOGICAL EFFECTS & CONSERVATION

While the logging industry is economically important, the ecological effects of logging can be devastating. Almost half of the Earth's original forest cover is gone, much of it destroyed after 1970. Since 1990, about one-third of forest around the world has been lost, with the vast majority of this loss occurring in the tropics. Logging for wood products is responsible for one-third of total global deforestation; other factors are land-clearing for agriculture, urban development, and road-building.

The early industrial approach to harvesting forests mainly involved clear-cutting with no reforestation. This practice created ecological devastation leading to erosion, desertification and local droughts, and forest-composition changes, all of which resulted in reduction of habitat for many flora and fauna species. It became clear that the forests were in many ways worth more than the logging products that could be extracted from them. Stakeholders outside the logging industry began to see the value of forests beyond the extractable forest products.

A deeper understanding of key ecological services provided by forests included carbon storage, regulation of river flows and sedimentation, habitat support, non-timber forest products (mushrooms, medicinal plants, craft woods, etc.), and a variety of aesthetic and cultural benefits. In many counties, logging companies must now follow regulations and management practices that carry out integrated resource management (IRM) plans.

During the past 150 years, pioneering individuals brought key legislation and approaches to the logging industry to help reduce the devastating effects of the industry and aid in sustainability. Before the 1950s, the logging industry was concentrated in temperate and boreal forests, and coping with the effects of unsustainable approaches to logging in these forests was key to the sustainability of the industry and the ecoservices provided by these forests. The earliest adoption of forest regulations and conservation was in California in the 1850s.

Conservation of some forested areas was implemented when sets of protected areas, such as national parks in the United States and Canada, were established. The logging companies began to see that the seemingly vast expanses of forest were not everlasting. They began to buy up forested land and moved to using harvest rotations in these temperate northern forests. By the mid-twentieth century, they used reforestation, silviculture, and plantations to ensure that product was available to be logged in the future.

Unfortunately, strategies such as reforestation and plantations were slow in terms of growth in the north. Northern forests containing jack, pine, and black spruce take 80 to 100 years to grow to maturity. Thus, some of the largest logging companies turned their attention to tropical areas for harvest and for plantation. Plantations located in tropical countries, such as those in Brazil and Indonesia, are likely to become much more prominent, as hardwoods are easier to plant here and grow at rapid rates.

Along with the logging industry, commercial banks, the United Nations, and other international bodies have seen plantations as a possible antidote to deforestation; they have largely supported this approach to a conservation-and-industry partnership. Biotechnology and hybridization have brought diverse commercial species to the logging industry to replace northern pulp and commercial timbers. One example in pulp supply is plantation-grown eucalyptus. Most species take five to seven years to grow to commercial size, and coppiced growth allows three crops from one root. Opponents see plantations as monocultures that provide little habitat support for the high biodiversity found in tropical settings.

HARVESTING APPROACHES

Logging companies, while historically built around clear-cutting, have to some measure changed their harvest approaches due to regulations and with an eye toward sustainability of forests. Harvest prescriptions are now set out to help mimic natural succession in different forest types. In even-aged stands, natural disasters such as fire and wind events are part of ecosystem cycling, in that most trees are killed off and grow up as one age class following the disaster. Newer forms of clear-cutting involve a reduction in the size of clear cuts, corridors for large terrestrial mammals, reserves on riparian areas, buffers around stick nests, and protection of forest reserves.

For temperate deciduous and most tropical forest types, which are mosaics of tree ages and species, single-tree selective logging can mimic natural gap formation.

Selective logging involves removing only a select proportion of the trees in a stand. Within selective logging, the approach has two main approaches. One involves a high grade or diameter cut. The loggers "take the best and leave the rest," resulting in leaving poor-genetic-legacy, diseased, and lower-valued trees in the stand. The second approach, also known as "best forest management practice," is the "worst first" approach. It removes lower-quality, diseased, and crooked trees. This method, in combination with attention to buffer zones around waterways and to preserving appropriate representation of all ages and classes of trees, including older trees with cavities for wildlife, allows for a more sustainable approach.

In tropical forests, harvest prescriptions and approaches can be controversial; it appears that logging in primary forests precipitates a significant loss of biodiversity. Selective logging is expanding rapidly throughout the tropics. In these settings, while selective logging is better than clear-cutting, selective logging may increase the vulnerability of forests to fire and future deforestation. Also, logging roads cut into remote forest areas typically grant easy access to developers, small farmers, hunters, and poachers. Selective logging in tropical forests is seen as degrading the forest, as the felling of a single large tree can bring down dozens of surrounding trees through the network of vines and lianas, creating large canopy openings and precipitating habitat changes by increased sunlight and drying winds. These types of effects are well-documented in the scientific study of forest fragmentation and edge effects.

One major problem in terms of regulating or creating sustainability within the industry is illegal logging. Its dramatic effects, especially in developing nations, include taking much-needed revenue away from governments, devastating effects on the livelihoods of forest-dependent people, and furthering corruption and civil conflict. At current rates of illegal logging, if all the timber involved had been harvested legally, governments could have earned up to $6.5 billion. The World Bank estimates that the value of the illegal trade is $10 billion to $15 billion per year.

INTERNATIONAL INITIATIVES & FUTURE MARKETS

Following the 1992 Earth Summit in Rio de Janeiro, Brazil, much-needed awareness was brought to the worldwide threats to forests, the implications of deforestation, and changing attitudes among the general public. As a result, attention was turned to furthering sustainability, keeping forests intact, and reducing the effects of climate change—as forests store 50 percent of the world's terrestrial carbon. The response by some of the largest companies within the logging industry was to partner with various groups, such as publicly recognized nongovernmental organizations, and to change their operations to be more sustainable through volunteer certification regimes.

There are more than 50 certification schemes in which the logging industry can participate, such as the Forest Stewardship Council and the Programme for the Endorsement of Forest Certification. These voluntary, third-party certifications are becoming more commonplace—but are not without criticism, as they are seen by some environmentalists as ways that the logging industry can manipulate the public by producing similarly named certifications with diluted standards.

Currently, the logging industry is moving into newer markets, such as turning waste into renewable energy. The industry also faces new challenges, such as the effect of invasive species and insects in the temperate forests of North America, urban sprawl, being called upon to mitigate various forms of pollution, and competition for land with mining and fossil-fuel extraction.

The logging industry has most recently turned its attention to the vast northern, or boreal, forests, where it is waging battles with government and environmental groups to obtain cutting rights. While both legal and illegal logging is found to be decreasing forest cover in tropical forest ecosystems, the boreal, temperate, and subtropical forests appear to be increasing in net forest cover due to reforestation, natural forest recovery, and establishments of forest plantations. The viability of these forest ecosystems remains at the heart of the logging industry's longevity—and that of forest habitats worldwide—into the twenty-first century and beyond.

FURTHER READING

Bacher, John. *Two Billion Trees and Counting: The Legacy of Edmund Zavitz.* Toronto: Dundurn Press, 2011.

Bryant, Dick, et al. *The Last Frontier Forests: Ecosystems and Economies on the Edge.* Washington, DC: World Resources Institute, 1997.

Diamond, Jared. *Collapse: How Societies Choose to Fail or Succeed.* London: Penguin Group, 2005.

Food and Agriculture Organization of the United Nations. "Forestry." 2012. http://www.fao.org/forestry/en.

Lawson, Sam and Larry MacFaul. *Illegal Logging and Related Trade: Indicators of the Global Response.* London: Chatham House, 2010.

Marchak, M. Patricia. *Logging the Globe.* Montreal: McGill-Queen's University Press, 1995.

M

Monoculture Farming

—D. R. Gossett

FIELDS OF STUDY

Plant Science; Crop Science; Botany; Agronomy; Agriculture; Biochemistry; Ecology; Entomology; Genetics; Soils Engineering.

SUMMARY

Almost all industrialized countries utilize monoculture as the major form of crop production. Monoculture is the most efficient means, in terms of both yield and cost, to produce food and fiber for large numbers of people, but some environmental concerns are associated with the practice. For example, monoculture increases the reliance on agricultural chemicals and decreases the genetic diversity of crop plants.

The continuing increase in the number of humans inhabiting the earth has placed greater and greater demands on agricultural systems to produce sufficient food and fiber to feed and clothe the population. In order to meet these demands, mechanized agriculture is utilized in countries where there is sufficient money and land to support it. Mechanized agriculture depends on the availability of large sections of relatively level land in order to facilitate the efficient operation of the machinery. In addition, mechanized agriculture relies on monoculture, the practice of planting the same crop on vast areas of land to increase the efficiency of planting, cultivating, and harvesting. Mechanized agriculture is practiced in most of Europe, many of the republics that once composed the Soviet Union, much of South America, almost all of North America, and a few other developed countries.

HISTORICAL PERSPECTIVE

Throughout most of human history, farms and ranches were owned by families who primarily practiced sustenance agriculture. Each farm or ranch produced a variety of crops sufficient to feed the resident family as well as a small excess that was sold for cash or bartered for other goods or services. Because almost all agricultural activities required human or animal labor, large families were desirable. Until the nineteenth century, agricultural tools such as plows were made of wood; however, the arrival of the Industrial Revolution changed the agriculture industry just as it did almost all other industries. Eli Whitney invented the cotton gin, a machine that could separate cotton fibers from the seeds of the plant, in 1793. Cyrus McCormick invented the mechanical reaper in 1833, and John Lane and John Deere began the commercial manufacture of the steel plow in 1837. Inventions such as these led the way for the development of many different types of agricultural machinery, which resulted in the mechanization of most farms and ranches. Most agricultural enterprises in the United States were mechanized by the early part of the twentieth century.

The Industrial Revolution also produced a significant change in the social nature of the United States. The nation went from being predominantly an agrarian society to being predominantly an urban society, as many people who had been involved in agricultural production left the farms to go to the cities to work in factories.

As the population continued to grow, the selection and production of higher-yielding crops became increasingly important, and the Green Revolution of

Farmers in Huambo who have received loans for production of improved varieties of Irish potato. USAID-supported agricultural programs have assisted farmers in growing various types of Irish potatoes, improving their crop production and their incomes. (C. Image by USAID Africa Bureau (Angolan Potato Farmers). Uploaded by Elitre.

the twentieth century helped to make this possible. Biological scientists supplied basic information that allowed agricultural scientists to develop new, higher-yielding varieties of numerous crops, particularly the seed grains that supply most of the calories necessary for maintenance of the world's population. These higher-yielding crop varieties resulted in tremendous increases in the world's food supply. The utilization of these new crop varieties, along with the increase in mechanization, also led to an increased reliance on monoculture.

ADVANTAGES

The major advantage to monoculture is increased efficiency, in terms of both crop yield and economics. Limited numbers of people are available for work as farm laborers, and this limited supply of manual labor resulted in the development of machinery that can be used to plant, cultivate, and harvest different crops. In many instances, different kinds of machines are necessary for different crops, so if farmers grow many different crops on numerous small plots of land, they need to change the machinery being used each time they handle a different crop. This requires large amounts of time and is expensive, in both labor and equipment costs. Farm machinery is very expensive, and few farmers can afford the equipment necessary to handle more than one or two different crops. If farmers intersperse different crops in one large field, they must also expend costly fuel for their machinery while they skip one part of the field to get to another. Many of these problems have been reduced through the use of monoculture. The farmer can buy fewer pieces of equipment, the equipment does not have to be changed out as often, and large tracts can be harvested at one time without the farmer's having to skip part of the field.

The use of mechanized monoculture tremendously increased the amount of agricultural produce available throughout the world during the twentieth century. Not only did the number of hectares under cultivation increase owing to the increased efficiency of machinery, but also the yield per hectare being farmed increased dramatically in many parts of the world. For example, corn yields almost doubled in the United States from 1970 to 2010. This increase came about for a variety of reasons, including improved farming methods, improved irrigation, new varieties of crops, the use of fertilizers and pesticides, and the availability of more efficient machinery, all of which developed in response to the increased reliance on monoculture. In the twenty-first century, the highly mechanized modern agricultural unit requires relatively few employees to produce large amounts of crops for an increasing number of consumers.

DISADVANTAGES

Although the practice of growing only one crop over a vast number of hectares has resulted in much higher yields, it has also caused some environmental problems. One of the problems associated with monoculture is the loss of genetic diversity. Farmers most often plant hybrid varieties of crops that have been developed to give high yields and to produce plants that are equal in size and other characteristics that provide uniformity in order to facilitate mechanized harvesting. When large numbers of farmers plant the same hybrid variety for a number of years, important genetic material that might provide pest resistance or some other important trait could be lost. It has been estimated that two-thirds of all seeds planted in developing countries belong to the same strains. This could prove disastrous in the future if these strains were to become infected with an uncontrollable pest. The use of the same hybrid by large numbers of farmers in a given area also results in pest control problems. Vast fields of the same crop attract large numbers of pests, including weeds, fungi, and insects. Monoculture is thus highly dependent on the use of pesticides to control these pests, and the long-term, large-scale application of pesticides could prove detrimental to other plants, animals, and humans.

The practice of monoculture generally means that the same crop may be planted year after year, with little or no crop rotation. As a rule, agricultural activity decreases the availability of a number of important plant nutrients found in the soil, and the lack of crop rotation can hasten the depletion of certain essential soil nutrients. This means that the farmer will have to apply fertilizer more often to restore soil fertility. Some of the nutrients in fertilizers, particularly nitrogen and phosphorus, can end up in nearby streams and lakes, where they contribute to water pollution.

In addition, the use of monoculture can often mean that large tracts of land remain uncovered between the harvesting of one crop and the planting of the next. When soil is left bare without any covering vegetation, it is subject to wind and water erosion, which results in the removal of the topsoil. Intensive mechanized monoculture can also result in the loss of organic matter from the soil. Soil suitable for growing plants contains organic matter that will remain in the soil if part of the crop is turned back under after it is harvested; however, in mechanized monoculture, the entire plant is often removed, thereby decreasing the organic matter content over time. Polyculture cropping systems (that is, the planting of several diverse crops) avoid some of the problems associated with monoculture, but the trade-off is lower overall agricultural productivity.

FURTHER READING

Enger, Eldon D., and Bradley F. Smith. *Environmental Science: A Study of Interrelationships.* 12th ed. Boston: McGraw-Hill Higher Education, 2010.

Hill, Marquita K. "Pesticides." In *Understanding Environmental Pollution.* 3d ed. New York: Cambridge University Press, 2010.

McBrewster, John, Frederic P. Miller, and Agnes F. Vandome, eds. *Industrial Agriculture: Factory Farming, Livestock, Aquaculture, Agribusiness, Monoculture, Agroecology, Organic Farming, Urban Agriculture.* Phoenix, Ariz.: Alphascript, 2009.

Miller, G. Tyler, Jr., and Scott Spoolman. *Environmental Science: Problems, Concepts, and Solutions.* 13th ed. Belmont, Calif.: Brooks/Cole, 2010.

Wojtkowski, Paul. *Landscape Agroecology.* Boca Raton, Fla.: CRC Press, 2003.

MORPHOGENESIS

—Charles R. Crumly

FIELDS OF STUDY

Evolution; Plant Science; Biochemistry; Genetics.

PRINCIPAL TERMS

bifurcation: the division of a Y-shaped and connected mesenchymal structure into a single proximal chondrogenic focus and two distal chondrogenic foci; this can lead to the formation of separate chondrification centers in a developing digit

chondrification: the process by which undifferentiated connective cells transform into chondrocytes (cells that make cartilage) and begin forming extracellular matrix

epithelium: the tissue that covers and lines all exposed surfaces of an organism, including internal body cavities such as the viscera and blood vessels

limb bud: thickened epithelial cells along the lateral body fold that are underlain by mesoderm, creating a paddle-shaped extension from the trunk

segmentation: the division of a structure into linearly arranged segments; it can lead to the formation of somites, or it can lead to the formation of separate chondrification centers in a developing digit

zone of polarizing activity (ZPA): a region at the posterior base of the limb bud that seems to influence the distal development of pattern in a developing limb

SUMMARY

Morphogenesis is the process, operating during the development of the organism and appearing through evolutionary history, by which patterns of form come into being in genetically controlled systems. By studying this multifaceted process, it is possible to understand underlying factors that regulate the appearance of patterns of structure in living organisms.

Morphogenesis is the process that leads to the appearance of form in specific patterns, through the differential reproduction, growth, and movement of cells and tissues and interactions between tissues. Such movements are controlled by a variety of factors, including cell adhesion molecules (CAMs), the nature of the extracellular matrix (ECM), and the size of the cells themselves.

Most multicellular organisms begin their lives as fertilized eggs. Although this seemingly simple cell is really quite highly organized, it is still relatively uncomplicated when compared to the structural complexity of the parent. As each egg cell divides, the daughter cells become structurally and functionally distinct. This is the beginning of the process of cell differentiation, which can lead to complex multicellular organisms. As new cells appear, grow, mature, and divide, there is a point at which the eventual fate of these cells is determined. That is, at some point the eventual location, structure, and function of the descendants of such cells are fixed. This fate determination frequently takes place very early in the development of an organism. After a particular fate is determined, it is the process of morphogenesis that allows the potential fate to be realized. Also through the process of morphogenesis, patterns of form develop that increase the structural and functional complexity of an organism.

THE RULES OF MORPHOGENESIS

Fate determination, morphogenesis, and pattern formation are not haphazard. In fact, these events are highly constrained, extremely stereotyped, and very predictable—so much so that models of morphogenesis have been formulated that rely on rules of development to establish the probabilities of particular forms developing. These rules take into account two types of conditions—initial and boundary conditions—that work together to produce pattern formation in a cascading process. Initial conditions are simply the cellular conditions that prevail in a certain part of an organism, such as cell size and number and the chemical makeup of cell membranes. Boundary conditions are the conditions that exist at boundaries of different types of tissue.

Only a few phenomena are responsible for setting the initial and boundary conditions within which a morphogenetic system operates. A variety of cellular phenomena are involved, including cell migration, cell division (producing new cell lineages), the rate of cell division, cell number, cell density, the orientation of daughter cells relative to parent cells, and even the places in which cell death occurs. Tissue-level phenomena also have an important impact on morphogenesis. In these cases, specific tissues possess certain properties and a limited ability to respond differently within the constraints allowed by those properties. For example, sheets of epithelial cells can become folded, but they seldom form a solid mass of cells. Finally, there are interactions that take place between different tissues. Among these interactions is induction, wherein one tissue will induce a specific response in an adjacent tissue. Examples of such induction are numerous and well known; they include the mesenchymal induction of dental epithelium during the formation of teeth.

LIMB MORPHOGENESIS

Although far from completely understood, one of the most thoroughly studied morphogenetic systems is the development of the tetrapod limb. (A tetrapod is a vertebrate with two pairs of limbs.) Because amphibians (such as frogs and salamanders) and chickens have large, easily obtainable eggs, they have been studied most. It will be useful to summarize the course of limb development, since it exemplifies many of the common features of morphogenesis and pattern formation.

Early in the formation of the limbs, a ridge forms along the flank of the embryo. Along this ridge, the apical ectodermal ridge (AER) develops as a thickened layer of epithelial cells. Undifferentiated connective tissue cells, called mesenchyme cells, accumulate underneath the AER and form the limb bud. These mesenchymal cells are derived from mesoderm that migrates into the limb bud following pathways of the ECM that are laced with CAMs, especially fibronectin. The AER is responsible for the formation of pattern in the proximodistal axis, which is a line from the base of a limb (the proximal end) to the outer end of the limb (the distal end).

Another focus of pattern formation also appears quite early in limb development; it is called the zone of polarizing activity (ZPA) and has influence over the anteroposterior axis of limb development. Together, the AER and the ZPA regulate a pattern of form that is shared by all tetrapods.

Under the influence of these pattern-generating centers, three types of mesenchymal chondrogenic foci will form: de novo condensations (which are unconnected), bifurcations, and segmentations. Segmentations and bifurcations both show recognizable connections, and they appear in a consistent and stereotyped fashion. Linear series of mesenchymal condensations showing connections with one another are called segmentations; a Y-shaped condensation with a single proximal focus connected to two distal foci is called a bifurcation.

In all tetrapods, a bifurcation leads to the formation of two skeletal elements in the region between the proximal bone in the arm (or leg) and the hand (or foot). In frogs and amniotes, the condensations at the base of the putative fourth digit, showing connection to the bone in the forearm (or lower leg), are the first to appear as the result of a bifurcation event. Each subsequent proximal digit condensation appears as a bifurcation. All distal condensations, after the proximal digital element forms, appear as the result of segmentation events. Thus, the development of limb elements is asymmetric and always emanates from the axis of the first digit to form chondrogenic foci.

The morphogenetic control of this pattern, which has been conserved throughout the long evolutionary history of tetrapods, is both simple and complex. It is simple in that only a few morphogenetic processes are responsible for generating the complex pattern of limb structure, so different in all the various tetrapods. It is complex in that so many phenomena can influence the ultimate structure of any particular limb. For example, the number of somites that contribute mesoderm to the limb bud mesenchyme can dictate the number of limb elements that will eventually form.

STUDYING MORPHOGENESIS

There are two general methods for studying morphogenesis. In the older—but still widely used—approach, tissues are removed from an organism in order to be examined under a microscope. Tissues are often stained (a variety of staining materials may be used) to make them more readily visible, and they are usually cut into thin slices called sections. The purpose is to determine the position of individual tissues. If carefully staged materials have been used, it is possible to get some ideas about whether cells have moved and, if so, approximately how far they have moved. There are problems with studying morphogenesis in this way: Details can be missed, or, more worrisome, preconceived notions can bias observations. For example, a small set of sections may support a theory, but, because thin sections of tissue can vary greatly, many sections of the material under study may contravene a pet theory of development. One could simply reject the many conflicting sections, claiming them to be poorly prepared or somehow damaged material. Indeed, many sections are so rejected. With the advent of transmission electron microscopy (TEM) and Scanning electron microscopy (SEM), more detailed observations could be made. New problems, however, accompany these methods. In TEM, the thinner sections which must be used increase the occurrence of variation, also increasing the chances of biased observations. SEM allows for the magnification of surfaces and a great increase in the depth of field, but it requires that materials be very carefully staged. In none of these methods is it possible actually to see individual cells move.

A second approach seeks to document cellular movements. Understanding such movements requires that individual cells be "marked." Marked cells are then grafted onto an unmarked host organism and followed. Both natural and artificial markers can be used. Pigment granules such as melanin and stained glycogen are examples of natural markers. These markers, however, can be affected by cellular activities, making it hard to follow the cells. Fortunately, there are other natural markers not so affected. For example, chick and quail cells stain differently, so the cells of one can be followed when they are transplanted into the other. One of the more frequently used artificial markers is tritiated thymidine. When it is introduced, this radioactive substance is permanently incorporated into the deoxyribonucleic acid (DNA) of the selected cells. Tritiated thymidine has several advantages: It works in any cell, it follows along with the cells, and it will be in all the offspring of the cell to which it was originally introduced. Unfortunately, as cells continue to divide, the concentrations of tritiated thymidine eventually approach undetectable levels.

Many systems have been studied using these methods, including the development of the tetrapod limb and vertebrate head, somite differentiation, and the appearance of the primary germ layers that eventually yield all other tissues.

Morphogenesis is the reflection of all the interactions that take place during the formation of a living organism and of the patterns of structure that characterize this organism. The study of morphogenesis is motivated by the desire to understand the appearance of patterns of structure. Such patterns are often conserved through evolutionary history. It is hoped that a clear picture of how these patterns are formed and regulated will show how the origin of novel morphologies is constrained. Thus, it will be possible to understand both the origin of novel forms and the maintenance of unchanging form.

FURTHER READING

Bard, Jonathan. *Morphogenesis: The Cellular and Molecular Processes of Developmental Anatomy.* New York: Cambridge University Press, 1992. A critical review of the physical mechanisms of morphogenesis. An appendix outlines three dozen areas where future research is needed. Clearly written, requiring only a minimal understanding of embryology. Extensive bibliography.

Bonner, John Tyler. *On Development: The Biology of Form.* Cambridge, Mass.: Harvard University Press, 1974. Updated from Bonner's earlier (1952) text on morphogenesis, this volume is also intended for the college-level student. Illustrations are rarely included. Endnotes and index.

De la Cruz, María Victoria, and Roger R. Markham. *Living Morphogenesis of the Heart.* Boston: Birkhäuser, 1998. An encyclopedic survey of the developmental anatomy of the heart, focusing on chicken embryos, with comparisons to other mammalian hearts. Written for medical professionals.

Edelman, Gerald M., and Jean-Paul Thiery, eds. *The Cell in Contact: Adhesions and Junctions as Morphogenetic Determinants.* New York: John Wiley & Sons, 1985. The subject of this book, intended for specialists, is the nature of the contacts that cells make and how they determine the fate of morphogenesis. Articles are contributed by researchers in this area and are grouped into five sections: membrane structure, determinants of animal form, cell adhesion molecules, extracellular matrix/cell-substrate adhesion, and specialized junctions. Most contributions are illustrated. Each article is followed by a literature cited section. The book ends with a subject index. Wessells (1977) is a more accessible college-level treatment of some of the same issues.

Hinchliffe, J. R., and D. R. Johnson. *The Development of the Vertebrate Limb: An Approach Through Experiment, Genetics, and Evolution.* Oxford, England: Clarendon Press, 1980. This volume is intended for specialists, but it can be understood by college-level students. It is a detailed and extensive treatment of limb development, divided into seven sections: the origin of the vertebrate limb, adaptation/diversity, descriptive and experimental embryology, regeneration, pattern, and mutants and the process of evolution. The well-illustrated text is followed by a long bibliography, an index, and plates.

Thomson, Keith S. *Morphogenesis and Evolution.* New York: Oxford University Press, 1988. This well-written, very short volume is intended for evolutionary biologists with a budding interest in development. Most of the examples chosen for discussion come from vertebrates. Nine chapters cover very broad topics including theories, the difference between pattern and process, pattern formation (both early and late), and the general properties of morphogenetic systems. The infrequent illustrations are all line drawings; the text is followed by references and a short index.

Trinkaus, John Philip. *Cells into Organs: The Forces That Shape the Embryo.* Englewood Cliffs, N.J.: Prentice-Hall, 1984. This exceptional book is intended as an upper-division college text that specialists might also find indispensable. It is an expanded version of an earlier edition of the same title. Fifteen chapters, illustrated by photographs and line drawings, extensively cover topics including cell movements, the structure of cell surfaces, cell adhesion, the environmental control of cell motility, mechanisms of morphogenetic movement, and the spread of cancer cells. Selected references follow each chapter. The author provides criticisms, suggests many avenues for future research, and notes many unsolved developmental problems.

Webster, Gerry, and B C. Goodwin. *Form and Transformation: Generative and Relational Principles in Biology.* Cambridge: Cambridge University Press, 2011. Print. Argues that the theory of Darwinian selection must be supplemented by a theory of biological form, based on a generative theory of organisms as developing and transforming entities of a distinctive type.

Wessells, Norman K. *Tissue Interactions and Development.* Menlo Park, Calif.: Benjamin/Cummings, 1977. This brief but detailed text is intended for undergraduate college students. Eighteen chapters cover the main processes of development, including morphogenesis. Emphasis is on how tissues interact to bring about morphogenesis. The text is well illustrated with line drawings and numerous photographs. Each chapter closes with a brief "Concepts" section that outlines important points, followed by a references-cited section. The last chapter is a handy summation of the major points discussed. Glossary and index.

Mountains

—*Jason A. Hubbart*

FIELDS OF STUDY

Geology; Environmental Sciences; Bioclimatology; Physical Geography; Ecology.

PRINCIPAL TERMS

asthenosphere: the region below the lithosphere where rock is less rigid than that above and below it

bioclimatic zone: a zone of transition between differing yet adjacent ecological systems

endemism: the occurrence of species only within narrow environmental ranges

local relief: the elevation difference between the lowest and highest points in an area

succession: directional change in communities of vegetation or animals

tectonic plate: tectonic plate theory suggests that the earth's surface is composed of a number of oceanic and continental plates which have the ability to move slowly across the earth's asthenosphere

SUMMARY

Mountains cover one fifth of the earth's terrestrial surface, and are one of the most extreme environments in the global ecosystem. They are globally significant landforms that function as storehouses for irreplaceable resources such as clean air and water, and biological and cultural diversity, as well as timber and mineral resources.

Mountains are the most conspicuous landforms on earth. They are found on every continent, and have been defined simply as elevated landforms of high local relief, with much of the surface in steep slopes, displaying distinct variations in climate and vegetation zones from the base to the summit. The earth's mountain ranges have been created by the collision of tectonic plates. Associated with many of these mountain ranges are volcanoes. If the solidified magma of a volcano builds up, it can become a mountain; likewise, if the collision involves two oceanic plates, a string of volcanic mountains, called an island arc, can form on the ocean floor.

MOUNTAIN HABITAT

Mountains are globally significant reservoirs of biodiversity. They contain rich assemblages of species and ecosystems. Because of the rapid changes in altitude and temperature along a mountain slope, multiple ecological zones are stacked upon one another, sometimes ranging from dense tropical jungles to glacial ice within a few kilometers. Many plant and animal species are found only on mountains, having evolved over centuries of isolation to inhabit these specialized environments. Mountains can also function as biological corridors, connecting isolated habitats or protected areas and allowing species to migrate between them. These extraordinary ecological conditions, coupled with many bioclimatic zones, have resulted in an extremely high number of ecological niches available for habituation in mountain ecosystems.

MOUNTAIN FAUNA

Because of the great diversity in habitats within mountainous regions, with each region showing a different combination of environmental factors, total mountain fauna is relatively rich and the variety of small communities very great, in spite of the general severity of the mountain environment as a whole. Likewise, this diversity has resulted in a wide range of endemic species that have evolved over centuries of isolation from other genetic material. Rocky Mountain National Park typifies this diversity as a home to 900 species of plants, 250 species of birds, and 60 species of mammals. Some are easily seen and others are elusive, but all are part of the ecosystem in the park. On a global scale, mountain fauna diversity includes many species of ungulates, including elk, bighorn sheep, moose, and deer. Also included in mountain communities are many species of rodents. Rodent species may include beaver, marmots, squirrels, and chipmunks. Other mammalian animal life includes bear, canids, including coyote and wolf, and many species of felids, such as mountain lions and bobcats. Mountain avian fauna includes many families of hummingbirds, bluebirds, hawks, falcons, eagles, and many more.

THREATS TO MOUNTAINS

Mountains are threatened in a variety of ways. There are constant threats from human activities, such as camping, hiking, and other recreational activities. Hikers create tracks in the soil that form erosion gullies. Likewise, hikers may trample on vegetation that has taken many years to grow. Commercial harvesting of trees in the lower forest zones of mountains is having an increasingly detrimental effect on biodiversity. Many countries have replanted indigenous trees with fast-growing coniferous trees, in an ill-fated effort to supply a growing human population with wood products. These hybrid forests are not nearly as beautiful as the native forests, but more to the point, they do not offer an environment conducive to the ecosystem that the native species supported. This problem creates a loss of wildlife, which becomes even more rare in these forests because of the decline of native vegetation. Global warming is another threat to mountain ecosystems. Snowlines are receding, and eventually, continued melting of glaciers and polar ice caps could lead to drying of major river systems which feed from them. Without question, human settlement and activities constitute the biggest threat to the mountain ecosystem.

Further Reading

Denniston, D. *High Priorities: Conserving Mountain Ecosystems and Cultures*. Worldwatch Paper 123. Washington, D.C.: Worldwatch Institute, 1995. This is a delightfully informative book. It is easy to read, with many relevant points regarding mountain ecosystem conservation.

Messerli, B., and J. D. Ives. *Mountains of the World: A Global Priority*. New York: Parthenon, 1997. This is a book for those readers who are more educated on the current environmental dilemmas concerning mountains and forestry. It is also a valuable source of general mountain ecosystem information.

Price, L. *Mountains and Man.* Berkeley: University of California Press, 1981. An easy-to-read, information-packed source concerning mountain-human interactions.

Sauvain, P. *Geography Detective: Mountains.* Minneapolis: Carolrhoda Books, 1996. This is an invaluable geographical perspective to mountains. It is loaded with easy-to-read information that is interesting and easily understood.

Stronach, N. *Mountains.* Minneapolis: Lerner, 1995. This book is filled with great information regarding general characteristics of mountains. The author also answers a wide array of questions regarding many mountain ecosystem specifics.

N

Nature Conservancy

—*Cynthia A. Bily*

FIELDS OF STUDY

Conservation; Environmental Sciences; Ecology.

SUMMARY

By purchasing land and by helping other landowners to develop conservation plans, the Nature Conservancy has been successful in protecting unspoiled areas from development and pollution.

The Nature Conservancy, based in Arlington, Virginia, traces its origins to a committee of the Ecological Society of America, which was founded in 1915. In 1946, the committee became the Ecologists' Union, which changed its name to the Nature Conservancy in 1950 and was incorporated as a nonprofit organization in 1951. The group, which included research scientists and conservation advocates, began acquiring land through donation and purchase and collaborating with other agencies to manage public lands to sustain biodiversity.

During the 1960's the Nature Conservancy helped create a new tool for encouraging conservation, the conservation easement, which enables private landowners to receive tax benefits for conserving their land and restricting development while retaining ownership. By 2010 the Nature Conservancy had more than one million members and branches in all fifty U.S. states, as well as in more than thirty other countries. It owned more than 405,000 hectares (1 million acres) of land in the United States and approximately the same amount total in other countries, and was involved with other landowners in hundreds of projects addressing the conservation of water, forests, and marine habitats, as well as issues related to climate change.

The Nature Conservancy partners with corporations, government agencies, and other environmental groups, as well as with private landowners. Most of its funding comes from private donations, although some of its experimental projects are supported by government grants. The organization has helped to create and preserve several important areas throughout the United States, including the Golden Gate National Recreation Area in California, the Tallgrass Prairie Preserve in Oklahoma, the Great Sand Dunes National Park and Preserve in Colorado, and Glacial Ridge National Wildlife Refuge in Minnesota. It also conducts programs in other countries and has helped organize debt-for-nature swaps, under which developing nations have some of their international debt canceled in exchange for conserving tracts of ecologically valuable land. In addition to working internationally on an increasingly large scale, the Nature Conservancy encourages its many members to participate in efforts such as the Plant a Billion Trees Campaign, which plants one tree in Brazil for every dollar donated, and its Adopt

Bison at the Tallgrass Prairie Nature Preserve.

an Acre program, which allows donors to designate particular habitats to be protected through their donations.

The Nature Conservancy is the wealthiest environmental group in the world, with more than one billion dollars in assets, and its large size and tremendous influence have created controversy. Critics have charged that some of the organization's acquisitions and logging, drilling, and mining projects, while intended to be models of sustainable development, have ignored the rights of indigenous peoples or threatened endangered species. The group's advocacy of integrated fire management has also been controversial. In addition, the Nature Conservancy has been criticized for acquiring private lands through charitable gifts and then reselling them against the wishes of donors, often at large profit to the U.S. federal government.

Further Reading

Birchard, Bill. *Nature's Keepers: The Remarkable Story of How the Nature Conservancy Became the Largest Environmental Organization in the World.* New York: John Wiley & Sons, 2005.

Brewer, Richard. *Conservancy: The Land Trust Movement in America.* Lebanon, N.H.: University Press of New England, 2004.

Novel Ecosystems

—Erika Marín-Spiotta

FIELDS OF STUDY

Ecosystems; Evolution; Environmental Sciences; Ecology.

SUMMARY

Human activities affect the distribution and abundance of species, often resulting in new combinations of species that did not occur previously in a given place. This change has unpredictable consequences for the biodiversity and ecology of a place.

Novel ecosystems are defined as ecosystems that have new assemblages or combinations of species resulting directly and indirectly from human action. They are called novel, or new, because species occur in combinations and in relative abundances unlike those that occurred previously within a given biome. Another term commonly used to describe novel ecosystems is *emerging ecosystems.*

Novel ecosystems are formed by changes in the physical and biological components of ecosystems; by species invasion of ecosystems; and/or by the abandonment of intensively managed ecosystems, such as agricultural or urban areas.

The distribution and abundance of species can fluctuate naturally. Changes in the physical environment through disturbances such as fires, landslides, floods, and volcanic eruptions cause the removal of some or all species. The opening of space and availability of resources allows for the colonization and reestablishment of species. In fact, disturbance can play an important role in maintaining species diversity in some ecosystems, by preventing the dominance of one species to the exclusion of others.

Ecological changes that occur after a disturbance are called *succession.* Primary succession occurs on a landscape that has been wiped clean of its original ecosystem, such as on a lava flow or a surface recently exposed by retreating glaciers. Secondary succession occurs when there are changes in species due to a disturbance that has affected but not removed the original ecosystem, as when a fire or a fallen tree in a forest opens space for plants to colonize.

On a global scale, changes in Earth's climate during the Pleistocene glacial and interglacial periods resulted in changing species assemblages. Species differ in their ability to adapt to changes in climate, and in their capacity to disperse or migrate in response to changes in their environment. By studying pollen records in lake cores and in sediments, scientists have found that past climatic changes resulted in new combinations of species through immigration of new species, emigration of species to a more suitable area, or extinction of those species that were unable to adapt to changing conditions. Paleoecologists call these communities no-analog communities because the resulting combinations of species have no analog in Earth's modern history.

Novel ecosystems differ from no-analog communities and from changes in species composition during natural succession in two ways: The changes in emerging ecosystems are driven by human activities which differ from natural past climatic changes and primary succession; the communities include new species that never co-existed there before, which differs from secondary succession. In addition, no-analog communities reconstructed from past environmental data are compared with modern systems, while novel ecosystems are compared with recent historical conditions.

While the emergence of novel ecosystems depends on some environmental change that is directly or indirectly caused by human activities, novel ecosystems—unlike agricultural ecosystems—do not require human management for their maintenance. Another important element of novel ecosystems is the unpredictability of the emerging species combinations, because they do not follow the well-studied patterns of species replacement during primary or secondary succession.

EFFECTS OF HUMAN ACTIVITY

Researchers are realizing that the influence of human activities extends all over the globe, which has led some to rename our current geologic epoch *Anthropocene*, for the Age of Humans. Human activities that change the landscape and species distributions include deforestation; conversion of natural forests, grasslands, and wetlands to agriculture (crops and grazing lands); and fragmentation of the landscape by urbanization. Through the domestication and trade of species, humans have deliberatively transported plant, animal, and microbial species around the world, with the result that many species now live in biomes where they never existed before and where they would not have been able to get to on their own.

Human activities have also caused the inadvertent movement of species around the world. The consequence is that many species are now finding themselves in new communities, interacting with species with which they have not co-evolved.

At the same time, humans have driven many species to extinction through overhunting, disease, habitat loss, and a breakdown of biological interactions required for the survival of a species. Through the fragmentation and degradation of landscapes, humans have created barriers to the natural dispersal or movement of some species from one place to another. All these activities result in communities becoming disassembled as individual species are affected in different ways. The reestablishment of the original species is prevented by changes in the physical environment—such as loss of soil nutrients or physical habitat—or by a reduction in the original species pool, such as through the loss of a seed bank or seed source.

EFFECTS ON ECOLOGICAL PROCESSES

Much recent research has focused on studying the role of novel ecosystems and their ability to restore forest and grassland cover to human-disturbed landscapes. Some novel forests harbor native species in the understory and can include a high biodiversity of plants and animals. However, because they are new ecosystems, they are less studied, and it is often unknown how changes in composition will affect important ecological processes.

Some ecological processes that are affected by which species inhabit an ecosystem are the amount of energy and food available for higher trophic levels through photosynthesis; plant-animal interactions, which can affect pollination, seed dispersal, and germination, as well as food-web dynamics; the breakdown of dead organic matter and recycling of important plant nutrients by bacteria, fungi, and invertebrates; and water availability for plant growth through species' effects on hydrology. The unpredictability of species combinations in novel ecosystems increases uncertainty about their continued ability to provide services that are important for humans and other organisms, such as habitat for biodiversity, watershed protection, regional climate stabilization, and carbon sequestration.

EXAMPLES OF NOVEL ECOSYSTEMS

Examples of novel ecosystems can be found around the world. Some are relatively recent and others have been around for a long time, because humans have been modifying their environment and moving species around for thousands of years. South Africa's fynbos shrublands are being invaded by pine species that are especially adapted to germinating rapidly after fire and thus preventing the reestablishment of the native shrub species. In Brazil, New Zealand, and parts of the United States, the introduction by humans of recurring fire and of intensive livestock grazing have caused the replacement of native forests by grasslands and shrublands.

Environmental pollution by nitrogen and other elements is changing soil chemistry in forests and grasslands in Europe and North America, allowing invasive species to outcompete those that were already present and that were better adapted to the old conditions.

In Puerto Rico, extensive deforestation for sugarcane, coffee, and cattle pasture at the beginning of the twentieth century has been reversed due to changes in economic policies. This Caribbean island has experienced a threefold increase in forest cover due to the large-scale abandonment of agriculture. The emerging forests on former agricultural lands have different combinations of tree species not found before on the island. Researchers are actively studying these subtropical forests to see how ecological processes are affected by the dominance of introduced species.

Examples also exist in the aquatic world. Human control of rivers for irrigation and flood prevention has altered water and sediment flows, changing the habitat of riverine species and facilitating the introduction of nonnative fish and aquatic plants. Many important estuaries, such as San Francisco Bay in California, are now completely dominated by new species combinations, due to the confluence of ships from all parts of the world that bring along hitchhikers.

CONSERVATION EFFORTS

Novel ecosystems are important because they bring attention to the extent of human influence on the environment. Many conservation practices attempt to restore ecosystems to their original condition. These efforts often fail because enough changes have happened to the ecosystem that the original species are not able to come back without intensive and expensive human assistance. Furthermore, many of these species are not able to survive and replace themselves without costly management. Often, information on the original species composition is lacking.

Novel ecosystems provide opportunities for scientists to study how ecosystems respond to environmental change. While those who value the preservation of wild nature devalue novel ecosystems because they often are dominated by nonnative, or exotic, species, others argue that novel ecosystems are nature's way of dealing with change, by creating new combinations of species that are better adapted to the transformed environment. Novel ecosystems on soils where nutrients have been severely depleted through intensive agriculture, or on soils contaminated by industrial pollution or overfertilization, for example, harbor species that are able to tolerate extreme conditions. Over time, these species can actually improve soil conditions and facilitate the establishment of other species, some of which might have been there before the environmental change.

CONTROVERSY

There is a vigorous debate in the ecological community on the role of species that have been introduced by humans, and what place they should have in different types of ecosystems. Similar questions arise about novel ecosystems that comprise numbers of these "unwanted" species. Some argue that these ecosystems are ecological disasters, unworthy of ecological study because they are "unnatural," and suggest that energy should be spent on studying and preserving the last wild places on Earth. Those who study novel ecosystems respond that more and more of our landscape contains these new combinations of species, and that it is important to understand how they work.

The topic of novel ecosystems provides a philosophical context for discussing the extent to which one species—humans—is modifying the planet's surface and atmosphere, and altering the composition of ecosystems and biomes at a much faster rate than past climatic or environmental changes. Novel ecosystems suggest that, in some places, irreversible changes in physical and biological conditions may have already occurred. Emerging ecosystems provide opportunities for ecologists to study how terrestrial and aquatic communities respond to disturbances caused by human activities, changes in nutrients, and species invasions. Global warming will further drive the emergence of new combinations of species.

Further Reading

Bridgewater, Peter, et al. "Engaging with Novel Ecosystems." *Frontiers in Ecology and Environment* 9 (2011).

Hobbs, Richard J., et al. "Novel Ecosystems: Theoretical and Management Aspects of the New Ecological World Order." *Global Ecology and Biogeography* 15 (2006).

Lugo, Ariel E. "The Emerging Era of Novel Tropical Forests." *Biotropica* 41 (2009).

Marris, Emma. "Ragamuffin Earth." *Nature* 460 (2009).

Seastedt, Timothy R., Richard J. Hobbs, and Katharine N. Sundig. "Management of Novel Ecosystems: Are Novel Approaches Required?" Ecological Society of America. http://www.esajournals.org/doi/abs/10.1890/070046.

Williams, John W. and Stephen T. Jackson. "Novel Climates, No-Analog Communities, and Ecological Surprises." *Frontiers in Ecology and the Environment* 5 (2007).

O

Organic Food Debate

—Angela Harmon

FIELDS OF STUDY

Agronomy; Agriculture; Food Systems; Farming; Ecology; Soil Science; Environmental Sciences; Environmental Studies; Botany.

PRINCIPAL TERMS

biological control: The use of natural agents, such as predators, parasites, and pathogens, to control pests.

bovine growth hormone: A modified hormone that is given to cows to increase production. This hormone is only legal in the United States, Mexico, and South America.

community-supported agriculture (CSA): Community members pre-pay local farmers for a portion of their crop, which they receive weekly.

compost: Accumulation of organic matter–including food waste, lawn clippings, and paper–that can be used to fertilize crops.

mitrates: A form of nitrogen (a nutrient) found in many foods. High levels of nitrates can cause "Blue Baby Syndrome," a cyanotic condition resulting from a decrease in hemoglobin capacity to transfer and carry oxygen.

SUMMARY

Organic food consists of any crop or animal product produced without the use of pesticides, manmade fertilizers, additives, or growth regulators. Genetically engineered foods, human sewage, and irradiation (the process through which radiation kills germs and bacteria in food) also are excluded from proper organic farming. Rather than synthetic chemicals, organic farmers rely on biological pest control, crop rotation, and animal manure.

Proponents of conventional (non-organic) farming argue that organic farming is not worth the time or cost and may not be any less contaminated or healthier than conventional foods. Organic products are usually more expensive than conventional foods due to the need for more intensive labor and hand weeding, and the cost of organic food for livestock. Though organic farmers do not use pesticides, chemicals may blow onto the site from adjacent farms. In addition, soils may be contaminated with pollutants from prior land use, such as heavy metals and polychlorinated biphenyls (PCBs), and may be taken up by the crops.

On the other hand, proponents of organic farming argue that organic farming reduces environmental impacts, produces healthier foods, and protects the soil. Field experiments over a number of years show a 97 percent reduction in pesticides in organic fields as opposed to conventional farms. The National Research Council (NRC) found that a majority of pesticides found in infants and children

Organic Farm Foods The packing facility of Organic Farm Foods. Image by David Luther Thomas CC-BY-SA-2.0.

originate from their diet, emphasizing the importance of consuming pesticide-free food during this vulnerable phase. Additionally, some organic food has more vitamins and less nitrates.

HISTORY

Traditional farming before the twentieth century was considered "organic" because it did not rely on synthetic chemicals. In the 1930s, British scientists Lady Eve Balfour and Albert Howard began to research the role of agricultural production in healthy food. In 1943, Howard advocated the recycling of all wastes by using animal waste and sewage on farms. Howard also worked to create the ideal compost heap by mixing plant and animal waste, proposing that healthy soil equals healthy livestock and crops.

With the widespread application of chemical pesticides and fertilizers in the mid-twentieth century, the separation between organic and conventional crops grew. This rift was apparent when even academics referred to organic farming as a cult and proponents as spreaders of "doom and gloom." In the 1960s and 1970s, the organic movement was accelerated by Rachel Carson's book *Silent Spring*, which emphasized the problems associated with pesticides. In addition, water pollution from farming and the energy crisis prompted farmers to use fewer chemicals.

In 1980, the U.S. Department of Agriculture (USDA) took an interest in organic farming, publishing a report titled *Report and Recommendations on Organic Farming*. This report gave organic farmers a voice. In 1981, the American Society for Agronomy argued that organic farming contributed to more sustainable agriculture. However, its voice was quickly silenced in 1981 when President Reagan buried the report and eliminated positions at the USDA related to organic farming. However, his move was too late as momentum for organic farming had increased and courses in organic farming had already been initiated at agricultural universities across the nation. As the environmental movement continued through the 1980s and 1990s, more emphasis was placed on maintaining biodiversity, animal welfare, and fair trade practices.

In the United States, the 1990 Organic Foods Production Act (OFPA) created national standards for marketing organic foods and was to be regulated by the USDA. The goal was to reduce or prevent water pollution by reducing over-fertilization with manure and increase natural soil fertility. In 2002, the USDA created National Organic Standards, overriding any state regulations and creating a labeling system. The labels include different levels such as "100 percent organic," which means the product must be made from only organic products, "organic" products that have at least 95 percent organic ingredients, and products "containing organic products" with at least 70 percent organic products.

ORGANIC FOOD TODAY

Food safety has been a major concern and may have inspired some consumers to switch to more naturally produced food. Since the late 1990s, the public has been concerned about potential issues related to food safety, including the use of radiation, genetically altering food, outbreaks of bovine spongiform encephalopathy (BSE, also known as "mad cow disease"), and food-borne illness due to *Escherichia coli* contamination.

However, organic food is not without problems. In the 2000s, *Salmonella* and *E. coli* outbreaks in organic foods such as spinach inspired concern regarding food safety. The risk of food contamination may increase when farmers use manure instead of synthetic fertilizers since many bacteria exist in manure. Manure has become more of a culprit in causing food-related illnesses, especially in vegetables eaten fresh rather than cooked. The bacteria in the manure may also live in soil for a while, increasing the risk. However, manure contamination may not be isolated to organic farming since the difference between manure use in organic and conventional farming is small. One difference is that chemical and antimicrobial washes are allowed in conventional but not organic farming. Recalls of organic foods suspected of bacterial contamination by *Salmonella* and *Listeria* accounted for between 5 and 7 percent of all food recalls during the first half of 2015, as compared to 1 and 3 percent in the preceding two years; however, industry advocates argue that the higher number of recalled organic units is proportional to organic foods' expanded market share and reflects improvements in food safety systems. In defense of organic farming, contamination may occur during any stage of transport from the farmer's field to the consumer's home.

A disk of Taza Salted Almond Chocolate showing the company's distinctive branding graphics. Image by Vistawhite (Own work) CC-BY-SA-3.0.

The debate on the health benefits of organic foods continues. A review of thirty-four studies of the vitamin content of organic foods produced during 1970s through the early twenty-first century found that many organic vegetables have the same or higher nutrient contents than do non-organic foods. In general, the amounts of calcium, magnesium, iron, and vitamin C were higher in organic foods, while nitrate concentrations were lower. Additionally, organic foods may counteract the effects of consuming conventional foods. A 2006 study by Emory University researchers found that organic diets protect against pesticide exposure in young children, reducing pesticides in the body by 50 to 80 percent.

Critics of organic farming cite many unknowns. Organic farming may enhance soil fertility and create an increase in potentially harmful minerals and phytochemicals in food. Little is known about the effects of phytochemicals on human health. Additionally, organic food is not guaranteed to be pesticide-free, and may have residues of pesticides from adjacent farms or previous farming practices. In addition, though farms are not certified organic for two to three years after organic practices have begun, the Organic Trade Association (OTA) states that the soil may still be contaminated for at least three years. Lastly, the appearance and taste may not differ between organic and conventional food products.

Organic foods have become more mainstream in recent years. Not only are they available at national health chains such as Whole Foods and Trader Joe's, but they are now commonplace at Wal-Mart and most grocery chains. In 2018, organic food was available in the majority conventional grocery stores and organic food sales accounted for over 5 percent of total U.S. food sales, according to the USDA Economic Research Service. A 2018 poll found that 82 percent of Americans were actively seeking to incorporate organic foods into their diets. The firm Research and Markets projected in 2015 that the global organic foods industry would enjoy a compound annual growth rate of more than 16 percent between then and 2020.

However, prices remain high, and families are faced with the decision of which organic foods to buy. In a previous report, the USDA Economic Research Service found that an average American household spent $4,000 more on organic than conventionally farmed food. In 2018, the Environmental Working Group (EWG) released a list of the most and least contaminated fruits and vegetables—The Dirty Dozen—as well as the least contaminated—The Clean Fifteen. This list can be used to decide which foods to buy organic foods when shopping on a budget. The most pesticide-laden products on the list are apples, followed by celery, sweet bell peppers, peaches, strawberries, imported nectarines, grapes, spinach, lettuce, cucumbers, blueberries and potatoes. Those with little-to-no pesticide residue include cabbage, onions, avocados, sweet corn, sweet peas, pineapples, mangoes, asparagus, kiwi, and grapefruit.

FURTHER READING

Alkon, Alison Hope. "The Socio-Nature of Local Organic Food." *Antipode* 45.3 (2013): 663–80. Print.

Anisimova, Tatiana, and Parves Sultan. "The Role of Brand Communications in Consumer Purchases of Organic Foods: A Research Framework." *Journal of Food Products Marketing* 20.5 (2014): 511–32. Print.

Bravata, Dena M., et al. "Are Organic Foods Safer or Healthier Than Conventional Alternatives?&amp" *Annals of Internal Medicine* 157.5 (2012): 348–66. *Academic Search Complete.* Web. 2 Jan. 2013.

Dimitri, C., and L. Oberholtzer. *Market-Led versus Government-Facilitated Growth: Development of the U.S. and EU Organic Agricultural Sectors.* Washington: WRS-05 -05, USDA, Economic Research Service, 2005. Print.

Givens, D.I., et al. *Health Benefits of Organic Food: Effects of the Environment.* X:CABI, 2009. Print.

Hagstrom, Jerry. "Briefings Planned For 'Know Your Farmer'." *CongressDaily* 10 May 2010: 10. Print.

Hauter, Wenonah. "Wal-Mart Squeezes Organics." *Progressive* 78.6 (2014): 18–20. Print.

Oz, Mehmet. "Give (Frozen) Peas A Chance And Carrots Too." *Time* 180.23 (2012): 38–46. Print.

Paarlberg, Robert. "Attention Whole Foods Shoppers." *Foreign Policy* 179 (2010): 1. Print.

Paumgarten, Nick. "Food Fighter." *New Yorker* 4 Jan. 2010: 36–47. Print.

Riffkin, Rebecca. "Forty-Five Percent of Americans Seek Out Organic Foods." *Gallup Poll Briefing* 7 Aug. 2014: 3. *Points of View Reference Center.* Web. 12 Nov. 2015.

Rogers, Heather. "Slowed Food Revolution." *American Prospect* 21.6 (2010). 12–16. Print.

Strom, Stephanie. "Recalls of Organic Food on the Rise, Report Says." *New York Times.* New York Times, 20 Aug. 2015. Web. 12 Nov. 2015.

Tranter, R.R., et al. "Consumers' Willingness-to-pay for Organic Conversion-grade Food: Evidence from Five EU Countries." *Food Policy* 34 (2009): 287–94. Print.

ORGANIC FOOD INDUSTRY

—Angela Harmon

FIELDS OF STUDY

Agronomy; Agriculture; Food Systems; Farming; Ecology; Political Action; Policy; Environmental Studies; Botany.

SUMMARY

The organic food movement began in response to the twentieth-century introduction of chemicals and pesticides to the food industry to increase yield. The organic food industry grew steadily in the United States after the passage of the Organic Foods Production Act of 1990, which helped set strict standards for organic food production. In the 2000s, many Americans, inspired by new books and films on food production and consumption, began seeking alternatives to the conventional food supply chain, including organic foods.

The Organic Foods Production Act of 1990 implemented national organic standards and established the National Organic Standards Board. The board makes recommendations to the National Organic Program (NOP) of the U.S. Department of Agriculture (USDA) to help regulate the organic food industry. The NOP, established in 2000, is in charge of developing and regulating organic food standards, which ensure that companies use the correct practices when growing and producing organic products. The NOP verifies that the players within the organic food industry—including farmers, ranchers, distributors, and processors—comply with these standards and become certified by accredited agents.

The organic food industry is heavily regulated by the USDA, which frequently inspects and audits organic companies to ensure that regulations are followed and organic products are labeled correctly. All organic products must carry the USDA organic seal, which verifies that the product is at least 95 percent

A sign outside of an organic apple orchard in Pateros, Washington reminding orchardists not to spray pesticides on these trees. Image by Christopher Thomas CC BY-SA 3.0.

organic or has been made with certified organic ingredients. The seal means the crops have been harvested without exposure to radiation, sewage sludge, synthetic fertilizers, synthetic pesticides, or genetically modified organisms. The seal also ensures that animals are raised according to standards that protect their health and well-being. Farmers raising livestock cannot use antibiotics or growth hormones and must feed the animals a 100 percent organic diet. The animals must also be provided with an outdoor pasture. Animals that have been cloned do not qualify as organic products.

LABELING

As awareness of food production methods grew in the 2000s, food labeling became an important issue. Consumers often had difficulty understanding the differences between the profusion of labels and health claims, especially those denoting organic and nonorganic food products. In addition to carrying the USDA organic seal, products from livestock—including meat, eggs, and milk—may also have other labels regulated by the USDA. Products that contain any of these various labels may or may not be organic. If a product is labeled "natural," for example, but does not contain a USDA organic seal, it is not an organic product. Products that are 70 percent organic can carry the label "made with organic ingredients" but cannot bear the USDA organic seal.

Animal products could bear a number of different labels in addition to or instead of organic. A product labeled "free range" means that the livestock was permitted unlimited access to indoor/outdoor areas and fresh food and water. "Cage-free" means the livestock was not kept in caged areas, but was permitted unlimited access to indoor areas. Products labeled "natural" do not contain any artificial ingredients and are minimally processed. This labeling applies to meat and egg products only. "Grass-fed" livestock are fed primarily grass. Unlike organic grass-fed livestock, these animals may be fed grass that has been treated with pesticides and may be given antibiotics and/or hormones. A "no-added hormones" label means that the livestock has been raised without the use of hormones or steroids. Other labels, such as "pasture-raised" and "humane," may or may not be regulated by non-USDA independent certifying organizations.

INDUSTRY GROWTH

The organic food industry grew steadily through the 2000s in the United States to become one of the fastest-growing segments in the food market. According to the Organic Trade Association, organic foods accounted for almost $14 billion in sales in 2005. By the following year, this number had risen to $17.7 billion. Sales continued to increase, even when the U.S. economy entered a recession during the late 2000s. In 2009, organic foods accounted for $26.6 billion in U.S. sales and $54.9 billion in global sales. The United States led the global market in organic food sales, with Germany and France not far behind. By 2014, the USDA estimated that organic products sold in the United States formed a $35 billion market, accounting for about 4 percent of all U.S. food sales.

The wider availability of organic foods at large grocery chains was cited as one of the main reasons for the continued increase. At first, organic foods could be found only at small specialty grocery stores and farmers markets. By 2016 organic food was stocked in three-fourths of all U.S. grocery stores as well as an additional twenty thousand or so specialty food stores, according to the USDA. People also began choosing more organic products because of an increased awareness of health issues, the environmental impact of food manufacturing, and food and animal safety concerns.

IMPACT

During the 2000s, people increasingly chose organic products because they believed these products were better for both their health and the environment. Organic foods do not contain potentially toxic chemicals such as pesticides, which can leave a residue on produce and allegedly affect its taste and appearance. However, the taste improvement and even the health benefits of organic foods have been contested by some scientific research, with skeptics suggesting that psychological factors cause perceived differences between organic and nonorganic products. Most experts agree that more research and data is needed to establish any advantage of disadvantage to organic food in these areas.

Organic foods also do not contain food additives, such as artificial sweeteners, colorings, flavorings, and preservatives. Organic farmers are prohibited from using chemical pesticides, which can wash from farmlands into waterways where they may contaminate the water supply and soil, kill wildlife, and destroy vegetation. Instead, organic farms use such practices as hand weeding and crop rotation, which are designed to benefit the environment and reduce pollution. Research suggest that in many cases the environmental benefits of organic agriculture are real. However, the size and complexity of the food supply and distribution chain in developed nations such as the United States raises many interrelated issues for people deciding between organic and nonorganic products. For example, organic crop yields may be lower, necessitating greater land use, while organic crops may require more water than drought-resistant genetically modified strains. These and similar factors can in some instances cancel out the benefits of the organic process. Additionally, while prices have fallen as organic agriculture has grown more common, in many cases organic foods are still more expensive than nonorganic counterparts, presenting questions of economic inequality and food access.

FURTHER READING

Dunn-Georgiou, Elisha. *Everything You Need to Know about Organic Foods.* New York: Rosen, 2002.

Fromartz, Samuel. *Organic, Inc.: Natural Foods and How They Grew.* Orlando: Harcourt, 2006.

Langley, Andrew. *Is Organic Food Better?* Chicago: Heinemann, 2009.

"Organic Certification." *United States Department of Agriculture Economic Research Service,* 19 Oct. 2016, https://www.ers.usda.gov/topics/natural-resources-environment/organic-agriculture/organic-certification/. Accessed 22 Nov. 2016.

"Organic Foods: Are They Safer? More Nutritious?" *Mayo Clinic.* Mayo Foundation for Medical Education and Research, 7 Sept. 2012. Web. 26 Nov. 2012.

"Organic Market Overview." *United States Department of Agriculture Economic Research Service,* 19 Oct. 2016, http://www.ers.usda.gov/topics/natural-resources-environment/organic-agriculture/organic-market-overview.aspx. Accessed 22 Nov. 2016.

Organic Trade Association. "Quick Overview: Organic Agriculture and Production." *Organic Trade Association.* Organic Trade Assn., 16 Feb. 2011. Web. 21 Nov. 2012.

ORGANIC FOOD MOVEMENT

—Melissa A. Barton

FIELDS OF STUDY

Agronomy; Agriculture; Food Systems; Farming; Ecology; Soil Science; Environmental Sciences; Environmental Studies; Botany.

SUMMARY

The organic food movement of the 1990's affected agriculture and grocery stores and launched a rapidly expanding industry.

The environmental movement of the 1960's and 1970's laid the foundation for the increased support for the organic food movement in the 1990's. Following grassroots advocacy for organic standards in the 1980's, the U.S. government passed the Organic Foods Production Act (OFPA), Title XXI of the Food, Agriculture, Conservation, and Trade Act of 1990. The purpose of the OFPA was to establish national standards for the marketing of certain agricultural products as organic; to assure consumers of a consistent organic standard; and to facilitate interstate commerce in organically produced food. The act established standards for organic products and criteria for certifying a farm or part of a farm as organic.

During the 1990's, the organic food industry experienced sustained growth. Major chains such as Whole Foods Market and Wild Oats Markets grew rapidly and incorporated smaller chains and independent organic grocery stores. These large organic grocery chains helped to bring the prices of organic food down, although it remained more expensive than conventional food in part because of increased production costs in some areas and higher expenses due to smaller-scale production, and because of the costs associated with governmental certification. These large chains marketed organic food to consumers as a moral and enjoyable alternative to conventionally grown food products.

COMMUNITY-SUPPORTED AGRICULTURE

Organic gardening and community-supported agriculture (CSA) increased in popularity during the 1990's. Home gardeners used organic techniques to grow fruits and vegetables as well as ornamentals. Home gardeners and some small organic farmers also focused on growing heirloom varieties.

Community-supported agriculture was a way for small farmers to have a successful closed market. In a CSA system, consumers subscribed to a weekly share of food products and accepted whatever was seasonal. Some CSA farms encouraged subscribers to work on the farm in exchange for part of membership costs. The system reduced risk to the farmer and provided consumers with a way to obtain fresh, local food. Not all CSA farms were organic, but many were, and CSA systems were closely tied to the local food movement, founded on similar principles.

CRITICISM & CONTROVERSY

Consumers wanted organic food for a variety of reasons. These included concerns about pesticide residues on conventionally grown plants and artificial hormones given to livestock; the belief that organic food tastes better; and a desire to lessen environmental impact. All of these reasons have been criticized as incorrect or unfounded, but evidence remains inconclusive.

The organic food movement has been criticized for being antitechnological and unsustainable. Some studies have shown organic farming to result in lower yield and higher impact than conventional farming; others have shown comparable or higher yields for organic techniques in certain parts of the world. Critics argue that conventional techniques produce more food per acre and have lower impact because they preserve more nonagricultural land; proponents of organic agriculture claim that small farms are more efficient overall, although they do not produce large volumes of single crops the way monoculture farms do.

Critics have also claimed that organic food is too expensive for low-income families and poorer countries and thus a luxury available only to the elite. Others contend that the cost of synthetic fertilizers and pesticides is prohibitively expensive to poorer countries and that application of organic farming methods would improve crop yield. The many variables involved in measuring agricultural yield make it difficult to determine definitively the effectiveness of organic farming as compared to conventional methods.

IMPACT

The organic food movement of the 1990's changed how many Americans viewed food and widely affected agricultural and marketing practices. The enormous success of organic food chains like Whole Foods Market and Wild Oats Markets proved that organic food could be profitable to retailers and producers as well as appealing to the general public.

The impact of the growth in the organic food industry reverberated throughout the decade and continued to arouse public debate in the twenty-first century. The increased popularity of organic agriculture brought up many issues, such as the relative impacts of different agricultural methods on the environment; food safety and public health; whether industrial organic agriculture is in keeping with the ideals of the organic food movement; and socioeconomic topics.

Further Reading

DeGregori, Thomas R. *Origins of the Organic Agriculture Debate.* Ames, Iowa: Blackwell, 2004. DeGregori discusses two historically contrasting views, the technological view that led to modern agriculture and the "vitalist" view that supports organic agriculture. The author is a proponent of responsible use of technology and argues that resources are not finite but rather created by technology.

Fromartz, Samuel. *Organic Inc.: Natural Foods and How They Grew.* Orlando, Fla.: Harcourt, 2006. Addresses the history of the organic movement and its connections to industry, assessing the compromises the movement has made to reach mainstream consumers.

Groh, Trauger, and Steven McFadden. *Farms of Tomorrow Revisited: Community Supported Farms, Farm Supported Communities.* Junction City, Oreg.: BioDynamic Farming and Gardening Association, 2000. Provides an introduction to, history of, and argument for community-supported agriculture.

Kristiansen, Paul, Acram Taji, and John Reganold. *Organic Agriculture: A Global Perspective.* Ithaca, N.Y.: Cornell University Press, 2006. The authors critically examine the successes and limitations of the organic agriculture movement worldwide, including in the United States.

Lipson, Elaine Marie. *The Organic Foods Sourcebook.* Chicago: Contemporary Books, 2001. Covers the history of the organic food movement, profiles influential people and companies, and provides a list of resources.

Norberg-Hodge, Helena, Todd Merrifield, and Steven Gorelick. *Bringing the Food Economy Home: Local Alternatives to Global Agribusiness.* London: Zed Books, 2002. Examines global food issues from multiple angles in an argument for local food.

ORGANIC FOOD MOVEMENT: NATIONAL STANDARDS FOR ORGANIC CERTIFICATION

—*Melissa A. Barton*

The U.S. Department of Agriculture's National Organic Program published a fact sheet in October, 2002, that explains the national standards for the certification of organic products laid out in the 1990 Organic Foods Production Act:

The national organic standards address the methods, practices, and substances used in producing and handling crops, livestock, and processed agricultural products. The requirements apply to the way the product is created, not to measurable properties of the product itself. Although specific practices and materials used by organic operations may vary, the standards require every aspect of organic production and handling to comply with the provisions of the Organic Foods Production Act (OFPA). Organically produced food cannot be produced using excluded methods, sewage sludge, or ionizing radiation.

CROP STANDARDS

The organic crop production standards say that:

Land will have no prohibited substances applied to it for at least 3 years before the harvest of an organic crop. The use of genetic engineering (included in excluded methods), ionizing radiation and sewage sludge is prohibited. Soil fertility and crop nutrients will be managed through tillage and cultivation practices, crop rotations, and cover crops, supplemented with animal and crop waste materials and allowed synthetic materials.

Preference will be given to the use of organic seeds and other planting stock, but a farmer may use non-organic seeds and planting stock under specified conditions. Crop pests, weeds, and diseases will be controlled primarily through management practices including physical, mechanical, and biological controls. When these practices are not sufficient, a biological, botanical, or synthetic substance approved for use on the National List may be used.

LIVESTOCK STANDARDS

These standards apply to animals used for meat, milk, eggs, and other animal products represented as organically produced.

The livestock standards say that:

Animals for slaughter must be raised under organic management from the last third of gestation, or no later than the second day of life for poultry. Producers are required to feed livestock agricultural feed products that are 100 percent organic, but may also provide allowed vitamin and mineral supplements. Producers may convert an entire, distinct dairy herd to organic production by providing

80 percent organically produced feed for 9 months, followed by 3 months of 100 percent organically produced feed. Organically raised animals may not be given hormones to promote growth, or antibiotics for any reason. Preventive management practices, including the use of vaccines, will be used to keep animals healthy. Producers are prohibited from withholding treatment from a sick or injured animal; however, animals treated with a prohibited medication may not be sold as organic. All organically raised animals must have access to the outdoors, including access to pasture for ruminants. They may be temporarily confined only for reasons of health, safety, the animal's stage of production, or to protect soil or water quality.

ORGANIC GARDENING & FARMING

—Dion Stewart & Toby Stewart

FIELDS OF STUDY

Agronomy; Agriculture; Ecology; Soil Science; Archeology; Insect and Pest Management; Ecosystem Management; Biotechnology; Environmental Sciences; Environmental Studies; Botany.

SUMMARY

Interest in organic farming has grown since the late twentieth century for a number of reasons, including concerns about the environmental effects of the chemical pesticides and fertilizers used by conventional large agricultural operations, and particularly about the possibility of negative health effects associated with ingestion of the food produced using such chemicals.

At the beginning of agricultural history, farmers believed that plants "ate the soil" in order to grow. During the nineteenth century, German chemist Justus von Liebig discovered that plants extract nitrogen, phosphorus, and potash from soil. His findings dramatically changed agriculture as farmers found they could grow crops in any type of soil, even sand and water solutions, if they added the right chemicals.

Diversified family farms eventually gave way to huge specialized operations; by the end of the twentieth century, less than 2 percent of the U.S. population was directly involved in crop production. Crop yields were raised with the use of chemicals, but farms and their soil and water were not being cared for as the unique ecosystems they are. The quality of the soil was ignored as chemicals were used to produce high crop yields to feed the 98 percent of the population not involved in farming. Over time, the organic quality of the soil was lost even though the chemicals remained in the soil.

Chemicals used in agriculture were also found to leach into the water supply. In 1988 the Environmental Protection Agency (EPA) found that groundwater in thirty-two U.S. states was contaminated with seventy-four different agricultural chemicals. Aside from the health consequences to humans, leaching causes once friable and fertile soils to turn hard and become nonproductive. Further, the use of chemical insecticides was proving to have toxic effects on both the foods grown and the farmworkers encountering them. According to the World Health Organization (WHO), cases of acute pesticide poisoning account for significant mortality worldwide, especially in developing countries where environmental regulations do not exist or are not enforced.

Organic farming gained popularity as some individuals attempted to return to diversified farming practices that emphasize working with nature to create a renewing, ecologically sound, and sustainable system of agriculture. Organic farmers are committed to crop-growing practices that are free of synthetic chemicals and genetic engineering. Soil must be free of chemicals for at least three years before products grown in it can be certified as organic. The organic certification requirements further state that both plants and livestock must be raised organically, without the use of chemicals, antibiotics, hormones, or synthetic feed additives. Organic farmers must comply with both organic regulations in their states and the 1990 federal Organic Foods Production Act (OFPA). In the late 1990s the U.S. Department of Agriculture, through the National Organic Standards Board, developed standards and regulations to ensure consistent national standards for organic products.

Organic cultivation of mixed vegetables on an organic farm in Capay, California. Image by Hajhouse (Own work).

SOIL FERTILITY

To produce chemical-free crops, organic farmers must rely on organic practices to ensure soil fertility and control unwanted plants and insects. Organic farmers build organic materials in the soil by adding green manure, compost, or animal manure.

"Green manure" is a term used for crops that are grown specifically to introduce organic matter and nutrients into the soil. These crops are raised expressly for the purpose of being plowed under rather than sold to consumers. Green manure crops protect against soil erosion, cycle nutrients from lower levels of the soil into the upper layers, suppress weeds, and keep much-needed nutrients in the soil rather than allowing them to leach out. Legumes are excellent green manure crops because they are able to extract nitrogen from the air and transfer it into the soil, leaving a supply of nitrogen for the next crop. Legumes have nitrogen nodules on their roots; when the legumes are tilled under and decompose, they add more nitrogen to the soil. As plants decay, they also make insoluble plant nutrients—such as carbon dioxide and acetic, butyric, lactic, and other organic acids—available in the soil.

Much of the organic material derived from green manure comes in the form of decaying roots. Alfalfa, one type of legume, sends its roots several feet down into the soil. When the alfalfa plants are turned, the entire root system decomposes into organic material, thus helping to improve water retention and soil quality at the same time. Some examples of legumes used for green manure are sweet clover, ladino clover, alfalfa, and trefoil. Nonlegumes or grasses used for green manure include rye, redtop, and timothy grass.

Greater soil fertility can be achieved if green manure is grazed by animals and the manure from those animals is deposited on the soil. Left on their own in a large field, cows, sheep, and horses will choose only their favorite bits of grasses to graze and thus will graze the field unevenly. Organic farmers use portable fencing to achieve strip rotational grazing, in which animals graze a strip of a field all the way down and then are moved to another strip of the field while the first pasture recovers. This process is repeated until the entire field has been grazed and fertilized. Also, when the grass is grazed, some of the roots die and rot to form a good amount of humus. This humus is the stable organic material that acts as a catalyst for allowing plants to find nutrients.

Compost can work for large or small farmers, but it is particularly useful to small farmers and gardeners who do not have enough land to grow green manure crops to plow under. Composting is a natural soil-building process that began with the first plants that existed on earth and continues as a natural process today. As falling leaves and dead plants, animals, and insects decompose into the soil, they form a rich organic layer. Farmers and gardeners can make compost by alternating layers of carbohydrate-rich plant cuttings and leaves, animal manure, and topsoil in a container or defined area and allowing the components to decay. The rich organic matter that results can then be added to the soil to be cultivated.

Animal manure is used as an organic fertilizer. Rich in nitrogen, the best animal manures come from animals that have a high amount of protein in their diets. For example, beef cattle being fattened for market have a higher level of protein in their solid waste than do dairy cattle that are producing milk for market. The application of manure to fields improves the structure of the soil, raises the organic nitrogen content, and stimulates the growth of soil bacteria and fungi necessary for healthy soil.

CROP ROTATION

Planting the same crop year after year on the same piece of ground results in depleted soil. Crops such as corn, tobacco, and cotton remove nutrients, especially nitrogen, from the soil. Crop rotation involves planting different crops in the same field every other year to keep soil fertile by adding nitrogen and achieving a balanced nutrient level. Planting a winter cover crop such as rye grass helps to protect the land from erosion, and when the crop is plowed under in the spring, it creates a nutrient-rich soil for the planting of a cash crop. Crop rotation also improves the physical condition of the soil because different crops vary in root depth, are cultivated differently, and respond to either deep or shallow soil preparation.

Monoculture farming puts a large source of the same crop in easy proximity to insects that are destructive to that particular crop. Insect offspring can multiply out of proportion when the same crop is grown in a given field year after year. Since insects are drawn to a particular home area, they will not be able to proliferate and thrive if the crop is changed every other year to a crop they do not eat. This is one of the reasons organic farmers rely heavily on crop rotation as one aspect of insect control. Rotating crops can also help control weeds. Some crops and cultivation methods inadvertently allow certain weeds to thrive. Farmers using crop rotation can incorporate successor crops that eradicate those weeds. Some crops, such as potatoes and winter squash, work as "cleaning crops" because of the different styles of cultivation that are used on them.

ORGANIC INSECT & WEED CONTROL

Advocates of organic farming methods assert that plants within a balanced ecosystem are resistant to disease and insect infestation and that plants stressed by unfavorable growing situations are much more susceptible to such problems. The whole premise of organic farming is that farmers should work with nature to help provide healthy, unstressed plants.

Rather than using chemical pesticides, which often create resistant generations of insects and thus the need for newer and stronger chemicals, organic farmers seek natural ways to diminish pest problems. They strive to maintain and replenish soil fertility to produce healthy plants that are resistant to insects. They also try to select plant species that are resistant to insects, weeds, and disease. Crop rotation, as noted above, is another method of keeping insect infestations down. Organic farmers may create barriers to insect pests by planting, in and around the crop field, rows of hedges, trees, or even plants that are not desirable to insects. They may also provide habitats for the insects' natural enemies: birds, beneficial insects, and garden snakes.

The ladybug and the praying mantis are two beneficial insects that can help rid a farm or garden of aphids, mites, mealy bugs, and grasshoppers. Trichogramma is a small wasp that will destroy moth eggs, squash borers, cankerworms, cabbage loopers, and corn earworm. Many beneficial insects are available to farmers and gardeners in large quantities through catalog or online retailers. Given that just one ladybug can consume fifty or more aphids per day and can lay up to one thousand fertile eggs, the overall cost of purchasing ladybugs is much lower than the cost of chemical insecticides.

Organic farmers and gardeners also often rely on so-called insecticide crops. Garlic planted near lettuce or peas will deter aphids. Geraniums or marigolds grown close to grapes, cantaloupes, corn, or cucumbers will deter Japanese and cucumber beetles. Herbs such as rosemary, sage, and thyme planted by cabbages will deter white butterfly pests. Potatoes will repel Mexican bean beetles if planted near beans, and tomatoes planted near asparagus will ward off asparagus beetles. Natural insecticides such as red pepper juice can be used for ant control, while a combination of garlic oil and lemon may be used against fleas, mosquito larvae, houseflies, and other insects.

Organic farming relies on the physical control of weeds, especially through the use of cutting (cultivation) or smothering (mulching and hilling). The cultivation method involves the shallow stirring of surface soil to cut off small developing weeds and prevent more from growing. This can be done with tractors, wheel hoes, tillers, or, for small gardens, hand hoes. Mulch is a soil cover that prevents weeds from getting the sunlight they need for growth. Mulching with fully biodegradable materials can help build soil fertility while controlling weeds. Plastic mulches can be used on organic farms as long as they are removed from the fields at the end of the growing or harvest season.

Organic farming offers a safe alternative to the use of synthetic chemicals in the production of food. Growing public awareness of the possible toxic effects of the traditional agricultural reliance on synthetic chemicals is reflected in a growing demand for organically produced foods, leading to significant growth in the organic food industry.

Further Reading

Coleman, Eliot. *New Organic Grower: A Master's Manual of Tools and Techniques for the Home and Market Gardener, 30th Anniversary Edition.* Chelsea Green, 2018.

Duram, Leslie A. *Good Growing: Why Organic Farming Works.* U of Nebraska P, 2005.

Fortier, Jean-Martin. *The Market Gardener: A Successful Grower's Handbook for Small-Scale Organic Farming.* New Society Publishers, 2014.

Fossel, Peter V. *Organic Farming: How to Raise, Certify, and Market Organic Crops and Livestock.* Voyageur Press, 2014.

Hamilton, Geoff. *Organic Gardening.* Rev. ed., Dorling Kindersley, 2008.

Koepf, Herbert H. *The Biodynamic Farm: Agriculture in the Service of the Earth and Humanity.* Anthroposophic Press, 2006.

Kristiansen, Paul, Acram Taji, and John Reganold, editors. *Organic Agriculture: A Global Perspective.* Cornell UP, 2006.

Lampkin, Nicolas. *Organic Farming.* Rev. ed., Diamond Farm Enterprises, 2002.

Paarlberg, Robert. "Organic and Local Food." *Food Politics: What Everyone Needs to Know.* Oxford UP, 2010.

"USDA Organic." *US Dept. of Agriculture,* 18 Oct. 2016, www.usda.gov/wps/portal/usda/usdahome?navid=organic-agriculture. Accessed 30 Nov. 2016.

P

Paleoecology

—Micah L Issitt

FIELDS OF STUDY

Agronomy; Agriculture; Ecology; Soil Science; Archeology; Ecosystem Management; Biotechnology; Environmental Sciences; Environmental Studies; Botany.

PRINCIPAL TERMS

actuopaleontology: the linguistic act of applying organismic features found on modern living animals to extinct organisms

brachiopod: any of a phylum of shelled marine animals that have formed valved shells superficially similar to such mollusk species as clams and oysters

foraminiferan: a single-celled aquatic animal with a shell composed of compacted minerals or grains of sand from surrounding sediment

paleoclimatology: the study of the climate of an extinct ecosystem as determined through geological evidence and the distribution of fossils

paleoecology: the study of the interrelationships between extinct animals and the paleoenvironments in which they lived

paleoenvironment: the extinct environment of a certain area during a chosen point in the history of the region

parsimony: the theory that, given unexplained phenomena, one should assume the simplest possible explanation for the observed data

substantive uniformitarianism: the theory that extinct animals would have been similar in both structure and behavior to living organisms, because the processes, rates of change, and the terrestrial materials have been relatively stable throughout the history of the earth

temperate zone: those intermediate latitude zones of the earth marked by seasonal changes in climate

SUMMARY

Paleoecology is a branch of ecology and paleontology focused on studying the interrelationships between extinct organisms and the paleoenvironments in which they lived. The study of paleoecology developed when geologists and paleontologists began using fossils and geological data to examine environmental conditions in ancient ecosystems. Paleoecology has helped scientists understand the nature of environmental changes occurring on Earth and has enabled scientists to better predict how conditions might continue to change if global warming continues.

DEVELOPMENT OF PALEOECOLOGY

Ecology is a branch of biological and physical science that studies the relationships among individual organisms, populations, and the environments in which they live. Paleontologists developed a branch of research known as paleoecology, which focuses on the interrelationships between extinct organisms and the paleoenvironments in which they lived.

Early paleontologists often made inferences about ancient environments from the types of fossil organisms found in various sediment samples. Paleontologists uncovering marine fossils from sediment in a modern desert, for instance, would likely infer that the sediment in the area was once part of a submerged ocean environment. Similarly, an abundance of fossils representing herbivorous organisms might indicate that there was once abundant vegetation in a certain fossil site.

American geologist Kirk Bryan (1888–1950) is often credited as one of the scientists responsible for the emergence of paleoecology as a distinct field of paleontological research. Bryan's research in the 1920's and 1930's focused on how to understand weather patterns in ancient environments by examining aspects of soil chemistry and the occurrence of fossilized pollen grains. Bryan went on to train a number of prominent students who later made major contributions to paleoecology and other branches of geological research.

Paleoecology utilizes data derived from the study of both paleozoology (the study of fossilized animal species) and paleobotany (the study of fossilized plant species) in an attempt to reconstruct assemblages of organisms that occupied ancient environments. In addition, paleoecologists utilize chemical analyses of soil and geological data regarding the movement of continents. Taken as a whole, these data allow paleoecologists to create models of extinct ecosystems that help them to understand how evolutionary drives and pressures create communities within specific environments. The study of these ancient ecosystems can provide insight for ecologists studying modern ecosystems and can help to create an understanding of how changes in environmental conditions alter the evolutionary path of an ecosystem.

THEORETICAL CONCEPTS OF PALEOECOLOGY

One of the most important theoretical underpinnings of paleoecology is the concept of substantive uniformitarianism, which is the theory that the physical and chemical properties of the earth have remained unchanged throughout the history of the planet. The concept of uniformitarianism was developed in the nineteenth century and popularized by scientists such as prominent Scottish geologist Charles Lyell and English scientist William Whewell, who believed that the properties and processes of the earth were universal constants that could be applied equally to both ancient and modern environments.

Since the nineteenth century, scientists have come to understand that the physical and chemical properties of the earth are not constant, and that they have changed considerably during the 3.4 billion years of life on the planet. Modern ecologists apply an altered version of uniformitarianism, which assumes that the properties of the earth have varied, but that they have remained relatively similar during most of life's history on the planet.

Applied to paleoecology, uniformitarianism holds that the materials that make up the earth and the forces that drive chemical reactions and geological processes are similar today to those active in ancient ecosystems. This assumption allows paleoecologists to tentatively apply modern theories about ecological relationships to ancient ecosystems. For example, paleontologists might assume that ancient microorganisms were able to gain energy by utilizing sulfur in the same way as modern sulfur-metabolizing bacteria, because the chemical reactions that drive the metabolism of sulfur functioned the same in the past as they do in the modern environment.

Similarly, the theory of actuopaleontology (also called analogy) holds that the features and behaviors of modern organisms are similar to the features and behaviors of ancient organisms. In general, actuopaleontology allows paleontologists to assume that extinct organisms behaved and functioned in similar ways to modern organisms. This applies not only to the structure and biological function of individual organisms but also to the formation of multispecies communities.

Therefore, when paleontologists observe a beak-like structure on a fossilized organism, they will often assume that the beak was used like the beaks of modern animals living in similar environments. Similarly, paleontologists often assume that organisms in ancient ecosystems filled the same roles as similar organisms in modern ecosystems of the same type. Paleontologists might therefore expect that organisms in an ancient marsh ecosystem would have displayed behaviors and environmental roles that are similar to organisms in modern marshes.

Research occasionally invalidates hypotheses developed using the principle of actuopaleontology. An extinct organism might, for instance, evolve a certain feature that is similar to features on modern organisms, but it might have used this feature for a different purpose. Similarly, ancient ecosystems that are superficially similar to modern ecosystems might be found to contain vastly different types of organisms with different sets of relationships not seen in modern ecosystems.

Whether or not to assume hypotheses based on the theories of actuopaleontology or uniformitarianism can be determined with the help of parsimony, defined as the idea that, given an unexplained phenomenon, it is best to assume the simplest possible explanation of the available data. Parsimony is a core concept that can be applied to all scientific thought because it discourages scientists from positing unlikely and overly complicated explanations.

Most scientists, for example, believe that life first originated on Earth about 3.4 billion years ago; however, some scientists have postulated that life came to Earth from elsewhere in the galaxy, carried on meteors from a distant planet. Although either explanation could be correct, the former is the simplest explanation in that it does not propose the extra assumption that life evolved elsewhere in the universe.

Applied to paleoecology, parsimony dictates that paleontologists utilize theories that encourage the simplest and most direct explanation. The principles of actuopaleontology and uniformitarianism are parsimonious, in that they assume that conditions in the past are similar to conditions in the present. While this may not always hold true, it is the simplest method for explaining similarities between extinct organisms and ecosystems and their modern counterparts.

PALEOCLIMATOLOGY

One of the most important steps to understanding extinct ecosystems is to determine the climate of the ecosystem in question. Geological data provide a variety of evidence regarding the climate of paleoenvironments. For instance, large deposits of coal are the result of decomposing woody plants, whose existence indicates the presence of a lush, humid forest system. By contrast, some ancient soils display characteristic signs of extended ice or frost coverage; these ancient soils include glacial tills, which are pulverized or striated rock segments that bear the scars of contact with moving glaciers. Rocks that show striations left by standing water as well as cracks resulting from heating and drying were exposed to an ancient monsoon system, in which hot, dry seasons alternated with periods of intense rainfall and flooding.

Fossils also can be used as indicators of climate conditions. Ecologists have noted that diversity of species is greatest near the equator and declines at higher latitudes; paleontologists have discovered the same trend in the diversity of fossil assemblages from different areas. This phenomenon can be partially explained by the warm and relatively stable climate near the equator, leading to the proliferation of plants and microorganisms that form the base of the food chain. Farther from the equator, the environment becomes more variable.

In the broad temperate zone, organisms are still plentiful, but diversity is lower overall than in equatorial zones. Here organisms are more generalized, filling multiple niches because food is less plentiful than in zones closer to the equator. The number of species continues to drop in the higher latitudes, where arctic weather conditions dominate and overall diversity is low.

By measuring the diversity of fossil species in a certain area, paleoecologists can begin to understand the relative climate of the region during the time that the fossils were deposited. By examining a diversity of fossils over time, paleoecologists also can begin to observe climate change in ancient environments.

An example of this approach is the research of paleontologist Warren O. Addicott, whose 1969 study of shallow-water mollusk fossils from the west coast of Mexico and the United States indicates that certain species of mollusk were gradually moving southward during the early portion of the Tertiary period (65 to 2.5 million years ago). Addicott believed that this movement indicates a pattern of global cooling that led to changing distributions of species requiring warm, tropical waters for feeding and breeding. Furthermore, Addicott's research also indicated that a pronounced period of global warming existed during the later part of the Tertiary, again demonstrated by the movement of mollusk species farther northward toward the upper edges of the United States.

Marine fossils are the most useful fossils for determining the specifics of ancient climate conditions. Brachiopods, which are members of a phylum of shelled, marine animals, first appear in the fossil record in the Cambrian period (542 to 488 million years ago) but are still represented by more than two hundred living species in modern marine environments. Brachiopods appear to be highly climate specific. Paleontologists have found that characteristic assemblages of brachiopod species appear at each climate zone. Because extinct brachiopod species appear to have changed little in millions of years of evolution, paleoecologists can examine the distribution of modern brachiopods and, by applying the theories of analogy, parsimony, and uniformitarianism, can assume that ancient brachiopods tended to exhibit the same climatic preferences as their modern counterparts. Paleontologists can therefore use the distribution of these characteristic brachiopod faunas to track changes in climate distribution in periods as distant as the late Cambrian.

Organisms that form shells are especially useful for studying marine climates because the process of shell formation is closely tied to the chemical composition of the marine environment. Oxygen isotopes occur naturally in marine environments, and they occur in multiple varieties. In warm environments, certain types of oxygen isotopes tend to disappear from the environment as they engage in chemical reactions and dissolve in the surrounding water. Therefore, in warmer waters, only the more stable types of oxygen isotopes remain. By examining the fossilized shells of ancient organisms, paleoecologists can measure the proportion of stable to unstable oxygen isotopes and can attempt to determine whether the organism lived in a warmer or cooler environment.

To determine climate in ancient terrestrial environments, paleoecologists often attempt to determine the closest living relatives of species represented in the fossil record of the paleoenvironment. For instance, paleontologists have used the distribution of fossil rodent species as an indicator of alternating periods of warm, productive climate and periods of cooler climate with less abundant food sources during the Pleistocene period (1.8 million to 10,000 years ago). In other cases, paleontologists have theorized that the presence of fossilized reptiles indicates a mean temperature in keeping with the temperature favored by similar-sized reptiles in the modern environment.

Paleontologists also have used the distribution of fossilized pollen grains to investigate climate change. Because plants produce pollen on an annual cycle, pollen grains may appear in layers in the sediment, with each layer representing a subsequent year in the history of the environment. Paleontologists can therefore retrieve a core sample of sediment from the environment of interest and then examine the distribution of pollen at each layer. In times of warmer weather and more rainfall, more pollen will appear in the sample, while cooler or drier temperatures will result in fewer pollen grains. By examining the distribution of pollen grains through longer periods, paleoecologists can begin to get an image of long-term trends in climate change.

PALEOECOLOGY OF THE PLEISTOCENE & HOLOCENE EPOCHS

Utilizing data from both geological sources and the distribution of fossil species, paleontologists have created a detailed model of the paleoecology of the Pleistocene period, when the earth experienced its most recent period of global cooling. The Pleistocene is characterized by extended glacial periods, when thick sheets of ice covered large portions of the temperate zone. These periods of intense cooling alternated with interglacial periods, when the climate grew warmer and the glaciers melted across much of the landscape.

Evidence for climate change in the Pleistocene has been derived partially from the distribution of fossil species that appear to be characteristic of certain climate zones. Single-celled organisms called foraminifera are some of the most important fossil organisms of the period. These tiny shelled microbes are associated with the growth of algae, and their distribution is tied closely to temperature and other climatic variables. The tiny shells of foraminiferans are abundant in many types of ocean sediment, and paleoecologists have used the proportion of oxygen isotopes, remaining from the pulverized shells of foraminiferans, to track shifting climatic patterns during the Pleistocene.

Paleontological knowledge of the Pleistocene ice age also has benefited from detailed investigations of fossilized pollen. The distribution of pollen grains shows distinct periods of increased productivity followed by periods of greatly reduced pollen production. Paleontologists have taken this as additional evidence that the Pleistocene experienced waves of glacial and interglacial conditions.

The Holocene epoch began approximately ten thousand years ago and continues into the twenty-first century. Paleoecologists believe that the Holocene represents an intermediate period between the last ice age of the Pleistocene and the next ice age. The climate has continued to fluctuate, and the earth experienced a period of cooling, beginning in the seventeenth century and ending at the beginning of the twentieth century. Sometimes called the little ice age, this period had lowered temperatures across most of the temperate zone, but it did not affect the globe as a whole.

Much evidence supports the assertion that human activity is accelerating the process of global warming across the planet, though scientists are not universally in agreement about the causes or scope of emerging climatic trends. Paleoecological research has helped scientists to understand how the environment will change if current trends in global warming continue. The study of ancient ecosystems indicates that further global warming will force many species to migrate to stay within comfortable climatic zones. Past periods of both global warming and cooling also have resulted in waves of extinction, during which thousands of species disappear from the earth as they are unable to adapt to changing environmental conditions.

FURTHER READING

Allmon, Warren, and David J. Bottjer. *Evolutionary Paleoecology.* New York: Columbia University Press, 2000. Thorough investigation of modern paleoecological principles with a focus on evolutionary theory and the evolutionary relationships existing in paleoenvironments. Focuses on speciation and other large-scale evolutionary changes that can be observed from aspects of the fossil record.

Benton, Michael J., and D. A. T. Harper. *Introduction to Paleobiology and the Fossil Record.* New York: Wiley-Blackwell, 2009. Introductory text covering a variety of topics in paleontological research. Chapter 4 covers theories in paleoecology and the reconstruction of ecological relationships from fossilized evidence.

Briggs, Derek E. G., and Peter R. Crowther, eds. *Paleobiology II.* New York: Wiley-Blackwell, 2001. Introduction to the study of paleobiology aimed at students of related disciplines. Contains a discussion of trace fossils, body fossils, and the fossilization process, and of methods used to evaluate biological activity from fossil remains.

Foote, Michael Joseph, et al. *Principles of Paleontology.* New York: Macmillan, 2007. Revised edition of a classic text on paleontology. Provides a basic introduction to field and laboratory methods used in the study of fossils and prehistoric life. Chapter 9 discusses the basic theories of paleoecology research.

Martin, Anthony J. *Introduction to the Study of Dinosaurs.* New York: Wiley-Blackwell, 2006. Introductory text covering a variety of concepts in paleontological research and in dinosaur behavior and classification. Includes a discussion of paleoecology in relation to dinosaur behavior.

Miller, William Charles. *Trace Fossils: Concepts, Problems, Prospects.* St. Louis, Mo.: Elsevier, 2007. Updated introduction to the history and modern methods used in trace-fossil research. Chapter 23 contains a discussion of using trace fossils to investigate paleoecology, including environmental conditions and animal behavior.

Shipman, Pat. *Life History of a Fossil: An Introduction to Taphonomy and Paleoecology.* Cambridge, Mass.: Harvard University Press, 1993. Thorough introduction to taphonomy and the use of fossils in reconstructing the paleoecology of extinct environments. Contains information on fossilized plants and animals and discussions of fossil assemblages and their use in reconstructing environmental variables.

Weishampel, David B., and Halszka Osmolska. *The Dinosauria.* Berkeley: University of California Press, 2004. Introduction to dinosaur species and the study of prehistoric environments. Chapter 26 contains a discussion of paleoecology in relation to dinosaur behavior, including detailed examinations of the ecology of herbivory and carnivory in dinosaurs.

PALEOSOLS

—*Michael M. Kimberley*

FIELDS OF STUDY

Agronomy; Agriculture; Ecology; Soil Science; Archeology; Ecosystem Management; Biotechnology; Environmental Sciences; Environmental Studies; Botany.

PRINCIPAL TERMS

caliche: a type of soil or paleosol that contains a high proportion of calcium carbonate, calcium sulfate, or both

clay mineral: a type of mineral that is the most common product of soil formation; it is composed of silicon, oxygen, hydrogen, usually aluminum, and possibly other elements

partial pressure: the proportion of a gas mixture (for example, the atmosphere) that a particular type of molecule (for example, carbon dioxide) represents

soil: all material that has been substantially altered at the earth's surface by interaction with the atmosphere, living things, or both, and that has not been laterally displaced subsequent to that alteration

soil horizon: a distinct layer in a soil

SUMMARY

Paleosols are ancient soils that have been buried. Although natural acids (largely dissolved carbon dioxide) are almost entirely supplied by the atmosphere during modern soil formation, that does not seem to have been the case with all paleosols, particularly the most ancient paleosols and those that are associated with ore deposits. Paleosols that were produced by ancient atmospheric gases may record the environmental conditions of the ancient earth.

ANCIENT SOILS

Soil is one of the best-known, yet least precisely defined, geologic entities. Many definitions would not exclude beds of graded sediment; other definitions would not exclude sediment that has experienced only mild physical and chemical alteration. Additionally, many soil science textbooks restrict "soil" to that lying within the depth of plant roots, but such a definition is inappropriate for Precambrian and Early Paleozoic soils, which formed before the evolution of rooted plants.

The rock record of the earth covers the past 4.28 billion years (out of a total earth history of 4.6 billion years). Throughout the known rock record, soil formation (weathering) has involved more energy than other geologic processes, such as mountain building. Mountain building is driven by the internal energy of the earth (for example, by natural radioactivity), whereas soil formation is driven by solar energy and by chemical reactions between acidic

Dessication crack preserved as a cast on the base of a sandstone, Inverness Formation (Pennsylvanian), Cape Breton Island, Nova Scotia. Image by Michael C. Rygel via Wikimedia Commons CC-BY-SA-3.0.

atmospheric gases and exposed rock. Solar energy fuels photosynthetic plants, which concentrate the most abundant atmospheric acid (carbon dioxide); subsequent decay of these plants releases concentrated carbon dioxide into soil waters, which thereby become acidic and dissolve minerals. The downward solar flux of energy, which enhances weathering, currently is about 7,500 times greater than the upward flux of internal energy. The internal energy of the earth has progressively decreased with time as radioactive elements decay; solar radiation has progressively increased. In addition, the input of solar energy has exceeded the output of internal energy by at least a factor of 1,000 since the beginning of the rock record.

The collective volume of soils produced throughout earth history may have been comparable to the present volume of the continental crust, but only an insignificant volume of these soils has become preserved by burial within the crust. Paleosols now constitute a smaller proportion of continental crust than most other well-known rock types. The proportion of paleosols appears to be particularly small in the oldest rocks. In these rocks, the preserved portion is so small and the conditions for preservation of paleosols appear to have been so peculiar that it is dangerous to make sweeping interpretations of ancient earth environments based on paleosols.

PALEOSOLS & SEDIMENTARY ROCK

Virtually all old soils have been eroded to become sediment rather than buried to become paleosols. Intense weathering causes a high proportion of rock to dissolve, and the remnant soil becomes rich in insoluble elements such as aluminum. Erosion of this soil produces aluminum-rich sediment, which usually accumulates in the shallow ocean as a clay-rich rock (mudrock), such as shale. Sedimentary rocks therefore may be environmental indicators, and the vastly greater volume of sedimentary rock has resulted in most interpretations of ancient environments coming from study of sedimentary rock rather than from study of paleosols. Environmental study of paleosols is less constrained than that of sedimentary rock, however, because the unweathered parent of the weathered material (soil) is observable beneath a paleosol, whereas the parent for a weathered sediment generally cannot be deduced with certainty; the parent rock for any given sediment may lie thousands of miles from where the sediment accumulates—for example, the Andean Mountain parent for much of the Amazon deltaic sediment.

Paleosols are known from all major divisions of earth history, from the Archean eon to the present epoch. The proportion of paleosols to other rock types roughly increases with time, consistent with the theory of continental growth through earth history; an increase in the area of exposed continents would lead to an increase in the volume of soil. Although the area of exposed continents probably was smallest in the Archean eon, well-preserved Archean paleosols are known from Canada and from South Africa, where Archean continental crust never has been so deeply buried that regional metamorphism could destroy evidence of the weathering processes that produced the paleosols. The area of exposed continents apparently increased dramatically from the Archean to the Proterozoic, so Proterozoic paleosols are correspondingly more abundant.

The role that plant life may have played in the weathering of these Proterozoic soils is unclear. Plants currently concentrate so much carbon in the upper portion of a typical soil that oxidation of this carbon provides more carbon dioxide to the underlying soil than does diffusion of carbon dioxide directly from the atmosphere. Much of this carbon dioxide comes from roots, but roots did not evolve until after the Proterozoic. Plant life in Precambrian (Archean plus Proterozoic) soils was limited to bacteria. Mats of photosynthesizing cyanobacteria (also called blue-green algae) lay at the surface, and other bacteria occurred deeper in the soil. Modern soil contains abundant bacteria below the surface, but the soil surface generally is occupied by complex photosynthesizing plants—for example, trees—instead of by cyanobacteria. A partial image of Precambrian soils may be obtained by studying cyanobacterial mats that grow in environments too harsh for more complex plants—for example, on salt flats and around geysers. These environments are not only harsh but also highly variable. The chemical composition of water on a modern salt flat may vary from hypersaline to nonsaline following a thunderstorm. The water temperature around a geyser may vary by several tens of degrees within a few minutes. The ability of cyanobacteria to withstand such variations may indicate that the environmental conditions of Precambrian soils either were consistently harsh or were more variable than those of more recent soils.

GEOGRAPHIC VARIATION

Ancient paleosols generally are too scarce for study of the geographic variation of soil types. Geographic variations in rainfall and temperature are the two prime variables that control the distribution of modern soils because the atmosphere mixes so rapidly that any local variation in the partial pressure of oxygen or carbon dioxide rapidly becomes globally homogenized. Molecular oxygen and carbon dioxide also would have been evenly distributed in ancient atmospheres, but their proportions of the atmosphere may have varied throughout earth history. The oldest paleosols therefore are studied to examine variation in the earth's atmosphere through earth history, whereas the youngest paleosols are studied to learn about geographic variation in climate. Paleosols and other indicators record dramatic variation in climates on the earth during the past 2 million years, related to the growth and melting of enormous continental glaciers.

Tropical climates produce distinct soils and thus distinct paleosols. Both wet and dry tropical climates characteristically produce yellow-to-red soils because of oxidation and retention of iron in the soil. Wet tropical conditions can produce soils that are so rich in aluminum that these soils may be mined profitably. In such soils, the more soluble elements have been leached away by groundwater draining lush vegetation. Decay of the vegetation produces acids that attack even quartz, leaving only aluminum-rich minerals. Aluminum-rich paleosols are among the oldest (Archean) paleosols on earth, so wet tropical conditions appear to have existed long ago, despite the fact that solar radiation should have been much smaller during the Archean eon.

Warm, dry climates may produce little soil of any kind. For example, more characteristic of the Sahara than the "sand seas" that are commonly illustrated in documentaries are vast stretches of bare rock. Caliche is a characteristic soil and paleosol produced in such a climate. Caliche generally forms by precipitation of calcium carbonate from upwardly moving groundwater as the groundwater approaches the earth's surface. The upward decrease in pressure may allow carbon dioxide to be released, just as carbon dioxide is released upon opening a bottle of soda pop. Release of carbon dioxide favors precipitation of calcium carbonate; this precipitation may be aided by evaporation of the groundwater—which points to a dry climate for the formation of an ancient caliche. A caliche paleosol generally may be interpreted to record an ancient dry climate, even if the precipitation of calcium carbonate was not related to evaporation within soil but was simply the result of release of carbon dioxide. The original excess of carbon dioxide could have been provided by the escape of gases to the earth's surface during metamorphism of carbon-bearing sedimentary rocks at great depth. The abundant infiltration of rainwater in a wet climate would dissolve calcium carbonate from soil, whatever its origin, so the preservation of calcium carbonate in a caliche paleosol generally records a dry climate, even if the precipitation of calcium carbonate were induced by deep crustal processes independent of climate.

ELEMENTAL COMPOSITION & MINERALOGY

Interpretation of the elemental composition and mineralogy of a paleosol generally is controversial, especially for paleosols older than 544 million years (Precambrian paleosols), because paleosols potentially have experienced substantial modification (diagenesis) after burial. One of the most consistent chemical peculiarities of Precambrian paleosols is that they contain extreme ratios of potassium to sodium, unlike modern clayey soils. No known weathering process could fractionate sodium from potassium so severely. Precambrian paleosols commonly contain more than ten times as much potassium as sodium, whereas these two elements generally behave similarly under modern weathering conditions. This potassium in Precambrian paleosols mostly occurs in fine-grained, aluminum-rich mica. The potassium either is a record of pervasive diagenetic alteration of Precambrian paleosols or indicates that, unlike modern soils, they did not form as a result of atmospheric acid-forming gases. The majority of Precambrian paleosols could represent alterations on the ancient land resulting from exhalation of acid-containing mud from deep in the earth. Although paleosols that are older than 544 million years generally have the greatest ratios of potassium to sodium, paleosols that are 245 to 544 million years old (Paleozoic paleosols) also are more potassium-rich than are modern soils. In these Paleozoic paleosols, the potassium-bearing mineral typically is a clay mineral called illite. In illite-bearing paleosols, some investigators attribute the high potassium content to peculiar weathering conditions, whereas others attribute it to precipitation of potassium from through-flowing groundwater long after burial of the soil.

FURTHER READING

Bloom, Arthur. *Geomorphology: A Systematic Analysis of Late Cenozoic Landforms.* 3rd ed. Long Grove, Ill.: Waveland Press, 2004. This college-level text covers the basics of geomorphology. Includes a section on the development of soils. Index and bibliography.

Chernicoff, Stanley, and Donna Whitney. *Geology: An Introduction to Physical Geology.* Upper Saddle River, N.J.: Prentice Hall, 2006. This is a good overview of the scientific understanding of the geology of the earth and surface processes. Includes a section on the development of paleosols. Includes a link to a Web site that provides regular updates on geologic events around the globe.

Kapur, Selim, and George Stoops, eds. *New Trends in Soil Micromorphology.* New York: Springer, 2010. This text is a compilation of articles suited for graduate students and researchers. Discusses the identification and analysis of paleosols, soil micromorphology of the Quaternary period. Articles address vertisols, andosols, and other paleosols.

Kimberley, M. M., and D. E. Grandstaff. "Profiles of Elemental Concentrations in Precambrian Paleosols on Basaltic and Granitic Parent Materials." *Precambrian Research* 32 (1986): 133–154. All papers in this issue are devoted to Precambrian paleosols. This particular paper compares Precambrian paleosols that developed on basalt with those that developed on granitic rocks.

Reinhardt, J., and W. R. Sigleo. *Paleosols and Weathering Through Geologic Time: Principals and Applications.* Special Paper 216. Boulder, Colo.: Geological Society of America, 1988. Several paleosols in the United States (from Pennsylvania, Kentucky, Georgia, Alabama, Colorado, and California) are described in this volume, along with field methods and theoretical models for the development of ancient paleosols.

Retallack, G. *Laboratory Exercises in Paleopedology.* Eugene: University of Oregon Press, 1985. Practical exercises are outlined for the study of paleosols. Examples of paleosols are described from several localities around the world.

__________. *Soils of the Past: An Introduction to Paleopedology.* 2d ed. New York: Blackwell Science, 2001. The best and most comprehensive coverage of this subject. This text is an excellent source of information on paleosols, discussing soil classification and formation, identifying paleosols, and their characteristics, and pointing out the factors involved in paleosol formation. The author covers the soil fossil records with discussion of ancient landscapes and ecology. Written for undergraduate students.

Rutter, Nat. "International Year of Planet Earth 4: Utilizing Paleosols in Quaternary Climate Change Studies." *Geoscience Canada* 36 (2009): 65–77. This article discusses paleosols as valuable tools for determining ancient environments. The author describes the process of identifying paleosols to determine past vegetation and climate. The article also provides an explanation of paleosol dating techniques. Best suited for researchers and graduate students.

Samama, J. C. *Ore Fields and Continental Weathering.* New York: Van Nostrand Reinhold, 1986. Some paleosols are ore deposits for aluminum or nickel; part of this book reviews the characteristics and metal-concentrating processes that produce such paleosols. Another part reviews a different class of paleosol-related ore in which subsurface holes in limestone coincidentally or subsequently become filled with metal-precipitating solutions. Research subsequent to this book has shown that the dissolution of ore-bearing limestone probably shortly precedes precipitation of the lead-zinc minerals and that both the limestone-dissolving and metal-precipitating fluids arise from deep in the earth.

Schaetzl, Randall J, and Michael L. Thompson. *Soils: Genesis and Geomorphology.* New York, NY: Cambridge University Press, 2016. Print. This text provides a large amount of fundamental content for studying soils. Part I discusses soil morphology and classification. Part II discusses the formation of soil and other soil dynamics. In the final part the authors discuss paleosol characteristics and reconstruction of environments through the study of paleosols.

Weil, Ray R, and Nyle C. Brady. *The Nature and Properties of Soils.* Harlow: Pearson Prentice Hall, 2017. Print. This advanced text requires some background in geology and geomorphology. However, it contains a wealth of information on the development of soils.

Wright, V. P., and Alfred Fischer, eds. *Paleosols: Their Recognition and Interpretation.* Princeton, N.J.: Princeton University Press, 1986. This volume emphasizes paleosols that occur outside the United States, particularly those in the British Isles, Spain, and New Zealand. Chapter 6 reviews selected paleosols in the western United States.

PASTURE LANDS

—José F. González–Maya and Eduardo Ponce

FIELDS OF STUDY

Agronomy; Agriculture; Ecology; Ecosystem Management; Biotechnology; Environmental Sciences; Environmental Studies; Botany.

SUMMARY

Pasture land is an anthropogenic biome represented by lands where predominant plants are grasses, grasslike plants, forbs, or shrubs. These lands are managed as grazing systems designed for the production of domestic animals for consumption, including production for meat, milk, and other major animal products. Since historical times these sites have been predominantly located in natural grasslands, savannas, steppes, wetlands, open and dense shrublands, tropical evergreen forests/woodlands, boreal evergreen forests/woodlands, boreal deciduous forests/woodlands, temperate needle-leaf evergreen forests/woodlands, some deserts, and tundra. For millennia, the widespread presence of human populations and their activities altered these ecosystems around the world, both intentionally and unintentionally. These effects started with low-impact practices like hunting and gathering, but through the centuries humans developed and controlled many tools that allowed them the permanent use of lands for agriculture, forestry, livestock, and settlements. These activities changed many natural processes and with them the composition and structure of flora and fauna, as well as the hydrologic and biochemical cycles at local, regional, and global scales.

"Today, we live at a time in which it is necessary to reconcile conservation strategies with livestock management strategies."

Pasture land is an anthropogenic biome represented by lands where predominant plants are grasses, grasslike plants, forbs, or shrubs. These lands are managed as grazing systems designed for the production of domestic animals for consumption, including production for meat, milk, and other major animal products. Since historical times these sites have been predominantly located in natural grasslands, savannas, steppes, wetlands, open and dense shrublands, tropical evergreen forests/woodlands, boreal evergreen forests/woodlands, boreal deciduous forests/woodlands, temperate needle-leaf evergreen forests/woodlands, some deserts, and tundra. For millennia, the widespread presence of human populations and their activities altered these ecosystems around the world, both intentionally and unintentionally. These effects started with low-impact practices like hunting and gathering, but through the centuries humans developed and controlled many tools that allowed them the permanent use of lands for agriculture, forestry, livestock, and settlements. These activities changed many natural processes and with them the composition and structure of flora and fauna, as well as the hydrologic and biochemical cycles at local, regional, and global scales.

The grazing activity of this herd of roughly 3,000 sheep, pasturing in the shadow of the Centennial Mountains in Dubois, Idaho, helps create a mosaic of different densities and ages of sagebrush crown. This in turn helps wildlife, such as the sage grouse, thrive. A wide variety of pasture lands are managed as grazing systems for livestock—natural grasslands, savannas, steppes, shrublands and even some deserts and tundras, to name a few. Image by U.S. Agricultural Research Service.

DEVELOPMENT OF PASTURE LANDS

Historically, people managed livestock to produce food, directly as meat and dairy products and indirectly as animal power that increased cropland production. In the eighteenth century livestock were managed at local scales because technology and transport facilities were poor, and because livestock depended on the availability of food resources in areas where disease constraints allowed these activities. For this reason, pasture lands were kept geographically close to human settlements and barely distributed around the world. However, during the following centuries there was a considerable increase in human population around the world, as well as a growth in technology development. Due to these factors, the demand for livestock products expanded pasture distribution in almost every corner of the planet in order to produce enough commodities to sustain the growing population. As a result, the location and extension of pasture lands were undergoing important shifts. The increase of this biome exerted an enormous pressure in all wild lands, principally in the drier biomes of the Americas, Australia, central Asia, and southern Africa, but also in the moister wooded biomes of sub-Saharan Africa, Central America, and Eurasia. As a result of the increase of human use of natural resources and appropriation of the net primary production, the global extension of pasture lands increased considerably since the Industrial Revolution, growing from 3 percent of the ice-free land cover before 1700 to 26 percent in 2000.

Because of these developments, pasture lands have now become one of the most extensive anthropogenic biomes, occupying more than 26 percent of the ice-free land in the world in an area around 13,513,576 square miles (35 million square kilometers), including large areas where there previously was little or no livestock grazing in North America, South America, and Australia. In many areas, pasture lands have expanded to occupy virtually all the land that can be grazed and for which there is no other activity. Except in bare areas located in dry or cold deserts, or in dense forests, pasture lands are present to some extent in all regions around the world. In North America a large extension of the continent has been considered as a pasture, except in areas such as the northern extreme in Canada and Greenland, where the extreme temperatures impedes this land use. Also the rough mountain systems of the Rocky Mountains in western North America, both the Sierra Madre Oriental and Occidental in Mexico, the Andes Range and Patagonia along the southwestern coast of South America, and the Brazilian highlands have been difficult places to sustain livestock production. The Amazon rainforest is also considered one of the few places in the world that has not been transformed to a pasture land. In Africa, the scenario is similar; most of the continent has been grazed except the Sahara Desert in the north and the tropical forest in central Africa. In Eurasia, the extreme weather conditions and complexity of the mountain system such as the Himalaya Range, the Gobi Desert, the Scandinavian and Siberian regions, as well as the tropical rainforests at the southeastern extreme of the Asiatic continent, have disabled these areas to use the land as pasture, while the rest of the world has virtually been transformed to pasture lands.

Pasture land productivity differs in different sites on the planet and is mainly defined by climatic and soil conditions. Managed grazing is preferentially employed in areas that are much drier and stabler than the biome mean, while regions with large interannual precipitation and temperature variation strongly affect vegetal production, grazing capacity, and human living conditions. Savannas, grasslands, shrublands, and deserts support the largest extent of managed pastoral systems; these dry-land biomes cover about 51 percent of the total land area of the Earth (around 25,888,845 square miles or 67 million square kilometers), and they support 78 percent of the global pasture lands of the Earth. Other biomes also support managed grazing systems but in less proportion. For example, temperate deciduous and temperate evergreen broadleaf forests and woodlands support grazing systems, and about 656,374 square miles (1.7 million square kilometers) of tropical evergreen broadleaf forests (corresponding to 10 percent of all tropical forests on the planet) have been cleared for managed grazing. Nowadays pasture land area is expanding annually in wild regions such as the Amazon Basin, Congo, and southeast Asia rainforest.

Besides savanna, shrubland and desert are the most-used biomes for grazing systems, although they are limited in their soil type and the fertility associated with it. For instance, many types of grassland are supported in soils composed of aridisols, entisols, and alfisols that are not fertile enough to sustain a grazing system for a long period of time. In the colder boreal biomes, grazing takes place preferentially on alfisols and spodosols but not on frozen gelisols, which are the most common soil order found in these regions, the reason why these lands are still under low grazing pressure. In humid tropical regions ultisols represent the best place to maintain a grazing system, while oxisols that are also well represented in the Amazon Basin, Congo, and southeastern Asia are nutrient-poor soils for managed grazing systems.

The five countries with the most pasture lands are Australia (1, 698,850 square miles or 4.4 million square kilometers), China (1,544,409 square miles or 4 million square kilometers), United States (926,645 square miles or 2.4 million square kilometers), Brazil (656,374 square miles or 1.7 million square kilometers), and Argentina (540,543 square miles or 1.4 million square kilometers). However, based on the proportion of total land area that a country uses for grazing, Mongolia, Botswana, and Uruguay are the most-affected countries, with 80 percent, 76 percent, and 76 percent of pasture land coverage, respectively. Countries with the highest stocking rates are Malaysia with 320 animal units per square kilometer, India with 272 animal units per square kilometer, North Korea with 213 per square kilometer, and Vietnam with 184 per square kilometer.

CONSEQUENCES OF PASTURE LAND EXPANSION & MANAGEMENT

Although managed grazing systems have been developed throughout the world for thousands of years, the spatial extension and intensity of grazing systems have considerably increased in the last three centuries. Pasture land development is influenced not only by a complex range of environmental conditions but also by many social and economic factors that have a strong influence on extension and pressure. Throughout human history, pastures have been managed under different conditions leaving the land in a broad number of states. Pasture lands have been managed in different ways according to property and access rights. Historically, three main types of land tenure have been recognized: private (owned by an individual or a company), communal, and public property. The interaction of this management generally is noxious for the ecosystem as a result of a complex and inadequate set of rules imposed by state and local officials who pretend to control the use of natural resources. This complexity generally leads to conflicts among stakeholders who claim access to pasture lands.

However, not all grazing systems or practices lead to these negative impacts. Domestic cattle (cows, horses, and goats) may have positive, neutral, or negative impacts on vegetation structure and composition, depending on the grazing intensity and the site. In areas where grazing intensity is light to moderate, the habitat of wild ungulates, the effect can be positive in stimulating the growth of some species and increasing diversity in the landscape. For the most part, however, where livestock is raised, pasture land is aimed toward the goal of producing the greatest possible number of animals. At these sites, generally bounded by fencing, cattle cause plants to recover. In these cases the rate of extinction of species tolerant of grazing increases dramatically, but there are also increases in the rate of colonization of grazing-tolerant species or those that are not consumed by livestock. Gradually the original vegetation (before introducing livestock) is replaced, and there comes a point where even when grazing decreases, the vegetation generally does not return quickly to its original state and can take many decades to return, or even not return to its original state. The impacts on soil structure are also notable. Much of the nutrients are either recycled naturally in the ecosystem, or removed and not redeposited in the place of origin, and the extensive areas without plant cover created by overgrazing promote soil erosion. After many years of inadequate, intense, chronic, and uncontrolled management, many pasture lands are now exposed to three main processes that have negative effects on ecosystems: desertification, woody encroachment, and deforestation.

The bioclimatic and edaphic conditions under which managed grazing occur have, to some extent, contributed to the development of these three processes. Desertification has occurred in arid regions of the southwestern United States, Australia, South Africa, and Argentina as a result of chronic grazing and pronounced climatic variability with special influence of drought periods. Woody encroachment has occurred in semi-arid to mesic environments as a result of large-scale grazing, fire suppression, and climatic variability. Deforestation continues to expand principally in the humid tropics in South America, Africa, and Asia in part because of grazing development on infertile soils that often cannot maintain large-scale livestock production.

As the human population grows and land scarcity increases, intensification and improvements in traditional animal production should be implemented to satisfy human demand for livestock products. Unfortunately, most pasture productivity is declining in many arid regions, and many grazing systems are disappearing in tropical forests. These are being replaced by more concentrated grazing systems that can lead to greater degradation of pasture land. In the northeastern United States, northwestern Europe, and densely populated areas of Asia, animal production has become mechanized and dependent on external fertilizer and feed inputs. Industrial meat production is growing rapidly, and in general in these systems animal numbers exceed the carrying capacity of the land, and waste is saturating the surrounding environment causing habitat fragmentation and eutrophication of freshwater and marine ecosystems.

Intensification of animal production and grazing systems is likely to continue, and it will require the expansion of pasture lands and improve livestock production through a supplemental feeding. In both cases, however, expansive management will cause greater degradation processes and will intensify environmental impacts over the land surface. Even more, the emergence of large economies such as China, India, Brazil, Russia, Mexico, South Korea, Indonesia, and Turkey as new centers of demand and production will potentially promote the expansion of pasture lands globally. Humid tropical ecosystems represent the only viable way to expand global grazing systems beyond their current geographic extent, but this expansion into marginal areas has already more or less reached the limits imposed by climate and soil factors, not to mention the degradation of some of the most biodiversity-rich areas in the world.

The success of pasture expansion into forest will depend mainly on local and regional policies in concerned areas. But since the prospect of expansion on pasture lands is limited, the intensification of pasture production on the most suitable land, and the loss of marginal pastures, is likely to continue. Climate change is also likely to alter pasture land systems. The impact on pastures will be greater than in other anthropogenic biomes. While the cropland biome can be more easily manipulated by irrigation or use of fertilizers, pasture lands depend largely on weather conditions. On dry pasture lands the impact is projected to be dramatic, reducing forage yields by as much as 25 percent by 2030, but on the other hand, pastures located in cold areas are expected to benefit from rising temperatures.

The management and long-term conservation of pasture lands requires a broad understanding of the patterns and processes that have determined their structure and composition, as well as the impact that humans have had on these lands for centuries. Today, we live at a time in which it is necessary to reconcile conservation strategies with livestock management strategies, and this will only be possible by integrating the knowledge of many disciplines. For this integration to occur it is essential to promote social and governmental participation in the design, dissemination, and implementation of best-management practices involving the conservation of ecological processes in the ecosystems in order to maintain their goods and services. It is necessary that management policies be based on sound scientific data to ensure better management of ecosystems, and that these policies integrate environmental and social needs focused on restoring and preserving the health of ecosystems in the long term.

Pasteurization & Irradiation

—James L. Robinson, PhD

FIELDS OF STUDY

Biology; Chemistry; Biotechnology; Environmental Sciences; Environmental Studies; Food Science.

SUMMARY

Pasteurization and irradiation are processes that partially sterilize food in order to make it safe to eat, without substantially altering its nutritional content, structure, and taste. Pasteurization uses mild heat treatment, whereas irradiation makes use of ionizing radiation. Both reduce the levels of pathogenic (disease-causing) and spoilage microorganisms to a level that renders the food safe to eat provided that it is stored appropriately for no longer than the prescribed time. Irradiation can also be used on fresh fruits and vegetables to kill insects and to retard biological processes, such as ripening. At high levels, irradiation will fully sterilize food, packing material, and disposable medical items.

DEFINITION & BASIC PRINCIPLES

Pasteurization is the process of using mild heat to treat food, whereas irradiation, sometimes called radiation pasteurization or cold pasteurization, uses ionizing radiation. The primary purpose of each is to destroy microorganisms that would be pathogenic to human consumers, without significantly changing the food's attributes. In addition, these processes can be used to destroy microorganisms or enzymes that spoil food, leading to a longer shelf life and less waste. Irradiation can also be used on fresh fruits and vegetables to kill insects and to delay germination, ripening, or sprouting.

Pasteurization, used primarily with liquid foods, such as milk, fruit juices, and beer, refers to heat treatments that do not exceed 100 degrees Celsius (C), whereas heat sterilization (such as canning) uses temperatures of 100 degrees C or higher. In general, high-temperature short-time (HTST) pasteurization destroys undesirable organisms while minimizing deleterious effects on the food. In milk, comparable killing of microorganisms can be achieved by conventional pasteurization at 63 degrees C for thirty minutes or by HTST at 72 degrees C for fifteen seconds or 88 degrees C for one second. Sterilization of milk at ultrahigh temperature (UHT) typically requires 138 degrees C for two seconds.

For irradiation, the ionizing radiation used are gamma rays (generated from the decay of radioisotopes cobalt 60 or cesium 137), X rays, and electrons, the latter two generated by machines for such purposes. The operators of equipment involving ionizing radiation need to be protected from its effects. The ability subsequently to pasteurize or irradiate food should not compensate for best practices to minimize contamination of food before treatment. Moreover, treated food also needs to be protected from subsequent contamination.

BACKGROUND & HISTORY

Thermal and nonthermal processes have long been used to ensure the safety and storage of food. Cooking and smoking food were practiced in prehistoric times and likely permitted a survival advantage. Subsequently, drying, salting, and pickling were also used.

Pasteurization was developed in 1862 by French scientists Louis Pasteur and Claude Bernard, initially for the preservation of beer and wine. The pasteurization of milk became widespread in the early 1900s.

X-rays were discovered by German physicist Wilhelm Röntgen in 1895, and radiation emitted from uranium and other radioactive elements was discovered by French physicists Henri Becquerel, Marie Curie, and Pierre Curie shortly thereafter. Patents were issued for food preservation using ionizing radiation in 1905 and for the use of X rays to destroy *Trichinella* in pork in 1921.

By 2005, food irradiation was widely used in Asia, less often used in America and Eastern Europe, and rarely used in Western Europe, where consumer resistance is high. The primary use of irradiation is for herbs, spices, and dry vegetables, followed by root crops (such as potatoes and garlic, to inhibit sprouting), grains and fruits, meat, and seafood. Food irradiation is strictly regulated within each country, and all irradiated foods must be labeled with words such as "Treated by irradiation" and the international Radura logo. Global trade would be enhanced by harmonization of national regulations.

HOW IT WORKS

PASTEURIZATION. Heat kills pathogenic and spoilage microorganisms by disrupting their cellular structure and metabolism. In pasteurization, sufficient heat is applied to the food being treated to kill undesirable organisms, but without damaging the food itself. In this balance, not all undesirable organisms are destroyed, but they are reduced to such a level that the product, if stored appropriately, will be safe for consumption until its use-by date.

Although some liquid foods, such as beer and fruit juices, may be pasteurized after filling containers (with warm water or steam applied to raise the temperature appropriately), most are pasteurized in a vat process or a continuous-flow process and then packaged. The vat (or batch) process involves heating in a well-agitated tank for the required time and temperature. The vat process is suitable for relatively small-scale operations.

The continuous flow process became possible when plate heat exchangers were developed in the late 1920s and has been enhanced by the development of concentric-tube heat exchangers. It is particularly well-suited to HTST and large-scale operations. In a typical system, the liquid to be pasteurized flows in a continuous tube from a cooled holding tank, through a preheater, through the heater (which heats the fluid to the required temperature), through a holding tube (whose size coupled with the flow rate determines the length of time that the liquid is held at the specified temperature), and then cooled down for storage. In practice, the preheater acts as a precooler, extracting heat from the heated liquid, before it is further cooled, permitting 85 to 90 percent of the heat to be reclaimed. The temperature in the holding tube must be monitored to ensure that the desired temperature has been maintained and, if not, the flow must be automatically diverted back to the starting tank.

In both vat and continuous-flow processes, the system must be thoroughly cleaned between uses, but a single use of the latter may last for many hours.

IRRADIATION. Radiation kills pathogenic and spoilage organisms, as well as insects. It also retards germination, ripening, and sprouting. It does so by disrupting cell structure, cell metabolism, and, most importantly, DNA molecules, preventing further growth and reproduction. Irradiation exerts its effects by direct action of the radiation or indirectly, principally via the radiolysis of water, which leads to the generation of highly reactive chemical species, such as hydroxyl radicals and hydrogen peroxide. Smaller organisms are more resistant than larger ones, for instance, viruses compared with bacteria. Spores of species such as *Clostridia* that cause botulism are more resistant than vegetative cells. Gram-negative bacteria, including primary food pathogens *Escherichia coli* and *Salmonella*, are more sensitive than gram-positive bacteria. Gamma rays are more effective than x-rays, which in turn are more effective than electrons. These differences relate to the penetrating power of the radiation, with electron irradiation only suitable for treatment of surfaces or thin packages.

In the process of irradiation, the product is brought in line with the source of radiation for the requisite period of time. Electrons or x-rays are generated by machines for these purposes and can be turned on and off as required. Because gamma rays result from radioactive decay, they cannot be turned on or off; when not needed, the sources of radioactivity are stored in a large water tank that absorbs the radiation. Irradiation with electrons and x-rays is well suited to a conveyor-belt system that brings the product into the radiation beam. With gamma rays, the use of an overhead rail system is preferred. The packages of product to be irradiated are suspended from that system and moved so that the package can be bombarded from various sides and angles to ensure uniformity of treatment. In all cases, a dosimeter (or dose meter) must be periodically included to ensure that the material has received the required dosage.

In food irradiation, sufficient radiation is applied to destroy undesirable organisms (including insects) or to inhibit a biological process, without adversely affecting the nutritive value and sensory characteristics of the food. As with pasteurization, organisms may not be completely eliminated but are reduced to a safe level provided the food is stored appropriately for no longer than the permitted time. Radiation is able to penetrate packaging materials, which reduces the risk of contamination after treatment. On the other hand, packaging materials can be affected by the ionizing radiation generating radiolysis that may migrate to the food and affect its taste. Careful choice of packaging material, as well as adhesives and printed material, must be made to avoid such problems. The fats in foods are susceptible to breakdown, forming products with unacceptable taste, but this effect is minimized by irradiating foods high in fat while frozen. Irradiation in the absence of oxygen minimizes the generation of byproducts that can affect the color and taste of the food.

APPLICATIONS & PRODUCTS

PASTEURIZED PRODUCTS. Pasteurization is typically applied to liquids—milk is the best-known example. Most milk consumed around the world is pasteurized or heat sterilized. Before the development of milk pasteurization, more than 25 percent of food-borne diseases were attributed to milk and milk products. Many microorganisms, including pathogenic ones, survive well in milk. Combined with aseptic packaging technology, pasteurization makes milk less prone to spreading disease and less perishable. Fruit juices and beers may be flash pasteurized (HTST) to minimize spoilage. A few wines are pasteurized; wines with less than 14 percent alcohol are sometimes pasteurized (or ultra-filtered) to stop any further fermentation. Flash pasteurization is also used to make wines acceptable to strict Orthodox Jews. Liquid eggs can be similarly pasteurized; those in shell can also be pasteurized in a series of warm-water baths.

Nonliquid pasteurized products include cheese, almonds, smokeless tobacco, crabmeat, bread, and ready-to-eat meals. Pasteurized cheese is made from pasteurized milk, so the liquid is treated in this process, although the cheese is subjected to heat treatments as well. Almonds can be pasteurized with a steam treatment, designed to kill any microorganisms on the outside of the nut; pasteurization of almonds can also refer to their treatment with propylene oxide, but that treatment is more appropriately called chemical fumigation. Smokeless (or chewing) tobacco is pasteurized by heating to 85 degrees C. Crabmeat labeled as pasteurized is heated to 113 degrees C for one minute in sealed cans or plastic containers. Because this temperature is higher than 100 degrees C, it is not properly termed pasteurization. Nevertheless, this treatment does not kill all pathogens present, it merely reduces them to a safe level, and the product must be stored at refrigerated temperatures for no longer than prescribed to ensure its safety. Bread and ready-to-eat meals are usually pasteurized by microwaves, which generate heat in the product being treated.

IRRADIATED PRODUCTS. The extent of irradiation of a food will vary according to the desired end point and nature of the substance being irradiated. The absorbed doses are expressed in units of gray (Gy) and kilogray (kGy). Low doses (less than one kGy) will inhibit sprouting of potatoes, garlic, onions and other root foods, disinfect insects on fruits, grains and dry foods, delay ripening of fresh fruits and vegetables, and inactivate parasites on pork and fresh fish. Medium doses (one to ten kGy) will extend shelf life of strawberries, mushrooms, fresh fish, and meat (if stored at between 0 and 4 degrees C), destroy parasites in meats, control molds on fresh fruit, and destroy pathogenic and spoilage microorganisms in spices, raw or frozen poultry, meat, and shrimp. High doses (greater than ten kGy) will sterilize herbs, spices, meat, poultry, seafood, food additives (such as enzymes and natural gums), packaging materials (such as wine corks), disposable medical items (such as syringes, tubing, and gloves), and hospital food (especially for immune-compromised patients). In the United States, the only exception to a maximum of thirty kGy is for sterilizing frozen packaged meats for National Aeronautics and Space Administration (NASA) space flights. When in space, American astronauts have been eating irradiated foods, such as beef, pork, smoked turkey, and corned beef, since the beginning of the space program. Interestingly, milk and milk products are not good prospects for irradiation because it generates undesirable flavors.

In irradiation, gamma rays are used more often than electrons or x-rays, largely because of their greater penetrating power. Irradiation of fresh fruits and vegetables to destroy any mature or immature insects obviates any need to quarantine these products. In the past, such disinfection was done with methyl bromide, but its use is being phased out because it is an ozone-depleting chemical. The costs of food irradiation include high capital costs and modest operating costs. Irradiation facilities must protect workers from the radiation used; this involves thick walls and fail-safe design that prevents accidental radiation exposure to employees when in operation. Several methods are available to determine if a food has been irradiated, but biological methods that enumerate dead and live microorganisms of the species of concern are particularly useful because they not only show how many survived, but the total burden before irradiation, providing a check on good handling practices before treatment.

SOCIAL CONTEXT & FUTURE PROSPECTS

Pasteurization is a well-accepted technology. Nevertheless, proponents of raw milk contend that pasteurization is unnecessary if milk is kept clean from the udder of the cow to the consumer and that it destroys some desirable components in milk. Replacing hand milking with milking machines and direct transfer of raw milk to a cooling tank have made contamination from microorganisms from the environment less likely but do not eliminate the risk. Recall that milk is a good medium for growing many microorganisms, including pathogens. With regard to components of milk that are destroyed in pasteurization, the only substantial loss is some enzymes, such as alkaline phosphatase. But, on eating, enzymes are inactivated and digested in the stomach and intestines of consumers and do not provide any demonstrable health benefits aside from being a source of amino acids. Public health and medical organizations promote the pasteurization of milk in the interest of food safety.

Irradiation is not as accepted, largely because of public concerns about nuclear radiation. Gamma irradiation does not increase radioactivity in food over what occurs naturally. Although it does produce radiolytic products in foods, animal testing indicates that irradiated food is safe and high doses of radiolytic products have no adverse effects. Public health and medical organizations, such as the American Medical Association, attest to the safety of irradiation in protecting the food supply. Public education at the point of sale has been demonstrated to be effective in overcoming resistance to irradiated foods.

FURTHER READING

Berk, Zeki. *Food Process Engineering and Technology.* Academic Press, 2009.

Fellows, P. J. *Food Processing Technology: Principles and Practice.* 3rd ed., CRC Press, 2009.

Komolprasert, Vanee, and Kim M. Morehouse, editors. *Irradiation of Food and Packaging: Recent Developments.* American Chemical Society, 2004.

Ortega-Rivas, Enrique, editor. *Processing Effects on Safety and Quality of Foods.* CRC Press, 2010.

Stewart, Eileen. "Food Irradiation: More Pros than Cons?" *The Biologist,* vol. 51, no. 2, 2004, pp. 91–96.

Stewart, Eileen. "Food Irradiation: More Pros than Cons? Part 2." *The Biologist,* vol. 51, no. 3, 2004, pp. 141–44.

PERMAFROST

—*C. Nicholas Raphael*

FIELDS OF STUDY

Climatogy; Soil Science; Ecology; Ecosystem Management; Environmental Sciences; Environmental Studies; Botany.

PRINCIPAL TERMS

periglacial: originally restricted to regions modified by frost weathering such as tundra

Pleistocene: the geologic era between 2.58 million and 11,000 years ago, characterized by extensive continental glaciation and a colder climate

subsidence: a downward vertical movement of the ground, commonly caused by land-cover changes of the surface

talik: a layer of unfrozen ground in permafrost; water under pressure frequently occurs in talik layers

thermokarst: a group of landforms in flat areas of degrading permafrost and melting ice

SUMMARY

Permafrost is a thermal condition existing in any type of earth material that is perennially frozen. It occurs in ground in which the temperature remains below 0 degrees Celsius for at least two consecutive years. Affecting about 25 percent of the earth's surface, the condition hampers construction in the far north in regions such as Siberia and Arctic Canada.

FORMATION & DISTRIBUTION

Permafrost, or permanently frozen ground, is a thickness of earth material such as soil, peat, or even bedrock, at a variable depth beneath the ground surface, in which a temperature below 0 degrees Celsius exists continuously for a number of years (from at least two years to several thousands of years). Permafrost is thus not a process or an effect, but rather a temperature condition of earth material on land or beneath continental shelves of the Arctic Ocean. Its occurrence is conditioned exclusively by temperature, irrespective of the presence of water or of the composition, lithology, and texture of the ground. However, unfrozen water horizons may exist within a body of permafrost as the result of various conditions such as pressure or impurities in the water.

Stone rings caused by permafrost in Spitzbergen, Norway. Image by Hannes Grobe 21:14, 26 October 2007 (UTC) (Own work) CC-BY-SA-2.5.

With regard to the climatic history of the earth, cold conditions and glaciers are the exception, not the rule. Because they are so rare, they are referred to as "climatic accidents." During the Pleistocene, advances of continental glaciers covered vast areas of Europe and North America. Glaciers continue to be significant landscape features, since they lock up as ice some 2.15 percent of the earth's waters. Permafrost is a relatively new condition introduced with the onset of the Pleistocene glacial events, and it will be maintained as long as the climate remains cold.

As the earth's climate has warmed during the past 20,000 years and the continental ice has receded, the location, distribution, and thickness of permafrost have changed. Today, the area underlain by permafrost is vast. It has been estimated that about 25 percent of the earth's surface is underlain by permafrost. One reason that vast areas of the earth possess permafrost is that the continents widen toward higher latitudes in North America and Eurasia. In the Northern Hemisphere, Mongolia, Russia, and China contain about 12.2 million square kilometers of land underlain with permafrost. North America and Greenland have 8.8 million square kilometers of permafrost. In the Southern Hemisphere, Antarctica accounts for 13.5 million square kilometers of permafrost. During the Pleistocene, perennially frozen ground had a different distribution. Permafrost produces soil and sediment structures that remain after the climate warms. Relict features such as ancient ice wedges can be identified, and former permafrost areas can be mapped. Evidence of preexisting permafrost conditions, and thus a record of climate change, has been extensively documented in Great Britain, southern Canada, and the central United States.

In terms of aerial distribution and continuity, permafrost may be classified as either continuous or discontinuous. At higher-latitude land circling the Arctic Ocean, the lateral extent of permafrost is uninterrupted nearly everywhere, except in areas of recent deposition and under large lakes that do not freeze to the bottom. Discontinuous permafrost includes areas of frozen ground separated by nonpermafrost areas. A transect from the Arctic Ocean southward, into Canada or Russia, reveals more and more permafrost-free areas. Farther south, in central Canada and central Russia, only patchy or sporadic zones of perennially frozen ground are to be found.

Permafrost is a unique condition because it is maintained very close to its melting point. Several thousand square kilometers are warmer than –3 degrees Celsius. If global warming continues, as many scientists predict, most discontinuous permafrost will eventually degrade. Russian scientists, in fact, have documented the northward retreat of permafrost near Archangel at a rate of 400 meters per year since 1837. In Canada, permafrost has retreated northward more than 100 kilometers since 1945, as ground temperatures have increased by 2 degrees Celsius in the northern prairie provinces. Changes in temperature may result from a local microclimate change, alteration of land cover, or changes in the atmosphere. Conversely, permafrost not only is degraded by global warming, but may also contribute to atmospheric change. Scientists suggest that thawing of Arctic terrain, which is rich in peat accumulation, may affect the chemistry of the atmosphere. It has been determined that one-fourth of the earth's terrestrial carbon is stored in organic matter in the permafrost and active layer. Long-term warming would release enormous quantities of greenhouse gases—methane and carbon dioxide in particular—accelerating global warming.

VERTICAL CHARACTERISTICS

A cross section of excavated earth in the high Arctic reveals the vertical characteristics of permafrost. The thickness of permafrost is variable because of differences in air temperatures, occurrence of water, composition and texture of earth materials, and many other factors. The permafrost in Siberia may be at least 5,000 meters thick, whereas in Alaska thicknesses of 740 meters have been recorded. Farther south, thicknesses decrease as discontinuous permafrost becomes more common. The thickness of permafrost is in part determined by a balance between the increase of internal heat with depth and the heat loss from the earth's surface. If the thickness and depth of permafrost remain unchanged for many years, the heat loss at the surface and the heat at the base of the frozen ground are in equilibrium, and a steady state is maintained. At the surface, the ground seasonally thaws and freezes. This section of the permafrost is known as the "active layer." Its thickness is controlled by numerous variables in addition to seasonal temperature variations, and it exhibits great variability in thickness. Vegetation, snow cover, albedo, water distribution, and human alteration of the surface can all act as insulators and affect the thickness of the active layer. However, in areas not disturbed by people, the ground will typically thaw to a depth of from 1 to 12 meters during the summer.

The deeper boundary separating the active layer from the permafrost is known as the "permafrost table" or "zero curtain." Within the permafrost, unfrozen lenses known by the Russian term *talik* occur. These unfrozen bodies frequently contain water and function as protected water sources or aquifers in the Arctic environment. Talik lenses may be isolated bodies completely enclosed in impermeable permafrost and may result from a change in the thermal regime in permafrost. Conversely, talik may be laterally extensive even in continuous permafrost and is frequently a valuable source of groundwater. Talik lenses are larger in discontinuous permafrost, and the active layer thickens at the expense of the permafrost.

Below the permafrost table, the perennially frozen ground has low permeability. Maps and photographs of the Arctic reveal many lakes, swamps, and marshes, suggesting high precipitation. However, evaporation rates are low because of the cold air temperatures, and standing water occurs because the ground is frozen at depth. Throughout the region, precipitation is probably less than 35 centimeters annually, equivalent to that of marginal deserts. The abundant water and poor drainage result from the occurrence of permafrost at depth, not to high atmospheric precipitation. As seasonal warming and cooling occur, the land surface expands and contracts, particularly in continuous permafrost areas. In essence, the lowering of the ground temperature causes thermal contraction of the ground, and vertical fissures or frost cracks occur. Water filling and freezing in these cracks creates ice wedges in the permafrost that widen with seasonal expansion and contraction. Ice wedges are often 3 to 4 meters wide and extend 5 to 10 meters into the perennially frozen ground below the active layer. In areas of degradation, ice wedges in the thawing permafrost begin to melt, creating thermokarst features characterized by numerous depressions similar to karst plains. As temperatures rise above freezing, the active layer extends deeper and deeper and eventually may intercept an ice wedge. If large ice wedges begin to thaw, the ground subsides, creating steep-sided conical depressions. The perimeter of a depression, often composed of very moist sediment, will slump into the depression, thus making it wider. Water then fills the depression, creating a thermokarst, or thaw lake. Averaging about 3 meters across, the lakes occur in clusters, are elliptical in shape, and are all oriented at right angles to the prevailing wind direction.

FURTHER READING

Anderson, Bjørn G., and Harold W. Borns, Jr. *The Ice Age World.* Oslo: Scandinavian University Press, 1994. The well-illustrated text details the Quaternary history of North America and northern Europe over the last 2.5 million years. Contains an extended glossary, references list, and index. For the general reader as well as the serious student.

Ballantyne, C. F., and C. Harris. *The Periglaciation of Great Britain.* New York: Cambridge University Press, 1994. A good upper-level description of permafrost in Britain. The discussion ranges from basic freeze-and-thaw processes to permafrost conditions and distribution in the geologic past.

Bloom, Arthur. *Geomorphology: A Systematic Analysis of Late Cenozoic Landforms.* 3rd ed. Long Grove, Ill.: Waveland Press, 2004. This college-level text covers the basics of geomorphology. Includes three chapters on glaciation, cold climates, and permafrost. Index and bibliography.

Chernicoff, Stanley, and Dona Whitney. *Geology: An Introduction to Physical Geology.* Upper Saddle River, N.J.: Prentice Hall, 2006. This is a good overview of the scientific understanding of the geology of the earth and surface processes. Includes a section on the development and effects of permafrost. Includes a link to a Web site that provides regular updates on geologic events around the globe.

Clark, M. J., ed. *Advances in Periglacial Geomorphology.* New York: Wiley InterScience, 1988. A series of state-of-the-art reviews of numerous processes and landforms in cold climates. Emphasizes processes of formation; very technical. Suitable for advanced earth science students as a reference and literature source.

Grosse, G., et al. "Why Permafrost Is Thawing, Not Melting." *EOS* 91 (2010): 87. Discusses terminology used in past and recent scientific communications and suggestions for future communications. The authors' recommendations provide the basis for an interesting debate over proper word usage to convey information beyond the scientific community. The article also provides a good description of the process of thawing, contrasting it to the physical process of melting.

Harris, S. A. *The Permafrost Environment.* Totowa, N.J.: Barnes & Noble, 1986. A primer on the consequences of construction in permafrost regions. Topics include the soil properties and methods of permafrost analysis. Written for contractors and environmental specialists.

Romanovsky, V., et al. "Permafrost Temperature Records: Indicators of Climate Change." *EOS, AGU Transactions* 83 (2002): 589–594. Discusses the changes to the ecosystem brought on by permafrost thawing. The article provides information on permafrost zones and the methods used to study their degradation.

Schuur, Edward A. G., et al. "The Effect of Permafrost Thaw on Old Carbon Release and Net Carbon Exchange from Tundra." *Nature* 459 (2009): 556–559. This article discusses the permafrost soils of the Arctic ecosystem. Provides a good understanding of how permafrost thaw alters the carbon cycle. Somewhat technical, so having a background in science is helpful for readers.

Williams, P. J., and M. W. Smith. *The Frozen Earth: Fundamentals of Geology.* New York: Cambridge University Press, 1991. A college-level textbook for science majors, emphasizing the physics of cold regions. Equations are used to illustrate significant physical ideas.

PLANT & ANIMAL INTERACTIONS

—*Thomas C. Moon*

FIELDS OF STUDY

Agronomy; Agriculture; Zoology; Ecology; Ecosystem Management; Biotechnology; Environmental Sciences; Environmental Studies; Botany.

PRINCIPAL TERMS

cellular respiration: the release of energy in organisms at the cell level, primarily through the use of oxygen

chlorophyll: one of several forms of photoactive green pigments in plant cells that is necessary for photosynthesis to occur

coevolution: a mutualistic relationship between two different organisms in which, as a result of natural selection, the organisms become interdependent

cross-pollination: the transfer of pollen grains and their enclosed sperm cells from the male portion of a flower to a female portion of another flower within the same species

food chain: a diagram illustrating the movement of food materials from green plants (producers) through various levels of animals (consumers) within natural environments

natural selection: the survival of variant types of organisms as a result of adaptability to environmental stresses

pistil: a female portion of a flower that produces unfertilized egg cells

stamen: a male portion of a flower that produces pollen grains and their enclosed sperm cells

SUMMARY

Interactions between plants and animals in natural environments often revolve around either food acquisition or pollination and seed dispersal. By studying such processes, scientists have discovered strategies to increase agricultural production and to duplicate naturally produced pesticides.

Bumblebee pollinating an azalea. Image by UpstateNYer (Own work) CC-BY-SA-3.0.

Ecology represents the organized body of knowledge that deals with the interrelationships between living organisms and their nonliving environments. Increasingly, the realm of ecology involves a systematic analysis of plant-animal interactions through the considerations of nutrient flow in food chains and food webs, exchange of such important gases as oxygen and carbon dioxide between plants and animals, and strategies of mutual survival between plant and animal species through the processes of pollination and seed dispersal.

Ecologists study both abiotic and biotic features of such plant and animal interactions. The abiotic aspects of any environment consist of nonliving, physical variables, such as temperature and moisture, that determine where species can survive and reproduce. The biotic (living) environment includes all other plants, animals, and microorganisms with which a particular species interacts. Certainly, two examples of plant and animal interactions involve the continual processes of photosynthesis and cellular respiration. Green plants are often classified as ecological producers and have the unique ability to carry out both these important chemical reactions. Animals, for the most part, can act only as consumers, taking the products of photosynthesis and chemically releasing them at the cellular level to produce energy for all life activities.

PLANT-ANIMAL MUTUALISM

One topic that has captured the attention of ecologists involves the phenomenon of mutualism, in which two different species of organisms beneficially reside together in close association, usually revolving around nutritional needs. One such example demonstrating a plant and animal association is a certain species of small aquatic flatworm that absorbs microscopic green plants called algae into its tissues. The benefit to the animal is one of added food supply; the adaptation to this alga has been so complete that the flatworm does not actively feed as an adult. The algae, in turn, receive adequate supplies of nitrogen and carbon dioxide and are literally transported throughout tidal flats in marine habitats as the flatworm migrates, thus exposing the algae favorably to increased sunlight. A similar example of mutualism has been reported by ecologists studying various types of reef-building corals, which are actually marine, colonial animals that grow single-celled green algae called zooxanthellae within their bodies. The coral organisms use the nutrients produced by these algae as additional energy supplies, enabling them to build more easily the massive coral reefs associated with tropical waters. In 1987, William B. Rudman reported a similar situation while researching the formation of such coral reefs in East African coastal waters. He discovered a type of sea slug called a nudibranch that absorbs green algae into its transparent digestive tract, producing an excellent camouflage as it moves about on the coral reefs in search of prey. In turn, the algae growing within both the coral and sea slugs receive important gases from these organisms for their own life necessities.

An example of plant-animal mutualism that has been documented as a classic example of coevolution involves the yucca plant and a species of small, white moth common throughout the southwestern United States. The concept of coevolution builds upon Charles Darwin's theories of natural selection, reported in 1859, and describes situations in which two decidedly different organisms have evolved into a close ecological relationship characterized by compatible structures in both. Thus, coevolution is a mutualistic relationship between two different species that, as a result of natural selection, have become intimately interdependent. The yucca plant and yucca moth reflect such a relationship. The female moth collects pollen grains bearing sperm cells from the stamens of one flower on the plant and transports these pollen loads to the pistil of another flower, thereby ensuring cross-pollination and fertilization. During this process, the moth will lay her own fertilized eggs in the flower's undeveloped seed pods. The developing moth larvae have a secure residence for growth and a steady food supply. These larvae will rarely consume all the developing plant seeds; thus, both species (plant and animal) benefit.

DEFENSIVE MUTUALISM

Although these examples demonstrate the evolution of structures and secretions that reflect mutual associations between plants and animals, other interactions are not so self-supportive. Plant-eating animals, called herbivores, have always been able to consume large quantities of green plants with little fear of reprisal. Yet, some types of carnivorous plants have evolved that capture and digest small insects and crustaceans as nutritional supplements to their normal photosynthetic activities. Many of these plants grow abundantly in marshlike environments, such as bogs and swamps, where many insects congregate to reproduce. Such well-known plants as the Venus's-flytrap, sundews, butterworts, and pitcher plants have modified stems and leaves to capture and consume insects and spiders rich in protein. On a smaller scale, in freshwater ponds and lakes, a submerged green plant commonly known as the bladderwort partially satisfies its protein requirements by snaring and digesting small crustaceans, such as the water flea, within its modified leaves.

A form of ecological interaction commonly classified as mimicry can be found worldwide in diverse environments. In such situations, an animal or plant has evolved structures or behavior patterns that allow it to mimic either its surroundings or another organism as a defensive or offensive strategy. Certain types of insects, such as leafhoppers, walking sticks, praying mantids, and katydids (a type of grasshopper), often duplicate plant structures in environments ranging from the tropical rain forests to the northern coniferous forests of the United States. Such exact mimicry of their plant hosts affords these insects protection from their predators as well as camouflage that enables them to capture their own prey readily. In other examples of mimicry, some insects will absorb unpalatable plant substances in their larval stages and retain these chemicals in their adult forms, making them undesirable to birds as food sources. The monarch butterfly demonstrates this type of interaction with the milkweed plant. The viceroy butterfly has evolved colorations and markings similar to those of the monarch, thereby ensuring its own survival against bird predators. Certain species of ambush bugs and crab spiders have evolved coloration patterns that allow them to hide within flower heads of such common plants as goldenrod, enabling them to grasp more securely the bees and flies that visit these flowers.

NONSYMBIOTIC MUTUALISM

Many ecologists have been studying the phenomenon known as nonsymbiotic mutualism: different plants and animals that have coevolved morphological structures and behavior patterns by which they benefit each other without necessarily living together physically. This type of mutualism can be demonstrated in the often unusual and bizarre shapes, patterns, and colorations that more advanced flowering plants have developed to attract various insects, birds, and small mammals for pollination and seed dispersal purposes. Pollination essentially is the transfer of pollen grains (and their enclosed sperm cells) from the male portion of a flower to the egg cells within the female portion of a flower. Pollination can be accomplished by the wind, by heavy dew or rains, or by animals, and it results in the plant's sexual production of seeds that represent the next generation of new embryo plants. Accessory structures, called fruits, often form around seeds and are usually tasty and brightly marked to attract animals for seed dispersal. Although the fruits themselves become biological bribes for animals to consume, often the seeds within these fruits are not easily digested and thus pass through the animals' digestive tracts unharmed, sometimes great distances from the original plant. Other types of seed dispersal mechanisms involve the evolution of hooks, barbs, and sticky substances on seeds that enable them to be easily transported by animals' fur, feet, feathers, or beaks to new regions for possible plant colonization. Such strategies of dispersal reduce competition between the parent plant and its new seedlings for moisture, living space, and nutritional requirements.

The evolution of flowering plants and their resulting use of animals in pollination and seed dispersal probably began in dense, tropical rain forests, where pollination by the wind would be cumbersome. Because insects are the most abundant form of animal life in rain forests, strategies based upon insect transport of pollen probably originated there. Because structural specialization increases the possibility that a flower's pollen will be transferred to a plant of the same species, many plants have evolved a vast array of scents, colors, and nutritional products to attract many insects, some birds, and a few mammals. Not only does pollen include the plant's sperm cells, but it is also rich in food for these animals. Another source of animal nutrition is a substance called nectar, a sugar-rich fluid often produced in specialized structures called nectaries within the flower itself or on adjacent stems and leaves. Assorted waxes and oils are also produced by plants to ensure plant-animal interactions. As species of bees, flies, wasps, butterflies, and hawkmoths are attracted to flower heads for these nutritional rewards, they unwittingly become agents of pollination by transferring pollen from male portions of flowers (stamens) to the appropriate female portions (pistils). Some flowers have evolved distinctive, unpleasant odors reminiscent of rotting flesh or feces, thereby attracting carrion beetles and flesh flies in search of places to reproduce and deposit their own fertilized eggs. As these animals consummate their own relationships, they often become agents of pollination for the plant itself. Some tropical plants such as orchids even mimic a female bee, wasp, or beetle, so that its male counterpart will attempt to mate with it, thereby ensuring precise pollination.

Among the bird species, probably the hummingbirds are the best examples of plant pollinators. Various types of flowers with bright, red colors, tubular shapes, and strong, sweet odors have evolved in tropical and temperate regions to take advantage of the hummingbirds' long beaks and tongues as an aid to pollination.

Because most mammals, such as small rodents and bats, do not detect colors as well as bees and butterflies do, flowers that use them as pollinators do not rely upon color cues in their petals but instead focus upon the production of strong, fermenting or fruit-like odors and abundant pollen rich in protein to attract them. In certain environments, bats and mice that are primarily nocturnal have replaced day-flying insects and birds in these important interactions between plants and animals.

EXPERIMENTS WITH PLANT-ANIMAL INTERACTION

Contemporary ecologists have gone beyond the purely descriptive observations of plant-animal interactions (initially within the realm of natural history) and have designed controlled experiments that are crucial to the development of such basic concepts as coevolution. For example, the use of radioactive isotopes and the marking of pollen with dye and fluorescent material in field settings have allowed ecologists to demonstrate precise distances and patterns of pollen dispersal. Ecologists and insect physiologists have cooperatively studied how certain insects, such as bees, are sensitive to ultraviolet light. When some flowers are viewed under ultraviolet light, distinct floral patterns become evident to guide these insects to nectar pollen sources. Through basic research, Carolyn Dickerman reported in 1986 that animal color preferences vary throughout the season. Insect pollinators, who must feed every day, will adapt to these changes by shifting their foraging behavior. Research in the field has demonstrated that some species of flowers, such as the scarlet gilia, will produce differently colored flowers to accommodate shifts in pollinator species. Early in the growing season, this plant will produce long, red, tubular-shaped flowers to attract hummingbirds. As the hummingbirds migrate, the flowers will later become lighter in hue and be pollinated primarily by nocturnal hawkmoths.

In the laboratory, ecologists and biochemists have cooperatively analyzed the chemical composition of plant secretions and products. The chemical analysis of nectar indicates great variation in composition, correlating with the type of pollinator. Flowers pollinated by beetles generally have high amino acid content. The nectar associated with hummingbird-pollinated flowers is rich in sugar. Pollen also varies widely in chemical composition within plant species. Oils and waxes are major chemical products in the pollen of plants visited primarily by bees and flies. For bat-pollinated flowers, the protein content is quite high.

Research has also successfully analyzed how certain plants have been able to develop toxins as chemical defenses against animals. These protective devices include such poisons as nicotine and rotenone that help prevent insect and small mammal attacks. A more remarkable group of protective compounds recently isolated from some plants are known as juvocimines. These chemicals actually mimic juvenile insect hormones. Insect larvae feeding on leaves containing juvocimines are prevented from undergoing their normal development into functional, breeding adults. Thus, a specific insect population that could cause extensive plant damage is locally reduced.

Ecological interactions between plants and animals are diverse and varied. These plant-animal interactions can be viewed as absolute necessities for developing food chains and food webs and for maintaining the global balances of such important gases as oxygen and carbon dioxide. The interactions can also be very precise, limited, and crucial for determining species survival or extinction. By analyzing varied plant-animal interactions, from the microscopic level to the global perspective, one can more fully appreciate all the ecological relationships that exist on the earth.

THE ECOLOGY OF INTERACTION

The ecological importance of plant-animal interactions cannot be stressed enough. Modern-day agriculture owes its existence to the activities of such insect pollinators as honeybees in regard to the production of domestic fruits, vegetables, and honey. It is becoming increasingly evident to many ecologists and forestry scientists how important certain bird species, such as blue jays and cedar waxwings, are in natural reforestation of burned and blighted areas through their seed dispersal strategies. The plant horticulture and floral industries also are developing an appreciation of specific plant-animal interactions that produce more viable natural strains of flowers and ornamental shrubbery. The study of natural chemical defenses produced by some plants against animal invasions is most promising. The renewed interest in earlier efforts to extract such plant products as nicotine, rotenone, pyrethrum, and caffeine may produce natural compounds that can be effective insecticides without the long-term, environmental hazards associated with such human-made pesticides as malathion, chlordane, and Dichlorodiphenyl-trichloroethane (DDT).

Finally, humankind is realizing that it is important to understand and protect certain plant-animal interactions associated with the tropical regions of the earth; otherwise, the global balance of oxygen and carbon dioxide could be seriously disrupted. Also, these tropical areas represent the last natural environments for the continuation of important plant species that produce secretions and products that have favorable medicinal qualities for humans and domestic livestock. By maintaining these populations and understanding how certain animals interact with them, humans can be guaranteed a viable supply of beneficial plant species whose medicinal values can be duplicated within the laboratory.

Further Reading

Abrahamson, Warren G., ed. *Plant-Animal Interactions.* New York: McGraw-Hill, 1989. Covers herbivore-plant interactions, carnivorous plant ecology and evolution, pollination and population dispersal agents, plant communities as habitats for animals, interactions in agroecosystems, and coevolution.

Barth, Frederick G. *Insects and Flowers: The Biology of a Partnership.* Princeton, N.J.: Princeton University Press, 1991. This beautifully illustrated book, rich in both color photographs and electron micrographs, describes all aspects of flower-insect interactions. Includes bibliography and index. The writing style is appropriate for both general readers and introductory-level biology students.

Dickerman, Carolyn. "Pollination: Strategies for Survival." *Ward's Natural Science Bulletin,* Summer, 1986, 1–4. This article traces the coevolution of flowering plants with their animal pollinators, especially the beetles, butterflies, and hawkmoths. An excellent introduction to both strategies of pollination and seed dispersal. The author also outlines some current research techniques and includes a simplified biological key that focuses upon flower structures that will help determine which animals pollinate them.

Howe, Henry F., and Lynne C. Westley. *Ecological Relationships of Plants and Animals.* Oxford Univ Pr, 2007. Print. Offers introductory students in botany, zoology, and ecology a comprehensive summary of both field and experimental studies on the ecology and evolution of plant and animal interactions. Excellent black-and-white photographs, illustrations, charts, and tables are found throughout. This textbook is written with clarity but is not only for the casual reader.

John, D. M., S. J. Hawkins, and J. H. Price, eds. *Plant-Animal Interactions in the Marine Benthos.* Oxford, England: Clarendon Press, 1992. The proceedings of a symposium on shallow-water marine plant-animal interactions.

Lanner, Ronald M. *Made for Each Other: A Symbiosis of Birds and Pines.* New York: Oxford University Press, 1996. Based on the author's research into the mutualism between Clark's nutcracker and several species of western white pines with wingless seeds. Accessible to the lay reader.

Meeuse, Bastian, and Sean Morris. *The Sex Life of Flowers.* New York: Facts on File, 1984. This well-written text includes beautiful photographs and illustrations, and it appeals to readers with varied scientific backgrounds. It describes the evolution of the flower as an organ of sexual reproduction and portrays selected animal adaptations in structure and behavior to accommodate pollination and seed dispersal. Includes a chapter-by-chapter listing of further reading sources. The photographs and writing are based upon the Public Broadcasting Service series entitled "Sexual Encounters of the Floral Kind."

Norstag, Knut, and Andrew J. Meyerriecks. *Biology.* 2d ed. Columbus, Ohio: Charles E. Merrill, 1985. A textbook written with exceptional clarity and organization. Discusses ecology and coevolution of plants and animals. Includes a glossary and index. Excellent photographs and diagrams. This is a college-level textbook but is also suitable for advanced high school students.

Rudman, William B. "Solar-Powered Animals." *Natural History* 96 (October, 1987): 50–53. A well-written article for all readers interested in plant-animal interactions, especially focusing upon mutualism involving marine corals, sea slugs, and their green algae allies. The undersea photography is excellent.

PLANT BIOTECHNOLOGY

—Robert A. Sinnott

FIELDS OF STUDY

Agronomy; Agriculture; Ecology; Ecosystem Management; Biotechnology; Environmental Sciences; Environmental Studies; Botany.

PRINCIPAL TERMS

genetic engineering: by direct manipulation of genetic elements (genes).

micropropagation: another biotechnology technique, is the production of many clonal plants using tissue culture methods.

microparticle bombardment: the introduction of foreign DNA constructs into plant cells by attaching the DNA to small metal particles and blasting the particles into plant cells using either a compressed air gun or a gun powered by a 0.22 caliber gun cartridge.

SUMMARY

Plant biotechnology may be defined as the application of knowledge obtained from study of the life sciences to create technological improvements in plant species. By this very broad definition, plant biotechnology has been conducted for more than ten thousand years.

BACKGROUND

The roots of plant biotechnology can be traced back to the time when humans started collecting seeds from their favorite wild plants and began cultivating them in tended fields. It appears that when the plants were harvested, the seeds of the most desirable plants were retained and replanted the next growing season. While these primitive agriculturists did not have extensive knowledge of the life sciences, they evidently did understand the basic principles of collecting and replanting the seeds of any naturally occurring variant plants with improved qualities, such as those with the largest fruits or the highest yield, in a process that we call *artificial selection.* This domestication and controlled improvement of plant species was the beginning of plant biotechnology. This very simple process of selectively breeding naturally occurring variants with observably improved qualities served as the basis of agriculture for thousands of years and resulted in thousands of domesticated plant cultivars that no longer resembled the wild plants from which they descended.

The second era of plant biotechnology began in the late 1800s as the base of knowledge derived from the study of the life sciences increased dramatically. In the 1860s Gregor Mendel, using data obtained from controlled pea breeding experiments, deduced some basic principles of genetics and presented these in a short monograph modestly titled *Versuch über pflanzen-hybriden* in 1865 (*Experiments in Plant-Hybridisation,* 1910). In this publication, Mendel proposed that heritable genetic factors segregate during sexual reproduction of plants and that factors for different traits assort independently of each other. Mendel's work suggests a mechanism of heritable factors that could be manipulated by controlled breeding of plants through selective fertilization; the work also suggests that the pattern of inheritance for these factors could be analyzed or, in some cases, predicted by the use of mathematical statistics.

These findings complemented the work of Charles Darwin, whose 1859 book *On the Origin of Species by Means of Natural Selection* expounds the principles of descent with modification and selection as the chief factor of evolutionary change. The application of these principles to agriculture resulted in deliberately produced hybrid varieties for a large number of cultivated plants via selective fertilization. These artificially selected hybrids soon began to benefit humankind with tremendous increases in both the productivity and the quality of food crops.

GENETIC ENGINEERING

The third era of plant biotechnology involves a drastic change in the way crop improvement may be accomplished, by direct manipulation of genetic elements (genes). This process is known as *genetic engineering* and results in plants that are called genetically modified organisms (GMOs), to distinguish them from plants that are produced by conventional plant-breeding methods. Genetically modified plants can contribute desirable genes from outside traditional breeding boundaries. Even genes from outside the plant kingdom can be brought into plants. For example, animal genes, including human genes, have been transferred into plants, a feat not replicated in nature.

PUBLIC CONCERN

It is perhaps this lack of natural boundaries for genetic exchange that seems so foreign to conventional scientific thought and that makes plant genetic engineering controversial. The thought of taking genes from animals, bacteria, viruses, or any other organism and putting them into plants, especially plants consumed for food, has raised a host of questions among concerned scientists and the public alike. Negative public perception of genetically modified crops has affected the development and commercialization of many plant biotechnology products, especially food plants. While there are dozens of genetically engineered plants ready for field production, public pressure has delayed the release of some of these plants and has caused the withdrawal of others from the marketplace.

This public concern also appears to have driven increased government review of products and decreased government funding for plant biotechnology projects in Europe. Negative public perceptions have not seemed to be as strong in Asia, since the pressures of feeding large populations tend to outweigh the perceived risks. The social climate of the United States toward biotechnology, although guarded, has appeared to be less apprehensive than that of most European countries. Therefore, many agricultural biotechnology projects have moved from European countries to U.S. laboratories.

ECONOMIC GOALS

To what end are humans genetically engineering plants? This is an essential question for researchers, executives of biotechnology companies, and consumers at large. Before addressing technical questions about how to apply biotechnology, the desired goals must be clearly defined. The general goals of plant biotechnology appear to be (1) economic improvement of existing products, (2) improvement of human nutrition, and (3) development of novel products from plants.

Economic improvements include increases in yield, quality, pest resistance, nutritional value, harvest ability, or any other change that adds value to an established agricultural product. Examples of this category include insect-protected tomatoes, potatoes, cotton, and corn; herbicide-resistant canola, corn, cotton, flax, and soybeans; canola and soybeans with genetically altered oil compositions; virus-resistant squash and papayas; and improved-ripening tomatoes. All of these examples were introduced to agriculture in the latter half of the 1990s, though some have been removed from the market.

NUTRITIONAL GOALS

Additionally, some products appearing in the scientific literature but awaiting commercialization have the potential to dramatically improve human nutritional deficiencies, which are especially prevalent in developing countries. These products include "golden rice," genetically modified rice that produces carotenoids, a dietary source of vitamin A. Golden rice has the potential to prevent vitamin A deficiency in developing countries, where this vitamin deficiency is a leading cause of blindness. The product has been met with resistance by anti-GMO activists and with caution by the World Health Organization.

Researchers are also using genetic engineering to increase the amount of the iron-storing protein ferritin in seed crops such as legumes. Iron deficiency, which affects 30 percent of the human population, can impair cognitive development and cause other health problems. This proposed enhancement of iron content in consumable plant products could help more than a billion people who suffer from chronic iron deficiency.

NOVEL PRODUCTS

Novel products include those not traditionally associated with plants and are limited only by imagination and available techniques. These include the production of plastics, vaccines, antibodies, human blood proteins, and new pharmaceuticals. One project has involved the production of hepatitis B vaccine in transgenic tomatoes. This project, which underwent clinical trials in the late 1990s, has the potential to provide a simple and inexpensive means of vaccinating people against hepatitis B. By oral administration of tomato juice containing the vaccine protein, humans are thought to develop an immune response that may protect them from infection by the hepatitis B virus. Hepatitis B has been epidemic in Asia and has increased at an alarming rate in the rest of the world. The disease ultimately causes liver disease, cancer, and death in nearly a million infected people worldwide.

PLANT TISSUE CULTURES

Central to plant biotechnology is the use of in vitro methods. Researchers use *plant tissue cultures,* for example, to grow plant cells on sterile nutrient media. Countless recipes for these nutrient media exist. The choice of which one to use is based on the plant species and the tissue type to be grown. All such media contain at least some of the important nutritional elements—such as nitrogen, potassium, calcium, magnesium, sulfur, phosphorus, iron, boron, manganese, zinc, iodine, molybdenum, copper, and cobalt, usually in the form of inorganic salts or as metal chelates—and an organic energy source, such as sucrose. The media may also contain vitamins, hormones, and other ingredients, depending on the intended use.

To initiate plant tissue culture, a piece of a living plant is excised and disinfected using a chemical disinfectant. This piece of plant tissue, called an *explant,* is placed on a sterile plant tissue culture medium to grow. Many plant tissues may be used to obtain explants for plant tissue culture, including those from leaves, petioles, shoots, tubers, roots, and meristematic regions. When an explant is placed in the sterile tissue culture medium, cells that are not terminally differentiated will grow and divide. If plant hormones are included in the recipe, the plant cells can be coaxed to develop into different types of tissues or organs. By using a succession of media containing different hormones, it is possible to regenerate whole plants from single cells. The choice of tissue used for the explant and the choice of hormones included in the tissue culture medium depend on the desired result.

MICROPROPAGATION

Micropropagation, another biotechnology technique, is the production of many clonal plants using tissue culture methods. By means of micropropagation, it is possible to generate many thousands of plant clones using tissue explants obtained from a single parent plant. The main advantage to micropropagation is the potential of producing thousands of exact copies of a plant with desirable traits. Micropropagation is especially important for rare plants, genetically engineered plants, and plants that have sexual reproductive problems. Many plant species are routinely propagated by micropropagation methods, including orchids, ferns, many flowering ornamentals, and vegetable plants.

STEPS IN GENETIC ENGINEERING

The first genetically engineered plants, tobacco plants, were reported in the scientific literature in 1984. Since 1984 there have been thousands of genetically engineered plants produced in laboratories worldwide. The process of genetically engineering a plant involves several key steps:

- isolating the genetic sequence (gene) to be placed from its biological source
- placing the gene in an appropriate vehicle to facilitate insertion into plant cells
- inserting the gene into the plant in a process known as *plant transformation*
- selecting the few plant cells that contain the new gene (transformed cells) out of all the plant cells in the explant
- multiplying the transformed cells in sterile tissue culture
- regenerating the transformed cells into a whole plant that can grow outside the tissue culture vessel

The gene or genes to be placed in the plant may be obtained from virtually any biological source: animals, bacteria, fungi, viruses, or other plants. Placing genes into an appropriate vehicle for transfer into a plant involves using various molecular biology techniques, such as restriction enzymes and ligation, to essentially "cut and paste" the gene or genes of interest into another DNA molecule, which serves as the transfer vehicle (vector).

PLANT TRANSFORMATION METHODS

Plant transformation with foreign genes may be accomplished by several proven methods, including bacteria-mediated transfer, microparticle bombardment, electroporation, microinjection, sonication, and chemical treatment.

By far, the most often utilized method of plant transformation involves the use of naturally occuring plant pathogenic bacteria from the genus *Agrobacterium.* In nature, this bacterium infects plants and transfers some of its own bacterial DNA into the plant. Through the action of proteins produced by the bacteria, bacterial DNA is made to integrate permanently into the plant's own genomic DNA. Expression of the bacterial

DNA in the plant causes the plant to produce unusual quantities of plant hormones and other compounds, called opines, which provide food for the bacteria. The unusual quantities of plant hormones around the infection site cause the plant cells to grow abnormally, producing characteristic tumors. Scientists have harnessed this pathogenic bacterium to insert genes into plants by deleting the bacterial genes that cause tumors in the plant and then inserting desirable genes in their place. When the modified *Agrobacterium* infects a plant, it transfers the desirable genes into the plant genome instead of causing tumors. The desirable genes become a permanent part of the plant genome, and expression of these genes in plant cells produces desirable products.

One major drawback of the *Agrobacterium* method is that insertion of bacterial DNA into the plant genome is essentially random. The gene may not be efficiently transcribed at its location, or the insertion of bacterial DNA may knock out an important plant gene by inserting in the middle of it—or both may occur. Therefore, the fact that a cell is genetically transformed does not guarantee that it will perform as desired.

Microparticle bombardment is the introduction of foreign DNA constructs into plant cells by attaching the DNA to small metal particles and blasting the particles into plant cells using either a compressed air gun or a gun powered by a 0.22 caliber gun cartridge. This is truly a "brute force" method of introducing DNA into a cell that inadvertently causes many lethal casualties among the bombarded plant cells. However, some plant cells blasted with the DNA-containing metal particles will recover and survive. The plant cells may express the DNA for only a short time (transient expression), because the DNA does not readily integrate into the plant genome, but occasionally the foreign DNA may spontaneously recombine into the plant genome and become permanent.

Other ways of introducing foreign DNA into plant cells include electroporation, microinjection, sonication, and chemical treatment. These methods are not used extensively, because they generally require the production of *protoplasts* (plant cells that lack their cell walls) from plant cells before transformation. To create protoplasts, the plant cell wall is removed by digestion with the enzymes cellulase and pectinase. Protoplasts are fragile structures, but the absence of a cell wall is desirable because it leaves only the plasma membrane as a barrier to foreign DNA entering a plant cell.

Electroporation uses very brief pulses of high-voltage electrical energy to create temporary holes in the plasma membrane through which the foreign DNA can pass. *Microinjection* involves physically injecting a small amount of DNA into a plant cell using a microscope and an extremely fine needle. *Sonication* uses ultrasonic waves to punch temporary holes in the plasma membrane; this method is therefore similar to electroporation. *Chemical treatment* involves the use of polyethylene glycol to render the plasma membrane permeable to foreign DNA.

All of the transformation procedures produce only a few transformed cells out of the millions of cells in an explant, so selection of transformed cells is essential.

SELECTION OF TRANSFORMED PLANT CELLS

Selecting the few transformed plant cells out of all the plant cells in an explant requires some advance planning. Most foreign DNA constructs introduced into a plant are designed and built to contain additional genes that function as selectable markers or reporter genes. *Selectable markers* include genes for resistance to antibiotics or herbicides. Plant cells containing and expressing these genes will be tolerant of antibiotics or herbicides added to the plant tissue culture media, while the nontransformed plant cells will be killed off. The surviving cells in the tissue culture media are mostly transformed.

Instead of selectable markers, *reporter genes* may be used. Reporter genes induce an easily observable trait to transformed plant cells that facilitates the physical isolation of these cells. Reporter genes include beta-glucuronidase, luciferase, and plant pigment genes. Beta-glucuronidase (commonly known as GUS) allows the plant cells expressing this gene to metabolize colorigenic substrates, while nontransformed plant cells cannot. To use this test, researchers treat a small amount of plant tissue with the colorigenic chemical substrate. If the cell turns color (blue) it is known to be transformed and expressing the GUS gene. If the cell does not turn color, it probably is not transformed. Another reporter gene is luciferase, an enzyme isolated from fireflies. Luciferase makes plant cells glow in the presence of certain chemicals if the gene is present; hence, transformed cells glow, whereas nontransformed cells do not glow. Plant pigment genes, such as anthocyanin pigment genes, occur naturally in plants and produce pigments that impart color to flowers. Inclusion of these pigment genes as reporter genes will allow transformed plant cells to be selected by their color. Transformed cells have color, while nontransformed cells remain colorless. Both selectable markers and reporter genes allow selection of cells into which genes have been successfully inserted and are operating properly.

REGENERATING WHOLE TRANSFORMED PLANTS

After successfully getting a gene construct into a plant cell and selecting the transformed cells, it is possible to get the plant cells to multiply in tissue culture. Also, by treating the plant cells with combinations of plant hormones, the cells are made to differentiate into various plant organs or whole plants.

For example, treating transformed plant cells with a high concentration of the plant hormone cytokinin causes shoots to develop. Transferring these shoots to another medium, one that is high in the plant hormone auxin, will cause roots to develop on the shoots. In this way a whole *transgenic plant* may be regenerated from transformed plant cells. Once a transformed plant is regenerated in tissue culture, the plant may be transferred to a climate-controlled greenhouse, where it can grow to maturity.

Future generations of transgenic plants may then be propagated sexually via seeds or asexually via vegetative propagation methods. Often transgenic plants must be grown in containment greenhouses to prevent accidental release into the environment. In such high-tech greenhouses, all factors contributing to optimal plant growth—lighting, temperature, humidity, nutrients, and other environmental conditions—are tightly controlled. Often hydroponic systems, which use a solution of plant nutrients as a growth medium in place of soil, are employed to control all aspects of plant nutrition.

Further Reading

Acquaah, George. *Principles of Plant Genetics and Breeding.* 2nd ed. Hoboken: Wiley, 2012. Print.

Barbosa-Cánovas, Gustavo, et al. *Global Issues in Food Science and Technology.* Burlington: Academic, 2009. Print.

Barnum, Susan R. *Biotechnology: An Introduction.* 2nd ed. Belmont: Wadsworth, 2005. Print.

Chrispeels, Maarten J., and David E. Sadava. *Plants, Genes, and Crop Biotechnology.* 2nd ed. Sudbury: Jones & Bartlett, 2003. Print.

Dixon, Richard A., and Robert A. Gonzales, eds. *Plant Cell Culture: A Practical Approach.* 2nd ed. New York: Oxford UP, 1994. Print.

Gamborg Oluf L., and Gregory C. Phillips, eds. *Plant Cell, Tissue and Organ Culture: Fundamental Methods.* New York: Springer, 1995. Print.

International Food Information Council Foundation. "Food Biotechnology Timeline." *Food Biotechnology: A Communicator's Guide to Improving Understanding.* Food Insight, n.d. PDF. Web. 24 Sept. 2013.

Kyte, Lydiane, and John G. Kleyn. *Plants from Test Tubes: An Introduction to Micropropagation.* 3rd ed. Portland: Timber, 1996. Print.

Owen, Meran R. L., and Jan Pen, eds. *Transgenic Plants: A Production System for Industrial and Pharmaceutical Proteins.* New York: Wiley, 1996. Print.

Pierce, Benjamin A. *Genetics: A Conceptual Approach.* 4th ed. New York: W. H. Freeman, 2012. Print.

"Plant Biotechnology Timeline." *National Institute of Food and Agriculture.* USDA, 18 Mar. 2009. Web. 24 Sept. 2013.

Thieman, William J., and Michael A. Palladino. *Introduction to Biotechnology.* 3rd ed. Boston: Pearson, 2013. Print.

Weasel, Lisa H. *Food Fray: Inside the Controversy over Genetically Modified Food.* New York: Amacom-American, 2009. Print.

TIME LINE OF PLANT BIOTECHNOLOGY

Year	*Event*
1838	German scientists Matthias Schleiden and Theodor Schwann present their cell theory: that all life-forms are made up of cells.
1858	Biologist Rudolf Virchow adds to the cell theory, proposing that "where a cell exists, there must have been a preexisting cell."
1902	Austrian botanist Gottlieb Haberlandt completes the cell theory with his idea of totipotency: cells must contain all the genetic information necessary to create an entire, multicellular organism. Therefore, every plant cell is capable of developing into an entire plant.

1939	Roger Jean Gautheret demonstrates the first successful culture of isolated plant tissues as a continuously dividing callus tissue.
1953	James Watson and Francis Crick make their landmark proposal for the double-helical structure of deoxyribonucleic acid (DNA), the molecule that carries genetic material.
1954	The first whole plant is regenerated, or cloned, from a single adult plant cell by W. H. Muir, Albert Hildebrandt, and Albert Riker.
1967	DNA ligase, the enzyme that joins DNA molecules, is discovered.
1970–1971	Daniel Nathans, Hamilton Smith, and Werner Arber discover restriction endonucleases.
1972	Researchers at Stanford University construct the first recombinant DNA molecules, and the following year DNA is inserted into *Escherichia coli* cells.
1980	The U.S. Supreme Court rules that genetically altered life-forms can be patented, allowing Exxon to patent a microorganism that eats oil.
1983	The National Institutes of Health permit scientists at the University of California at Berkeley to release genetically engineered bacteria designed to retard frost formation on crop plants.
1983	The U.S. Patent Office begins to issue a series of patents for genetically modified plants.
1984	The Plant Gene Expression Center, a collaborative effort between academia and the U.S. Department of Agriculture, is established to research plant molecular biology, sequence plant genomes, and develop genetically modified plants.
1985	Field testing of plants genetically modified to resist plant pathogens and disease vectors begins.
1986	The first release of a genetically modified crop, genetically engineered tobacco plants, is approved by the Environmental Protection Agency.
1987	Calgene receives a patent for a DNA sequence that extends the shelf life of tomatoes.
1987	Advanced Genetic Sciences Inc. field-tests a recombinant organism designed to inhibit frost in strawberries.
1990	Calgene conducts field experiments with herbicide-resistant cotton plants.
1990	At the Plant Gene Expression Center, biologist Michael Fromm announces the use of a high-speed "gene gun" to transform corn. Gene guns are used to shoot genetic material directly into cells via DNA-coated microparticles.
1993	Kary Mullis wins the Nobel Prize in Chemistry for the development of polymerase chain reaction technology, a technique he invented in 1981 for quickly multiplying DNA sequences in vitro.
1993	The U.S. Food and Drug Administration (FDA) announces its finding that genetically modified foods are not "inherently dangerous" and not in need of regulation. The following year, the FDA approves the first genetically modified whole food crop, Calgene's Flavr Savr tomato (later removed from the market).
1996	The genomes of *Saccharomyces cerevisiae* (baker's yeast), with twelve million base pairs, and of ancient archaea cells, which live near thermal vents at the ocean bottom, advance biologists' understanding of the evolution of life.
1996	GM seeds of soybean, corn, canola, tomato, cotton, and potato are planted in the United States, Canada, Mexico, Australia, Argentina, and China.
1998	Virus-resistant papaya is planted in Hawaii.
1999	Swiss and German scientists develop golden rice, a grain fortified with betacarotene.
2000	Researchers complete the full genomic sequence for the model flowering plant *Arabidopsis thaliana.*
2001	A transgenic tomato that can thrive in salty conditions is developed by American and Canadian scientists.
2002	Researchers complete the genomic sequence for rice, *Oryza sativa.*
2008	Biotech sugar beets are commercialized.
2011	"High-oleic" soybean varieties, high in monounsaturated fats, become available in the United States. More GM foods are submitted for government review, including low-acrylamide potatoes and non-browning apples.
2012	A reported 420.8 million acres of biotech crops are grown in twenty-eight countries. More than 90 percent of the 17.3 million farmers growing these crops live in developing countries.

Plant Breeding & Propagation

—*Martha A. Sherwood, PhD*

FIELDS OF STUDY

Agronomy; Agriculture; Ecology; Ecosystem Management; Environmental Sciences; Environmental Studies; Botany.

SUMMARY

The science of plant breeding and propagation is the controlled, systematic identification and multiplication of useful plant varieties. It is critical to human survival. Of all discoveries furthering the advance of civilization, improvements in food production have arguably been the most significant. Breeding food crops for higher productivity, improved nutritional content, and greater resistance to stress and disease; preserving rare species and reintroducing them into the natural environment; and creating plant varieties that act as factories for complex pharmaceuticals are among the most active fields for the plant breeder. Genetic engineering techniques are revolutionizing the industry.

DEFINITION & BASIC PRINCIPLES

Plant breeding and propagation science encompasses any systematic attempt to create or identify and select useful varieties of plants and multiply them for commercial production. Historically, the raw materials came from naturally occurring variations. Using increased knowledge of genetics and the mechanisms of inheritance, modern plant breeders have been able to crossbreed strains to produce hybrids with specific suites of characters. Breeders may use radiation or chemical mutagens to increase variability. Beginning in the 1980's, genetic engineering has enabled plant scientists to insert specific genes into genomes of cultivated plants.

Having produced and selected a desirable strain, the breeder must then produce sufficient numbers of plants to test for trait stability and determine optimal conditions for commercial production. Unlike annuals that produce abundant seed, perennials with long generation times, such as forest trees, are a challenge to the plant breeder. For species such as conifers, advances in tissue-culturing techniques have transformed production processes in both the development and the marketing phases.

The advent of genetic engineering in agriculture has produced a number of hotly debated issues concerning the safety, long-term environmental impact, and wisdom of introducing nonplant genes into human dietary staples on a global scale. This debate has slowed the spread of this technology.

BACKGROUND & HISTORY

Humans have been selecting and propagating useful varieties of plants for more than 10,000 years. The domestication of wheat, rice, and barley in the Old World and maize and potatoes in the New World was critical to the development of the earliest civilizations. When plant breeding became systematic enough to constitute science is difficult to pinpoint. Roman agricultural science, which drew heavily on Greek and Mesopotamian antecedents, included the deliberate introduction and propagation of novel plant varieties. Contemporary Chinese were at least as advanced as the Romans in agronomy and horticulture, and the sophisticated agricultural techniques and variety of cultivars in pre-Columbian Mexico and Peru argues for the existence of a scientific approach in those cultures.

Agricultural science came into renewed prominence in the eighteenth century as educated landowners brought the tools of the Enlightenment to bear. Private botanical gardens became establishments for introducing, testing, and propagating exotic species and varieties of food and ornamental plants. By midcentury, breeders recognized that flowering plants did have a sexual cycle and that the principles of animal breeding applied to them.

Beginning with the concept of genes pioneered by Gregor Mendel in the nineteenth century, the science of genetics progressed steadily through the twentieth century, giving plant breeders increasing understanding of the processes underlying their work. Plant breeders have contributed a great deal to genetics in general.

HOW IT WORKS

Conventional Plant Breeding. The plant breeder seeks to combine the desirable traits from different strains into a single variety. Sources for the genes producing these traits include existing commercial cultivars, field collections or germ plasm banks, closely related wild plants, and spontaneous or induced mutations. Under carefully controlled conditions, the researcher cross-pollinates two varieties, plants the resulting seed, and evaluates the progeny for the desired combination of characteristics. To produce uniform seed of the resulting hybrid for sale, a company will develop breeding stocks of the two parent strains, crossing them in production plots. A gene for male sterility is often incorporated into one of the parents, preventing self-pollination.

Tremendous improvements in methods of vegetative propagation have enabled it to be used for commercial production of perennials with a generation time of many years. There have also been significant advances in methods for evaluating desirable characteristics at an earlier stage in a plant's life cycle. If the biochemical basis for a mature trait is known, precursors can often be detected in very young seedlings. Alternatively, if a mature trait is closely linked genetically to a trait that is expressed earlier in the life cycle, the presence of that marker can be used for screening.

Many plants form viable interspecific crosses. Often these are sterile, which results in seedless fruits. Seedless varieties must either be hybrids of seed-producing parents or propagated vegetatively. Sometimes breeders use colchicine to induce polyploidy, resulting in a sexually reproducing interspecific hybrid with a double chromosome complement.

Genetic Engineering. Certain viruses and virus-like plasmids invade cells and attach their DNA to that of the host. Any DNA attached to the virus or plasmid will also be incorporated. Plant genetic engineering uses a plasmid from a phytopathogenic bacterium, *Agrobacterium tumefaciens*, to transform plant cells. Because transformation rates are very low, geneticists incorporate an antibiotic-resistance gene to facilitate screening. Plant cells in undifferentiated tissue culture are exposed to transformed *Agrobacterium* plasmids and transferred to an antibiotic-containing medium. The surviving cells are then grown out as plants.

In theory, the ability to synthesize any biologically produced compound can be transferred to any plant species by this method. In practice, the absence of activators or presence of inhibitory genes stymies many attempts.

Once engineered into a plant variety, a gene propagates normally from generation to generation and, under field conditions, into other populations of the same species. A gene for herbicide resistance, introduced into a crop to facilitate management by chemical means, can backfire if it spreads to closely related weeds. Of serious concern is so-called terminator technology, a sterility gene introduced into genetically engineered seed by seed companies to prevent patent infringement.

Modern Methods of Plant Propagation. A key feature of modern plant propagation is the use of explants and their proliferation under sterile conditions to produce large numbers of genetically identical, pathogen-free plant starts. Undifferentiated plant cells are grown in liquid or solid media promoting rapid growth. To produce a mature plant, subcultures are then subjected to growth regimes promoting differentiation and maturation. For biopharmaceutical production, plant cells may be grown indefinitely in liquid-culture bioreactors and harvested in the same manner as microbial populations, often heterotrophically.

APPLICATIONS & PRODUCTS

Seed. The primary use of plant breeding is to improve the productivity, cultivation characteristics, stress tolerance, and nutritional content of food crops, particularly cereal grains. The green revolution stressed productivity under modern agricultural methods. Later, more attention was paid to tailoring crops to specific environmental and social conditions. An example is yellow rice, genetically engineered for high vitamin A content, to combat vitamin A deficiency in rice-dependent populations.

Preservation of Biodiversity. Tissue culture methods for propagation have been a great boon for the proliferation of medicinal plants, heritage varieties, and stock for habitat restoration. Orchids and cycads, which are difficult to propagate from seed, are under tremendous pressure from collectors who encourage poaching in nature preserves. Modern nursery propagation helps protect these vulnerable species.

Molecular Farms. The use of genetically engineered plants to produce exotic organic compounds is still in the development phase but shows promise. Researchers have successfully produced a strain of *Arabis* (a mustard) that synthesizes hirudin, an anticlotting agent, from leeches. Molecular-farmed pharmaceuticals of plant origin are safer than those extracted from animals or human plasma. Another promising line of development is edible vaccines—food plants synthesizing proteins that provoke an antigen response to a particular disease-causing bacterium.

Reforestation. With tissue culture and rapid mass multiplication of stocks of woody plants, reforestation following disturbance has become more targeted. The forester can readily access seedlings adapted to the site, with built-in insect and disease resistance. Species such as the American chestnut, eliminated from most of its original range by blight in the early 1900's, are being successfully reintroduced as highly selected resistant strains.

Virus-Indexed Plants. Production of disease-free stock of potatoes and bananas has always been a challenge. The ability to produce commercial quantities of plant starts from a small amount of sterilized tissue greatly reduces the spread of viruses and other pathogens in these vegetatively propagated crops.

SOCIAL CONTEXT & FUTURE PROSPECTS

Advances in plant breeding have provoked many controversies, the most pressing of which are the contribution of improved crop varieties to unsustainable population growth, ownership of rights to germ plasm and to the products of genetic engineering, the safety of transgenic plants as human food, the risks of the unplanned spread of modified genes, and the adverse effects on traditional agriculture in developing countries. Concern over safety has led to a patchwork of national laws that inhibit but fail to effectively regulate global commerce in the products of genetic engineering.

The relationship of increased crop productivity to exponential population growth, already noted by William Malthus in 1798, was a feature of the green revolution in the twentieth century. Gains from improved technology or new crops are quickly canceled out unless population growth slows. Rural populations may actually end up worse off because improved technology favors large-scale operations and displaces farmers. Institutes in developing countries, such as the International Maize and Wheat Improvement Center (Centro Internacional de Mejoramiento de Maíz y Trigo, or CIMMYT), are working to ensure that the new round of agricultural technology will have a more beneficial human impact.

New varieties of plants under patent can be propagated only under license from the originator, which prevents farmers from saving seed. This increases costs and forces small operations to use older, less productive strains. When genetically engineered strains cross with crops in adjacent fields, the farmer of the traditional crop may be unable to sell it for human consumption. There are conflicting claims over the ownership of rights to improved varieties whose base stock came from field collections in developing nations.

If these objections can be overcome, the potential for the new technology is tremendous. The combination of greater productivity and improved nutritional content in staple crops could be of immense benefit. Incorporating resistance to disease and stress should lower the need for pesticides and herbicides and make farming on marginal lands more environmentally friendly. Using genetically engineered green plants to produce pharmaceutical and other complex organic chemicals ought to lower their cost and increase availability. The ability to propagate a species from vegetative portions of a few individuals and reintroduce it into its native habitat will undoubtedly be a boon in preserving endangered species, restoring native vegetation, and promoting sustainable ecologically friendly landscaping. As long as science and the public do not lose sight of potential social costs and maintain adequate safeguards, plant breeding will continue to serve humankind well.

Further Reading

Hrazdina, Geza, ed. *The Use of Agriculturally Important Genes in Biotechnology.* Washington, D.C.: NATO Scientific Affairs Division, 2000.

Kalloo, G., and J. B. Chowdhury, eds. *Distant Hybridization of Crop Plants.* Berlin: Springer-Verlag, 1992.

Liang, George H., and Daniel Z. Skinner. *Genetically Modified Crops: Their Development, Uses, and Risks.* New York: Food Products Press, 2004.

Parekh, Sarad R. *The GMO Handbook: Genetically Modified Animals, Microbes, and Plants in Biotechnology.* Totowa, N.J.: Humana Press, 2010.

Slater, Adrian, Nigel W. Scott, and Mark Fowler. *Plant Biotechnology: The Genetic Manipulation of Plants.* New York: Oxford University Press, 2008.

PLANTATION FORESTS

—David R. Coyle & Kamal J. K. Gandhi

FIELDS OF STUDY

Ecology; Ecosystem Management; Environmental Sciences; Environmental Studies; Forestry.

SUMMARY

The world's population just exceeded 7 billion people, and the demand for products derived from wood for this number of people simply cannot be attained by harvesting from natural forests. To augment our needs, we have turned trees into a crop plant by creating plantation forests around the world. Plantation forests, also called "tree farms," are grown on every continent except Antarctica and have been planted for many centuries. Plantations are grown almost exclusively for economic purposes, while benefits to wildlife and the environment are much less emphasized. As natural forests have shrunk in size, plantations forests have increased in area over time.

Plantation forests share many characteristics with more traditional agricultural crops in North America. Plantation forests are generally monocultures, consisting of a single genotype or species of tree. Occasionally, plantation forests have multiple genotypes of a single species, creating closely related polycultures. Tree species used are generally fast-growing, early successional species such as pine (*Pinus* spp.), poplars (*Populus* spp.), or eucalypts (*Eucalyptus* spp.). Regardless of the number of genotypes within a plantation forest, the treatment of these areas is generally the same. Trees are planted at uniform spacing, which is often determined by the size of the equipment and machines used to plant, apply pesticides, and harvest the stand. Trees almost always receive fertilization and receive irrigation in some drought-prone areas. Pest and weed control is often administered through the use of chemicals, but sometimes this can be accomplished through silvicultural and biological control methods.

Plantation forests contrast greatly with natural forests, which can have many different tree species. Natural forests often arise through succession, whereas plantation forests are planted, harvested after several years, and replanted, over and over again. Natural forests have many different-aged trees, while trees in plantation forests are all the same age. Natural forests have a diverse forest floor that supports woody vines and shrubs, herbaceous plants, many different arthropods, and even mammals. Plantation forests are generally kept weed- and insect-free, which in turn greatly reduces the attractiveness to most mammals, with the exception of deer, which may feed in plantation forests when the trees are young. To that end, however, most plantation forest managers implement deer-exclusion fences to eliminate deer browse damage. Natural forests do not receive irrigation or fertilization. Rather, they are dependent on nature to provide necessary water and nutrients for tree growth. Plantation forests, however, may be irrigated and fertilized in an effort to maximize tree growth and production.

PURPOSE & LAND AREA

Plantation forests are typically grown for industrial (48 percent as of 2000) and nonindustrial purposes (26 percent), with another 26 percent as unspecified uses. In terms of industrial use, plantation forests focus on roundwood production that is classified as either industrial roundwood (e.g., sawlogs, veneer logs, plywood, and pulpwood) or wood fuel (e.g., wood used for cooking, heating, or power) production. Current worldwide harvests for industrial roundwood production are around 1.8 billion cubic meters (more than 63.5 billion cubic feet). Even though plantation forests constitute about 5 percent

A plantation of eastern cottonwood (Populus deltoides) in the southeastern United States. Eastern cottonwood is a valuable timber species that is highly suitable for plantation management. Image by David R. Coyle.

of total forests in the world, they may contribute up to two-thirds of wood production in some areas. In recent years, there has been a tremendous push to develop technology to make liquid biofuels (ethanol) using wood residues from plantation forests, as well as to use woody biomass along with coal in cofiring power plants. In the next decade, this technology may lead to widespread use of forest biomass for transportation and energy needs. Nonindustrial use of plantations includes providing ecological services such as wind protection, soil and water conservation, carbon sinks, and enhancement of barren and urban landscapes.

The area occupied by forest plantations varies considerably among continents. In North America, forest plantations cover an area over 72,587 square miles (188,000 square kilometers)—that's an area larger than Kentucky and South Carolina combined. Poplars (*Populus* spp. and hybrids) are grown primarily in the Pacific Northwest and the Great Lakes regions of the United States, with smaller areas in the northeast and southeast. Most plantations are grown on 8–12 year rotations, with the majority of wood going for pulp. However, some plantations in the Pacific Northwest are growing trees longer, with a targeted end use as lumber. Trees are generally planted at 2.5-by-3-meters (8.2-by-9.8-feet) spacing and are not thinned prior to harvest. When harvested, the entire site is clear-cut. Trees for the next rotation in the plantation forest are planted between the rows of harvested stumps, so by the time the second generation is ready to harvest, the old stumps have biodegraded. Willows (*Salix* spp.) are grown in the northeast, especially in New York State. These willow plantations are grown for 2–3 years and harvested up to four times before being replanted—a technique called "coppicing." In this case, the biomass produced is used for energy, not pulp or lumber, and may be cut and made into large bales with modified agricultural equipment. In addition to hardwoods, several coniferous species, including pine and spruce (*Picea* spp.) trees are grown in the southeastern and northeastern United States and in parts of Canada. Conifer species are slower growing than poplars and require 1–2 thinning cuts after initial planting. Thinning cuts allow more space to grow for the trees that are left on site. Trees removed in the thinning are generally smaller and used for pulp, while the larger, older trees (up to 30 years old for *Pinus* spp. and older for *Picea* spp.) are used for lumber.

PLANTATION FORESTS IN ASIA

Asia has the greatest area of forest plantations in the world, with over 250,194 square miles (648,000 square kilometers) planted—an area nearly the size of Texas. This region, along with Australia, has the highest rate of forest plantation establishment in the world, with Japan, China, and India having the largest amount of forest plantation land area. Over 45 percent of Japan is classified as forest plantation, with trees such as Japanese cedar (*Cryptomeria japonica*), Japanese cypress (*Chamaecyparis obtusa*), Japanese larch (*Larix leptolepis*), and pine species. Japan uses its forest plantation trees for pulp and lumber products. Several pine and spruce species are grown in China, as well as hardwood species such as eucalyptus, paulownia (*Paulownia* spp.), birch (*Betula* spp.), locust (*Robinia* spp.), and poplars. Trees are harvested in as little as 6 years, and are given intensive nutrient, weed, and pest management. These trees are used for all types of wood products, including pulp for paper and lumber. China alone accounts for nearly three-fourths of the world's poplar production. India grows teak (*Tectona* spp.), eucalyptus, and poplar in its forest plantations. Teak is planted at wide spacing, with one thinning occurring when trees are around 33 feet (10 meters) tall, and another when trees are 99 feet (20 meters) tall. Branches are often removed on the lower portions of the stem, especially when teak trees are young. Teak is generally harvested by 30 years of age, but some plantations up to 60 years old do exist.

PLANTATION FORESTS IN EUROPE

Europe as a whole has over 106,564 square miles (276,000 square kilometers) of forest plantations, which is an area slightly larger than Colorado. Sweden, Finland, and other Nordic and Baltic countries have many acres of willow plantations. In these coppice systems, the trees are cut back after the first year to promote multiple sprouts. Biomass is harvested every 3–5 years thereafter for another 3–4 rotations. Plantations are fertilized, harvested during dormant periods, and treated much like traditional agricultural crops. Sweden is widely regarded as the world authority on willow coppice systems, as it has conducted extensive research and field trials on silviculture and harvesting methods. Pine and eucalypt plantation forests are found in Spain, while poplars are grown in France. Plantation forests in Europe are primarily used for biomass and energy, with a smaller portion going toward pulp and lumber.

PLANTATION FORESTS IN AFRICA

With the exception of South Africa, there is little reliable information regarding forest plantations in Africa. Forest plantations cover an area almost as large as Louisiana (over 50,579 square miles, or 131,000 square kilometers), and are primarily planted with teak, pine, eucalypts, acacia (*Acacia* spp.), and types of rubber tree (*Hevea* spp.). As countries in Africa continue to develop, the demand for wood products is growing. A large portion of forest plantations are found in western Africa, where their primary uses are for fuelwood and environmental stability (e.g., preventing sand dunes). The eastern portion of Africa, unfortunately, has been one of the most unstable regions in the world in the last several decades, and likewise forest plantations in this region have been neglected and are impacted by pests and disease in addition to a general lack of silviculture. Central Africa has a low proportion of forest plantations, mostly pines and eucalypts. Here, the wood is used for fuel and lumber. The majority of forest plantations are in southern Africa, specifically South Africa. Eucalypts and pines are most often planted and used as fuelwood, pulp, or lumber for construction. South Africa is widely recognized as a world leader in forest plantation management and is able to supply the majority of the country's wood needs through its plantation forests.

PLANTATION FORESTS IN SOUTH AMERICA

Eucalypts and pines are the most commonly grown tree species in South America, which has over 43,629 square miles (113,000 square kilometers)—an area almost the size of Ohio—planted in forest plantations. Brazil, Chile, Argentina, and Uruguay have the most land area, while other countries have much less area in forest plantations. Brazil is one of the most active producers of pulpwood in the world. South American forest plantations are used for not only pulp and lumber but also for exports of wood chips and logs, as well as for biomass and cofiring plants. It is illegal to clear native forests in Chile, making the renewable wood from forest plantations even more valuable. In addition to pines and eucalypts, poplars and willows are grown on a smaller scale, especially in Argentina.

PLANTATION FORESTS IN AUSTRALIA

Australia (and Oceania) has over 14,672 square miles (38,000 square kilometers) of plantation forests, or an area the size of Maryland and Delaware combined. In Australia, plantation forests are mostly located in the southeastern area and Tasmania. The predominant tree species planted are pine, teak, mahogany (*Sweitenia* spp.), and eucalypts, with one native conifer, hoop pine (*Araucaria cunninghamii*) also used. These species are used for pulp and paper products, saw timber for building and furniture, and wood products such as particle board and plywood. Plantation forests are an important component of forestry in this region and account for a great deal of wood products. Forest plantations are managed very intensively, with pest and weed control. Australia and New Zealand are regarded as leaders in the development of plantation forest silviculture, and this is reflected in the high economic impact that this sector has on the local economy.

CRITICISMS OF PLANTATION FORESTS

The main criticism of plantation forests is that they are monocultures that are even-aged with trees grown primarily for quick wood and fiber production. Hence, plantation forests have a much lower diversity of plant species and other biota that those of non-plantation forests. As monocultures, the trees may be more susceptible to outbreaks of insects and diseases because the trees may represent a single genetic stock and pests can move easily across the homogenous landscape. Plantation forests typically tend to be dominated by nonnative tree species that may have adverse ecological impacts. For example, eucalyptus trees that are widely used in plantation forests have a higher requirement for water than native species and show allelopathy where they produce compounds that inhibit the growth of other plants.

There have been cases where nonnative trees (e.g., pine trees planted in the Southern Hemisphere) have escaped plantations and have invaded natural ecosystems with significant alterations to native forest structure and composition. Nonnative forest pests have frequently been introduced to plantations, from where they can establish and spread into the natural forests. Because fertilizers and insecticides are often used in plantation forests, this may cause chemical pollution in the general area. Forests have been cleared to create plantations in areas where indigenous people used to live, thus causing their displacement. Further, there is a concern that if significant biofuel technology is developed, it could lead to large-scale replacement of natural forests by plantation forests, especially in tropical areas.

BENEFITS OF PLANTATION FORESTS

There could be many ecological benefits to plantation forests, especially if plantations are managed with appropriate silvicultural methods and sustainable forest practices. Plantation forests have the maximum benefit when they are planted on degraded lands due to soil erosion after intensive agriculture and mining activities. In some instances, plantation trees have been used to competitively exclude nonnative weeds from landscapes, so that forests could be regenerated to a natural state over years. Planted trees that are hardier than native tree species may prevent further soil loss, rehabilitate the land, and assist with watershed protection. Afforestation of these areas has led to higher sequestration of carbon stocks, which is one of the reasons why many tropical countries are creating plantation forests. It is also argued that the use of existing plantation forests to the maximum capability may reduce the need to cut down natural forests for fiber production in the future. If managed properly, plantations can continue to provide a renewable source of wood for a long time. Since plantations are more productive than natural forests, more wood and fiber can be produced using less area. Sewage and industrial waste has been sprayed in plantations where trees break it down and absorb it, leading to a more environmentally friendly way of disposing waste. Certain types of trees in forest plantations, especially poplars, can also be used for phytoremediation, or removing pollutants from the soil by drawing them up through the roots and accumulating pollutants in tree tissues. Those trees can then be removed and disposed, taking the pollutants with them.

Plantation forests are excellent areas in which to conduct research. Because of the relatively sterile environment compared to natural forests (e.g., no ground vegetation, high survival of trees, low genetic diversity), plantation forests are well suited for experiments on tree physiology, tree growth and nutrient dynamics, and interactions with pests and diseases. By using plantation forests, scientists can isolate specific issues and adequately test them without lots of confounding effects that are often present in natural forests. Thus, plantation forests may not only produce a high amount of wood quickly on a small land area, but they also provide scientists with sites to conduct research to improve forest productivity. Research and production forest plantations contribute to the local economy, especially in the third world countries through providing livelihoods, poverty alleviation, and food security.

THE WORLD'S "WOOD-BASKETS"

Overall, plantation forests that are present in most countries around the world are considered "wood-baskets." The area occupied and the importance of plantation forests are increasing over time. Adverse effects of plantation forests include changes in the original forest structure, composition, and processes, with subsequent impacts on native species. The beneficial effects of plantation forests include production of wood and fiber that could be sustainable and renewable, some ecological values especially in degraded and urban areas, and a potential source of income to the local communities. A major challenge for plantation foresters is how to optimize production and efficiency with minimal impacts by insects and disease, while maintaining and enhancing local biodiversity. Management of plantation forests in the future is likely going to be heavily affected by changing technology, policy, and the value any society places on these highly modified habitats.

POLLUTION EFFECTS ON ANIMAL LIFE

—Ming Y. Zheng

FIELDS OF STUDY

Ecology; Ecosystem Management; Environmental Sciences; Environmental Studies; Zoology.

SUMMARY

Pollutants in soil, water, and atmosphere have created enormous problems for the living world. Destroyed habitats and polluted food sources and drinking water for animals have caused deformations in animal growth, development, and reproduction, as well as a shortening of life span, all of which contribute to an accelerated decrease in biodiversity and the extinction of more species. Realization of the severity of pollution is the first step toward seeking a long-term remedy.

During the last decade of the twentieth century, the environmental problems predicted by environmental scientists decades previously began to be aggravated in a variety of ways. These included population explosion, food imbalances, inflation brought about by scarcity of energy resources, acid rain, toxic and hazardous wastes, water shortages, major soil erosion, the depletion of the ozone layer, and greenhouse effects. As a result of pollution, decreases in biodiversity and the extinction of both plant and animal species have accelerated. The burning and cutting of thousands of square miles of rain forest not only destroyed habitats for numerous animal species but also caused irreversible damage to ecosystems and climate. The recurring drought and famine in Africa are testimony to the negative effects of human activity. The well-being of animals as well as humans will not be protected against the ecological consequences of human actions by remaining ignorant of those actions. In order to take effective measures to reduce pollution and protect natural resources and the environment, people must first recognize these problems. The ignorance and inaction of ordinary citizens will lead to disastrous consequences for the environment, threatening humanity's very existence.

SOURCES & TYPES OF POLLUTION

Industrialization and the expansion of the human population have left relatively few places on earth undisturbed. In simple terms, human interference in natural ecosystems is the single most important source of pollution. First, heavy dependence on fossil fuels for energy, and on synthetic chemicals and materials, helped to dump millions of metric tons of nonnatural compounds and chemicals into the environment. Among such chemicals and compounds are fertilizers, insecticides, fungicides, and herbicides, and home products. Application of excess chemical fertilizers to soil hampers natural cycling of nutrients, depletes the soil's own fertility, and destroys the habitat for thousands of small animals residing in the soil. Farm runoff carries priceless topsoil, expensive fertilizer, and animal manure into rivers and lakes, where these potential resources become pollutants. In the city, water pours from sidewalks, rooftops, and streets, picking up soot, silt, oil, heavy metals, and garbage. It races down gutters into storm sewers, and a weakly toxic soup gushes into the nearest stream or river. Many of these chemicals also seep into the ground, contaminating groundwater.

Plants and factories manufacturing these chemical products are another source of pollutants and contamination. Burning fossil fuels releases greenhouse gases, carbon oxides, and methane. Coupled with deforestation in many regions of the world, carbon dioxide concentration in the atmosphere has steadily climbed, from 290 parts per million in 1860 to more than 395 parts per million (ppm) in 2013, more than a 36 percent rise due to industrialization. That same year, atmospheric carbon dioxide surpassed 400 ppm for the first time in Earth's history. In 2016, the average concentration of carbon dioxide was 402.9 ppm, and in May 2017, it reached 409 ppm.. The resultant global warming will have far-reaching effects on plants, animals, and humans in ways still not understood.

Acid rain, a result of overcharging the atmosphere with nitric oxides and sulfur dioxide (two gases also released by the burning of fossil fuels), has increased the acidity of soil and lakes to levels many organisms cannot survive. In 1991, the U.S. National Acidic Precipitation Assessment Program reported that 5 percent of New England lakes were acidic, some to such levels that they were no longer able to support fish. This happens when much of the food web that sustains the fish is destroyed. Clams, snails, crayfish, and insect larvae die first, then amphibians, and finally fish. The detrimental effect is not limited to aquatic animals. The loss of insects and their larvae and small aquatic animals has contributed to a dramatic decline in the population of black ducks that feed on them. The result is a crystal-clear lake, beautiful but dead.

Another serious problem created by the chemical industry is ozone depletion. Chlorofluorocarbon (CFC) compounds contain chlorine, fluorine, and carbon. Since their development in the 1930s, these compounds were widely used as coolants in refrigerators and air conditioners, aerosol-spray propellants, agents for producing Styrofoam, and cleansers for electronic parts. These chemicals are very stable and for decades were considered to be safe. Their stability, however, turned out to be a real problem. In gaseous form, they rise into the atmosphere, where the high energy level of ultraviolet (UV) light breaks them down, releasing chlorine atoms, which in turn catalyzes the breakdown of ozone to oxygen gas. As a result of the decline of ozone and the depletion of the ozone layer, UV radiation rose by an average of 8 to 10 percent per decade between 1979 and 1992. This depletion of the ozone layer poses a threat to humans, animals, plants, and even microorganisms.

Pollution in the form of plastic waste harms marine habitats and wildlife, such as when marine animals eat or become tangled in such debris. Plastic waste makes up 60 to 80 percent of marine debris, and the demand for plastic bags, bottles, and containers is growing. According to PlasticsEurope, the global production of plastics reached 322 million tons in 2015, a significant increase from 230 million tons a decade earlier. And, according to a report released by the Ellen MacArthur Foundation in January 2016, plastic is produced at twenty times the rate that it was in 1964. Of the plastic produced, 40 percent winds up in landfills and only 5 percent are recycled effectively.

POLLUTION EFFECTS OF CHEMICALS

The degradation of air, land, and water as a result of the release of chemical and biological wastes has wide-ranging effects on animals. On a large scale, pollution destroys habitats and leads to population crashes and even the extinction of species. Hazardous chemicals introduced into an environment sometime render it unfit for life, as in Love Canal, New York, and Times Beach, Missouri. At the individual level, pollution causes abnormalities in growth, development, and reproduction. Hazardous chemicals, introduced either intentionally (such as fertilizers, herbicides, and pesticides) or through neglect (such as industrial wastes), have a variety of detrimental, sometimes devastating effects on animals. They affect the metabolism, growth and development, reproduction, and average life span of many species.

In the 1940s, the new insecticide dichlorodiphenyltrichloroethane (DDT) was regarded as a miracle. It saved millions of lives in the tropics by killing the mosquitoes that spread deadly malaria, and it saved millions more lives by destroying insect pests that fed on crops, thus increasing crop yields. This miraculous pesticide, however, turned out to be a long-lasting nemesis to the environment and many species of wildlife. In the mid-1950s, the World Health Organization (WHO) used DDT on the island of Borneo to control malaria. DDT entered food webs through a caterpillar. Wasps that fed on the caterpillar were destroyed, and then gecko lizards that ate the poisoned insects accumulated high levels of DDT in their bodies. Both geckos and the village cats that ate the geckos died of DDT poisoning. As the cats died, the local rat population exploded. The village was then threatened with an outbreak of plague, carried by the uncontrolled rats. Meanwhile, in the United States, ecologists and wildlife biologists during the 1950s and 1960s witnessed a stunning decline in the populations of several predatory birds, especially fish eaters such as bald eagles, cormorants, ospreys, and brown pelicans. The population decline drove the brown pelican and bald eagle close to extinction. In 1973, the U.S. Congress passed the Endangered Species Act, which banned the use of DDT, and the once-threatened species have since somewhat recovered.

Although DDT has been banned in much of the world, there is a growing concern over the effects of a number of similar compounds known as xenoestrogens or environmental estrogens. These compounds interfere with normal sex-hormone functions by mimicking the effects of the hormone estrogen or enhancing estrogen's potency. High levels of xenoestrogens, such as dioxin and polychlorinated biphenyls (PCBs), in the Great Lakes have led to a sharp decline in populations of river otters and a variety of fish-eating birds, including the newly returned bald eagles. These chemicals are also the cause of deformed offspring or eggs that never hatch. In 1980, a spill of dichlorodiphenyldichloroethylene (DDE), a common by-product of DDT, in Florida's Lake Apopka led to a 90 percent drop in the birth rate of the lake's alligators.

EFFECTS OF AIR POLLUTION

Air pollution leads to acid rain and the greenhouse effect, as well as damage to the ozone layer. Acid rain falls in areas that are great distances from the source of the acids, destroying forests and lakes in sensitive regions and causing fish populations to dwindle or disappear entirely. Data collected from the Adirondack lakes and Nova Scotia rivers since the 1980s clearly show declines in acid-sensitive species. Similar results were obtained by analyzing fish population and water acidity in Maine, Massachusetts, Pennsylvania, and Vermont. The effects of acid rain on other animals are indirect, either through the dwindling fish population, thus eliminating a food source, or through stunted forest growth, thus disturbing other animals' habitats.

The effect of global warming on the animal kingdom is also a serious and complex issue. As global temperature rises, ice caps in polar regions and glaciers melt, ocean waters expand in response to atmospheric warming, and the sea level elevates. The expected sea-level rise will flood coastal cities and coastal wetlands. These threatened ecosystems are habitats and breeding grounds for numerous species of birds, fish, shrimp, and crabs whose populations could be severely diminished. The Florida Everglades will virtually disappear if the sea level rises two feet. The impact of global warming on forests could be profound. The distribution of tree species is exquisitely sensitive to average annual temperature, and small changes could dramatically alter the extent and species composition of forests, and thus the population distribution of animals. A rise may make temperatures unsuitable for some species, hence reducing biodiversity.

The effect of the depleted ozone layer on animals is yet to be fully understood. It is known that the high energy level of ultraviolet (UV) radiation can damage biological molecules, including the genetic material deoxyribonucleic acid (DNA), causing mutation. In small quantities, UV light helps the skin of humans and many animals produce vitamin D, and causes tanning. However, in large doses, UV light causes sunburn and premature aging of skin, skin cancer, and cataracts, a condition in which the lens of the eye becomes cloudy. Due to UV radiation's ability to penetrate, even animals covered by hair and thick fur cannot escape from these detrimental effects. UV damage caused by depleted ozone costs U.S. farmers billions of dollars annually in reduced crop yields. All who depend on forestry and agriculture may bear a much higher cost if the emission of pollutants that destroy ozone is not regulated soon.

POSSIBLE REMEDIES

The various types of pollution all have serious effects on the plant and animal species that share this planet. It is all too easy to document the impacts of pollution on human health and ignore its effects on the rest of the living world. Any possible remedies to alleviate these problems should start with education at an individual as well as a global level. The tasks seem to be insurmountable, and no organization or no country can do it alone. It takes willingness to accept short-term inconvenience or economic sacrifice for the benefit of the long run.

Synthetic chemical pollutants that are poisoning both people and wildlife could be largely eliminated without disrupting the economy, as reported in a study published in 2000 by the Worldwatch Institute, a Washington, DC–based environmental organization. The report presents strong evidence from three sectors that are major sources of these pollutants—paper manufacturing, pesticides, and polyvinyl chloride (PVC) plastics—to show that nontoxic options are available at competitive prices. Agricultural pollution can be mitigated, significantly reduced, or virtually eliminated through the use of proper regulation and economic incentives. Farmers from Indonesia to Kenya are learning how to use less of various chemicals while boosting yields. Since 1998, all farmers in China's Yunnan Province have eliminated their use of fungicides while doubling rice yields by planting more diverse varieties of the grain.

Other efforts have already led to positive changes. In 1989, the U.S. government established the Acid Rain Program in an effort to reduce sulfur dioxide and nitrogen oxide emissions. By 2007, emissions of sulfur dioxide had been reduced by 40 percent. Also in 1989, the Montreal Protocol went into effect; this international treaty, which as of 2012 has been signed by 197 nations, calls for the complete elimination of CFCs.

In most, if not all, cases, the question is not whether it is possible to alleviate pollution of the environment; rather, it is whether we realize the urgency and are willing to sacrifice to do it. For the common well-being of generations to come, better approaches have to be taken to preserve the environment and biodiversity.

FURTHER READING

Bondareff, Joan M., Maggie Carey, and Carleen Lyden-Kluss. "Plastics in the Ocean: The Environmental Plague of Our Time." *Roger Williams University Law Review*, vol. 22, no. 2, 2017, pp. 360–383. *Legal Source*, search.ebscohost.com/login.aspx?direct=true&db=lgs&AN=124086721&site=eds-live. Accessed 5 Feb. 2018.

Brown, Lester R. *State of the World, 2000.* New York: Norton, 2000. Print.

Elliott, John E., Christine A. Bishop, and Christy A. Morrissey, eds. *Wildlife Ecotoxicology: Forensic Approaches.* New York: Springer, 2011. Print.

"Graphic: The Relentless Rise of Carbon Dioxide." *Global Climate Change*, NASA, 26 Jan. 2018, climate.nasa.gov/climate_resources/24/. Accessed 5 Feb. 2018.

Hill, Julia Butterfly. *The Legacy of Luna: The Story of a Tree, a Woman, and the Struggle to Save the Redwoods.* San Francisco: Harper, 2000. Print.

Johnson, Arthur H. "Acid Deposition: Trends, Relationships, and Effects." *Environment* 28.4 (1986): 6+. Print.

Kendall, Ronald J., et al., eds. *Wildlife Toxicology: Emerging Contaminant and Biodiversity Issues.* Boca Raton: CRC, 2010. Print.

Lindsey, Rebecca. "Climate Change: Atmospheric Carbon Dioxide." *Climate.gov*, National Oceanic and Atmospheric Administration, 17 Oct. 2017, www.climate.gov/news-features/understanding-climate/climate-change-atmospheric-carbon-dioxide. Accessed 5 Feb. 2018.

Lippmann, Morton, ed. *Environmental Toxicants: Human Exposures and Their Health Effects.* 3rd ed. Hoboken: Wiley, 2009. Print.

Sampat, Payal. *Deep Trouble: The Hidden Threat of Groundwater Pollution.* Washington: Worldwatch, 2000. Print.

Sindermann, Carl J. *Coastal Pollution: Effects on Living Resources and Humans.* Boca Raton: Taylor, 2006. Print.

PONDS

—Hannah Bement

FIELDS OF STUDY

Ecology; Ecosystem Management; Environmental Sciences; Environmental Studies; Hydroclimatology; Physical Geography.

SUMMARY

Because vernal ponds tend to dry up annually, they are generally fishless, which allows for the safe development of amphibians and insect larvae that would normally be preyed upon.

Ponds are small lentic or still bodies of freshwater. There is much disagreement about how to define a pond, as well as the proper method of distinguishing ponds from lakes. Some defining methods involve size requirements. For example, some areas of the United States define a pond as having surface areas of less than 10 acres. Other methods specify a maximum depth, stating that they must be as shallow as 12 to 15 feet (3.66 to 4.57 meters), or shallow enough that light can penetrate to the bottom uniformly across the water body so that plants may root all the way through the pond. This shallow depth also means that ponds do not have temperature layers or stratification caused by heat convection as lakes do; however, they may have thermally driven microclimates. Another criterion is that ponds must be a quiet body of water, lacking wave action on the shoreline, though they may have wind-driven currents. Regardless of these criteria, many ponds have been given the misnomer of lake and vice versa. Ponds can form in any depression in the ground that collects and retains enough groundwater or precipitation to pool. These depressions can result from a wide range of natural geological and ecological events. For example, kettle ponds were formed when glaciers retreating across North America and Europe ground out deep potholes in the landscape, which are now filled from underground aquifers. Many pond-forming activities are now constrained by human activities, such as filling or irrigation.

Ponds are small lentic or still bodies of freshwater. There is much disagreement about how to define a pond, as well as the proper method of distinguishing ponds from lakes. Some defining methods involve size requirements. For example, some areas of the United States define a pond as having surface areas of less than 10 acres. Other methods specify a maximum depth, stating that they must be as shallow as 12 to 15 feet (3.66 to 4.57 meters), or shallow enough that light can penetrate to the bottom uniformly across the water body so that plants may root all the way through the pond. This shallow depth also means that ponds do not have temperature layers or stratification caused by heat convection as lakes do; however, they may have thermally driven microclimates. Another criterion is that ponds must be a quiet body of water, lacking wave action on the shoreline, though they may have wind-driven currents. Regardless of these criteria, many ponds have been given the misnomer of lake and vice versa. Ponds can form in any depression in the ground that collects and retains enough groundwater or precipitation to pool. These depressions can result from a wide range of natural geological and ecological events. For example, kettle ponds were formed when glaciers retreating across North America and Europe ground out deep potholes in the landscape, which are now filled from underground aquifers. Many pond-forming activities are now constrained by human activities, such as filling or irrigation.

VERNAL PONDS

Some ponds are temporary, only holding water for part of the year, and are known as ephemeral or vernal pools. Most vernal ponds fill with winter rains and snowmelt, meaning that they are at their peak level in the spring, hence the term *vernal*, meaning spring. Because these ponds tend to dry up annually, they are generally fishless, which allows for the safe development of amphibians and insect larvae that would normally be preyed upon. However, some species of fish, called killifish, have adapted to this environment by laying eggs that can survive partial dehydration. There are several species of organisms that rely upon these pools and are considered indicator species by scientists trying to identify pond types. Aquatic invertebrates and amphibians can serve as indicators of pond health and type because they have specific habitat requirements and different species respond differentially to changes in the biological, chemical, and physical makeup of a pond. Vernal pools are important breeding sites for frogs and toads, such as wood frogs (*Rana sylvatica*) and spadefoot toads (*Scaphiosus*), and mole salamanders like spotted salamanders (*Ambystoma maculatum*) and marbled salamanders (*Ambystoma opacum*). They are also habitats for daphnia and fairy shrimp, small invertebrates that survive the periodic dry periods by laying resting eggs in the dried mud of the pond bottom, which remain dormant until the pond refills with water in the spring. Some eggs can remain dormant for decades, ensuring the continuation of the species life cycle, even through years of drought.

POND SEASONALITY

Most ponds retain water year round, and in temperate climates freeze partially or completely during the winter season. Aquatic organisms tend to tolerate a narrow temperature range, and have a variety of methods for dealing with seasonal changes. Cold-blooded animals, like reptiles and amphibians, can change their metabolism and reduce their activity levels to preserve energy stores. Turtles and frogs will often burrow in the mud of the pond or surrounding area and hibernate through the winter. Some frog species, such as the wood frog, can survive being partially frozen because of the presence of cryoprotectant chemicals that act like antifreeze in their blood and body tissues. Many insects also burrow and hibernate or lay winter eggs that will hatch in the spring. Fish and snails move to deeper waters and remain less active beneath the ice. Some aquatic plants form winter buds that drop to the bottom of the pond and germinate in the spring. Zooplankton, like rotifers or cladocerans, produce winter eggs, whereas protists and annelids protect themselves by forming cysts around their bodies.

NUTRIENT CYCLING IN A CLOSED ECOSYSTEM

Many ponds have no surface outflow and are spring-fed, making them self-contained ecosystems. All ecosystems are made up of interactions among the biotic, or living, plants, animals, and microorganisms relating among themselves and to the abiotic, or non-living, physical and chemical factors of the environment. In ponds, the abiotic factors include: water, carbon dioxide, oxygen, phosphoric salts, amino acids, and nitrogen. Like most closed ecosystems, ponds function through the interlocked relationship of two food chains cycling these nutrients through their environment. The first food chain consists of light-dependent primary producers or autotrophic organisms, and the organisms that eat them, called heterotrophs. Autotrophic organisms produce complex organic compounds from simple inorganic molecules. Though a few autotrophs can do this by using chemicals like hydrogen sulfide, most ecosystems, including ponds, derive their energy from the sun and are dominated by photosynthetic autotrophs. Photosynthesis involves the reduction of carbon dioxide molecules through the addition of oxygen, which creates chemical energy and sugars that are used by organisms to create biomass or growth.

The primary producers in ponds are phytoplankton, periphytic algae, submerged plants, floating plants, emergent plants, and shore plants. Phytoplankton, whose name means "wandering plants," are microscopic algae that float in open water, giving it a green appearance. Periphytic algae are microscopic algae that attach to substrates, making pond rocks a slimy, greenish brown. This algae helps to produce oxygen at the bottom of ponds, which is important for decomposers found in the depths. Submerged plants grow with their entire structure completely under the water. Floating plants are rooted at the bottom of ponds, but extend to float on the surface. Emergent plants are rooted in shallow water and have their stems and leaves above the water most of the time, whereas shore plants are completely above the water most of the time.

The energy stored in the algae and plant biomass is moved up the food chain by grazing organisms. These organisms are called consumers, or heterotrophs, which include animals, bacteria, and fungus. Ponds are filled with tiny zooplankton, microscopic animals that eat phytoplankton or other small zooplankton. They include single-celled animals like the foraminifera, tiny crustaceans such as copepods, or immature stages of larger animals. Algae and zooplankton are eaten by macroinvertebrates, larger animals that have no backbone, including thousands of species of pond dwelling insects.

Plants, zooplankton, and microinvertebrates are then eaten by larger vertebrates, animals having a backbone, which include fish, frogs, salamanders, turtles, snakes, water fowl, wading birds, and semi-aquatic mammals like muskrats, beaver, and otters, as well as raccoons. Consumers that eat only plants, herbivores, are considered primary consumers. Consumers that eat primary consumers are called secondary consumers, and can be strict carnivores, eating only animals; or omnivores, which eat both animals and plants. Tertiary consumers are considered apex or top predators, and can feed on both primary and secondary consumers. An example of a tertiary predator in a pond is the otter, which may eat both herbivores, such as some fish, as well as carnivores like frogs.

When the nutrients in a pond cycle up through the food chain as organisms are eaten by larger and larger animals, they are lost to the surrounding environment only when animals leave the pond through external predation or migration. The rest of the nutrients are cycled back down to the abiotic pool of nutrients by a second food chain. The second food chain is made up of non-light dependent organisms that feed on and break down the detritus from the first food chain. These detritivores are bacteria, fungi, and other organisms that feed on the waste products of the producers and consumers. In turn, the detritivores break the organic waste back into their inorganic components, which the primary producers, plants and algae, can then again use in the process of photosynthesis. In this way, the nutrients are cycled continuously through the pond ecosystem.

POND HABITATS

Ponds are made up of four distinct habitat zones: the shoreline, the surface film, the open water, and the bottom water. The surface film and the open water make up the pelagic or upper habitats of the pond. The surface film refers to the surface of the pond on which insects, such as water striders, and free-floating organisms range. It also includes the water just below the surface, as well as the underside of floating plants.

The open water is home to large, free-swimming organisms like fish, and drifters like plankton. The bottom water and shoreline make up the benthic or lower level habitats in the pond. The bottom water habitat varies depending on the pond depth. Quiet, standing water ponds often have muddy and silted bottoms into which crayfish, mayfly nymphs, and other macro- and microorganisms burrow. The shoreline is considered the area of the pond where water meets land, and is dominated by rooted plants and air-breathing organisms that vary depending on the substrate type. This is the richest part of the pond in terms of biodiversity because of the dense plant life, which is used for foraging and breeding by many species.

Though not strictly considered part of a pond, the area of land around a pond is essential to the health of the pond and the organisms within it. It is called a buffer because it often serves to protect or buffer the pond from exposure to harmful external forces such as pollution. Wide vegetated buffers around ponds provide essential nesting sites, winter habitat, and escape cover for wildlife. Species of amphibians like the spotted salamander are especially dependent on large and healthy buffer zones around ponds because, though they are dependent on the ponds to lay their eggs and for their larvae to develop, the adults only remain at the pond long enough to breed. Afterward, they return to the surrounding terrestrial habitats to live and feed for the rest of the year and over winter. Reptiles like snakes and turtles may depend on the pond for food and cover, but they require the surrounding area to lay their eggs buried on dry land.

Without a healthy buffer zone of at least 50 feet, many species cannot survive, despite the presence of a pond. A vegetated buffer also controls soil erosion and increases the amount of water filtration through the soil. It helps to improve water quality, filtering out runoff and pollutants like fertilizers. In addition, buffers can isolate ponds from intensive wading from livestock, such as cattle, which can disturb pond vegetation, increase water turbidity (the amount of suspended particles in the water), accelerate bank erosion, and increase nitrogen levels in the water.

POND EUTROPHICATION

Life in ponds depends on a pool of nutrients that includes nitrogen and phosphorous. The amounts of these nutrients present in a pond can limit the amount and type of organisms that can survive in a pond. The rate of release of these compounds into the pond from the surrounding environment can vary depending on the temperature cycle and seasonality of the pond, as well as the amount of sunlight it receives and the regional climate. Sometimes, ponds receive excess amounts of nitrogen and phosphorous through both natural and artificial means. Natural cases of increased nutrient intake occur when nutrients accumulate through naturally occurring processes, such as depositional environments wherein sediments are laid down or where ephemeral or seasonal nutrient flows occur. Artificial nutrient enrichment occurs from human activities such as untreated sewage, fertilizer runoff, or allowing livestock direct access to ponds and rivers, into which they expel fecal matter.

Spikes in nitrogen and phosphorous levels can lead to eutrophication, which is the increase of plant biomass in a body of water. This process usually favors simple phytoplankton or algae over more complex plants, and often leads to algal blooms.

The excess nutrients allow these photosynthetic species to convert even more sunlight into biomass, meaning that they can increase their populations rapidly. Algal blooms can coat water surfaces or make water cloudy, often turning ponds a shade of green, yellow, brown, or red depending on the species of algae. This overgrowth limits the amount of sunlight that can reach bottom-dwelling organisms, and can cause drastic changes in the amount of dissolved oxygen in the pond.

When photosynthesis occurs, oxygen is released from the breakdown of carbon dioxide and water. In a healthy pond, more oxygen is made during the day than can be consumed by respiring organisms and builds up in the pond. This oxygen reserve is then used by the pond organisms at night, without which they would otherwise suffocate. During an algal bloom, large excesses of oxygen are made during the day, but the unnaturally large populations of phytoplankton continue to undergo cellular respiration at night and can deplete oxygen stores within a pond. This oxygen depletion leads to hypoxic or oxygen-poor conditions in the pond, which can cause large die-offs of fish and other animal populations, especially immobile bottom dwellers. Through such die-offs, eutrophication can cause large changes in the species composition and richness of a pond. Eutrophication in a pond can also threaten species through toxic effects. When large amounts of certain species of algae die and are eaten, they can release both neuro- (affecting the brain) and hepato- (affecting the liver) toxins. These toxins can cause the death of many animals, including species of zooplankton. Livestock can also become sick from ingesting toxic algal blooms, and the toxins can even pose a threat to human health.

Laws have been successful in reducing some instances of eutrophication by regulating the discharge and treatment of sewage, thereby reducing a source of nutrient inflow to groundwater sources. However, there is need for regulation dealing with the use of agricultural fertilizer and animal waste. Fertilizer and pesticides that are sprayed on crops often end up washed by rains into nearby water bodies, especially when applied liberally. Reducing the amount, frequency, and timing of application of agricultural chemicals could help to reduce the occurrence of eutrophication. Vegetated buffers can again serve the health of a pond by helping to filter out chemicals that would otherwise wash into ponds, as well as keeping livestock from directly accessing water bodies.

POND POLLUTION

Other types of pollution can also be harmful to ponds, changing the chemical makeup of the pond and affecting organisms. Foamy areas, fluorescence, or distinctive odors can signal pollution in a pond, but can also be misleading. Foam can be produced in ponds by harmless inorganic salts or some species of phytoplankton. Fluorescence and distinct odors can be caused by certain protozoan species. Carbon dioxide levels in water can be effected by pollution and are important to determining the pH of a pond, which can determine which types of life can survive in that environment. Agricultural pesticides and synthetic estrogens have been found contaminating ponds in agricultural and urbanized environments where frogs and fish are experiencing external and reproductive deformities, including the feminization of males. Ongoing research and policy development are needed to determine the effects of pollution on pond ecosystems and the best ways to protect water bodies like ponds from contamination.

Poultry Farming

—Tyler Biscontini

FIELDS OF STUDY

Ecology; Ecosystem Management; Animal Husbandry; Farming.

SUMMARY

Poultry farming is the process of raising commercial birds for their meat or eggs. Modern poultry farming is a long, complicated process, often involving several separate facilities for different stages of processing. It also involves the use of specialized breeds of poultry bred to produce the most meat per animal.

To begin the process, chicks are hatched at a facility. They are raised until they hit market weight at another facility, and are often processed at a third facility. The poultry is then shipped to stores in specialized refrigerated trucks. In many cases, a single company owns all these facilities. This allows the company to control every step of the poultry farming process.

BACKGROUND

Poultry refers to domesticated fowl raised for food. The most common poultry species are farmers were making a career out of raising poultry. They realized that certain breeds of chickens produced better eggs while other breeds produced higher quality meat. Soon farmers began raising chickens for either eggs or meat instead of simply harvesting chickens for meat toward the end of their lives. This marked the beginning of true commercial chicken breeds.

Within the next decade, poultry farmers began to dress and clean poultry before sale. Prior to this, if people wanted to purchase poultry from a farmer, they would purchase a live animal and process it themselves. Purchasing cleaned and processed poultry greatly reduced the consumer's work.

Approaching the Eggmobile, at Polyface Farms in Virginia. Image by Jessica Reeder, CC BY-SA 2.0.

Once vendors began selling preprocessed poultry, the government began certifying meats. This made consumers aware of the quality of chicken for sale, even if they knew very little about poultry. It also guaranteed that all meat sold to consumers was safe to eat.

During the 1960s and 1970s, the poultry farming became significantly more industrialized. Egg-laying hens were kept in wire cages, causing their eggs to immediately fall into a collection chamber. This enabled eggs to be immediately processed and sent to stores in the shortest time possible. Similarly, the processing of meat chickens became increasingly automated. Automation provided consumers with a consistent, regular product.

INTRODUCTION

Modern poultry farming is almost entirely industrialized. Many steps and facilities are involved in the poultry-farming process. Smaller firms often deal with only one level of the process, while many larger firms control every step of the process. This allows larger firms to attain the strictest quality control possible.

Poultry farming begins with primary breeders, farmers who breed poultry to produce the highest quality products. Modern chicken breeds are separated into egg-laying chickens and meat chickens. Egg-laying breeds are bred to produce the greatest quantity of high-quality eggs, while meat chickens are bred to grow extremely large with more muscle than the average chicken.

Primary breeders have crafted many varieties of chickens to suit each purpose. The Cornish Cross, White Cornish, and White Plymouth Rock are commonly raised for meat. They grow more quickly than other chickens, allowing them to reach an appropriate weight for processing at a rapid pace. White Leghorns and Rhode Island Reds are the leading varieties used in egg production.

Primary breeders sell the chicks of these varieties to breeders, who use them to create populations for poultry farms. Breeders force the chickens and turkeys to reproduce rapidly rate, quickly creating and sustaining a new population. They then carefully raise the chicks to a specific weight.

Approaching the Eggmobile, at Polyface Farms in Virginia. Image by Jessica Reeder, CC BY-SA 2.0.

During this process, the chicks are fed by feed mills. Instead of buying poultry food from a third party, most large-scale poultry farmers own and operate their own feed mills. These mills take raw materials purchased by the farmers and turn it into an efficient food source for the poultry. In most cases, feed mills create a different type of feed for each stage of the poultry's growth. The feed recipes are scientifically formulated to help poultry grow as large as possible.

Once the chicks reach a stage of development specified by the poultry farmer, they are shipped to growout farms, which are specialized facilities owned by independent contractors. These contractors raise the poultry until it grows large enough to sell at market. In most cases, the poultry farmer provides the growout farm with any supplies necessary for this process.

When the poultry is fully grown, it is shipped to a processing plant, which prepares the poultry for sale. Workers clean the meat and portion it into whole animals or parts. While the process is often entirely automated, most processing plants have to abide by strict government regulations and pass periodic inspections. This ensures that the processing is humane, and the meat is both clean and properly packaged. It also ensures that the meat is truly of whatever quality is advertised to the consumer.

After its time in the processing plant, some meat is shipped directly to stores. However, other meat is shipped to post-processing plants. At these plants, the poultry is cooked, marinated, breaded, or prepared in other ways. Once the process is concluded, the ready-for-sale poultry can be shipped to stores.

Poultry is shipped under carefully controlled conditions. It is packaged in safe, sealed containers to limit contamination. It is then loaded into refrigerated trucks and shipped across the country to be sold in supermarkets.

Further Reading

"A Look at the Poultry Industry – How Does Chicken Get on Your Plate?" *Farmaid.org*, 2015, https://www.farmaid.org/issues/industrial-agriculture/a-look-at-the-poultry-industry-how-does-chicken-get-on-your-plate/. Accessed 26 May, 2017.

"An Introduction to Poultry and Eggs Farming." *Department of Agriculture, Environment and Rural Affairs*, 2017, https://www.daera-ni.gov.uk/articles/ introduction-poultry-and-eggs-farming. Accessed 26 May, 2017.

"History of Poultry Production." *USPoultry.org*, https://www.uspoultry.org/educationprograms/PandEP_Curriculum/Documents/PDFs/Lesson2/HistoryofPoultryProductionver3Pres.pdf. Accessed 26 May, 2017.

"Insider Info." *CFNC.org*, 2017, https://www1.cfnc.org/. Accessed 26 May, 2017.

"Introduction to the Poultry Industry." *USPoultry.org*, https://www.uspoultry.org/educationprograms/PandEP_Curriculum/Documents/PDFs/Lesson1/IntroductiontothePoultryIndustryver2.pdf. Accessed 26 May, 2017.

"Large-Scale Pastured Poultry Farming in the U.S." *UW-Madison Center for Integrated Agricultural Systems*, 2003, https://www.cias.wisc.edu/large-scale-pastured-poultry-farming-in-the-us/. Accessed 26 May, 2017.

"Poultry." *Farms.com*, 2017, http://www.farms.com/poultry/. Accessed 26 May, 2017.

"Poultry Farming." *Roy's Farm*, 2017, http://www.roysfarm.com/poultry-farming. Accessed 26 May, 2017.

PREDATION

—*Lawrence E. Hurd*

FIELDS OF STUDY

Ecology; Ecosystem Management; Climatology; Bioclimatology; Climate Modeling; Environmental Sciences; Environmental Studies.

PRINCIPAL TERMS

competition: the interaction among two or more organisms of the same or different species that results when they share a limited resource

food web: the sum total of the feeding relationships (links) between trophic levels in ecosystems

functional response: the rate at which an individual predator consumes prey, dependent upon the abundance of that prey in a habitat

mimicry: the resemblance of one species (the model) to one or more other species (mimics), such that a predator cannot distinguish among them

numerical response: the abundance of predators dependent upon the abundance of prey in a habitat

population regulation: stabilization of population size by factors such as predation and competition, the relative impact of which depends on abundance of the population in a habitat

trophic level: a level at which energy is acquired in a food web—the herbivore level obtains energy from plants; the carnivore level from herbivores and other carnivores

SUMMARY

The relationships among predators and their prey in natural communities are varied and complex. These interactions provide clues as to how natural populations regulate one another, as well as to how to preserve and manage exploited populations more successfully.

Predation is an interaction between two organisms in which one of them, the predator, derives nutrition by killing and eating the other, the prey. Obvious examples include lions feeding on zebras and hawks eating rodents, but predation is not limited to interactions among animals. Birds that feed on seeds are legitimate predators since they are killing individual organisms (embryonic plants) to derive energy. There are a number of species of carnivorous plants, such as sundews and pitcher plants, that capture and consume small animals to obtain nitrogen in habitats lacking sufficient quantities of that nutrient. Most animals that feed on plants (herbivores) do not kill the entire plant and therefore are not really predators. Exceptions to this generalization are some insects that reach infestation levels, such as gypsy moths or locusts, and can kill the plants upon which they feed. The majority of herbivore-plant associations, however, are more properly described as parasite-host interactions in which the host plant may suffer damage but does not die.

A pelican nesting in the grass. Image by US Fish and Wildlife Service.

There are special cases in which parasitism and predation may be combined. One of these is the interaction between parasitoid wasps and their hosts, usually flies. Adult female parasitoids attack and inject eggs into fly pupae (the resting stage, during which fly larvae metamorphose into adults), and the larvae of the wasp consume the fly. The adult parasitoid is therefore a parasite, while the larval wasp acts as a predator.

PREDATOR-PREY INTERACTIONS

Predator-prey interactions can be divided into two considerations: the effects of prey on predators, and the effects of predators on their prey. Predators respond to changes in prey density (the number of prey organisms in the habitat) in two principal ways. The first is called numerical response, which means that predators change their numbers in response to changes in prey density. This may be accomplished by increasing or decreasing reproduction or by immigrating to or emigrating from a habitat. If prey density increases, predators may immigrate from other habitats to take advantage of this increased resource, or those predators already present may produce more offspring. When prey density decreases, the opposite will occur. Some predators, which are known as fugitive species, are specialized at finding habitats with abundant prey, migrating to them, and reproducing rapidly once they are established.

Cape May warblers are good at finding high densities of spruce budworms (a serious pest of conifers) and then converting the energy from their prey into offspring. This strategy allows the birds to persist only because the budworms are never completely wiped out; they are better at dispersing to new habitats than are the birds.

The second response of predators to changing prey density is called functional response. The rate at which predators capture and consume prey depends upon the rate at which they encounter prey, which is a function of prey density. If the predator has a choice of several prey species, it may learn to prefer one of them. If that prey is sufficiently abundant, this situation results in a phenomenon known as switching, the concentration by the predator on the preferred prey. It may entail a change in searching behavior on the part of the predator, such that former prey items will no longer be encountered as frequently.

Animals have evolved a number of defense mechanisms that reduce their probability of being eaten by predators. Spines on horned lizards, threatening displays by harmless snakes, camouflage of many cryptic animals, toxic or distasteful chemicals in insects and amphibians, and simply rapid movement—all are adaptations that may have evolved in response to natural selection by predators. A predator that can learn to prefer one prey item over another is smart enough to learn to avoid less desirable prey. That capability is the basis for a phenomenon—known as aposematism—among potential prey species which are toxic and/or distasteful to their predators. Aposematic organisms advertise their toxicity by bright coloration, making it easy for predators to learn to avoid them, which in turn saves the prey population from frequent taste-testing. Many species of insects are aposematic. Monarch butterflies are bright orange with black stripes, an easy signal to recognize. They owe their toxicity to the milkweed plant, which they eat as caterpillars. The plant contains cardiac glycosides, which are very toxic. The monarch caterpillar is immune to the poison and stores it in its body so that the adult has a high concentration of it in its wings. If a bird grabs the butterfly in flight, it is likely to get a piece of wing first, and this will teach it not to try orange butterflies in the future.

Some potential prey species that are not themselves toxic have evolved to resemble those that are; these are called Batesian mimics. Viceroy butterflies, which are not toxic, mimic monarchs very closely, so that birds cannot tell them apart. One limit to Batesian mimicry is that mimics can never get very numerous, or their predators will not get a strong enough message to leave them alone. Another kind of mimicry involves mimics that are as toxic as their models. The advantage with this type, Müllerian mimicry, is that the predator has to learn only one coloration signal, which reduces risk for both prey populations. In this relationship, the mimic population does not have to remain at low levels relative to the model population. A third type of mimicry is more insidious—aggressive mimicry, in which a predator resembles a prey or the resource of that prey in order to lure it close enough to capture. There are tropical praying mantids that closely resemble orchid flowers, thus attracting the bees upon which they prey. There are some species of fireflies that eat other species of fireflies, using the flashing light signal of their prey to lure them within range.

THE CHOICE OF PREY

What determines predator preference for prey? Since prey are a source of energy for the predator, it might be expected that predators would simply attack the largest prey they could handle. To an extent, this choice holds true for many predators, but there is a cost to be considered. The cost involves the energy a predator must expend to search for, capture, handle, and consume prey. In order to be profitable, a prey item must yield much more energy than it costs. Natural selection should favor reduction in energetic cost relative to energetic gain, the basis for optimal foraging theory. According to this theory, many predators have evolved hunting strategies to optimize the time and energy spent in searching for and capturing prey. Some predators, such as web-building spiders and boa constrictors, ambush their prey. The low energetic cost of sit-and-wait is an advantage in environments that provide plentiful prey. If encounters with prey become less predictably reliable, however, an ambush predator may experience starvation. Spiders can lower their metabolic energy requirements when prey is unavailable, whereas more mobile predators, such as boa constrictors, can simply shift to active searching. Probably because of the likelihood of facing starvation for extended periods of time, ambush predation is more common among animals that do not expend metabolic energy to regulate their body temperatures (ectotherms) than among those that do (endotherms). Some predators, such as wolves and lions, hunt in groups. This allows them to tackle larger (more profitable) prey than if they hunted alone. Solitary hunters generally have to hunt smaller prey.

Natural communities consist of food webs, constructed of links (feeding relationships) among trophic levels. Each prey species is linked to one or more predators. Most predators in nature are generalists with respect to their prey. Spiders, snakes, hawks, lions, and wolves all feed on a variety of prey. Some of these prey are herbivores, but some are themselves predators. Praying mantids eat grasshoppers (herbivores), but they also eat spiders (carnivores) and each other. Thus, generalist predators have a bitrophic niche, in that they occupy two trophic levels at the same time.

PREDATORY RELATIONSHIPS & POPULATION FLUCTUATION

It is an open question whether predators and prey commonly regulate each other's numbers in nature. There are many examples of cyclic changes in abundance over time, in which an increase in prey density is followed by an increase in the numbers of predators, and then the availability of prey decreases, also followed by a decrease in predators. Are predators causing their prey to fluctuate, or are prey responding to some other environmental factor, such as their own food supply? In the second case, prey may be regulated by food, and in turn, may be regulating predators, but not the reverse.

Predators can sometimes determine the number of prey species that can coexist in a habitat. If a predator feeds on a prey species that could outcompete (competitively exclude) other prey species in a habitat, it may free more resources for those other species. This relationship is known as the keystone effect. Empirical studies have indicated that the number of prey species in some communities is directly related to the intensity of predation (numerical and functional responses of predators) such that at low intensity, few species coexist because of competitive exclusion; at intermediate intensity, the diversity of the prey community is greatest; and at high intensity, diversity decreases because overgrazing begins to eliminate species. This intermediate predation hypothesis depends upon competition among prey species, which is not always the case.

STUDYING PREDATION

The central question in the study of predation is: To what extend do predators and their prey regulate one another? Most studies suggest that predators are usually food limited, but the extent to which they regulate their prey is uncertain. It is one thing to observe predators in nature and another to assess their importance to the dynamics of natural communities. Like other aspects of ecology, studies of predation can be descriptive, experimental, and/or mathematical.

At the descriptive level, characteristics of both predator and prey populations are assessed: rates of birth and mortality, age structure, environmental requirements, and behavioral traits. Qualitative and quantitative information of this type is necessary before predictions can be made about the interactions between predator and prey populations. General lack of such information in natural ecosystems is largely responsible for failures at biological control of pests and management of exploited populations.

Experimental studies of predation involve manipulation of predator and/or prey populations. A powerful method of testing the importance of predation is to exclude a predator from portions of its accustomed habitat, leaving other portions intact as experimental controls. In one such experiment, excluding starfish from marine intertidal communities of sessile invertebrates resulted in domination by mussels and exclusion of barnacles and other attached species; in the absence of the predator, one prey species was capable of competitively excluding others. This keystone effect depends upon two factors—that the prey assemblage structure is determined by competition and that the predator preferentially feeds on the species that is the best competitor in the assemblage. Clearly, not all food webs are likely to be structured in this way.

Another method of experimental manipulation is to enhance the numbers of predators in a community. For complex natural communities, both additions and exclusions of predators have revealed direct (depression of prey) and indirect (enhancement) effects. Since generalist predators are bitrophic in nature, they may interact with other carnivores in such a way as to enhance the survival of herbivores that normally would fall victim. In one experiment, adding praying mantids to an insect community resulted in a decrease in spiders and a consequent increase in aphids, normally eaten by these spiders. Such results are not uncommon and contribute to the uncertainty of prediction. Mathematical models have been constructed to depict predator-prey interactions in terms of how each population affects the growth of the other. The simplest of these models, known as the Lotka-Volterra model for the mathematicians who developed it, describes a situation in which prey and predator populations are assumed to be mutually regulating. This model, which was developed for a single prey population and single predator species, has been modified by many workers to provide more realism, but it is far from predicting many competitive situations in complex natural communities.

As with the rest of modern ecology, these different approaches must be blended in order to build a robust picture of how important predators are in natural ecosystems. This knowledge would allow more successful prediction of the outcomes of human intervention and more intelligent management of exploited populations. Predation is a key interaction in natural ecosystems; understanding the nature of this interaction is central to any understanding of nature itself.

FURTHER READING

Allen, K. R. *Conservation and Management of Whales.* Seattle: University of Washington Press, 1980. A fascinating account of the history and methods of human exploitation of a poorly managed resource. Requires no special training to read and appreciate. Diagrams and references provided.

Crawley, M. J., ed. *Natural Enemies: The Population Biology of Predators, Parasites, and Diseases.* Boston: Blackwell Scientific Publications, 1992. Covers evolution, morphology, and population dynamics, the biology of predator-prey interactions, and predator-prey coevolution.

Ehrlich, Paul R., and Anne H. Ehrlich. *Extinction: The Causes and Consequences of the Disappearance of Species.* New York: Random House, 1995. This is a thoughtful and lucid treatment of one of the most crucial of environmental topics. Paul Ehrlich is one of the best known American environmentalists and is also a highly respected researcher in evolutionary ecology. As such, his point of view is not to be taken lightly. Illustrated with references.

Fabre, Jean Henri. *The Life of the Spider.* Translated by Alexander Teixeira de Mattos. New York: Dodd, Mead, 1916. The famous French naturalist provides some interesting insights into the natural history of this important group of predators. Written for lay readers and amateur naturalists, this work is both informative and entertaining. Considered to be a classic.

Gause, G. F. *The Struggle for Existence.* Reprint. New York: Dover, 1971. This is a reprint of the classic 1934 edition. Gause blends mathematical insights with simple laboratory experiments in competition and predation. This book can be read by those with a basic understanding of algebra. Diagrams and references.

Hassell, Michael P. *Arthropod Predator-Prey Systems.* Princeton, N.J.: Princeton University Press, 1978. This is a book for serious students of mathematical ecology who have had some college calculus. Among topics covered are basic models of predator-prey dynamics, functional responses, and theoretical considerations of biological control. Diagrams and references provided.

Krebs, J. R., and N. B. Davies, eds. *Behavioral Ecology: An Evolutionary Approach.* 4th ed. Boston: Blackwell Scientific Publications, 1997. A textbook for upper level undergraduates or graduate students.

Levy, Charles Kingston. *Evolutionary Wars: A Three-Billion Years Arms Race.* New York: Basingstoke, 2000. Traces the coevolution of predators and prey from the earliest days of animal life on earth. Focuses on animals' natural "weapons systems."

Mowat, Farley. *Never Cry Wolf.* Boston: Atlantic-Little, Brown, 1963. An intimate natural history of wolves of the North from the standpoint of a man who lived in close proximity to them. Mowat reveals details about the natural history of these diminishing predators in an entertaining and sympathetic manner. Written in narrative form for a general readership.

PROTOZOA

—Robert T. Klose

FIELDS OF STUDY

Conservation Biology; Biology; Chemistry; Ecology; Environmental Studies.

PRINCIPAL TERMS

autotroph: any organism capable of synthesizing its own food by using either solar or chemical energy

ciliate: a protozoan that uses short, hairlike structures called cilia for locomotion

flagellate: a protozoan that uses long, whiplike structures called flagella for locomotion

heterotroph: any organism that must consume other organisms or organic substances to obtain nutrition

nucleus: a cellular organelle that coordinates the cell's activities and contains the genetic information

organelle: any of several structures found inside an individual cell, analogous to the organs of multicellular organisms

phagocytosis: obtaining food by engulfing it

pseudopods: cytoplasmic extensions of a protozoan's body, used for locomotion and the engulfing of food

unicellular: consisting of only one cell

SUMMARY

The protozoa are a heterogeneous group of organisms the only common characteristic of which is that they are all unicellular. Otherwise, they are extremely diverse in their structures and habits: Some are animal-like and others plantlike; some are free-living, while others are parasitic and cause pernicious diseases.

The protozoa are a vast assemblage of extremely diverse organisms whose only common characteristic is that they are all unicellular. They were previously classified as members of the animal kingdom but are now recognized as a distinct group and occupy a kingdom of their own, Protista.

Protozoa have been known only since the seventeenth century, when Dutch naturalist Antoni van Leeuwenhoek observed them with simple microscopes of his own making. Leeuwenhoek's discovery was accidental; he was simply curious about the microscopic nature of such common substances as rainwater and scrapings from his mouth and teeth. His observations of small, active creatures associated with these substances were the first recorded descriptions of microscopic life. Although Leeuwenhoek made no effort to classify protozoa (which he called "little beasties"), his observations were later confirmed by scientists who recognized the importance of his discoveries.

PROTOZOAN DIVERSITY

The diversity of protozoan types cannot be overemphasized. Though some have a close evolutionary relationship, others have followed very different paths. Some protozoa are animal-like, obtaining energy by consuming food from external sources, and others resemble plants in that they contain chlorophyll and use the energy of the sun to manufacture food.

Protozoa occur in salt water and freshwater, in the soil, and inside larger organisms. In short, they are wherever moisture is present. Most are motile, although some are sessile (attached), and others, like volvox, form colonies. The great majority of protozoa are microscopic, although the very largest are just visible to the naked eye. The blood parasite Anaplasma is only one-sixth to one-tenth the size of a human red blood cell, while the ciliate Spirostomum (a freshwater protozoan that uses tiny hairlike cilia for locomotion) is up to three millimeters long and can be seen without a microscope.

An individual protozoan, because it consists of only one cell, does not contain any organs. Instead, it harbors analogous structures called organelles, which enable it to carry out its life functions. Central among these organelles is the nucleus, which contains the genes and serves as a sort of control center for cell activities. The cells of protozoa contain from one to many nuclei, depending on the species. Other organelles that distinguish the protozoa are those used for locomotion. These include whiplike flagella, hairlike cilia, and flowing extensions of the cell called pseudopodia, or false feet, which are generated in the direction of movement. Locomotory organelles are important in the classification of the protozoa. Many protozoa, especially freshwater species, possess an organelle called the contractile vacuole, which serves to pump excess water out of the cell. This organelle is especially important in freshwater protozoa, which tend to accumulate water by osmosis because their bodies contain more dissolved substances than the water in which they live. As water flows into these protozoa, the contractile vacuole absorbs it, expands, and finally contracts to flush the water through the cell membrane and into the surrounding environment.

Protozoa can be divided into two groups based on how they get their food: autotrophic and heterotrophic. Autotrophic protozoa contain chlorophyll and are able to convert solar energy into usable energy. The freshwater Euglena are autotrophs. Heterotrophic protozoa must capture their food, either funneling it through mouthlike openings or, like the amoeba, engulfing food particles with their pseudopods. This latter method of obtaining food is called phagocytosis.

Gas exchange in protozoa occurs through simple diffusion of gases across the cell membrane. Some protozoa live in environments such as stagnant water or the guts of animals where little oxygen is present. These protozoa are called facultative anaerobes, which means they can live with or without oxygen.

PROTOZOAN REPRODUCTION

Most protozoa reproduce asexually: a single parent produces offspring that are identical genetic copies of itself. Asexual reproduction in protozoa is accomplished in two ways, through fission and budding. In fission, the parent cell simply divides into two offspring cells of equal size. In budding, the offspring is a smaller cell that grows off the larger parent cell. Some protozoa carry out sexual reproduction, in which two parent cells release smaller sex cells called gametes. Two gametes fuse and grow into a new protozoan that is genetically different from either parent.

Conjugation is an interesting process that occurs in ciliate protozoa and may be linked to the reproductive process. It involves the temporary linking of two cells, followed by an exchange of nuclei, after which the cells separate. However, no offspring are formed. It is thought that conjugation is a means of renewing the nuclei of certain protozoa so that they may continue to reproduce asexually. Another process that many protozoa undergo is called encystment, in which the protozoa secrete a thickened sheath around themselves and enter a period of dormancy. Encystment enables protozoa to survive adverse environmental conditions such as drought or cold. When conditions become favorable again, the protozoa emerge from their cysts and resume their normal activities.

PROTOZOAN PHYLA

The kingdom Protista is subdivided into five groups, or phyla: Mastigophora, Sarcodia, Sporozoa, Cnidospora, and Ciliophora. The phylum Mastigophora contains the flagellates: protozoa that move by means of one or more whiplike flagella. The mastigophorans are of two types: phytoflagellates, which contain chlorophyll and are primarily autotrophic; and zooflagellates, which are primarily heterotrophic. Many zooflagellates are parasites of insects and vertebrates. The flagellates of the phylum Mastigophora show such a high degree of structural diversity that it is impossible to describe a typical representative. However, among the more interesting of the phylum members are the dinoflagellates, which occur in both marine and freshwater environments. They have two flagella, one of which points away from the cell and the other of which lies transversely around the cell. Many dinoflagellates contain a golden-brown photosynthetic pigment called xanthophyll and are therefore autotrophic, but others are unpigmented and are heterotrophic. These protozoa are encased in a cellulose sheath, or pellicle, which is often highly ornamented with spikes, hooks, or other prominences. Certain marine species are responsible for red tides, a population explosion that is responsible for massive fish kills and causes paralytic poisoning in humans who consume shellfish taken from red tide waters.

The protozoa in the phylum Sarcodina move about and engulf their food by means of pseudopodia. Amoebas belong to this phylum, as well as many marine, freshwater, and terrestrial species. The members of this phylum are the simplest protozoa and have few organelles. Some have exquisitely crafted external skeletons. The radiolarians, which are restricted to the marine environment, have beautifully sculptured exoskeletons made of silicon dioxide or strontium sulfate. The skeleton is riddled with tiny pores through which the living organism within is able to extrude thin filaments of cytoplasm for purposes of feeding, locomotion, and anchorage. Radiolarians exist in great numbers in the surface waters of the ocean and, upon dying, sink to the bottom where in some areas their skeletons form a large proportion of ocean bottom sediments.

Two phyla are closely related: the Sporozoa and Cnidospora. These are parasitic protozoa collectively referred to as sporozoans because some members of both groups go through sporelike infective stages. The phylum Sporozoa contains parasites that infect cells of the intestine and blood and are responsible for causing malaria in humans. These sporozoans belong to the genus Plasmodium, a group containing more than fifty species, four of which infect humans. All require a mosquito for transmission of the disease-causing protozoan. When the mosquito bites a person, the parasite is introduced into the blood with the saliva of the insect. From the blood, the protozoan travels to the individual's liver, where it reproduces, then escapes to invade red blood cells. Once in the red blood cells, it consumes hemoglobin, reproduces, and escapes to invade additional red blood cells. This cycle of invasion and escape is responsible for the debilitating chills and fever characteristic of malaria. The phylum Cnidospora contains amoeboid parasites of fishes and insects. This phylum's members are separated from those in Sporozoa based on differences in the nature of the infective spores.

The largest phylum is Ciliophora, which has more than eight thousand species. Of all the phyla, this one has the most homogeneous members as far as their evolutionary relationships are concerned. They all appear to have arisen from a common ancestor. Members of this phylum are commonly referred to as ciliates because they use cilia for locomotion and for sweeping food particles into the cytostome (cell mouth). They are also characterized by the presence of two nuclei, a small micronucleus and a larger macronucleus whose sole purpose is the synthesis of deoxyribonucleic acid (DNA, the genetic material). Although some ciliophorans are sessile, or attached, many are nimble, quick, and carnivorous and are therefore perhaps the most animal-like of all the protozoa. They reproduce either asexually by fission or through sexual conjugation, but they never release free gametes into the surrounding environment. Ciliophorans are common in fresh and salt water. Some are parasitic.

PROTOZOOLOGY

Although Antoni van Leeuwenhoek had little sense of what he was looking at when he first observed protozoa in the seventeenth century, he used the same tool that later scientists did, the microscope. Protozoology—the study of organisms too small to be observed with the naked eye—has become an area of specialization within microbiology.

Protozoa are best examined while alive, as specimens that have been killed, fixed, and stained often become distorted. Some of the smaller species can be viewed under the cover slip of a microscope slide;

others, especially the larger specimens, must be viewed under less restrictive conditions: in a petri dish with a dissecting microscope or under a raised cover slip with corners that have been supported with paraffin or some other material.

One of the challenges of studying protozoa with the microscope is the speed with which they move and the tedium of keeping track of them by constantly adjusting the position of the slide. For this reason, a solution of methyl cellulose is often added to the slide preparation to increase the viscosity of the water and slow down the organisms.

As many protozoa are almost transparent, it is necessary to highlight their features by using dyes or stains that will not kill them. Such substances, called vital stains, render various structures in the protozoa visible under the microscope. For example, the vital stain janus green b selectively stains mitochondria, the energy-producing organelles, while neutral red has an affinity for vacuoles, the organelles that contain engulfed food particles. In studying protozoan structure, scientists use an assortment of vital stains and regular stains, which are used when protozoa are killed and preserved on a slide. Electron microscopy, used to examine the extremely fine structure of protozoa, has special stains and fixatives of its own for maintaining the structure of these organisms under the sustained electron bombardment required to illuminate the specimen.

Protozoa are ideal subjects of study for cytologists, geneticists, and developmental biologists because they reproduce asexually and therefore provide the researcher with an endless succession of clones, or genetically identical individuals. Scientists can subject these clones to various environmental conditions and observe the effects in their offspring.

The primary method of maintaining protozoa in a laboratory is a culture, which usually consists of one species of protozoa and the bacterium on which it feeds. The object is to exclude other organisms that might harm the protozoan. As toxic waste products build up in the culture dish, a small number of protozoa are transferred to a clean culture medium. This procedure is called subculturing and continues as long as the culture is maintained. There are two types of cultures: monoxenic and axenic. In monoxenic cultures, a single species of protozoa is maintained with a single species of bacteria. In axenic cultures, a single species of protozoa lives in isolation from any other organisms. The axenic culture is of value in certain types of nutritional studies.

Parasitic protozoa can be examined separate from their hosts in a laboratory but only when kept under conditions that provide them with a chemical and physical environment that will keep them alive for as long as necessary. For example, some parasitic protozoa of warm-blooded animals must be kept in a warm culture medium to prevent their being distorted by cooling.

THE BASE OF THE FOOD CHAIN

Protozoa are of first and foremost importance to ecology. These organisms make up the lowest links of the food chain that eventually leads to humans. Autotrophic protozoa are called producers because, like plants, they are able to use the energy of sunlight to manufacture food substances. Other protozoa are heterotrophs, or consumers, and must take food from their environments. Dinoflagellates are a group of autotrophs with a notorious role in the marine environment. These organisms, which are responsible for red tides, can pass their potent toxins up the food chain to cause paralytic shellfish poisoning in humans. The toxic materials they release into the ocean during periods of intense reproductive activity also lead to massive fish kills, which mean losses for the fishing industry.

Foraminifers are a type of shelled amoeba in the phylum Sarcodina. Some foraminifers existed millions of years ago, when the earth's fossil fuel deposits were being generated. Therefore, the presence of their fossil remains in drill cores helps indicate where oil is likely to be found.

An interesting relationship exists between protozoa and termites. Cellulolytic (cellulose-breaking) protozoa inhabit the guts of termites and are the organisms actually responsible for digestion of the wood that termites eat. Without these protozoa, the termites would not be able to feed on wood and would no longer be able to damage homes and other structures.

PROTOZOAN INFECTIONS

In people, protozoa cause numerous infections, generally limited to four sites in the body: the intestines, the genital tract, the bloodstream and tissues, and the central nervous system (brain and spinal cord). Not all parasitic protozoa cause death in humans, but they are responsible for untold misery in wide areas of the world, especially underdeveloped countries where preventive medicine is unknown and too expensive for these societies to employ.

The two most common protozoan infections are malaria and amoebic dysentery. The malarial parasite is transmitted through the bite of the Anopheles mosquito and eventually destroys the body's red blood cells, causing anemia and recurrent debilitating chills and fever. Malaria is a major health problem in parts of Africa, Asia, and Central and South America. More than one hundred million people are afflicted at any given time, and about one million die of the disease each year. Amoebic dysentery flourishes in overcrowded, unsanitary conditions. The disease gradually erodes the inner lining of the intestines, causing ulcers. Amoebic dysentery is entirely avoidable when food and water are handled carefully to prevent their contamination or are purified if contamination is suspected.

Other protozoa of medical importance are Trichomonas vaginalis, which parasitizes the genital tract; Giardia lamblia, a parasite of the small intestine; and Toxoplasma gondii, which infects the fetus and can cause congenital abnormalities.

Protozoa play an important role in the degradation of both human and industrial wastes in the environment. In sewage treatment, for example, anaerobic protozoa (those that flourish in the absence of oxygen) and bacteria play a role in the degradation of raw sewage. After they have done their work, the effluent is passed to a set of tanks with aerobic (oxygen-requiring) protozoa for further processing. Autotrophic protozoa have been used to deal with industrial wastes containing high levels of nitrates and phosphates. The settling tanks are illuminated to promote the growth of the autotrophs, which absorb and metabolize the industrial chemicals as part of their normal life processes.

Further Reading

Barnes, Robert D. Invertebrate Zoology. 6th ed. Philadelphia: Saunders College Publishing, 1993. This general invertebrate zoology textbook contains a well-organized chapter on protozoa that is detailed yet accessible and has numerous illustrations. Indispensable for those seeking a comprehensive yet concise understanding of protozoa.

Hegner, Robert W. Big Fleas Have Little Fleas: Or, Who's Who Among the Protozoa. New York: Dover, 1968. A lively and lighthearted look at protozoa. This book is a rarity: fact-based science that also functions as literature. Highly recommended as an introduction to protozoa.

McKane, L., and J. Kandel. Microbiology: Essentials and Applications. New York: McGraw-Hill, 1996. This microbiology text presents a comprehensive picture of protozoa, from their evolution through their classification to the human diseases they cause. Contains some beautiful photomicrographs of protozoa feeding, regulating water balance, and reproducing.

Margulis, L., et al. Illustrated Glossary of Protoctista. Boston: Jones and Bartlett, 1993. A handsome reference volume, nicely organized, easy to use, and illustrated with line drawings. Contains a detailed glossary of protozoology terms, making it a good companion to other books on the subject. For the advanced student.

Morholt, E., and Paul F. Brandwein. A Sourcebook for the Biological Sciences. San Diego, Calif.: Harcourt Brace, 1980. The information in this extremely useful reference book is arranged cookbook-style. The entries are brief, practical, and readable, describing procedures for the collection, staining, observation, and preservation of specific protozoa. What is unique about this book is that it prefaces the practical how-to entries with well-written background information.

Roberts, Larry S., et al. Foundations of Parasitology. 6th ed. Boston: McGraw-Hill, 2000. The sections on protozoa are limited to those species that parasitize human hosts. The text is densely written, emphasizing scientific terminology, classification, and the fine structure of these organisms. For the reader with a serious interest in this topic.

Whitten, R., and W. Pendergrass. Carolina Protozoa and Invertebrates Manual. Burlington, N.C.: Carolina Biological Supply, 1980. A short, attractive, ready-reference guide to the culture and observation of protozoa by a biology supply house. Entirely accessible to the interested middle school or high-school student. Has large, colorful illustrations, recipes for media, and suggestions for further reading.

R

Rain Forests

—Nancy Farm Männikkö

FIELDS OF STUDY

Conservation; Ecology; Foretry; Environmental Studies.

PRINCIPAL TERMS

herbivore: a plant-eating animal

temperate zone: the regions of the earth north of the Tropic of Cancer but south of the Arctic Circle, or south of the Tropic of Capricorn but north of the Antarctic Circle

tolerance: ability of a tree to grow in the shade of other trees

tropics: the region of the earth lying between the Tropic of Cancer and the Tropic of Capricorn

SUMMARY

A forest growing in a region that receives over one hundred inches of rain annually is considered to be a rain forest. Rain forests can be found in both tropical and temperate climates, and are noted for their remarkable biodiversity. Thousands of different animal and plant species can be found within only an acre or two of a rain forest.

Rain forests are forests found in regions of the world that receive large amounts of precipitation annually. Rain forests present an incredibly diverse range of habitats, as they exist both at low elevations and high in mountain ranges. Many unusual and seldom-seen creatures inhabit the world's rain forests, including spiders so large they eat small birds and colorful but highly poisonous tree frogs. The enigmatic sloth, an animal that spends its entire life hanging upside down from tree limbs and moving so slowly that moss grows on its fur, is found in the rain forests of South America.

Although tropical rain forests such as those in the Amazon River drainage system of South America are perhaps the best known, rain forests do exist in temperate regions as well. Olympic National Park in the state of Washington preserves a temperate climate rain forest, while much of the coast of British Columbia and southeastern Alaska also receives well over one hundred inches of rain annually. The primary difference between temperate and tropical rain forests is that in a temperate rain forest, often one or two species of trees will become dominant. In the coniferous rain forest of the Pacific Northwest, for example, Douglas fir and western red cedar are the dominant species, while other trees are found in much smaller numbers. In a tropical rain forest, in contrast, several hundred species of trees may grow side by side within a very small geographic area. The

View of the south side of the island. Dominica features lush mountainous rain forests, and is the home of many rare plant, animal, and bird species (including the Sisserou Parrot featured on its flag). (Courtesy of the CIA World Fact Book). © CIA World Fact Book.

majority of trees found in tropical rain forests tend to be broad-leaved, such as the rubber tree, while temperate rain forests are dominated by conifers. The leaves of many plants in rain forests have a waxy texture or come to a point to help shed water more quickly and prevent the growth of fungi or mold.

CHARACTERISTICS OF RAIN FORESTS

Although rain forests are remarkably diverse, they do share a few characteristics in common. The abundant moisture in a rain forest gives the woodland a lush, fertile appearance. This is particularly true in tropical regions. Even in the understory, close to ground level where light is limited, vegetation may be dense. This appearance of fertility is often deceptive. Dead plant matter decays rapidly in a tropical forest, but the nutrients are used quickly by the numerous competing plants. In addition, the trees in tropical forests are evergreen, which means the litter that does fall to the forest floor does so irregularly, unlike temperate broadleaf forests, where trees lose their leaves annually as the seasons change. Leaves will remain indefinitely on tropical species, such as fig and rubber trees, which is one reason small specimens of these trees are popular as houseplants. As a consequence of this lack of mulch, topsoil is often thin and the root systems of the trees are quite shallow.

One reason tropical rain forests are evergreen is that in the tropics there is little seasonal variation in the hours of daylight. The closer to the equator a forest lies, the less change there is from season to season. In temperate climates, many plants have evolved to bloom, set seeds, or lose their leaves based on the number of hours of sunlight available each day. As the seasons change annually, plants bloom in the spring or early summer; fruit ripens in the fall, photosynthesis slows, and leaves change color and die. In the tropics, where the number of hours of daily sunlight never varies, plants follow a different schedule. Many tropical plants bear new flowers and mature fruit simultaneously. The evergreen foliage and continuous supply of certain fruits has led to the adaptation of some animals to a very restricted diet: koalas, for example, which feed exclusively on eucalyptus leaves, or parakeets that eat only figs. Exceptions to this pattern are the forests where rainfall is seasonal, such as regions of the world like southeast Asia, where much of the rain comes in the form of annual monsoon storms. In those cases, flowering and setting fruit will coincide with the seasonal rains.

FOREST ZONES

A rain forest can be divided into four zones, each of which has its own distinct characteristics. The lowest level, the forest floor, is often dark and gloomy. Little sunlight penetrates to this level, and there is little air movement. Numerous insects, such as beetles, cockroaches, and termites, live in the decaying litter and provide food for larger animals and birds. Many of the insects, birds, reptiles, and amphibians that live in the lower levels of the rain forest are brightly colored. Scientists speculate that the animals have evolved in this fashion to more easily attract potential mates. Other scientists believe that colors warn potential predators to stay away. In either case, the vivid colors make the animals more easily seen in what is otherwise a dark environment.

Just above the forest floor is the understory. Many of the plants in the understory have large, dark leaves to maximize their light-collecting ability. Because there is little natural air movement within the lower levels of a rain forest due to the canopy blocking any natural breezes, the flowering plants in the understory often have strongly scented or vividly colored flowers to help attract insects or birds to assist with pollination. Lizards, snakes, amphibians such as tree frogs and salamanders, small birds, and mammals as large as the jaguar all call the understory home. The plants found only in the understory seldom exceed fifteen to twenty feet in height. The coffee shrub is an example of a small, shade-tolerant, tropical tree. Until horticulturalists developed strains of coffee for use in plantations where the coffee bushes are the only plants grown, coffee grew naturally in the understory of tropical forests.

The densest layer of plant life is the canopy. High above the rain forest floor, the branches of mature trees form a dense intertwined zone of vegetation extending up as much as 150 feet above the ground. Numerous plants sprout in the crotches of trees, where debris may collect. Tree limbs are festooned with vines and mosses, and bromeliads and orchids grow on the rough bark of tree trunks. Even other trees may start their life cycles a hundred feet above the ground: The strangler figs of Borneo are a relatively shade-intolerant species. A fig seed that lands and sprouts on the ground will probably not survive due to low light levels on the forest floor. Strangler figs have adapted so that their seedlings do best high in the canopy. The figs begin life in the crotches of other trees. The roots of a young fig will gradually creep down the trunk of the tree on which it sprouted. Over time, the strangler fig's roots will completely encircle the host tree, as well as penetrating the forest floor. The fig thrives, but the host tree dies, choked off by the strangler fig. Primates such as gibbons, orangutans, and lemurs spend much of their lives in the canopy, feeding on the fruit of trees such as the strangler fig, as do the sloth and other herbivorous mammals.

The emergent layer of the rain forest consists of the tallest trees, some of which exceed two hundred feet in height. The tops of these trees provide a habitat for large, predatory birds, such as eagles, as well as being home to assorted snakes, monkeys, and other animals. Every layer of the rain forest teems with life, and often what can be found at ground level gives no hint of the diversity that exists two hundred feet above in the tree tops.

RAIN FOREST CONSERVATION

Many of the trees found in rain forests are valued for their commercial use as lumber, while others have been exploited for their fruits or other products, causing much habitat loss. Tropical hardwoods, such as teak and mahogany, for example, have long been used in construction and in furniture. Teak resists rotting and as a result is often used for products that are going to be exposed to the weather, such as garden furniture. Due to teak's desirability as lumber, timber companies are increasingly planting it in plantations for a sustainable yield rather than relying solely on natural forests as a source. Logging of rain forests, such as those in Sumatra, Indonesia, has also been done to supply the paper and pulp industries.

Activists hoping to preserve the tropical rain forest have encouraged indigenous peoples to collect forest products, such as nuts or sap, as a way to create a viable economy while at the same time discouraging industrial clear-cutting of the forest. Native people tap rubber trees in Amazonia, for example, to collect latex. Rubber trees are native to the rain forests of South America, although they are also grown in plantations in other tropical regions of the world, such as Southeast Asia.

The biggest threat to the world's rain forests may not come from commercial logging, however. In many regions of the world, rain forests have fallen victim to population pressures. Forests continue to be clear-cut for agricultural use, as in Brazil and Indonesia, even when the farmers, ranchers, and plantation owners know the exposed soil's fertility will be quickly exhausted. Such incursions have been made to promote global soy, palm oil, and beef production. In some cases, the cleared land becomes an arid wasteland as the tropical sun bakes the soil too hard to absorb rain water. In others, the land is farmed for a year or two and then abandoned. Given enough time, the rain forest may regenerate, but the process will take hundreds of years. Mining of precious metals, coal, and petroleum represent yet another threat, from both the land clearance itself and the risks of contamination from extraction-related chemicals.

Tropical rain forests cover less than 5 percent of the earth's land area, and an estimated 2 percent are formally protected. Most remaining tropical rain forests are found in developing countries. Such countries in particular have struggled to implement and enforce viable legal protections. These countries are often poorer than their subtropical and temperate counterparts, and their governments often feel pressure to exploit their rain forests for economic development. Corruption, dubious land deals and permits, and government loopholes or policy reversals contribute to ongoing rain forest destruction. Nongovernmental agencies and local indigenous people often decry these conditions and attempt to hamper illegal clearances.

FURTHER READING

Bowman, David M. J. S. *Australian Rainforests: Islands of Green in a Land of Fire.* New York: Cambridge University Press, 2000. Looks at Australia's rain forests, which exist as patches separated from each other by vegetation from other environments, and discusses the ecosystems unique to Australia.

Durbin, Kathie. *Pulp Politics and the Fight for the Alaska Rain Forest.* Corvallis: Oregon State University Press, 1999. Fascinating study of the Tongass National Forest in southeast Alaska.

Goldsmith, Frank Barrie. *Tropical Rain Forest: A Wider Perspective.* New York: Chapman & Hall, 1998. A good overview of tropical rain forests globally. This book helps the reader gain an appreciation of the incredible diversity of tropical forests.

Killman, Wolf, and Lay Thong Hong. "Rubberwood: The Success of an Agricultural By-Product." *Unasylva* 51 (2000): 66–72. Looks at the rubber tree and its uses once the tree no longer produces harvestable latex. An interesting, focused discussion of one specific type of tropical tree.

Maser, Chris. *Forest Primeval: The Natural History of an Ancient Forest.* Reprint. Corvallis: Oregon State University Press, 2001. An in-depth study of the old-growth forests of the Pacific Northwest. Provides insight into temperate rain forests and how they differ from tropical.

Mysteries of the Rain Forest. Pleasantville, N.Y.: Reader's Digest, 1998. A lavishly illustrated book meant for the general reader. Deals with the rain forests as ecosystems and provides information on plants, animals, and indigenous peoples of the rain forests of the world.

Schiffman, Richard. "Amazon Rainforest under Threat as Brazil Tears Up Protections." *New Statesman,* 26 Apr. 2017, www.newscientist.com/article/2129024-amazon-rainforest-under-threat-as-brazil-tears-up-protections. Accessed 31 Jan. 2018.

Tricart, Jean. *The Landforms of the Humid Tropics, Forests, and Savannas.* Translated by Conrad J. Kiewiet de Jonge. New York: St. Martin's Press, 1972. Discusses the different ecosystems found in the tropics and helps the reader to understand what distinguishes one type of forest from another.

Vidal, John. "Q&A: Rainforests." *The Guardian,* 26 July 2007, www.theguardian.com/environment/2007/jul/26/conservation.endangered-habitats. Accessed 31 Jan. 2018.

Vidal, John. "'The Sumatran Rainforest Will Mostly Disappear within 20 Years.'" *The Observer,* 26 May 2013, www.theguardian.com/world/2013/may/26/sumatra-borneo-deforestation-tigers-palm-oil. Accessed 31 Jan. 2018.

Rainwater Harvesting

—*Robin L. Wulffson*

FIELDS OF STUDY

Conservation Biology; Ecology; Environmental Studies.

SUMMARY

Rainwater harvesting includes collecting and storing rainwater to meet freshwater needs. Increasing urbanization throughout the planet requires an ever-increasing supply of water. Rainwater is a free resource that can be collected through a variety of relatively inexpensive methods, and rainwater harvesting can provide water for drinking, irrigation, and the refilling of aquifers. Rainwater harvesting is a key component to low-impact development (LID) and offers a boost or an alternative to traditional water management systems. Bioretention, permeable pavement surfacing, and green roofs are additional complimentary LID strategies related to rainwater harvesting.

Because many natural water sources around the world have been depleted or even completely exhausted, rainwater harvesting is becoming increasingly important to the well-being of some populations. Rainwater can be collected from roofs or from areas at ground level. Rainwater collected from roofs is sometimes potable (fit for drinking) without any processing, but collected rainwater usually must undergo some processing before it is suitable for human or animal consumption. Harvested rainwater is often suitable for irrigation, flushing toilets, and laundering clothes without processing, however. Harvested rainwater can serve to supplement available water sources; in some areas, collected rainwater is the only readily available source of water. For example, small islands with high annual rainfall may be sustained totally by rainwater harvesting. In urban areas, rainwater harvesting can supplement city water supplies and reduce the likelihood of flooding.

METHODS

Rainwater harvested from roofs can be channeled into storage tanks through systems of gutters and pipes. The gutters must have an incline so that water does not stand in them and must be strong enough to handle peak flow. Pollutants that may be found in rooftop water include pesticides, dust particles, animal and bird feces, minerals, and dissolved gases such as carbon dioxide, sulfur dioxide, and nitrous oxide. The level of pollutants is commonly highest in water from the first rainfall after a dry spell; such water must therefore be discarded or put through some form of decontaminating treatment, such as boiling or chemical treatment.

Ground collection of rainwater involves a number of techniques that can improve the runoff capacity of a land area: clearing or altering vegetation, increasing the land slope, and decreasing soil permeability (penetration) through compaction or the application of chemicals (or both). Ground collection methodologies provide opportunities to collect water from larger surface areas than rooftop collection allows. The placement of a subsurface dike can obstruct the natural flow of groundwater through an aquifer (underground water channel), raising the groundwater level as well as increasing the amount of water stored in the aquifer (this process is known as groundwater recharge).

Harvested water is commonly collected in tanks; these tanks must be covered to prevent evaporative loss, algae growth, and mosquito breeding. The damming of small creeks and streams, including those that contain water only during heavy rains, is an inexpensive method of storing water during dry spells. Sometimes, however, large portions of water stored in this way are lost through infiltration into the ground. Usually this water is unfit for human consumption; thus it is often solely used for irrigation.

WORLDWIDE PRACTICES & REGULATIONS

Rooftop rainfall harvesting is widespread in China and Brazil, where the water is consumed by humans and animals; it is also used for crop irrigation and replenishment of groundwater stores. In India, where urbanization is growing at a rapid rate, rooftop harvesting is also widespread; in this nation it represents an essential component of the water supply. In island nations such as Bermuda and the U.S. Virgin Islands, rainwater harvesting is commonly practiced. In rural portions of New Zealand, the only source of water for all household activities is the rooftop collection of rainwater. In the United Kingdom, it is a common practice to collect rainwater for irrigating gardens.

In some parts of the world, no statutes are in place to regulate rainwater harvesting or the allowable uses of the water harvested. In the United States, regulations vary among state and local governments. California, for example, had no state laws related to rainwater harvesting before 2012, but some local governments in the state created regulations. San Francisco, for instance, allows the use of rainwater for nonpotable purposes such as toilet flushing and irrigation. In 2012, the state legislature passed the Rainwater Capture Act, authorized expanded rainwater harvesting in the light of severe droughts, water shortages, and high storm waters in the state. California's neighbor to the north, Oregon, which receives high levels of rainfall annually, has established statutes regulating rainwater harvesting, and the city of Portland specifically permits the use of rainwater for human consumption by request. Likewise, harvested rainwater may be used for potable purposes in Texas and Ohio. For a long time Colorado had laws prohibiting rainwater harvesting; the argument for these statutes was that harvesting could infringe on the water rights of those living downstream. The laws were changed after a 2007 study found that in an average year, almost all of the precipitation that fell in the southern suburbs of Colorado never reached a stream; it either evaporated or was absorbed by plants. Two years later, a pair of state laws relaxed requirements for the use of harvested rainwater. In addition, several states, including Arizona, Rhode Island, Texas, and Virginia, offer tax credits to those who install rainwater collection equipment.

Further Reading

Allen, Laura. *The Water-Wise Home: How to Conserve, Capture, and Reuse Water in Your Home and Landscape.* North Adams, MA: Storey Publishing, 2015.

Grant, Gary. *The Water Sensitive City*. Chichester, UK: John Wiley & Sons, Ltd., 2016.

Lancaster, Brad, and Joe Marshall. *Rainwater Harvesting for Drylands and Beyond.* 2 vols. Tucson, AZ: Rainsource Press, 2013.

National Conference of State Legislatures. "State Rainwater: Graywater Harvesting Laws and Legislation." *NCSL.org*, September 1, 2013.

Rossmiller, Ronald Leroy. *Stormwater Design for Sustainable Development.* New York: McGraw-Hill Education, 2014.

Tandlich, R. *Novel Approaches to Rainwater Harvesting and Sanitation in Developing Countries*. New York: Nova Publishers, 2017.

Zhu, Qiang, John Gould, Yuanghong Li, and Chengxiang Ma. *Rainwater Harvesting for Agriculture and Water Supply*. Singapore: Springer, 2016.

RIVER VALLEYS

—Michael W. Mayfield

FIELDS OF STUDY

Geology; Agriculture; History; Conservation.

PRINCIPAL TERMS

aggradation: the process by which a stream elevates its bed through deposition of sediment

base level: the theoretical vertical limit below which streams cannot cut their beds

fluvial: of or related to streams and their actions

prokaryotic: an upland area between valleys

stream equilibrium: a state in which a stream's erosive energy is balanced by its sediment load such that it is neither eroding nor building up its channel

underfit stream: a stream that is significantly smaller in proportion than the valley through which it flows

valley: part of the earth's surface where stream systems are established; it includes streams and adjacent slopes

SUMMARY

The valleys in which streams flow are produced by those streams through long-term erosion and deposition. The landforms produced by fluvial action are quite diverse, ranging from spectacular canyons to wide, gently sloping valleys. The patterns formed by stream networks are complex and generally reflect the bedrock geology and terrain characteristics.

FORMATION OF RIVER VALLEYS

River valleys consist of valley bottoms and the adjacent valley sides. Between valleys are undissected uplands known as interfluves. Valley floors may be quite narrow, as in the case of the Black Canyon of the Gunnison River, or they may be quite wide, as in the case of the Hwang Ho or the Brahmaputra. Similarly, valley sides may have very gentle to rolling slopes, or they may be nearly sheer, as in the case of the Arkansas River's Royal Gorge. In many areas, the interfluves are simply divides between adjacent valleys, but on tablelands such as the Colorado Plateau, they may be tens of kilometers wide in places.

River channels and river valleys are products of the streams that flow through them. As a stream erodes a channel for itself in newly uplifted terrain, it eventually carves a valley whose form is determined by the erosive power of the stream, by the structural integrity of the rock and debris of the valley walls, by the length of time that the stream has been operating on its surroundings, and by past environmental conditions. These past environmental conditions are attested by stream channel and valley profiles that have not entirely erased landforms produced during the most recent episodes of glaciation and climate change. The valley of the Mississippi River, for

A river gorge at Fall Creek, Ithaca, Thompkins County, New York. The gorge wall shows erosion partially controlled by joints. Image by U.S. Geological Survey.

example, is formed in the complex deposits of what was a much larger, more heavily laden glacial meltwater stream that existed only 15,000 years ago. River channels and valleys may be referred to as palimpsests, a term originally used to describe parchment manuscripts that had been partly scraped clean, then reused. On the fluvial (river-carved or river-deposited) landscape, previous landscape elements are seen, just as old words show through on a recycled piece of parchment.

In many parts of the world, streams are found flowing through valleys that appear to have been formed by a far larger stream. The valley width, amplitude of meanders, and caliber of coarse sediment are proportional to far larger stream courses. Such streams are said to be underfit, and the valleys are largely remnant features from times of wetter and cooler climates that accompanied glaciation, or the streams were glacial meltwater channels during deglaciation.

The fact that stream channels and valleys are not chance features on the landscape was first noted by British geologist John Playfair in 1802. Playfair suggested that instead of being isolated features on the landscape, streams are part of well-integrated networks. More important, these networks are finely adjusted to the landscape and to one another in such a way that tributaries almost always join the main trunk stream at the same level as that stream. Streams that must plunge over waterfalls to join a larger stream are quite rare. Such discordant streams are found primarily in recently glaciated terrain, where the stream-valley system has not yet become fully adjusted, such as in Yosemite Valley. This remarkable consistency of stream accordance is the strongest evidence indicating that streams carve their own valleys.

Streams develop into network patterns that are strongly influenced by bedrock structure. Where the bedrock is relatively uniform and without strong joints and faults, a dendritic pattern of drainage develops. In this pattern, a stream system is branched like a tree. On inclined plains such as the Atlantic coastal plain, the stream pattern is often parallel, with major streams flowing directly down the topographic slope to the sea. Inclined mountain systems such as the Sierra Nevada also produce parallel drainage. Where structural folding of the terrain has produced linear ridge and valley topography, main trunk streams occupy the linear valleys and are quite long, with short, steep tributaries feeding them off the flanks of the hills or mountains. In areas such as the Great Valley of Virginia, this type of pattern has developed and is known as trellis drainage. Volcanoes often produce a radial pattern of drainage. Mount Egmont in New Zealand is often cited as an example.

BASE LEVEL

The depth of a river valley is a function of the height of the land above base level, the length of time that has passed since the stream began to erode, the resistance of the bedrock to erosion, the load of sediment that the stream is carrying, and the spacing of adjacent streams. "Base level" refers to the theoretical lower limit to which a stream can cut. Ultimate base level for all streams is sea level, but local base level is significantly higher for many streams. Streams require a minimum slope to transport their sediment to the sea; this limits the depth to which a stream may cut. The upper section of the Hwang Ho near Tibet flows at elevations of more than 3,000 meters, but the river must still flow more than 4,000 kilometers to the sea. In order to carry its heavy load of sediment over that distance, the river must maintain its channel at a rather high elevation.

Valleys that are deep and relatively narrow are called canyons. Valleys that are especially narrow are called gorges. Streams require time to erode great canyons, although the rate of cutting can be quite rapid compared with most geologic processes. The process of downwearing of interfluves is far slower, so young (recently uplifted) landscapes produce the deepest canyons. As downcutting by the main stream ceases, upland weathering and erosion lower the local relief. The deepest, narrowest canyons are eroded in strong, homogeneous rock such as granite and quartzite. The Royal Gorge of the Arkansas River and the Colca Gorge of Peru are excellent examples; both are cut in recently uplifted masses of resistant rock formations. The Colca Gorge is far narrower than is the Grand Canyon and, at more than 3,000 meters, it is nearly twice as deep.

As a mountain mass or tableland is uplifted, streams often develop that flow directly down the initial slope of the land. Such streams are called consequent streams because their course is a direct consequence of the terrain slope. Because streams seek the path of least resistance, their courses often follow the outcrop pattern of weak rocks such as shale, producing what is known as a subsequent stream pattern. In many cases, however, streams seem to ignore the terrain and structural slope of the land entirely, flowing through mountains of quite resistant rock. An excellent example is the Black Canyon of the Gunnison River in Colorado, where the river carves a deep canyon through a high plateau, with its channel cut in resistant gneisses and igneous intrusives. What makes this so surprising is that much lower terrain, underlain by thick sequences of weak shale, lies only 3 kilometers to the west of the head of the gorge, which would seem to provide a much easier path to the sea. The reason for this course—and many other anomalous stream courses—is that the course of the Gunnison was established prior to the uplift of the plateau. As the land rose beneath the river, it maintained its position, carving an ever-deepening gorge in the rocks as they rose. Far older and more extensive is the anomalous course of the New River, which cuts directly across the structure of the Appalachians, flowing a great distance to the Ohio River rather than the shorter, more direct route to the Atlantic. Once again, this river is one that was established prior to the uplift of the land in which it is now entrenched—in this case, the ancient Appalachians. Thus, the river is rather ironically named, considering that its course is older than the Appalachians.

At any given time, a stream may either erode or aggrade (build up) its bed. This vertical change in stream profile is determined by many factors internal to the stream, as well as outside environmental factors. Over geological time, the tendency of streams is to erode their beds deeper and deeper. Over shorter time intervals, however, a stream may reach a state of equilibrium between its erosive energy and the load of sediment that the stream is carrying. A delicate balance is maintained within the stream channel. If stream energy is increased by an increase in flood volume or frequency, for example, the stream is likely to respond by eroding its bed. Conversely, if stream power remains constant and the sediment load is significantly increased, the stream will respond by aggrading. This channel aggradation increases the channel slope, giving the stream more energy to transport the sediment load, and a new equilibrium is reached.

Further Reading

Bloom, Arthur. *Geomorphology: A Systematic Analysis of Late Cenozoic Landforms.* 3rd ed. Long Grove Ill.: Waveland Press, 2004. This college-level text covers the basics of geomorphology. Includes three chapters on fluvial systems and the development of river valleys. Index and bibliography.

Chorley, R. J., S. A. Schumm, and D. E. Sugden. *Geomorphology.* New York: Methuen, 1985. An advanced text in geomorphology that is best used as a reference. Provides detailed citations of the important scholarly works on each subject addressed and is quite thorough. Many diagrams, graphs, and line drawings supplement the material, but there are few photographs. Suitable for college students who have already gained some background in geomorphology.

Dobrazanska, Halina, Erzsebet Jerem, and Tomasz Kalicki, eds. *The Geoarchaeology of River Valleys.* Budapest: Archaeolingua, 2004. A collection of papers that discuss the development of societies along rivers and the ways in which river valleys influence culture. Also covers anthropogenic effects on river valley evolution.

Fletcher, Charles. *Physical Geology: The Science of Earth.* Hoboken, N.J.: John Wiley & Sons, 2011. This text deals with all aspects of physical geology. It is intended for undergraduate students but accessible to the interested reader.

Hunt, C. B. *Natural Regions of the United States and Canada.* San Francisco: W. H. Freeman, 1974. This book is primarily a physiography text, aimed at describing and explaining the pattern of landforms across North America. The author produced it in the light of "the general public's interest in the natural environment and the need for an authoritative account of it in language as nontechnical as possible." In both regards, the book is quite successful. Well illustrated with diagrams, line drawings, and photographs. Introductory chapters provide an explanation of the general phenomena, and the remaining chapters provide a very thorough, understandable coverage of the regional expression of landforms. Suitable for high school students and above.

McKnight, T. L., and Darrel Hess. *Physical Geography: A Landscape Appreciation.* 9th ed. Englewood Cliffs, N.J.: Prentice-Hall, 2007. This physical geography text provides very thorough but understandable coverage of the concepts of stream networks, stream channel history, and fluvial dynamics. Richly illustrated with color diagrams and photographs. Suitable for high school students and above.

Ollier, Cliff, and Colin Pain. *The Origin of Mountains.* London: Cambridge University Press, 2000. This text encompasses the geographic, geological, and geophysical concepts of mountains. Topics include fault-block mountains, fold belts and folding, gravity structures, rift valleys and river valleys, plateaus and mountain erosion. Begins with a general overview of subduction and plate tectonics. Mountain ranges discussed in detail include the Appalachians, Andes, Cascades, Himalaya, and the Rockies.

Restrepo, Juan D., Sergio A. Lopez, and Juan C. Restrepo. "The Effects of Geomorphic Controls on Sediment Yield in the Andean Rivers of Colombia." *Latin American Journal of Sedimentology and Basin Analysis* 16 (2009): 79–92. A technical and detailed study on the sedimentation within the Andean basin. Compares Colombian runoff of the Pacific system and the Caribbean system. Hydroligic, climatic, and morphometric data were collected to build a model for predicting sediment yield. A background in geology or hydrology is recommended.

Skinner, Brian J., and Stephen C. Porter. *Physical Geology.* New York: John Wiley & Sons, 1987. A consistently excellent text in introductory physical geology, richly and generously illustrated in full color. Very thorough and quite extensive (more than seven hundred pages). Chapter 11, "Streams and Drainage Systems," is one of the best chapters. Suitable for advanced high school students and above.

ROOTS OF TOBACCO FARMING IN NORTH AMERICA

—*Diana M. Casey*

FIELDS OF STUDY

Plant Science; Crop Science; Botany; Agronomy; Agriculture; History; Soils Engineering.

KEY FIGURES

John Rolfe (1585–1622), English-born colonial official and first North American tobacco farmer

Sir Walter Ralegh (c. 1552–1618), English explorer who popularized tobacco in England

James I (1566–1625), king of England, r. 1603–1625, and king of Scotland as James VI, r. 1567–1625

SUMMARY

Tobacco, already gaining popularity in Europe, was introduced as a commercial crop in Jamestown, Virginia. The New World plant's cultivation corresponded with the rise of tobacco culture in Europe, provided America with a fundamental economic underpinning by which plantation owners gained immense fortunes, and aided in the destruction and displacement of Native Americans.

SUMMARY OF EVENT

Tobacco was first introduced into Europe from North America before its establishment as a crop in Virginia. Native Americans consumed wild tobacco, which was plentiful in the Americas, rather than cultivating tobacco fields. The growing European desire for the New World plant, or "weede," as it was initially referred to, literally saved the lives of the struggling British colonists and provided a means for their colonies to grow and prosper. While tobacco was widespread during the sixteenth and early seventeenth centuries, it was not until the second half of the seventeenth century that the price of tobacco fell drastically and tobacco became truly popular. The addictive desire for tobacco in Europe and eventually the rest of the world brought about enormous social, physical, and economic change.

It is difficult to imagine Europeans' initial response when they first viewed pipe smoking or snuff inhalation. They simply had no referential context whatsoever by which to understand the practice, so it is hardly any wonder such phrases as "drinking smoke" or "drinking fog," were invented for lack of better terms. Native Americans viewed tobacco as a medicine and used it also within social frameworks to bring people together in peace and camaraderie. With electrifying speed, Europeans, and soon after the rest of the world, borrowed American cultural and social practices surrounding the use of tobacco.

Tobacco, which gained an early reputation as a medicinal panacea, was introduced to France in 1556, Portugal in 1558, Spain in 1559, and finally to England in 1565. The explorer Christopher Columbus is actually credited with bringing tobacco to Europe at the end of the 1400', but it did not become popular until the middle of the sixteenth century, when diplomats like France's Jean Nicot, for whom nicotine is named, began to use it as a curative. In France, where tobacco could only be purchased with a prescription, Queen Catherine de Médicis

was so favorably impressed when it reportedly cured her headaches that she decreed that it be called Herba Regina, the "queen's herb."

After tobacco use moved beyond the medical arena, a culture of smoking, initially based on Native American social practices, sprung up in Europe. Pipes were ritually passed around from person to person in communal camaraderie by Native Americans to pledge an oath or ratify a treaty. This practice inspired the phrase "peace pipe." When Hernán Cortés landed in Mexico (1519), the natives offered him tobacco as "a pledge of peace and good will." Similarly, in early seventeenth century Europe, tobacco was "drunk" from a communal pipe that was handed from man to man around the table. In time, tobacco etiquette demanded that people of both genders cultivate the proper manner of smoking, which included a delicate hand.

Both genders and all ages smoked, including children, and, while smoking was the English nicotine delivery method of choice during the Stuart and Cromwellian eras, like many other imported French court manners, inhaling snuff gained in popularity after the Restoration of Charles II. Among aristocrats, it was common to have a different, enormously expensive snuff box for every day of the week.

In addition to the tobacco culture—smoking, snuff taking, and their accompanying mannerisms and accoutrements—tobacco also had a physical influence on seventeenth century bodies. Its ability to diminish the appetite led it to be used often in place of food. As a result, a reduction in the intake of meat, milk, and cheese accompanied Europeans' consumption of tobacco and allowed workers to work longer hours. Tobacco also played a role in major military conflicts, permitting soldiers to fight longer by increasing energy while keeping their bodies relaxed. In fact, seventeenth century European soldiers carried pipes in their caps, and tobacco constituted part of their daily rations.

The famous pipe smoker Sir Walter Ralegh is credited with popularizing tobacco in England during the reign of Queen Elizabeth I (r. 1558–1603). In 1587, Ralegh established a settlement on Roanoke Island in Virginia. After its failure, he attempted to colonize Guiana, rumored to be the mythical El Dorado, but instead of gold, he returned to England with tobacco.

The first successful commercial tobacco crop was cultivated in Virginia in 1612 by Englishman John Rolfe, secretary and recorder of the fledgling Jamestown Colony. Rolfe, who is also well known for marrying the Powhatan princess Pocahontas in 1614, obtained tobacco seeds in the Caribbean and planted them in Virginia in lieu of the more bitter native tobacco. This act literally saved the starving colonists. Two years later, he sent to England 4 hogsheads (252 dry gallons) of tobacco, weighing about 2,600 pounds (1,180 kilograms). The colony exploded with the growing and selling of tobacco. In the Chesapeake region of Virginia, tobacco was used as a medium of exchange by some Native Americans: It became a form of currency for the colonists as well, with everything bought and sold with tobacco.

Concern about his subjects' health had led the English king James I to issue *A Counterblaste to Tobacco* (1604), condemning tobacco as a "heathenish" poison. However, the tobacco market continued to expand, and, by 1617, Virginia had shipped around 10 tons (9,100 kilograms) of tobacco to England. The European demand for tobacco grew at a bewildering rate. In 1603, England had imported a total of only 25,000 pounds (11,300 kilograms) of the plant, but by 1640, tobacco had risen to first place among London's imports. In 1700, the nation imported almost at 38 million pounds (17 million kilograms) and in 1771, over 100 million pounds (45 million kilograms).

SIGNIFICANCE

The introduction of tobacco crops as a commercial crop in seventeenth century North America had far-reaching social, physical, and economic significance. Its cultivation spurred the developing tobacco culture in Europe and throughout the world. Tobacco joined sugar and other plantation crops to form the heart of the colonial American economy, binding the colonies to developing world markets and providing an ever-greater impetus to import slaves to work the plantations. Historians agree that tobacco made more fortunes than all the silver in North and South America.

Although mild when compared with today's hard narcotics, nicotine's advance into Britain during the seventeenth and eighteenth centuries marked a major step in the development of global, imperialist capitalism. Tobacco, as an ever-increasing commodity, ensured the ongoing development of Virginia and provided an economic incentive for further expansion and settlement of the New World.

Furthermore, tobacco greatly influenced the displacement of Native Americans. Since it is a crop that is exhausting to the soil, virgin land must be cultivated continuously to keep up production. Thus, more and more land was needed to fulfil the rapidly expanding worldwide desire for tobacco. In order to meet the need for new territory, settlers took advantage of the Native Americans: By trading for mere pots and pans, the Indians lost control of their lands. The European and colonial craving for tobacco therefore played a major role in imperial acquisition and colonial expansion.

FURTHER READING

Gately, Iain. *Tobacco: A Cultural History of How an Exotic Plant Seduced Civilization.* New York: Grove Press, 2003. Traces the history of tobacco from pre-Columbian America through the litigation of the 1990'. Fascinating account of tobacco in literature, film, and society.

Goodman, Jordan. *Tobacco in History: The Cultures of Dependence.* New York: Routledge, 1994. A far-reaching analysis of the culture and business of tobacco. Explores the chemical addictive nature of tobacco, details the introduction of tobacco to Europe, and examines the role of government and the enormous economic impact worldwide. Provides a forty-page bibliography.

Middleton, Arthur Pierce. *Tobacco Coast: A Maritime History of the Chesapeake Bay in the Colonial Era.* Baltimore: Johns Hopkins University Press, 1984. Scholarly history of tobacco's role in the Chesapeake Bay economy. Discusses how the early colonists thrived on the cultivation of tobacco and the plant's increasing importance as the area's primary economic mainstay.

Schivelbusch, Wolfgang. *Tastes of Paradise: A Social History of Spices, Stimulants, and Intoxicants.* Translated by David Jacobson. New York: Pantheon Books, 1992. Although this book deals primarily with coffee, tea, and alcohol, it contains a highly informative section on the history of tobacco and its cultural and economic impact.

RURAL AREAS

—Krishna Roka

FIELDS OF STUDY

Plant Science; Crop Science; Botany; Agronomy; Agriculture; Biochemistry; Ecology; Entomology; Genetics; Landscape Architecture; Soils Engineering.

SUMMARY

Rural areas, in addition to farms and farmers, contain forests, minerals, water, wildlife, and landscapes, and are the scenes of various transformation and development schemes.

Rural areas, regardless of where they are, consist of farms, forests, villages, ranches, small towns, and absence of busy roads or other large infrastructure. These are frequently food producing areas that contain such vital resources as water, minerals, fossil fuels, forests, and wildlife. Rural areas are valued for their scenic landscapes and natural environment, which people also use for various recreational activities.

Many indigenous groups with unique cultures, beliefs, and lifestyles tend to inhabit rural areas. Currently, pastoral communities are considered critical to ensuring global food demands and energy security. Rural areas are adaptable to climate change, can help preserve biodiversity, and conserve indigenous cultures. Therefore, many countries are working to preserve them.

DEFINITION

The word *rural* is complex and its strict definition contested, making it difficult to draw a line between urban and rural areas. Derived from the Latin word *rus,* which means *open space,* it is synonymous with the words *country* and *countryside* in English-speaking countries. *Rural* also simply means places that have few people or cities. The term is a subjective state of mind for some and an objective quantitative measure for others. The most generalized definition of rural areas, one that's been used for hundreds of years, is that they are not urbanized.

The U.S. Census Bureau's classification of *rural* includes all territory with a population density of less than 500 people per square mile (2.6 square kilometers). Other countries use different measurements to differentiate rural areas from urban areas.

People working in a rice paddy in a rural area of Nepal. Rates of urbanization vary a great deal between developed and developing countries: while about 79 percent of Americans live in urban areas, by contrast, in Nepal, a developing country, only 15 percent are city dwellers. Image by World Bank/Curt Carnemark.

TRANSFORMATION

Rural areas have been transforming rapidly, especially since humans settled and adopted agriculture. It was around settled areas, mostly surrounding river basins, that several human civilizations evolved. Over time, as civilizations evolved and cities were established as centers of commerce and administration, rural areas became thought of simply as sources of food.

The Industrial Revolution of the nineteenth century accelerated the creation of cities throughout the world. Since then, cities have grown, attracting millions of migrants from rural areas. Even though the world has become increasingly urbanized with more people living in metropolitan areas, rural communities still play a significant role in sustaining the urban population. The rate of urbanization varies by country, as developed countries are more urbanized than developing nations. In the United States, for example, about 79 percent of Americans live in cities—but in Nepal, only 15 percent do. The type of urbanization also varies, but rural areas today, except in a few countries, generally have access to technology, transportation, commerce, and communication.

HYBRID SPACES

The rapid transformation of both rural areas and urbanization is blurring the distinction between the two, resulting in hybrid spaces, which have characteristics of both types of communities. Hybrid spaces include networks of agricultural production that combine human, nonhuman, and technological components, and they have processes of development for social or economic transformation purposes.

Some of these areas include recreational activities that depend on complex technologies and on particular landscapes and climates, transportation patterns comprising human-machine relations, and hybrid networks of local and extralocal agencies. The pervasive influence of urbanization and adoption of nonagricultural economic activities, urban cultural practices, and consumption patterns have made *rural* a hybrid concept.

In North America, some such areas are known as *suburban* or *exurban,* and they are characterized by rural landscapes and settlements that are tightly linked to urban labor markets and service centers. In recent years, real estate developers have attracted urban residents to suburbs by invoking images of an idyllic rural lifestyle, proximity to nature, solidarity and community spirit, high-quality public services, and cultural diversity. Such development also has raised concerns about environmental sustainability and inefficient use of resources.

CHARACTERISTICS OF RURAL AREAS

Despite rural-urban overlap, rural areas have several distinct characteristics, chiefly the closer attachment of people to nature. Often, rural areas are regarded as the last frontier, full of wilderness and open spaces. People who live in rural areas are viewed as being simple, hard-working, innocent, virtuous, peaceful, and living in harmony with nature. Also, rural societies are considered to have tightly connected structures of families and communities. Over time, these characteristics have been idealized and mythologized through art, literature, media, and movies.

The expansion of urban areas in the United States and Europe led to the growth of this romanticized version of rural areas. During the colonial period, Europeans exported their version of rural areas to colonies worldwide. They even tried to convert the existing landscape to reflect their ideas of countryside. British rulers, for example, established rural estates to re-create English country gardens in India, South Africa, and the West Indies. The transportation of rural ideas was not all one-way, but also flowed back from the colonies to Europe. Several crops were introduced to the Old World from the New World, most notably potatoes, beans, maize, and spices.

ECONOMIC FACTORS

Economically, rural areas are attached to natural resources, mostly agriculture and extractive industries such as mining, fishing, and forestry. In the past, these activities were the chief source of food and income for rural families. Even today, millions of people in developing countries depend on subsistence agriculture to support their families. This dependence on traditional agriculture is considered the primary reason for nondevelopment of rural areas. However, such theories are contested. As cities grew and became associated with advanced economy, rural areas became synonymous with backward economies plagued by poverty, traditionalism, and lack of opportunities.

Several factors contributed to the underdevelopment of rural areas. Historically, the abundance of resources in rural areas resulted in the exploitation of those resources through farming, forestry, mining, quarrying, fishing, hunting, and energy production. The growth of cities, industries, and expansion of a capitalist global economy facilitated the extraction of resources from rural areas. In the process, rural people abandoned traditional agriculture and became part of the global economic system by working for companies and corporations. This transformation brought drastic changes to rural economies.

When the natural resources were exhausted, rural dwellers, instead of going back to traditional occupations, looked for other means of income. Many bucolic communities attracted industries at the cost of environmental and cultural degradation. In the United States, some rural communities now depend on such enterprises as waste facilities and penitentiaries for employment and revenue.

In pursuit of a better life and economic opportunities, thousands of people migrated to urban areas. Many of the families remaining in the rural areas, especially areas that have transitioned away from fully agriculture-based communities, now depend on government subsidies to fulfill their basic needs.

A river gorge at Fall Creek, Ithaca, Thompkins County, New York. The gorge wall shows erosion partially controlled by joints. Image by U.S. Geological Surve

As the cities grew in size and population, the press of urbanization weighed heavily on urban dwellers, causing some to move to the countryside. In recent years, the urban-to-rural shift has in a sense become a major industry: rural tourism.

Rural communities all over the United States have adopted tourism as their major source of income. Rural tourism is flourishing in places that contain unique culture, biodiversity, and quaint, natural settings. Increasingly, however, rural areas are becoming economically vulnerable as they become dependent on a single industry or activity. The dependent economy, along with other factors, has made rural areas lag urban or suburban areas in development scale. As a result, rural development is a major objective for governments in both developed and developing countries.

RURAL DEVELOPMENT

Rural development strategies differ from country to country. The strategies depend on the nature and availability of natural and social capital that can be mobilized for development.

Conventional approaches to rural development were based on the concept of modernization, which assumes the underdevelopment of rural areas. It assumed that underdevelopment could be overcome with sustainable economic growth that eventually would improve the living conditions of rural people and raise those areas to national standards of development.

Over the past several decades, many governments have been trying to update and improve living conditions in rural communities through enhanced agricultural services, economic conditions, infrastructure, and social values. Agricultural modernization involves mechanized farming practices, application of agrochemicals, and adoption of biotechnology. Economic modernization often entails diversifying rural economies by establishing small industries. Infrastructure updates include expansion of roads, electrical lines, telecommunications, and increased housing options. Social modernization promotes modern education, challenges superstitions and traditional beliefs, and raises awareness about civic rights.

After 1950, this four-pillar process became the dominant rural development paradigm in many parts of the world. Despite decades of investment, however, many rural areas have not been successfully developed along these lines. Since the early 1980s, criticism of this model has focused on environmental degradation versus sustainability, and the drawbacks of dependency on transnational banks and corporations.

A number of alternative paradigms have been proposed and applied. These approaches center on sustainable development, community participation, involvement of nongovernmental organizations, and environmentally friendly approaches. However, there is no universal model to achieve these goals, which have become vague and broad.

FUTURE OF RURAL AREAS

The future of rural areas depends on sociopolitical decisions and environmental changes associated with climate change. Policies of local, national, and international agencies on rural areas will influence the appearance of rural landscapes, the structure of their economies, the pattern of their settlement, and the dimension of the populations. The future also depends on the status of biodiversity, construction of infrastructure, commercialization of rural resources, and the quality of life of rural people.

Policies can have positive or negative effects. The greatest threat to rural areas, however, is climate change. On the positive side, rising global temperatures could increase the yield of a few food crops such as wheat, rice, corn, and sugarcane. The negative effects of rising temperatures include possible water shortages, reduced soil fertility, pest mobility, and increased frequency of flooding and wildfires. Such changes will disrupt the human-nature relationship in rural areas.

Rural tourism, which depends on the natural environment, will be hit hard by climate change as non-agricultural economic activities are forced to relocate, putting millions of people at risk. The response will be migration. Thousands of people have already moved out of rural areas and are living in urban slums or in neighboring countries as climate refugees. Rural areas, despite their crucial role in the urban world, have never been so vulnerable in recent history.

History also suggests that rural areas have the capacity to fight back through resiliency and adoption of new techniques. The fight against climate change has begun, for example, as rural areas from Africa to Asia are responding to the challenges facing them by adopting new techniques to conserve water, growing drought-resistant crops, and planting trees.

Bucolic communities also are diversifying their economies and moving away from dependence on a single activity. There are programs being implemented to raise awareness about the impact of climate change and ways to reduce carbon footprints. These efforts will take time to yield results. In the meantime, rural areas depend more than ever on urbanites to reduce emission of hydrocarbon gases.

Further Reading

Halseth, Greg, Sean Markey, and David Bruce. "The Next Rural Economies: Constructing Landesa." Rural Development Institute. http://www.landesa.org.

Shucksmith, Mark, David L. Brown, Sally Shortall, and Jo Vergunst. *Rural Transformations and Rural Policies in the U.S. and UK.* New York: Routledge, 2012.

Woods, Michael. *Rural.* New York: Routledge, 2011.

S

Savannas & Animal Life

—*Mary Hurd*

FIELDS OF STUDY

Zoology; Biology; Ecology; Environmental Science.

PRINCIPAL TERMS

carnivores: flesh-eating animals
herbivores: plant-eating animals
marsupials: animals having a pouch on the abdomen for carrying their young
ruminants: grass-eating animals that chew again food that has been swallowed
ungulates: hoofed animals

SUMMARY

Savannas are grasslands in tropical regions near the equator that support a larger number of plant-eating animals than any other area in the world.

Savannas are vast open spaces on which grow a large variety of plant life. Savannas usually endure long periods of drought that are punctuated by one or two rainy seasons. When the rains come, large herds of animals, mostly ungulates, make their annual journey along centuries-old migration routes from the river valleys where they have spent the dry season, to the fresh grass on the savanna. Although the principal vegetation is grass, trees or tall bushes appear occasionally on the landscape or along streams.

The huge African savanna is ancient, having probably evolved about sixty-five million years ago. Other areas are newer, some having been created by humans when forests were cleared to accommodate farming. Savannas also exist in South America, largely in Venezuela, and in northern Australia.

PLANT-EATING ANIMALS

The African savanna varies from very dry regions to areas of swamp, lake, and woodland, and can support the largest variety of herbivores in the world. Many animals are capable of living together because most of them have their own specific feeding habits. The hippopotamus, reedbuck, and waterbuck remain near the water, while various gazelles prefer dry areas, receiving moisture from plants. While the zebra chooses open grassland, the wildebeest (gnu), giraffe, and antelope are equipped to forage in the bush and also the woodland by virtue of their long snouts for gathering leaves and stems. All parts of trees and shrubs provide food; while some animals feed on the tough outer parts of grass, others eat the tender, fresh foliage or leaves of wild flowers.

The herbivores found on the African savanna are also the world's largest land animals. African elephants live in grassland, bush, and forest, in mountainous country and near lakes. Every day, elephants eat vast quantities of grass, leaves, twigs and bark, sometimes destroying trees and helping to create savannas. Elephant herds incorporate smaller groups of four to twenty elephants, led by the older females.

The white rhinoceros, weighing more than three tons, is one of Africa's rarest animals. The herds are composed of small family groups of one male, one or two females, and several young. Its smaller relative, the black rhinoceros, exists more abundantly.

Browsing on leaves and branches, the black rhinos are protected by their size and horns and tusks. Humans are their only real enemies. Many of the grass eaters on the African savanna are antelope, which belong to a suborder of animals called ruminants. Equipped with complex stomachs, ruminants eat food which passes from the mouth, through the several chambers of the stomach, and back again to the mouth. The slow process of rumination, or chewing the cud, provides the animal with more safety from predators as chewing can be accomplished later in a less dangerous place than grazing.

In the savanna of northern Australia are found marsupials—kangaroos, koalas, and wombats. Marsupials also live on the South American savanna, along with the armadillo. Other particular species found in these geographically isolated areas are long-legged, flightless birds—the ostrich, rhea, and the emu.

PREDATORS & SCAVENGERS

When the sun's energy is converted by plants into food for herbivores, or primary consumers, then the predators and scavengers, or animals who live by preying upon other animals, become secondary consumers in the food chain. Big cats, such as the lion, cheetah, and leopard, are powerful animals who stalk and run down their prey. Lions function in teams, with females assuming most of the work.

African wild dogs are smaller animals that have strong jaws, sharp teeth, and a keen sense of smell. They live and hunt in packs made up of six to twenty members. Other predators, such as the hyena and the jackal, kill their prey and feed at night off the carcass of the animal; powerful, far-sighted vultures and marabou stork feed by day, each eating a different part of the carcass. Smaller scavengers—crows, ravens, rats, and insects—move in, helping to dispose of dead bodies that might carry disease organisms. Smaller predators of the African savanna include the desert lynx, which pursues the smaller antelope, and the black spotted serval, which hunts ground squirrels, large rats, and guinea fowl. Other predators are the genet and the fox. Puff adders and cobras poison their victims and are themselves attacked by the mongoose and the secretary bird. The aardvark hunts ants and termites at night.

Various species of monkeys, baboons, and vervets have adapted to living on the savanna by living mostly on the ground, trying to avoid predators. In the Australian savanna, the kangaroos are preyed upon by the dingoes, or wild dogs.

FURTHER READING

Alden, Peter, et al. *National Audubon Society Field Guide to African Wildife.* New York: Chanticleer Press, 1995. Visually organized mammal guide with striking photographs of birds, mammals, reptiles; contains a wealth of information on more than 850 species, including animal behavior, habitats, park, reserves.

Combes, Simon. *Great Cats: Stories and Art From a World Traveler.* Shelton, Conn.: Greenwich Workshop Press, 1998. Beautifully photographed and sketched; detailed narratives of real-life adventures with large cats. One section focuses upon lions, cheetahs, and leopards of African savanna, their habitats, behavior and feeding habits.

Miius, S. "When Elephants Can't Take It Anymore." *Science News* 155 (May 29, 1999): 341–342. Article reflects research on the decline of the elephant population in Africa. Maintains that when human population reaches a certain threshold, elephants disappear rapidly; the main threat to elephant survival is land-use clashes with humans.

Sherr, Lynn. *Tall Blondes: A Book About Giraffes.* Kansas City, Mo.: Andrews McMeel, 1997. The chapter "Where the Giraffes Are" is concerned with the exploitation and extermination of giraffes in Africa at the end of the nineteenth century and the eventual creation of wildlife laws that prevent slaughter of giraffes.

Silver, Donald M. *African Savanna.* New York: McGraw-Hill, 1997. One volume in a series of science and nature books for children called *One Small Square.* Includes stunning full-color illustrations of habitat and its creatures; also contains experiments and activities.

SCIENCE OF CLONING

—*Michael A. Buratovich, PhD*

FIELDS OF STUDY

Genetics; Biology; Chemistry.

SUMMARY

Cloning is any type of biological reproduction that produces offspring that are genetically identical to their parents. Cloning occurs naturally, since many organisms routinely reproduce through natural cloning processes. Artificial cloning technologies include molecular cloning, which reproduces large quantities of discrete segments of DNA; reproductive cloning, which uses assisted reproductive technologies to produce animals that share the same desirable genetic characteristics as another living or previously existing organism; and therapeutic cloning, which uses the same techniques as reproductive cloning but instead derives useful cell lines from cloned embryos.

DEFINITION & BASIC PRINCIPLES

Cloning is a means of producing biological organisms, cells, or DNA molecules that are genetically identical to their progenitors. There are natural forms of cloning and three main types of artificial cloning: molecular, reproductive, and therapeutic cloning.

Natural mechanisms of cloning occur in organisms such as bacteria that simply split or fragment into identical copies of themselves. In other organisms, reproductive cells, or gametes, undergo a process called parthenogenesis, in which they initiate development without the benefit of fertilization. Cloning is uncommon in mammals, but rarely, early mammalian embryos undergo a form of cloning called twinning, in which the embryo splits into two embryos, which develop into genetically identical twins.

Molecular cloning, also known as recombinant DNA technology or DNA cloning, involves the transfer of an isolated fragment of DNA from an organism of interest to a host cell that replicates it. Such isolated DNA fragments are known as cloned DNA or genes.

Reproductive cloning uses assisted reproductive technologies to generate animals with the same nuclear genome as another animal. The particular procedure used during reproductive cloning is called somatic cell nuclear transfer (SCNT). Cloned embryos are gestated in the womb of a surrogate mother until they come to term. Cloned organisms are not genetically modified organisms but are simply produced through a type of assisted reproduction.

Therapeutic cloning uses the same procedures as reproductive cloning; however, instead of transferring the cloned embryo into the womb of a surrogate mother, the embryo is further manipulated in the laboratory to make cell cultures of embryonic cells for basic or clinical research.

BACKGROUND & HISTORY

Sea urchins were the first animal cloned in the laboratory. In 1894, Hans Dreisch isolated sea urchin embryo cells and watched them develop into small, separate larvae. In 1902, Hans Spemann used the same procedure of embryo splitting to isolate cells from salamander embryos, which also developed into identical adult salamanders. In 1903, U.S. Department of Agriculture employee Herbert Webber coined the word "clon" for asexually produced cells or organisms, which later evolved into "clone." This term comes from the Greek *klon*, which means "trunk" or "branch." Horticulturists have used this term for more than a century, since an entire new plant can grow from a cutting, resulting in a plant that is genetically identical to the plant from which the cutting was taken.

In 1928, Spemann cloned salamanders by transferring the nucleus, the subcellular compartment that houses the chromosomes, from one salamander embryo into the egg of another. Since Spemann's seminal experiments, scientists have adapted nuclear transfer technology to clone other organisms. In 1952, frogs were cloned, and in 1963, the Chinese embryologist Tong Dizhou cloned a carp to produce the first cloned fish. During the 1980s and 1990s, sheep, cows, and mice were cloned. However, all these animals were cloned by using nuclei from embryos. In 1996, Ian Wilmut and his team at the Roslin Institute in Edinburgh, Scotland, cloned a sheep from an adult cell, demonstrating that adult cells could serve as the source of genetic material for animal clones. This technological feat was followed by the cloning of goats, mules, gaurs (an endangered species), horses, pigs, mouflons (a wild sheep), mice, rats, dogs, cats, water buffalos, camels, rabbits, deer, wolves, and African wildcats, and even embryos from nonhuman primates and humans.

HOW IT WORKS

Molecular Cloning. To clone a gene, the DNA of the model organism is selectively fragmented by enzymes called restriction endonucleases (REs) and inserted into another piece of DNA called a cloning vector. Cloning vectors are either small circles of DNA called plasmids, bacterial viruses, or bacterial or yeast artificial chromosomes. They ferry the DNA fragments from the genome of the model organism into a host cell (either a bacterium or yeast). This population of host cells collectively carries the entire genome of the model organism in small fragments, and is called a gene library.

To isolate a gene from a gene library requires a probe, which is a fragment of DNA or RNA of any length that has a sequence that is complementary to the sequence of the gene that is to be isolated. Probes can be made synthetically or can come from the genes of closely related organisms. By screening the gene library with the probe, the gene of interest is cloned, which simply means to isolate it from all the other sequences found in the genome of the model organism.

Alternatively, scientists can synthesize small strands of DNA called primers, whose sequences are complementary to different locations in the gene. These primers can be used to specifically amplify the gene from the library by means of a Polymerase chain reaction (PCR). A polymerase chain reaction makes large quantities of the gene of interest from a very small amount of starting material, and the amplified DNA can also be cloned into a cloning vector or analyzed directly.

Reproductive Cloning. To clone an animal, mature eggs are isolated from females of the animal species that is to be cloned. The egg is enucleated by piercing it with a microscopically narrow (0.0002-inch-wide) glass tube that is used to vacuum out the egg nucleus. The enucleated egg is fused with a cell from the body of the animal to be cloned and activated with either chemicals or an electric current. This procedure is called somatic cell nuclear transplantation (SCNT).

After activation, the egg divides and grows like a newly formed embryo. However, if the animal is a mammal, the embryo can survive only for a limited period of time before it must implant into the inner layer of the mother's womb. Therefore, a surrogate female from the same species of the animal to be cloned, or a closely related species, is made pseudopregnant by feeding her hormones, and the embryo is released into her receptive womb where it then implants. Barring any technical or biological mishap, the cloned embryo will develop, and the process will result in a live birth.

Therapeutic Cloning. To make embryonic cell cultures, cloned embryos are made by means of somatic cell nuclear transplantation. They are then either disassembled in the laboratory and used to establish embryonic cell cultures or gestated in a surrogate mother to the fetal stage, at which time the fetus is aborted, and cells from the fetus are used to establish fetal cell cultures.

By culturing specific cells from cloned embryos, scientists can make embryonic stem cell (ESC) cultures. During mammalian development, two distinct cell populations form after the first few days of embryonic development. The trophoblast, or the flattened, outer layer of cells, will eventually form the placenta and its associated structures. The inner cell mass (ICM) is the round, inner clump of cells that develop to form the embryo proper and a few structures associated with the placenta. If ICM cells are isolated and cultured on feeder cells, a layer of nondividing skin cells that secrete a cocktail of growth-promoting chemicals, the ICM cells will grow and spread over the surface of the culture dish. Such a culture is an embryonic stem cell culture, and these cells are pluripotent, which means that they can differentiate into any cell type in the adult body.

APPLICATIONS & PRODUCTS

Molecular Cloning. Organisms that express cloned genes make many useful pharmaceuticals such as human insulin, growth hormone, clotting factors, fertility drugs, and vaccines. Cloned genes are also used to genetically screen individuals for genetic diseases. Pharmacologists even use cloned genes for pharmacogenetics, which screens patients for the presence of gene variants that can profoundly affect the efficacy and toxicity of particular drugs. This allows clinicians to tailor treatment to the exact genetic makeup of the patient to maximize treatment efficacy and minimize side effects. Such a strategy is called personalized medicine. Cloned genes are also used in gene therapy, which delivers cloned genes into the bodies of patients who suffer from genetic diseases in an attempt to cure them. Patients with cancer and inherited deficiencies of the immune system, blindness, and blood-based defects have been treated with gene therapy protocols.

In agriculture, the introduction of cloned genes into plants that are used as food crops has generated transgenic crops. These crops display several advantageous traits: reduced dependence on agrochemical applications (for example, Bt-corn and herbicide-resistant crops), increased nutritional value (for example, Golden Rice), increased resistance to environmental stresses, and reduced spoilage (for example, the Flavr Savr tomato).

Despite safety and ethics concerns regarding genetically modified organisms (GMOs)in agricultural products, the International Service for the Acquisition of Agri-Biotech Applications (ISAAA) reported in 2013 that genetically modified (GM) crops were planted in twenty-seven countries by approximately 18 million farmers worldwide. Over 60 percent of the world's population live in the twenty-seven countries that are planting GM crops. In 2006, U.S. government statistics showed that 87 percent of the global genetically modified crops were grown in developed countries. By 2013, however, ISAAA reported that Latin American, Asian, and African farmers grew 54 percent of the global GM crops compared to the 46 percent grown in developed countries worldwide. This trend continued into 2016, when the ISAAA reported increased soybean plantings in countries such as Argentina and Brazil. Corn, soybeans, cotton, alfalfa, and canola were the major crops, often modified for insect resistance. Rice has been genetically enhanced for more iron and vitamins to alleviate malnutrition in Asia. Other plants have been modified to survive weather variances.

Reproductive Cloning. When farmers identify food animals with desirable traits, they typically breed those animals as much as possible to improve the genetic quality of their herds and flocks. However, such prize animals inevitably die. Propagating these animals by reproductive cloning and mating them to as many animals as possible preserves the exceptional genetic content of a prize animal and allows it to produce far more offspring. This significantly raises the genetic quality of the flock or herd, and commercial dissemination of such cloned animals to other farmers raises the overall genetic quality of food animals. Reproductive cloning also eliminates the need for artificial insemination, which is often expensive and inconvenient.

Cloning effectively maintains high-quality animal stocks. Reproductive cloning of only the healthiest and most productive animals increases their numbers and improves the gene pool (sum total of genetic diversity) and overall health of food animals. This results in safer and healthier food and reduces the use of growth hormones, antibiotics, and other chemicals in the raising of animals.

In the field of conservation biology, the numbers of endangered species are often increased by captive breeding programs. However, not all endangered species can effectively breed in captivity. Reproductive cloning can aid in the preservation of those organisms that do not reproduce in captivity. Cloning can also resurrect genetic material from dead animals and potentially expand the gene pool of endangered species. In 2001, scientists at the University of Teramo, Italy, cloned the European mouflon, an endangered sheep, from cells sampled from a dead animal. When combined with other reproductive technologies, cloning can help save endangered species.

Cloned animals also serve as excellent research models. Because each cloned animal is genetically identical, experiments on cloned animals are devoid of differences caused by heterogeneous genetic backgrounds. Genetic manipulation of cloned animals allows researchers to modify genes of interest and more completely analyze their contribution to development and disease. Modifying particular genes of cloned animals also generates model systems for particular genetic diseases. Cloned, transgenic mice and cloned knockout mice, which have had a specific gene inactivated, are examples of the vast usefulness of such model systems.

Of enormous interest is modifying the genomes of cloned animals so that they can produce clinically and pharmaceutically significant products. By genetically modifying pigs, it is possible to make cloned pigs that contain organs that are fit for transplantation into humans (xenotransplantation). Also, producing antibodies, clotting factors, or even vaccines in the blood or milk of farm animals provides a means to mass-produce potentially expensive pharmaceutical agents at a fraction of the normal cost. This process is called pharming.

Therapeutic Cloning. Therapeutic cloning has tremendous potential for numerous clinical applications. Embryonic stem cells (ESCs) made from therapeutic cloning procedures are pluripotent. Therefore, injured, diseased, or failing tissues or organs could potentially be replaced by tissues or organs manufactured from embryonic stem cells in the laboratory or fetal cells from cloned fetuses. Furthermore, embryonic stem cells made from cloned embryos, or any tissues or organs fashioned from these cells, would not be regarded by the patient's body as foreign. Experiments in laboratory animals have shown that such scenarios are possible. Therapeutic cloning, coupled with embryonic stem cells technology, could christen a new era of regenerative medicine.

Embryonic stem cells from cloned embryos have toxicological applications. Toxicologists typically use laboratory animals or cultured cells to gauge the biological effects of natural or industrially produced molecules on human beings. Unfortunately, laboratory animals show limited utility as a model for human toxicology, and cultured cells do not represent the response of an organ or tissue to foreign molecules. Furthermore, neither of these model systems can assess the individual responses people will have to such molecules, because the genetic variation between individual humans causes differential responses to drugs, toxins, or environmental pollutants. However, cultured embryonic stem cells from cloned embryos can test the biological effects of drugs or environmental pollutants on cells made from a specific person. In addition, because these cells can be differentiated into various tissues and even organs, they can be used to evaluate the individual and tissue-specific responses people might have to particular drugs or pollutants.

SOCIAL CONTEXT & FUTURE PROSPECTS

Despite the reservations of some people, cloning is a part of everyday life. Many of the foods Americans consume contain some genetically engineered products. Physicians prescribe medicines, give vaccines, and apply other biological products made by genetically engineered microorganisms on a quotidian basis.

Nevertheless, many people have raised concerns over cloning technologies. First, conservation biologists have suggested that cloning endangered species does not address the habitat destruction and environmental degradation that pushed these species to near extinction in the first place. Second, cloning only makes one species and does not re-create an ecosystem. For example, cloning cannot recapitulate a coral reef or an old growth forest. Thus, many argue that it is the wrong solution for the problem.

Genetically modified organisms have become the focal point of concern for several environmental activism groups. Such groups oppose GMOs because they believe that the cloned genes inserted into them can spread to other species and cause severe environmental disruption and that genetically engineered foods have not been sufficiently tested and are potentially dangerous to human health.

The most contentious aspect of cloning technologies is human genetic engineering and reproductive cloning. Until the cloning of the sheep Dolly in 1997, it was thought that adult specialized cells could not be made to revert to nonspecialized cells that can give rise to any type of cell. However, Dolly was created from a specialized adult cell from a ewe's udder. U.S. president Bill Clinton asked the National Bioethics Advisory Commission to form recommendations about the ethical, religious, and legal implications of human cloning. In June 1997, the commission concluded that attempts to clone humans are "morally unacceptable" for safety and ethical reasons. There was then a moratorium placed on using federal funds for human cloning. In January 1998, the U.S. Food and Drug Administration (FDA) declared that it had the authority to regulate human cloning and that any human cloning must have FDA approval.

In 2013, the science of cloning experienced another major breakthrough when researchers at the Oregon Health & Science University successfully cloned human embryonic stem cells. Using nuclear transfer, the scientists implanted the skin of a fetus into the nucleus of an egg before prompting division, creating a ball of cells that includes embryonic stem cells with the same genetic material as the original skin cell. The main purpose of the research was not to further the prospect of human cloning, however, but to establish a process to consistently clone stem cells for use in the treatment of diseases and other human ailments.

China, a country that has always striven to become a global leader in the advancement of cloning technology, announced in late 2015 that one of its major companies, BoyaLife, would be opening a facility designated to mass commercialization of cloned cattle largely in response to the country's increased need for beef, which represented a marked departure from other countries' focus on cloning solely for the purpose of breeding. Earlier that year, the European Parliament had voted to institute a ban on the cloning of any farm animals, their offspring, and derivative products.

Transhumanists are some of the most energetic proponents of human cloning and genetic enhancement. As a movement, Transhumanism regards infirmity, disease, aging, and death as undesirable and unnecessary and views science and technology as the means to defeat human limitations. Transhumanists' main argument for human cloning is that reproductive freedoms extend to everyone, and therefore, every human being has an inherent right to clone himself or herself.

Opponents of human cloning object to the manufacturing of human beings. Cloned children are made to be identical to someone else and therefore will always live in the shadow of the original person and never be completely the person they choose to be. These unreasonable expectations can psychologically damage them and violate their human dignity and individuality. Cloning would also alter the concept of human nature and therefore undermine the very foundation of liberal democracy.

In the future, the argument and debate over cloning will not dissipate, but cloning research will certainly advance and provide more and more examples of the utility of this remarkable technology.

Further Reading

Alexander, Brian. *Rapture: A Raucous Tour of Cloning, Transhumanism, and the New Era of Immortality.* Basic, 2004.

Fukuyama, Francis. *Our Posthuman Future: Consequences of the Biotechnology Revolution.* Picador, 2003.

Hanson, Charles, et al. "Transplantation of Human Embryonic Stem cells onto a Partially Wounded Human Cornea In Vitro." *Acta Ophthalmologica,* vol. 91, no. 2, 2013, pp. 127–30.

"ISAAA Brief 46–2013: Executive Summary." *ISAAA,* 25 Mar. 2014, www.isaaa.org/resources/publications/briefs/46/executivesummary/. Accessed 28 July 2014.

Jensen, Eric A. *The Therapeutic Cloning Debate: Global Science and Journalism in the Public Sphere.* Ashgate, 2014.

Lynch, Colum. "UN Backs Human Cloning Ban." *Washington Post,* 9 Mar. 2005, www.washingtonpost.com/wp-co/hotcontent/index.html?section=nation/specials/science/cloning. Accessed 28 July 2014.

Mitchell, C. Ben, et al. *Biotechnology and the Human Good.* Georgetown UP, 2007.

Park, Alice. "Scientists Report First Success in Cloning Human Stem Cells." *CNN,* 16 May 2013, www.cnn.com/2013/05/15/health/time-cloning-stem-cells. Accessed 24 Oct. 2016.

Phillips, Tom. "Largest Animal Cloning Factory Can Save Species, Says Chinese Founder." *Guardian,* 24 Nov. 2015, www.theguardian.com/world/2015/nov/24/worlds-largest-animal-cloning-factory-can-save-species-says-chinese-founder. Accessed 24 Oct. 2016.

"Pocket K No. 16: Biotech Crop Highlights in 2015." *ISAAA,* June 2016, isaaa.org/resources/publications/pocketk/16/default.asp. Accessed 24 Oct. 2016.

Prado, José Rafael, et al. "Genetically Engineered Crops: From Idea to Product." *Annual Review of Plant Biology,* vol. 65, 2014, pp. 769–90.

Semple, Kirk. "UN to Consider Whether to Ban Cloning of Human Embryos." *The New York Times,* 3 Nov. 2003, www.nytimes.com/2003/11/03/world/un-to-consider-whether-to-ban-some-or-all-forms-of-cloning-of-human-embryos.html. Accessed 28 July 2014.

Shanks, Pete. *Human Genetic Engineering: A Guide for Activists, Skeptics, and the Very Perplexed.* Nation, 2005.

Silver, Lee. *Challenging Nature: The Clash Between Biotechnology and Spirituality.* Harper Perennial, 2006.

Silver, Lee. *Remaking Eden: How Genetic Engineering and Cloning Will Transform the American Family.* Harper Perennial, 2007.

Wilmut, Ian, et al. *The Second Creation: Dolly and the Age of Biological Control.* Farrar, 2000.

SEA SHEPHERD CONSERVATION SOCIETY (SSCS)

—Adam B. Smith

FIELDS OF STUDY

Physical Science; Conservation; Biology; Chemistry; Environmental Studies.

SUMMARY

The Sea Shepherd Conservation Society's use of direct action and confrontation to disrupt fishing, seal hunting, and whaling operations have been the subject of controversy, but the group's activities have nonetheless focused public and policy attention on overexploitation of the seas.

In 1977 Paul Watson, a founding member of the environmental organization Greenpeace, left that organization after a disagreement with other members and founded the Earthforce Society, the group that later became the Sea Shepherd Conservation Society (SSCS). One of the first interventions carried out by Earthforce was the ramming of a vessel hunting whales in contravention of the worldwide whaling moratorium, an incident that set the tone for subsequent campaigns. Other actions the group has carried out since then have included marking baby seals with an organic dye so their pelts would be worthless to hunters, exposing illegal shark finners, capturing and destroying drift nets, and harassing the Japanese whaling fleet in the Southern Ocean Whale Sanctuary.

Bateau Sea Shepherd Conservation Society à l'Armada Rouen 2013. Image by Copyleft (Own work) [CC-BY-SA-3.0.

SSCS has been controversial since its inception. Detractors often label it a vigilante organization (a label SSCS proudly adopts) or an ecoterrorism group, whereas supporters note that SSCS enforces marine conservation laws and treaties that would otherwise go unenforced. On occasion SSCS has partnered with governments unable to police their own waters for illegal fishing, but it also operates on its own in national waters and on the high seas. SSCS cites Article 21 of the United Nations World Charter for Nature as the legal grounding for its activities, although legal scholars note that the organization's interpretation is probably outside the spirit of the text. Despite efforts by governments of targeted operations to disenfranchise SSCS by impounding its vessels, arresting members, and protesting to the nations in which SSCS branches are located, the organization remains largely unhindered. In fact, it has often used its encounters with law-enforcement agencies and upset citizenry to increase its profile in the news media.

SEED BANKS

—*Amy Sisson*

FIELDS OF STUDY

Agronomy; Agriculture; Soil Science; Physical Science; Conservation; Biology; Chemistry; Ecology; Environmental Studies.

SUMMARY

Seed banks are repositories for the genetic material contained within seeds of agricultural and other plant species. Because modern agricultural practices focus on only a limited number of crops, seed banks have been established to help promote biodiversity and protect the world's future food supply by preserving the seeds, or genetic material, of thousands of plant species that might otherwise become extinct. Seed banks also provide a backup of genetic plant material in the event of wars, accidents, or environmental disasters on a local or global scale.

Modern-day seed banks, also referred to as gene banks, are based on an agricultural tradition that has persisted for thousands of years: saving seeds from one season to ensure a supply that can be used to plant crops the next season. Seeds have also been passed down from one generation of farmers to the next, but as agricultural mass production has increased around the world, thousands of plant species have been irrevocably lost simply because no farmers or agricultural corporations chose to grow them over a period of time.

By the early twenty-first century, approximately 1,400 organizations identifying themselves as seed or gene banks were in operation worldwide. These banks can range from small, informal facilities that specialize in certain types of seeds to larger organizations such as the Seed Savers Exchange, a nonprofit organization established in 1975 to preserve not only the genetic but also the cultural and historical heritages of plants. Located on 360 hectares (890 acres) in Decorah, Iowa, the Seed Savers Exchange maintains has more than 13,000 members globally, distributing more than 20,000 varieties of fruit, flower, vegetable, and herb seeds. This group goes beyond the activities traditionally conducted by seed banks in that it makes many uncommon seed varieties available for sale, and it plants its seed stock on a rotating basis in order to generate new seeds and ensure that the plant species remain part of the planet's active ecosystem.

One of the world's most impressive seed banks is the vast Svalbard Global Seed Vault, a facility opened by the Norwegian government in 2008 with assistance from the Global Crop Diversity Trust. Built inside a mountain on an island within the Arctic Circle, this high-tech seed bank has the capacity to store 4.5 million seed samples; it maintains a constant interior climate in spite of temperature variations in the surrounding permafrost. The low temperatures in the vault along with lowered oxygen levels reduce metabolic activity and slow seed aging. Nicknamed the "Doomsday Vault" by the media, it is designed to withstand nuclear warfare and terrorist attacks as well as natural disasters such as flooding, and is therefore considered to be a global "insurance policy" in the event of agricultural disaster. The Svalbard Global Seed Vault is actually a bank for banks; most of its "depositors" are smaller seed banks around the world that are located in areas that are geographically or politically less stable. It thus represents a vital step in efforts to protect the earth's plant biodiversity. The Global Crop Diversity Trust has the further goal of providing funding and other support for multiple crop repositories around the world.

In 2015 researchers were forced to withdraw samples of seeds from the vault originally contributed by a regional seed bank near Alleppo, Syria, to replenish seed losses resulting from the ongoing civil war in the country. In 2016 melting permafrost near the Svalbard Global Seed Vault entrance caused minor seepage flooding at the entrance, but water froze long before entering the seed storage portions of the vault.

FURTHER READING

Fenner, Michael, and Ken Thompson. "Soil Seed Banks." In *The Ecology of Seeds*. New York: Cambridge University Press, 2005.

Fowler, Cary. "The Svalbard Global Seed Vault: Securing the Future of Agriculture." Global Crop Diversity Trust. 2008.

Fowler, Cary. "The Svalbard Seed Vault and Crop Security." *BioScience* 58, no. 3 (2008): 190–191.

Fry, Carolyn, Sue Seddon, and Gail Vines. *The Last Great Plant Hunt: The Story of Kew's Millennium Seed Bank.* Richmond, Surrey: Kew, 2011.

Murphy, Janice. *Seed Banks: Types, Roles and Research.* Hauppauge, New York: Nova Science, 2016.

Rosner, Hillary. "The Gatherers." *Popular Science,* January, 2008, 60–64, 66, 91.

Shea, Neil. "Norway's Ark." *National Geographic,* June, 2007, 14–21.

Vernooy, Ronnie. *Community Seed Banks: Origins, Evolution and Prospects.* London: Routledge, Taylor & Francis, 2015.

SEGMENTED WORMS

—D. R. Gossett

FIELDS OF STUDY

Agronomy; Agriculture; Soil Science; Physical Science; Biology; Chemistry; Ecology; Environmental Studies.

PRINCIPAL TERMS

hermaphroditic: a condition where both male and female reproductive organs are located in the same animal

nephridia: excretory tubules specialized for excretion and osmoregulation, with an external opening and with or without an internal opening

parapodia: paired lateral projections on each side of most segments, variously modified for locomotion, respiration, or feeding

septa: a thin, muscular tissue used to separate segments into a series of ringlike cavities

setae: a needlelike, chitinous structure of the integument, used for locomotion

SEGMENTED WORM FACTS

Segmented worms are a class (Annelida) whose body parts are arranged in a longitudinal series of rings, or segments. Many are hermaphroditic.

CLASSIFICATION

Kingdom: Animalia
Subkingdom: Bilateria
Phylum: Annelida
Classes: Polychaeta (with parapodia and numerous setae), Archiannelida (small marine annelides with simple body), Oligochaeta (parapodia absent and few setae), Hirudinea (leeches, parapodia and setae absent)
Orders: Polychaeta—Errantia (palolo worms), Sedentaria (lugworms); Oligochaeta—Lumbricus (earthworms); Hirudinea—Rhynchobdellida (no jaws, pharynx eversible, colorless blood), Gnathobdellida (three jaws, red blood)

GEOGRAPHICAL LOCATION: Found all over the world

HABITAT: Polychaeta—mostly marine, found near the shore or on the bottom of shallow areas, with a few species in living in brackish water or freshwater; Archiannelida—marine; Oligochaeta—freshwater and terrestrial forms found burrowing in soil or leaf mold; Hirudinea—freshwater and terrestrial environments often found attached to the body of a host

GESTATIONAL PERIOD: Varies among species, but most species lay eggs within a few days after fertilization; eggs usually hatch within a few days to a few weeks after being deposited

LIFE SPAN: Varies among species; can be as short as a year in some polychaetes and up to several years in some earthworms

SPECIAL ANATOMY: Elongated, metameric (segmented) bilateral invertebrates with appendages (parapodia) and chitinous setae in many species but lacking in Hirudinea; possess a true coelom (body cavity lined with epithelial tissue) divided by septa with a closed circulatory system, complete digestive system and an excretory system (nephridia) in each segment; leeches have specially adapted mouth parts for attaching to the body of the host

GENERAL CHARACTERISTICS OF THE ANNELIDS

The Annelida are segmented worms in which the body wall, coelom (body cavity), epidermis, circular muscle, longitudinal muscle, and peritoneum are arranged into a longitudinal series of rings or segments. True segmented animals exhibit metamerism, a repetition of a structure or organ from segment to segment. Each segment has the same fundamental structures as all the others. With the exception of the digestive system, the major organ systems of the annelids are metameric in structure. Young annelids generally have few segments, but as they grow, new segments are formed by the division of the terminal segment. Annelids represent the most highly organized animals capable of complete regeneration.

The mouth lies between the first and second segments and forms one segment called the prostomium. In leeches, the mouth contains suckers for attaching to the body of a host. The brain originates in the prostomium and develops a pair of circumpharyngeal nerve rings that reach around the pharynx to form the ventral nerve cord, which appears as a chain of ganglia, one pair in each segment. In the Polychaetes, a pair of swimming or crawling parapodia are located on most of the segments. Both the Polychaetes and Oligochaetes contain external setae to assist in locomotion.

The annelid body is covered with a thin cuticle. Each segment has a ring of circular and longitudinal muscles that contract to either elongate or shorten the segment. A spacious coelom, divided by septa, lies between the body wall and an internal digestive tract. The coelom is filled with fluid and serves as a hydrostatic skeleton in all annelids except the leeches. The coelom also contains the circulatory and excretory systems. A system of large vessels (hearts) pump blood through a ventral vessel into capillary beds that invade all of the tissues. The blood is returned to the hearts via the dorsal vessel. Each segment, except the first and last, contains a pair of nephridia, which collect wastes and deliver them to the outside.

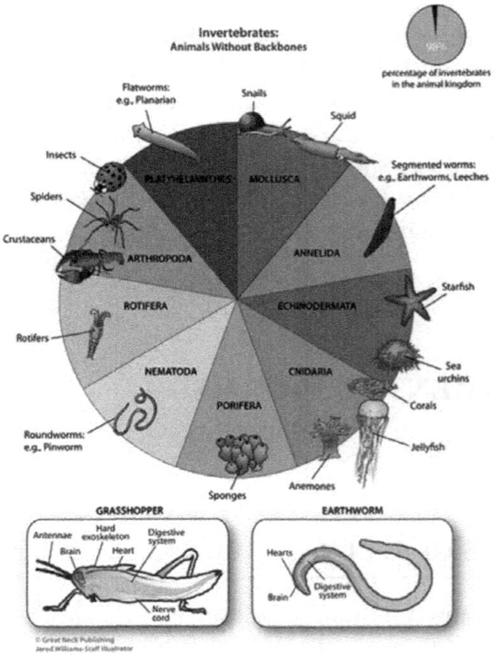

Invertebrates: Invertebrates are animals without backbones. Most animals on Earth are invertebrates. Examples include snails, worms, jellyfish, spiders, and insects. © EBSCO.

REPRODUCTION IN THE ANNELIDS

In many species of polychaete worms, fertilization is external and takes place in the open sea water. The palolo worm provides a good example of polychaete reproduction. Through most of the year, the worms exist as sexually immature animals called atokes, but during the breeding season, the posterior segments develop gonads, and the coelom becomes filled with gametes. On the night of breeding, individuals back out of their holes, and the posterior portion, called an epitoke, breaks free. The epitokes swim to the surface for a few minutes and burst, shedding eggs or sperm and leaving a rapidly disintegrating body. Reproduction in the palolo worms is tied to an annual cycle designating the month, a lunar rhythm designating the day, and a diurnal cycle designating the hour of reproduction. Over 90 percent of the population breeds within a single two-hour period of the entire year.

Oligochaete worms are hermaphroditic, with a pair of testes in the tenth and eleventh segments and a pair of ovaries in the thirteenth segment. During copulation, two worms exchange sperm. Once the eggs are fertilized, the clitellum, a swollen glandular region of the epidermis, secretes a membrane that slips forward along the body so that the eggs are laid directly into it as it passes. Finally, the cocoon slips off the head, and the eggs develop into tiny worms, which later emerge from the cocoon. The reproductive system of leeches is very similar to that of the Oligochaetes.

PRINCIPAL TERMS

hermaphroditic: a condition where both male and female reproductive organs are located in the same animal

nephridia: excretory tubules specialized for excretion and osmoregulation, with an external opening and with or without an internal opening

parapodia: paired lateral projections on each side of most segments, variously modified for locomotion, respiration, or feeding

septa: a thin, muscular tissue used to separate segments into a series of ringlike cavities

setae: a needlelike, chitinous structure of the integument, used for locomotion

FURTHER READING

Conniff, R. "The Little Suckers Have Made a Comeback." *Discover* 8 (1987): 84–94. An interesting article that describes the medical uses of leeches in microsurgery.

Dales, R. P. *Annelids.* 2d ed. London: Hutchinson, 1967. An older but excellent and concise description of the major groups of Annelids.

Hickman, C. P., L. S. Roberts, and A. Larson. *Integrated Principles of Zoology.* Boston: McGraw-Hill, 2001. An introductory zoology text that gives a very good review of the characteristics of flatworms.

Kingman, J., and P. Kingman. "The Dance of the Luminescent Threadworms." *Underwater Naturalist* 22 (1993): 36. An excellent description of the reproductive swarming of these Polychaete worms off the coast of Belize during the full moon of July and August.

Lent, C. M., and M. H. Dickinson. "The Neurobiology of Feeding in Leeches." *Scientific American* 258 (1988): 98–103. A description of how feeding behavior in leeches is controlled by the nervous secretion of serotonin.

SLASH-AND-BURN AGRICULTURE

—*Oghenekome U. Onokpise*

FIELDS OF STUDY

Agronomy; Agriculture; Soil Science; Physical Science; Conservation; Biology; Chemistry; Ecology; Environmental Studies.

SUMMARY

Among the negative environmental impacts of the widespread use of slash-and-burn agriculture are habitat fragmentation, air pollution, and soil erosion.

Slash-and-burn agriculture has been practiced for many centuries among people living in tropical rain forests. Initially, this farming system involved small populations. Therefore, land could be allowed to lie fallow for many years, leading to the full regeneration of secondary forests and hence restoration of the ecosystems. During the second half of the twentieth century, however, several factors led to drastically reduced fallow periods. In some places the use of such fallow systems ended, resulting in the transformation of forests into shrub and grasslands, negative effects on agricultural productivity for small farmers, and disastrous consequences for the environment.

Among the factors responsible for reduced or nonexistent fallow periods during the 1980's and 1990's were increased population in the Tropics, increased demand for wood-based energy, and, perhaps most important, increased worldwide demand for tropical commodities, especially products such as palm oil and natural rubber. These factors helped industrialize slash-and-burn agriculture, which had been practiced for centuries by small farmers. Ordinarily, when small farmers engage in slash-and-burn farming, they are able to control their fires, which might be compared in size to small forest fires triggered by lightning in

Slash and burn shifting cultivation, Thailand. Image by mattmangum CC BY 2.0.

the northwestern or southeastern United States. However, the continued reduction in fallow periods, coupled with increased burning by subsistence farmers and large agribusiness, especially in Asia and Latin America, resulted in increased negative impacts on the environment.

Although slash-and-burn agriculture seldom takes place in temperate regions, some agricultural burning occurs in the Pacific Northwest of the United States, where it has been estimated that three thousand to five thousand agricultural fires are set each year in Washington State alone. These fires also create problems for human health and the environment.

HABITAT FRAGMENTATION

One of the most easily recognizable results of slash-and-burn agriculture is habitat fragmentation, which leads to significant loss of the vegetation needed to maintain effective gaseous exchange in tropical regions and throughout the world. For every hectare of land lost to slash-and-burn agriculture, ten to fifteen times that amount of land is fragmented, resulting in the loss of habitat for wildlife, plant species, and innumerable macro- and microorganisms yet to be identified. This also creates problems for management and wildlife conservation efforts in parts of the world with few or no resources to feed their large populations. The recognition of the problems caused by habitat fragmentation led to intensive discussions on global warming. While slash-and-burn agriculture by itself is not completely responsible for global warming, the industrialization of the process could make it a significant component of the problem, as more and more vegetation is fragmented.

HUMAN HEALTH

The impacts of slash-and-burn agriculture on human health and the environment are best exemplified by the impacts related to the 1997 Asian fires that resulted from such practices. Monsoon rains normally extinguish the fires set by farmers, but a strong El Niño weather phenomenon delayed the expected rains in 1997, and the fires burned out of control for months. Thick smoke caused severe health problems. It has been estimated that more than twenty million people in Indonesia alone were treated for asthma, bronchitis, emphysema, and eye, skin, and cardiovascular problems as a result of the fires. Similar problems have been reported for smaller agricultural fires.

Three major problems are associated with air pollution: particulate matter, pollutant gases, and volatile organic compounds. Particulate compounds of 10 microns or smaller that are inhaled become attached to the alveoli of the lungs and can result in severe illness. Research conducted by the U.S. Environmental Protection Agency (EPA) and the University of Washington found that death rates associated with respiratory illnesses increase when fine-particulate air pollution increases. Pollutant gases such as carbon monoxide, nitric oxide, nitrogen dioxide, and sulfur dioxide become respiratory irritants when they combine with vapor to form acid rain or fog. Until the Asian fires, air pollutants stemming from the small fires of slash-and-burn agriculture that occur every planting season often went unnoticed. Thus it is likely that millions of people in the Tropics experience environmental health problems because of slash-and-burn agriculture that are never reported.

SOIL & WATER QUALITY

The loss of vegetation that follows slash-and-burn agriculture causes an increased level of soil erosion. The soils of the humid tropics create a hard pan underneath a thick layer of organic matter. Therefore, upon the removal of vegetation cover, huge areas of land become exposed to the torrential rainfalls that occur in these regions. The result is severe soil erosion. As evidenced by the impact of Hurricane Mitch on Honduras during 1998, these exposed lands can give rise to large mudslides that can lead to significant loss of life. While slash-and-burn agriculture may not be the ultimate cause of sudden mudslides, it contributes to the problem by predisposing the land to erosion.

Associated with erosion is the impact of slash-and-burn agriculture on water quality. As erosion continues, sedimentation of streams increases. This sedimentation affects stream flow and freshwater discharge for catchment-area populations. Mixed with the sediment are minerals such as phosphorus and nitrogen-related compounds that enhance algal growth in streams and estuaries, which depletes the supply of oxygen that aquatic organisms require to survive. Although fertility is initially increased on noneroded soils, nutrient deposition and migration into drinking-water supplies continues to increase.

CONTROLLING SLASH-AND-BURN AGRICULTURE

Given the fact that slash-and-burn agriculture has significant effects on the environment, not only in regions where it is the mainstay of the agricultural system but also in other regions of the world, environmental activists and scientists have encouraged the exploration of different approaches to agriculture in the regions using the slash-and-burn system. Because slash-and-burn agriculture has evolved to be a socioculturally accepted way of making a livelihood, to be successfully implemented any recommended changes to this practice must be consistent with the ways of life of peoples who have minimal resources for extensive agricultural systems.

Among the alternatives are new agroecosystems, such as agroforestry systems and sustainable agriculture systems, that do not rely so much on the slashing and burning of forestlands. These systems allow for the cultivation of agronomic crops and livestock within forest ecosystems, thus protecting soils from being eroded. Another approach involves the education of small rural farmers, absentee landlords, and big agribusiness concerns in developing countries so that they understand the environmental impacts of slash-and-burn agriculture. Small rural farmers do not have the resources to renovate forestlands that have been slashed and burned for farming, but the big businesses that profit from such farming can organize restoration of the original ecosystems; this kind of restoration has been undertaken in many of the world's developed nations.

Further Reading

Dean, Warren. *With Broadax and Firebrand: The Destruction of the Brazilian Atlantic Forest.* Berkeley: University of California Press, 1997.

Mazoyer, Marcel, and Laurence Roudart. *A History of World Agriculture: From the Neolithic Age to the Current Crisis.* New York: Monthly Review Press, 2006.

Palm, Cheryl A., et al., eds. *Slash-and-Burn Agriculture: The Search for Alternatives.* New York: Columbia University Press, 2005.

Terborgh, J. *Diversity and the Tropical Rain Forest.* New York: Scientific American Library, 1992.

Vandermeer, John H. *The Ecology of Agroecosystems.* Sudbury, Mass.: Jones and Bartlett, 2011.

Soil

—*Judith J. Bradshaw-Rouse*

FIELDS OF STUDY

Agronomy; Agriculture; Soil Science; Physical Science; Conservation; Biology; Chemistry; Ecology; Environmental Studies.

SUMMARY

Soil is a product of the physical and chemical breakdown of the Earth's surface into small fragments, including sand, silt, and clay. Soil is also the product of organic matter decomposition—the composting of dead plant and animal debris. Soils are classified on the basis of soil profile and soil formation. They can be grouped according to a number of characteristics, including agronomic use, color, organic matter content, texture, and moisture condition.

BACKGROUND

Typical soil is about 45 percent minerals and about 5 percent organic matter. The other 50 percent of soil consists of pores that hold either water or air. The liquid portion of soil contains dissolved minerals and organic compounds, produced by plants and microorganisms. The gases found in soil often are the same as those found in the air above it. Soil can support plant life if climate and moisture are suitable. It is a changing and dynamic body, adjusting to conditions of climate, topography, and vegetation. In turn, soil influences plant and root growth, available moisture, and the nutrients available to plants. While "the soil" is a collective term for all soils, "a soil" means one individual soil body with a particular length, depth, and breadth.

SOIL PROFILE

In a typical soil, the top layer is usually dark with decomposing organic matter; the layers below are sand, silt, clay, or some combination of the three. Soil scientists classify soils on the basis of soil profile and soil formation.

Typically the top soil layer is called the O horizon, or organic matter horizon. It has rotten logs, leaf litter, and other recognizable bits of plants and animals. Underneath the O horizon is the A horizon. It is characterized by thoroughly decomposed organic matter. Water passing through the A horizon carries clay particles and organic acids through it into the B horizon. Clay or organic substances passing into the B horizon glue soil particles together, forming soil aggregates. Soil aggregates—granular, columnar, and so on—are indicators of a mature, healthy soil. The lowest level of the soil profile is the C horizon. It contains bedrock or soil parent material that shows little or no evidence of plant growth or soil formation.

SOIL FORMATION

Soil formation takes hundreds, even thousands, of years. Parent material, climate, organisms, topography, and time all contribute. Sources of parent material include igneous, sedimentary, and metamorphic rocks (fragments of which may be deposited by water, wind, and ice), and plant and animal deposits.

Soil formation is the result of the physical, chemical, and biochemical breakdown of parent material. It also reflects the processes of weathering and change within the soil mass. Many substances are added to soil—rain, water from irrigation, nitrogen from bacteria, sediment, salts, organic residues, and a variety of substances created by humans. However, many substances are also removed from the soil—water-soluble minerals, clay, plants, carbon dioxide, and nitrogen. Other transformations also are occurring: Organic matter is decomposing, and minerals are solubilizing and changing chemical form. Clays and soluble salts that move along with the soil water cause color and chemical changes in the soil.

The Twelve Soil Orders in the U.S. Classification System

Soil Order	Features
Alfisols	Soils in humid and subhumid climates with precipitation from 500 to 1,300 millimeters (20 to 50 inches), frequently under forest vegetation. Clay accumulation in the B horizon and available water most of the growing season. Slightly to moderately acid soils.
Andisols	Soils with greater than 60 percent volcanic ash, cinders, pumice, and basalt. They have a dark A horizon as well as high absorption and immobilization of phosphorus and very high cation exchange capacity.
Aridisols	Soils that exist in dry climates. Some have horizons of lime or gypsum accumulations, salty layers, and A and slight B horizon development.
Entisols	Soils with no profile development except a shallow A horizon. Many recent river floodplains, volcanic ash deposits, severely eroded areas, and sand are entisols.
Gelisols	Soils that commonly have a dark organic surface layer and mineral layers underlain by permafrost, which forms a barrier to downward movement of soil solution. Common in tundra regions of Alaska. Alternate thawing and freezing of ice layers results in special features in the soil; slow decomposition of the organic matter due to cold temperatures results in a peat layer at the surface in many gelisols.
Histosols	Organic soils of variable depths of accumulated plant remains in bogs, marshes, and swamps.
Inceptisols	Soils found in humid climates that have weak to moderate horizon development. Horizon development may have been delayed because of cold climate or waterlogging.
Mollisols	Mostly grassland soils, but with some broadleaf forest-covered soils with relatively deep, dark A horizons, a possible B horizon, and lime accumulation.
Oxisols	Excessively weathered soils. Oxisols are over 3 meters (10 feet) deep, have low fertility, have dominantly iron and aluminum oxide clays, and are acid. Oxisols are found in tropical and subtropical climates.
Spodosols	Sandy leached soils of the cool coniferous forests, usually with an organic or O horizon and a strongly acidic profile. The distinguishing feature of spodosols is a B horizon with accumulated organic matter plus iron and aluminum oxides.
Ultisols	Strongly acid and severely weathered soils of tropical and subtropical climates. They have clay accumulation in the B horizon.
Vertisols	Soils with a high clay content that swell when wet and crack when dry. Vertisols exist in temperate and tropical climates with distinct dry and wet seasons. Usually vertisols have only a deep self-mixing A horizon. When the topsoil is dry, it falls into the cracks, mixing the soil to the depth of the cracks.

The Twelve Soil Orders in the U.S. Classification System.

Parent material is a primary determinant of soil type or soil classification. All soils at the lowest category of soil classification are distinct if the parent material differs. The differences in parent materials—weathering rates, the plant nutrient content, and soil texture resulting from parent material breakdown—contribute to the formation of distinctive soils. For example, sandstone yields sandy soil with low fertility.

EFFECTS OF CLIMATE

Soils slowly change color and density as a result of wetting and drying, warming and cooling, and freezing and thawing. During weathering—the rubbing, grinding, and moving of rocks by water, wind, and gravity—rocks are split into smaller and smaller fragments. Soil is composed of fragments 2 millimeters or less in diameter.

The expansion force of water as it freezes is sufficient to split minerals. However, water also is involved in chemical weathering—solution, hydrolysis, carbonation, reduction, oxidation, and hydration. A simple example of solution, the dissolving of minerals in liquid, is the dissolving of salt in water. The salts then move along with the liquid. In hot arid climates, salts can move to the surface as water evaporates, creating salt flats. In wetter climates, salts can move through the soil, depleting it of necessary plant nutrients and contaminating groundwater.

Hydrolysis is the splitting of a water molecule to form hydroxides and soluble hydroxide compounds, such as sodium hydroxide. Hydration is the addition of water to minerals in rock. When a mineral such as hematite (an oxide of iron) hydrates, it expands, softens, and changes color. Carbonation is the reaction of a compound with carbonic acid, a weak acid produced when carbon dioxide dissolves in water. Water often contains carbonic acid and other organic acids produced by organic matter decomposition. These acids increase the power of the water to disintegrate rock. Oxidation is the addition of oxygen to a mineral, and reduction is the removal of oxygen from a mineral.

Climate also influences soil formation indirectly through its action on vegetation. Soils in arid climates have sparse vegetation, less organic matter, and little soil profile development. Wet soil, however, usually has thick vegetation and high organic matter, and therefore a deep soil profile.

BIOLOGICAL WEATHERING

Biological weathering is a combination of physical and chemical disintegration of rocks to produce soil. The roots of plants can crack rocks and break them apart. Plant roots also produce carbon dioxide, which combines with water to produce carbonic acid. Carbonic acid dissolves certain minerals, speeding the breakdown of parent material and chemically changing the soil.

Plants and animals also add humus (organic matter) to soil, increasing its fertility and water-holding capacity and speeding rock weathering. Animals such as earthworms, ants, prairie dogs, gophers, and moles also contribute to soil aeration and fertility by mixing the soil. In areas where animal populations are large, they can influence both the formation and destruction of soil.

TOPOGRAPHY

The shape of the land is referred to as its topography. Each landform—valleys, plains, hills, and mountains—is covered with a crazy quilt of different soil types. For example, the steep sides of the Sandia Mountains near Albuquerque, New Mexico, which are severely eroded by wind and summer rains, contain a variety of soil types—forest soils, sandy soils, and rocky soils. Sand, silt, and clay eroded from the mountains and nearby extinct volcanoes combine in the moist and fertile Rio Grande Valley. The valley has deep sandy soils, layered sand and clay soils, and soils eroded by flash floods.

Soils located in similar climates that develop from similar parent material on steep hillsides usually have thin A and B horizons because less water moves through the soil. Similar materials on shallower slopes allow more water to pass through them. Topography and climate work together either to allow or to prohibit plant growth and organic matter deposition. Without moisture, plants cannot grow to impede soil erosion, and soil development is slow. With moisture, plants can grow, hold the soil in place, add organic matter to the soil, and speed soil development.

The age of a soil may be reckoned in tens, hundreds, or thousands of years. Under ideal conditions, a soil profile may develop in two hundred years; however, under less favorable conditions soil development may take several thousand years.

SOIL CLASSIFICATION

Scientists have identified and classified soils for hundreds of years. Soils can be grouped according to agronomic use, color, organic matter content, texture, moisture condition, and other characteristics. Each of these groupings serves a particular purpose. U.S. soil scientists adopted a system of soil classification on January 1, 1965, that was based on the knowledge they had about soil genesis, morphology, and classification. The U.S. system is divided into six categories: order, suborder, great group, subgroup, family, and series. (Soil taxonomy is patterned after the worldwide system of plant and animal taxonomy, which contains phylum, class, order, family, genus, and species.)

Changes to the system have proceeded through a number of major revisions or approximations. The system can be used to classify soils anywhere in the world. Soil classification is based on similar physical, chemical, and mineralogical properties. The minimum volume of soil that scientists consider when they classify a soil is the pedon, which can range from 1 to 10 meters square and is as deep as roots extend into a soil.

The U.S. soil classification system recognizes twelve soil orders. The differences among orders reflect the dominant soil-forming processes and the degree of soil formation. Each order is identified by a word ending in "-sol." Each order is divided into suborders, primarily on the basis of properties that influence soil genesis, are important to plant growth, and reflect the most important variables within the orders. The last syllable in the name of a suborder indicates the order. An example is "aquent," meaning water, plus "-ent," from "entisol." Suborders are distinctive to each order and are not interchangeable between orders.

Each suborder is divided into great groups on the basis of additional soil properties and horizons resulting from differences in soil moisture and soil temperature. Great groups are denoted by a prefix that indicates a property of the soil. An example is "psammaquents" ("psamm" referring to sandy texture and "aquent" being the suborder of the entisols that has an aquic moisture regime). Soil scientists have identified about 190 great groups in the United States. Great groups are distinguished on the basis of differing horizons and soil features. The differing soil horizons include those with accumulated clay, iron, or organic matter and those hardened or cemented by soil cultivation or other human activities. The differentiating soil features include self-mixing of soil because of clay content, soil temperature, and differences in content of calcium, magnesium, sodium, potassium, gypsum, and other salts.

Each great group is divided into three subgroups: a typic subgroup, an intergrade subgroup, and an extragrade subgroup. The typic subgroup represents the central spectrum of a soil group. The intergrade subgroup represents soils with properties like those of other orders, suborders, or great groups. The extragrade subgroup represents soils with some properties that are not representative of the great group but do not indicate transitions to any other known kind of soil. Each subgroup is identified by one or more adjectives preceding the name of the great group.

Families are established within a subgroup on the basis of physical and chemical properties and other characteristics that are important to plant growth or that are related to the behavior of soils that are important for engineering concerns. Among the properties and characteristics considered are particle size, mineral content, temperature regime, depth of the root zone, moisture, slope, and permanent cracks. A family name consists of the name of a subgroup preceded by terms that indicate soil properties. About forty-five hundred families have been identified in the United States.

SOIL TEXTURE, STRUCTURE & CONSISTENCY

Soil texture is determined by the percent of sand, silt, and clay in a soil sample. Most fertile or productive soils have a loam texture, or about equal amounts of sand, silt, and clay, and a high organic matter content (about 5 to 10 percent). Soil texture determines the water-holding and nutrient-holding capacity of a soil. Thus, clay soils have a high nutrient-holding capacity, but they waterlog easily. Sandy soils have a lower nutrient-holding capacity but dry out easily. Farmers base their plans of how to fertilize and irrigate their crops partly on the texture of the soil.

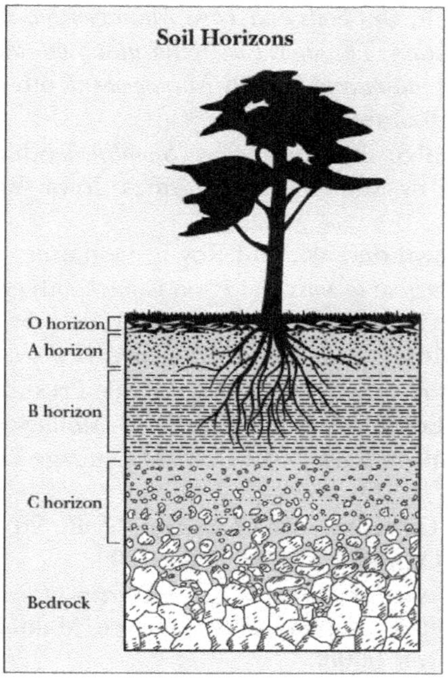

Soil Horizons.

Soil structure refers to how soil particles are glued together to form aggregates. During soil formation, soil particles are glued together with clay, dead microorganisms, earthworm slime, and plant roots, and they form air and water channels. Plants need these channels so they can absorb nutrients, water, and air. Soil structure may be destroyed when farmers cultivate wet or waterlogged soils with heavy farm machinery; destroying soil structure makes a soil unsuitable for plant growth.

Soil consistency is the "feel" of a soil and the ease with which a lump can be crushed in one's fingers. Common soil consistencies are loose, friable, firm, plastic, sticky, hard, and soft. Clay soils, for example, are sticky or plastic when they are wet, but they become hard or harsh when they are dry. The best time to work a clay soil is when it is soft or friable. Sandy soils, on the other hand, do not become plastic or sticky when they are wet or hard or harsh when they are dry. They have a tendency to stay loose, which makes them easier to work. Loam and silt loam soils are intermediate in behavior. When farmers are trying to determine whether to work the soil or wait for better soil moisture conditions, they usually check the soil consistency.

SOIL AERATION & SOIL MOISTURE

Soil aeration relates to the exchange of soil air with atmospheric air. Growing roots need oxygen and are constantly expiring carbon dioxide. Unless there is a continuous flow of oxygen into soil and carbon dioxide out of the soil, oxygen becomes depleted. When their oxygen supply is cut off, the roots will die.

Soil moisture refers to water held in soil pores. A plant draws water from soil the same way a child draws water from a cup with a straw. When the cup is full, it is easy for the child to draw up the water, but as the cup empties, the child must work harder to get water. Similarly, plants draw water from soil easily when the soil has plenty of water. As the soil dries, however, plants must work harder to pull water out of the soil until they reach a wilting point.

SOIL FERTILITY

Plants absorb many of the nutrients they need from soil, including phosphorus, potassium, calcium, magnesium, sulfur, boron, chlorine, cobalt, copper, iron, manganese, molybdenum, and zinc. They may obtain carbon, hydrogen, and nitrogen from the air and water.

Soil testing services give farmers specific fertilizer and lime recommendations based on soil texture and chemical analysis. Farmers use soil tests to determine if their soil has enough essential nutrients for a crop to grow. The absence of one essential nutrient can limit overall crop growth. Nitrogen, phosphorus, and potassium are commonly applied to the soil as commercial fertilizer and manure. Calcium and magnesium are applied as lime, which is also used to reduce the acidity of soil and to increase the solubility of some minerals. Manure and other organic matter added to soils increase water-holding and nutrient-holding capacity and therefore boost crop yields.

Agricultural extension services offer guidelines for the maximum amounts of manure, sewage sludge, fertilizer, and other chemicals that farmers should apply to soils. Farmers are encouraged to apply nitrogen fertilizer in small applications at times when plants are growing rapidly. This soil management practice decreases deep percolation losses that could pollute groundwater.

With an understanding of soil characteristics, farmers and gardeners can learn to manage a wide variety of soils. Some soils are naturally fertile and need few amendments to promote high crop yields. Other soils, whether because of their parent material or climate, are naturally infertile and might best be used for purposes other than agriculture. Like the water and the air, the soil is a crucial natural resource. From an airplane, all soils look about the same, but from an ant's view, soils are all different. Differences in soil type make huge differences to plants, animals, humans, and the environment.

Further Reading

Ashman, M. R., and G. Puri. *Essential Soil Science: A Clear and Concise Introduction to Soil Science.* Malden, Mass.: Blackwell, 2002.

Brady, Nyle C., and Ray R. Weil. *The Nature and Properties of Soils.* 14th ed. Upper Saddle River, N.J.: Prentice Hall, 2008.

Davies, Bryan, David Eagle, and Bryan Finney. *Resource Management: Soil.* Brighton, Ont.: Diamond Farm Book, 2001.

Donahue, Roy L., Roy Hunter Follett, and Rodney W. Tulloch. *Our Soils and Their Management: Increasing Production Through Environmental Soil and Water Conservation and Fertility Management.* 6th ed. Danville, Ill.: Interstate, 1990.

Esch, Neal S., et al. *Soil Science Simplified.* 5th ed. Illustrated by Mary C. Bratz. Ames, Iowa: Blackwell, 2008.

Miller, Raymond W., and Roy L. Donahue. *Soils: An Introduction to Soils and Plant Growth.* 6th ed. Englewood Cliffs, N.J.: Prentice-Hall, 1990.

Paul, Eldor A., ed. *Soil Microbiology, Ecology, and Biochemistry.* 3d ed. Boston: Academic Press, 2007.

Plaster, Edward J. *Soil Science and Management.* 5th ed. Clifton Park, N.Y.: Delmar Cengage Learning, 2009.

Sposito, Garrison. *The Chemistry of Soils.* 2d ed. New York: Oxford University Press, 2008.

White, Robert E. *Principles and Practice of Soil Science: The Soil as a Natural Resource.* 4th ed. Malden, Mass.: Blackwell, 2006.

Soil Chemistry

—Richard W. Arnseth

FIELDS OF STUDY

Agronomy; Agriculture; Soil Science; Physical Science; Conservation; Biology; Chemistry; Ecology; Environmental Studies.

PRINCIPAL TERMS

cation: an element that has lost one or more electrons such that the atom carries a positive charge

clay mineral: a group of minerals, commonly the result of weathering reactions, composed of sheets of silicon and aluminum atoms

hydroxl: a combination of an oxygen and hydrogen atom, which behaves as a single unit with a negative charge

lysimeter: a simple pan or porous cup that is inserted into the soil to collect soil water for analysis

stable isotopes: atoms of the same element that differ by the number of neutrons in their nuclei yet are not radioactive

weathering: reactions between water and rock minerals at or near the earth's surface that result in the rock minerals being altered to a new form that is more stable

SUMMARY

Soils are complex chemical factories. Regardless of the type of soil, chemical processes such as plant growth, organic decay, mineral weathering, and water purification are ongoing processes.

Soils are complex chemical factories. Regardless of the type of soil, chemical processes such as plant growth, organic decay, mineral weathering, and water purification are ongoing processes.

WEATHERING PROCESS

Soil chemistry has been studied as long as there has been sustainable agriculture. Although they did not recognize it as such, those first successful farmers who plowed under plant stalks, cover crops, or animal wastes were actively managing the soil chemistry of their fields. These early farmers knew that to have productive farms in one location season after season, they had to return something to the soil. It is now understood that soil chemistry is a complex of chemical and biochemical reactions. The most obvious result of this complex of reactions is that some soils are very fertile whereas other soils are not. Soil itself is a unique environment because all of the "spheres"—the atmosphere, hydrosphere, geosphere, and biosphere—are intimately mixed there. For this reason, soil and soil chemistry are extremely important.

Soil chemistry begins with rock weathering. The minerals making up a rock exposed at the earth's surface are continually bathed in a shower of mildly acidic rain—not polluted rainwater but naturally occurring acid rain. Each rain droplet forming in the atmosphere absorbs a small amount of carbon dioxide gas. Some of the dissolved carbon dioxide reacts with the water to form a dilute solution of carbonic acid. A more concentrated solution of carbonic acid is found in any bottle of sparkling water.

Most of the common rock-forming minerals, such as feldspar, will react slowly with rainwater. Some of the chemical elements of the mineral, such as sodium, potassium, calcium, and magnesium, are very soluble in rainwater and are carried away with the water as it moves over the rock surface. Other chemical elements of the mineral, such as aluminum, silicon, and iron, are much less soluble. Some of these elements are dissolved in the water and carried away; most, however, remain near the original weathering, where they recombine into new, more resistant minerals. Many of the new minerals are of a type called clays.

Clay minerals tend to be very small crystals composed of layers of aluminum and silicon combined with oxygen or hydroxyl ions. Between the layers of aluminum and silicon atoms are positively charged ions (cations) of sodium, potassium, calcium, and magnesium. The cations hold the layers of some clays together by electrostatic attraction. In most cases, the interlayer cations are not held very tightly. They can migrate out of the clay and into the water surrounding the clay mineral, to be replaced by another cation from the soil solution. This phenomenon is called cation exchange.

The weathering reactions between rainwater and rock minerals produce a thin mantle of clay mineral soil. The depth to fresh, unweathered rock is not great at first, but rainwater continues to fall, percolating through the thin soil and reacting with fresh rock minerals. In this way, the weathering front (the line between weathered minerals and fresh rock) penetrates farther into the rock, and the overlying soil gets thicker.

BIOLOGICAL PROCESSES

Throughout the weathering process, biological processes contribute to the pace of soil formation. In the very early stages, lichens and fungi are attached to what appear to be bare rock surfaces. In reality, they are using their own acids to "digest" the rock minerals. They absorb the elements of the mineral they need, simultaneously extending fine filaments into the rock for attachment, and the remainder is left to form soil minerals. As the soil gets thicker, larger plants and animals begin to colonize it. Large plants send roots down into the soil looking for water and nutrients. Some of the necessary nutrients, such as potassium, are available as exchangeable cations on soil clays or in the form of deeper, unweathered minerals. In either case, the plant obtains the nutrients by using its own weathering reaction carried on through its roots. The nutrient elements are removed from minerals and become part of the growing plant's tissue.

Without a way to replenish the nutrients in the soil, the uptake of nutrients by plants will eventually deplete the fertility of the soil. Nutrients are returned to the soil through the death and decay of plants. Microorganisms in the soil, such as bacteria and fungi, speed up the decay. Since the bulk of the decaying plant material is found at the surface (the dead plant's roots also decay), most of the nutrients are released to the surface layer of the soil. Some of the nutrients are carried down to roots deep in the soil by infiltrating rainwater. Most of the nutrients, however, are removed from the water by the shallow root systems of smaller plants. The deeper roots of typically large plants can mine the untapped nutrients at the deep, relatively unweathered soil-rock boundary.

The soil and its soil chemistry are now well established, with plants growing on the surface and their roots reaching toward mineral nutrients at depth. Water is flowing through the soil, carrying dissolved nutrients and the soluble by-products of weathering reactions. Yet the soil continues to evolve even after it is well established. Eventually, when the downward migration of the weathering front matches the rate at which soil is eroded from the surface, the soil reaches its maximum evolution. In tropical and subtropical climates, the end stage of soil evolution is very deep soil in which most of the nutrients are gone. Virtually all nutrients are contained in the vegetation, and people often assume, falsely, that such soils are extremely productive.

STUDY OF SOIL WATER

The study of soil chemistry is concerned with the composition of soil water and how that composition changes as the water interacts with soil atmosphere, minerals, plants, and animals. Soil and its chemistry can be studied in its natural environment, or samples can be brought into the laboratory for testing. Some tests have been standardized and are best conducted in the laboratory so that they can be compared with the results of other researchers. Most standardized tests, such as measurements of the soil's acidity and cation-exchange capacity, are related to measurements of the soil's fertility and its overall suitability for plant growth. These tests measure average values for a soil sample because large original samples are dried and thoroughly mixed before smaller samples are taken for the specific test.

Increasingly, soil chemists are looking for ways to study the fine details of soil chemical processes. They know, for example, that soil water chemistry changes as the water percolates through succeeding layers of the soil. The water flowing through the soil during a rainstorm has a different chemical composition from that of water clinging to soil particles, at the same depth, several days later. Finally, during a rainstorm, the water flowing through large cracks in the soil has a chemical composition different from that of the same rainwater flowing through the tiny spaces between soil particles.

SAMPLING TECHNIQUES

Soil chemists use several sampling techniques to collect the different types of soil water. During a rainstorm, water flows under the influence of gravity. After digging a trench in the area of interest, researchers push several sheets of metal or plastic, called pan lysimeters, into the wall of the trench at specified depths below the surface. The pans have a very shallow V shape. Soil water flowing through the soil collects in the pan, flows toward the bottom of the V, and flows out of the pan into a collection bottle. Comparing the chemical compositions of rainwater that has passed through different thicknesses of soils (marked by the depth of each pan) allows the soil chemist to identify specific soil reactions with specific depths.

After the soil water stops flowing, water is still trapped in the soil. The soil water clings to soil particles and is said to be held by tension. Tension water can spend a long time in the soil between rainstorms. During that time, it reacts with soil mineral grains and soil microorganisms. Tension water is sampled by placing another type of lysimeter, a tension lysimeter, into the soil at a known depth. A tension lysimeter is like the nozzle of a vacuum cleaner with a filter over the opening. Soil chemists actually vacuum the tension water out of the soil and to the surface for analysis.

DETERMINATION OF ISOTOPIC COMPOSITION

Nonradioactive, stable isotopes of common elements are being used more often by soil chemists to trace both the movement of water through the soil and the chemical reactions that change the composition of the water. Trace stable isotopes behave chemically just the way their more common counterparts do. For example, deuterium, an isotope of hydrogen, substitutes for hydrogen in the water molecule and allows the soil chemist to follow the water's movements. Similarly, carbon-13 and nitrogen-15 are relatively uncommon isotopes of biologically important common elements. Using these isotopes, soil chemists can study the influences of soil organisms on the composition of soil water. Depending on what the soil chemist is studying, the isotope may be added, or spiked, to the soil in the laboratory or in the field. Spiking allows the movement of the isotope to be tracked by following the unusual concentration of the isotope. Alternatively, naturally occurring concentrations of the isotope in rain or snowmelt may be used. Regardless, soil water samples are collected by one or more of the lysimeter methods, and their isotopic composition is determined.

THE SOIL CHEMICAL FACTORY

The wonderful interactions of complex chemical and biochemical reactions that are soil chemistry are one indication of the uniqueness of planet Earth. Without the interaction of liquid water and the gases in the atmosphere, many of the nutrients necessary for life would have remained locked up in rock minerals. Thanks to weathering reactions, the soil chemical factory started to produce nutrients, which resulted in the exploitation of the soil environment by millions of organisms. The processes involved in soil chemistry—from weathering reactions that turn rock into new soil to the recycling of plant nutrients through microbial decay—are vital to every human being. Without fertile soil, plants will not grow. Without plants as a source of oxygen and food, there would be no animal life.

Because of the complex chemical interrelationships that have developed in the soil environment, it may seem that nothing can disrupt the "factory" operation. As more is understood about soil chemistry and the ways in which human activities stress soil chemistry, it is apparent that the factory is fragile. Not only do humans rely on soil fertility for their very existence, but they also are taking advantage of soil chemical processes to help them survive their own past mistakes. Soil has been and continues to be used as a garbage filter. Garbage, whether solid or liquid, has been dumped on or buried in soil for ages. Natural chemical processes broke down the garbage into simpler, less toxic or unsanitary forms and recycled the nutrients. When garbage began to contain toxic chemicals, those chemicals, when in small quantities, were either destroyed by soil bacteria or firmly attached to soil particles. The result is that water—percolating through garbage, on its way to the local groundwater, stream, or lake—does not carry with it as much contamination as one might expect. Soil chemistry has, so far, kept contaminated garbage from ruining drinking water in many cases. There are well-known cases, however, where the volume and composition of waste buried or spilled were such that the local soil chemistry was overwhelmed. In cases of large industrial spills, or when artificial chemicals are spilled or buried, the soil needs help to recover. The recovery efforts are usually very expensive but, faced with the possible long-term loss of large parts of the soil chemical factory, humankind cannot afford to neglect this aspect of the environment.

FURTHER READING

Albarede, Francis. *Geochemistry: An Introduction.* 2d ed. Cambridge, Mass.: Cambridge University Press, 2009. An introduction for students looking to gain knowledge in geochemistry. Covers basic topics in physics and chemistry, including isotopes, fractionation, geochemical cycles, and the geochemistry of select elements.

Berner, Elizabeth K., and Robert A. Berner. *The Global Water Cycle: Geochemistry and Environment.* Englewood Cliffs, N.J.: Prentice-Hall, 1987. Designed to teach environmental geochemistry to college students. Covers the role of water in soil chemistry. Emphasizes the hydrologic cycle. Discusses the unique properties of water and the water cycle. Features chapters on rainwater and soil water. Includes an extensive bibliography of more technical references.

Bohn, Heinrich L., Rick A. Myer, and George A. O'Connor. *Soil Chemistry.* 3d ed. New York: John Wiley & Sons, 2001. Fairly widely used college text on soil chemistry. Introductory chapter and chapter on weathering are accessible to any reader with a college background. Provides an annotated bibliography at the end of each chapter for further reading.

Brill, Winston. "Agricultural Microbiology." *Scientific American* 245 (September, 1981): 198. Discusses the link between soil chemistry and microbiological processes. Emphasizes nitrogen biochemistry and genetic engineering. Several figures nicely illustrate the complexity and cyclic nature of soil nitrogen chemistry. Appropriate for those with a background in the terminology of microbiology.

Evangelou, V. P. *Environmental Soil and Water Chemistry: Principles and Applications.* New York: Wiley, 1998. Examines soil and water chemistry, the pollution levels of water and soils, and the chemical processes used to determine those levels.

Lloyd, G. B. *Don't Call It Dirt.* Ontario, Calif.: Bookworm Publishing, 1976. Provides a backyard gardener's point of view on soil fertility, microorganisms, and the interrelationships between soil moisture and soil processes. Information is pertinent and accessible to the general reader.

McBride, Murray B. *Environmental Chemistry of Soils.* New York: Oxford University Press, 1994. Focuses on the chemical make-up and properties of soils in different regions. Written for the college-level reader.

Millot, Georges. "Clay." *Scientific American* 240 (April, 1979): 108. Emphasizes the chemistry and industrial uses of clay minerals. Offers a brief discussion of weathering as the source of clays and as part of the geologic cycling of clays. Discusses how weathering reactions vary in different climates and yield different clay minerals in the local soils. Provides several spectacular scanning electron microphotographs of representative clay types. Accessible to the general reader.

Randolph, John. *Environmental Land Use Planning and Management.* Washington, D.C.: Island Press, 2004. Describes basic principles and strategies of land-use planning and management. Discusses various land features, types, and environmental issues, such as soils, wetlands, forests, groundwater, biodiversity, and runoff pollution. Provides case studies and specific examples.

Sparks, Donald S., ed. *Soil Physical Chemistry.* 2d ed. Boca Raton, Fla.: CRC Press, 1999. Introduction to soil chemistry. Offers a clear look into the practices, procedures, and applications of the field. Appropriate for the careful reader.

Sposito, Garrison. *The Chemistry of Soils.* 2d ed. New York: Oxford University Press, 2008. Provides a comprehensive discussion of soil composition. Describes chemical makeup and basic soil components. Discusses physical and chemical characteristics and processes within soil types. Includes a list of references for further reading.

Storer, Donald A. *The Chemistry of Soil Analysis.* Middletown, Ohio: Terrific Science Press, 2005. Contains valuable information on common tests conducted in soil chemistry. Designed for practical use by students, environmental laboratory technicians, agronomists, archaeologists, and engineers. Describes specific analyses and techniques in a manner accessible to those with a limited background in geology.

Tan, Kim Howard. *Principles of Soil Chemistry.* 4th ed. Boca Raton, Fla.: CRC Press, 2011. Expanded in this third edition. Written as an introduction to soil chemistry; explains the processes and application of the science in an easy-to-understand manner, accompanied by plenty of illustrations. Includes a twenty-five-page bibliography, as well as an index.

Walther, John Victor. *Essentials of Geochemistry.* 2d ed. Burlington, Mass.: Jones & Bartlett Publishers, 2008. Contains chapters on radioisotope and stable isotope dating and radioactive decay. Geared more toward geology and geophysics than chemistry; provides content on thermodynamics, soil formation, and chemical kinetics.

SOIL CONSERVATION

—*Gordon Neal Diem*

FIELDS OF STUDY

Agronomy; Agriculture; Soil Science; Physical Science; Conservation; Biology; Chemistry; Ecology; Environmental Studies.

SUMMARY

By using agricultural methods that protect soil from degradation, farmers can reduce erosion, prevent soil particles from contributing to air and water pollution, and improve crop production.

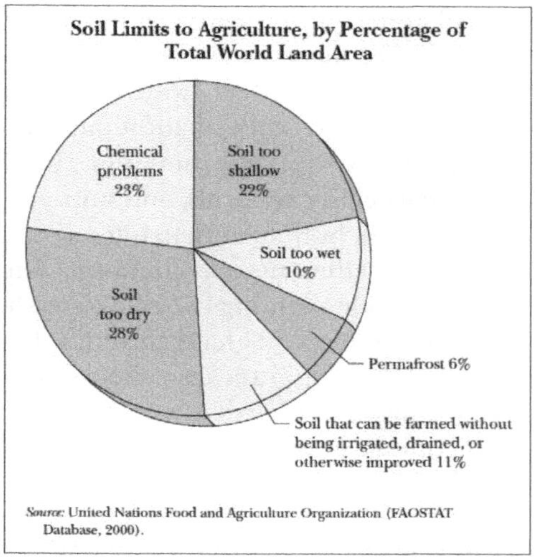

United Nations Food and Agriculture Organization (FAOSTAT Database, 2000). Image by United Nations Food and Agriculture Organization.

According to the United Nations Environment Programme, by 2014 approximately 23 percent of the earth's usable land was degraded, a situation that poses a threat to agricultural production around the world. The introduction of minerals, metals, nutrients, fertilizers, pesticides, bacteria, and pathogens suspended in topsoil runoff into waterways is a significant source of water pollution and is a threat to fisheries, wildlife habitats, and drinking-water supplies. The introduction of soil particles into the air through wind erosion is a significant source of air pollution. Soil conservation is the effort by farmers and other land users to prevent the loss of topsoil from wind erosion, water erosion, desertification, and chemical deterioration such as the buildup of salts and fertilizer acids.

The Industrial Revolution of the nineteenth century and the population explosion of the twentieth century encouraged people to till new land, cut down forests, and disturb soil for the expansion of towns and cities. The newly exposed topsoil quickly succumbed to erosion from rainfall, floods, wind, ice, and snow. The Dust Bowl, which occurred in the Great Plains in the United States during the 1930's, is one example of the devastating effects of wind erosion.

Hugh Hammond Bennett, often called the father of soil conservation, lobbied for congressional establishment of the Soil Erosion Service, which was formed in the U.S. Department of the Interior in 1933, and the establishment of voluntary Soil Conservation Districts in each state. Bennett was named the first chief of the renamed Soil Conservation Service, now part of the Department of Agriculture, in 1937 (in 1994 the name of the agency changed again, to the Natural Resources Conservation Service). On August 4, 1937, the Brown Creek Conservation District in Bennett's home county, Anson County, North Carolina, became the first Soil Conservation District in the United States. Local landowners voted to establish the district by three hundred to one, proving that farmers were concerned about soil conservation. A reporter for the *Charlotte Observer* newspaper sought out the one negative voter, and after the program was explained to him, he changed his opinion. By 1948 more than 2,100 districts had been established nationwide; this number grew to 3,000 by 2012. The districts were eventually renamed Soil and Water Conservation Districts.

The Food Security Act of 1985 authorized the Conservation Reserve Program to take out of production any land deemed to be highly susceptible to erosion; it also required farmers to develop soil conservation plans for any remaining susceptible land. The Natural Resources Conservation Service estimated that with such soil conservation measures, the loss of topsoil in the United States was cut nearly in half, reduced from 1.6 billion tons per year to 0.9 billion tons. The European Community and Australia also adopted soil conservation measures during the 1990's.

Soil conservation practices include covering the soil with vegetation, reducing soil exposure on tilled land, creating wind and water barriers, and installing buffers. Vegetative cover slows the wind at ground level, slows water runoff, protects soil particles from being detached, and traps blowing or floating soil particles, chemicals, and nutrients. Because the greatest wind and water erosion damage often occurs during seasons in which no crops are growing or natural vegetation is dormant, soil conservation often depends on permitting the dead residues and standing stubble of the previous crop to remain in place until the next planting time. In forested areas, annual tree foliage loss serves as a natural ground mulch. Farmers can also reduce erosion by planting grass or legume cover crops until the next planting season for their primary crops or as part of a crop rotation cycle or no-till planting system.

Modern no-till and mulch-till planting systems reduce soil exposure to wind and rain. No-till systems leave the soil cover undisturbed before planting; crop seeds are inserted into the ground through narrow slots in the soil. In mulch-till planting, a high percentage of the dead residues of previous crops are retained on the soil's surface when a new crop is planted.

The ways in which crops are planted can also help to reduce erosion. Row crops can be planted at right angles to the prevailing winds and to the slope of the land in order to absorb wind and rainwater runoff energy and trap moving soil particles. Crops may be planted in small fields to prevent the avalanching caused by an increase in the amount of soil particles transported by wind or water as the distance across bare soil increases. As the amount of soil moved by wind or water increases, the erosive effects of the wind and water also increase. Smaller fields reduce the length and width of unprotected areas of soil.

Wind and water barriers include tree plantings and crosswind strips of perennial shrubs and tall grasses, which act as windbreaks, slowing wind speeds at the surface of the soil. The areas protected by such windbreaks extend for ten times the height of the barriers. In alley cropping, which is used in areas of sustained high winds, crops are planted between rows of larger mature trees. Contour strip farming on slopes, planting grass waterways in areas where rainwater runoff concentrates, and planting grass field borders 3 meters (10 feet) wide on all edges of cultivated or disturbed soil are additional methods for reducing wind speed and rainwater runoff and trapping soil particles, chemicals, and nutrients.

Soil conservation buffers work to filter agricultural runoff to remove sediments and chemicals. Riparian buffers are waterside plantings of trees, shrubs, and grasses, usually 6 meters (20 feet) in width. Riparian buffers planted only in grass are called filter strips. Grassed waterways, field borders, water containment ponds, and contour grass strips are other types of soil conservation buffers.

FURTHER READING

Blanco, Humberto, and Rattan Lal. *Principles of Soil Conservation and Management.* New York: Springer, 2008.

Field, Harry L., and John B. Solie. "Erosion and Erosion Control." In *Introduction to Agricultural Engineering Technology: A Problem Solving Approach.* 3d ed. New York: Springer, 2007.

Plaster, Edward. *Soil Science and Management.* 5th ed. Clifton Park, N.Y.: Delmar, 2008.

Schwab, Glen, et al. *Soil and Water Conservation Engineering.* 5th ed. Clifton Park, N.Y.: Delmar, 2005.

SOIL CONTAMINATION

—John P. DiVincenzo

FIELDS OF STUDY

Agronomy; Agriculture; Soil Science; Physical Science; Biology; Chemistry; Ecology; Environmental Studies.

SUMMARY

Soils contaminated with high concentrations of hazardous substances pose potential risks to human health and the earth's thin layer of productive soil.

To be productive, soil depends on bacteria, fungi, and other microbes to break down wastes and release and cycle nutrients that are essential to plants. Healthy soil is essential for humankind's ability to grow enough food for the world's increasing population. Soil also serves as both a filter and a buffer between human activities and natural water resources, which ultimately serve as the primary source of drinking water. Soil that is contaminated may cause water pollution through the leaching of contaminants into groundwater and through runoff into surface waters such as lakes, rivers, and streams.

The U.S. government has tried to address the problem of soil contamination by passing two landmark legislative acts. The Resource Conservation and Recovery Act (RCRA) of 1976 regulates hazardous and toxic wastes from the point of generation

Soil contamination caused by underground storage tanks containing tar. Encountered during remediation works at a disused gasworks. Image by Dumelow (Own work) CC-BY-SA-3.0.

to disposal. The Comprehensive Environmental Response, Compensation, and Liability Act (CERCLA) of 1980, also known as Superfund, identifies past contaminated sites and implements remedial action.

Soils can become contaminated by many human activities, including fertilizer or pesticide application, direct discharge of pollutants at the soil surface, leaking of underground storage tanks or pipes, leaching from landfills, and atmospheric deposition. Additionally, soil contamination may be of natural origin. For example, soils with high concentrations of heavy metals can occur naturally because of their close proximity to metal ore deposits. Common contaminants include inorganic compounds such as nitrate and heavy metals (for example, lead, mercury, cadmium, arsenic, and chromium); volatile hydrocarbons found in fuels, such as benzene, toluene, ethylene, and xylene (BTEX) compounds; and chlorinated organic compounds such as polychlorinated biphenyls (PCB) and pentachlorophenol (PCP).

Contaminants may also include substances that occur naturally when these appear in concentrations that are elevated above normal levels. Examples are substances such as nitrogen- and phosphorus-containing compounds, which are often added to agricultural lands as fertilizers. Since nitrogen and phosphorus are typically the limiting nutrients for plant and microbial growth, their accumulation in the soil is usually not a concern, but the leaching and runoff of these nutrients into nearby water sources is a problem, as it may lead to oxygen depletion of lakes. Furthermore, nitrate is a concern in drinking water because it poses a direct risk to human infants (blue-baby syndrome).

Contaminants may reside in the solid, liquid, and gaseous phases of the soil. Most will occupy all three phases but will favor one phase over the others. The physical and chemical properties of the contaminant and the soil will determine which phase the contaminant favors. The substance may preferentially adsorb to the solid phase. This may include either the inorganic (mineral) or the organic (organic matter) fraction of the soil. The attraction to the solid phase may be weak or strong. The contaminant may also volatilize into the gaseous phase of the soil. If the contaminant is soluble in water, it will dwell mainly in the liquid-filled pores of the soil.

Contaminants may remain in soils for years or wind up in the atmosphere or nearby water sources. Additionally, contaminating compounds may be broken down or taken up by the biological component of the soil. This may include plants, bacteria, fungi, and other soil-dwelling microbes. The volatile compounds may slowly move from the gaseous phase of the soil into the atmosphere. The contaminants that are bound to the solid phase may remain intact or be carried off in runoff attached to soil particles and flow into surface waters. Compounds that favor the liquid phase, such as nitrate, will either wind up in surface waters or leach down into the groundwater.

Metals display a range of behaviors. Some bind strongly to the solid phase of the soil, while others easily dissolve and wind up in surface water or groundwater. PCBs and similar compounds bind strongly to the solid surface and remain in the soil for years. These compounds can still pose a threat to waterways because, over long periods of time, they slowly dissolve from the solid phase into the water at trace quantities. Fuel components favor the gaseous phase but will bind to the solid phase and dissolve at trace quantities into the water. However, even trace quantities of some compounds can pose serious ecological and health risks. When a contaminant causes a harmful effect, it is classified as a pollutant.

One of two general approaches is used in cleaning up a contaminated soil site: treatment of the soil in place (in situ) or removal of the contaminated soil to another location for treatment (ex situ). In situ methods, which have the advantage of minimizing exposure pathways, include biodegradation, volatilization, leaching, vitrification (glassification), and isolation or containment. Ex situ methods generate additional concerns about exposure during the process of transporting the contaminated soil. Ex situ options include thermal treatment (incineration), land treatment, chemical extraction, solidification or stabilization, excavation, and asphalt incorporation. The choice of method depends on the quantity and type of contaminants and the nature of the soil.

FURTHER READING

Connell, Des W. "Soil Contamination." In *Basic Concepts of Environmental Chemistry.* 2d ed. Boca Raton, Fla.: CRC Press, 2005.

Pierzynski, Gary M., J. Thomas Sims, and George F. Vance. *Soils and Environmental Quality.* 3d ed. Boca Raton, Fla.: CRC Press, 2005.

Sparks, Donald L. "Soil Decontamination." In *Handbook of Hazardous Materials,* edited by Morton Corn. San Diego, Calif.: Academic Press, 1993.

Testa, Stephen M. *The Reuse and Recycling of Contaminated Soil.* Boca Raton, Fla.: Lewis, 1997.

SOIL DEGRADATION

—Judith J Bradshaw-Rouse

FIELDS OF STUDY

Agronomy; Agriculture; Soil Science; Physical Science; Biology; Chemistry; Ecology; Environmental Studies.

SUMMARY

Soil degradation is a decline in soil quality, productivity, and usefulness because of natural causes, human activities, or both. Degradation may be caused by unfavorable alterations in one or all of a soil's physical, chemical, and biological attributes.

BACKGROUND

In 1992, for the first global study of soil degradation, the World Resources Institute in Washington, D.C., reported that 1.2 billion hectares of land worldwide had been seriously degraded from World War II to the present. It also stated that 9 million hectares of once usable land could no longer support crops.

NATURAL PROCESSES & HUMAN ACTIVITIES

Of the total lands lost to soil degradation, almost two-thirds is in Asia and Africa; most of the loss is attributable to water and wind erosion resulting from agricultural activities, overgrazing, deforestation, and firewood collection. There are also seriously degraded soils in Central America, where degradation is caused primarily by deforestation and overgrazing. In Europe, industrial and urban wastes, pesticides, and other substances have poisoned soils in much of Poland, Germany, Hungary, and southern Sweden. In the United States, the U.S. Department of Agriculture estimates that a quarter of the nation's croplands have been depleted through deep plowing, removal of crop residue, conversion to permanent pasture, and other conventional agricultural practices. Although unwise management practices contribute significantly to soil degradation, soil degradation also involves three natural soil processes: physical, chemical, and biological degradation.

Soil Limits to Agriculture, by Percentage of Total World Land Area

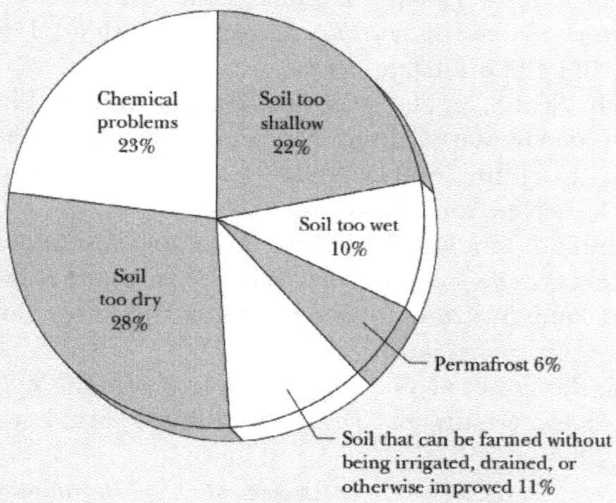

Source: United Nations Food and Agriculture Organization (FAOSTAT Database, 2000).

Soil Limits to Agriculture, by Percentage of Total World Land Area. Image by United Nations Food and Agriculture Organization (FAOSTAT Database, 2000).

PHYSICAL SOIL DEGRADATION

Physical soil degradation involves deterioration in soil structure, leading to compaction, crusting, accelerated erosion, reduced water-holding capacity, and decreased aeration. Soil compaction is the compression of soil particles into a smaller volume. Excessively compacted soil suffers from poor aeration and reduced gas exchange, which can restrict the depth of root penetration. Soil compaction also causes accelerated runoff and erosion of soils.

Crusting is the formation of a hard layer a few millimeters or a few tens of millimeters thick at the soil surface. Crusts affect drainage, leading to waterlogging at the soil surface and to salinity or alkalinity problems. Once crusts called "duricrusts" form, soil moisture recharge declines, and vegetation cannot root. Sheet and gully erosion increases as the land fails to absorb precipitation. Hard layers can also form below the cultivation depth and are called hardpans (other names are plow soles, traffic pans, and plow pans). These compacted layers can restrict root growth, making crops and trees vulnerable to drought and lodging (falling over).

CHEMICAL DEGRADATION

Chemical degradation comprises changes in soil's chemical properties that regulate nutrient availability. Nutrient depletion is the major factor in chemical soil degradation. Soil nutrient depletion may be caused or exacerbated by many factors, including monocropping, leaching of nutrients, and salt buildup.

A historic example of nutrient depletion is the depletion of soils in the southeastern United States by the growing of cotton. As late as 1950, "King Cotton" was the most valuable farm commodity produced in Alabama, Arkansas, Georgia, Louisiana, Mississippi, South Carolina, Tennessee, and Texas. In the eighteenth and nineteenth centuries, the growing of cotton ruined soil fertility as it spread westward from the Atlantic to the Texas panhandle. Cotton growth without regard to topography in hilly regions contributed to soil erosion. topsoil was eventually removed from many fields, which further depleted nutrients. One reason that peanuts became a major crop in the South is that they were nitrogen-fixing plants that could grow in soils depleted of nitrogen by cotton.

Nutrient leaching is another problem. Continuous irrigation can leach nutrients and cause salt buildup in soils where drainage is poor. Leaching can move essential but soluble nutrients past the root zone deeper into the soil and into groundwater. In addition, the water used to irrigate soil often contains salts that can accumulate to toxic levels and inhibit plant growth where evaporation occurs readily. Thick crusts of salt on farmland in Pakistan, Australia, Ethiopia, Sudan, and Egypt have made soil unfit for crops.

Laterization refers to the product and process of wetting and drying that leads to the irreversible consolidation and hardening of aluminum- and iron-rich clays (plinthitic materials) into hardpans, sometimes of great thickness. In Greek *plinthos* means "brick." Laterization is particularly common in the humid and subhumid tropics.

BIOLOGICAL DEGRADATION

The loss of organic matter and soil nutrients needed by plants can occur in any environment, but it is most dramatic in hot, dry regions. Organic matter is important in maintaining soil structure, supporting microorganisms, and retaining plant nutrients. Because organic matter is near the soil surface, it is generally the first soil component to be lost. Organic matter may be lost through brushfires, stubble-burning, overgrazing, or the removal of crops, fodder, wood, and dung. Loss of organic matter can be accelerated when soil moisture is reduced, when soil aeration is increased, or both. For example, peat soils that are drained decompose rapidly and subside. In drier climates, the loss of organic matter reduces the soil's moisture-holding capacity and lowers soil fertility, which leads to lower crop yields and thus to less organic matter being returned to the soil.

Tropical forests such as those of the Amazon basin in South America seem lush, so people widely assume tropical soils to be fertile and high in organic matter. Although tropical forests do produce considerable organic matter, the amount that stays in the soil is surprisingly small, and the soils actually have low nutrient levels. Soil microorganisms in the rain forest break down the organic matter and release nutrients that are absorbed by growing plants. However, warm temperatures and high rainfall cause accelerated nutrient loss if plants are absent. Nutrients that would buffer the pH of the soil are lost. Consequently, the clearing of rain forests exposes the soil to erosion, leaching, acidification, and rapid nutrient depletion.

Further Reading

Ashman, M. R., and G. Puri. *Essential Soil Science: A Clear and Concise Introduction to Soil Science.* Malden, Mass.: Blackwell, 2002.

Brady, Nyle C., and Ray R. Weil. *The Nature and Properties of Soils.* 14th ed. Upper Saddle River, N.J.: Prentice Hall, 2008.

Braimoh, Ademola K., and Paul L. G. Vlek, eds. *Land Use and Soil Resources.* London: Springer, 2008.

Davies, Bryan, David Eagle, and Bryan Finney. *Resource Management: Soil.* Brighton, Ont.: Diamond Farm Book, 2001.

Donahue, Roy L., Roy Hunter Follett, and Rodney W. Tulloch. *Our Soils and Their Management: Increasing Production Through Environmental Soil and Water Conservation and Fertility Management.* 6th ed. Danville, Ill.: Interstate, 1990.

Esch, Neal S., et al. *Soil Science Simplified.* 5th ed. Illustrated by Mary C. Bratz. Ames, Iowa: Blackwell, 2008.

Gerrard, John. "Soil Degradation." In *Fundamentals of Soils.* New York: Routledge, 2000.

Hannam, Ian, and Ben Boer. *Legal and Institutional Frameworks for Sustainable Soils: A Preliminary Report.* Gland, Switzerland: International Union for Conservation of Nature, 2002.

Lal, Rattan, et al. *Soil Degradation in the United States: Extent, Severity, and Trends.* Boca Raton, Fla.: Lewis, 2004.

_______, eds. *Methods for Assessment of Soil Degradation.* Boca Raton, Fla.: CRC Press, 1998.

Miller, Raymond W., and Roy L. Donahue. *Soils: An Introduction to Soils and Plant Growth.* 6th ed. Englewood Cliffs, N.J.: Prentice-Hall, 1990.

Paul, Eldor A., ed. *Soil Microbiology, Ecology, and Biochemistry.* 3d ed. Boston: Academic Press, 2007.

Plaster, Edward J. *Soil Science and Management.* 5th ed. Clifton Park, N.Y.: Delmar Cengage Learning, 2009.

Raman, Saroja. "Soil Degradation and Its Impact on Agriculture." In *Agricultural Sustainability: Principles, Processes, and Prospects.* New York: Food Products Press, 2006.

Singer, Michael J., and Donald N. Munns. "Soil Degradation." In *Soils: An Introduction.* 5th ed. Upper Saddle River, N.J.: Prentice Hall, 2002.

Sposito, Garrison. *The Chemistry of Soils.* 2d ed. New York: Oxford University Press, 2008.

White, Robert E. *Principles and Practice of Soil Science: The Soil as a Natural Resource.* 4th ed. Malden, Mass.: Blackwell, 2006.

Zimmerer, Karl S. "Environmental Discourses on Soil Degradation in Bolivia: Sustainability and the Search for Socioenvironmental 'Middle Ground.'" In *Liberation Ecologies: Environment, Development, Social Movements,* edited by Richard Peet and Michael Watts. 2d ed. New York: Routledge, 2004.

University of Michigan, Global Change Program. Land Degradation. ttp://www.globalchange.umich.edu

SOIL EROSION

—Alan Brown

FIELDS OF STUDY

Agronomy; Agriculture; Soil Science; Physical Science; Biology; Chemistry; Ecology; Environmental Studies.

PRINCIPAL TERMS

granules: small grains or pellets
porosity: the ability to admit the passage of gas or liquid through pores
rills: small rivulets in channels
suspension: a condition in which particles are dispersed through a supporting medium
understrata: material lying beneath the surface of the soil

SUMMARY

Soil erosion, which plays an important role in shaping the landscape, is a process that takes extended periods, although people have greatly accelerated it by removing vegetation. Controlling soil erosion is essential to maintaining the world's food supply.

SHAPING THE LANDSCAPE

Soil erosion is a natural process whereby rock and soil are broken loose from the earth's surface at one location and moved to another. Erosion creates and transforms land by filling in valleys, wearing down mountains, and making rivers appear and disappear. Generally, this process takes thousands or even millions of years.

The active forces of erosion are water and wind. Water erosion involves the movement of soil by the action of rainwater and melted snow moving rapidly over exposed land surfaces. The principal types of water erosion are splash erosion and gully erosion.

SPLASH EROSION

Splash erosion is the removal of thin layers, one drop at a time, from land surfaces. Fine-grained soil such as silt loams, fragile sandy soil, and all soils deficient in organic matter are especially vulnerable. Most troublesome are lands that tend to slope, lands subjected to heavy rainfall, and lands composed of shallow surface soils overlying dense clay subsoil.

When rainwater is absorbed into the ground, small sieve-like openings are made in the soil. Eventually, fine particles carried by the water plug the openings, thereby causing the rainwater to flow off the land. As the soil is softened by the rainfall, clods, lumps, and granules break down and form a pasty mass, which resists penetration by rainwater. As a result, the runoff increases and forms a relatively impervious, skin-like film sometimes referred to as puddled soil. As rain continues, the abrasive force of the rain results in cutting, which penetrates the skin-like layer and starts trenching.

Soil erosion. Image by Богдан Малько (Panoramio) CC-BY-SA-3.0.

When splash erosion reaches an advanced stage, it is referred to as rill erosion. Rill erosion can be defined as localized small washes in channels that can usually be eliminated by ordinary plowing. Rilling is the most common form of erosion on soft, freshly plowed soils that are high in silt content where the slopes are deeper than 4 or 5 percent.

There are several specialized forms of splash erosion. Pedestal erosion occurs when a stone or tree root protects easily eroded soil from splash erosion. Consequently, isolated pedestals containing resistant material are left standing. This type of erosion occurs after several years, primarily on bare patches of grazing land, and often can be seen on a small scale on pebbly ground after a rain. The erosion patterns formed in highly erodible soils are called pinnacle erosion. This erosion is usually found in deep vertical rills in the sides of gullies. When these gullies cut back and join, they form pinnacles. When pinnacle erosion is present, reclamation is difficult. Another specialized form of splash erosion is piping, which is associated with the formation of continuous pipes or channels underground. Piping usually occurs in soil types that are subject to pinnacle erosion. It involves water penetrating the soil surfaces and moving downward until it reaches a less permeable layer. The fine particles of the more porous soil may be washed out if the water flows over the less permeable layer through an outlet. The more rapid lateral flow increases the sideways erosion, causing the entire surface flow to disappear down a vertical pipe. The water then flows underground until it reappears in the side of a gully. Once pipe erosion starts, it cannot be controlled; this has been a factor in some dam collapses. Fortunately, pipe erosion is restricted to the "bad lands." The last type of splash erosion is slump erosion. It is prominent in areas of high rainfall with deep soils. Slumping can become the chief agent in the development of gullies, probably as a result of flood flow in channels. Riverbank collapse and coastal erosion are the other main cases of slumping.

GULLY EROSION

The second principal type of water erosion is gully erosion. This type of erosion occurs in places where runoff from a slope is sufficient in volume and velocity to cut deep trenches. It also takes place in areas where concentrated water continues cutting in the same groove long enough to form deep rills. Gullies usually begin in slight depressions in or below fields where water concentrates, in ruts left by farm machinery, in livestock trails, or along furrows between crop rows. Gullies ordinarily carry water only during or following rains or melting snows. Most gullies cannot be removed by normal plowing because of their size; in fact, some of them take the form of huge chasms 15–30 meters deep. A field gully is a channel at least 45 centimeters wide and between 25 and 30 centimeters deep; a woodland channel, by comparison, is considered to be a channel deep enough to expose the main lateral roots of trees. Woodland gullies develop in wooded areas that received water from cultivated slopes immediately above or on closely cut, severely burned-over woodland.

There are two main types of gullies. V-shaped gullies, which have sloping sides with narrow bottoms, are the normal type of gully. Gullies that have straight, more or less vertical sides with broad bottoms are U-shaped gullies. They are less winding than the V-shaped gullies because the soft materials at the base of the sides tend to give way to the impact of currents, and irregularities in the channel walls are planed away. Usually, the presence of gullies means that the land has been overused or abused. U-shaped gullies are usually the more serious because the soft, unstable materials commonly found in their lower depths, such as sand, loose gravel, or soft rocks, are easily cut out by floodwaters. This cutting near the bottom causes the banks to split off from above in great vertical blocks. When the caved-in material is washed out of the gully, the trench is left box-shaped. Most gullies tend to branch out as they grow, but the tendency is more serious in the U-shaped gullies. They are the most difficult to control because of the instability of the understrata.

Waterfall erosion is an important form of gully erosion because it does so much damage. It is caused by water cascading over the heads and sides of gullies, over dams, and over terraces whose channels have been filled with the debris of erosion. This type of erosion is most commonly associated with flooding.

WIND EROSION

The second active force of soil erosion is wind erosion. Wind erosion can occur in places where water erosion is also active, but it reaches its most serious proportions on both level and sloping areas during dry times. Like water erosion, wind erosion usually proceeds very slowly; however, wind erosion is likely to increase rapidly in relatively flat and gently undulating treeless regions, like the Great Plains. When grass is plowed up, the cultivated soil becomes much less cohesive. Organic material that normally collects under a cover of grass and that serves as a bonding agent when grass is present disappears by decay and oxidation when the grass is gone. After periods of drought, the soil turns into a dry powdery mass, which is easily swept up by the wind and lifted into the pathways of high air currents, which carry it hundreds and, at times, thousands of kilometers. Coarser, heavier particles, known as ground drift, are blown along near the surface of the ground and pile up in drifts about houses, fences, farm implements, and clumps of vegetation. The fine materials that are blown away are dust; the coarser materials that are left behind are sand. The susceptibility of soils to wind erosion depends on the size of the particles and on the content of the organic matter. Coarse sands are more likely than heavy clays to blow away immediately after plowing. Ironically, the finer-textured soils, especially those of granular structure, show the greatest resistance. In fact, they sometimes remain undisturbed through years of cultivation until their organic material is disrupted.

The massive removal of soil particles through the action of the wind takes the form of dust storms. Early in 1934, a dust storm originating in the Texas-Oklahoma Panhandle covered a vast territory extending eastward from the Rocky Mountains to several hundred kilometers over the Atlantic. This one storm deprived the Great Plains states of 200 to 300 million tons of soil.

There are essentially five different types of wind erosion, although there is some overlapping, and several of the processes occur simultaneously. Detrusion is the wearing away of rocks and soil formations by fine particles carried away in suspension. This process often carves large rocks in deserts into grotesque shapes or streamlined hills, called yardangs. Abrasion occurs close to the ground where the moving particles are larger and bound over the surface. The removal of very fine particles, carried off in suspension, is efflation, and the rolling away of large particles is extrusion. The removal of particles of intermediate size bouncing downwind is known as effluxion, and the bouncing process is called saltation. The latter has nothing to do with salt but is derived from the Latin word for "leap."

SEVERITY OF EROSION

The severity of erosion is determined by a great many factors; one of the least controllable factors is climate. The United States has more erosion problems than England because it lacks England's gentle rains and mists. Many areas in the United States have rainstorms of sufficient strength and intensity to erode many centimeters of soil from a field not protected by vegetation within minutes. For example, rain-induced erosion may be severe in the Corn Belt (roughly, the states of Indiana, Illinois, Missouri, Nebraska, and Kansas) because rains frequently occur in these states in June when the plants have not matured enough to protect the soil; in arid areas of the western United States, serious water erosion may exist because there is not enough rain to establish a protective ground cover for the infrequent rains.

Soil type and topography also influence the severity of erosion. Soils in the United States range from poorly drained and saturated to very dry, from sandy to clayish, from acid to alkaline, and from shallow to deep. They also vary in slope characteristics, porosity, organic content, temperature, and the capacity to supply nutrients. Land is better for crops if it is nearly level, with just enough slope for good drainage. About 45 percent of the cropland in the United States falls into this category.

A farmer's cropping practices can also greatly influence the severity of erosion. Erosion is likely to occur when fields are plowed in the conventional way: in straight rows regardless of the topography, with all plant cover and crop residue removed. Erosion can be retarded if the farmer uses the best land for crops and puts any other land to different uses. After that, any number of conservation practices, such as strip cropping, contour planting, crop rotation, terracing, and various conservation tillage practices, can be introduced.

MEASURING EROSION

Soil losses per acre are measured by either the Universal Soil Loss Equation (USLE) or the Wind Erosion Equation (WEE). These formulas have been developed from field experiments in various parts of the country. Both equations measure the tons of each soil type that are lost annually through the action of climate, cropping systems, management practices, and topography.

The use of these equations involves a number of limitations. Although both the USLE and the WEE measure the movement of the soil, they do not reveal the distance the soil has traveled or where the soil was deposited. Thus, the USLE could overestimate the severity of the erosion. Another drawback to the USLE is its failure to measure losses resulting from snowmelt. A revised version of the USLE, the RUSLE, has been developed to overcome these limitations.

The soil losses measured by the RUSLE or WEE are average figures taken from measurements over an extended period. They are usually reported in tons per acre. The computed losses are often connected with soil-loss tolerances, or T-values. These figures are the maximum soil losses that can be sustained without adversely affecting productivity. Some erosion is natural even without human disturbance. As long as soil evolution keeps pace with erosion, the soil loss is sustainable. Excess soil erosion is frequently defined as amounts greater than T-values. The U.S. Department of Agriculture has assigned T-values that usually range from 1 to 5 tons per acre, depending on the properties of the soil. Unfortunately, the validity of these numbers in representing maximum sustainable soil losses is doubtful. A soil loss of 5 tons per acre per year translates into a net loss of 2.54 centimeters (1 inch) of soil every thirty years. Even in fairly steep topography, natural long-term erosion rates are only a few millimeters in thirty years. T-values have been set too high on some soils to assure long-term maintenance of the soil. T-values are also limited in their value because they do not reflect the impact of technology on crop yields.

Erosion can also be studied according to how it will be affected by different types of rain and how it will vary for different types of soil. Therefore, the amount of erosion that occurs depends on a combination of the ability of the soil to withstand rain and the power of the rain to cause erosion. This relationship between factors can be expressed in mathematical terms. Erosion is a function of the erosivity (of the rain) and the erodibility (of the soil); that is, erosion equals erosivity multiplied by erodibility.

For given soil conditions, one rainstorm can be compared quantitatively with another, and a numerical scale of values of erosivity can be created. For given rainfall conditions, one soil condition can be compared quantitatively with another, and a numerical scale of values of erodibility can be created. Erodibility of the soil can be subdivided into two parts: the inherent characteristics of the soil (that is, mechanical, chemical, and physical composition) and the way the soil is managed. Management may, in turn, be subdivided into land management and crop management.

COSTS & BENEFITS OF EROSION

Soil erosion can be both beneficial and harmful. Erosion benefits people by contributing to the formation of soil by breaking up rocks. Erosion also creates rich, fertile areas as it deposits soil at the mouths of rivers and on the floors of valleys. The Nile River valley in Egypt has been sustained for thousands of years by sediments eroded from farther upstream. Erosion is also important from the standpoint of aesthetics. The Grand Canyon, for example, was created over millions of years through the eroding action of the Colorado River.

Yet soil erosion is one of the leading threats to the food supply, as it robs farmland of productive topsoil. Soil scientists estimate that it takes nature between three hundred and one thousand years to produce about 2.5 centimeters of topsoil, although humans can replace topsoil at a much faster rate. Still, considering that about 70 percent of the United States is subject to erosion, the prospect of building productive soil becomes overwhelming. Once the topsoil has been removed, a heavy layer of clay often remains, which may not contain enough porosity to support a good crop.

Crops can also be damaged by the loss of nutrients to erosion. Most serious are losses of nitrogen, sulfur, and phosphorus. In addition to the natural nutrients taken from the soil, synthetic fertilizers and chemicals are often washed from fields and into lakes and rivers, thereby contributing to pollution.

Finally, erosion that forms deep gullies presents problems to farmers. Gullies caused by flowing water are often too deep for farm machinery to cross. Unable to accommodate tractors and other farm equipment, fields riddled with gullies face ruin.

FURTHER READING

Agassi, Menachem, ed. *Soil Erosion, Conservation, and Rehabilitation.* New York: Marcel Dekker, 1996. Assesses soil erosion processes through an examination of rainfall and water runoff. Deals with methods used in soil conservation and rehabilitation programs. Illustrations, index, and bibliography.

Batie, Sandra S. *Soil Erosion: Crisis in America's Croplands?* Washington, D.C.: Conservation Foundation, 1983. Attempts to structure an improved public policy for soil conservation. Assesses the extent and effects of erosion and explains the techniques for reducing soil erosion in language that can be easily understood by the nonscientist. Contains photographs and diagrams.

Bennett, Hugh Hammond. *Elements of Soil Conservation.* 2d ed. New York: McGraw-Hill, 1955. Although some of the information is dated, the chapter describing the processes and effects of erosion is up-to-date. Contains many photographs. Also includes a bibliography of books, pamphlets, and films concerning soil conservation.

Boardman, John, and David Favis-Mortlock, eds. *Modelling Soil Erosion by Water.* New York: Springer-Verlag, 1998. Covers the effects of climatic factors and climatic changes on soil erosion. Discusses possible climatic causes and effects and uses computer-simulated erosion models to illustrate complex processes. Technical at times.

Cornforth, Derek H. *Landslides in Practice.* Hoboken, N.J.: John Wiley & Sons, 2005. Explains how and why landslides occur. Describes the resulting changes to a landscape and prevention methods. Compares various soil types and slope stability, and examines a number of case studies. Provides the tools and techniques needed to evaluate landslide potential and determine preventative strategies.

Hudson, Norman. *Soil Conservation.* 3d ed. Ames: Iowa State University Press, 1995. Covers the history of soil conservation and demonstrates how the methods of soil conservation practiced in the United States can be applied in developing countries. Adopts an engineering approach to explain soil conservation. Includes many charts, illustrations, and photographs.

Lal, Rattan, ed. *Soil Quality and Soil Erosion.* Boca Raton, Fla.: CRC Press, 1999. Based on papers presented at a 1996 symposium on soil and water conservation. Offers an examination of the quality, management, and erosion of soils and water. Deals with methods of determining those levels.

Morgan, Royston Philip Charles. *Soil Erosion and Conservation.* 3d ed. Malden, Mass.: Blackwell Science, 2005. Explains the procedures and protocol used in determining the rates and effects of soil erosion. Describes how the application of these techniques is used in soil conservation programs. Illustrations, maps, thirty-page bibliography, and index.

Newson, Malcolm. *Land, Water and Development: Sustainable and Adaptive Management of Rivers.* 3d ed. London: Routledge, 2008. Presents land-water interactions. Discusses recent research, study tools and methods, and technical issues, such as soil erosion and damming. Appropriate for undergraduates and professionals. Covers concepts in managing land and water resources in the developed world.

Norris, Joanne E., et al., eds. *Slope Stability and Erosion Control: Ecotechnological Solutions.* New York: Springer, 2010. Discusses erosion processes, mass wasting, and landslides as they occur on varying slopes. Evaluates vegetation as a solution to slope instability. Cites many studies but is not overly technical. Appropriate for engineers, landscape architects, ecologists, foresters, and undergraduate students.

Simms, O. Harper. *The Soil Conservation Service.* New York: Praeger, 1970. Primarily a record of the activities of the U.S. Soil Conservation Service. Discusses the methods that the government recommends for preventing soil erosion. Appropriate for the general reader.

Stallings, J. H. *Soil: Use and Improvement.* Englewood Cliffs, N.J.: Prentice-Hall, 1957. Focuses on proper land management. Explains past and present methods of soil conservation in terms appropriate for the general reader.

Troeh, Frederick R., and Louis M. Thompson. *Soils and Soil Fertility.* 5th ed. New York: Oxford University Press, 1993. Introductory textbook in soils for students of agriculture. Covers every aspect of soil, including the physical properties of soil and soil conservation. Highly readable and well supplemented with charts, illustrations, and diagrams, although the reference sections are dated.

U.S. Soil Conservation Series. *Our American Land: Use the Land, Save the Soil.* Washington, D.C.: Government Printing Office, 1967. Promotes the need for soil conservation by providing statistical evidence of damage that has been done in the United States by erosion. Also recommends proper land management techniques. Appropriate for the general reader.

SOIL FORMATION

—*Scott F. Burns*

FIELDS OF STUDY

Agronomy; Agriculture; Soil Science; Physical Science; Biology; Chemistry; Ecology; Environmental Studies.

SUMMARY

A soil is formed when decomposed organic material is encompassed into weathered mineral material at the earth's surface. The climate, the organisms living in the soil, the type of parent material, the local topography, and the amount of time the soil has been developing all influence the resulting soil characteristics. Changes in the soil arise from the main processes of soil formation: addition, removal, transfer, and transformation of parts of the soil.

STAGES OF FORMATION

Soil formation and development are continuing processes. A soil is a layer of unconsolidated weathered mineral matter on the earth's surface that can support life. It is initially formed when the remains of plants are mixed into weathered mineral fragments. Soil characteristics can change, however, for soil develops over time as the environment dictates.

Many processes contribute to the two stages of soil formation, which may grade indistinctly into each other. The first stage is the accumulation of rock fragments known as parent material. These sediments may have formed in place from the physical and chemical weathering of hard rocks below or by the accumulation of sediments moved by wind, water, gravity, or glaciers. The second stage is the formation of soil horizons, which may either follow or occur simultaneously with the first stage. The near-surface horizon, where materials are removed, is called the A horizon. Some of the removed material accumulates in the subsurface, in a zone called the B horizon. If the soil has an organic-rich top layer, that horizon is called the O horizon. In some cases, the A horizon has so much iron removed that it looks bleached. A very light zone at the base of the A horizon is called the E horizon. Unmodified parent material is termed the C horizon.

The characteristics of the horizons and their degree of differentiation depend upon the relative strengths of four major processes in the soil: weathering, transfers, gains, and losses of the soil constituents. These processes have been observed directly in laboratory experiments but also are implied by relating physical, chemical, and morphological properties in different parts of the soil profile.

TRANSFORMATIONS & TRANSFERS

Weathering is the fundamental process of soil formation, as it transforms soil constituents into new chemicals. The primary minerals found in the initial sediments and rocks are decomposed by physical and chemical agents into salts that are soluble in water and new minerals. Clay is a very important new mineral formed from the breakdown of primary minerals by the agents of water and air. Calcite, or calcium carbonate, is another mineral formed through weathering, especially of rocks such as basalt. Iron and aluminum hydroxides form a third major group of weathering-produced minerals. Organic matter is added to the soil when roots die and vegetation residues fall onto the surface. These are transformed by microorganisms into a mixture of chemically stable, organic substances called humus.

The transfer of soil constituents is generally in a downward direction. Rainwater is the major transporting agent. As it moves through the soil, it picks up salts and humus in solution and clays and humus in suspension. These materials are usually deposited in another horizon when the water is withdrawn by roots or evaporation, or when the materials are precipitated as a result of differences in pH (degree of acidity) or salt concentration. The clays normally coat the exteriors of sediment grains and line the interiors of pores. This transfer of materials through solution is commonly called leaching and requires good drainage. The soil profile often becomes vertically zoned according to the solubility of the dissolved substances. In forest soils, leaching generally removes iron and aluminum, leaving behind a bleached-looking zone. This process is called podzolization. In arid soils, calcium minerals precipitate in the subsoil to form calcite or gypsum, a deposit called caliche.

GAINS & LOSSES

Gains consist of additions of new materials to the soil. Wind can deposit fine sands and silts if there is a source nearby. Chemical reactions during weathering add oxygen and water to the soil minerals. Groundwater brings new minerals into the soil pores. Floods can add new sediment to the soils in a floodplain. The largest addition to soils, however, comes from organic matter from the decomposition of plant material. Annual additions of new organic material between 255 and 22,900 kilograms per hectare have been estimated for soils in different environments ranging from deserts to tropical rain forests.

Losses from the soil profile generally occur when there is so much water moving through the system that the dissolved and suspended materials exit at the bottom of the profile and move downslope and eventually enter a stream. Soil erosion also occurs on the surface by water runoff, especially where there is minimal vegetation to control the sediment. Soil erosion is one of the major environmental problems facing the world today. Oxidation of plant matter balances some of the additions of new organic material.

CLIMATE

After describing many soil profiles around the world and classifying them, scientists concluded that this weathered material at the earth's surface was quite variable and differed from environment to environment. In 1980, Hans Jenny summarized the work of early researchers and listed the five main soil-forming factors that created this variation: climate, organisms, parent material, topography, and time. The rates and degrees of these processes in different soils depend upon these five factors. These factors are interdependent, and, therefore, very different soils may be formed from the same parent material.

Of the five factors, climate has probably the greatest impact on soil development. This dominance is best shown on a large-scale geographic basis, in which the distribution of soils on soil maps is related to vegetation type, which is in turn related to climatic zones. Precipitation provides the water for the transfer processes in the soils. In dry climates, where little water is moving through the system, calcium carbonate and salts build up in the soils. In humid climates, the abundance of water moving through the systems allows leaching and clay movement processes to dominate in the soils. Temperature is the second important element of climate, as it controls the rates of chemical weathering reactions in the soil. In colder environments, soil weathering and development are slower because weathering reactions are not active all year round. Some extreme climatic conditions also can control soil development, such as permafrost, which limits soil depth and drainage. Overall, climate controls vegetation, which also controls soil development.

ORGANISMS & PARENT MATERIALS

Living organisms, primarily vegetation, control the distribution and types of humus that are formed in soils. In turn, humus controls some of the important soil-forming processes. Grassland vegetation produces humus that is highest in mineral matter and not very acid, so resulting soils have thick A horizons that are high in organics. Conifer forests produce humus that is low in mineral constituents and highly acidic. These forest soils are dominated by leaching and the production of O, E, and B horizons (B horizons are rich in iron and aluminum). Deciduous forest soils are located somewhere between the conifer forest and the grassland soils in characteristics. Therefore, it is rare to find white, bleached E horizons under grassland vegetation, and vice versa—it is rare to find thick A horizons under conifer forest vegetation. Through its leaves and root systems, vegetation also protects the soil from erosion and, therefore, promotes soil development.

The parent material is the rock or sediment deposit from which the soil has developed. The composition of the original minerals controls the complexity of the soil processes by determining the weathering by-products, and controls the rate of these processes by porosity. A parent material of pure quartz sand would undergo minimal weathering, producing a simple soil profile. A parent material of wind-deposited silt, or loess, produces a more complex soil, as it has many minerals initially that can weather to by-products such as clays, salts, carbonates, and oxides. A soil produced on granite would yield a fairly deep, sandy soil, whereas a soil produced on basalt would yield a fairly thin, clay-rich soil. A soil developed in shale would be clay-rich, whereas one developed on sandstone would be sandy. Soils formed on rocks rich in iron tend to produce very red soils. Soils formed on limestone are difficult to acidify and podzolize because the parent material neutralizes those chemical processes. The porosity, or the amount of pore space, of the parent material gives ready access of gases and liquids to weathering. In porous, or permeable, soils, weathering and transfers of soil constituents are more rapid than in soils with low permeability.

TOPOGRAPHY & TIME

Topography is the geometric configuration of the soil surface and is defined by the slope gradient, orientation, and elevation. This factor essentially controls the microclimate. A slope modifies rainfall by allowing rapid runoff on a steep slope or ponding in low places. Slope orientation modifies soil temperatures by the differences at which the sun's rays strike the soil. In the northern latitudes, southfacing slopes are warmer and drier than those on north-facing slopes. In mountain ranges, rainfall increases and temperature decreases with increases in elevation.

Time is the factor of the duration of the processes working in the soil. All the characteristics in the soil take time to form. The A horizon forms most rapidly, taking as little as 10 years with the aid of humankind to more than 1,000 years in a cold and dry arctic or alpine environment. B horizons take longer to form because minerals weather slowly. In the dry western parts of the United States, discoloration from the weathering of iron-rich minerals appears at about 1,000 years, but the strongest red colors require 100,000 years of development in the same environment. In the same part of the United States, initial detection of clay movement into the B horizon in a sandy parent material occurs after about 10,000 years of development, with maximum development taking more than 100,000 years. Calcium carbonate deposition in arid regions of the world takes between 100,000 and 500,000 years to produce a strongly cemented horizon. It is estimated that it takes more than 1 million years to produce a highly weathered lateritic soil (decayed iron-rich rock that is red in color) in the tropics.

Soils change with time just as humans change as they grow. Young or poorly developed soils in a grassland environment have only A and C horizons. With time and the movement of iron oxides and clay into the lower parts of the profile, a faint B horizon develops between the A and C horizons. An old or well-developed soil in this environment would have a thick A horizon and a thick B- horizon with abundant clay moved into it, all overlying a C horizon.

SOIL LIQUEFACTION

—John F. Shroder Jr.

FIELDS OF STUDY

Agronomy; Agriculture; Soil Science; Physical Science; Biology; Chemistry; Ecology; Environmental Studies.

PRINCIPAL TERMS

bearing capacity: the ability of granular soil materials to support the weight of building structures

clay minerals: the diverse group of very finely crystalline mineral structures that are predominately composed of silicon, aluminum, and oxygen and that have very different water retention capabilities

cohesion: molecular attraction by which the particles of a body are united throughout the mass, whether alike or unlike

electrostatic charge: the fundamental atomic force in which objects that have a similar electric charge

Earthquake damage in Anchorage, Alaska, on March 27, 1964. Much of the damage was the result of liquefaction of an unstable layer of clay a few meters below the surface. Image by U.S. Geological Survey.

repel each other whereas those with unlike charges attract each other

flocculation: the sedimentation process by which a number of individual minute suspended particles are held together in clotlike masses by electrostatic forces

remolding: the property of some sensitive clays upon disturbance to reorient their particles, which softens them, and to flow in a liquid form

sensitive (quick): describes fine-grained deposits that are characterized by considerable strength in the undisturbed condition, but upon disturbance their ability to support themselves declines dramatically

van der Waals force: a weak electrostatic attraction that arises because certain atoms and molecules are distorted from a spherical shape so that one side of the structure carries more of the charge than does the other

SUMMARY

Soil liquefaction is the group of processes by which otherwise solid soil particles are shaken apart by earthquakes or collapse away from edge-to-edge contact with one another and become temporarily supported by the pore water contained within them. The resulting fluid mush can allow buildings to sink; this phenomenon is responsible for considerable loss of property and life.

TYPES OF LIQUEFACTION FAILURE

Soil liquefaction is the abrupt and temporary change of seemingly solid soil materials into liquid mush. The soil can lose all of its cohesion and bearing capacity in a matter of moments. Once the process begins, most commonly when an earthquake strikes, ground that had supported a hospital or a high-rise apartment can suddenly become a fluid into which the buildings sink like rocks into quicksand. Anything built upon such materials can slip or sink into the new liquid. Buried gasoline or septic tanks can suddenly become buoyant and float to the surface.

A number of different sorts of liquefaction failure are recognized; several occur on land and a few are subaqueous. Quick condition failure (not to be confused with sensitive or quick-clay landslips) is the complete loss of bearing capacity caused by liquefaction of sands and silts so that structures sink or rise in material that appears otherwise solid. Flow landslips, or flowslides, can occur on moderate slopes on land or beneath water and involve either sands or clays. These types tend to retrogress, or work their way backward, slice by slice as the material beneath them liquefies. Lateral spread landslips can occur on gentle, or nearly flat, slopes. Many lateral spreads are in quick clays. In extreme cases, the liquefaction slides result from spontaneous subaqueous liquefaction that propagates, or spreads, in all directions.

WATER-SATURATED SAND LIQUEFACTION

Two classes of soil materials can liquefy: water-saturated sands and silts and the unusual sensitive, or quick, clays. In the first example, layers of loosely packed, well-sorted, fine- to medium-grained sands and coarse silts are subject to liquefaction where groundwater tables are within 10–15 meters of the ground surface. When this water-saturated sediment starts to shake apart in an earthquake or other vibration, the grains temporarily lose rigid contact with one another and collapse inward. Much of the pore water (water in the gaps between soil grains) is then superfluous but does not escape at once, so that the rearranged grains cannot fit close to one another. The particle weight is thereby transferred to the pore water, its pressure increases (which reduces friction between the grains), and the soil becomes liquid for a short time.

When water-saturated sands liquefy during the course of an earthquake, sand boils erupt muddy water and sand from ground fractures and turn the surface into a quagmire. Sand boils tend to be roughly circular in plan and can have a depressed center like a volcano. After a very large earthquake, sand boils may form as much as several hundred meters across and several meters high. Their presence in an area is a reliable sign of past earthquake activity.

Water-saturated earth-fill dams are also subject to soil liquefaction, with obvious disastrous consequences. For example, during the 1971 San Fernando Valley earthquake in California, the Lower Van Norman dam collapsed. The 40-meter-high dam had been built with a core of clay surrounded by a fill of water-saturated sand. After about twelve seconds of strong shaking, a large, wedge-shaped segment of this water-saturated sand fill liquefied. Eight large blocks on the upstream side of the dam slid into the reservoir as parts of the sediment fill liquefied. Before the quake, the crest of the dam was a safe 10 meters above the water level of the reservoir. Afterward, however, only a thin barrier of 1.5 meters was left above the water level. Had this minor amount of remaining dam surface moved down but a fraction, the dam would have quickly failed, because water would have rushed down and eroded the downstream side. The 80,000 people who lived directly below the reservoir thus escaped disaster by a slim margin.

QUICK-CLAY LIQUEFACTION

In the quick-clay type of soil liquefaction, because of a rather special geologic history, certain clays develop the ability to liquefy, or to become quick, as with quicksand. Two main theories have been advanced to account for the unusual distribution of these quick clays. The saltwater theory holds that sensitive clays are first formed where glaciers erode very fine-grained clay mineral platelets from the soils and bedrock over which they ride. Where the weight of such glaciers depresses the surface of the land close to the sea, the small platelets can be deposited directly into the salty waters.

The small clay mineral platelets of a sensitive clay carry a negative charge, and as they settle in ocean water they tend to pick up positively charged sodium ions from the salt in solution. These oppositely charged, mutually attractive forces act as a glue to cause the particles to clump together into a sort of coagulated honeycomb structure. The platelets thus develop an edge-to-edge, or "house of cards," structure held together by the electrostatic forces of the salt ions attached to the platelet edges.

Glacial marine clays may remain quite stable as long as the salt water remains in the pore spaces of the open card-house structure. These deposits can later be raised above sea level by rebound of the land following removal of the glacier ice load. Then the salty pore waters may be flushed out of the clays by the fresh waters that normally occur above sea level. Removal of the mutually attractive electrochemical charges then sets the stage for eventual liquefaction.

A different theory explains the freshwater type of quick clays. In this case, glacial erosion also provides extremely fine particles of other than clay minerals. Instead, the fine particles (mostly quartz minerals) are so small that the normally weak van der Waals attraction is sufficient to hold them together. When the particles are small enough, the ratio of these weak attractions to the weight of the particles is greater, and the material will be cohesive, strong but brittle, and sensitive. When the cohesive bonds are broken through shaking, the short-range forces are ineffective, and a total loss of strength results. If there is sufficient pore water, the material will liquefy and flow on gentle slopes.

An example of a well-documented failure in quick clay is the St. Jean-Vianney failure of May 4, 1971, in Quebec, Canada. This landslip was located in the middle of a much larger prior failure that was at least five hundred years old. At about 7:00 P.M., instability first began with no warning on the steep west bank of the small Petit Bras tributary to the Saguenay River. The first actual failure, however, did not occur until 10:15 P.M., when, in fifteen minutes, the failure surface retrogressed 150 meters west into the slope. The resulting debris moved into the river and blocked both the river valley and the opening to the failure. The dam blocked the outflow of the liquefied material from the crater until the pressure became too high. Finally, the dam burst and allowed the remolded fluid to flow down the river valley with a wavefront about 18 meters high and traveling about 26 kilometers per hour. The mass carried with it thirty-four houses, one bus, and an undetermined number of cars. Thirty-one lives were lost as well.

SOIL STRENGTH

The strength, or load-carrying capacity, of all soils varies considerably. In addition, the strength of any specific soil can vary under different conditions of moisture and density. Sensitive soils have natural water contents already above the liquid limit, which is the moisture content at which a soil passes from a plastic to a liquid state. Plasticity is a characteristic of clayey soils that allows them to be squeezed and easily deformed without disintegration, whereas being above the liquid limit, as the name implies, allows the soil to flow easily and remold itself.

Sensitive clays are generally considered to be thixothropic in the sense that strength diminishes upon disturbance and is regained when disturbance ceases. Once collapse of the card-house matrix is initiated and flow begins, the clays seem to be remolded by lining up of the platelets and particles in a parallel or linear fashion. Remolded clays slip easily over one another in the watery mixture. Once these instabilities develop, they can spread rapidly throughout an area of retrogressive failure. The regaining of strength is not well understood but is thought to result from the gradual rearrangement of particles into positions of increasing mechanical stability under the action of new bonding forces. Perhaps the compaction and gradual expelling of excess water from the moving mass also are responsible for some restoration of strength. The artificial addition of new salts to such materials can be used in certain cases to regain strength and thereby stabilize hazardous areas.

REMOLDING

The typical lateral spread and flow landslips in sensitive clays have certain characteristics that make them distinctive. The processes are essentially a gravitational remolding that transforms the clay into a viscous slurry, or mixture of liquid and solid. Overlying sediments can break into strips or blocks that then become separated. The cracks between the blocks fill with either soft material squeezed up from below or detritus (loose material resulting from disintegration) from above.

Such failures commonly start in the lower part of a slope as a result of local oversteepening through stream or other erosion. After the initial slip, the failure spreads retrogressively, slice by slice, farther and farther into the bank. Movement generally begins quickly, without appreciable warning, and proceeds with rapid to very rapid velocity. A large, bowl-shaped crater commonly results.

SOIL LIQUEFACTION STUDY

Analysis of soil liquefaction is not extensive, because the phenomenon has been known as an important hazard for a relatively short period of time and is an unusual event in any case. The rarity of the phenomenon means that scientists have had trouble installing instruments and making measurements of the ground in places that would eventually liquefy and provide a well-documented record. In most cases, studies had to be done on ground that had liquefied in the past but had since become stable or in the rather artificial situations of the laboratory.

After soil liquefaction was first discovered following catastrophic ground failure, interviews with survivors provided most of the early information. Borehole cores and measurements of displaced ground showed types of sediments and characteristic depths and amounts of water involved. In many engineering applications, the undisturbed cores are subjected to various stresses to determine their behavior when shaken or loaded. Stirring of undisturbed quick clays to a liquid state in the lab, followed by resolidification upon addition of salt, is a means of analyzing the condition. Such studies allow the hazardous condition of certain areas to be represented on maps of liquefaction susceptibility that indicate places in which the phenomenon is likely to occur.

Success in observing soil liquefaction in action has occurred where instruments were previously installed in the ground to measure pore-water pressures at various depths during carefully measured earthquakes. Surprisingly, and in contrast to laboratory experiments in which liquefaction occurs at the same time as does strong shaking, pore-water pressures rose only slowly as the shaking intensified, and sandy layers completely liquefied only well after strong earthquake motions had ceased. The delay appears to occur as uneven pore-water pressures are redistributed in the ground.

EVER-PRESENT EARTHQUAKE HAZARD

The significance of soil-liquefaction potential is enormous. Many of the world's major cities are partly built upon weakly consolidated, water-saturated sediments. Understanding the mechanisms of liquefaction is also important in analyzing earthquakes long past; old sand boils have been used to date and to estimate magnitudes of prehistoric earthquakes. Major earthquakes in Niigata, Japan, and Anchorage, Alaska, brought about a heightened appreciation of the importance of soil liquefaction as a general geologic process. Following these events, geologists, engineers, and planners have come to recognize soil liquefaction as an ever-present earthquake hazard and have associated it with almost all major earthquakes since.

In June 1964, for example, a magnitude 7.3 earthquake occurred 55 kilometers from the city of Niigata on Japan's west coast. In 50 seconds of shaking, the city of 300,000 was subjected to dramatic soil liquefaction that affected thousands of dwellings and industrial structures. Much of the city was built originally upon sand deposits about 100 meters thick along the Shinano River and upon younger lowland sediments and reclaimed riverfront land. During the earthquake, subsurface sand and water flowed up and out of cracks in the ground. The liquefaction caused major destruction of highways, bridges, railroads, utilities, oil refineries, and harbor facilities. There were 3,018 houses destroyed outright and 9,750 damaged moderately or severely because of cracking and unequal settlement of the ground; much of this damage occurred on the newly reclaimed areas. In a most spectacular occurrence, a number of large apartment buildings, which had been designed to be earthquake-resistant, tipped over on their sides to settle at angles of as much as 80 degrees, though the structures themselves remained intact. People were able to escape by walking down the sides of the buildings. Several of these apartment houses were later jacked up, reinforced, and opened for reoccupation.

The Good Friday (March 27) earthquake of 1964 in Alaska was, at 8.4–8.6 magnitude, the largest ever recorded in North America. The Turnagain Heights landslide in Anchorage, Alaska, took place in flat terrain along the steep coastal bluffs that border the Knick Arm of the Cook Inlet there. Before the earthquake, the bluffs rose steeply some 30–35 meters above sea level. Marine clays and silts with layers and lenses of sand of the Bootlegger Cove Clay were exposed in the bluffs. A 6-meter thickness of sand and gravel on the flat terrace above had provided an apparently fine place to build homes. During the 1964 earthquake, giant blocks of Bootlegger Cove Clay and the overlying sands and gravels were set in motion toward the sea as the sand layers in the clay formation were liquefied. The first movements of the blocks began about two minutes after the onset of intense earthquake shaking. In the next five minutes, the previously flat terrain was transformed into a jumble of blocks capped by tilted trees and broken buildings. Seventy-five homes and three lives were lost in the breakup of the ground. More than seven huge blocks became widely separated as they moved toward the sea.

One positive result of soil liquefaction is the tendency for earthquake shaking to be significantly reduced, as liquids do not support shear stress. Once the soil liquefies due to shaking, shear waves are not transferred to buildings at the ground surface.

INCIDENCES OF SUBAQUEOUS FAILURE

In a subaqueous example of failure, the sand beds along the coast of Zeeland in the Netherlands periodically liquefy. The coast is located on a thick layer of fine quartz sand that consists of rounded quartz grains. The slope of the beach is only about 15 degrees. Once every few decades, especially after exceptionally high spring tides, the structure of the sand breaks down beneath a short section of the coastal belt. The sand flows out and spreads with great speed in a fan-shaped sheet over the bottom of the adjacent body of water. The tongue of such a flowslide is always much broader than is the source; the flowslides themselves commonly have surface slope angles of as small as 3–4 degrees. Such a failure occurred at Borssele in 1874 and involved nearly 2 million cubic meters of sand.

A liquefaction process also seems to be significant in the fine silts and clays of the Mississippi Delta region, particularly in the formation of collapse depressions and elongate flows. In these sediments, the pore-water pressures are extremely large, and the pore spaces also contain large amounts of methane gas from decaying organic matter. Following initial failure of these materials, softening of the highly pressured clay/water/gas system causes remolding and strength loss similar to a type of liquefaction or quick behavior. Collapse of offshore structures and sinking of pipelines and seafloor monitoring equipment vertically into the sediment may result. Increased pore pressures during storms also cause bottom movement, collapse, and indications that the sediment can become active as a fluid. On very low-angle slopes (0.1–0.2 degree), distinct collapse depressions are formed, whereas on slightly steeper slopes (0.3–0.4 degree), more elongate flows can result.

Further Reading

Das, Braja M., and G. V. Ramana. *Principles of Soil Dynamics.* Stamford: Cengage Learning, 2011. Provides background information on physics of vibrations, elastic waves, stress waves, and earthquakes. Chapter 10 covers aspects of soil liquefaction procedures and tests.

Dennen, W. H., and B. R. Moore. *Geology and Engineering.* Dubuque, Iowa: Wm. C. Brown, 1986. This volume is one of the most nontechnical but accurate books available on general geotechnical subjects. The sections on soil liquefaction and related phenomena are not long but are easy to understand.

Doyle, Hugh A. *Seismology.* New York: John Wiley, 1995. A good introduction to the study of earthquakes and the earth's lithosphere. Written for the layperson, the book contains many useful illustrations.

Holzer, T. L., T. L. Youd, and T. C. Hanks. "Dynamics of Liquefaction During the Superstition Hills, California, Earthquake." *Science* 244 (April 7, 1989): 56–59. This article reports the details of the first record of a natural liquefaction event after an earthquake. An array of instruments had been installed up to 12 meters deep that recorded excess pore pressures generated once horizontal ground acceleration from the earthquake exceeded a certain threshold value.

Idriss, I. M., and R. W. Boulanger. *Soil Liquefaction During Earthquakes.* Earthquake Engineering Research Institute, 2008. This text covers the basic principles of liquefaction behavior of soils and analysis methods.

Lade, Poul V., and Jerry A. Yamamuro. "Evaluation of Static Liquefaction Potential of Silty Sand Slopes." *Canadian Geotechnical Journal* 48 (2011): 247–264. The article compares sand types and their liquefaction potential.

Lundgren, Lawrence. *Environmental Geology.* 2d ed. Englewood Cliffs, N.J.: Prentice-Hall, 1998. A general book, with several excellent sections and photographs of soil liquefaction phenomena. Not too difficult or technical for the average interested reader.

Penick, J. L., Jr. *The New Madrid Earthquakes.* Rev. ed. Columbia: University of Missouri Press, 1981. This well-written account covers the most intense earthquakes ever to strike the North American continent and details their effects upon people, animals, waterways, and land. Some of the largest sand boils from soil liquefaction ever recorded are described here. The vivid description of the devastation wrought upon the face of the land gives a picture of the dramatic changes caused by the upheaval of natural forces.

Pipkin, Bernard W., and Richard J. Proctor. *Engineering Geology Practice in Southern California.* Belmont, Calif.: Star Publications, 1992. This book provides a detailed description of geological engineering in areas prone to earthquakes, including attempts to prepare for the destructive effects of soil liquefaction. Illustrations and bibliographic references.

Plescan, Costel, and Ancuta Rotaru. "Aspects Concerning the Improvement of Soils Against Liquefaction." *Bulletin of the Polytechnic Institute of Iasi* (2010): 39–45. Presents options for improving foundation soils to protect against soil liquefaction. Discusses jet grouting as an effective preventative method.

Plummer, Charles C., and Diane Carlson. *Physical Geology.* 12th ed. Boston: McGraw-Hill, 2007. A college-level introductory geology textbook that is clearly written and wonderfully illustrated. An excellent sourcebook of basic information on geologic terminology and fundamentals of geologic processes. An excellent glossary.

Spangler, M. G., and R. L. Handy. *Soil Engineering.* New York: Harper & Row, 1982. This book is a detailed engineering text with many equations. Nevertheless, readers will find the discussions on soil liquefaction useful even if they disregard the numerical equations.

Terzaghi, K., and R. B. Peck. *Soil Mechanics in Engineering Practice.* New York: John Wiley & Sons, 1948. This text is the classic work on soil liquefaction. Although it includes many equations, the authors have also provided plentiful written descriptions that are easy to understand.

SOIL MANAGEMENT

—*Judith J. Bradshaw-Rouse & Mark S. Coyne*

FIELDS OF STUDY

Agronomy; Agriculture; Soil Science; Physical Science; Biology; Chemistry; Ecology; Environmental Studies.

SUMMARY

Soil management refers to the collection of tillage, conservation, and cropping practices that are used to preserve soil resources while optimizing soil use.

DEFINITION

Soils are managed differently depending on their intended use. Soil management groups are soil types with similar adaptations or management requirements for specific purposes, such as use with crops or cropping rotations, drainage, fertilization, forestry, highway engineering, and construction. For agriculture, soil management includes all tillage and planting operations, cropping practices, fertilization, liming, irrigation, herbicide and insecticide application, and other treatments conducted on or applied to the soil surface for the production of plants.

Crop consultant draws a soil sample early in the crop year to test nitrogen availability in the soil. Image by Tim McCabe, USDA Natural Resources Conservation Service.

INTRODUCTION

The most basic aspect of soil management is the way in which it is cultivated or tilled for crop growth. Tillage is the mechanical manipulation of the soil profile to modify soil conditions, manage crop residues or weeds, or incorporate chemicals for crop production. Tillage can be exhaustive or minimal. Conventional tillage uses multiple tillage operations to bury existing crop residue and prepare a uniform, weed-free seedbed for planting. This method breaks up soil aggregates in the process and destroys soil structure. Consequently, it can result in excessive wind and water erosion. conservation tillage, or minimum tillage, involves soil management practices that leave much more crop residue on the soil surface and cause much less soil disruption. As a result, the soil is less susceptible to erosion, and the plant residue acts as a mulch to protect the soil surface from the destructive impact of rainfall as well as to reduce evaporation. No tillage, or chemical tillage, is a soil management practice adapted to sloping soils in which herbicides rather than tillage are used to control weeds, while the disruption of soil structure is limited to a narrow slit in the soil surface in which the seeds are planted.

Soil management extends to the way in which soils are manipulated. Terraces, for example, are raised horizontal strips of earth constructed along the contour of a hill to slow the movement of downward-flowing water. Tile drains are perforated ceramic or plastic pipes buried in poorly drained soils that act as underground channels to carry water away, lower the water table, and allow a soil to drain faster after rainfall. The benefit of managing potentially erodible soils on hill slopes as permanent pastures has gained increasing recognition. Likewise, the value of retaining wet soils as wetlands has been acknowledged. Wetlands provide wildlife habitat, assist in flood control, and act as buffers to protect surface waterways from nutrient and soil runoff from cultivated fields.

Soil management also involves the addition of chemicals to soil: lime to make acid soils more neutral; fertilizers to increase the nutrient level; herbicides and insecticides to control weed and insect pests; and soil conditioners to improve soil aggregation, structure, and permeability. A growing technology is the use of mobile global positioning system (GPS) units attached to the equipment that applies these chemicals to soil. Called "site specific management," it uses computer technology to regulate chemical addition based on the exact position in a field and previous yield or fertility maps that indicate whether the soil needs to be amended. The goal of site specific management is to optimize chemical use and profit while minimizing potential chemical loss to other environments by only applying the chemicals to areas where they are needed.

SOIL PROFILES

—Scott F. Burns

FIELDS OF STUDY

Agronomy; Agriculture; Soil Science; Physical Science; Biology; Chemistry; Ecology; Environmental Studies.

PRINCIPAL TERMS

horizon: a layer of soil material approximately parallel to the surface of the land that differs from adjacent related layers in physical, chemical, and biological properties

leaching: the dissolving or removal of soluble materials from a soil horizon by percolating water

sediment: rock fragments such as clay, silt, sand, gravel, and cobbles

structure: the arrangement of primary soil particles into secondary units called peds

texture: the relative proportions of varying sediment sizes in a soil

weathering: the mechanical disintegration and chemical decomposition of rocks and sediments

SUMMARY

A soil profile is the vertical section of a soil extending through its horizons into the unweathered parent material. Scientists use the different arrangements of horizons to give classification names to the soil profiles, which can be used to interpret landscape history and determine future land use.

HORIZON NOMENCLATURE

The term "soil" has many definitions. To the soil scientist, it is anything that will support plant growth. To the civil engineer, it is an unconsolidated surficial material that can be penetrated by a shovel. To the geologist, it is an unconsolidated layer of weathered mineral matter arranged in layers at the earth's surface. A soil profile is a vertical arrangement of soil horizons down to and including the parent material in which the soil has developed. Two sets of horizon nomenclature are used to describe soil profiles: symbols for field descriptions and diagnostic horizon names for classification. The field description

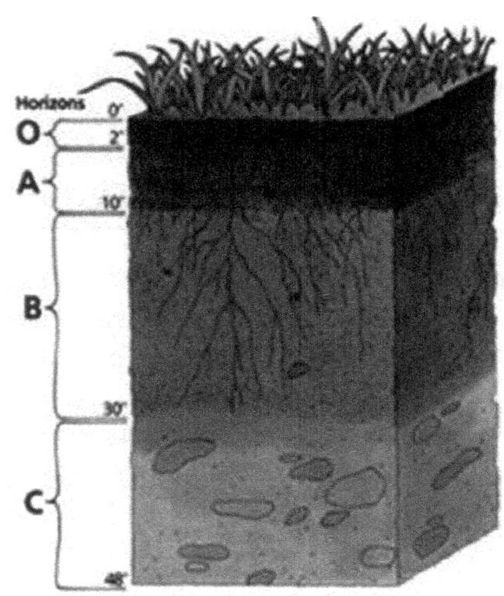

Graphic of a soil profile showing O, A, B, and C horizons.

symbols have been in use since the nineteenth century, whereas the diagnostic horizon names came into use only in the 1960's.

The field description symbols used to identify various kinds of horizons were developed in Russia in the 1880's. There, soil scientists such as V. V. Dokuchayev applied the letters A, B, and C to the main horizons of the black soils of the steppes of Russia. The A horizon was designated the zone of maximum organic material accumulation, the C horizon was the unaltered parent material, and the B horizon was the layer in between the A and the C. These concepts spread to the rest of the world, with the B horizon concept being modified to be a zone of the accumulation of iron, aluminum, and clays that had moved down the profile. Horizons may be prominent or so weak that they can be detected only in the laboratory; they can be thick or thin. The use of the A, B, and C master horizons has remained the backbone of horizon symbols. Three additional master horizons, however, have also been included: O, for organic-rich layers; E, for layers that have been intensively leached; and R, for rock. Soil terms like "podzol" (from *pod*, meaning "underneath," and *zola*, meaning "ash") and "chernozem" (from *cherny*, meaning "black," and *zemlya*, meaning "earth") are still used and reflect the Russian roots of soil classification.

Since the 1930's, many sets of horizon symbols have been developed around the world. With the United States and many other countries adopting a horizon nomenclature similar to that produced by the Food and Agricultural Organization (FAO) of the United Nations in the early 1980's, the world has come closer to a universally accepted set of horizon symbols.

REFINING HORIZON NOMENCLATURE

There are three kinds of symbols used in soil horizon descriptions: capital letters, lowercase letters, and Arabic numerals. The capital letters describe the master horizons, the lowercase letters depict some specific characteristic of a master horizon, and the numeral characterizes a further subdivision or parent material layering. An example of a horizon symbol might be Bg1.

The "g" describes a "gleyed" horizon, or one where the iron has been removed during soil-forming processes, or saturation with stagnant water in a high water table has preserved the reduced state of the iron, giving the layer a neutral or gray color. This letter also can describe a soil with red and gray specks from the oxidized and reduced forms of iron. A horizon that is more than 90 percent cemented is given the "m" designation. Horizons with abundant silica are described with a "q" and abundant sodium with an "n." If the horizon is very densely packed, which commonly occurs in silt-rich soils, the "x" designation is used. A "b" denotes a buried soil horizon. A "c" signifies that weathering has formed concretions, or nodules, in the soil.

Transitions between master horizons are commonly found in nature. Where the properties of both horizons are mixed in the same layer, both capital letters are used, with the first letter denoting the horizon whose properties dominate. The term BA horizon would be a transition between the A and B horizons, where the B horizon characteristics prevail. Where the horizon has distinct parts of both horizons, the two capital letters are separated by a slash mark (A/B, E/B, B/C).

ORGANIC SOIL HORIZONS

The O horizon is a master horizon in which there is an accumulation of mainly organic matter that overlies a mineral soil. It is dominated by fresh or partly decomposed organic litter such as twigs and needles, and many times the original form of most of the vegetative matter is visible to the naked eye. This horizon contains between 20 and 30 percent organic material, sometimes more depending upon the clay content of the underlying mineral horizons. Three subdivisions of this master horizon are based on the amount of decomposition in the organic material and range from the Oi (least decomposition) to the Oe (intermediate decomposition) and the Oa (most decomposition).

The A horizon is a master horizon in which decomposed organic matter, humus, is mixed with mineral sediments. The organic matter content is not great enough to be classified as an O horizon. This horizon is generally a surface horizon, except when located below an O horizon. Because of the presence of organic material, this layer can be darker than underlying horizons. The organic material is assumed to be derived from the decomposition of plant and animal remains deposited on the surface. This layer is the zone of maximum biological activity. A subdivision is the Ap, or "plow horizon," where the A horizon has been disturbed by cultivation.

MINERAL SOIL HORIZONS

The E horizon is the master horizon commonly located below the O or A horizons in forest environments where leaching is dominant in the soil profile. Abundant water passing through the O horizon may become very acidic; this water leaches iron, aluminum, and organics from the A horizon as it passes to the lower part of the profile. The remaining mineral matrix is light-colored because of the color of the primary mineral sand grains that remain. Sometimes the horizon is almost pure white. For many years, this horizon was called an A2 horizon but has been removed from the category, as it is not a zone of organic material accumulation. The E horizon sometimes is light enough to look like ash, explaining the Russian term "podzol"; the process that forms such soils is still called "podzolization."

The B horizon is the master zone of accumulation of materials that have been moved down by water from the O, A, and/or E horizons. These suspended materials include clay, iron, aluminum, and organic matter. The soil horizon shows little or no evidence of the original sediment or rock structure. Several kinds of B horizons are recognized depending upon the materials moved into them and, of course, not all horizons are present in any given soil. A Bh horizon is an accumulation of abundant organic material and therefore is very black in color. In a Bs horizon, the deep red color depicts the accumulation of abundant iron and aluminum (the latter of which is colorless). A Bk horizon has an abundant amount of calcium carbonate and a white color throughout. A Bt horizon includes a large amount of clay that has moved down from the horizons above. Clay films are noted on the sediments, and the soil samples are very sticky. A By horizon is an accumulation of gypsum. A Bz horizon contains salts more soluble than gypsum. A Bw horizon is a weakly developed B horizon in which a reddish color has developed through weathering, but few or no apparent materials have been moved into it from above.

The C master horizon is the subsurface layer of partially weathered parent material. Soil-forming processes, in which particles and chemicals move downward from the O, A, E, and B horizons, have not affected this horizon, yet weathering of the sediments has slightly changed the color of the parent material. If the horizon material has the structure of the parent rock but is weathered enough to get a shovel through it, the horizon is called a Cr horizon; geologists also refer to this horizon as a saprolite. The R master horizon is the unweathered, consolidated bedrock underlying the soil.

SOIL TAXONOMY

A second set of diagnostic horizon nomenclature was developed in the latter half of the twentieth century in the United States, using a new system called Soil Taxonomy. Information from diagnostic surface horizons (epipedons), subsurface horizons, soil-moisture, and temperature regimes together are used to determine the soil classification. Soil Taxonomy is based mainly on measurable soil properties based in seventeen epipedons and subsurface horizons rather than on theories of soil formation, which form the basis for older classification systems. The classification system is very exact and offers exotic combinations of Latin and Greek syllables, such as Pachic Cryumbrepts and Natraqualfic Mazaquerts. Each syllable is a descriptive term for some aspect of the soil. For example, "natraqualfic" means sodium-rich (natr-), wet (aq-), and having a B horizon rich in iron and aluminum (alfic). Field observations must be supplemented by laboratory analyses of properties plus climatic data. Similar approaches using this type of horizon nomenclature have also been developed in Canada and by the FAO (the UN Food and Agriculture Organization). In the United States, soils are classified into twelve soil orders on the basis of their horizon materials and structures. Most soil types are defined by the distinctness of their horizons and degree of evolution, but some, like soils developed from peat, volcanic ash, swelling clays, and on permafrost, are defined by those special characteristics.

Soil Taxonomy describes four different epipedons. The Mollic epipedon is a thick, dark-colored surface horizon rich in mineral nutrients. The Umbric epipedon is similar to the Mollic visually but is low in mineral nutrients. The Ochric epipedon is a thin surface horizon that does not meet the requirements of the Mollic or Umbric. The Histic epipedon is basically an O horizon.

The additional Soil Taxonomy diagnostic horizons are subsurface horizons. The Albic horizon is basically an E horizon, as it has been highly leached. An Argillic horizon is a B horizon that has been enriched significantly with clay from higher in the profile. The Natric horizon contains a high abundance of salt. The Spodic horizon is a B horizon with abundant iron, aluminum, and/or organics that have been moved from above. The Cambic horizon is a B horizon that is slightly enriched in red color but, unlike the Spodic and Argillic horizons, lacks an accumulation of products from above. The Oxic horizon is a B horizon with an abundance of highly weathered minerals. A Calcic horizon has an abundance of calcium carbonate in it. A Petrocalcic horizon is a cemented Calcic horizon, often called caliche. The Gypsic horizon contains an abundance of gypsum. A Petrogypsic horizon is a cemented Gypsic horizon. A Duripan is cemented with silica. A Fragipan horizon is very dense and is many times formed in silt particles.

STUDYING A SOIL

The first step in the study of a soil is producing a description of the soil profile. A scientist studying a particular soil exposes a soil profile by either digging a soil pit or finding a road cut. First, the soil horizons are delineated (A, B, C, and so on). Second, each horizon is characterized by thickness, color, presence of films, and boundaries. The horizon texture (the relative amount of sediment sizes) is recorded as well. For example, a sandy loam soil has an abundance of sand with some silt in it. The structure of that horizon, or how the sediments are put together, is described as well. Third, samples from each horizon are gathered to be analyzed in the laboratory. Finally, an approximate classification is made based on the minimal information available in the soil pit.

The soil samples are returned to the laboratory for a chemical and physical characterization. The pH (degree of acidity) and amounts of chemical nutrients are generally determined using pH meters and titrations. The exact texture is determined using sieves and water-settling columns. The percentage of organic matter is calculated using titration. Additional information can be obtained using an X-ray machine to determine the types of clays in the soil.

Once the field and laboratory data have been assembled, diagnostic horizons can be selected and the exact classification is determined using Soil Taxonomy. Twelve major classification groupings, or orders, are used in this system. Entisols are young soils that have no B horizon, whereas Inceptisols have a weak B horizon. In humid climates the major mature soil types are Mollisols, Alfisols, and Spodosols. The Mollisols are soils with a Mollic epipedon and are often grassland soils. Alfisols have an Argillic horizon and abundant nutrients; they are often deciduous forest soils. Spodosols are highly leached soils that have Spodic (E) horizons and are common in conifer forests. Ultisols are highly weathered soils of warm climates, and Oxisols are extremely weathered soils of the tropics. Aridisols are arid climate soils. Histosols are peaty soils that have thick Histic epipedons. Vertisols are high in clays that shrink and swell with seasonal moisture variation and exhibit extensive cracking during dry seasons. The Andisols are volcanically produced soils. Gelisols are soils formed above permafrost. Using the proper classification, scientists can interpret the soil's age, future land use, the crops it can best support, and its past history.

A FUNDAMENTAL TOOL

For years, the soil profile has been the basic descriptive tool of scientists. This cross-section constitutes the format from which scientists make their soil classifications. Without a proper soil profile description, one cannot perform the soil classification required before environmental interpretations and soil maps can be produced. The symbols that are used to describe the horizons of the soil profile may have changed, but the basic A-B-C concept that was developed during the nineteenth century has not changed much. Different systems of description of the soil profile and the resulting soil classification systems are present around the world, but with time they are becoming more similar.

In the United States, the Soil Taxonomy system has been officially used since 1975. This new classification offers a very exact hierarchical system similar to what biologists use to classify plants and animals. The twelve major orders are divided into suborders, great groups, subgroups, and finally soil series. The subgroup name, such as "Typic Paleudult," is similar to the genus/species designation of a plant or animal.

Soil Taxonomy is the most comprehensive soil classification system in the world and therefore is used more than any other approach. It is such an exact system that one can give a scientist familiar with the taxonomy a subgroup name and he or she can construct a complete soil profile description very close to the actual one without even seeing the soil. Soil classification may be the most difficult classification problem in science, as a large number of factors must be considered and a wide variety of client groups want to apply the results to very different problems.

FURTHER READING

Al-Rawas, Amer Ali, and Mattheus F. A. Goosen, eds. *Expansive Soils: Recent Advances in Characterization and Treatment.* New York: Taylor & Francis, 2006. Discusses the classification of expansive soils. Provides information on swelling and consolidation, and the effects these processes have on buildings. Discusses how site characteristics and stabilizing agents can alter expansion processes.

Birkeland, P. W. *Soils and Geomorphology.* 3d ed. New York: Oxford University Press, 1999. One of the best books written on soils from the perspective of a geologist. Offers many examples from around the world. Covers the newly adopted soil profile description symbols. Discusses Soil Taxonomy in the early chapters. Suitable for university-level readers; early chapters are appropriate for high school-level readers.

Buol, S. W., F. D. Hole, and R. J. McCracken. *Soil Genesis and Classification.* 5th ed. Ames: Iowa State University Press, 2003. An update of a classic text on soil science. New to the edition is an update on the new soil horizon symbol nomenclature. Includes an in-depth discussion on soil development. Suitable for university-level readers.

Carr, Margaret, and Paul Zwick. *Smart Land-Use Analysis: The LUCIS Model.* Redlands, Calif.: ESRI Press, 2007. Discusses the use of ArcGIS software to analyze land suitability. Provides a step-by-step guide to ArcGIS software. The LUCIS model is explained with case studies to provide examples of potential conflicts of land use.

Idriss, I. M., and R. W. Boulanger. *Soil Liquefaction During Earthquakes.* Oakland, Calif.: Earthquake Engineering Research Institute, 2008. Covers the basic principles of the liquefaction behavior of soils and analysis methods.

Jenny, Hans. *Factors of Soil Formation: A System of Quantitative Pedology.* New York: Dover, 1994. Describes the factors involved with variations in soils. Suitable for university-level readers. Illustrations, index, and bibliography.

__________. *The Soil Resource: Origin and Behavior.* New York: Springer-Verlag, 1980. A rewrite of the classic book from 1941. Describes variations in soils. Suitable for university-level readers.

Kapur, Selim, and George Stoops, eds. *New Trends in Soil Micromorphology.* New York: Springer, 2010. Compiles articles suited for graduate students and researchers. Discusses the identification and analysis of paleosols, and soil micromorphology of the Quaternary period. Addresses vertisols, andosols, and other paleosols.

Landa, Edward R., and Christian Feller, eds. *Soil and Culture.* New York: Springer, 2010. Provides a full history of the relationship between humans and the earth. Organized into parts discussing soil's function in art, philosophy, culture, and health. Provides interesting information from an anthropological perspective.

Paton, T. R., G. S. Humphreys, and P. B. Mitchell. *Soils: A New Global View.* New Haven, Conn.: Yale University Press, 1995. Aimed at readers at the university level. Offers many good examples of soils and soil formation from around the world. Color illustrations, maps, bibliography, and index.

Singer, M. J., and D. N. Munns. *Soils.* 6th ed. Upper Saddle River, N.J.: Prentice Hall, 2005. Includes well-explained soil profile descriptions and classification. Suitable for high school-level readers.

Soil Science Society of America. *Glossary of Soil Science Terms.* Madison, Wis.: Author, 2001. Defines soil science terms; written by soil science professionals.

Sparks, Donald S., ed. *Soil Physical Chemistry.* 2d ed. Boca Raton, Fla.: CRC Press, 1999. Offers a clear look into the practices, procedures, and applications of the field.

Sposito, Garrison. *The Chemistry of Soils.* 2d ed. New York: Oxford University Press, 2008. Provides a comprehensive discussion of soil composition. Describes chemical makeup and basic soil components. Discusses physical and chemical characteristics and processes within soil types. Concludes with a list of references for further reading.

Tan, Kim Howard. *Principles of Soil Chemistry.* 4th ed. Boca Raton, Fla.: CRC Press, 2011. Written as an introduction to soil chemistry. Explains the processes and application of the science in an easy-to-understand manner accompanied by plenty of illustrations. Includes a twenty-five-page bibliography, as well as an index.

U.S. Department of Agriculture, Soil Survey Staff. *Keys to Soil Taxonomy*. Honolulu: University Press of the Pacific, 2005. A difficult but important text on the official classification system of the United States.

SOIL SALINIZATION

—Robert E. Carver

FIELDS OF STUDY

Agronomy; Agriculture; Soil Science; Physical Science; Biology; Chemistry; Ecology; Environmental Studies.

SUMMARY

The salinization of soil on agricultural lands can result in poor plant growth and reduced crop yields, and attempts to reduce salinity through increased irrigation can lead to contamination of water supplies.

In the process of soil salinization, water-soluble salts build up in the part of the soil known as the root zone, where the soil comes into contact with the roots of plants, blocking the movement of water and nutrients into plant tissues. Soil salinization rarely occurs naturally. Rainwater is virtually free of dissolved solids, but surface waters and groundwater contain significant quantities of dissolved solids, ultimately produced by the weathering of rocks. Evaporation of water at the land surface results in an increase in dissolved solids in soil that may adversely affect the ability of plant roots to absorb water and nutrients.

In arid regions, evaporation of soil water potentially exceeds rainfall. Shallow wetting of the soil followed by surface evaporation lifts the available dissolved solids to near the surface of the soil. The near-surface soil therefore becomes richer in soluble salts. In natural arid areas, soluble salts in the subsurface are limited in quantity because rock weathering is an extremely slow process, and degrees of soil salinization detrimental to plants are uncommon.

The irrigation of arid-climate soils with surface water or groundwater provides a constant new supply of soluble salt. As the irrigation water evaporates and moves through plants to the atmosphere, the dissolved solid content of the soil water increases. Eventually, the increase in soil salt will inhibit or stop plant growth. It is therefore necessary to apply much more water to fields in arid climates than the amount required for plant growth, so that the water flushes salts away from the plant root zone. If the excess water drains easily to the groundwater zone, however, the groundwater becomes enriched in dissolved solids, which may be detrimental.

If the groundwater table is near the surface, or if impermeable soil zones are close to the surface, overirrigation will not alleviate the problem of soil salinization. Alleviation of this condition requires the installation of subsurface drains to carry the excess soil water and salts to a surface outlet. The problem with this method is that disposing of the salty drain water is difficult. If the drain water is released into surface streams, it degrades the quality of the stream

Soil Salinity. Image by R.J.Oosterbaan at nl.wikipedia.

water, adversely affecting downstream users. If the water is discharged into evaporation ponds, it has the potential to seep into the groundwater zone or produce a dangerously contaminated body of surface water, as occurred at the Kesterson National Wildlife Refuge in California, where concentrations of the trace element selenium rose to levels that interfered with the reproduction of resident birds.

Further Reading

Blanco, Humberto, and Rattan Lal. *Principles of Soil Conservation and Management.* New York: Springer, 2008.

Vengosh, A. "Salinization and Saline Environments." In *Environmental Geochemistry,* edited by Barbara Sherwood Lollar. Oxford: Elsevier, 2005.

Soil Science

—*Domingo M. Jariel Jr., PhD*

FIELDS OF STUDY

Agronomy; Agriculture; Soil Science; Physical Science; Biology; Chemistry; Ecology; Environmental Studies.

SUMMARY

Soil science is the multidisciplinary study of soils, which are composed of mineral, water, air, and organic matter. Soil supports the growth of terrestrial plants, controls the percolation of water, recycles elements and organic molecules, provides habitats for biota, alters dust and gaseous components in the atmosphere, and stabilizes the foundations of buildings and constructions.

DEFINITION & BASIC PRINCIPLES

Soil science is the study of the physical and chemical properties of soils. Soil is the end product of the weathering of rocks and minerals and the decomposition of living organisms. Soil solids include minerals (inorganic particles) and humus (organic matter). As soil solids cling together, pore spaces, or voids, are formed between inorganic and organic materials to accommodate air and water. A typical loam soil contains 45 percent minerals, 5 percent humus, 25 percent water, and 25 percent air. Among the three inorganic particles (sand, silt, and clay), pure sands have the largest particle size, the lowest surface area per unit volume, and the lowest percent of pores. Sands have the largest pores, which allow faster water percolation and drainage. Clay particles are the smallest and have the highest surface area per unit volume and the highest percent of pore spaces. Therefore, clays have the greatest water-holding and element-adsorbing capacity among inorganic particles. Organic humus is the lightest and has the greatest percent of pore spaces among the soil solids. Addition of humus increases the retention of water and elemental nutrients in sandy soils and improves the drainage of water in clay soils.

Soils have net negative charge because the surfaces of soil particles have more negative sites than positive sites. Negative sites attract positively charged ions (cations) of nutrient elements. The attracted cations are loosely adsorbed and can be easily replaced by other cations. For example, application of ammonium fertilizers in the soils releases ammonium cations (NH_4^+), which can replace other adsorbed cations, such as the cations of potassium (K^+), hydrogen (H^+), magnesium (Mg^{2+}), calcium (Ca^{2+}), or sodium (Na^+). This chemical property of the soils plays an important role in crop production and in retaining chemical wastes and other pollutants.

BACKGROUND & HISTORY

Edaphologists, who studied soil as a medium for plant growth, could be considered the pioneers of soil science. The next group of scientists, the pedologists, studied soil as a geologic entity, examining its origin, formation, morphology, and classification.

Edaphology research began in the early 1600's, when Dutch scientist Jan van Helmont conducted an

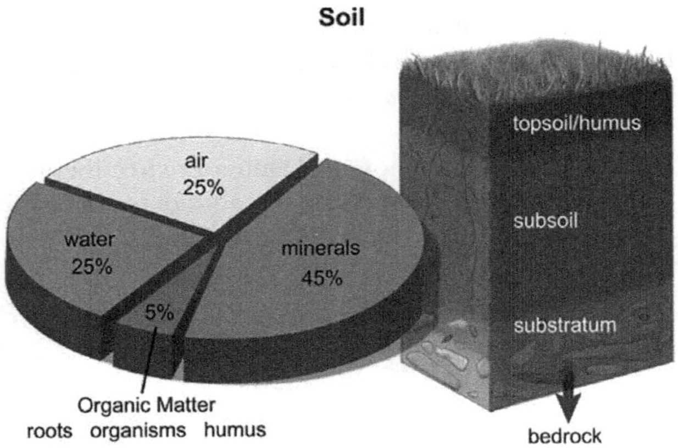

experiment involving soil. He planted a five-pound willow tree in 200 pounds of soil and let it grow for five years. During those years, the plant received only rainwater, no nutrients or fertilizers. At the end of the experiment, the weight of the soil had decreased while that of the plant increased. Helmont concluded that the elements from the soil had contributed to the weight increase of the willow tree. In the 1800's, German chemist Justus von Liebig postulated that the growth of plant is limited by sixteen essential nutrient elements, a number that scientists later increased to seventeen.

In the early 1870's, Russian geomorphologist Vasily Vasilyevich Dokuchayev, the father of pedology, conceived of soils as natural bodies, each with its own specific characteristics resulting from a unique process of development. In the 1890's, Dokuchayev created a classification system for Russian soils. In the 1920's, the Russian technique was adapted by soil scientist C. F. Marbut, who applied it to the soils of United States. In 1941, Swiss-born scientist Hans Jenny published *Factors of Soil Formation: A System of Quantitative Pedology*. His work summarized and illustrated investigations into soil science and became a standard work in the field. In 1965, an official soil classification system was completed for the U.S. National Cooperative Soil Survey; this was later published in 1975 as *Soil Taxonomy: A Basic System of Soil Classification for Making and Interpreting Soil Surveys.*

HOW IT WORKS

Soil Formation. It takes hundreds of year for the soils to develop from parent materials. The first step in soil formation is the weathering of rocks and minerals by physical, chemical, or biological processes. Over time, the consolidated rocks turn into unconsolidated soil materials lying on top of the existing bedrock. Some unconsolidated materials can be carried by wind, water, gravity, or a glacier and then deposited as sediment in different locations. Bedrock and transported sediments become the parent materials of inorganic (mineral) soils. Some land areas, which are cold or inundated with stagnant waters, are subject to faster accumulations of plant and animals residues. The decomposed residues serve as the parent materials of organic soils.

As the soil develops on top of the parent materials, a soil profile is formed with distinct layers of soil horizons. Soil taxonomists use the unique properties of horizons in a profile as the basis for soil classification. Soils are classified into twelve orders.

Physical Properties. The physical properties of soil (such as texture, density, and structure) affect their role in agriculture, the environment, and engineering. Soil texture represents the percentages of sand, silt, and clay particles. The greater the proportion of clay, the more the soil responds to compaction, which increases bulk density, decreases porosity, and reduces aeration. As a result, compacted clays prohibit root penetration but provide better foundation for engineering projects.

Chemical Properties. The chemical properties of soil—cation exchange capacity (CEC), pH (acidity-alkalinity), and nutrient availability—depend solely on microscopic soil colloids, clays, and humus. Each colloid possesses more negatively charged surfaces than positively charged surfaces. The total amount of cations that can be adsorbed by the soil colloids is called the cation exchange capacity. Exchangeable cations held by soil colloids are not easily removed by percolating water but can be easily removed by root absorption and replaced by other cations. Other adsorption reactions bind the ions more snugly, preventing them from leaching into the environment.

The cation exchange capacity influences the development of acidic or alkaline soils. In rainy locations, H^+ ions from rain replace most of the basic cations (K^+, Ca^{2+}, Mg^{2+}, Na^+) on the negative sites. As rainwater percolates through the soil, the basic cations leach with water to the deeper horizons, leaving the top soil acidic with pH lower than 7 because of high H^+ accumulation in the exchange sites. In arid regions, alkaline soils with pH greater than 7 are found because basic cations are retained in the exchange sites.

Soil Biota. Based on their feeding habits, soil-dwelling organisms (plants, animals, fungi, protists, bacteria, and archaea) play a role in the formation of soil, contribute to the accumulation of organic matter, and recycle nutrients in the soil. Large animals (such as badgers, mice, and prairie dogs) dig underground tunnels and aerate the soil. Earthworms strengthen the soil structure. As they burrow, they ingest plant and animals residues, including clays, and excrete them as granular aggregate. Most soil fungi (molds and mushrooms) are decomposers. They use the nutrients and organic molecules from the tissues of dead and living plants and animals. Protists include plant-like (algae), animal-like (protozoa), and fungus-like (slime molds) organisms. Algae, like plants and cyanobacteria, are not decomposers but rather primary producers because they are photosynthetic organisms. Protozoa (such as amoeba) are unicellular organisms that ingest small living organisms. Most bacteria and archaea, which decompose the tissues of all dead organisms, are found in the lowest part of the food chain.

APPLICATIONS & PRODUCTS

Medium for Plant Growth. Soil anchors the roots of a plant (or shoot) and holds its above-ground parts in place. Pore spaces between soil solids supply oxygen to the roots of established plants or to germinating seeds during respiration. Soil pores also store water, which is supplied to the roots for absorption or to the seeds for imbibition. Moisture in the pores minimizes soil temperature fluctuations in the rhizosphere. As a result, soil serves as an insulator that protects plant roots from extremely hot and cold conditions.

All higher plants require seventeen essential elements to complete their life cycle. Three elements—carbon (C), hydrogen (H), and oxygen (O)—mostly come from air and water. The remaining fourteen elements—nitrogen (N), phosphorus (P), potassium (K), calcium (Ca), magnesium (Mg), sulfur (S), boron (B), copper (Cu), chlorine (Cl), iron (Fe), manganese (Mn), molybdenum (Mo), nickel (Ni), and zinc (Zn)—are supplied by the soil. Macronutrients (N, P, K, Ca, Mg, and S) are the elements needed by the plants in large amounts. Therefore, incorporation of these nutrients in the soil (as fertilizers) is often necessary. Micronutrients (B, Cu, Cl, Fe, Mn, Mo, Ni, and Zn) are required by plants in small amounts, such that the application of fertilizers containing micronutrients is usually not necessary.

Nutrient elements from the soil are absorbed by plant roots in the form of negatively charged ions (anions) or positively charged ions (cations). The forms of ions easily absorbed by plant roots are anions—nitrate (NO_3^-), hydrogen phosphate ($H_2PO_4^-$), sulfate (SO_4^{2-}), hydrogen borate ($H_4BO_4^-$), Cl^-, and molybdate (MoO_4^{2-})—on the positive sites and cations—NH_4^+, K^+, Ca^{2+}, Mg^{2+}, Cu^{2+}, Fe^{2+}, Mn^{2+}, Ni^{2+}, and Zn^{2+}—on the negative sites.

Soil amendments increase cation exchange capacity, improve the water-holding capacity and drainage of the soil, and supply more nutrients to the plants. Commercial potting mix for ground and container gardening contains a combination of fertilizer, perlite, vermiculite, and compost. Synthetic soils, which are made of potting mix plus additional amendments such as sludges, hygdrogels, and organic mulches, are commercially manufactured in large quantities and used for reclamation or remediation of abandoned mine sites, barren soils, or unproductive wetlands.

Water Storage and Filtration System. Water is stored in the soil, but some water travels through soil pores and moves to streams, rivers, lakes, and aquifers. Contaminated waters that are spilled onto soils undergo purification and clarification processes that remove toxic ions and impurities before the cleansed water enters waterways. This filtration mechanism relies on the pores and exchange sites of the soil. As water moves through the soil, large molecules of pollutants are trapped within the microscopic pores, allowing only small molecules of water to pass through. Soil further removes ions of toxic wastes and heavy metals by electrostatic reaction with the ion exchange sites of the soil colloids.

Not all soils can store and filter water effectively. Sandy soils make ineffective water storage and filtration systems because of their large pores. Compacted clays have tiny pores that become impermeable to water. Torrential rains that pour into denuded mountains with impermeable clays result in flash floods of muddy water.

Nutrient and Waste Recycling. Soils are the major repositories of wastes created by humans.

The capacity of soil to recycle nutrients from wastes depends on the living organisms that inhabit the soils. Organic wastes are decomposed into valuable humus by fungi and microbes. Mineral nutrients from humus are used by plants and animals. Plants acquire carbon as carbon dioxide from the atmosphere and release oxygen into the atmosphere through photosynthesis. Animals release carbon dioxide into the atmosphere through respiration. When plants and animals die and decompose, mineral nutrients are deposited in the soil.

Atmosphere Modification. Exposed soils that are dry and poorly structured are susceptible to erosion by wind, which can carry and disperse the soil particles as dust in the atmosphere. Suspended dust decreases the clarity of the air, increases the chance of developing respiratory problems, and cools the atmosphere and the ground. Soils that are covered with vegetation are less affected by erosion.

Soil alters the amounts of ammonia (NH_3) and nitrogen gas (N_2) in the atmosphere. Decomposition of organic matter by bacteria and fungi in the soil generates NH_4^+ through ammonification. Some NH_4^+ is absorbed by plants and some volatilizes into the atmosphere as ammonia (NH_3). Soil bacteria convert NH_4^+ into nitrate (NO_3^-) by nitrification. Some nitrate is assimilated by plants, and some is denitrified by soil bacteria into nitrogen gas (N_2). Some N_2 in the soil pores is fixed by bacteria that live in the roots of the plants. Some atmospheric N_2 is converted into NO_3^- by lightning. Whatever form of nitrogen is in the soil, nitrogen is subject to rapid loss. For this reason, soil nitrogen is the element most needed by plants. Nitrogen is often replaced by the addition of nitrogen-containing fertilizers or compost to soil.

Medium for Engineering Projects. Soil mechanics, an examination of the architecture and physical properties of the soil, determines which types of soil are best suited for construction. Soil, as compacted solid ground, serves as a stable and firm foundation on which to build roads, bridges, houses, towers, and other infrastructure. Most structures are erected on the soil surface, and many engineering projects (such as underground tunnels) involve digging deeper in the soil. Soil stability is based on its compressibility. Sands are difficult to compress, but compressed clays form a very stable compacted material. However, clays that shrink and form deep cracks when dry and expand when wet (such as montmorillonite) are not suitable sites for buildings. Highly weathered clays such as kaolinite do not shrink and swell, and are appropriate types of clays for engineering projects.

Soils are extensively used as materials in building houses. Walls of African huts are made of mud, and walls of modern Western houses use brick (baked clay), hollow cement blocks (a mixture of sand and cement), silica glass (made from sand), and other soil-cement mixtures. Green roofs are those that are partly or totally covered with soil and vegetation. They insulate the structures they cover and help to lower the air temperature within cities.

SOCIAL CONTEXT & FUTURE PROSPECTS

Soils are highly used and abused natural resources. For example, farmers who employ conventional crop production methods cultivate the soils each year, and if they routinely apply more than the recommended amounts of nutrient fertilizers, this could lead to overfertilization. In rainy areas, the excess nutrients leach down to the aquifer and run off horizontally, polluting the surface water and groundwater. In arid regions, the excess nutrients facilitate the development of saline soils. Annual cultivation of soil removes organic matter and exposes the soil to wind and water erosion. Many modern farmers use windbreaks and no-till farming to reduce soil erosion.

The U.S. government has implemented many programs to combat soil problems. For example, the Natural Resources Conservation Service planted grasses on barren soils in Texas to counteract wind erosion. As a result, dust particles in the air were reduced drastically. Bioremediation projects operated by the Environmental Protection Agency have been able to identify plants that can extract toxic heavy metals from the soil and water. Although technologies are being continually developed to increase crop production and to improve soil and water quality, ultimately, it is up to humans to ensure that environmental quality is maintained for future generations.

FURTHER READING

Jariel, Domingo M., et al. "Plant and Soil Nutrient Concentrations in a Restored Cajun Prairie." *Proceedings of the Twenty-first North American Prairie Conference* 21 (2010): 95–107.

Lal, Rattan, and Manoj K. Shukla. *Principles of Soil Physics.* New York: Marcel Dekker, 2004.
Liu, Cheng, and Jack Evett. *Soil Properties: Testing, Measurement, and Evaluation.* 6th ed. Upper Saddle River, N.J.: Pearson/Prentice Hall, 2009.
Sposito, Garrison. *The Chemistry of Soils.* 2d ed. New York: Oxford University Press, 2008.
Sylvia, David M., Jeffrey J. Fuhrmann, Peter G. Hartel, and David A. Zuberer, eds. *Principles and Applications of Soil Microbiology.* 2d ed. Upper Saddle River, N.J.: Pearson/Prentice Hall, 2005.

SOIL TESTING & ANALYSIS

—*Panagiotis D. Scarlatos*

FIELDS OF STUDY

Agronomy; Agriculture; Soil Science; Physical Science; Biology; Chemistry; Ecology; Environmental Studies.

SUMMARY

Understanding soil properties is crucial to a large variety of agricultural, geological, and soil science applications as well as to geotechnical, environmental, and foundation engineering projects. Soil testing can be conducted in situ or at the laboratory.

BACKGROUND

Soil is the top part of the Earth's crust; it comprises loose, unconsolidated materials that have been formed by physicochemical weathering and disintegration of rocks. Soils are usually composed of mixtures of inorganic minerals, organic matter (humus), water, and air. Inorganic soils are mainly classified as coarse (granular) soils and fine (cohesive) soils. Coarse soils include boulders, cobbles, gravel, and sand. Fine soils include silts and clays that have a mean particle diameter of less than 0.062 millimeter. The properties of coarse soils are affected mainly by mechanical forces such as gravity, buoyancy, drag, and inertia. In addition to the mechanical forces, fine soils are affected by electrochemical phenomena such as Van der Waals forces (weak molecular forces based on electric polarization) and electric surface forces.

Depending on the purpose of the study, there are a number of in situ and laboratory soil tests. In general, soil testing and analysis can be categorized as physical or chemical. Some of the tests are focused directly on soil properties, while others investigate the behavior of the soil relative to the water content.

PHYSICAL TESTING & ANALYSIS

Physical testing investigates the physical properties and behavior of soils. The most fundamental level of physical testing includes estimation of specific gravity, determination of moisture content, sieve analysis, and hydrometer analysis.

Specific gravity is defined as the ratio of "unit weight of soil solids only" to "unit weight of water." For sandy soils the specific gravity ranges from 2.63 to 2.67, while for clays it is between 2.67 and 2.90. The specific gravity of organic soils is less than 2.0. Moisture content is defined as the ratio of "the weight

A soil scientist inspects a field of canola plants, which will be used for cleansing soils rich in selenium. Image by United States Department of Agriculture.

of water present in a soil sample" to "the weight of dry soil." Moisture content is about 15 to 20 percent for coarse soils and 80 to 100 percent for clays. Organic soils may have moisture content in excess of 500 percent.

In order to determine the grain-size distribution of coarse soils, the soil sample is subjected to sieve analysis. This analysis is conducted by using a stack of sieves that have a decreasing sieve diameter from the top to the bottom. The sieves are usually made of woven wires. After the sieves are shaken for a specified period of time, the amount of sediment retained at each sieve is weighted. Based on these data, each particle diameter is plotted in terms of the percentage of material that is finer than this particular diameter. For the plot, a semilogarithmic paper is used. For the determination of particle size distribution of fine soils, the hydrometer analysis is applied. The lower limit of the particle diameter that can be detected by this analysis is 0.001 millimeter. The hydrometer measures the particle diameter indirectly. This analysis is based on the principle that the hydrometer will be subject to higher buoyancy forces in a well-mixed water-sediment system. However, as the suspended solid particles settle, the density of the water-sediment mixture decreases, and the hydrometer tends to sink. The grain-size distribution is integral to studies involving sediment transport, protection against erosion, and soil contamination.

ENGINEERING TESTING & ANALYSIS

Tests that are most useful in geotechnical and foundation engineering are those used to define the so-called Atterberg limits for clay soils: liquid limit, plastic limit, and shrinkage limit. The liquid limit is the moisture content that causes clay soils to behave as viscous liquids. Liquid limit is estimated by counting the number of "blows" required (using a Casagrande device) to close a groove made in the soil sample. The plastic limit of moisture content is the point at which the soil sample will turn from a plastic state to a semisolid state. The plastic limit is determined by the ability of the sample to roll and form threads 3.18 millimeters in diameter without crumbling. When clay soils are losing moisture, their volume generally decreases. The moisture content, expressed as a percentage, at which the soil volume ceases to decrease is called the shrinkage limit.

To determine the performance of soils under various loading conditions a number of soil tests are necessary. These tests include the Proctor compaction test, the unconfined compaction test, the shear test on sands, the consolidation test, and the triaxial tests in clay. The Proctor compaction test identifies the optimum moisture content that would result in a maximum dry unit weight of soil. This information is applied to the design of airports, highways, and structural foundations. The unconfined compaction test determines the stress-strain relation of a soil specimen and is useful in studies regarding retaining walls and landslides.

The shear test in sands is used to estimate the ability of sands to sustain shear loading. It is directly related to its angle of internal friction. Quantification of the time-dependent settling of saturated clays subject to increased loading is achieved by the consolidation test. Finally, the triaxial tests in clays define the general stress relationships for unconsolidated undrained, consolidated drained, or consolidated undrained specimens.

Permeability tests are integral to hydrogeological studies. They define the flow of water through a soil sample under either a constant hydraulic head or a falling hydraulic head. Permeability tests are indicative not of soil porosity but of the connectivity among the pores and their ability to form conduits that allow the water to flow freely through the soil.

ENVIRONMENTAL TESTING & ANALYSIS

Soils and sediment can be subjected to contamination by a variety of pollutants. Fine sediments and organic soils, because of their large specific surface, show a particular affinity to adsorb or absorb chemicals in dissolved or particulate form. The most common chemical testing of soils involves estimation of the pH value; carbonate, chloride, sulfate, and organic content; and total dissolved solids. The organic content can be defined easily through a loss-on-ignition test. The total dissolved solids can be estimated by evaporation.

FURTHER READING

Budhu, Muni. *Soil Mechanics and Foundations.* 2d ed. Hoboken, N.J.: Wiley, 2007.

Das, Braja M. *Soil Mechanics Laboratory Manual.* 5th ed. Austin, Tex.: Engineering Press, 1997.

Day, Robert W. *Soil Testing Manual: Procedures, Classification Data, and Sampling Practices.* New York: McGraw-Hill, 2001.

Head, K. H. *Soil Technicians' Handbook.* New York: Halsted Press, 1989.

Kézdi, Árpád. *Soil Testing.* Vol. 2 in *Handbook of Soil Mechanics.* New York: Elsevier Scientific, 1974–1990.

Liu, Cheng, and Jack B. Evett. *Soil Properties: Testing, Measurement, and Evaluation.* 6th ed. Upper Saddle River, N.J.: Prentice Hall, 2008.

Mudroch, Alena, and José M. Azcue. *Manual of Aquatic Sediment Sampling.* Boca Raton, Fla.: Lewis, 1995.

SOIL TYPES

—Alan Brown

FIELDS OF STUDY

Agronomy; Agriculture; Soil Science; Physical Science; Biology; Chemistry; Ecology; Environmental Studies.

PRINCIPAL TERMS

A horizon: the surface soil layer; also known as topsoil

belt: a geographic region that in some way is distinctive

B horizon: the soil layer just beneath the topsoil

leach: to remove, or be removed from, by the action of a percolating liquid

monolith: a large soil profile created by dovetailing monolith containers together and combining small, individual profiles

pedologist: a soil scientist

polypedons: bodies of individual kinds of soil in a geographic area

profile: a representative soil sample containing different layers of soil

sesquioxide: an oxide consisting of three oxygen atoms for every two atoms of other elements, the most common being aluminum (Al_2O_3) and iron (Fe_2O_3) sesquioxides

zone: an individual soil group within a horizon

SUMMARY

The classification of soils is of crucial importance, the taking soil surveys a primary task of the U.S. Department of Agriculture. Knowledge of soil types indicates the crops that are best suited for a particular soil. Soil classification also provides farmers with the most appropriate methods for preventing soil erosion.

THREE BASIC TYPES

Because soils consist of particles varying greatly in size and shape, specific terms are required to give some indication of their physical properties. Soils can be classified according to different characteristics. The simplest way to classify soils is by their texture. Three broad yet fundamental groups of soils are now recognized by pedologists: sands, clays, and loams. All class names that have been devised over the years are based on these three groups. The sand group includes all soils, of which sand makes up more than 70 percent of the material by weight. In contrast to the heavier groups of soils, which are stickier and more clayey by nature, this group is characteristically sandy in texture. Two specific classes within the sand group are recognized: sand and loamy soil. Soils included in the clay group must consist of at least 35 percent clay and in most cases not less than 40 percent. The names are "sandy clay," "silty clay," or simply "clay," which is the most common of all. Sandy clays often contain more sand than clay; similarly, the silt content of silty clays usually exceeds that of the clay fraction itself. The loam group, which is the most important for agriculture, is more difficult to explain. An ideal loam is a combination of sand, silt, and clay particles. It also exhibits light and heavy properties in roughly equal proportions. In most cases, the quantities of sand, silt, or clay require a modified class name. For example, a loam in which sand is present is classified as a sandy loam.

Qualifying attributes such as stone, gravel, and various grades of sand must be considered when placing some soils in classes. For example, one refers to stony-clay loams, gravelly sands, or fine-sandy loams. Pedologists also classify soils according to their composition. A great soil group consists of many soil types whose profiles have major features in common. Every soil type has the same numbers and kinds of definitive horizons, but they need not be expressed in the same profile to the same degree. For example, the Fayette, Dubuque, Downs, and Quandahl soils are all members of a single great soil group.

Geographic belts marked by certain combinations of great soil groups can be shown on maps of small scale. A schematic soil map of the world consists of six broad belts. One belt consists of rough landscapes, such as mountains, in which many of the soils are stony, shallow, or both. The patterns of soil types in these areas are especially complex. The other five broad belts have simpler patterns, but each includes a number of great soil groups. The soil types within a single farm, for example, commonly represent two or more great soil groups. The thousands of soil types in the United States are classified in twelve major soil orders. Collectively, the groups have wide ranges in their characteristics. They also vary greatly in such qualities as fertility, ability to hold available moisture, and susceptibility to erosion.

PODZOLIC & LATOSOLIC SOILS

Although terms like "podzol" and "chernozem" have long been superseded by more precise classifications, they still are useful as broad labels to describe major groups of soils.

Podzolic soils dominate a broad belt in the higher latitudes of the Northern Hemisphere and some smaller areas in the southern half of the world. They include the brown podzolic soils, gray-brown podzolic soils, and gray wooded soils. These groups were formed under forest vegetation in humid, temperate climates. The B horizons of these soils vary. Some podzolic soils have B horizons that are composed primarily of humus, sesquioxides, or both; others have B horizons that are mainly accumulations of clay. Podzolic soils are more strongly weathered and leached than many of the other soil groups. Because they are usually acidic, low in bases such as calcium, and low in organic matter, levels of fertility are moderate to low. As a group, however, these soils are responsive to scientific management. According to Soil Taxonomy, the system used in the United States, these soils are mostly termed spodosols if they show strong leaching, or alfisols if they show B horizons rich in iron and aluminum.

The equatorial belts of Africa and South America are dominated by latosolic soils, which are termed oxisols. These soils are also dominant in the southeastern parts of Asia and North America, in northeastern Australia, and in the larger islands of the western Pacific Ocean. Latosolic soils include the great soil groups known as laterites, reddish-brown lateritic soils, yellowish-brown lateritic soils, red-yellow podzolic soils, and several kinds of latosols. Red-yellow podzolic soils are so named because they share features with both the latosolic and podzolic groups, but they are more closely related to the latosols. Latosolic soils have been formed under forest and savanna vegetation in tropical and subtropical to fairly dry climates. Although these soils do not extend into arid regions, they may be found in alternately wet and dry areas with low rainfall. The latosols are the most strongly weathered soils in the world. As a result of the large amounts of iron oxides formed through intense weathering, the profiles commonly display red and yellow colors. Except for a darkened surface layer, most of these soils lack distinct horizons; therefore, the profile may remain unchanged for many feet. Although supplies of plant nutrients are usually low, these soils contain a high capacity to fix phosphorus in unavailable forms. Most of these soils are resistant to erosion and easily penetrated by plant roots. In fact, some tree roots in southeastern Brazil have extended more than 60 feet below the surface layer. The moisture capacities in latosolic soils are mostly moderate to high, although they are low in some. Without the benefit of modern science and industry, productivity is normally low.

CHERNOZEMIC & DESERTIC SOILS

Chernozemic soils, termed mollisols in the United States, have formed under prairie or grass vegetation in climates that are humid to semiarid, temperate to tropical. Although these soils are most extensive in temperate zones, some large areas exist in the Tropics. Chernozemic soils include the great soil groups known as chernozems, prairie soils (or brunizems), reddish prairie soils, chestnut soils, and reddish chestnut soils in temperate regions. In tropical and subtropical regions, these soils are often known as black cotton soils, grumusols, regurs, and dark clays. The A horizons of chernozemic soils are normally very thick, fertile, and slightly weathered. The B horizons are usually much less distinct. The A horizons of chernozemic soils are high in organic matter and nitrogen in temperate zones but not in tropical and subtropical zones. These soils are less acidic and higher in bases and plant nutrients than podzolic or latosolic soils; therefore, they are among the most fertile soils in the world, producing about 90 percent of the world's grain. It is no exaggeration to say that temperate chernozemic soils (mollisols) feed the human race. Within the United States, the chernozemic soils form the heart of the Corn Belt and wheat-producing regions. Because the soils extend from the margins of humid zones into semiarid zones, production varies with seasonal weather. Chernozemic soils in tropical and subtropical zones are not suitable for cultivation because they are high in clay, are plastic, and subject to great shrinking and swelling. Handling these soils is difficult.

Desertic soils, or aridosols, have formed alternately under shrub or grass vegetation in arid climates ranging from hot to cold. These soils are commonly found in the great deserts of Africa, Asia, and Australia, and in the smaller deserts of North and South America. They include the great soil groups known as desert soils, red desert soils, sierozems, brown soils, and reddish-brown soils. Desertic soils have been slightly weathered and leached because of the shortage of moisture. The lack of adequate rainfall also limits plant growth; consequently, these soils are low in organic matter and nitrogen. Limited rainfall is also reflected in the shallow profiles. Most of these soils feature faint horizons. The A horizon is lighter in color than the B horizon because it has commonly lost carbohydrates and perhaps some bases and clay. The slightly darker B horizon has some accumulation of clay but is very low in organic matter. Levels of nutrients other than nitrogen are usually moderate to high in these soils. Available moisture capacities vary, depending on the thickness of the profiles and textures of horizons. Land management can have a tremendous impact on these soils. Often the B horizon is cemented by calcite in a deposit known as caliche.

SOIL SAMPLES

The most simple and common method for determining the class name of a soil is by feel. In fact, rubbing the soil between the thumb and fingers can probably tell a scientist as much about a soil as any other superficial means. To estimate plasticity more accurately, it is helpful to wet the sample; this technique causes the soil to "slick out," thereby giving the observer a good idea of the amount of clay present. There is ample information available in the way the soil particles feel: sand particles are gritty; silt has a floury or talcum-powder feel when dry and is only moderately plastic or sticky when wet; and silt and clay often take the form of clods. Wetting the soil and rolling it into a rope or flattening it into a ribbon are simple but effective ways to gauge relative amounts of silt and clay.

A surveyor can learn more about the nature of soils by using simple tools. The only tool on which the fieldworker can absolutely rely is the spade. Although many spades tend to compact the soil layers somewhat, the gouge or cheese-sampler types of spade obtain samples of near perfection. Good samples that reveal the differentiation of horizons can also be taken with a butcher's knife. A horizontal cut is made between the horizons and two vertical cuts are made in the face. Individual elements can then be picked out and used as representative zone samples.

A permanent record of the soil character can be taken by means of a monolith. A monolith container is made of steel and is 66 centimeters long. Since these containers are made to dovetail into each other, soil profiles can be taken at any depth. After the sample has been taken, a fixing solution is applied to preserve it. As soon as the sample has been "set," it can be erected in a vertical position so that the soil layers can be examined in their natural order. Individual soil fragments taken by hand can also be preserved. The specimen is first air-dried and then oven-dried at 50 degrees Celsius until it reaches the same temperature throughout. While it is still warm, the specimen is immersed in a bath containing a solution of 3 percent cellulose acetate in acetone until bubbles cease to appear. It is then allowed to drain and dry on a sieve. Specimens fixed in this manner may be handled without breaking down.

A field map illustrating the soil types present in a large area can also be made by taking samples. A surveyor bores a hole with an auger until sure that a representative sample of the site has been acquired. After describing the profile in a notebook and then marking the site with a triangle and number, the surveyor moves to a neighboring site and continues taking samples at regular intervals.

AERIAL PHOTOGRAPHS & DIAGRAMS

Much can also be learned about soil characteristics from the "layout" of fields in aerial photographs. These are especially useful for examining soils in more inaccessible areas, such as vast forests. Aerial photographs show only the external or environmental characteristics of soil. With experience, however, scientists can deduce from the study of these features a good idea of what lies underneath. Aerial photographs can generally be studied for soil recognition without the aid of sophisticated photographic apparatus. Quality prints at a scale of about 15 centimeters to 1.5 kilometers usually give enough definition for general purposes. Later, simple stereoscopic studies of "stereo pairs" of contact prints provide three-dimensional detail.

Probably the most accurate method of determining the class of soils has been devised by the U.S. Department of Agriculture. This method involves the creation of a diagram that can be obtained through mechanical analysis. The diagram reveals that a soil is a mixture of different sizes of particles and that a close correlation exists between particle size distribution and the properties of soils. A surveyor can readily check the accuracy of class designations once mechanical analyses of the field soils under scrutiny are available.

The U.S. Conservation Service groups soils into eight land-capability classes. Soils in classes I to III are generally suitable for cultivation. Soils in classes IV to VIII are severely limited in their utility for farming.

FARMING APPLICATIONS

Knowledge of soil types is essential for effective farming. Farming methods have to be tailored according to the type of soil that is being cultivated. As people from the eastern United States moved to the prairies and the plains, for example, the farming practices that they brought with them had to change according to the different soils types that they encountered: chernozem, prairie, and chestnut soils. Even today, farming practices are still undergoing change because no one knows just what type of farming is best suited to the chestnut and brown soils.

Failure to adapt farming practices to the soil type can have a devastating impact on a community. A decline of productivity often occurs because of the inability of people in some regions to establish an appropriate type of farming. This is an especially severe problem with latisols (oxisols), which look superficially rich but cannot sustain intensive agriculture. People may attempt to force a pattern of use on soils that are not suited to it, as on the reddish-chestnut soils of the southern plains in the United States. Soil erosion and soil depletion, which often occur because of inappropriate practices, are symptoms of bad relationships between people and soil. It follows that not only do people live differently on different soils, but they also must live differently to live at all. This fact may be illustrated by civilizations that largely developed on one particular type of soil: the Egyptians on the alluvial soils along the Nile River; the ancient Greeks and Romans on the terra rossa ("red earth") soils around the Mediterranean Sea; the Arabians on the soils of the deserts; and the Westerns on the gray-brown podzolic of western Europe, the northeastern United States, eastern Canada, and parts of Australia, New Zealand, and South Africa. Not only were the types of crops upon which these cultures relied determined in great part by the soil type, but important artifacts were as well, such as the kinds of pottery that they used.

FURTHER READING

Al-Rawas, Amer Ali, and Mattheus F. A. Goosen, eds. *Expansive Soils: Recent Advances in Characterization and Treatment.* New York: Taylor & Francis, 2006. Discusses classification of expansive soils. Provides information on swelling and consolidation and the effects these processes have on buildings. Discusses how site characteristics and stabilizing agents can alter expansion processes.

Bear, Firman E. *Earth: The Stuff of Life.* 2d ed. Norman: University of Oklahoma Press, 1990. Explains how soil came into being; primarily a treatise on soils in relation to the growth of plants, animals, and humans, as well as a plea for the sensible conservation of soil. Recommended for the general reader.

Birkeland, P. W. *Soils and Geomorphology*. 3d ed. New York: Oxford University Press, 1999. One of the best books written on soils from the perspective of a geologist. Offers many examples from around the world. Covers the newly adopted soil profile description symbols in the early chapters. Includes a very elementary discussion of soil taxonomy in the early chapters. Suitable for university-level readers, but the early chapters are easily understood by high-school-level readers.

Brady, Nyle C., and Ray R. Weil. *The Natures and Properties of Soils*. 14th ed. Upper Saddle River, N.J.: Prentice Hall, 2007. The most complete book on soil science available. Covers soil composition, formation, and reactions, and soil's effect on the world's food supply. Contains ample maps, diagrams, and photographs, as well as a complete glossary. For college-level students.

Buol, S. W., F. D. Hole, and R. J. McCracken. *Soil Genesis and Classification*. 5th ed. Ames: Iowa State University Press, 2003. An update of a classic soil-science text. Includes an update on the new soil-horizon symbol nomenclature. Includes an in-depth discussion on soil development. Suitable for university-level readers.

Carr, Margaret, and Paul Zwick. *Smart Land-Use Analysis: The LUCIS Model*. Redlands, Calif.: ESRI Press, 2007. Discusses the use of ArcGIS software to analyze land suitability. Provides a step-by-step guide to using ArcGIS software. Explains the LUCIS model with case studies to provide examples of potential conflicts of land use.

Clarke, G. R. *The Study of Soil in the Field*. 5th ed. Oxford, England: Clarendon Press, 1971. A comprehensive introduction to soil science. Includes such areas as soil site characteristics, sampling, surveys, and evaluation. Charts and diagrams clarify some of the esoteric terminology. Includes a short section of references. For college-level students.

Kellogg, Charles E. *Our Garden Soils*. New York: Macmillan, 1952. Intended as a manual for gardeners. Explains the composition and characteristics of various types of soil. Introduces the general reader to the terminology that scientists use to describe soils.

__________. *The Soils That Support Us*. New York: Macmillan, 1956. A comprehensive and easy-to-read introduction to soil science, with an emphasis on the relationship between people and soil. Includes photographs, charts, maps, and several useful appendices. For high school and college students.

Landa, Edward R., and Christian Feller, eds. *Soil and Culture*. New York: Springer, 2010. Provides a full history of the relationship between humans and the earth. Organized into parts discussing soil's function in art, philosophy, culture, and health, among others. Provides interesting information from an anthropological perspective.

Paton, T. R., G. S. Humphreys, and P. B. Mitchell. *Soils: A New Global View*. New Haven, Conn.: Yale University Press, 1995. Aimed at university-level readers. Easily comprehensible and offers many good examples of soils and soil formation from around the world. Color illustrations, maps, bibliography, and index.

Retallack, G. J. *Soils of the Past*. 2d ed. New York: Blackwell Science, 2001. An excellent source of information on paleosols. Discusses soil classification and formation, identifying paleosols and their characteristics, and the factors involved in paleosol formation. Covers soil fossil records with discussion of ancient landscapes and ecology.

Simonson, Roy W. "What Soils Are." In *Soil*, edited by the U.S. Department of Agriculture. Washington, D.C.: Government Printing Office, 1957. Explains soil formation and composition in a somewhat technical but very comprehensive way. Includes a chart and diagram. For the college-level student.

Sparks, Donald S., ed. *Soil Physical Chemistry*. 2d ed. Boca Raton, Fla.: CRC Press, 1999. An excellent introduction to soil chemistry. Offers a clear look into the practices, procedures, and applications of the field and is easily understood by the careful reader.

Troeh, Frederick R., and Louis M. Thompson. *Soils and Soil Fertility*. 5th ed. New York: Oxford University Press, 1993. Appropriate for an introductory college course in soils for students of agriculture. Covers every aspect of soil, including the physical properties of soil and soil conservation. Highly readable and well supplemented with charts, illustrations, and diagrams.

Soilborne Illness & Disease

—*Elizabeth D. Schafer, PhD*

FIELDS OF STUDY

Agronomy; Agriculture; Soil Science; Physical Science; Biology; Chemistry; Ecology; Environmental Studies.

SUMMARY

Soilborne illnesses and diseases are caused by numerous microorganisms and parasites that live in soils. Soil serves as an ecosystem for diverse microbes that perform various roles and that range from useful organisms in biological and geological processes to dangerous transmitters of diseases. In May, 2001, a resolution of the World Health Assembly of the United Nations emphasized the need for increased medical intervention to minimize the occurrence of soil-transmitted diseases. The next year, the United Nations reiterated the goal of preventing soilborne worm infestations.

DEFINITION

Soilborne illnesses and diseases are caused by numerous microorganisms and parasites that live in soils. Soil serves as an ecosystem for diverse microbes that perform various roles and that range from useful organisms in biological and geological processes to dangerous transmitters of diseases. In May, 2001, a resolution of the World Health Assembly of the United Nations emphasized the need for increased medical intervention to minimize the occurrence of soil-transmitted diseases. The next year, the United Nations reiterated the goal of preventing soilborne worm infestations.

CAUSES

Soilborne, or soil-transmitted, diseases are common, especially in tropical regions, and affect more than two billion people globally in the early twenty-first century, according to the World Health Organization (WHO). In 2010, more than one-half million children internationally suffered sicknesses, particularly helminthiasis infections, contracted from soil. Although most people survive these illnesses, sources indicate approximately 12,000 to 135,000 people die yearly from soilborne helminth infections. Scientists estimate that the life spans of people infected with the most prevalent soilborne diseases and parasites are reduced by 43.5 million life years, more than the lifespan reductions of measles and malaria and exceeded only by tuberculosis.

Most soilborne illnesses are transmitted from soil to humans through the pathogens (such as bacteria, fungi, protozoa, and worms) that are shed in fecal material, which contaminate soil. These pathogens infect people who eat the plants grown in the contaminated soil or who drink water polluted by that soil. Human and nonhuman animals infected with soil-transmitted illnesses perpetuate the cycle when their feces contact soil. Bacteria and viruses associated with enteric diseases, specifically infections in the gastrointestinal tract, often are transmitted to humans in soils that have been used to bury sewage. Landfills that do not have procedures to control leaching enable bacteria and viruses to contaminate soil.

Helminthiasis is a frequently diagnosed disease reported worldwide that is transmitted by contact with soil. Eggs from worms, or helminths, are dormant in soil until they enter human or animal

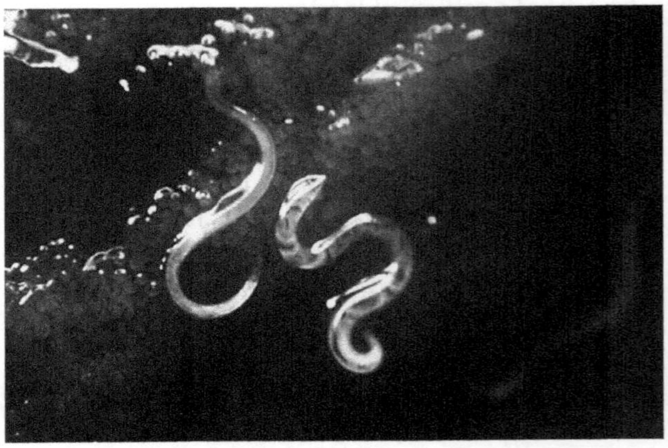

Ancylostoma caninum, a type of hookworm, attached to the intestinal mucosa. Image by Joelmills at en.wikipedia.

bodies, move through those bodies as larvae, and mature within three months into worms that infest the intestines. Worm infestation is often chronic, as parasites can live several years in their hosts. Soils host numerous types of worms (also referred to as nematodes), including roundworms, hookworms, and whipworms, which transmit diseases to humans.

The roundworm *Ascaris lumbricoides* infects humans with ascariasis. According to WHO, more than one billion humans are infected with this worm. Soils host these parasites' eggs. Roundworms are prolific, producing approximately 200,000 eggs daily. Humans are exposed to these immature worms by touching contaminated soil or by swallowing dirt or food that has not been adequately cleaned. Inside human bodies, roundworm eggs hatch into larvae that then invade essential body parts, including nerves and organs.

Other worms transmitted through soil include the hookworms *Necator americanus* and *Ancylostoma duodenale*. Those larvae enter skin from soil, infecting people with ancylostomiasis. WHO states approximately 740 million people have hookworms. The disease trichuriasis results from eggs from the whipworm, *Trichuris trichiura*, contaminating soil. About 795 million people have this infection, according to WHO.

Humans also are infested with the threadworm *Strongyloides stercorali*, which causes the disease strongyloidiasis as it moves from soil into the feet. Through worm eggs shed in fecal material, infested domestic pets expose humans to parasites and diseases associated with them, including visceral larva migrans and toxoplasmosis infections caused by *Toxoplasma gondii.*

As a cause of anthrax infections, dormant *Bacillus anthracis* bacteria spores often exist in soil for long durations, ranging from several years to decades. Growing grass blades transport anthrax spores from soil when grass-grazing animals ingest them. Associated with listeriosis, the bacterium *Listeria monocytogenes* is frequently found in soils and manure, contaminating livestock and plants used as food sources. The *Brucellosis* bacterium, often associated with livestock, can enter bodies through exposure to touching or breathing dust. The *Leptospira interrogans* bacterium, which causes leptospirosis, is present in soil and muddy areas where rodent urine containing that pathogen soaks into the ground. The *Acinetobacter baumannii* bacterium, which lives in soil, causes acinetobacter.

Most soil-transmitted disease deaths involve the pathogen *Clostridium tetani*, responsible for the deaths of approximately 450,000 infants and 50,000 adults each year. Scientists state that this bacterium's spores can be dormant in soil for almost one-half century and can still infect humans. *C. botulinum* bacteria spores also remain dormant in soils for extended times, causing botulism infections that are spread through foods. Soils in tropical areas often host the bacteria *Burkholderia pseudomallei* and *B. mallei*, which infect people with melioidosis and glanders when they touch or inhale soil or eat foods cultivated in contaminated fields. Nocardiosis infections occur when people breathe *Nocardia* bacteria found in dust or when contaminated soil contacts a person's skin injury.

Rarer, and often more deadly, soil-transmitted diseases occur when *Chromobacterium violaceum* bacteria infect humans by spreading from skin openings through the circulatory system and attacking organs simultaneously, preventing their function. Although Legionnaires' disease is usually transmitted through air or water, the Centers for Disease Control and Prevention (CDC) in 2000 reported cases in which the bacterium *Legionella longbeachae* infected people who had touched potting soil. Scientists have linked poliovirus 1 to soil, noting that the virus can survive more than three months underground and contaminate crops.

Fungi in soils, including the fungi *Mucorales*, *Aspergillus*, *Fusarium*, and *Blastomyces dermatitidis*, also transmit diseases. *Cryptococcus neoformans* can infect people who inhale dust containing it, leading to cryptococcal meningitis. Histoplasmosis is another soilborne illness, this time caused by *Histoplasma*

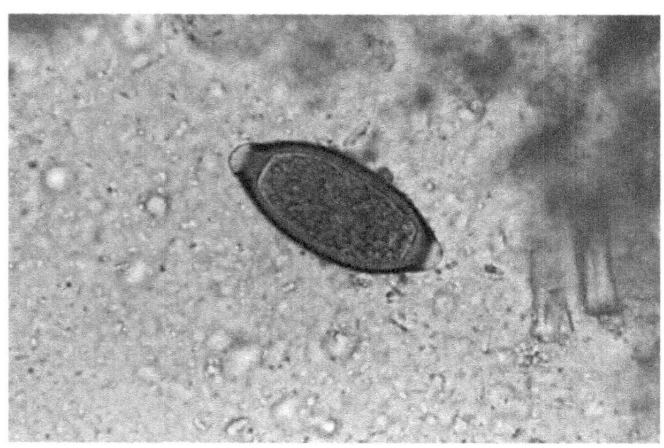

Photomicrograph showing an egg of the whipworm Trichuris trichiura, causative agent of trichuriasis.

fungus, which exists in soils as mold and affects the lung. Fungal *Coccidioides immitis* spores in soil cause coccidioidomycosis infections.

RISK FACTORS

Socioeconomic factors increase the risk of a person being infected with soilborne diseases. Poor sanitation exposes humans to soils contaminated with microorganisms and parasites. Areas without hygienic toilets or other devices to contain human wastes contribute to soil-transmitted infestations. Inadequate sewage systems result in fecal material being present near houses, schools, and other buildings. Rain washes these soils into water supplies, often rivers and ponds, which people use as a source for drinking water and use for bathing, swimming, fishing, and cleaning of cooking and eating implements.

Impoverished populations often lack access to sufficient preventive medical care. Illiteracy and restricted educational opportunities result in people not having information on preventing soilborne diseases or on treating those conditions if infected. Many people who suffer soil-transmitted diseases do not consume diets with nutritional foods that provide vitamins and minerals. They often eat foods that have not been thoroughly cooked or cleaned, and they do not have access to pasteurized products and boiled water. Pregnant women and people with weak immune systems or other health problems are extra-susceptible to soilborne illnesses and suffer miscarriages and mortality caused by infections.

Populations living in areas where crops are fertilized with feces or irrigated with polluted water risk being infected by diseases transmitted by soil, either by consuming foods, especially vegetables and fruits, grown in those soils or by working in contaminated fields. Runoff from pastures can contaminate communities with pathogens associated with livestock manure. Risks associated with contacting contaminated soils increase with people's proximity to landfills or other sites where sewage is stored in soil and where pathogens seep into the ground.

Dangers associated with exposure to soil depend on how accessible a person's skin is to contacting soils directly. People who work outdoors , performing landscaping, agricultural, forestry, sewage, or other jobs that involve contact with soil are at high risk of acquiring soil-transmitted illnesses. Hikers and others participating in outdoor recreation come into contact with soils and the pathogens they host. Children's risk is increased because they often play in dirt that can be contaminated. People with the disorder called pica (in which they consume, among other non-nutritive substances, soil) are especially vulnerable to soilborne illnesses.

Weather can intensify the occurrence of soilborne diseases. Floods, for example, can force to the surface underground soil layers and the pathogens they host. Violent storms such as hurricanes and typhoons can move soil and pathogens great distances. Soilborne diseases identified as potential biological weapons also present public safety concerns. Terrorists have threatened to use anthrax spores secured from soil sources. The CDC designated *B. pseudomallei* and *B. mallei* to be bioterrorism agents because of their universal availability in soils.

SYMPTOMS

Various symptoms are exhibited by people infected with soil-transmitted diseases. Probably the most obvious symptom associated with soilborne illness is the shedding of parasites while vomiting or during a bowel movement. People infected with soil-transmitted helminths often suffer gastrointestinal pain and swollen stomachs. Other organs occasionally swell. Infected people frequently experience diarrhea, nausea, bloody bowels, and vomiting, and they can become anemic.

Soil-transmitted infections usually cause people to become weak and listless. Fevers, rashes, headaches, and stiffness are common symptoms too. Infected persons often are not strong enough to attend school or perform labor. Some people with soilborne illnesses exhibit impaired cognition, experiencing problems with memory and language functions. Long-term symptoms include decreased mental and physical development in children. Lung damage from soilborne illness is often exhibited through pneumonia, coughing, or asthma. Although most adults do not exhibit the symptoms of toxoplasmosis infection, children who were infected in utero develop symptoms as they mature. Health problems associated with toxoplasmosis include blindness, deafness, and retardation.

Illnesses associated with soil often weaken immune systems, causing people to develop infections and conditions unrelated to soil. Soil-transmitted diseases are sometimes described as food-borne illnesses, even though pathogens such as *Escherichia coli* O157:H7 and *Salmonella*, which are associated with food poisoning, are present in soil that contaminates foods. Poor agricultural yields, damaged plants, and ill livestock may indicate the presence of pathogens and parasites in soil.

SCREENING & DIAGNOSIS

Medical professionals evaluate those with soil-transmitted illnesses according to conditions and physical ailments unique to each person. Examinations usually begin with recording an infected person's medical history and asking where the person lives and if they have traveled to other areas. This geographical information helps clinicians determine the most likely soil-transmitted disease infected the patient.

Health care workers assess various tissue samples to diagnose soilborne diseases associated with microorganisms. Specimens acquired for analysis frequently include blood, sputum, urine, feces, skin, bone marrow, and cerebrospinal fluids. Blood tests reveal the presence of antibodies to bacteria. X rays and biopsies are often used to detect fungal soil-transmitted diseases such as cryptococcosis. Some physicians utilize magnetic resonance imaging or computed tomography scans to assess damage by mucormycosis and other microbe infections.

For diagnosis of helminth infections, patients provide fecal samples for laboratory analysis to detect evidence of parasites and to examine worms. Technicians utilize several methods to evaluate specimens, including formalinethyl acetate sedimentation and Kato-Katz fecal-thick smear to count worm eggs. Medical personnel use imaging procedures and tools, such as endoscopy and ultrasonography, to determine any internal damage to organs and intestines.

TREATMENT & THERAPY

Persons diagnosed with soil-transmitted diseases undergo various methods to treat their infection. Many treatments focus on removing worms and include tablets composed of benzimidazole anthelmintics. The drugs most frequently dispensed include albendazole and mebendazole in doses of 400 to 500 milligrams (mg) for children two years of age and older and for adults, including women who have reached their second trimester of pregnancy. Toddlers younger than two years of age receive 200 mg. Amounts of the drug praziquantel are determined by measurements of the patients' height with a dose pole. Helminth-infected persons often receive iron and vitamin A supplements. People at risk of being reinfected receive additional drug doses at later times.

Researchers are developing new pharmaceuticals and methods to control soil-transmitted parasites that have become resistant to standard treatments. Other drugs sometimes used include levamisole, pyrantel pamoate, nitazoxanide, and tribendimidine. Many soilborne infections caused by microbes are treated with antibiotics. These drugs, however, often cannot defeat pathogens that become resistant to antibiotics. Medical researchers have tested the use of recombinant larval antigen ASP2 to create a hookworm vaccination. Scientists have investigated incorporating outer membrane proteins in a leptospirosis vaccine, reporting successes in 2010 in strengthening immunities in test animals.

PREVENTION & OUTCOMES

Most people contact soil daily. Interaction with soil varies depending on a person's activities, exposure to the outdoors, and dietary habits. People can minimize the possibility of being infected by avoiding areas most likely to host parasites and pathogens. One should wash his or her hands, feet, or any bare skin that has been in contact with soil. Wearing gloves while gardening lessens hazards associated with handling soils. Wounds, cuts, abrasions, cracks, and other skin damage should be covered with bandages. Shoes prevent soilborne illnesses from being transmitted through the soles of the feet. People should avoid inhaling dust. Masks help block spores in areas where fungi thrive in soil.

Soil contaminant hazards can be minimized through purifying water supplies, cleaning unsanitary sites, and providing sanitary toilet facilities. Other preventive measures include washing soil from raw foods harvested from gardens or bought at markets. Cooking meats thoroughly to destroy parasites helps reduce risks associated with consuming livestock products that might have been infected while animals grazed on forage that grew on contaminated soil.

The transmission of diseases associated with soil can be prevented by not eating soil. Crops fertilized with raw waste or irrigated with wastewater should not be consumed. Agricultural laborers should avoid handling those contaminated soils. People should regularly deworm domesticated pets and not touch any fecal material.

Some schools, particularly in tropical regions or developing countries, sponsor programs to deworm students. Health care personnel provide children medical treatments to purge and prevent further helminth infections. Educational presentations teach children and adults hygienic behavior and discourage contact with hazardous soils. Several charities and shoe manufacturers distribute free shoes in impoverished communities where residents are at risk of contracting soilborne diseases.

Further Reading

Albonico, Marco, Dirk Engels, and Lorenzo Savioli. "Monitoring Drug Efficacy and Early Detection of Drug Resistance in Human Soil-Transmitted Nematodes: A Pressing Public Health Agenda for Helminth Control." *International Journal of Parasitology* 34, no. 11 (2004): 1205–1210.

Albonico, Marco, et al. *Preventative Chemotherapy in Human Helminthiasis.* Geneva: World Health Organization, 2006.

Ambrosioni, Juan, Daniel Lew, and Jorge Garbino. "Nocardiosis: Updated Clinical Review and Experience at a Tertiary Center." *Infection* 38, no. 2 (2010): 89–97.

Baumgardner, Dennis J. "Soil-Related Bacterial and Fungal Infections." *Journal of the American Board of Family Medicine,* vol. 25, no. 5, pp. 733–44.

Bethony, Jeffrey, et al. "Soil-Transmitted Helminth Infections: Ascariasis, Trichuriasis, and Hookworm." *The Lancet* 367 (May 6, 2006): 1521–1532.

Brooker, Simon, et al. "Global Epidemiology, Ecology, and Control of Soil-Transmitted Helminth Infections." In *Global Mapping of Infectious Diseases,* edited by Simon I. Hay, Alastair Graham, and David J. Rogers. Amsterdam: Elsevier, 2006.

De Siqueira, Isadora Cristina, et al. " *Chromobacterium violaceum* in Siblings, Brazil." *Emerging Infectious Diseases* 11, no. 9 (2005): 1443–1445.

Jeffery, Simon, and Wim H. van der Putten. "The Ecology of Soil-Borne Human Diseases." *Microbiology Today,* Nov. 2016, www.microbiologysociety.org/all-microsite-sections/microbiology-today/index.cfm/article/7A5364C5-F6CC-4CFA-8835962649F50F22. Accessed 17 Nov. 2016.

"Legionnaires' Disease Associated with Potting Soil—California, Oregon, and Washington, May-June, 2000." *Journal of the American Medical Association* 284, no. 12 (September 27, 2000): 1510.

Santamaría, Johanna, and Gary A. Toranzos. "Enteric Pathogens and Soil: A Short Review." *International Microbiology* 6, no. 1 (2003): 5–9.

World Health Organization. "Soil-Transmitted Helminthiasis: Number of Children Treated 2007–2008: Update on the 2010 Global Target." *Weekly Epidemiological Record* 85 (April 16, 2010): 141–148.

Strip Farming

—Gordon Neal Diem

FIELDS OF STUDY

Agronomy; Agriculture; Ecology; Environmental Studies.

SUMMARY

Strip farming helps to alleviate several environmental problems associated with large-field agriculture at the same time it improves crop production: It reduces soil erosion from wind and rain; traps minerals, metals, fertilizers, pesticides, bacteria, and pathogens before they leave cultivated fields; and provides habitat for wildlife.

The origins of strip farming can be traced to the enclosure movement of post-medieval Great Britain. Landlords consolidated the small, fragmented strips of land farmed by tenant peasants into large block fields in an effort to increase agricultural production to meet the demands of growing human populations. A peasant plot was typically 0.4 hectare (1 acre) in size: 201 meters (220 yards), or 1 furlong (the distance a team of oxen could plow before resting), in length and 20 meters (22 yards) in width (the total area being the amount of land one team of oxen could plow in one day). After enclosure, fields were 40.5 hectares (100 acres) or more in size. Larger fields were more productive, but the soil in such fields was also more exposed to wind and water erosion and nutritional exhaustion.

As agricultural production gradually shifted to new lands in the Americas and colonial Africa, farmers continued to use large-field farming techniques and developed large-field plantations. By the early twentieth century, all readily tilled lands had been opened by the plow and were suffering the effects of water and wind erosion. Strip farming, also known as strip cropping, was developed as a soil conservation measure during the 1930's. During the 1960's strip farming became an important tool to prevent water and air pollution and improve wildlife habitats.

Wind erosion begins when wind velocity at 0.3 meters (1 foot) above soil level increases beyond 21 kilometers (13 miles) per hour. Wind moves soil by saltation and surface creep. In saltation, small particles are lifted off the surface, travel ten to fifteen times the height to which they are lifted, then spin downward with sufficient force to dislodge other soil particles and break earth clods into smaller particles. Surface creep occurs when particles too small to be lifted move along the surface in a rolling motion. The wider the field, the greater the cumulative effects of saltation and surface creep, leading to an avalanche of soil particles across the widest fields even during moderate wind gusts.

Water erosion begins when rain or flowing water detaches and suspends soil particles above the surface and transports them downslope by splash or runoff. Water ice crystals expand, then contract when melted, dislodging soil particles and making them available for both water and wind erosion. Water also leaches nutrients and chemicals from the soil, causing the soil to experience both nutrient loss and an increase in salts and acids.

The U.S. Department of Agriculture computes annual soil loss from agricultural and developed land using the formula $A = RKLSCP$. In this formula, A equals annual soil loss, R equals the amount of rainfall on the plot, K equals the erosion factor for the type of soil on the plot, L equals the length of the slope on which the plot is located, S equals the angle of the slope, C equals the type of crop or soil cover on the plot, and P equals the presence of management conservation practices such as buffers, terraces, and strip farming. Soil loss tolerance—that is, the amount of soil that can be lost without any reduction in productivity—is computed for each plot. Farmers and developers reduce soil losses to tolerance levels by reducing soil exposure to wind and rain and by utilizing conservation practices such as strip farming.

An aerial photograph in the vicinity of Jefferson, Monroe County, Wisconsin, USA in 1957, showing strip farming, a planting method designed to control erosion. In the foreground is St. Mary's Church. Image by E. W. Cole.

Strip farming reduces field width, thus reducing erosion. Large fields are subdivided into narrow cultivated strips. The planting of crops along the contour lines around hills is called contour strip cropping. Another variation is field stripping, or the planting of crops in strips across the tops of predominant slopes. Crops are arranged so that a strip planted with hay plants (which include grasses, herbaceous plants, and legumes, such as clover or alfalfa) or a strip of close-growing small grain (such as wheat or oats) is alternated with a strip of cultivated row crop (such as tobacco, cotton, or corn). Rainwater runoff or blown dust from the row-crop strip is trapped as it passes through the subsequent strip of hay plants or grain, thus reducing soil erosion and pollution of waterways. Contour or field strip cropping can reduce soil erosion by 65 to 75 percent on a 3 to 8 percent slope.

Cropping in the strips is usually rotated each year. In a typical four-strip field, each strip will be cultivated with a cover crop for one or two years, grain for one year, and row-crop planting for one year. Each strip benefits from one or two years of nitrogen replenishment from nitrogen-fixing cover crops such as alfalfa, and each strip benefits from one year of absorbing nutrient and fertilizer runoff from the adjacent row-crop strip.

Strip widths are determined by the slope of the land: The greater the slope, the narrower the strips. In areas of high wind, the greater the average wind velocity, the narrower the strips. The number of grass or small-grain strips must be equal to or greater than the number of row-cropped strips.

Terraces are often constructed to reduce the slope of agricultural land. To reduce erosion, at least one-half of the land between terrace walls is cultivated with grass or a close-growing crop. Diversion ditches are often used to redirect water from its downhill course across agricultural land. These ditches usually run through permanently grassed strips, through downhill grass waterways constructed across the width of the strips, and through grassed field borders surrounding each field.

Further Reading

Gliessman, Stephen R., and Martha Rosemeyer, eds. *The Conversion to Sustainable Agriculture: Principles, Processes, and Practices.* Boca Raton, Fla.: CRC Press, 2010.

Hart, John Fraser. *The Rural Landscape.* Baltimore: The Johns Hopkins University Press, 1998.

Koepf, Herbert H. *The Biodynamic Farm: Agriculture in the Service of the Earth and Humanity.* Hudson, N.Y.: Anthroposophic Press, 2006.

Troeh, Frederick R., and Louis M. Thompson. *Soils and Soil Fertility.* 6th ed. Malden, Mass.: Blackwell, 2005.

—*Toby Stewart and Dion Stewart*

Sustainable Agriculture

FIELDS OF STUDY

Agronomy; Agriculture; Conservation Biology; Evolution; Ecology; Environmental Studies.

SUMMARY

Most twentieth-century agricultural practices were based on continued economic growth. This practice demonstrated dramatic increases in production but had negative impacts on the environment through the losses of plant and animal habitats, depletion of soil nutrients, and pollution of water supplies. The concept of sustainable development focuses on the use of renewable resources and working in harmony with existing ecological systems. The World Commission on Environment and Development described sustainable development as the ability "to meet the needs of the present without compromising the ability of future generations to meet their own needs." Practitioners of sustainable agriculture strive to manage their agricultural activities in such a way as to protect air, soil, and water quality, as well as conserve wildlife habitats and biodiversity.

Sustainable agriculture tries to match crops and livestock to the topography, soil characteristics, and climatic conditions that exist in a given region. The crops selected for cultivation must be well suited to the existing soil and site conditions and should also be resistant to known pests in the area.

PROBLEMS CAUSED BY AGRICULTURE

Water pollution is one of the most damaging and widespread effects of modern agriculture. The runoff from farms accounts for more than 50 percent of sediment damage to natural waterways, and the chemicals and nutrients associated with this runoff in the United States cost billions of dollars per year to clean. Heavy application of nitrogen fertilizers, insecticides, and herbicides has raised the potential for groundwater contamination. Feedlots that concentrate manure production lead to further groundwater contamination. Several of the most commonly used pesticides have been detected in the groundwater of at least one-half of U.S. states. In addition, monoculture, or the growing of a single crop on an area of land, requires heavy reliance on agricultural chemicals, because the practice of growing the same crop repeatedly depletes the natural organic nutrients that were formerly rich in North American topsoils.

Research has found that many of the chemical agents, pesticides, fertilizers, plant-growth regulators, and antibiotics used in agriculture have ended up in the food supply. These chemicals can be harmful to humans at moderate doses, and chronic effects can develop with prolonged exposure at lower doses. Further, widespread pesticide use has been shown to result in severe stress in other animal populations, including bees. Pesticide use is often followed by occurrences of secondary pest outbreaks and resurgence of pests that have developed resistance to the pesticides previously used.

Because of these growing problems, many American farmers have turned to the practices associated with sustainable agriculture. The U.S. government offers guidance for this transition through the Food, Agriculture, Conservation, and Trade Act of 1990. This legislation describes sustainable agriculture as agriculture that, through an integrated system of plant and animal production practices, can, over the long run, meet human food and fiber needs, enhance environmental quality and natural resources, make the most efficient use of nonrenewable resources, maintain economic viability of farm operations, and enhance the quality of life for farmers and society.

WATER CONSERVATION & SOIL CONSERVATION

In the western United States, the ability to irrigate crops is an especially important factor in agriculture, as much of the region is naturally arid or semiarid. In California, the limited supply of surface water has caused overdraft of groundwater and the consequent intrusion of salt water, which causes the permanent collapse of aquifers. In order to counteract such negative effects, sustainable farmers in California have introduced improved water conservation and storage methods, selecting drought-resistant crop species, using reduced-volume irrigation systems, and managing crops to reduce water loss. Drip and trickle irrigation methods also dramatically reduce water usage and water loss while helping to avoid such problems as soil salinization.

Farmers can temporarily manage salinization and contamination of groundwater with pesticides, nitrates, and selenium by using tile drainage to remove water and salt. This method often has adverse effects on the environment, however. Long-term solutions include conversion from the planting of row crops to production of drought-tolerant forages and the restoration of wildlife habitats.

One of the most important aspects of sustainable agriculture is soil conservation. In order to prevent excessive erosion, farmers might leave grass strips in the waterways of their fields to capture soil that begins to erode. A field with a 5 percent slope has three times the water runoff volume of a field with a 1 percent slope and eight times the soil erosion rate. Contour plowing, which involves plowing across a hill rather than up and down the hill, helps capture overland flow and reduce water runoff. Contour plowing is often combined with strip farming, where different kinds of crops are planted in alternating strips along the contours of the land. As one crop is harvested, another is still growing and helps recapture wind- and waterborne soil while preventing runoff from flowing quickly downhill. In areas of heavy rainfall, sustainable farmers might construct tiered ridges to trap water and prevent runoff. This method involves a series of ridges that are constructed at right angles to one another to direct runoff and slow it down so that the water has a chance to soak into the soil.

Another farming method that contributes to soil conservation is terracing, in which the land is shaped into level shelves of earth to hold in water and soil. Soil-anchoring plants are grown on the edges of the terraces to provide further stability for the soil. Terracing is costly, but it can make enable farmers to grow crops on steep hillsides that they otherwise could not use for production. To protect fragile or unstable soil on sloping sites or along waterways, farmers may need to plant perennial species of grasses every year.

LIVESTOCK & ANIMAL MANURE

Ruminant animals (sheep, cattle, and goats) can be raised on rangeland, pasture, cultivated forage, cover crops, shrubs, weeds, and crop residues. The breeds that have lower growth and milk production potential can adapt better to environments with sparse or seasonal forage. By growing row crops on level soil and growing pasture on steeper slopes, farmers can help reduce soil erosion; growing pasture and forage crops in rotation can help improve soil quality. When ruminants and other farm animals are allowed to graze in pasturelands, those fields are fertilized naturally. Farmers can direct how their fields are fertilized with animal manure by using portable fencing to keep livestock grazing in one area or strip of pasture all the way down before moving the animals to another strip of the field while the first strip of pasture recovers.

Sustainable farmers also use so-called green manure—crops that are raised specifically to be plowed under—to introduce organic matter and nutrients into the soil. Green manure crops help protect against erosion, cycle nutrients from lower levels of the soil into the upper layers, suppress weeds, and keep nutrients in the soil rather than allowing them to leach out. Legumes such as sweet clover, ladino clover, and alfalfa are excellent green manure crops. They are able to extract nitrogen from the air into the soil and leave a supply of nitrogen for the next crop that is grown. Some crops, such as beans and corn, can cause high soil erosion rates because they leave the ground bare most of the year. One way sustainable farmers combat this is by leaving crop residues on the land after harvest. Residues help reduce soil erosion and even excessive soil temperatures in hot climates. Many farmers choose to use cover crops rather than residue crops. The decision about which cover crop to grow is based on the farm's geographical location and the purpose of the crop: to control erosion, to capture nitrogen for the soil, to release nitrogen to the crop, or to improve soil structure and suppress weeds.

COVER CROPS

Cover crops such as hairy vetch or clover are well suited to the needs of later crops with high nitrogen requirements, such as tomatoes or sweet corn. Both of these cover crops decompose and release nutrients into the soil within one month. To fight erosion, a farmer might choose a fast-growing cover crop, such as rye. Rye provides abundant ground cover and an extensive root system that can prevent soil erosion and capture nutrients. Alfalfa, rye, or clover can be planted after harvest to protect the soil and add nutrients and can then be plowed under at planting time to provide a green manure for the crop. Cover crops can also be flattened with rollers, and seeds can be planted in their residue. This gives the new young plants a protective cover and discourages weeds from overtaking them. Use of natural nitrogen also reduces the risk of water contamination by agricultural chemicals.

Sustainable agriculture emphasizes the use of reduced tillage systems, which are intended to disturb the soil as little as possible when preparing it for planting. Minimum-till farming involves using the disc of a chisel plow to make a trench in the soil where seeds are planted. Plant debris is left on the surface of the ground between the rows, which helps prevent erosion. Conservation tillage, or conser-till farming, uses a cutting tool (a coulter) attached to a plow to open slots in the ground just wide enough to insert seeds without disturbing the soil. No-till planting involves drilling seeds into the ground directly through any ground cover or mulch.

CROP ROTATION & MONOCULTURE

Planting the same crop every year on a given field can result in depleted soil. In order to keep the soil fertile, practitioners of sustainable agriculture rotate nitrogen-depleting crops (such as sweet corn, tomatoes, and cotton) from year to year with legumes, which add nitrogen to the soil. By planting winter cover crops, such as rye grass, farmers can protect their land from erosion. Such cover crops will, when plowed under, provide nutrient-rich soil for the planting of cash crops. Crop rotation also improves the physical condition of the soil because the different crops vary in root depth and in the ways they are cultivated.

In nature, various kinds of plants grow in mixed meadows, and this helps them to avoid insect infestations. The agricultural practice of monoculture places great quantities of the food of choice in easy proximity to insect predators. Insect populations that feed on particular crops can multiply more than they otherwise would when those crops are grown in the same fields year after year. Since most insects are instinctively drawn to the same home area every year, they cannot proliferate and thrive if their crops of choice are not in the same fields every year.

In addition to helping farmers use fewer pesticides, crop rotation helps to control weeds naturally. Some crops and cultivation methods inadvertently allow certain weeds to thrive. Sustainable farmers often incorporate into their crop rotations successor crops that eradicate weeds. Some crops, such as potatoes and winter squash, work as cleaning crops because of the different style of cultivation that is used on them. Pumpkins planted between rows of corn will help keep weeds at bay.

INTEGRATED PEST MANAGEMENT

Most sustainable farmers control insect pests through the practice known as integrated pest management (IPM). In IPM, each crop and its pests are evaluated as an ecological system. A plan is developed to manage the damage that pests can do through the use of particular cultivation techniques, biological methods, and chemical methods at different timed intervals. IPM techniques are effective, profitable, and safe and have been adopted for crops such as tomatoes, citrus, and apples.

The goal of IPM is to keep pest populations below the size where they can cause damage to crops. Fields are monitored to gauge the level of pest damage. If farmers begin to see crop damage, they put cultivation and biological methods into effect to control the pests; physical techniques such as vacuuming bugs off crops are also used. IPM encourages the growth and diversity of beneficial organisms that enhance the defenses and vigor of plants. Small amounts of pesticides are used only if all other methods fail to control pests. It has been found that IPM, when done properly, can reduce inputs of fertilizer, lower the use of irrigation water, and reduce preharvest crop losses by 50 percent. Reduced pesticide use can cut the costs of pest control by 50 to 90 percent and can increase crop yields without increasing production costs.

Further Reading

Chiras, Daniel D. "Creating a Sustainable System of Agriculture to Feed the World's People." *Environmental Science.* 9th ed. Jones, 2013.

Conway, Gordon R., and Edward B. Barbier. *After The Green Revolution: Sustainable Agriculture for Development.* Routledge, 2013.

Francis, Charles A., Raymond P. Poincelot, and George W. Bird, eds. *Developing and Extending Sustainable Agriculture: A New Social Contract.* Haworth, 2006.

Hand, Carol, and Chad Kruger. *Sustainable Agriculture.* Abdo, 2016.

Gliessman, Stephen R., and Martha Rosemeyer, editors. *The Conversion to Sustainable Agriculture: Principles, Processes, and Practices.* CRC, 2010.

Koepf, Herbert H. *The Biodynamic Farm: Agriculture in the Service of the Earth and Humanity.* Hudson: Anthroposophic, 2006.

Lyson, Thomas A. *Civic Agriculture: Reconnecting Farm, Food, and Community.* Lebanon: Tufts UP, 2004.

Norton, Tomas, Brijesh K. Tiwari, and Nick Holden. *Sustainable Food Processing.* John Wiley, 2014.

Obach, Brian K. *Organic Struggle: The Movement for Sustainable Agriculture in the United States.* MIT P, 2015.

Owings, Lisa. *Sustainable Agriculture.* Minneapolis: ABDO, 2013.

Troeh, Frederick R., and Louis M. Thompson. *Soils and Soil Fertility.* 6th ed. Malden: Blackwell, 2005.

Wheeler, Tim, and Joachim Von Braun. "Climate Change Impacts on Global Food Security." *Science,* vol. 341, no. 6145, 2013, pp. 508–13.

Symbiosis

—Sneed B. Collard

FIELDS OF STUDY

Conservation Biology; Evolution; Ecology; Environmental Studies.

PRINCIPAL TERMS

commensalism: symbiotic associations based chiefly on some form of food sharing, which may also involve shelter, protection, or cleaning

host: by convention, the larger of two species involved in a symbiotic association

intermediate host: an animal species in which nonsexual developmental stages of some commensals and parasites occur

mutualism: a type of commensalism in which both symbiotes benefit from the association in terms of food, shelter, or protection

parasite: a symbiote that must live in intimate contact with its host to survive; a parasite may be pathogenic or beneficial to the host

parasite mix: all the individuals and species of symbiotes living in a host concurrently

reservoir host: a host species other than the one of primary interest in a given research study

symbiosis: all forms of evolved, nonaccidental, nontrivial, interspecies associations, excluding predator-prey relationships

symbiote: a species involved in any form of symbiotic association with another species

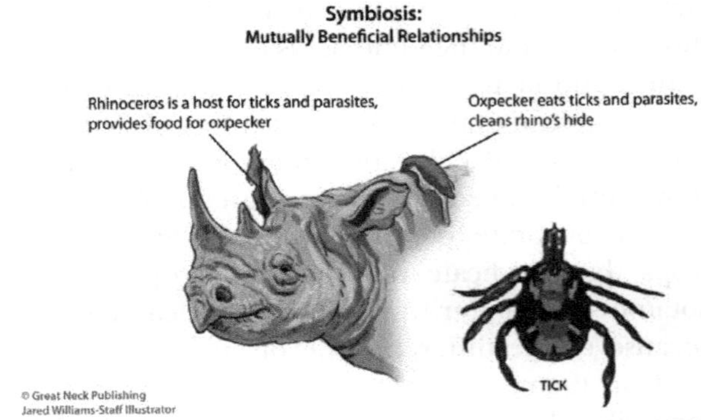

Symbiosis occurs among living things who live in the same environment and whose presence benefits one another. © EBSCO.

SUMMARY

All animals live in close association, or symbiosis, with other species. Most symbioses are based on nutritional interrelationships involving competition or cooperation. Some animals cannot survive without their symbiotic partners, while others are harmed or killed by them.

Understanding the ways in which different species of animals interact in nature is one of the fundamental goals of biology. Predator-prey relationships, competition between species for limited resources, and symbiosis are the major forms of species interactions, and these have profoundly influenced the diversity and ecology of all forms of life. Significant advances have been made in understanding how organisms interact, but in studies of symbiosis (which literally means "living together") one finds the most complex, interesting, and important examples of both cooperation and exploitation known in the living world.

Symbiosis involves many types of dependent or interdependent associations between species. In contrast to predator-prey interactions, however, symbioses are seldom rapidly fatal to either of the associating species (symbiotes) and are often of long duration. With the exception of grazing animals that do not often entirely consume or destroy their plant "prey," most predators quickly kill and consume their prey. While a predator may share its prey with other individuals of the same species (clearly an example of "living together"), such intraspecific behavior is not considered to be a type of symbiosis. Fleas, some ticks, mites, mosquitoes, and other bloodsucking flies are viewed as micropredators rather than parasites.

All organisms are involved in some form of competition. The abundance and availability of environmental resources are finite, and competition for resources occurs both between members of the same species and between individuals and populations of different species. When the number of individuals in a population increases, the intensity of competition for limited food, water, shelter, space, and other resources necessary for survival and reproduction also increases. Thus, competition plays a major role in populations of free-living animals (those not inhabiting the body of other organisms) and in populations living on or in other animals. For example, both tapeworms and whales must compete for resources, and both have evolved habitat-specific adaptations to accomplish this goal. Whales compete with whales, fishes, and other predators for food; tapeworms compete with tapeworms and other symbiotes (such as roundworms) for food and space; and tapeworms and whales compete with each other for food in the whale's gut.

"Symbiosis" is a term used to describe nonaccidental, nonpredatory associations between species. When used by itself, the term "symbiosis" does not provide information on how or why species live together, or the biological consequences of their interactions. Recognizably different forms of symbioses all have one or more characteristics in common. All involve "living together"; most involve food sharing; many involve shelter; and some involve damage to one or both symbiotes.

HOSTS & SYMBIOTES

Host species may be thought of as landlords. Hosts provide their symbiotes (also called symbionts) with transportation, shelter, protection, space, some form of nutrition, or some combination of these. Host species are generally larger and structurally more complex than their symbiotes, and different parts of a host's body (skin, gills, and gut, for example) may provide habitats for several different kinds of symbiotes at the same time. The three primary categories of symbiosis most commonly referred to in popular and scientific works are commensalism, mutualism, and parasitism.

Symbiotes that share a common food source are known as commensals (literally, "mess-mates"). In the usual definition of commensalism, one species (usually referred to as the commensal, although both species are commensals) is said to benefit from the relationship, while the other (usually referred to as the host) neither benefits nor is harmed by the other. Adult tapeworms which live in the intestinal tracts of vertebrate hosts provide a classic example of commensals. Adult tapeworms share the host's food, usually with little or no effect on otherwise healthy hosts. As in all species, however, too large a tapeworm population may result in excessive competition, lower fitness, or disease in both the host and the tapeworms. For example, the broad fish tapeworm of man, *Diphyllobothrium latum*, may cause a vitamin B12 deficiency and anemia in humans when the worm burden is high. In addition to tapeworms, many human symbiotes called "parasites" are, in fact, commensals.

External commensals (those living on the skin, fur, scales, or feathers of their hosts) are called epizoites. A good example of an epizoite is the fish louse (a distant relative of the copepod), which feeds on mucus of the skin and scales of fishes. Another type of commensalism is called phoresis (phoresy), which involves passive transportation of the commensal (phoront) by its host. Examples of phoreses include barnacles carried by whales and sea turtles, and remoras (sharksuckers), which, in the absence of sharks, may temporarily attach themselves to human swimmers. In inquilinism, the transported commensal (inquiline) shares, or more accurately, steals, food from the host, or may even eat parts of the host. Perhaps the best-known inquilines are the glass- or pearlfishes, which take refuge in the cloacae of sea cucumbers and often eat part of the host's respiratory system. A unique type of commensalism, known as symphilism, is found in certain ants and some other insects (hosts) which "farm" aphids (symphiles) and induce them to secrete a sugary substance which the ants eat.

MUTUALISM

The most diverse type of commensalism is mutualism. In some works, particularly those dealing with animal behavior, mutualism is used as a synonym of symbiosis; hence, the reader must use caution in order to determine an author's usage of these terms. As used here, mutualism is a special case of commensalism, a category of symbiosis. The relationship between mutuals may be obligatory on the part of one or both species, but it is always reciprocally beneficial, as the following examples illustrate. Some species of hermit crabs place sea anemones on their shells or claws (sea anemones are carnivores which possess stinging cells in their tentacles). Hermit crabs without anemones on their shells or claws may be more vulnerable to predators than those with an anemone partner. Hermit crabs, which shred their food in processing it, lose some of the scraps to the water, which the anemones intercept, and eat. Thus, the crab provides food to the anemone, which in turn protects its provider. Such relationships, which are species-specific, are probably the result of a long period of coevolution.

A different type of mutualism, but one having the same outcome as the crab-anemone example, is found in associations between certain clown fishes and sea anemones. Clown fishes appear to be fearless and vigorously attack intruders of any size (including scuba divers) that venture too close to "their" anemone. When threatened or attacked by predators, these small fishes dive into an anemone's stinging tentacles, where they find relative safety. Anemones apparently share in food captured by clown fishes, which have been observed to drop food on their host anemone's tentacles.

Cleaning symbiosis is another unique type of mutualism found in the marine environment. In this type of association, marine fishes and shrimp of several species "advertise" their presence by bright and distinctive color patterns or by conspicuous movements. Locations where this behavior occurs are called "cleaning stations." Instead of being consumed by predatory fishes, these carnivores approach the cleaner fish or shrimp, stop swimming, and sometimes assume unusual postures. Barracudas, groupers, and other predators often open their mouths and gill covers to permit the cleaners easy entrance and access to the teeth and gills. Cleaners feed on epizoites, ectoparasites, and necrotic tissue that they find on host fishes, to the benefit of both species. Some studies have shown that removal of cleaning symbiotes from a coral reef results in a significant decrease in the health of resident fishes.

PARASITISM

Parasitism is a category of symbiosis involving species associations that are very intimate and in which competitive interactions for resources may be both acute and costly. The extreme intimacy (rather than damage) between host and parasite is the chief difference between parasitism and other forms of symbiosis. Parasites often, but not always, live within the cells and tissues of their hosts, using them as a source of food. Some types of commensals also consume host tissue, but in such cases (pearl fishes and sea cucumbers, for example) significant damage to the host rarely occurs. Commensalism is associated with nutritional theft.

Some, but not all, parasites harm their hosts, by tissue destruction (consumption or mechanical damage) or toxic metabolic by-products (ammonia, for example). Commonly, however, damage to the host is primarily the result of the host's own immune response to the presence of the parasite in its body, cells, or tissues. In extreme cases, parasites may directly or indirectly cause the host's death. When the host dies, its parasites usually die as well. It follows that the vast majority of host-parasite relationships are sublethal. A number of parasites are actually beneficial or crucial to the survival of their hosts. The modern, and biologically reasonable, definition of parasitism as an intimate type of symbiosis, rather than an exclusively pathogenic association between species, promotes an ecological-evolutionary understanding of interspecies associations. Most nonmedical ecologists and symbiotologists agree that two distinct forms of intimate associations, or parasitisms (with many intermediate types), occur in nature. The most familiar are those involving decreased fitness in humans and in their domestic animals and crops.

Among animal parasites, malarial parasites, hookworms, trypanosomes, and schistosomes (blood flukes) cause death and disease in millions of people each year. The degree to which these parasites are pathogenic, however, is partly the result of preexisting conditions of ill health, malnutrition, other diseases, unsanitary living conditions, overcrowding, or lack of education and prevention. Parasites which frequently kill or prevent reproduction of their hosts do not survive in an evolutionary sense, because both the parasites and their hosts perish. Both members of intimate symbiotic relationships constantly adapt to their environments, and to each other. Over time, evolutionary selection pressures result in coadaptation (lessening of pathogenicity) or destruction or change in form of the symbiosis.

Nonpathogenic or beneficial host-parasite associations are among the most highly evolved of reciprocal interactions between species. The extreme degree of intimacy of the symbiotes (not lack of pathogenicity) distinguishes this type of parasitism from mutualism. Parasitic dinoflagellates (relatives of the algae that cause "red tides") are found in the tissues of all reef-building corals. These photosynthetic organisms use carbon dioxide and other waste products produced by corals. In turn, the dinoflagellates (*Symbiodinium microadriaticum*) provide their hosts with oxygen and nutrients that the corals cannot obtain or produce by themselves. Without parasitic dinoflagellates, reef-building corals starve to death. Similar host-parasite relationships occur in termites, which, without cellulose-digesting parasitic protozoans in their gut, would starve to death.

STUDYING SYMBIOSIS

Early studies of symbiosis focused primarily on discovery and description of commensals and parasites found in humans and their domestic food animals. Malarial parasites and some of the important trematodes (flatworms known as flukes) and nematodes (roundworms) of humans were described by the ancient Greeks, and references to the guinea worm (*Dracunculus medinensis*) are found in the Bible, where it is called "the fiery serpent." Some of the dietary conventions or laws observed in modern cultures have the side effect of preventing harmful symbioses, although it is still debated whether these proscriptions actually have their basis in early observations. It is widely known, for example, that pork products, if eaten at all, should never be consumed without thorough cooking. Swine are intermediate hosts for two very pathogenic human parasites, the trichina worm (the nematode *Trichinella spiralis*) and the bladderworm (the infective stage of the tapeworm *Taenia solium*). Much research in parasitology involves the description of symbiotes, particularly those of potential medical, veterinary, or agricultural importance.

The life cycles of many commensals and parasites are extremely complex and often involve two or more intermediate hosts living in different environments, as well as free-living developmental stages. Except for symbiotes of medical importance, relatively few complete life cycles have been worked out. Knowledge of life cycles remains as one of the most important areas of research in parasitology and is usually the phase of research following the description of new species.

Scientists have long recognized that "chemical warfare" (antibiotics, antihelminthics, insecticides) against microbial and animal parasites, and their insect and other vectors, provides only short-term solutions to the control or eradication of symbiotes of medical importance. Research attempts are being made to find ways of interrupting life cycles, sometimes with the use of parasites of other parasites. This research requires sophisticated ecological and biochemical knowledge of both the host-parasite relationship and the parasite mix. Studies of the parasite mix are ecological (parasite-parasite and host-parasite competition), immunological (host defense mechanisms and parasite avoidance strategies), and ethological (host and symbiote behavioral interactions) in nature. Investigators involved in this kind of research must be well trained in many of the biological disciplines, including epidemiology (the distribution and demographics of disease).

Immunology is the most promising modern research area in parasitology. Not only have specific diagnostic tests for the presence of cryptic (hidden or hard to find) parasites been developed, but also vaccines may be discovered that can protect people from such destructive protozoan diseases as malaria. Malaria has killed more humans than any other disease in history, and it currently causes the death of more than one million people, and lowers the quality of life for millions of others, each year.

THE INTERRELATIONSHIP OF SPECIES

All species are involved in complex interrelationships with other species that live in or on their bodies, or with which they intimately interact behaviorally or ecologically. Such interactions may play a minor role in the life and well-being of one or both of the associates, or they may be necessary for the mutual survival of both. In relatively few symbiotic relationships, one or both species may suffer damage or death. Pathogenic associations are relatively rare, because disease or death of one symbiote generally results in corresponding disease or death of the other. Such relationships, which cannot persist over evolutionarily long periods of time, may nevertheless cause catastrophic loss of life in nonadapted host populations.

Domestic animals cannot live in some parts of the world, such as the central portion of Africa, because they have little or no resistance to parasites of wild species, which are the normal hosts and are not harmed. Native species have coadapted with the parasites. This situation presents a moral dilemma to humans. In the face of human needs for space and other resources, should native animals be displaced or killed? Or should human populations proactively slow their reproductive rates? History shows that humanity has often chosen to take the former course.

The common view that animals which live in other animals are degenerate creatures that take advantage of more deserving forms of life is understandable but inaccurate. Symbiotes are highly specialized animals that do not live cost-free, or always to the detriment of their hosts. Symbiotic relationships between species have vastly increased the diversity, complexity, and beauty of the living world.

Further Reading

Boothroyd, John C., and Richard Komuniecki, eds. *Molecular Approaches to Parasitology.* New York: Wiley-Liss, 1995. Print.

Caullery, Maurice. *Parasitism and Symbiosis.* London: Sidgwick and Jackson, 1952. Print.

Limbaugh, Conrad. “Cleaning Symbiosis.” *Scientific American* 205 (1961): 42–49. Print.

Margulis, Lynn. “Symbiosis and Evolution.” *Scientific American* 225 (1971): 48–57. Print.

Margulis, Lynn, and Dorion Sagan. *Slanted Truths: Essays On Gaia, Symbiosis, and Evolution.* New York: Copernicus, 1997. Print.

Noble, Elmer, Glenn Noble, Gerhard Schad, and Austin MacGinnes. *Parasitology: The Biology of Animal Parasites.* 6th ed. Philadelphia: Lea & Febiger, 1989. Print.

Toft, Catherine Ann, Andre Aeschlimann, and Liana Bolis, eds. *Parasite-Host Associations: Coexistence or Conflict?* New York: Oxford UP, 1991. Print.

Whitefield, Philip. *The Biology of Parasitism: An Introduction to the Study of Associating Organisms.* Baltimore: University Park, 1979. Print.

Zann, Leon P. *Living Together in the Sea.* Neptune City, N.J.: T. F. H., 1980. Print.

Zinsser, Hans. *Rats, Lice, and History.* Reprint. New York: Bantam, 2000. Print.

T

Temperate Coniferous Forests

—*Jacqueline E. Mohan*

FIELDS OF STUDY

Biology; Forestry; Zoology; Biochemistry; Chemistry; Physical Science; Animal Science; Ecology.

SUMMARY

Trees with needlelike leaves, known as "conifers," are an ancient plant group, with fossils that go back 300 million years to the late Carboniferous Period when much of Earth's coal was formed. Coniferous trees are generally more tolerant of drought and fire than are deciduous trees, although some species do inhabit moist forest ecosystems. Coniferous forests of the temperate zone mainly grow in North America, where they occur in the Pacific Northwest, the Rocky Mountains, as stands of eastern hemlock in the northeast (currently disappearing due to an invasive insect pest), and in the Coastal Plain of the southeast. Very small coniferous temperate forests also grow in Japan, Argentina, and Chile. At the highest elevations in the southern Appalachian Mountains in North Carolina coniferous spruce-fir forests occur, but these are ecologically more similar to northern boreal forests than to temperate coniferous forests. With the exception of cypress, which grows primarily in southeastern swamps, the tree species of coniferous forests are almost always evergreen as opposed to deciduous trees, which lose their leaves every autumn. Fire is an important feature of coniferous forest ecosystems, and without fire most of these ecosystems would eventually develop, or "succeed," into different forest types. Other important environmental determinants of these forests are the amount of soil moisture and, in the Rocky Mountains, elevation. As winters become warmer and growing seasons are lengthened in duration, insect pests and disease pathogens are increasing the mortality of many temperate conifer tree species.

Trees with needlelike leaves, known as "conifers," are an ancient plant group, with fossils that go back 300 million years to the late Carboniferous Period when much of Earth's coal was formed. Coniferous trees are generally more tolerant of drought and fire than are deciduous trees, although some species do inhabit moist forest ecosystems. Coniferous forests of the temperate zone mainly grow in North America, where they occur in the Pacific Northwest, the Rocky Mountains, as stands of eastern hemlock in the northeast (currently disappearing due to an invasive insect pest), and in the Coastal Plain of the southeast.

The Hall of Mosses in Olympic National Park's Hoh Rainforest in Washington. The humid, moist conditions encourage the exuberant growth of ferns, mosses, and lichen on tree trunks and branches. The climate in the temperate evergreen forests of the Pacific Northwest is mild and very moist; they even receive enough rainfall to be called temperate rainforests. There are long intervals between fire events, making these forests very unique. Image by Wikimedia/Konrad Roeder.

Very small coniferous temperate forests also grow in Japan, Argentina, and Chile. At the highest elevations in the southern Appalachian Mountains in North Carolina coniferous spruce-fir forests occur, but these are ecologically more similar to northern boreal forests than to temperate coniferous forests. With the exception of cypress, which grows primarily in southeastern swamps, the tree species of coniferous forests are almost always evergreen as opposed to deciduous trees, which lose their leaves every autumn. Fire is an important feature of coniferous forest ecosystems, and without fire most of these ecosystems would eventually develop, or "succeed," into different forest types. Other important environmental determinants of these forests are the amount of soil moisture and, in the Rocky Mountains, elevation. As winters become warmer and growing seasons are lengthened in duration, insect pests and disease pathogens are increasing the mortality of many temperate conifer tree species.

PACIFIC NORTHWEST CONIFEROUS FORESTS

The beautiful temperate evergreen forests of the Pacific Northwest hug the coast of the Pacific Ocean, where climates are mild and very moist. These forests grow within 37 to 74 miles (60 to 120 kilometers) of the coast and receive enough rainfall to be called "temperate rainforests." Massive evergreen conifers are important for ecosystem functioning, wildlife, recreation, and local and regional economies. Giant trees include Douglas fir, western hemlock, western red cedar, Sitka spruce, and coast redwood. These trees occur in diverse tree communities with multiple dominant species and are generally very large and very long-lived. Some of these species, notably the famous redwood (*Sequoia sempervirens*), include the oldest trees in the world. These moist, mild forests with very long intervals between natural fire events (several hundred years) are unusual on the planet, as most such ecosystems are dominated by deciduous tree species, not conifers. The humid, moist conditions result in exuberant growth of ferns as well as mosses and lichen on tree trunks and branches, leading to names like "Hall of Mosses" and "Elvin Forests" for these temperate rainforest ecosystems.

Major environmental changes over the landscape ("gradients") that explain differences between forest types in the Pacific Northwest track changes in latitude, distance from the coast, and elevation. The two most important environmental factors explaining differences in forest types are soil moisture and temperature. In coastal regions, temperatures are mild and there are many cloudy days. Differences between day and night temperatures, and differences in temperatures from one day to the next, are mild, making these climates very moist and stable. Most of the precipitation occurs from October through March. As the landscape position moves from south to north, increasing in latitude, the temperature generally decreases and the precipitation increases.

The main forest type in the Pacific Northwest is dominated by Douglas fir and western hemlock and often includes western red cedar and other species of primarily coniferous trees. Forests of these massive trees range from sea level up to 2,296–3,280 feet (700–1,000 meters), with Douglas fir trees that may be 1,000 years old.

ROCKY MOUNTAIN CONIFEROUS FORESTS

The eastern slope of the Rocky Mountains rises dramatically from the relatively flat expanse of the Great Plains, reaching peaks of 14,436 feet (4,400 meters) in elevation. Not surprisingly, the elevation where an ecosystem occurs most strongly determines what type of ecosystem it will be. As elevation increases, temperatures decline and precipitation generally increases, as do solar radiation, wind, and snow depth. Although there are important differences in temperature and moisture between, for example, north- and south-facing slopes at a given elevation, lower-elevation climates are generally warmer and drier, whereas higher-elevations sites are colder and moister. The lowest elevations of the Rockies are dominated by grasslands and mixed oak deciduous forests, whereas the highest elevations are evergreen boreal forests grading into alpine meadows. Temperate coniferous forests dominate large expanses of the midslopes of these mountains. Woodlands comprised of juniper and pinyon pine tree species occur upslope from the oak forests, and heavy grazing by livestock has increased the abundance of juniper trees over this landscape by eating grass and reducing the grass-competitive shading of juniper seedlings. Juniper-pinyon grade upslope to Ponderosa pine woodlands and forests and then further upslope to forests dominated by Douglas fir, an important tree for use as a Christmas tree in the United States, that grow at about 5,410 to 8,900 feet (1,650 to 2,700 meters) in elevation. The highest elevation temperate deciduous forest is dominated by lodgepole pine, with grades into spruce-fir boreal forests at higher elevations.

EASTERN HEMLOCK FORESTS OF NORTHEASTERN NORTH AMERICA

Eastern hemlocks are "climax" species in forests, meaning that they do not require a disturbance event such as fire or the death of canopy trees creating a light gap in order to successfully reproduce and replace themselves in the forest canopy. In fact, hemlocks produce such dense shade that few if any other tree species are able to successfully regenerate under hemlock stands; instead, seedlings of other tree species generally die within a few months or years of germinating. Hence, despite having a dense canopy of evergreen needles, there is little plant ground cover or undergrowth in dark hemlock stands.

Hemlock stands do, however, have some unique hydrological features that enable them to form temporary freshwater "vernal pools" in the spring that are essential for populations of amphibians including frogs, toads, and salamanders. Because these pools are temporary, they do not support populations of fish, which could otherwise eat tadpoles and other offspring that inhabit the pool. And dense, evergreen hemlock forests provide essential wildlife habitat in the winter months, helping to protect white-tailed deer, wild turkey, and other animals from cold winter winds and snow. Hemlock trees are very slow growing and extremely long-lived, but unfortunately they are currently being decimated by an introduced species called the hemlock woolly adelgid.

SOUTHEASTERN COASTAL PLAIN CONIFEROUS WOODLANDS & FORESTS

The Coastal Plain is a flat, sandy region of the southeastern United States with a large amount of biodiversity and also one of the fastest-growing human populations in the nation. Three main types of temperate coniferous forests occur in the southeastern Coastal Plain, and differences among them depend largely on soil moisture and associated fire frequency. The first is the longleaf pine forest, which once accounted for 36 million hectares (90 million acres) of the landscape from North Carolina to Florida to eastern Texas, and occurs in various vegetative structures from quite wet to very dry soil moisture conditions. The vast majority of the former longleaf pine system has been converted to agriculture and forest plantations, making natural longleaf pine forests one of the most endangered ecosystems in the United States, with only 2 percent still remaining.

Wet longleaf pine woodlands contain some of the highest numbers of species ever recorded at small spatial scales, including wild orchids and carnivorous plants like sundews, pitcher plants, and Venus flytraps. Like other coniferous forest systems, longleaf pine forests and savannas depend on regular fire events to remain healthy. The second type of temperate coniferous forest in the southeast is very wet bald cypress and pond cypress swamp forests, or "bayous," which can have standing water several meters deep. These wet forests are less prone to fire, but during especially dry periods or when humans alter groundwater aquifers, these forests, too, can burn. Unlike most other conifer trees, cypress trees lose their needles in the autumn, and these swamp forests are important for freshwater fisheries, including crayfish and shrimp. The third type of temperate coniferous forest in the southeast is the sand pine woodland, or "scrub," that occurs on very deep, very infertile, and extremely dry white sands in central Florida. Many of these woodlands have been converted to citrus plantations. Sand pine scrub woodlands are the habitat of rare animal species, such as the gopher tortoise and scrub jay, as well as 12 types of endangered plant species, including wildflowers.

FIRE IN CONIFEROUS FOREST ECOSYSTEMS

Fire, set either by natural lightening strikes or by human land managers, is an important feature of almost all coniferous forests, with the exception of those dominated by eastern hemlock. Evidence of past fires is found in fire scars on tree trunks and in charcoal layers in soils. There is a major difference between ground- or surface-level fires, which burn less intensely and generally do not kill large trees, and crown or canopy fires, which do. When fires are suppressed by people, these coniferous forests can develop into deciduous hardwood-dominated stands or change into overly crowded tree conditions that burn much more intensely with more extreme damage than natural fires when a burn finally does occur. The latter happened in the Yellowstone fire event of 1988, which formed the largest recorded fire in the history of Yellowstone National Park.

Fires had been suppressed for decades, and normally widely spaced trees became crowded by smaller, younger trees that would have otherwise been burned as seedlings when ground-level fires occurred naturally. Instead, due to fire elimination, these younger trees were allowed to grow taller and when fires did ignite, they formed a "ladder effect" and the flames burned up into the large trees, resulting in a canopy fire. Thus land managers now set controlled, or "prescribed," fires to mimic natural fire frequencies and promote the health of these ecosystems.

In the wet and foggy Pacific Northwest, forests have fire intervals of 500–800 years as evidenced by fire scars and soil charcoal. When a fire does occur, it is a catastrophic fire burning up the giant canopy trees as fuel. In contrast, other temperate coniferous forests typically have more frequent and less severe fire events. Douglas fir forests of the Rocky Mountains have typical fire frequencies of 50–100 years, while the more arid Ponderosa pine forests burn more frequently every 40 years or so. Natural fire frequencies can be as often as every few years in moist longleaf pine forests and woodlands of the southeastern Coastal Plain. Florida sand pine scrub systems are so dry that plant cover is extremely limited and fuel that forms from dead plants and leaves takes a long time to accumulate. Hence, fire intervals in these systems may be 20–50 years, whereas nearby moist longleaf pine woodlands could burn every few years. The cones containing the seeds of sand pine are serotinous, meaning they require the high heat of being burned in a fire before they can open and release the seeds to produce the next generation of trees.

Frequencies and sizes of natural and accidental arson-caused fires have been going up in the United States over the last several decades, largely as a result of earlier spring snowmelt and higher spring and summer temperatures. For example, the National Interagency Fire Center finds that the size of the area burned by an average fire in the United States was about 47 acres between 1987 and 2000, and increased to 100 acres during 2001–07.

INSECT PESTS & DISEASE PATHOGENS

Like fire, climate is another important controller of the insect pests and pathogens that shape the health of all ecosystems, including temperate coniferous forests. With warming winter temperatures, disease pathogens and insects like pine bark beetles as well as human pets' outdoor flea populations are better able to survive and even increase their distribution over regions like mountaintops that were formerly too cold. This is occurring now as the mountain pine bark beetle that is causing over 90 percent tree mortality rates in western lodgepole pine stands has been able to cross over the Rocky Mountain Continental Divide and now is killing jack pine trees in eastern North America. The near synchronous death of these standing trees forms fuel for extreme canopy-level fires, further hastening change in these forest ecosystems and having major ecological and economic consequences. Longer growing seasons also enable insects and pathogens to produce more generations per year, which exacerbates negative effects on tree health. For example, the hemlock woolly adelgid beetle, which invaded the east coast of North America from Asia, is wiping out trees and stands of eastern hemlock. In New England, this insect generally produces two generations per year and takes 15 years or so to kill the trees. In the warmer climates of the southern Appalachian Mountains, this insect can produce three generations per year and typically kills mature trees in just four years.

TEMPERATE DECIDUOUS FORESTS

—Alexis S. Reed

"The four distinct seasons and the changes that broad-leaved trees experience during the seasonal cycle are the most defining characteristics for the temperate deciduous forests."

FIELDS OF STUDY

Biology; Forestry; Zoology; Biochemistry; Chemistry; Physical Science; Animal Science; Ecology.

SUMMARY

Temperate deciduous forests exist in the mid-latitude regions around the globe, including North America, Europe, Asia, and southern South America. Four seasons characterized by cold, relatively dry winters and warm, wetter summers are found in these forests. The cycle of seasons influences the shedding of leaves by deciduous tree species. These forests provide important resources to civilizations around the world, but many modern activities and urbanization threaten to decrease the extent of these forests. Conservation efforts focusing on increasing management and awareness of temperate deciduous forests are working around the world to preserve these ecosystems in the face of global changes in climate and land use.

Temperate deciduous forests exist in the mid-latitude regions around the globe, including North America, Europe, Asia, and southern South America. Four seasons characterized by cold, relatively dry winters and warm, wetter summers are found in these forests. The cycle of seasons influences the shedding of leaves by deciduous tree species. These forests provide important resources to civilizations around the world, but many modern activities and urbanization threaten to decrease the extent of these forests. Conservation efforts focusing on increasing management and awareness of temperate deciduous forests are working around the world to preserve these ecosystems in the face of global changes in climate and land use.

Temperate deciduous forest regions are characterized by relatively warm summers and cold winters with growing seasons ranging from 140 to 200 days with four to six frost-free months. Annual temperatures in temperate deciduous forests range from minus 22 degrees F to 86 degrees F (minus 30 degrees C to 30 degrees C) with an average of 50 degrees F (10 degrees C), and precipitation ranges from 29 to 59 inches (750 to 1,500 millimeters) per year and mostly occurs during warm months. Temperate forest species typically thrive on neutral to slightly basic soils, and soils are typically very fertile due to the continuous decomposition of newly deposited litter such as falling leaves and woody debris. The four distinct seasons that occur in all temperate deciduous forests trigger the characteristic shedding of leaves by deciduous, broad-leaved species and subsequent regrowth of leaves during spring. The cyclical adaptation of shedding leaves during the cold, drier periods allows plants to survive harsh winters in a rather dormant stage and reanimate during more favorable spring conditions.

Black bears are adapted to temperate conditions with sharp claws for climbing trees. As omnivores, they eat vegetation and other forest creatures. Thick fur protects them against cold temperatures. Image by U.S. Forest Service.

DISTINGUISHING CHARACTERISTICS

The four distinct seasons and the changes that broad-leaved trees experience during the seasonal cycle are the most defining characteristics for the temperate deciduous forests. The changes experienced physiologically by trees are unique across the seasons and represent a significant adaptation to the harsher conditions experienced by trees. For comparison, the trees in warm, wet climates such as the tropics never experience a dormant period in which leaves are shed. The shedding of leaves by trees in the autumn decreases water loss due to transpiration, a side effect of photosynthesis. When leaves are shed, water loss is decreased and photosynthesis is in turn decreased. Since new carbohydrates are no longer produced by photosynthesis, trees enter a stage of torpor, or dormancy, in which most physiological functioning ceases. Growth of woody material stops as well.

Leaves experience a change of color in autumn triggered by a change in the allocation of sugars and chemical components and a withdrawal of chlorophyll from the leaves. Trees are able to draw nutrients from the leaves back into woody tissues for storage over winter. For example, nitrogen, an often limiting nutrient in many forests, can be translocated, or transferred, from the leaf back into the tree prior to leaf shedding. Once leaves are shed, trees enter dormancy for the winter, and transpiration and photosynthesis cease for the most part. Thick exterior bark allows trees to withstand cold winter temperatures. Leaves and flowers reemerge in the spring when growing conditions become more favorable for growth, and photosynthesis resumes. In the spring, large conducting elements comprising the tree's vascular system (i.e., phloem and xylem) reestablish as well, allowing for transport of water from the roots to the treetops and transfer of sugars and carbohydrates generated by photosynthesis from the leaves to the rest of the tree for use in growth and storage. These conducting elements comprise the annual growth rings of trees, and thus a new, annual ring is added each year to deciduous species as trees exit dormancy and resume growth and photosynthesis each spring.

The most common deciduous tree species that exist in temperate forests belong to the oak (*Quercus*), ash (*Fraxinus*), maple (*Acer*), beech (*Fagus*), hickory (*Carya*), and chestnut (*Castanea*) families. Biodiversity is high; along with canopy-dominant tree species, understory shrubs, herbs, and flowering plants are also important for these ecosystems. Closed-canopy forests typically contain shade-tolerant tree species reestablishing in the forest understory. When forest stands reach maturity, canopies are referred to as "closed," signifying a relative closure in leaf space and a decrease, but not total elimination, of the sunlight penetrating the canopy. There are also several different kinds of plants that exist in the shady conditions of the forest floor where only small amounts of sunlight penetrate, such as shrubs, flowers, and mosses.

GLOBAL DISTRIBUTION & STRUCTURE

In terms of global distribution, temperate deciduous forests are found in the eastern half of North America in latitudes between 35 and 48 degrees north and in central Europe in latitudes between 45 and 60 degrees north. Deciduous forests reach farther north in Europe than in other regions of the Northern Hemisphere as a result of the Atlantic conveyor belt carrying warmer waters from the mid-Atlantic and Caribbean up around Great Britain, resulting in warmer, temperate climates in higher, northern latitudes. The deciduous forests of Asia are located in southwest Russia, Japan, Korea, and eastern China. In South America, southern Chile and the midcoastal regions of Paraguay contain deciduous forests. Also in the Southern Hemisphere, New Zealand and southeastern Australia have extensive deciduous forests. While all these regions are characterized by four distinct seasons over the year, there are unique characteristics of each region attributed to changes in soil texture, nutrients, topography, and common species of flora and fauna. For example, the deciduous forests of the eastern United States contain interspersed coniferous trees, especially in northern regions and areas of high elevation such as the Appalachian Mountains, with species indigenous only to North America.

Temperate deciduous forests cover large areas in the Northern and, to a lesser degree, Southern Hemispheres. Further categorization of these forest communities can be done by assessing the repeating units of dominant tree species in a stand. For example, some of the most common forest communities in the United States can be described by eight categories: mixed mesophytic, Appalachian oak, hemlock-white pine-northern hardwoods, oak-hickory, maple-basswood, beech-maple, oak-pine, and southern pine. Mesophytic signifies an environment receiving a moderate amount of moisture.

The forest structure in temperate deciduous regions is separated into five zones based on vegetation height. The tree stratum zone comprises the first zone and ranges in height between 60 and 100 feet (10 to 30 meters). The tree stratum zone contains trees such as oak, beech, maple, chestnut hickory, elm, basswood, linden, walnut, and sweet gum trees. Smaller tree and juvenile saplings are in the second zone. The shrub zone, or third zone, consists of short shrubs and flowering plants such as rhododendrons, primrose, bluebells, painted trillium, azaleas, mountain laurel, and huckleberries. The herb zone is the fourth zone, and it contains plants close to the ground such as herbal plants like ferns and perennial forbs that blossom in spring. The ground zone contains low-to-the-ground growth forms such as lichens, true mosses, and club mosses. The canopy in most deciduous temperate forests contains openings and gaps created by fallen trees and low tree densities that allow some light to penetrate to the forest floor, thus allowing for a great diversity in flora, which in turn supports a great diversity in fauna.

Animals found in temperate deciduous forests must adapt to the seasonal cycles, so many hibernate or migrate when freezing temperatures arrive. Some animals are adapted to the harsher winter months and can withstand the high risk of predation and cold temperatures during the winter. For example, black bears are adapted to temperate conditions. They have sharp claws for climbing trees and, as omnivores, they eat other vegetation and prey on other forest creatures. While their thick fur protects them from the cold temperatures of winter, they do hibernate during the time of hardest freeze to avoid food shortage. Other common forest animals are deer, gray squirrels, opossums, mice, and raccoons. Amphibians such as salamanders and frogs, reptiles such as snakes and turtles, and a multitude of insect species including beetles and mosquitoes also contribute to the diversity of fauna found in deciduous temperate forests. A huge variety of bird species including robins, eagles, cardinals, chickadees, broad-winged hawks, various owls and woodpeckers, hummingbirds, and wild turkeys also are found in these forests.

TIMBER & RESOURCE MANAGEMENT

Temperate deciduous forests have acted as important sources of timber historically and continue to provide important resources presently; thus, these areas have often been settled and used for resources. Less than one-quarter of old-growth deciduous forests still exist globally due not only to timber harvesting but also due to land clearing for agriculture. For example, many European colonists in the northeastern United States cleared the deciduous forests for agriculture. Upon finding more arable, or easier to plow, soils toward the western United States, colonists allowed many former agricultural fields in the northeast to return to forested states. Prior to European arrival, forests in the northeastern United States contained higher species diversity and greater natural variation in forest stands, which were influenced by natural and human disturbances such as hurricanes and clearing of small areas for native villages. This led to trees of varying age, density, size, and species across a wide range of forested sites.

With the arrival of Europeans, forests diversity decreased and tree ages and densities became more homogenous, or had decreases in variety, due in part to the influences of larger-scale hunting, trapping, and clear-cutting by Europeans beginning around 1820. Through the passage of about 30 years, European agricultural efforts in New England notably declined and many of the fields returned to forested states. Early forest regrowth was primarily represented by white pine, but conifers have since been replaced by the diversity of broadleaf species that currently comprise eastern forests. Forest regrowth in the northeast United States has continued for the most part, and the deciduous forests of New England represent secondary succession forest growth, or rather the forest has regrown after the original forests were cleared for timber or agriculture. In another example, forests in China have been harvested for over 4,000 years, thus most of the forests existing in the country are human-made.

Secondary succession and current forest management of temperate deciduous forests create some ecosystem problems; because most trees are of the same age, or even-aged, stands are denser than historic stands, and species are often selected for their commercial value instead of to maintain historic biodiversity with a wide variety of species. To contrast current temperate deciduous forests, it is important to recognize some of the valuable characteristics of old-growth deciduous temperate forests. Old-growth forests are more complex forest structures characterized by trees representing a variety of ages, large fallen trees and woody debris on the forest floor, and multiple layers in the canopy representing differences in tree size. Forest regrowth typically decreases the complexity of forest communities by decreasing the variety of age, layers, species, and habitats available.

The ecosystem services provided by temperate deciduous forests, or the commercial or aesthetic value that is placed on nature, are not limited to timber products such as wood or paper. Forests provide oxygen as a by-product of photosynthesis, decrease atmospheric carbon dioxide through sequestration, provide food including mushrooms and game, create recreational spaces and natural beauty, and purify the water that many communities around the globe rely upon. The change in the autumn colors of foliage in deciduous trees leads to great beauty. This can be recognized among the wide array of tours and recreational opportunities in the fall in New England. Tourists flock to these regions to view the autumn colors among the maples, and Arcadian National Park in Maine is flooded with guests enjoying the reds, yellows, and oranges that flame through the leaves during the fall. Not only does this represent a great aesthetic value, but also an important economic aspect for the region.

Forests are often managed to maintain high densities of valuable tree species such as red oak, and stands are often even-aged because most trees within a stand would have been established at the same time following forest clear-cutting. High densities of trees make stands more vulnerable to the negative impacts of drought as competition between neighboring trees creates more demand for limited water resources. This makes trees more vulnerable to drought-related mortality or secondary stresses such as insect or disease pathogen attacks. An example of this can be observed in the oak-hickory stands of northwest Arkansas. An oak decline event in the early twentieth century that resulted in high mortality of red oak species was linked to increased vulnerability of trees to insect infestation following a cycle of droughts. The mortality event was also related to the unnaturally high densities of red oak species that existed in the area. As a commercially valuable species, the even-aged stands were managed for high timber production, and these high densities increased moisture stress during drought. The increased stress may have made trees more vulnerable to insect attacks as more resources had to be allocated to accessing water (i.e., roots) than to defensive chemicals. Decreased diversity of species also poses a threat to forest ecosystems.

For example, eastern U.S. forests were dominated by chestnut trees in the early 1900s until the chestnut blight, an introduced fungal pathogen, mortally impacted this key species and left the canopy in many forest stands open to other species such as maples and sycamores. A pathogen decimating an entire forest species is rare, but as introduced pathogens become more common, it is vital for continued ecosystem function to maintain a wide variety of codominant tree species within forest canopies. The management practices that create even-aged, high-density forest stands with relatively low diversity found in many temperate deciduous forests make these regions more vulnerable to the impacts of disturbances such as hurricanes, windfall events, and disease pathogens.

ENVIRONMENTAL CHALLENGES & FOREST CONSERVATION

Temperate deciduous forest habitats are also threatened by timber harvesting, acid rain, and invasive plants and animals worldwide. Nonnative, invasive species may compete for food and habitats, potentially threatening the native plant and animal species. Because temperate deciduous forests provide rich areas of flora and fauna diversity along with globally important ecosystem services, conservation of these regions is important. Recycling programs and awareness programs sponsored by organizations such as the U.S. Environmental Protection Agency (EPA) and the European Union (EU) European Commission have led to increased recycling rates over the past several decades. Acid rain impacts leaf photosynthesis and soil nutrient balances and has been recognized as a serious problem in many areas such as central Europe. Evidence of leaf bleaching and slowed growth resulting from acid rain deposition in forest stands has been linked to vehicle fossil-fuel emissions through decades of research. Programs in the early 1990s through the EPA and the EU to decrease fossil-fuel emissions of sulfur by-products, which cause sulfuric acid deposition in rain, have helped decrease the impacts of acid rain on forest communities up to 40 percent since the 1970s. Another threat to temperate deciduous forests is land use change. Continued clearing for agriculture and commercial mining threatens forests globally. For example, in Germany, much of the historic deciduous forest land has been strip-mined for metal and minerals, decreasing the range of forests in the region.

The large amount of standing biomass existing in temperate deciduous forests represents a significant contribution to the carbon balance of the Earth. The carbon stored in the woody biomass, roots, and soil of these forests is a large component of the global carbon balance, which is a biogeochemical cycle of exchanges among the biological components, soil, rocks, atmosphere, and water cycles of the Earth. The carbon cycle is vital for recycling and reuse throughout the biosphere and all organisms globally.

Forests in general store up to 86 percent of global carbon stocks. Disturbance of soils along with burning or decay of woody biomass releases this stored carbon back to the atmosphere. Because trees convert carbon dioxide into carbohydrates during photosynthesis, releasing oxygen, they are an important component of global atmospheric carbon dioxide levels. The drawdown of global atmospheric carbon dioxide levels can be observed in deciduous forests during leafing-out in the spring. Given that deciduous forests are most extensive in the Northern Hemisphere due to larger land masses, the Northern Hemisphere spring dominates the drawdown trend, most famously observed by the Keeling curve, which measures global carbon dioxide levels.

Related to changes in global carbon dioxide levels, the recognized increases in global temperatures also threaten forest biomes through a decrease in moisture availability and increases in secondary stress factors such as invasive insects and pathogens. Climate change impacts may also alter the species habitats, composition, and adaptations of forest stands, leading to changes in forest structure or function. For example, the impacts of global climate change have already been noted in hardwood forests in Ohio, where decreases in annual snowfall, important for soil water recharge, and increased temperatures in winters have placed carbohydrate and water constraints on some forest tree species. With global increases in disturbances and climate impacts, conservation efforts in forests are an ever-increasing concern for many U.S. national organizations such as the Sierra Club and Audubon Society and for international organizations such as the Commonwealth Forestry Association.

TEMPERATE GRASSLANDS

—Christina Alba

FIELDS OF STUDY

Agricultural Geography; Biology; Zoology; Biochemistry; Chemistry; Physical Science; Statistics; Botany; Insect And Pest Management;Animal Science; Ecology.

SUMMARY

Several characteristics of the temperate grassland biome have contributed to its historic and current importance in shaping people's livelihoods and customs, as well as local, regional, and now global economies. In particular, the location of temperate grasslands at mid-latitude regions ranging from 30 degrees to 60 degrees north and south of the equator is conducive to human settlement. Large swaths of these grasslands cover parts of North America, South America, Eurasia, and southern Africa, while less extensive grasslands occur in Australia and New Zealand. An agreeable climate, highly fertile soils, and the availability of forage for livestock provided incentive for a diversity of people from these regions to settle in the temperate grassland zone. The many names used to describe temperate grasslands, including *prairie* in North America, *steppe* in Eurasia, *pampas* in South America, and *veld* in South Africa, reflect how different cultures identify portions of this expansive biome. Given the history of human settlement in temperate grasslands, one of the main threats to their integrity is land-use change. Prudent stewardship is required so that the healthy functioning of this biome, which is highly productive and of great economic importance, can be maintained over the long term.

Temperate grasslands are largely defined by their lack of woody vegetation. They are typically dominated by perennial grasses and sedges (graminoids) that are well adapted to semi-arid to subhumid climates with marked extremes in temperature (both among seasons and over the course of a day) and seasonal precipitation. These grasslands also contain a variety of nongraiminoid herbaceous plants called forbs, which, although less common than the dominant graminoids, are the main contributors to plant diversity in these systems. Forbs often flower in synchrony following monsoonal moisture events, giving rise to a colorful palette of blooms that cover vast expanses of land. In some cases small shrubs are present, for example, where the local topography and soil texture favor them. More generally, plant community composition varies in relationship to many factors other than climate including edaphic (soil) conditions, fire regimes, and grazing by wild and domesticated ungulates (hoofed mammals).

Temperate grasslands are broadly considered to be water-limited in the sense that woody vegetation is typically not supported. However, there is substantial variation in the amount and timing of precipitation among regions, which gives rise to distinct differences in grass canopy height and species composition. For example, in North America, as one moves from the subhumid central lowlands in the east to the semi-arid western extent of the Great Plains, precipitation becomes more limiting. This gradient in precipitation is mirrored by a transition from tall-grass prairie in the east, to short-grass prairie in the west, with midgrass prairie occupying the area in between. In contrast, the gradient of precipitation availability in the Eurasian steppe runs from north to south, with grasses and forbs becoming shorter, less abundant, and less diverse as one moves south and precipitation becomes more limiting.

The Badlands National Park in South Dakota is home to bison, one of the quintessential symbols of the North American plains. Temperate grasslands like this attract human settlements for abundant hunting or livestock grazing. Image by National Park Service.

PHOTOSYNTHETIC GRASS PATHWAYS

An important distinction among grasses is whether they have a C_3 (when carbon dioxide breaks up into a three-carbon compound) or C_4 (when carbon dioxide breaks up into a four-carbon compound) photosynthetic pathway. The designation as a C_3 or C_4 grass indicates whether the first product of carbon fixation during photosynthesis is a three-carbon or four-carbon molecule. Several differences in the C_3 and C_4 photosynthetic pathways result in plants of each type achieving optimal photosynthetic rates under different climatic conditions. As a result, the distribution of C_3 and C_4 grasses varies strongly in relationship to water availability and temperature among regions. The C_3 grasses, often referred to as "cool season" grasses, dominate in areas with relatively cool temperatures and where precipitation is high during the winter months. These grasses germinate in early spring and achieve maximum growth when temperatures are still relatively low (approximately 20 degrees C or 68 degrees F).

In contrast, the C_4, or "warm season" grasses, dominate in areas where both temperature and average annual rainfall are higher than where C_3 grasses dominate, and where the majority of precipitation occurs during the summer. Germination of C_4 grasses is delayed until the summer months, and maximum growth occurs at quite high temperatures (approximately 35 degrees C or 95 degrees F). The distribution of C_3 and C_4 plants in relation to their climatic requirements results in gradients in their relative abundance at large spatial scales. For example, temperate grasslands in the southeastern United States are dominated by C_4 grasses, while C_3 grasses become increasingly dominant to the north and west. Differences in the chemical composition of C_3 and C_4 grasses (for example, the ratio of carbon to nitrogen in their tissues) affect ecosystem-level properties such as rates of litter decomposition and nutrient cycling.

Grasses also exhibit a number of morphological (structural) characteristics that adapt them to the precipitation and temperature extremes to which they are exposed, as well as to fire and grazing by herbivores. In particular, grasses have several characteristics that help them cope with stressful conditions, including vegetative (asexual) reproduction; buds (new growth) that occur at or below the soil surface; cell division (new growth) that occurs at the base of leaf blades rather than at the tip; and fibrous root systems that are concentrated in the shallow layers of the soil profile. Vegetative reproduction occurs when individual plants send out specialized structures such as stolons or rhizomes in order to establish a new individual. Eventually the new individual breaks off from the mother plant, forming an independent, genetically identical individual (or clone). This type of reproduction is beneficial to plants growing in harsh environments because it requires less time and energy than sexual reproduction, which depends on pollination and the formation of seeds. Once grasses have established, new buds are often produced at or below the soil surface, which protects this young and delicate tissue from fire and grazing herbivores. Additionally, cell division typically occurs at the base of leaf blades, rather than at the tip, as is common in other types of plants. This allows for quick tissue regeneration, even if the majority of a leaf is lost to herbivory. Grasses also have fibrous root systems that extend horizontally from the plant through the shallow layers of the soil profile. The fibrous roots, which are finely branched and have a high surface area, allow grasses to take maximum advantage of periodic monsoonal rains that are short in duration and infiltrate only the shallow soil layers.

SOIL DIVERSITY IN TEMPERATE GRASSLANDS

There is a great diversity of soils that underlie temperate grasslands. This diversity reflects variation in the many factors that shape soil properties, including climate, parent material (for example, bedrock), topography, and the plant and microbial (that is, bacterial and fungal) communities present in a given area. Temperate grassland soils do share some basic characteristics, however. Importantly, they tend to be extremely fertile, and because of this they have largely shaped patterns of human settlement following the discovery of agriculture. They are typically high in dissolved nutrients (for example, nitrogen, phosphorous, and potassium) and organic matter, making them amenable to crop production. The top soil layers are often dark and rich in color due to a well-developed layer of humus (organic matter). High evaporation rates and capillary action prevent the leaching of nutrients to deep soil layers, where they would be inaccessible by plant roots. Additionally, the dense fibrous root systems of native grasses effectively bind nutrients, which keep them from leaving the system, and they prevent soil erosion by anchoring the soil and trapping moisture. The texture of grassland soils, which is defined by the particle-size distribution, varies from clays to loams to sands (with clays being composed of the smallest particles and sands of the largest particles). Soil texture is important because it influences water infiltration and percolation rates, as well as water-storage capacity, all of which directly affect plant distributions.

All of the attributes discussed so far, including abiotic factors such as temperature, precipitation, and soil characteristics, and biotic factors such as plant community composition and plant canopy height, affect nutrient cycling dynamics in grasslands. In addition, microbes, which are the main decomposers of organic matter, are important in shaping nutrient cycles. Nutrients such as carbon and nitrogen move through different "compartments" of the ecosystem, including the atmosphere; living organisms such as plants, animals, and microbes; and the soil. They exist in different forms depending on which compartment they are in. Often they enter the system as inorganic molecules (for example, gaseous carbon dioxide [CO_2] or nitrogen gas [N_2] from the atmosphere) and are transformed into usable (organic) forms that can be assimilated by plants and animals. For example, during photosynthesis, plants provide the crucial service of "fixing" CO_2 into organic forms that can be used by herbivores. In turn, most plants rely on microbes that transform inorganic nitrogen into organic forms that can be used to make the proteins that catalyze photosynthesis. Given that a diversity of species shape nutrient cycling dynamics in natural grasslands, modifying native plant and soil communities (for example, by converting native grasslands to single-species crops) can have marked impacts on nutrient cycling regimes.

FAUNA OF TEMPERATE GRASSLANDS

The vegetation of temperate grasslands in large part determines the abundance and composition of the associated fauna. Grasses support several characteristic mammal assemblages, most notably herds of grazing ungulates as well as burrowing rodents and lagomorphs (for example, rabbits and hares). Bison (*Bison bison americana*) and pronghorns (*Antilocapra americana*) represent the dominant large ungulates of the North American Great Plains, while wild horses (*Equus przewalskii* and *E. gmelini*), asses (*E. hemionus* and *E. kiang*), and saiga antelope (*Saiga tatarica*) were prevalent on the Eurasian steppe. Other examples of grassland ungulates include the pampas deer (*Ozotoceros bezoarticus celer*) of South America and the sable antelope (*Hippotragus niger*) of the South African veld. Vast herds of these animals historically populated grassland ecosystems worldwide, acting as keystone species that shaped the abundance, distribution, and adaptive characteristics of the plants on which they grazed. While some of these species, such as the pronghorn, are still widespread, most of them have been hunted to near extinction or widely displaced due to habitat loss. For example, in a few short decades, European settlers of the American west hunted an estimated 70 million bison to near extinction.

Many reasons underlie this tragic slaughter including the attractiveness of the skins for commercial sale, the perception that bison competed with cattle for forage, and the desire of the U.S. government to pressure Native American tribes who relied on bison for food and clothing. In many cases, individuals of these once prolific keystone ungulates exist only in nature reserves or as managed semi-natural populations (as with the American bison and the pampas deer). Instead, domesticated ungulates such as cattle or sheep number in the millions in many temperate grasslands and have in some places virtually replaced the native herbivore assemblages. Unfortunately, this shift also has negative implications for many of the large carnivores that historically fed on native ungulates.

Fossorial rodents and lagomorphs have also been important in shaping temperate grasslands, due in large part to the effect that their burrowing has on soil characteristics and plant distributions. The natural disturbances created by these ground-dwelling mammals, including the black-tailed prairie dog (*Cynomys ludovicianus*) of North America and marmots (*Marmota sibirica*) and social voles (*Microtus socialis*) of the Eurasian steppe, help to maintain plant and animal biodiversity in grassland systems. In lieu of other protective structures such as trees or caves, burrows provide shelter to a large variety of these relatively small, herbivorous animals. Burrow excavation and maintenance act to mix the soil layers, often times redistributing deeper soils to the surface and reducing soil compaction. Burrows also serve as habitat "microsites" that are different from the surrounding intact grassland. For example, some forb species recruit almost exclusively on disturbed soil created by burrowing, while insects and spiders often prefer the cooler, more humid habitat provided by underground burrows. In many cases, burrowing rodents live socially, in groups of related individuals that share tasks such as raising offspring and keeping watch for predators. Social living concentrates the number of individuals occupying a given area, leading to pronounced effects on habitat structure and function. For example, the preference of black-tailed prairie dogs for native grasses gives co-occurring forbs a competitive advantage, eventually leading to pronounced differences between plant community composition on and off prairie dog colonies.

IMPACT OF HUMAN ACTIVITIES

The history and future of temperate grasslands cannot be fully considered without understanding how human activities have altered this biome over time. Because of intensive land-use change, temperate grasslands today have been mostly altered from their pristine state. One of the earliest impacts humans had on grassland ecosystems occurred in the veld of southern Africa, where people used fire as a tool in hunting game. Fires were set to corner large herbivores, thereby facilitating their capture, or to create flushes of nutritious new plant growth that would attract herds of grazers. However, because periodic fires are a natural occurrence in most temperate grasslands, the use of fire by early peoples may not have had a large negative impact (although human-caused fires likely differed in frequency and intensity from natural fires). Indeed, in wetter regions, periodic fire, along with sustained grazing, may actually be required to prevent the encroachment of woody vegetation.

A more challenging issue in terms of land-use change, at least in regard to the maintenance of native biodiversity and ecosystem health, is the conversion of millions of acres of native grassland to cropping systems or for use as intensive livestock operations. The spread and intensification of agriculture, in large part due to its mechanization over the last 200 years, is associated with marked increases in human populations as well as a more sedentary lifestyle. Because agriculture-based societies have access to a stable food supply, they have largely circumvented the need to migrate in search of the plants and animals they historically used for food or clothing. Unfortunately, the modern agricultural practices that became widespread in the early twentieth century tend to be largely unsustainable, having led to the degradation of grassland habitats worldwide. For example, soil erosion due to poor farming practices and overgrazing of livestock has threatened the integrity of temperate grasslands in North America, Eurasia, and South America.

A classic example of the hardship associated with unsustainable farming practices is provided by the 1930s Dust Bowl era of the American west. After the Civil War, the government provided incentive for large numbers of settlers from Europe and the eastern United States to homestead the Great Plains. During this time, when precipitation levels happened to be much higher than usual, agriculture became widespread and intense, eventually replacing millions of acres of native grasses with cultivated crops. Many circumstances, including overplowing and severe drought, eventually led to extreme soil erosion across the American and Canadian Great Plains. The unprecedented loss of topsoil rendered the land useless for farming, forcing countless farmers to abandon their homesteads for lack of food and livelihood. The devastating ecological and social impacts that resulted are captured by pictures of denuded landscapes and massive dust clouds rolling across the plains and engulfing abandoned homesteads. Some of the dust clouds traveled far enough to darken the skies of far-away cities like Chicago and Boston.

The issue of soil erosion is not unique to the North American prairie. For example, as sedentary agriculture replaced nomadic grazing in northern China, overgrazing led to erosion and salinization (an increase in salt content) of the soil. The same issue arose in the Argentine pampas following the introduction of horses, cattle, and sheep and the advent of sedentary agriculture in areas that experienced periodic drought. Fortunately, an increased awareness of the biological properties of these systems has led to the implementation of sound policy, which has to some extent reversed the degradation associated with early intensive agriculture.

Indeed, the concept of sustainable agriculture is gaining traction today as different stakeholders acknowledge that maximizing agricultural production at the expense of ecosystem functioning and socioeconomic justice is shortsighted. However, the movement toward more sustainable practices must be balanced with the obligation to feed and cloth an ever-growing human population that has come to rely in large part on industrial-scale agriculture. Several organizations have begun to advocate a responsible transition from industrial agriculture back to smaller family-owned farms. They argue several benefits for such a transition including a reinvigoration of local farming economies; an incentive to maintain the natural resources (for example, water and healthy soil) that maintain profitable farms over the long term; a reduction in harmful pollutants due to large-scale application of chemical pesticides and fertilizers; and healthier foods derived from organic farming practices. Such organizations include the National Sustainable Agriculture Coalition in the United States and the European Initiative for Sustainable Development in Agriculture. There are also agencies that specifically advocate sustainable ranching practices, for example, by adjusting stocking rates (that is, the number of animals per unit area) to levels that are sustainable given the various ecological characteristics of the land. In fact, agricultural land is coming to be viewed as an important resource in the face of increasing urbanization worldwide. Given that much of the world's agricultural land is concentrated in the temperate grassland biome, long-term solutions for its sustainable use are needed.

TROPICAL & SUBTROPICAL GRASSLANDS

—Pedro Maria de Abreu Ferreira & Gerhard Ernst Overbeck

FIELDS OF STUDY

Agricultural Geography; Biodiversity; Forestry; Biology; Zoology; Ecology.

SUMMARY

Grasslands can be defined as ecosystems dominated by grasses (Poaceae family) or graminoids (grass-like plants, usually monocots, for example, from the Sedge family), with few or no trees. There are two principal ways that can be used to classify the vegetation of a defined area: in terms of floristic composition or in terms of vegetation physiognomy. The floristic classification is based on the presence of a set of species (that is, composition), which will be used to define this area as a floristic unit. The physiognomic classification, in contrast, takes into account parameters of the entire species set at any given point in time, such as height, estimated cover, and importance of different growth forms (that is, vegetation structure). The definition of grasslands presented above contains both floristic (large predominance of one botanical family, Poaceae) and physiognomic (comparatively low vegetation height due to dominance of herbaceous species, high grass/tree ratio, and growth form, considering the grassy habit as predominant) elements.

However, it is widely accepted that the worldwide distribution of ecosystems, including grasslands, is mainly related to climate conditions, and this can be used for a classification of grass-dominated ecosystems. David J. Gibson, for example, uses the Köppen climate classification system to define different grassland ecosystems around the globe.

Grasslands occur in tropical wet, dry, and desert climates, as well as in subtropical, temperate, and alpine climates. These different climates encompass a large variation in factors such as soil type, mean temperature, rainfall, and moisture. Considering that different species have different adaptations and strategies to endure different environmental conditions, this climatic variation is reflected in differences of composition and structure among different grassland ecosystems.

The frequency and abundance of tree species is an important characteristic for grassland classification—especially in tropical and subtropical grasslands. This is chiefly due to the conceptual overlapping between two ecosystem types in which grasses dominate: grasslands and savannas. Following the classical definition by Heinrich Walter, savanna ecosystems are tropical formations with the presence of grasses and woody plants, but do not include open grasslands without a woody component. Pragmatically, savannas can be considered as (tropical) grassland ecosystems with more than 10 percent tree cover, usually with tussock grasses and shrubs shaping the lower layer of the vegetation. This tree cover may be relatively uniform and dense (as in some parts of the Brazilian Cerrado) or clumped, shaping a parklike landscape (as in some parts of Africa). Grasslands, on the other hand, present no or less than 10 percent of tree or shrub cover. In many regions around the world, savannas and grasslands form mosaiclike patterns in the landscape, and the limits between both formations are neither clear nor stable. In this article we will focus on grassland ecosystems. However, the grassland component of savanna ecosystems will be briefly considered.

Impalas graze on the new grass sprouting under fire-blackened woody vegetation in Kruger National Park, South Africa. In August 2011, one year after a prescribed burn to kill off woody shrubs and trees that had been encroaching on the area where wildlife graze, the tropical grassland area had completely regenerated. Fire disturbance is sometimes necessary for the existence of grassland species, which show a number of adaptations to this process. Image by NASA.

NATURAL GRASSLANDS WORLDWIDE: A BRIEF EVOLUTIONARY HISTORY

Grasslands cover 15.84 million to 21.62 million square miles (41 to 56 million square kilometers), which represents 31 to 43 percent of the world's surface (excluding permafrost areas). During past geological eras, under different climates and land mass distribution, grassland-covered areas where even larger. Although the first grasses grew under forest cover or near forest borders, current evidence indicates that expansion of grasslands and diversification of grasses took place under conditions of increased aridity. The emergence of considerable shrub or tree layers in grasslands (as we see in tropical savannas today) occurred much later, when shifts in temperature and moisture allowed the establishment of such growth forms.

Fossil records provide evidence that large grazing animals and fire were important factors in the evolutionary history of grasslands. Altogether, climatic and, in some cases, edaphic conditions, grazing, and fire are constraining forces that have shaped grassland ecosystems as we see them today. In consequence, many species in the extant grassland flora show adaptations to avoid or resist drought and disturbances such as grazing and fire.

Extant grasslands may be natural or secondary, that is, the consequence of habitat conversion by human populations. Most grassland ecosystems present in highland montane climates are natural and largely determined by climate. Temperate grasslands in central and western Europe are considered secondary, being mostly linked to historical forest cleaning and subsequent mowing, burning, and grazing. Eurasian steppes and South American and African grasslands are considered natural (or climatogenic). In places where present climatic conditions are suitable for forest establishment, open grasslands gradually gave place to savannas (for example, Africa and central Brazil), shrublands or forests (southern Brazil and Africa). Some grasslands in tropical regions are clearly secondary, for example, cultivated pastures used for cattle grazing in tropical rainforest regions, originating from destruction of natural forests. This type of grassland will not be discussed in this article.

GRASSES: THE DETERMINANTS OF GRASSLAND'S PHYSIOGNOMY

The grass family does not only give the name to the vegetation type but also dominates most grassland systems. Current estimations of species richness in Poaceae range from 7,500 to 11,000 species, distributed in 600 to 700 genera in all continents and environments. The importance of this family to humankind is unquestionable, since all important cereal crops and sources of forage are grasses. Moreover, almost half of the Earth's surface is covered by landscapes (natural or human-driven) chiefly characterized by grasses.

Grass species are known to thrive under a wide amplitude of ecological conditions; grasses can be found from wet tropical climates to deserts. Many grasses can resist disturbance events such as complete burning or removal by grazing, or manage to resprout immediately. These abilities are consequences of evolutionary adaptations within the family, such as reduction in many vegetative and reproductive structures, growth forms adapted to disturbance (buds protected by leaf sheaths), or high importance of vegetative propagation in many grass species. Further, different photosynthetic pathways allow for good performance under different climatic conditions.

Reduction and simplification in aboveground structures directly influence effectiveness of biomass allocation: Grasses spend less energy and grow faster than trees, for example. The trade-off for this reduction seems to be clear: Grasses are apparently more fragile than trees, because they lack features such as secondary thickening in cell walls. However, grasses compensate for this with astonishing resistance to drought, resprouting ability after disturbances or unfavorable seasons, and ability to reach faraway sites by wind dispersal.

A considerable share of the resistance and resilience of most grasses can be attributed to their growth form. Meristematic tissues of grass shoots ("tillers") in most grasses are located near, at, or below the surface of the soil, providing protection to disturbances. Each tiller comprises photosynthetic leaves and (potentially) reproductive structures of the species. Most grass species can produce many tillers in a single growing season. This feature leads to the formation of dense tufts and multiple flowering culms in some species. Growing tillers may shape various patterns depending on the species, but there are two generally recognized growing forms: caespitose growth (erect tussocks) and rhizomatous growth (prostrate). The production of new tillers, which may be nearly unlimited in perennial species, allows the species to expand its cover in the landscape, "moving" to new sites with potentially better resources without the need of a full seed-dependent cycle of colonization and establishment.

Grasses can be divided according to their photosynthetic pathways: C_3 (when carbon dioxide breaks up into a three-carbon compound) species are cool season grasses, usually with short life cycles, and C_4 (when carbon dioxide breaks up into a four-carbon compound) species are warm season grasses, usually with long life cycles. The two pathways are named after the first stable product of photosynthesis, consisting of three carbons in C_3 grasses and four in C_4 grasses. The C_4 pathway is optimized considering carbon assimilation and enhances the plant's water use efficiency, which is advantageous in hot and dry environmental conditions. As a consequence, there are ecological differences between C_3 and C_4 grasses: The first have more efficient physiological processes under temperate conditions, whereas the latter are more efficient under tropical conditions, in which temperatures are higher throughout the year.

THE ROLE OF DISTURBANCE IN GRASSLAND ECOSYSTEMS

Disturbance can be defined as the partial or total removal of biomass from an individual plant. On the other hand, on the level of the plant community, grazing or fire can be considered normal or, in case of climate favorable to forest development, even necessary factors for the existence of this vegetation type. The plant species present in a community subject to recurrent disturbances show a number of adaptations to these processes, such as high regeneration ability. Fire may cause local removal of individuals and even species, but prevents shrub encroachment and keeps at bay the process of forest expansion over grasslands that takes place in ecosystems under moister conditions. Grazing may negatively affect grassland diversity when management is not adequate, and overgrazing may lead to decreased soil cover, erosion, and species removal. However, well-managed cattle-raising may be one of the most ecologically sustainable economic activities in grassland ecosystems, since native grass and legume species may be maintained and used as the primary source of forage.

DIFFERENCES BETWEEN TROPICAL & TEMPERATE GRASSLANDS

In comparison with their tropical and subtropical counterparts, temperate grasslands suffer colder winters and milder summers, larger amplitudes in temperatures around the year, a shorter period of vegetation growth, and much more frequent frost events. Annual rainfall is usually lower in temperate grasslands than in tropical and subtropical grasslands. However, many (but not all) tropical and subtropical grasslands show great seasonality in rainfall: a pronounced dry season, when mean rainfall may be less than 2.36 inches (60 millimeters), followed by a similarly pronounced wet season. Soils in temperate grasslands are usually deep and fertile, but plant growth is nutrient-limited because much of the soil nitrogen is inaccessible to the roots.

In addition, tropical and subtropical grasslands receive much more solar radiation throughout the year, which implies consequences concerning species composition. In tropical grasslands, C_4 grasses are dominant, and the contribution of C_3 grasses is usually reduced. However, the further south or north of the equator (that is, passing first to subtropical and then to temperate grasslands), the contribution of C_3 species enhances, considering both composition (more C_3 species) and structure (higher abundances of C_3 species).

With few exceptions (for example, the Indian subcontinent), natural grassland vegetation can be found—even though not always as a dominant vegetation type—in tropical and subtropical regions around the world. In the following we will briefly present the different vegetation types, beginning with those under a tropical climate.

SOUTH AMERICA

The South American llanos (Spanish for "plains") follow the valleys of the Orinoco River in Venezuela and Colombia. The region is dominated by savannas, but patches of typical grasslands, with almost no trees, also occur. They can be divided into different local vegetation physiognomies, classified mostly according to soil water availability. The lower layer of the vegetation is dominated by arrow-grasses (*Trachypogon* spp.) in some places, and three-awn grasses are also common (*Aristida* spp.). Legume species are abundant, enhancing forage value. Grasslands and savannas are grazed by cattle and the native *Hydrochoerus hydrochaeris* (capybara or chigüiro), the largest living rodent in the world. The llanos are threatened by land conversion to crops and monospecific plantations of exotic trees.

The Brazilian Cerrado is a biome that covers about 772,204 square miles (2 million square kilometers) in South America. Although the biome as a whole can be considered a tropical savanna, its landscape actually consists of mosaics of tree-free grasslands, of different savanna physiognomies (from scattered to dense tree cover), and of gallery forests. The disturbance regime (fire and grazing) and, to some extent, soil properties shape the distribution between savanna and grasslands. Typical grasslands are locally classified according to the presence of woody elements: *campos sujos* contain some scattered shrubs and trees, whereas *campos limpos* are dominated by grasses (common genera are *Echinolaena, Elyonurus, Paspalum, Trachypogon,* and *Tristachya*), without woody species. The Cerrado is considered to be one of the biodiversity hot spots of the world; the biome as a whole and the grassland areas are threatened by land conversion to establish pastures with nonnative species, uncontrollable fires in tree-dense savannas, and invasion of exotic species (*Brachiaria* spp.)

Further to the south, the flooding grasslands of the Pantanal cover 54,054 square miles (140,000 square kilometers) of Brazilian, Bolivian, and Paraguayan territory. Large areas of grasslands and forests that grow along river courses are yearly flooded. Natural grasslands are mostly used for cattle-raising, sometimes with native grasses as the main source of forage. Among these grasses, especially *Paspalum almum* and *P. plicatulum* present high forage value. The fauna of the Pantanal is outstandingly rich (for example, 650 birds, 1,100 butterflies, 80 mammals, and more than 250 fishes). Hunting is a major threat for this ecosystem since it has historically depleted populations of large predators such as the jaguar (*Panthera onca*). The invasion of the exotic tree *Vochysia divergens* (cambará) threatens the grasslands, and is kept at bay with fire (which is also considered a threat, mostly to animals) and selective logging.

The Campos of southern Brazil are under subtropical climates, and harbor approximately 2,500 nonwoody plant species, many of which are endemic. At higher altitudes (greater than 700 miles or 1,127 kilometers), these grasslands occur in mosaics with Atlantic and Araucaria forests, and *Andropogon lateralis* (caninha grass) is the dominant grass over large areas. At lower altitudes, grassland vegetation is dominant, and forests are mostly restricted to water courses. In these lower altitudes many grasses with high forage values are found, such as *Paspalum notatum, P. nicorae, P. pumilum,* and *Axonopus* spp. Different classifications for local grassland physiognomies exist, mostly related to soil and climatic factors. The southernmost areas of the Campos stretch south to temperate areas and the pampa biome (or Rio de la Plata Grasslands) of southern Brazil, Uruguay, and Argentina. Extensive cattle-raising is historically the most important economic activity in the Campos, and land conversion to crops and exotic tree plantations are the most pressing threats to this ecosystem.

AFRICA

Grasslands and savannas are the dominant landscapes in large areas of Africa. Atlantic forests mostly follow the equator line and the Zaire River, in central Africa, and are surrounded by woody savannas. Further north toward the Sahel and then the Sahara Desert, and east toward the Horn of Africa, trees become rarer and more scattered in the landscape, and the woody savanna gives way to typical grassland. Transitional ecosystems such as arid grasslands, shrublands, and open-canopy woodlands also occur, and vegetation physiognomy is largely determined by water availability. African grasslands and savannas are known for their large migratory herds of wildebeest (*Connochaetes taurinus*), zebra (*Equus burchelli*), and eland (*Taurotragus oryx*); by the Thompson's gazelle (*Gazella thomsonii*), buffalo (*Syncerus caffer*), and topi (*Damaliscus korrigum*); and by their famous predators, the lion (*Panthera leo*) and the cheetah (*Acinonyx jubatus*). There are also giraffes (*Giraffa camelopardalis*) and African elephants (*Loxodonta africana*), mostly in *Acacia* savannas and in the Serengeti volcanic grasslands. Threats to African grasslands include overgrazing, conversion to croplands, and uncontrollable wildfires. Trophy-hunting is also a problem, and has greatly depleted populations of mammals such as the white and black rhinos (*Ceratotherium simum* and *Diceros bicornis*).

The velds ("fields" in Dutch) are grasslands and savannas from the plateaus of South Africa and Zimbabwe. Locally, these ecosystems are divided into high velds (above 1,400 miles, or 2,253 kilometers in altitude, cooler, and with high rainfall) and low velds (below 700 miles or 1,127 kilometers in altitude, hot and dry). The dominant grasses are *Cymbopogon plurinodis, Diheteropogon filifolius, Heteropogon contortus, Themeda triandra, Brachiaria serrata, Digitaria eriantha,* and *Setaria flabellata.* In heavily grazed areas, this composition changes with more forbs, replacement of some dominant grasses, and tree encroaching.

ASIA, EURASIA & AUSTRALIA

Most tropical and subtropical grasslands in southern and southeastern Asia are secondary, being the result of abandoned crop areas that were originally rainforests. Large areas of midaltitudes (300 to 700 miles, or 483 to 1,127 kilometers) are dominated by *Imperata cylindrica*, whereas *Arundo madagascariensis* is the dominant grass in higher areas (greater than 900 miles or 1,148 kilometers). The larger continuous grass-dominated ecosystem in the region is the *Sehima nervosum–Dichanthium annulatum* grasslands in India. In some tropical areas trees (for example, *Pinus merkusii, Eucalyptus alba,* and *Casuarina junghuhniana*) form a rather continuous layer, characterizing savanna-like landscapes.

In Australia, climate conditions and soil fertility define different grassland ecosystems. A mosaic of savannas and grasslands are distributed in a belt around the arid center of the continent. Typical savannas, with a continuous tree layer, are usually dominated by eucalypts (*Eucalyptus* spp.) or wattles (*Acacia* spp.). In low-fertility soils, there are tall-grass grasslands dominated by *Themeda triandra, Schizachyrium fragile,* and *Sorghum* spp. Additionally, tropical and subtropical tall-grass formations with *Heteropogon contortus* and *H. triticeus,* as well as midgrass grasslands with *Aristida* spp., *Dichanthium sericeum, Bothriochloa decipiens, B. bladhii,* and *Chloris* spp., can be found. Fire is a natural disturbance and is also used as a management practice to prevent forest expansion. In some areas, absence of fire for long periods leads to closed eucalyptus forests.

Natural grasslands can also be found in the tropics and subtropics as "islands" of often relict vegetation scattered in forest ecosystems. These grasslands usually occur at higher altitudes, where temperatures are lower throughout the year, and are mostly maintained in the present climate by fire and grazing. Such formations are common in the Brazilian Atlantic Forest, in hilltops of mountain ranges above 800 miles altitude (1,287 kilometers), and are locally known as *campos rupestres.* Similar altitude grasslands occur in mountain ranges of Africa and Asia, for example, on so-called inselbergs.

LAND USE & IMPACTS ON CONSERVATION

Most grasslands originally occupied landscapes with a rather gentle topography (except in altitude grasslands), and in many cases were associated with deep and fertile soils (although grasslands may also occur on shallow, litholic soils). Also, it is easier to remove the original vegetation and plow the land in grasslands than in forests. The combination of these factors led to the conversion of many grassland ecosystems to crop production throughout the tropics and subtropics, which has led to species and habitat loss. In the past decades, natural grasslands in the tropics and subtropics have been increasingly converted to planted forests, often monocultures of exotic species, mainly to produce paper, resins, and firewood, again leading to pronounced biodiversity losses and landscape changes.

Grasslands are historically related to pastoral activities, dating back to the Neolithic. Today, large areas of natural grasslands in the tropics and subtropics still support herds of cattle, sheep, and horses. In many cases, the raising of grazing animals relies solely (or mainly) on native forage, which usually maintains the local biodiversity (both vegetal and animal) in good conservation status.

Abandonment, or nonmanagement, may have negative effects on natural grasslands. The absence of fire and grazing, the main nonclimatic factors that have influenced the origin, evolution, and maintenance of grassland ecosystems worldwide, may promote the dominance of tall, coarse, caespitose grasses, causing diversity losses in the plant community. This process may be followed by shrub encroachment, which leads to out shading of herbaceous plants. Under tropical and subtropical climates, where moisture conditions are suitable to forest establishment, this shrub encroachment may be followed by forest expansion and complete loss of the original natural grassland ecosystem.

FUTURE OF TROPICAL & SUBTROPICAL GRASSLANDS

As human population grows, so does the demand for food. In many regions of the world, grasslands are the primary targets for conversion into croplands. The conversion rates exceed legal protection in grasslands more than in any other inland ecosystem: approximately 1 hectare of natural grassland is protected for every 10 hectares lost.

A less obvious but nonetheless serious threat to grassland biodiversity is the suppression of management and disturbance. In many tropical and subtropical grasslands, current climate conditions are suitable for development or expansion of shrubland or forest vegetation, and grazing and fire impede these successional processes. If they are suppressed, the grassland community will, in the long run, be converted into a community dominated by woody species, losing its original biodiversity and ecological properties. Conservation efforts that focus on natural grasslands therefore must take management into account.

Because of human-induced climatic change, both the global temperature and atmospheric carbon dioxide concentration are expected to rise at unprecedented rates. These changes most likely will directly affect tropical and subtropical grasslands, where C_3 (temperate grasses and most forbs, shrubs, and trees) and C_4 species (tropical grasses) coexist. The effects of climate change on grassland distribution and on the distribution and abundance of C_3 and C_4 grasses has become an important research topic.

As in many other ecosystems, invasion by alien species also threatens grasslands. In African and Australian grasslands and savannas, invasive nonnative species increase the amount of flammable biomass in natural grasslands, enhancing the intensity of fire events and thus having an impact in the community at the local scale (for example, local extinctions).

Also, many grass species that are first introduced outside their native range as forage may become dominant and may lead to losses in biodiversity and productivity. In southern Brazil, the African grasses *Eragrostis plana* and *Brachiaria* spp. that present no invasive behavior in their natural environment become immensely aggressive, outcompeting and excluding virtually any native grassland species.

No vegetation type is stable, and changes are not necessarily negative, both from conservationist and agricultural perspectives. However, as natural grasslands constitute the basis for livestock production and harbor important—and often neglected—parts of the world's biodiversity, however, management decisions should consider possible changes and aim at the stabilization of the important ecosystem functions and services provided by grassland vegetation.

TROPICAL RAINFORESTS

—Hannah Bement

FIELDS OF STUDY

Agricultural Geography; Biodiversity; Forestry; Biology; Zoology; Ecology.

SUMMARY

Tropical rainforests are forest biomes with an average rainfall between 67 inches (170 centimeters) and 400 inches (1,016 centimeters). The World Wildlife Fund (WWF) defines them as a type of tropical wet forest, tropical moist broad-leaf forest, or lowland equatorial evergreen rainforest. Tropical rainforests are found in the equatorial zone between the Tropics of Cancer and Capricorn, roughly between latitudes 28 degrees north or south of the equator, where the direct angle of the sun rays causes high rates of evaporation. As water vapor evaporates from the forests, it rises, cools, and falls about an eighth-inch per day, giving these forests their name. Average humidity is usually around 77–88 percent. Mean monthly temperatures exceed 64 degrees F (18 degrees C) throughout the entire year. Tropical rainforests are found in five major geographical areas: Central and South America in the Amazon River Basin; Africa in the Zaire Basin; a small amount in West Africa and East Madagascar; in Indo-Malaysia on the west coast of India, Assam, southeast Asia; as well as New Guinea and Queensland, Australia.

Tropical rainforests are forest biomes with an average rainfall between 67 inches (170 centimeters) and 400 inches (1,016 centimeters). The World Wildlife Fund (WWF) defines them as a type of tropical wet forest, tropical moist broad-leaf forest, or lowland equatorial evergreen rainforest. Tropical rainforests are found in the equatorial zone between the Tropics of Cancer and Capricorn, roughly between latitudes 28 degrees north or south of the equator, where the direct angle of the sun rays causes high rates of evaporation. As water vapor evaporates from the forests, it rises, cools, and falls about an eighth-inch per day, giving these forests their name. Average humidity is usually around 77–88 percent. Mean monthly temperatures exceed 64 degrees F (18 degrees C) throughout the entire year. Tropical rainforests are found in five major geographical areas: Central and South America in the Amazon River Basin; Africa in the Zaire Basin; a small amount in West Africa and East Madagascar; in Indo-Malaysia on the west coast of India, Assam, southeast Asia; as well as New Guinea and Queensland, Australia.

There are four main types of rainforest; lowland equatorial evergreen rainforest, moist deciduous and semi-evergreen seasonal forests, montane rainforests, and flooded forests. Lowland equatorial evergreen rainforests have high rainfall throughout the year. They form a belt around the equator of the globe, covering the Amazon Basin of South America, the Congo Basin of central Africa, Indonesia, and New Guinea. Moist deciduous and semi-evergreen seasonal forests have high rainfall overall but have a distinct warm summer wet season and a cooler winter dry season. Some species of trees in this type of rainforest even drop their leaves during the winter dry season to conserve water. These forests make up parts of South America, Central America and the

The forest anole Anolis nitens lives in leaf litter under the canopy of the Amazon rainforest. Tropical rainforests host an enormous diversity of animal life, many of which also live in the canopy. Image by National Science Foundation.

Caribbean, coastal west Africa, India, and Indochina. Montane rainforests are also sometimes called cloud forests because of their almost constant shroud of fog. These forests are found in cooler-climate mountain areas, depending on the latitude, between elevations of about 4,920 and 18,827 feet (1,500 and 3,300 meters). Flooded forests are caused by seasonal flooding of tropical rivers. There are seven types of flooded forests recognized in the Tambopata Reserve in Amazonian Peru alone.

It is estimated that more than half of all plant and animal species are found in this biome and there are likely millions of species yet to be discovered. Many of these species are, or could be, important sources of food and medicine. Tropical rainforests produce about 40 percent of the Earth's oxygen and sequester carbon from the atmosphere, which helps to mediate climate change. Due to heavy logging and clearing for agriculture throughout the twentieth century, tropical rainforests now cover less than 6 percent of the Earth's land surface. They are threatened globally by deforestation and fragmentation for agriculture and urbanization, mining, drilling, and climate change.

LAYERS OF THE TROPICAL RAINFOREST

Tropical rainforests are typically made up of four layers or zones typified by the structure of the vegetation. These different zones form habitats for a variety of animal species. Though this stratification is not always clear-cut, these four layers are generally found at different heights in the forest: starting at the forest floor, the lower canopy, the upper canopy, and ending at the emergent layer.

The forest floor of the tropical rainforest is completely shaded from the dense forest canopy except for clearings, formed by fallen trees, where the sun can reach through the opening. Only about 1 percent of the sunlight that strikes the upper canopy reaches the forest floor. Except for dense growth near riverbanks, swamps, and clearings, most areas are so shaded that very little can grow, leaving a relatively open forest floor that allows for easy movement of large animals. Litter falls to the ground from the trees above and then is broken down by decomposers like termites, earthworms, and fungi. The constant heat and humidity accelerates the decomposition process. This leads to a very short nutrient cycle wherein organic matter is consumed almost as quickly as it is deposited, leaving only a thin layer of poor quality soil.

Also called the lower canopy, the understory is composed of 60-foot trees, the trunks of upper canopy trees, shrubs, and plants. Epiphytes, mosses, and lichens grow on the trunks and branches of trees. This area is dense and humid, with very little air movement. This area receives slightly more light than the forest floor, 5 percent of the light that strikes the upper canopy, but is mostly still in constant shade.

The canopy, found above the understory, is also called the upper canopy and it is made up of 60- to 130-foot trees. The trees in this layer compete and grow taller to reach the sunlight and block light access to the layers below. They form the roof of the rainforest. Leaves on these tallest trees often have "drip spouts" that allow rainwater to run off them to keep them dry and light and to prevent mold growth. Most of the animals in the rainforest live in this layer, where there is lots of food available from fruiting trees. Many animals that live in the upper canopy never have to go to the ground, getting most of their water from their food or rain pooled in the trees.

The emergent layer is the only one truly unique to the tropical rainforest, as the other layers are found in temperate rainforests as well. This layer is made of widely spaced trees that grow above the upper canopy. These trees, called emergents, are 100 to 240 feet (30 to 73 meters) tall with umbrella-shaped crowns. Emergent trees are exposed to extreme drying winds and heat. Many have adapted small pointed leaves that they sometimes drop during the dry season in monsoon areas to conserve water. Due to the shallow soil, these giant trees have shallow roots and must grow large buttress structures that can grow up to 35 feet wide for support. The emergent layer is home to animals like the crowned eagle (*Stephanoaetus coronatus*), king colobus (*Colobus polykomos*), and large flying fox (*Pteropus vampyrus*).

DISTURBANCE & SUCCESSION

Succession is the ecological process that changes biotic community structure over time. This generally trends toward a more stable, diverse community structure after the initial disturbance. Disturbances can be a natural phenomenon or human-caused. Natural phenomena include hurricanes, volcanoes, river movements, or fallen trees. Such natural disturbances are well documented in the fossil record to encourage speciation and endemism. Fallen trees, often caused by windstorms, are especially important disturbances in tropical rainforests. They open clearings in the dense canopy, allowing light to reach the forest floor where seedlings are waiting to spring into growth to take advantage of the gaps. These saplings might otherwise never be able to establish themselves and mature. About 70 percent of seedlings depend on such gaps for germination or growth past seedling size.

SOIL ECOLOGY

As previously mentioned, soil quality in tropical rainforests is typically poor due to the short nutrient cycle of the ecosystem. High rates of precipitation also contribute to the leaching or loss of nutrients from the system. Soil types throughout rainforests are, however, variable and some areas are very fertile. Each area's soil type is influenced by that region's climate, vegetation, topographic position, parent material, and age. There are two main classes of soil: ultisols and oxisols. Ultisols are defined as well-weathered, acidic red clay soils. They have low levels of calcium and potassium. Their high clay content helps them to retain water. Oxisols are also acidic, reddish soils but are old and highly weathered and leached because of good drainage, leaving them nutrient-poor. The red color in both of these soil types originates from the high heat and moisture environment, which forms oxides of iron and aluminum. These oxides are then insoluble in water and not readily taken up by rainforest plants.

The soils of the eastern and central Amazon rainforests and those of southeast Asia are generally old and mineral-poor. The nutrient flood plains of the western Amazon rainforests in Ecuador and Peru and volcanic areas of Costa Rica, southeast Asia, Africa, and Central America have young, mineral-rich soil. These soil characteristics influence the ecological processes of the regions in question. For example, primary productivity, or wood production, is higher in the western Amazon region than in the eastern Amazon oxisols. Above-ground productivity is closely linked to soil types and mineral contents. While the physical properties of a forest, like disturbance regimes, control tree turnover rates, the chemical properties such as nitrogen and phosphorous levels control the growth rates of trees. Phosphorous, potassium, calcium, and magnesium are all needed by growing trees and generally come from the weathering of rocks. Increased rates of decomposition result from increased phosphorous content in the soils. Soils that are limited in phosphorous content limit the growth of trees and subsequently the ability of a forest to uptake and sequester carbon.

The Amazon River Basin contains more types of plant and animal life than any other biome. The second-most biodiverse rainforest is found in southeast Asia. African tropical rainforests have the lowest variety of all the rainforests. In areas of high diversity, tropical rainforests can harbor between 40 and 100 species in just two and a half acres.

PLANT BIODIVERSITY

Seventy percent of all plants in tropical rainforests are trees. There are more kinds of trees in this biome than in any other area. For example, one area in South America was found to have between 100 and 300 tree species in just two and a half acres (one hectare) of forest. Trees in the rainforest are often straight and unbranched due to the need to grow fast and tall to compete for light and space. They tend to have smooth, thin bark because they do not need protection from cold temperatures or to retain water. This smooth bark also deters parasitic epiphytic plants. Due to this similarity in adaptations of rainforest trees, it is often difficult to identify species by their bark and leaves and they can only be differentiated by their flowers, which are often highly specialized to different pollinators.

Many plants are adapted to help them deal with the intense water inputs to the system. They have developed structures like drip tips and grooved leaves with oily coatings that shed water so that it does not weigh them down, which might cause them to break. The trees give off water through their pores, called stomata, which further increases the humidity of these forests. This transpiration can account for as much as 50 percent of the precipitation in a rainforest. Some upper canopy trees have small, dark green, leathery leaves to reduce water loss in the direct sun but the same trees may produce large leaves lower in the canopy to funnel off water and increase their light-catching surface.

Competition for sunlight and space has also led to many plant adaptations. Plants and trees in the lower canopy have large leaves to absorb as much sunlight as possible. Some leaves even turn to track the sun as its angle changes throughout the day. There are plants that specialize in growing on tall trees in order to gain access to the light. These plants are called epiphytes and derive their water and nutrients from the air and rain as well as from debris that accumulates around them. Epiphytes include orchids and bromeliads.

There are over 2,500 species of vine that grow in the tropical rainforests. Lianas are long-stemmed, woody vines that start out as small shrubs and then send out tendrils to grab onto sapling trees. As the tree grows upward, the liana vine grows with it to reach the sunlight of the canopy. Lianas will even grow from one tree to another to reach the best source of light and can make up about 40 percent of the plant mass in the rainforest canopy. The rattan vine similarly has spikes on its underside that point backward to grab onto saplings. Strangler vines use trees as support and as they grow thicker they can actually strangle their host tree, which will die and rot away, leaving the "hollow" tree form of the vine.

Tall rainforest trees develop wide buttress roots or a system of stilt roots to support them in shallow soil, but these structures have other benefits for the trees as well. Buttress roots aid in the competition for and efficient uptake of nutrients and water. The roots help to collect leaf litter and funnel nutrient-rich rain water as it drains down the trunk and holds it in pools that can be absorbed as needed. The large root structures increase the surface area of the tree, which increases gas exchange. They also help to reduce soil erosion that would wash away needed nutrients.

Throughout tropical rainforests, tree species are not often found in monoculture, or large areas of only one species. This spacing is due to high seed dispersal from fruit-eating animals as well as the highly competitive nature of tree establishment in the forest. This well-spaced biodiversity ultimately helps to prevent the spread of infections among the trees and decreases mass die-offs. Many tree species also have staggered blooming and fruiting periods, which helps to provide a year-round source of food to rainforest animals.

ANIMAL BIODIVERSITY

Tropical rainforests are home to a huge diversity of animal life. Some forests support ungulates like okapi (*Okapia johnstoni*), tapir (*Tapirus* spp.), or Sumatran rhinoceros (*Dicerorhinus sumatrensis*). Others have apes like the western lowland gorilla (*Gorilla gorilla*). There are many predators in rainforests, such as leopards (*Panthera pardus*), piranha, poison dart frogs, ring-tailed coati (*Nasua nasua*), boa constrictor (*Boa constrictor*), and many insects, which make up the largest group of animal species in tropical rainforests. Many species are adapted for a specialist life in the treetops and have bright colors, sharp patterns, loud vocalizations, fruit-heavy diets, and structures to aid in climbing. For example, New World monkeys such as the spider monkey (*Ateles* spp.) all have prehensile tails that help them to climb in their canopy home. Other treetop specialists include the three-toed sloth (*Bradypus* spp.) which may only climb to the forest floor once a month to defecate, the kinkajou (*Potos flavus*), harpy eagle (*Harpia harpyja*), macaws, and other parrots.

Tropical rainforests have existed for hundreds of millions of years. Most that remain today are fragments of forest that covered the Mesozoic era Gondwana supercontinent. When this land mass separated, it resulted in a great loss of amphibian diversity but the consequent drier climate led to the diversification of reptiles. Many amphibians (frogs, toads, salamanders, newts, and wormlike caecilians) and reptiles (snakes, lizards, turtles, tortoises, and crocodiles) are found in current-day tropical rainforests. They form an important part of the rainforest ecosystem, helping to cycle nutrients through the food web. The most common types of amphibian in the rainforest are frogs. More than 1,000 species of frog are found in the Amazon Basin alone. Many rainforest frogs are quite different from temperate frogs. The high humidity of the rainforest allows them to stay moist outside bodies of water and many are adapted to live high in the trees. These species also often lay their eggs in moist leaf litter or puddles of water formed in bromeliads high in the canopy.

THEORIES FOR THE ORIGIN OF BIODIVERSITY

There are two main theories for the origin of the great biodiversity of the tropical rainforests: the interspecific competition hypothesis and the Pleistocene refugia hypothesis. The interspecific competition hypothesis states that in tropical rainforests there is a high density of species with similar niches, or the way that they respond to distributions of resources and competitors within an ecosystem. Since there are limited resources available, if two species share a niche and are therefore directly competing for resources, then one of those species must eventually lose the competition and find a new niche or go extinct. This pressure to find a new niche is called niche partitioning. Niche partitioning can involve utilizing a different habitat, food source, or behavior, such as eating the same food but at a different time of day.

The Pleistocene refugia hypothesis was formulated by Jurgen Haffer in 1969. He theorized that the present diversity was a product of rainforest patches being separated by stretches on nonrainforest vegetation during the last glacial period. These remaining rainforest patches served as refuges for the species of the rainforest and over time the species in each of these patches changed, causing speciation, or the creation of new species. At the end of the glacial period when humidity increased and the rainforest patches reconnected, the new rainforest stretch was populated by a larger diversity of species. This theory has been met with much debate and skepticism based on evidence that places rainforest speciation before the Pleistocene glacial period.

IMPORTANCE TO HUMANS: CARBON SEQUESTRATION, MEDICINE & FOOD

Tropical rainforests provide many important services to humans, including oxygen production, carbon sequestration, and sources of food and medicine. It is estimated that more than half of all plant and animal species are found in tropical rainforests and there are likely millions of species yet to be discovered. Many of these species are, or could be, important sources of food and medicine. Tropical rainforests produce about 40 percent of the Earth's oxygen and sequester carbon from the atmosphere, which helps to mediate climate change.

Carbon flux is the exchange of carbon dioxide between the atmosphere and a reservoir, also known as a sink. The net primary productivity is the amount of carbon that is retained in plant biomass over time. It is also the gross primary productivity minus carbon that is released through autotrophic respiration. This metric is used to measure the amount of carbon in a tropical rainforest. Tropical rainforests are important as a carbon sink because of their above- and below-ground biomass, the amount of carbon they store, and the rate of fixation by photosynthesis. Tropical rainforests absorb about 4.8 billion tons of carbon dioxide per year.

One-quarter of all medicines originate from tropical rainforest plants. For example, curare is made from a tropical vine (*Chondrodendron tomentosum*) that is used as an anesthetic and muscle relaxer, and quinine comes from the chinchona tree (*Cinchona* spp.) and is used to treat malaria. More than 1,400 plants are thought to contain properties that could potentially cure cancer. Rainforest frogs have been especially important for pharmaceutical discoveries because of the unique chemicals found in their skin. These chemicals, which the frogs use for defense or healing, have been found to act as powerful painkillers, muscle relaxants, and offer possible cures for cancer.

Many cultivated foods and spices originate from the tropical rainforest biome. Over 250 types of fruit have been harvested from the rainforests, including banana, mango, and papaya as well as yam, coffee, chocolate, macadamia, and sugarcane. In New Guinea alone, 251 edible fruit trees have been found and most of these in the last few decades. These crops are now mostly grown on plantations but much of their genetic diversity is maintained through pollination from wild stock. This introduction of new genes helps the crops to avoid disease and pest damage.

VARIOUS THREATS TO TROPICAL RAINFORESTS

Most of the original extant of tropical rainforests has been lost due to clearing of forests for logging, mining, drilling, and agriculture. The ecosystem of the forest biome has also been affected by poaching by humans, the introduction of invasive species and diseases, and climate change.

Deposits of precious metals such as gold and diamonds and fossil fuels such as oil and natural gas have instigated forest clearing to allow access to these resources. Such mining and drilling is often given priority over environmental concerns to ensure economic growth in developing countries. For example, after decades of mining, Ghana has only 12 percent of its original tropical rainforest left intact. Mining and drilling activities are also often associated with air and water pollution that contaminate rainforest ecosystems.

Much of the tropical rainforests worldwide have been converted to open farm land. Then thin, infertile soil is quickly leached of all minerals from farming and is quickly eroded by heavy rains. This poor productivity has led to a technique of slash-and-burn agriculture. Once one patch of land has been used up, farmers push farther in to the forest, which is cleared and then burned to add the organic matter to the soil.

The abandoned patch of forest is often completely eroded of soil, leaving only hard red clay behind, making it difficult for new trees to grow. However, this fallow period after agriculture can allow secondary forest to regrow and slowly replenish the soil layer. Fertilizers and pesticides used by farmers to boost the nutrient content of the rainforest soil often run off and contaminate soil and rivers.

The delicate ecosystem of the tropical rainforest biome can be affected by many anthropogenic activities other than forest clearing. Poaching can remove species that are important to the system's functioning. Staggering amounts of wildlife are taken from rainforests every year for food and medicinal uses. In the Brazilian Amazon alone, 9.6 to 23.5 million mammals, birds, and reptiles are harvested annually. It is estimated that this number might be six times higher in Africa due to the demand for meat in poor and rural areas.

Invasive species are plants and animals that, when introduced to a new region, adversely affect the ecosystems therein. Invasive species compete with native species and can bring with them new parasites and diseases. For example, the brown tree snake (*Boiga irregularis*) was accidentally brought to Guam in a ship's cargo and has since extirpated, or caused the local extinction of, most of the native forest vertebrate species. Invasive diseases can be detrimental to tropical rainforest ecosystems as well. Such a disease is caused by the chytrid fungus (*Batrachochytrium dendrobatidis*), which is thought to have caused the extinction of over 100 species of frogs in the past few decades. Frog species in the tropical rainforests are especially hard hit because of the constant moisture and temperatures.

Tropical rainforest ecology is very dependent on climate, with many of its processes driven by the characteristic constant high heat, humidity, and rainfall. Therefore, this biome is very sensitive to changes in precipitation and temperature regimes such as those that might accompany global climate change. The climate in these equatorial forests is controlled by bands of clouds called the intertropical convergence zone (ITCZ) that forms near the equator. This zone is created by the convergence of trade winds from the Northern and Southern Hemispheres. Changes in the position of this convergence zone throughout the year cause slight seasonality in tropical rainforests, leading to wet and dry seasons in some forests.

Some regions have already experienced strong warming trends with an increase in average temperature of almost half a degree Fahrenheit (0.26 degree C). These regions have also experienced a decrease in rates of precipitation. For example, tropical rainforests in Asia have suffered from increased dry season intensity, with longer periods of drought conditions. However, there has thus far been little change in the tropical rainforests of Amazonia. El Niño Southern Oscillation events also drive rainforest climate variability between years. These events are likely to increase in duration and intensity with anthropogenic warming. These altered climate regimes may cause stress and increased mortality of trees within the tropical rainforest biome.

TROPICAL SAVANNAS

—*Alex W. Kisingo, Henry K. Njovu & Gideon Alfred*

FIELDS OF STUDY

Agricultural Geography; Forestry; Biology; Zoology; Ecology.

SUMMARY

Tropical regions contain higher species diversity than any other biomes of the world. This is attributed to their extensive area that creates many habitat types, their comparatively stable and warm climate with higher primary productivity, their long growing season, and the long time since they underwent major perturbations. Tropical savannas denote communities with an expanse of grass layer and scattered woody vegetation (shrubs and trees). They are found in tropical and subtropical regions, shaped by a succession of dry and wet seasons that differentiate them from other biomes. Tropical savannas cover about 10,656,420 square miles (27.6 million square kilometers), approximately 12.5 percent of the global land mass. They comprise roughly 40 percent of Africa, and they are also represented in parts of Australia, South America, and Asia.

Tropical regions contain higher species diversity than any other biomes of the world. This is attributed to their extensive area that creates many habitat types, their comparatively stable and warm climate with higher primary productivity, their long growing season, and the long time since they underwent major perturbations. Tropical savannas denote communities with an expanse of grass layer and scattered woody vegetation (shrubs and trees). They are found in tropical and subtropical regions, shaped by a succession of dry and wet seasons that differentiate them from other biomes. Tropical savannas cover about 10,656,420 square miles (27.6 million square kilometers), approximately 12.5 percent of the global land mass. They comprise roughly 40 percent of Africa, and they are also represented in parts of Australia, South America, and Asia.

One of the views of the extensive, tree-dotted savanna in Kenya's Samburu National Reserve. African savannas, which are grassland-dominated vegetation mixed with scattered and isolated trees, are rich in animal diversity and home to many species of mammals, reptiles, birds, and insects. This abundance supports traditional pastoralists and bushmen while also attracting worldwide tourism. Image by National Science Foundation.

Thus, savanna ecology is influenced by both grasses and woody plants, generally warmer temperatures throughout the year, and two distinct dry seasons. Summers are wet, hot, and humid with most rainfall during this period 15–25 inches (381–638 millimeters), while the dry season lasts for a longer period of the year (up to eight months) with warmer temperatures. However, the amount of rainfall and humidity in the savannas are not sufficient to support forest growth. Furthermore, high evaporative demands as a result of intense illumination cause savannas to experience a net water deficit for most of the year. Tropical savannas are maintained by complex and dynamic interactions among climatic factors, topography, edaphic factors, geomorphology, herbivory (grazing and browsing), fire, and human activities.

Savannas are generally categorized based on their canopy cover, spatial layout of woody plants, and stature. One of the resultant categories is *savanna grassland* consisting of sparsely scattered trees or shrubs. On the other hand, *savanna parkland* comprises discrete patches of woody vegetation interspersed over a continuous herbaceous plant layer. Due to various abiotic influences such as soil, altitude, and topography, savannas are interrupted by bands or areas of woodlands, forests (riparian, groundwater, or montane forests) or wetlands forming what are popularly called *savanna landscapes.*

Tropical savannas are known to be subclimax communities that are maintained by the soil characteristics (edaphic subclimax savannas), grazing (biotic subclimax savannas), and fire (pyric subclimax savannas). Termites are especially abundant in tropical savannas of the world, and their tall termitaria are conspicuous elements of the savanna landscape. Termites are important in soil formation and their termitaria provide shelter for other animals. Termites are also important food sources for anteaters (endemics of the neotropical zoogeographic province) and aardvarks and pangolins in African savannas.

Tropical savannas are inhabited by the world's proportionately large and rapidly growing human population, and a majority of their rangelands are dominated by livestock. Large population size associated with the prevalence of poverty increases the demands for immediate environmental resources. Overharvesting of timber and nontimber products, introduction of exotic species of plants and animals, overstocking, deliberate massive killing of wildlife, and fire suppression in some areas have significantly degraded tropical savannas, changing their scenery. There is likelihood that with climate change impacts, tropical savanna might be completely replaced by a different landscape.

AFRICAN TROPICAL SAVANNAS

African savannas are grassland-dominated vegetation mixed with scattered and isolated trees found between latitudes 15 degrees north and 30 degrees south and longitudes of 15 degrees west and 40 degrees east. Savannas in Africa are found in several countries including Guinea Bissau, Sierra Leone, Liberia, Ghana, Somalia, Nigeria, Ethiopia, Ivory Coast, Benin, Togo, Central African Republic, Cameroon, Chad, Tanzania, Sudan, Democratic Republic of Congo, Ethiopia, Kenya, Angola, Malawi, Burundi, Uganda, Rwanda, Zimbabwe, Zambia, Mozambique, Botswana, Madagascar, and South Africa. The savannas occupy areas such as the Maasai steppe ecosystems in northern Tanzania and southern Kenya within the Somali-Maasai regional center of endemism. Savannas in Africa also include the widely spread miombo (mopane) woodlands spanning from southwestern Tanzania to Zambia, Malawi, Mozambique, Zimbabwe, parts of Botswana and Namibia, and Angola. Miombo is an extensive tropical savanna falling mostly under the Zambezian regional center of endemism, and mostly characterized by such plant species in the genera *Isoberlina, Jubernadia,* and *Brachystegia.*

FLORA & FAUNA

In terms of plant species, savannas in Africa are characterized by such species as *Acacia,* miombo/ mopane, candelabra trees, baobabs (*Adansonia digitata*), sausage trees (*Kigelia africana*), amarula plants (*Sclerocarya birrea*), *Combretum* species, star grass, elephant grass, Bermuda grass, and many more.

African savannas are known for their diversity of both invertebrates and vertebrates. Millions of animals are known to coexist and are spatially or temporally specialized to environmental resources in savannas. There are carnivores, including cats (lions, leopards, cheetahs, serval cats, caracals, and civets); dogs such as wild dogs and jackals, hyenids (aardwolves, spotted and striped hyenas); and mustelids (honey badgers). Herbivores include elephants, hippopotamuses, kudus (greater and lesser), giraffes, elands, topis, sable antelopes and roan antelopes, hartebeests, black and white rhinoceroses, wildebeests, gazelles, impalas, gerenuks, aardvarks, waterbucks, bushbucks, reedbucks, buffalo, zebras, oryx, dik diks, lemurs, and sunis. These herbivores utilize different parts and types of plants, thus reducing competition, and hence increasing coexistence between them.

There are also a large number of birds such as ostriches, vultures, chaffinch, doves, pigeons, eagles, shrikes, starlings, shoebills, storks, ducks, hammerkops, hornbills, and many others. Reptiles include snakes (venomous and nonvenomous), crocodiles, lizards, and tortoises. There are also amphibians such as frogs and toads, and many invertebrates, with insects accounting for the majority. Biota in African savannas interacts in a complex way and produces a complex food web. Some animal species such as lions, elephants, beetles, and termites are keystone species maintaining diversity of the biomes. For example, elephants play the roles of ecosystem engineers, maintaining and creating vegetation structure, which increases usable habitat for other species.

ECONOMY & CULTURE

Wildlife abundance and the multitude of culture from the people residing in these areas make them attractive destinations for cultural tourism. Bushmen (San in Namibia and South Africa, Baswara in Botswana, and Hadzabe in Tanzania) and pastoralists (Maasai, Tutsis, and Ankole) persistently maintain their traditional ways of life. Other communities are composed of agropastoralists (peasants and small-scale livestock keepers). Preferred types of crops are maize, beans, peas, wheat, millet, cassava, coconuts, groundnuts, sugarcane, and cashew nuts, and the common livestock are an indigenous breed of cattle, sheep, and goats.

THREATS TO AFRICAN SAVANNAS

The majority of community livelihood in African savannas relies on the use of the immediate environmental resources for survival. People in these areas lack alternative livelihoods, which places direct pressure on natural biological resources, posing a greater threat to their survival. Indigenous trees are unsustainably harvested for construction, fuelwood, and charcoal production, while animals are illegally overhunted or captured for food and trade to satisfy the desires of the growing population in nearby towns and abroad for food and trophies. The tragedy of the ivory trade in recent years has increasingly jeopardized the survival of elephants.

Human-wildlife conflicts are prevalent in the African savannas. Conflicts between pastoralists on one side and African wild dogs and cheetahs on the other side have driven these wildlife species to the danger of extinction. On the other hand, contact between wildlife and domestic animals creates an outbreak of zoonotic diseases such as rabies, canine distemper, and anthrax, claiming the life of many wildlife, domestic animals, and human beings. Trampling by moving domestic animals creates loose and bare soils, which erode easily during rainy seasons but also suppress fire recurrence. The growing human population requires large areas for expansion of settlement and agriculture; this creates encroachment into protected areas and blocks wildlife migratory corridors and dispersal areas. Ultimately areas with wild animals (for example, protected areas) remain as ecological islands in the sea of human-modified landscapes.

Climate change and associated impacts invariably affect tropical African savannas resulting in changed climatic patterns in some areas. Frequent drought occurrences in the Horn of Africa (Kenya, Sudan, Ethiopia, and Somalia) dramatically reduce food production for humans and kill livestock (a major asset for most people) as a result of the loss of pasture and water.

CONSERVATION STATUS

African savannas comprise major protected areas of the world, differing in their conservation status, ranging from fenced national parks such as Kruger National Park in South Africa to the free-ranging wildlife in the major natural ecosystems such as the Serengeti in Tanzania. Other notable protected areas falling under the African savannas include Selous, Rungwa, and Moyowosi Game Reserves; Ruaha and Mikumi National Parks in Tanzania; Maasai Mara, Amboseli, and Tsavo National Parks in Kenya; Luangwa Valley in Zambia; Gonarezue in Zimbabwe; and the Great Limpopo Transfrontier Park straddling the borders between Mozambique, Zimbabwe, and South Africa. These areas preserve pristine habitats for wildlife and play a crucial role as representative areas for biodiversity conservation in the region annually visited by millions of tourists. Many countries in this region are members of international conservation treaties regulating management of biological resources. These treaties include the International Union for Conservation of Nature (IUCN), Convention on International Trade in Endangered Species of Wild Fauna and Flora (CITES), Convention on Biological Diversity (CBD), and the Ramsar treaty for wetland management.

SOUTH AMERICAN TROPICAL SAVANNAS

Tropical savanna also extends to South America in llanos and cerrado, but with relatively few species existing. Savannas occupy almost 965,255 square miles (2.5 million square kilometers) of Brazil, Colombia, and Venezuela, an area equivalent to one-quarter the size of Canada. Animals dispersing from the neighboring biomes play a key role in maintaining diversity and persistence of this region.

LLANOS IN SOUTH AMERICA

Flooded annually by the Orinoco River, the llanos of the Orinoco Basin of Venezuela and Columbia contain plants adapted to growing for long periods in stagnant waters. They span in a northeast direction from the foothills of the Andes Mountains in Colombia along the Orinoco River to its delta in the Atlantic Ocean. The vegetation in these areas is dominated by grass species in the genera *Melinis, Panicum, Homolepis, Scleria, Paspalum, Trachipogon, Gymnopogon,* and *Axonopus* with shrub species in the genus *Casia.* Fauna in the llanos includes capybara and marsh deer (mammal species adapted to a semi-aquatic life). The area also harbors the Orinoco crocodile, an endangered endemic crocodile species. Other animals include the world's extant largest boa anaconda, Orinoco turtle, giant armadillo, giant otter, black and chestnut eagle, swallow-tailed kite, broad-winged hawk, neotropical migratory shorebirds, and many species of cat fishes.

BRAZILIAN CERRADO

The Brazilian Cerrado is an extensive open woodland savanna of short twisted trees composing 21 percent of the country. The Cerrado savanna extends from the Amazonian forest margin to outlying areas in the southern states of Sao Paulo and Parana, to Bolivia and Paraguay. The Cerrado is a moister tropical savanna with rainfall ranging from about 31 to 79 inches (800 to 2,000 millimeters), with vegetation less adapted to waterlogging compared to its northern counterparts, the llano of Venezuela and Columbia. The Cerrado has several endemic plants and animals, and is a biodiversity hot spot consisting largely of savanna ecosystem. Plants in the Cerrado are estimated at around 5,000 species, with dominant grassy level vegetation five to seven times the tree and shrub layer. Similar to other savanna, it is characterized by remarkably long dry seasons supporting drought and fire-adapted plant species. Common genera are *Anisacanthus, Dicliptera, Geissomeria, Hygrophila, Jacobinia, Justicia, Lophoslachys, Mendoncia, Poikilancathus, Ruellia, Stauurogyne, Stenandrium,* and *Thunbergia.*

The Cerrado is characterized by indigenous animal species such as edentates (tamanduas, anteaters, and armadillos), marsupials (opposums), latyrrhine monkeys (marmosets, howlers, and capuchins), rodents (agoutis, pacas, capybaras, and many mouse-sized species) and maned wolf. The Cerrado has great diversity of animal life, though not in comparison to its counterpart in Africa.

ECONOMY, CULTURE & THREATS IN THE CERRADO

In the past, the Cerrado domain was sparsely populated by backwoodsmen and Indians, typically Brazilian countrymen, while llanos were shielded from human settlement and other threats. In most cases the countrymen kept livestock, cultivated crops, and utilized natural resources on a subsistence scale. In the past three decades, neoliberal changes in government policies have commercialized agriculture in the Cerrado leading to encroachment of many areas for farming. Crops like soya, maize, rice, and cassava are commonly preferred. Invasive plant species have increased in number as a result of tree plantation and livestock fodder improvement. Construction of Brazil's new capital city (Brasilia) and its associated development also created demand for the growing urban population, which increased pressure on natural resources. With livestock numbering around 48 million, it is estimated that 35 percent of the Cerrado has been destroyed.

Another threat to Brazilian savannas comes from the Brazilian steel industry, which solely depends on charcoal as source of energy. This places a remarkably high pressure on trees and is the second-largest threat after agriculture. The industry gets 71 percent of its charcoal from native trees. On the other hand, expansion of agriculture has led to dramatic changes in water use in llanos, diverting water to irrigation areas and drying up wetlands. Crops commonly cultivated are palm oil and rice, leading to fragmentation of the once continuous ecosystem. Oil exploration and extraction together with construction of dikes and water channels exacerbate habitat fragmentation. Furthermore, agrotoxins poison the ecosystem, leading to fatal consequences.

CONSERVATION STATUS IN THE CERRADO

The value of the biodiversity-rich Cerrado and llano has always been undermined. When compared to other biomes, Cerrado is considered to be poor, and not even considered by the Brazilian constitution as a national heritage. Although only 1 percent of the area is currently conserved, local conservation organizations collaborating with the World Wildlife Fund (WWF) are advocating for conserving and rescuing the biome. On the other hand, llanos have been poorly represented in protected area systems. Again, WWF, working with a local conservation agency, has proposed more than 63,000 acres in Colombia to be the first llano protected area. Increasing private areas for conservation and collaboration with the U.S. Fish and Wildlife Service are other efforts that are in place to rescue these ecosystems and their wildlife populations.

AUSTRALIAN TROPICAL SAVANNAS

Australia's tropical savannas are landscapes of dense grass and scattered trees covering approximately one-quarter of the continent, about 733,594 square miles (1.9 million square kilometers). They cross the continent from Rockhampton on the east coast, across the Gulf of Carpentaria, to the Top End of the Northern Territory, and over to The Kimberley in western Australia. Australian savannas fall within the tropical latitudinal ranges 10 degrees to 20 degrees south with medium to low mean annual rainfall (16 to 47 inches or 400 to 1,200 millimeters). Climates are characterized by two distinct seasons: the wet and the dry, with high daytime temperatures and high evapotranspiration rates. The dry season lasts five to six months, usually from May to October. The wet season lasts five to six months from December to March. Australia's tropical savannas are also referred to as the *monsoonal tropics* and the *wet-dry tropics*. The soils of Australia's tropical savannas are generally poor, with the exception of the southeast sector of the tropical savannas, which is more fertile.

AUSTRALIAN FAUNA & FLORA

Australian tropical savanna is one of the major biodiversity reservoirs which is less intensively developed compared to the temperate areas of the world, and this accounts for its relatively intact habitat compared to that of southern Australia. The savanna is home to hundreds of species of native plants, mammals, birds, reptiles, and amphibians and tens of thousands of different species of invertebrates. Many species in these groups are found nowhere else in the world. This part of Australia also has World Heritage areas like Kakadu National Park and Purnululu National Park.

Australian savannas are dominated by marsupials, mammals whose young are born undeveloped. These animals include the echidna, eastern gray kangaroo, the koala agile and whiptail wallabies, wallaroos, possums, gliders, the northern quoll, and the golden bandicoot. These animals live in or near the few trees in this area, utilizing them for shade, food, and water. Reptiles dominate in the other parts of the tropical savannas. The saltwater crocodile, which is found here, is the world's largest reptile, growing up to 26 feet (8 meters long). Associated with these are many species of birds, amphibians, and invertebrates, many of which are endemic to the region.

Open woodlands with grassy understory are the major vegetation type in Australia's tropical savannas. *Eucalyptus* trees of restricted height (49 to 82 feet or 15 to 25 meters) and open canopy dominate Australia's tropical savanna woodlands in areas receiving over 23 inches (600 millimeters) mean annual rainfall, with acacia or *Casuarina* species inhabiting the drier, and less fertile, areas. *Eucalyptus* species grow in warm climates and are the most important tree for timber, oil, and gum in Australia.

AUSTRALIAN SOCIOECONOMIC ACTIVITIES

Australian savannas have a rich culture from the Aborigine people, who have a long association with the land and much traditional knowledge of land management. The didgeridu wind instrument and the band Yothu Yindi come from the tropical savannas. Tourists and locals also value the Australian tropical savannas for their wild and spectacular natural features. The main areas of employment include the livestock, conservation, tourism, mining, and horticultural industries. Until a few decades ago, livestock was the main economic base of the tropical savannas, accompanied by lesser but more intensive uses such as mining, agriculture, or urban development.

More recently, however, mining and tourism have become the dominant economic industries. Like many other savanna ecosystems, Australian savannas are home to a pastoral industry that includes some of the most extensive operations in Australia and manages the largest area of land of any group in the northern part of the country. The livestock industry is the major user and manager of land in the tropical savannas. It contributes hundreds of millions of dollars to regional economies across northern Australia and provides employment for thousands, both directly and indirectly via service industries

Australian tropical savanna is experiencing significant population pressure and land-use changes and is vulnerable to rapid land degradation. Other threats include habitat fragmentation, deforestation, climate change and variability, as well as frequent fires.

ASIAN TROPICAL SAVANNAS (TERAI-DUAR SAVANNA)

Asian savannas are estimated to cover an area of 13,400 square miles (34,706 square kilometers) spreading through Nepal, India, and Bhutan. These savannas are dominated by subtropical grasslands and shrublands in the Terai-Duar Savanna. The Terai or "moist land" is actually the belt of marshy grassland, savanna, and forest located at the base of the Himalayas, about 15.5 miles (25 kilometers) wide, but it is very long. The area is mainly found in the semi-arid and semi-humid climate regions of subtropical and tropical latitudes. Rainfall ranges between 12 and 60 inches (300 and 1500 millimeters) annually, with climate significantly influenced by the topography, especially the great complex of mountains that flanks the Himalayas.

The Asian savannas have originated over generations from woodland ecosystems through deforestation, abandoned cultivation, and burning and are maintained at a subclimax by repeated grazing and burning. These mixed forces have led to the formation of mosaic types of savanna communities, depending upon the age, mode of origin, and the intensity of biotic disturbance. Savanna communities at different seral stages tend to differ in their biotic composition (that is, species composition, productive potential, and nutrient cycling).

ASIAN FAUNA & FLORA

The Asian savannas are home to diverse and abundant invertebrates and vertebrates. These include diverse amphibians, birds, and reptiles such as the gharial crocodile (*Gavialis gangeticus*). There are also mammals, including tigers (*Panthera tigris*), pygmy hogs (*Sus salvinius*), swamp deer or barasinghas (*Cervus duvauceli*), and the greater one-horned rhinoceros (*Rhinoceros unicornis*). Asian savannas are home to three endemic birds: the spiny babbler (*Turdoides nipalensis*), gray-crowned prinia (*Prinia cinereocapilla*), and Manipur bush-quail (*Perdicula manipurensis*).

There are a number of plant species in the region including wild sugarcane (*Saccharum spontaneum*), 10-foot tall tropical reeds (*Phragmitis kharka*), and silky kangaroo grass (*Themeda villosa*). Smaller species include cranberry grass (*Imperata cylindrical*) and six-weeks three-awn (*Aristida adscensionis*). Many of the grass species found in the Terai-Duar Savanna serve as fodder for livestock and wildlife species such as elephants and rhinoceroses, as well as cover for the predatory species such as tigers.

Agriculture is the principal economic activity in the region. Other major employment sectors include livestock keeping, nature-based tourism, cultural tourism, and natural resources. The main threats to the region include poaching for wildlife species, overgrazing, clearing for cultivation, population growth, irrigation projects, and water diversion.

TROPICAL SEASONAL FORESTS

—*Clara B.*

FIELDS OF STUDY

Agricultural Geography; Forestry; Biology; Zoology; Ecology.

SUMMARY

When most people think of a tropical forest they envision a jungle that is hot, humid, dark, noisy, and filled with strange, possibly dangerous animals and plants. In fact, many different types of forests can be found in zones surrounding the equator, and these forests vary in their physical characteristics (average rainfall and soil types) and biological characteristics (plant and animal species). Although all forest types exhibit some degree of temporal variation, tropical rainforests are characterized by less severe cyclic variations compared to seasonal forests (Table 1). In tropical seasonal forests (Figure 1), temporal patterns of temperature and rainfall, as well as plant and animal activities, are correlated within and between years, requiring organisms to shift their responses to changing conditions, to escape harsh conditions, or to minimize potential threats to survival and reproduction with intermediate traits.

When most people think of a tropical forest they envision a jungle that is hot, humid, dark, noisy, and filled with strange, possibly dangerous animals and plants. In fact, many different types of forests can be found in zones surrounding the equator, and these forests vary in their physical characteristics (average rainfall and soil types) and biological characteristics (plant and animal species). Although all forest types exhibit some degree of temporal variation, tropical rainforests are characterized by less severe cyclic variations compared to seasonal forests (Table 1). In tropical seasonal forests (Figure 1), temporal patterns of temperature and rainfall, as well as plant and animal activities, are correlated within and between years, requiring organisms to shift their responses to changing conditions, to escape harsh conditions, or to minimize potential threats to survival and reproduction with intermediate traits.

For example, the root systems of plants inhabiting seasonal environments must have the capacity to regulate water and nutrient absorption in both wet and dry conditions. Animals also must accommodate to seasonal changes since resources required for survival and reproduction, particularly food and mating sites, are not continuously available. If a population of a given species is sufficiently diverse, evolutionary processes are expected to favor traits minimizing negative consequences to life history (patterns of mortality and reproduction) occasioned by variability in resource dispersion and quality over time and space.

Scientists utilize a variety of research methods to evaluate the nature of variability in tropical seasonal forests and the effects of these variations on plant and animal populations. One of these techniques utilizes a multiyear series of rainfall data yielding an unbroken sequence of points, each representing total monthly rainfall. A quantitative test, time-series analysis, is used to produce a model summarizing past, current, and estimated future rainfall events. This statistical approach describes features of the rainfall series that may be similar within and between years and that may serve as predictable environmental stimuli eliciting compensatory physiological and behavioral responses in organisms.

Thus, individuals in populations located in tropical seasonal regimes survive or die depending upon the success or failure of their phenotypic traits. In the real world, however, reoccurring cycles of rainfall or other environmental stimuli will not be perfectly correlated over time within and between cycles, exposing organisms to some degree of unpredictability and risk. Since the response of plants and animals inhabiting tropical seasonal forests will not reliably fit many challenges presented by changing conditions, the survival and reproductive success of individuals in populations will depend upon phenotypic flexibility. In tropical seasonal forests, then, developmental, physiological, and behavioral features are expected to represent "best-of-a-bad-job," rather than optimal, responses.

Forests, their canopies, and other characteristics are fundamental to the processes, functions, stability, maintenance, and persistence of ecosystems. Locally and globally, the physical and biological elements of forests filter and cool the atmosphere, absorb and deflect sound and light, provide refuge for animals and plants, and buffer the potentially damaging effects of temperature (desiccation), rainfall (erosion), and wind (tornadoes may destroy forests and extirpate species). Detritus (falling leaves) and decomposing trees produce humus, the primary building block of soil. Trees also produce large quantities of oxygen and absorb carbon dioxide, processes critical for regulation of the negative consequences of global climate and other anthropogenic effects.

The varieties of tree species grouped into different tropical seasonal forest types (habitats) are determined by the range of local (food) and global (climate) conditions in given environments that change across time and space. Local (niche) and global factors responsible for habitat characteristics may be classified as physical (abiotic) or biological (biotic). Abiotic factors include all inorganic properties of forest landscapes such as altitude, geological and geographic formations, and waterways. Biotic factors include all organic features of forest landscapes, for example, distribution, abundance, type, and composition of plant and animal species.

CENTRAL AMERICAN TROPICAL DRY FORESTS

The tropical dry forests of Central America can be used as prototypes for tropical seasonal forests worldwide. These landscapes are characterized by clearly defined wet and dry seasons of about six months duration each and by two easily distinguishable habitat types, deciduous and riparian (riverine). Gordon Frankie reported that Costa Rican riparian and deciduous forests exhibit flower and fruit activity primarily during the dry season, November through April. In the deciduous forest, leaf fall is synchronized for most trees during the early to mid-dry season. In both habitats, soil, altitudinal gradients, and local disturbances (e.g., tree-fall gaps) are important determinants of plant growth, distribution, and abundance patterns. Climate, particularly rainfall, is the most significant predictor of seasonality.

Most trees in the riparian forest retain their leaves throughout the year, displaying a phenological (tree activity) pattern similar to wet forest sites in Costa Rica. Riparian habitat, with higher humidity and greater proportion of evergreen vegetation, is most likely characterized by a higher level of primary productivity compared to deciduous habitat, though few empirical studies exist to test this assumption. Riparian habitat is also likely to exhibit greater complexity and stability resulting in greater resilience when confronted with environmental perturbations (for example, drought and hurricanes).

FRUITING PATTERNS

For more than 30 years, Theodore Fleming, a biologist at the University of Miami, has conducted research on plant-animal interactions in many Latin American tropical seasonal forests, including those in Costa Rica. During that time, he has served as a faculty member and administrator with the Organization for Tropical Studies (OTS), an internationally recognized consortium of universities in the United States and Costa Rica whose primary goals are to facilitate scientific initiatives in tropical ecosystems and to train future tropical ecologists. Comparing fruiting patterns over time and space in New World (Central and South America) and Old World (Asia and Africa) tropics, Fleming and his colleagues demonstrated that seasonal unpredictability of fruiting patterns is greater in Old World forests. This finding has important implications for differential physical (soil composition and temperature) and biological (species composition, abundance, survival, and reproductive strategies) features. The differential environmental patterns documented by Fleming's group are expected to have significant consequences for patterns of mortality and survivorship of individuals, populations, and species inhabiting tropical forests west and east of the Atlantic Ocean.

TABLE 1. WHAT IS A TROPICAL SEASONAL FOREST?

Component	*Definition*
Tree	A long-lived woody plant occurring in a variety of types with a minimum height of 5 meters and distinctive characteristics (such as leaves and average circumference and height)
Forest	Land with tree-crown (canopy) layer greater than 10 percent in an area greater than 0.5 hectare
Natural forest	Forest composed of indigenous (native) trees that have not been deliberately planted or replanted
Seasonal forest	Forest in changing environments (for example, summer, fall, winter, spring, or wet season, dry season)
Tropics	The tropics include regions between 20 degrees north and 20 degrees south of the equator with average temperatures varying more over the short-term than average monthly temperatures vary over the long-term. Sea-level temperatures in the tropical zone are usually greater than 20 degrees C (60 degrees F), and forests in the tropical zone are characterized by variations in rainfall.
Tropical seasonal forest	Tropical seasonal forest environments are distributed worldwide in North America, Central and South America (Latin America), Asia, Africa, and Australia. They are generally found between 10 degrees north and 10 degrees south of the equator and differ from other tropical forests by rainfall patterns.

The presence of temporal variation in tropical seasonal forests is generally problematic from the organisms' point of view, since their biological processes, from molecular and cellular to phenotypic levels, must be sufficiently flexible to endure both wet and dry conditions. Although plants and animals experience disadvantages associated with inhabiting tropical seasonal forests, a major advantage is that time-varying patterns associated with seasonality yield environmental predictability, providing clear cues and signals about changing conditions (for example, changes in sunlight, humidity, or rainfall patterns). The discrete stimuli signaling wet and dry seasons in time-varying environments modify physiological and behavioral processes of organisms, allowing adjustments to the stress of different temporal features (for example, abundance and distribution of nutrient and water resources).

Ecologists studying plants and animals inhabiting tropical seasonal forests measure patterns of mortality to evaluate the success of organisms' responses to temporally varying regimes. On the whole, these studies show that many organisms display a trade-off between adult mortality and mortality of immature age-sex classes, and adult survivorship is commonly favored in seasonal forests if survival and reproduction exhibit certain features (extended life span). Ecological studies have shown that this life-history trade-off results from uncertainty and risk associated with reproducing in seasonally changing habitats in which, compared to wet tropical forests, cues and signals predicting the initiation and termination of seasons are highly variable. These challenges may stress developmental, physiological, and behavioral processes influencing lifetime reproductive success of individuals in a population. Statistical analyses of time-varying features are predictable on average, but the onset, duration, and termination of temporal patterns within and between years are usually difficult for the perceptual mechanisms of organisms to assess. Ambiguity and error resulting from between-cycle variability usually leads to significant mortality of one or more age-sex classes in a population.

MODES OF REPRODUCTION

All other things being equal, plant and animal populations inhabiting seasonal environments may respond to cycle variability with one of six modes of reproduction, and it is important to remember that each life-history strategy entails both costs and benefits for the organisms displaying it. One mode is characterized by brood parasitism whereby some bird species in tropical seasonal forests are parasitized by other species of birds that lay clutches in nests of the parasitized (host) species. The parasitic species destroys some proportion of the host's clutch, resulting in partial or complete loss of the host's seasonal reproductive investment. Parasite-host associations are highly specialized with respect to the species involved, and parasites gain an advantage over their hosts by laying their eggs prior to hatching of host eggs. Though this life-history pattern appears to be a reproductive dead-end for hosts, the lifetime reproductive benefits to hosts must have been higher than alternative patterns available over evolutionary time. In New World (neotropical) tropical seasonal forests, ground cuckoos (*Tapera naevia, Dromococcyx pavoninus, D. phasianellus*) are common brood parasites.

A second reproductive pattern is displayed when plants and animals specialize to the most common type or subtype of tropical seasonal forests, such as riparian or deciduous habitat. Specialists are usually characterized by a diet limited in food types and a tolerance for a narrow range of environmental conditions. Of approximately 230 primate species, my research suggests that 18 species are extremely specialized. Of these 18, however, most are exclusive to rainforests, two prefer wet forests (*Hylobates hoolock, H. syndactylus*), while the geographic distributions of five species are limited to or include large areas of seasonal forests (*Propithecus diadema, Nasalis larvatus, Presbytis pileata, P. potenziani, Gorilla gorilla*). It is interesting to note that the diets of the previous seven species include significant proportions of leaves, an observation worthy of further study since a folivorous diet is often thought to buffer primates from environmental stressors because, all other things being equal, leaves are continuously available. The neotropical plant *Tabebuia chrysantha* (araguaney) displays a life-history mode specialized to optimize lifetime reproductive success in deciduous habitat, and, like other specialist plants, produces flowers and fruits in dry season, the period most beneficial to the survival of new growth. A specialist strategy may be beneficial when conditions favor the specialized traits; however, specialists may be highly vulnerable if environmental stimuli change significantly (changes in rainfall patterns incompatible with a root system's particular functions).

In a third case, some species exhibit a phenotypic compromise or mix of traits advantageous to an intermediate fit across changing conditions (across both deciduous and riparian habitats in Central American environments). These generalist phenotypes are the most common ones displayed by plants and animals in tropical seasonal environments. All other things being equal, generalists thrive in a wide range of environmental conditions and utilize a variety of resources (for example, food items, resting sites). Compared to other life-history modes, generalist strategies may seem intuitively to be least costly to lifetime reproductive success; however, generalists may be spread too thin in time and space to respond successfully to many stressful environmental events, such as climate change or habitat destruction. Recent scientific evidence supports this proposition. The mantled howler monkey (*Alouatta palliata*) inhabits a broad range of forest types, including tropical seasonal forests, throughout the forests of Middle America and the Pacific coast of northern South America. Mantled howlers prefer a diet of flowers, fruit, and new leaves. Kathryn Stoner and other investigators have shown that mature leaves serve not only as a fallback food when preferred items are unavailable to howler monkeys but also as a buffer against environmental stress if distribution, abundance, or quality of preferred plant dispersion changes in time or space. The *Alouatta* genus is extremely successful biogeographically as the most widely distributed neotropical primate genus. However, the most well-known and widely ranging generalist species among mammals is the omnivorous *Homo sapiens*.

A fourth reproductive pattern displayed by some tropical seasonal forest species entails specialization on one of the environment's morphs (for example, dry or wet conditions). A number of these taxa are migrating species such as cockatoos (*Calyptorhynchus* spp.) that move to and from rainforests and tropical seasonal forests to maintain a relatively unchanging exposure to a tolerable range of humidity. This strategy permits the species to escape the negative effects of the least favorable suite of conditions to the more favorable suite of conditions (favorable for breeding or preferred food items).

A fifth life-history strategy observed in tropical seasonal forests is mast fruiting, whereby individuals of a given plant species coordinate flower and fruit production, usually in accordance with the highest availability of pollinators. In the tropical seasonal forests of Central America, *Manilkara* spp. display mast fruiting. A sixth life-history pattern observed by a relatively few tropical seasonal forest plant species such as *Pithecellobium saman* entails the production of flowers and fruit biannually. This mode of reproductive allocation may represent a lifetime energy-conservation strategy for some large trees. For each of the six life-history modes, the evolutionary and ecological strategies characteristic of specific populations depend upon a suite of factors. Among the most important factors are ancestral and derived traits of a species, as well as differential life-history metrics describing populations, such as generation time (T), rate of intrinsic increase (r), and net reproductive rate (R_0).

THREATS TO TROPICAL SEASONAL FORESTS

The tropical seasonal forest biome buffers the tropical rainforest biome and the tropical savanna biome. Thus, tropical seasonal forests are of critical importance for the preservation of biodiversity in tropical zones worldwide. Habitat destruction and other anthropogenic effects threaten evolved (for example, cooperative) and by-product (for example, competitive) associations within and among species. Local (patch or niche) and global (climate) effects interact, potentially inducing ecosystem perturbation and instability, in some cases leading to a cascade of network changes, including loss of species and ecosystem collapse.

Climate change and other factors varying in frequency, rate, duration, intensity, and quality contaminate, modify, and extirpate biogeochemical environmental components and attendant processes in tropical seasonal forests. The severity of anthropogenic outcomes continues to increase, with the potential to drive inherently resilient ecosystems beyond their capacities to accommodate and adapt to changing regimes. Jean Carlos Santos and his colleagues studied the research and conservation status of the Brazilian Caatinga, a tropical seasonal forest. These investigators showed that tropical seasonal forests are typically resilient and, thus, of particular scientific interest for a general understanding of factors associated with developmental, physiological, and phenotypic as well as biogeochemical buffering in complex systems. Santos's group highlighted the ecological, social, and economic importance of tropical seasonal forests. However, using tropical dry forests as an assay, these authors found that only 1 percent of dry forests in Central and South America, and only 5 percent of these forests worldwide, are protected.

These Brazilian scientists also found that tropical seasonal forests have been relatively ignored in the scientific literature, with tropical humid forests receiving the most research and conservation focus. Tropical seasonal forests are frequently located in areas having few academic resources, are poorly represented in the scientific literature, and are rarely highlighted as hot spots. Tropical seasonal forests require attention from professional, nonprofessional, and student conservationists, who can apply their knowledge, skills, and labor in ongoing initiatives from community to international levels in order to understand, address, and manage crises attendant to changes in tropical ecosystems. As two high-profile ecologists stated in the early 1990s: "On one point there is no argument: tropical forests are, indeed, in trouble." The worldwide status of tropical seasonal forests deserves treatment equal to that received by humid tropical ecosystems.

TUNDRA (ANIMAL LIFE)

—Mary Hurd

FIELDS OF STUDY

Agricultural Geography; Biology; Zoology; Biochemistry; Chemistry; Physical Science; Statistics; Animal Science; Ecology.

SUMMARY

The tundra, a characteristically treeless area of shrubs and patchy grass, lies above 60 degrees north latitude and covers about 10 percent of the earth's surface. It is inhabited by large numbers of a small variety of animals.

From March to September, the tundra, which covers about one-tenth of the earth's surface, is warmed by sunshine, with the sun visible for days or weeks. Over much of the tundra, from September to March the sun does not rise at all, with temperatures of –40 to –50 degrees Fahrenheit common during the dark months.

Basically, two kinds of tundra exist: the Arctic tundra, which follows the coniferous forest belt and covers the northernmost landmass of Europe, Asia, North America, Greenland, and Iceland; and alpine tundra, found on mountain slopes above the tree line in temperate areas.

Permafrost is a consistent feature of the Arctic tundra, as well as of the poorly developed tundra of the Antarctic. The summer thaw in the Arctic tundra extends to a depth of six to twelve inches, with most plant root depth and burrowing of animals limited to this thawable area. The tundra permafrost retards drainage, thereby causing boggy, saturated lowland during the summer thaw. Alpine tundra is drier because of drainage.

YEAR-ROUND INHABITANTS

Some hardy animals have learned to cope with the cold. Musk oxen, bulky animals with shaggy protective coats of hair, live along the shores of the Arctic Ocean and in other desolate areas of the tundra, where darkness continues for months and the temperature never rises above zero. These herbivores do not seek shelter even in the coldest weather, but search for frozen twigs or grass beneath the snow. Also capable of enduring the cold winter is the arctic fox, whose thick coat of fur turns white in winter and provides camouflage from his enemies. Foxes feed on arctic hares, birds, and lemmings, small rodents that dig beneath the snow. Lemmings remain active all winter, living under the snow and feeding upon grass and roots. Frequently, they reproduce under the snow.

Ptarmigans spend winters on the tundra, having feathers on their feet to aid traveling in the snow.

Tundra. Image by U. S. Fish & Wildlife Service.

Living in the driest areas of the tundra, they eat berries and tender leaves in the summer and rely on frozen vegetation during the winter. Their feathers turn white in winter, helping camouflage them from their fiercest enemy, the snowy owl. The owl, a bird of prey, moves southward into the forest in winter when the food supply becomes scarce in the tundra. Owls feed upon lemmings, small birds, insects, and arctic hares that are twice their size.

Some alpine mammals, such as marmots and ground squirrels, hibernate, eating large amounts of vegetation in summer and early fall before hibernation begins. Other small animals, including rabbits, forage as they can for winter feeding. Foxes range over alpine tundra in winter.

MIGRATORY INHABITANTS

As summer approaches and the days slowly grow longer and warmer, the frozen tundra begins to sprout grass, leaves, and wildflowers. The plants which have adapted to the tundra's short growing season and shallow topsoil above the permafrost are unusually small and provide food for animals. Huge herds of caribou migrate north from the forest to roam across the tundra, feasting on tender, young vegetation. The tundra wolf, or arctic wolf, a strong, fast animal endowed with exceptional hearing, vision, sense of smell, and endurance, follows the caribou into the tundra. Living and hunting in family groups or packs, the wolves feed off young, sick, and old members of the caribou herd.

After hibernating in the forest for the winter, grizzly bears move north to enjoy the berries and plants of the summer. These blond-colored bears also eat fish, lemmings, and carrion (remains left over by other animals). Polar bears, who prefer to eat meat, live near the Arctic Ocean to enjoy the seals, walruses, and fish. Occasionally, they move inland and devour berries, carrion, and other tundra animals.

Many birds also move north from the warm southern areas to feed, nest, and raise their young. Swans, ducks, and geese converge on the ponds, while gulls and terns occupy the tundra's beaches, and falcons and eagles soar at great heights above the tundra seeking their prey. In summer on the tundra, birds hatch as food becomes plentiful. As new vegetation is abundant and fish are plentiful in ponds and lakes, there are large numbers of insects on the tundra. Butterflies, moths, and bees abound, and flies and mosquitoes swarm after the caribou herds, providing food for the birds and fish.

Alpine animals, such as mountain sheep, ibex, wildcats, and many birds, who are not equipped for year-round alpine life, migrate south into more temperate forest environments in winter, returning to higher regions in the summer.

FURTHER READING

Johnson, Rebecca L. *A Walk in the Tundra.* Minneapolis: Carolrhoda Books, 2000. Newest entry in *Biomes of North America* series, including a map and a resource section. Includes much information about the habitat, the flora and fauna, and color photography. Intended for juveniles.

Sayre, April Pulley. *Tundra.* Frederick, Md.: Twenty-first Century Books, 1995. Part of the series *Exploring Earth's Biomes.* Aimed at young readers. Presents an overview and variety of facts and activities about the ecological community of the tundra. Encourages young readers to read further and communicate with pertinent ecological organizations.

Shepherd, Donna W. *Tundra.* New York: Franklin Watts, 1997. Clear, succinct introduction to the tundra that surveys climate, landforms, seasonal change, plants, animals, and people; includes color photographs and maps. Intended for juvenile readers.

Walker, Tom. *Caribou: Wanderer of the Tundra.* Portland, Oreg.: Graphic Arts Center, 2000. Using vivid photography, Alaskan wildlife photographer and author follows large herds of elusive caribou through their life cycle; includes natural history information and personal observation.

Zwinger, Ann H., and Beatrice E. Willard. *Land Above the Trees: A Guide to American Alpine Tundra.* Boulder, Colo.: Johnson Books, 1996. Interesting, well-written coverage of Colorado Rockies, California Sierra Nevada, Olympia Mountains, White Mountains, and Cascades. Discusses natural science and ecology.

TUNDRA ECOSYSTEMS

—Zhiqiang Chen

FIELDS OF STUDY

Ecology; Environmental Studies; Soil Science.

PRINCIPAL TERMS

coniferous forest: great northern forests, mostly evergreen, that end where the tundra begins

herbivores: plant-eating animals

lichens: organisms formed by algae and fungi that are a source of food for tundra animals

migratory animals: animals that move from one place to another for feeding or breeding

permafrost: a permanently frozen layer below the earth's surface

temperate region: a mild climatic area between the Tropic of Cancer and the Arctic Circle

SUMMARY

A circumpolar band of roughly 5 million acres of treeless country called the arctic tundra lies between the forests to the south and the Arctic Ocean and the polar ice caps to the north. Smaller yet ecologically similar regions found above the timberline on high mountains are called alpine tundra. The Russian word *tundra* came from the term *tunturi*, meaning "treeless plain." This name gives a brief but effective description of the plant species that grow on the northern flanks of the latitudinal tree line in the Northern Hemisphere, and above the altitude of the vertical tree line. Based on the temperature and environment, tundra can usually be categorized as short-term (hours, days, or a half month), seasonal (a half month to several months), and permafrost (years to thousands of years). Geographically, short-term, seasonal frozen soil and permafrost areas occupy about 50 percent of land area on Earth, while the permafrost area can take up to 25 percent of the land area. Nearly half of the territory of the former Soviet Union and Canada is permafrost, and 85 percent of the land in Alaska is permafrost. In addition, even near the equator, permafrost is found on the peak of Mount Kilimanjaro.

Among many factors that affect the distribution of ecosystem types, the temperature is a very important one. The higher the temperature is, the more energy is available for plants to use in photosynthesis and to be stored in the biosphere. When we go forward north from the equator, the temperature generally decreases. Also, as altitude increases, the temperature declines. So there is a general similarity between latitude and altitude in determining the distribution of ecosystems. Bearing this in mind, arctic tundra and alpine tundra are described and compared below.

In tundra, cold is the number-one limiting factor. Temperature often ranges from 60 degrees F (15.5 degrees C) in the summer to minus 130 degrees F

A caribou antler rests among wildflowers in the arctic tundra in the Bering Land Bridge, Alaska. In Alaska, 85 percent of the land is permafrost. Some of the estimated 48 animal species include arctic fox, ermine, and grizzly bear. Image by National Park Service.

(minus 90 degrees C) in the winter. Precipitation is also low, but water is not a limiting factor because of the very low evaporation rate in the tundra region. The growth of woody plants is essentially restricted due to the lack of warm summer at high latitude and altitude. Therefore, vascular plants that can survive in these cold environments must complete their annual growth rather quickly during the short but cool summer, and need to have the ability to endure the long yet extremely cold winter. Consequently, the species diversity in the tundra is lower compared to most other biomes. In fact, tundra can be considered as a cold marsh or a wet grassland that is frozen for a portion or even most of the year.

TUNDRA TEMPERATURE & SOIL

The tundra regions at high latitude areas are dominated by cold, dry, polar wind for most of the year. Air with limited moisture only enters the tundra region in summer when the sea ice begins to melt down. The typical summer temperature is usually below 50 degrees F (below 10 degrees C), and the temperature stays above the freezing point for only two to six months in a year. In the short summer, the sun remains above the horizon for most of the day, and does not even set for several weeks at extreme latitudes. But in the winter, the sun rises only briefly or not at all. Alpine tundra is similar to arctic tundra in terms of having a rather short growing season but a long and cold winter. Otherwise, the climates are quite different. Alpine tundra has less extreme photoperiods, more precipitation (mainly as snow), stronger winds, and greater day-night temperature fluctuations. Therefore, in contrast to arctic tundra, the weather in alpine is very variable. In addition, alpine tundra differs from arctic tundra in having higher ultraviolet radiation and lower oxygen concentration.

In addition to temperature, soil is an important environmental factor for the formation and function of ecosystems. Many types of soil are present in tundra areas, ranging from moist deep peats to coarse rocky materials where only mosses and lichens are able to grow. Usually only the very top layer of soils can thaw every year. The length and depth of thawing depends primarily on the amount of solar radiation received. However, factors that affect surface thermal properties, including soil color and moisture, vegetation cover, snow depth, and so on, will also play a role. The thawing layer can range from about three feet (one meter) to only a few centimeters in depth. In most areas, lower layers can remain frozen throughout the year. Consequently, drainage is usually poor and many soils remain waterlogged in summer. In contrast, alpine tundra soils generally lack permafrost but are thinner and immature, mostly classified as cryic (cold) entisols or inceptisols. Thus, they are better drained than arctic tundra soils. Because of the good drainage, late-summer drought can occur, limiting plant growth. Therefore, rocky sites with little plant cover are fairly common in alpine tundra. Solifluction (also known as soil fluction), a type of mass wasting in which waterlogged sediment moves slowly downslope over impermeable material (in tundra's case, a deeper frozen layer or bedrock), occurs in both arctic and alpine tundra, especially in alpine tundra because of the slope. Solifluction tends to produce an undulating landscape.

TUNDRA FAUNA & FLORA

To survive in extremely cold environments, plants must have special morphological and physiological characteristics. In arctic tundra, the dominant plants are dwarf shrubs and perennial herbs. These plants grow very close to the ground surface from which they get more energy during the rather short growing season. But in extreme environments, other than lichens and mosses, very few plants can survive.

These plants usually have the ability to go dormant during severe drought or cold and resume growth when conditions become better. Some plants can even survive under continued snow cover for years. In addition, plants must be able to tolerate short and cool summers when temperatures are only slightly above the physiological limit for plant growth. They have to grow fast because the summer is very short in tundra areas. Also, many tundra plants have perennating organs underground to help them survive in extreme habitats.

Trees are almost entirely lacking except in very limited areas near rivers, or on sheltered slopes, where the unfrozen soil can be relatively deep. The herb layer is in fact common in tundra areas and there are quite a few different species of plants, especially grasses and sedges. Surprisingly, the lichen layer covering exposed rocks is better developed than in any other biome. Alpine tundra is often similar to arctic tundra in terms of plant species. However, there are exceptions. For example, the tundra of the Sierra Nevada shares only 15 to 20 percent of its plants species with the arctic tundra.

Therefore, the ecosystems in arctic and alpine are similar, and both can be placed into the category of "tundra." However, there are many obvious differences between them. Therefore, we must formally call them arctic tundra and alpine tundra separately to show the differences.

Although extremely cold, there are animals in the tundra area. It has been suggested that approximately 48 species of land mammals are found on arctic tundra, but surprisingly the abundance of each species is not low. Most noticeable mammals and birds in arctic tundra are arctic fox, caribou, ermine, grizzly bear, harlequin duck, musk ox, polar bear, and snowy owl. There are not many species of insects in the tundra, and the commonly found ones are black flies, deer flies, and mosquitoes. The surviving strategy of mosquitoes is to keep themselves from freezing by replacing the water in their bodies with glycerol, a chemical that works like an antifreeze and allows them to stay alive during cold winter. Also, it should be mentioned that arctic and alpine tundra share few animals.

In addition to the commonly recognized arctic tundra and alpine tundra, Antarctic tundra occurs on Antarctica and on several Antarctic and sub-Antarctic islands. Most areas in Antarctica are very cold and dry, and most of Antarctica is covered by ice fields, so plants are not supported. However, limited areas of Antarctica, particularly the Antarctic Peninsula, have coverage of rocky soil that can support vegetation. Compared to the arctic and alpine tundra, the Antarctic tundra has much less mammal population, largely because of its physical separation from other continents. Seabirds and sea mammals such as penguins and seals have their habitats nearby the shore. Some small mammals such as cats and rabbits occasionally present in Antarctic tundra, and are usually considered to be introduced by humans to some sub-Antarctic islands.

TUNDRA & GLOBAL WARMING

Soils in tundra are considered to have the highest carbon density (21.8 per square meter) among all biomes, and the total soil carbon in tundra ecosystem is also the highest (191.8 times 10^{15} grams) among all specific ecosystems. In fact, tundra is recognized as one of Earth's three major carbon dioxide sinks. A carbon dioxide sink is a living entity (basically biomass) that takes in more carbon dioxide than it releases. Therefore, when discussing tundra, one thing cannot be neglected in this century: global warming. Global warming began to draw researchers' attention in the 1990s. It is a function of the greenhouse effect and caused by many kinds of greenhouse gases, such as carbon dioxide, methane, ozone, and nitrous oxides, of which carbon dioxide is thought to be the most important one. Carbon dioxide (CO_2) levels in the atmosphere are increasing, which may be mainly due to continued use of fossil fuels. It is estimated that during the twenty-first century, the carbon dioxide concentration could double and, consequently, simulations from general circulation models suggest that global average temperature would rise between 1.5 and 4.5 degrees C (34.7 and 40.1 degrees F). The gradual warming is accompanied by the melting of inland glaciers. One fundamental concern is that higher temperature in tundra and taiga, in which around 30 percent of the total soil-bound carbon found on our planet's surfaces are stored, will accelerate the degradation of peatlands, layers of permafrost, and forest litter, and thus lead to a large release of CO_2 and methane, exceeding the speed of biomass production. The second fear may be that a hotter, drier climate will bring more natural fires to areas in tundra where dead vegetation and peat have accumulated, which will also increase the amount of CO_2 in the atmosphere. In addition, not only the density of vegetation but also the composition of plants will be affected by climate change. Animal communities will be affected indirectly through their food plants.

From another viewpoint, the melt water from permafrost provides a direct source of water for plants in a severe drought summer, and keeps surplus water in the soil until the next summer. Global warming may disturb this permafrost system and, consequently, plants in tundra ecosystems could be negatively impacted. Also, freezing and subsequent thawing of soil in tundra may cause an increase in microbial activity, which is an important factor to keep the soil ecosystem active. Unfortunately, climate change might break the balance. However, climate warming, which brings increased availability of nitrogen and longer growing seasons, will likely increase biomass production in tundra ecosystems; this is what we should not ignore.

U

U.S. Hardwood Forests, Central

—Alexis S. Reed

FIELDS OF STUDY

Ecology; Forestry; Environmental Studies; Agronomy.

SUMMARY

These forests are biologically rich, even after two centuries of continually expanding habitat fragmentation.

The central hardwood forests of the United States are characterized by temperate broadleaf (angiosperms) and mixed broadleaf-coniferous (angiosperm-gymnosperm) stands. They cover more than 100 million acres (40 million hectares) in all. The boundaries of the ecoregion are rather ambiguous, but the commonly accepted range occurs along transition zones or ecotones, with several other bioregions, with the northern hardwood-conifer forest forming a northern border, the eastern and southern boundaries covered by the southeastern-pine forest, and the tall-grass prairie region bordering the west.

First referenced by ecologist E. Lucy Braun in 1950, the definition of Central Hardwood Forest has developed over time and pertains to the forest communities of deciduous and coniferous trees existing mainly on the floodplains of the Ohio-Mississippi River confluence area in the east-central United States, and extending into the more rugged topographic regions of the Ozark Plateau and some hill-and-valley regions in the western parts of Tennessee and Kentucky.

A hillside hardwood forest in the southeastern United States carpeted with white atamasco lilies (Zephyranthes atamasca) in spring. Central hardwood forests like these cover about 100 million acres (40 million hectares) of the United States and are one of the country's most diverse regions for flora. Image by U.S. Fish and Wildlife Service/Pete Pattavina.

The region today consists of some of the most diverse floral habitats in North America, and the climate is temperate. The Central U.S. Hardwood Forests biome covers approximately 116,000 square miles (300,000 square kilometers).

BIODIVERSITY

Before European contact, the regions consisted of both open oak savannas and forests dominated by oak and hickory species. Oaks in the intact remnants here include red (*Quercus rubra*), white (*Q. alba*), and black oak (*Q. velutina*); prevalent hickories are shagbark (*Carya ovata*) and bitternut hickory (*C. cordiformis*).

Most of these native forests and historic oak savannas have been disturbed, with more than 95 percent ecosystem alteration in some regions, and great swaths have been replaced by agriculture, urbanized areas, or introduced species. The oak savanna regions, of which only about 0.2 percent remains, are fragmented and heavily interspersed with other native vegetation, led by flowering dogwood (*Cornus florida*), sassafras (*Sassafras albidum*), and hop-hornbeam or ironwood (*Ostrya virginiana*). The ecoregion toward its northern boundary contains patches of open prairie. The riparian zones, bottomlands, and marsh areas feature tulip (*Liriodendron tulipfera*), sweetgum (*Liquidambar styraciflua*), and American elm (*Ulmus americana*) as the main tree types.

Along with native flora, the fauna that exist in hardwood forests in the east-central United States are quite diverse and include eastern gray squirrels, opossums, chipmunks, raccoons, deer, and American black bear among major mammals. An array of bird species includes warblers, vireos, wood thrush, sapsuckers and woodpeckers, tanagers, hawks, eagles, owls, and black-billed cuckoo, among many others. Also present are large populations of moths and butterflies, snakes and tortoises, and salamanders and frogs.

Policies of fire suppression implemented in the 1930s precipitated a major decline event in this biome, altering the composition of flora and fauna. With fire no longer keeping savannas in an open state, closed-canopy forests were allowed to develop in many regions. Shade-tolerant species such as maples and yellow poplar now dominate the formerly shade-intolerant composition of oaks; thus, decreases in oak regeneration and establishment have been noted. Invasive vegetation—such as kudzu, garlic mustard, and privet—threaten forest understory species throughout the Central Hardwood Forests biome.

While most forest stands of this region are predominantly angiosperms, softwood conifers are found throughout. A common conifer across the central hardwoods is the red cedar (*Juniperus virginiana*), which has adapted to disturbed sites and does well in drier conditions. Other common conifers are more often found in xeric, or low-moisture, locations in the Ozark Mountains of southern Missouri, and in some of the hillside zones on the eastern fringes of the biome; these include shortleaf and Virginia pine (*Pinus echinata* and *P. virginiana*).

THREATS

Much damage has already been done to this biome. Forest-clearing for pasture, cropland, monoculture plantation forests, urban and suburban development, and transportation infrastructure have combined to destroy the majority of the original habitat area. Native species in the fragmented stands that remain must contend with continued countercyclical fire suppression regimes, human overuse for recreation, climate change that is leading to a shift in habitat by latitude and elevation, and the persistent onslaught of invasive species. Insects and disease also exact a toll more readily than in the past, as some parts of the complex web of defenses have vanished along with the flora and fauna that once provided them.

There are some generally small reserve areas throughout the biome. Among them: Flat Rock Cedar Glades and Barrens, in central Tennessee, with 846 acres (342 hectares) protecting such species as leafy prairie-clover (*Dalea foliosa*), Boykin's milkwort (*Polygala boykinii*), and wavy-leaf purple coneflower (*Echinacea simulata*); Hoosier National Forest, in Indiana, with 202,000 acres (82,000 hectares) protecting oak and hickory stands, along with pileated woodpecker, wild turkey, white-tailed deer, and fox; and Shawnee Hills National Forest, in Illinois, with 280,000 acres (113,000 hectares) protecting cherrybark oak (*Quercus pagoda*), cerulean warbler (*Setophaga cerulea*), northern copperbelly watersnake (*Nerodia erythrogaster neglecta*), and bigclaw crayfish (*Orconectes placidus*).

FURTHER READING

Fralish, James S. *The Central Hardwood Forest: Its Boundaries and Physiographic Provinces.* St. Paul, MN: U.S. Forest Service, 2003.

Illinois Endangered Species Protection Board (IESPB). *Checklist of Endangered and Threatened Animals and Plants of Illinois.* Springfield, IL: IESPB, 1999.

Shawnee National Forest. *Biological Evaluation of Regional Forester Sensitive Species.* Washington, DC: Shawnee National Forest, U.S. Forest Service, 2004.

UPWELLING ECOSYSTEMS

—*Alexandra M. Avila*

FIELDS OF STUDY

Ecology; Environmental Studies; Agronomy.

SUMMARY

"Upwelling is the process in which surface currents move away from one another, or when winds push surface waters away from the shoreline."

Upwelling ecosystems occur where cold, often nutrient-rich waters from the ocean depths rise to the surface. They can also occur in freshwater bodies such as lakes. Upwelling ecosystems are considered some of the most productive ecosystems in the world. Upwelling is the process in which surface currents move away from one another, or when winds push surface waters away from the shoreline. This draws deeper water upward to replace the surface water. This process of the rising of cold nutrient-rich waters is essential for productivity in marine and freshwater ecosystems. The importance of the nutrient-rich waters has a direct tie to the food web. This is because most primary productivity in the oceans occurs in surface waters, but most of the organic matter can be found at the bottom of the sea. The reason for this is because almost all food chains begin with photosynthesis, and photosynthesis occurs in sunlight. Marine plants, therefore, will only live near the surface where light can penetrate, keeping herbivores and their predators up near the surface. But once these organisms die or shed scales, eggs, leaves, shells, or feces, they sink down toward the seafloor and decompose.

Upwelling thus brings all these organic nutrients back up to the surface water. With these nutrients, the phytoplankton and seaweed population grow, forming the basis of many food webs in marine and freshwater ecosystems. These blooms form the ultimate energy base for large animal populations higher in the food chain, including marine and freshwater fish, mammals, and seabirds. The raising of benthic nutrients to the surface waters occurs in regions where the flow of water brings currents of differing temperatures together. There are at least five types of upwellings: coastal upwelling, large-scale wind-driven upwelling in the ocean interior, upwelling associated with eddies, topographically associated upwelling, and broad-diffusive upwelling in the ocean interior.

This satellite image illustrates how the convergence of the Oyashio and Kuroshio Currents affect phytoplankton. When currents with different temperatures and densities collide, they create eddies. Phytoplankton growing there become concentrated and create visible swirls consistent with the water movement, seen above. Image by NASA.

COASTAL UPWELLING

Winds affected by the rotation of the Earth create a phenomenon known as the Coriolis effect. Along a coastline oriented north–south, like much of the west coast of the United States, winds that blow from the north tend to drive ocean surface currents to the right of the wind direction, thus pushing surface waters offshore. As surface waters are pushed offshore, water is drawn from below to replace them; these are coastal upwellings. Coastal upwelling usually occurs in the subtropics along the continental coasts, where prevailing trade winds drive the surface water away from shore, drawing deeper water upward to take its place. Because of the abundance of krill and other nutrients in the colder waters, these regions are rich feeding grounds for a variety of marine species, including whales, sharks, and many other fish species, as well as avian species (coastal birds). Some of the most well-known examples of this type of upwelling include: the Canary Current in Northern Africa, the Benguela Current in Southern Africa, the California Current off of the west coast of the United States, the Humboldt Current in South America, and the Somali Current off the western side of India. All of these currents support major fisheries. Along the west coast of southern Africa, the Benguela upwelling distinguishes itself from other upwelling systems by the intrusion of warm waters at its two boundaries: to the north, via the Angola Benguela front, and to the south, via the Agulhas Current that terminates the western boundary current of the Indian Ocean.

To spawn, sardines (*Sardinops sagax*) and anchovies (*Engraulis encrasicolus*) migrate toward the Agulhas Bank, where the warm waters transported by the Agulhas Current create a highly stratified environment. Eggs and larvae are then rapidly transported northward by a coastal current. After several days, they reach the upwelling region of the east coast of Africa, where certain numbers are transported out to the open sea by wind-driven currents, a source of mortality. Other sources of early life cycle mortality must not be overlooked, in particular, those linked to starving larvae and predation of eggs and larvae. A week after hatching, larvae must feed. As their swimming abilities are limited, they need not only a high density of suitably sized prey (mainly microzooplankton), but also optimal turbulence conditions. Excessive turbulence disperses plankton aggregations and reduces capture. Using more precise modeling of currents and production should shed new light on these issues. Although primary productivity of the Benguela system seems globally in excess, deficits may occur in certain spatiotemporal strata, such as that of the Agulhas Bank or the offshore feeding area during the reproductive season. In these conditions, feeding oases may be found within the large offshore eddies caused by the meeting of different currents and water masses. The low productivity of the Agulhas Bank creates intense food competition among the millions of tons of spawners that congregate there during the spawning season. This encourages parental cannibalism and the predation of eggs and larvae by other species. Because of the absence of monthly spatialized data on fish distribution, modeling of this phenomenon remains difficult but efforts at quantification are possible.

The Benguela ecosystem can be divided into two subsystems, the northern Benguela (southern Angola and Namibia) and the southern Benguela (South Africa and Namibia), separated by the permanent upwelling cell of Luderitz, the strongest in the world. In South Africa, historical fisheries data suggest a stock collapse of sardines in the late 1960s, followed by slow recovery, whereas the first level of collapse was noted in Namibia during the same period, but worsened in the late 1970s. The biomass of the Namibian stock has remained at a very low level. Scientists are not sure if these collapses are because of an environmental process (upwelling) or overfishing.

The structure and ecosystem dynamics of the southern Benguela have been undergoing progressive changes for over 30 years, whereas the northern Benguela ecosystem has been completely "reorganized" following a "regime shift." Today, the northern Benguela is dominated by jellyfish and fish feeding on detritus (gobies). Pelagic fishes have become low in abundance, and fisheries for most commercial species (hake and horse mackerel) are threatened.

UPWELLING IN THE OCEAN INTERIOR

Upwelling can also occur in the middle of the oceans where cyclonic circulation is relatively permanent or where southern trade winds cross the equator. The churning of a cyclone eventually draws up cooler water from lower layers of the ocean, which, in turn, causes the cyclone to weaken. Another very important example of upwellings that occur in very special habitats are upwellings that occur off the shore of islands, ridges, or seamounts because of a deflection of deeper currents. Just like other upwelling habitats, this provides a nutrient-rich area in otherwise low productivity ocean areas. Upwelling in seamounts has made them a very unique habitat with much endemism, and can be considered a biodiversity hot spot. A specific example includes upwellings around the Galapagos Islands with the Cromwell Current, and the Seychelles Islands, both which have major pelagic fisheries. In the Galapagos, the Crowmwell Current is a very important feeding ground for the endemic Galapagos penguin, which heavily depends on the smaller fish that feed on the plankton caused by the upwelling. This upwelling is also very important for the marine iguana, also endemic to the Galapagos Islands. When this upwelling fails to occur during an El Niño year, there is great devastation to the populations of these endemic animals that depend so heavily on it.

Another type of powerful upwelling occurs around Antarctica. The deepwater in the global conveyer belt flows across the seafloor all the way from the North Atlantic before it reaches Antarctica, so it has plenty of time to fill up with nutrients. Antarctica has many surface currents flowing around it, which stir up the waters, including the southern parts of the Pacific, Atlantic, and Indian Ocean gyres and the Circumpolar and Subpolar Currents. Also causing upwelling in this region are the strong polar winds that usually blow offshore.

FRESHWATER UPWELLING

Upwelling in freshwater ecosystems tends to be seasonal. It is most likely to occur in lakes in temperate regions. During the summer, the lake surface is heated with infrared radiation from the sun, causing a thermal stratification (also known as thermocline) in the lake. It is because of this stratification that upwelling does not occur. By the time fall comes around, the stratification breaks down, and the temperature of the epilimnion (surface layers) becomes the same as the temperature at the hypolimnion (deeper layers). This means that the thermocline shifts and allows upwelling to occur; also, stronger seasonal winds aid the process of upwelling. When winter comes, the surface water of the lake freezes and upwelling or mixing does not occur. As the ice begins to melt during the spring, lakes become isothermal, which means that the epilimnion and the hypolimnion have the same temperature because of the wind action that brings deeper waters to the surface as a result of mixing.

CANARIES CURRENT

In the Canaries Current, there are three major ecosystems: the northern Moroccan coast with seasonal upwellings in the summer; the south Moroccan and north Mauritanian coast with permanent upwellings (Sahara Desert); and the south Mauritanian and Senegalese coast, with boreal upwellings. The southern part of this ecosystem is characterized by high seasonal variability, alternating between an ecosystem under tropical influence in the summer and a coastal upwelling ecosystem in the winter. This alternation is accompanied by a migration of certain tropical species (tuna and tuna-like fishes) up to 20 degrees north during summer, and by a southward extension of the habitat of temperate species such as the sardine *S. pilchardus* during the winter. The Canaries Current system has wide continental shelves in the south, whereas the eastern regions are generally characterized by narrow continental shelves because of their young geological age. Principal spawning areas have been linked to regions with wide shelves. This association would be the result of physical processes developing over wide and shallow continental shelves, which would limit exchanges between the coast and pelagic areas.

The Canaries Current ecosystems were dominated by large demersal fish, which were rapidly overexploited. In the southern part of the area, the adaptability of artisanal fisheries took advantage of the seasonal migration of numerous species (a balance between seasonal upwelling and the warm season dominated by tropical influences). Over a longer period of time, West African fisheries have been spared the sudden collapses observed in other systems. A lower fishing exploitation rate until the 1980s and high seasonal fluctuations could have contributed to this relative resilience. Three countries experienced rapid and unexpected growth of octopus stocks in the 1970s.

Because of its commercial value, octopus fishing has become one of the essential components of West African fisheries. This shift in species was interpreted as the result of the absence of top-down control, following the overexploitation of demersal species that favored the development of short-lived prey species, such as the octopus (as well as shrimp and pelagic fishes). Further to the north, the reasons that have lead to the southern displacement of seiner fleets exploiting sardines remain largely unknown.

HUMBOLDT CURRENT

The Humboldt Current system, with its permanent upwellings off the coast of Peru and its seasonal upwelling along the coast of Chile, is by far the most productive in fish landings. With less than 1 percent of the world's ocean surface, it provides 15 to 20 percent of world marine catches (up to nearly 20 million tons per year for Peru and Chile combined). A second particularity lies in the presence of a very intense, extensive shallow zone that is very low in oxygen. A final particularity is its position under the direct influence of the El Niño Southern Oscillation (ENSO) mechanism. Alternate upwellings of nutrient-poor and nutrient-rich waters off the coast of Ecuador and Peru are associated with El Niño and La Niña episodes in the tropical Pacific. During El Niño, the pycnocline is so deep that the upwelled waters come from the nutrient-poor waters above the pycnocline. In extreme cases, nutrient-deficient waters coupled with overfishing cause fisheries to collapse, bringing about severe, extended economic impacts.

Coastal upwelling ecosystems, like the west coast of the United States, are some of the most productive ecosystems in the world and support many of the world's most important fisheries. Although coastal upwelling regions account for only 1 percent of the ocean surface, they contribute roughly 40 to 50 percent of the world's fisheries landings (this refers to the part of the fish catch that is put ashore. Frequently, landings provide the only record of total catch).

In upwelling systems, large variations in fish recruitment in fisheries appear to be from fluctuations in mortality during the early life-history stages of the fish, and thereby must be essentially associated to climatic variability. Survival through these stages is essentially linked to hydrodynamic structures which, in certain spatiotemporal strata, favor the retention, enriching, and concentration of plankton and ichthyoplankton. Fluctuations in abundance of pelagic fish stocks reflect significant changes in the structure and functioning of upwelling ecosystems.

High mortality rates have been observed at upper trophic levels (birds, marine mammals, and large predatory fish) in response to the diminished abundance of prey. Effects at lower trophic levels may also bring to light evidence of reduced predation by pelagic fishes on planktonic species, bringing about in turn changes throughout the entire trophic web. An example is the change in relative abundance of two species of anchovy associated with the El Niño phenomenon. Long-term fluctuations in dominant species have been observed in most upwelling ecosystems, such as the alternating dominance of sardines and anchovies in the Humbolt, Benguela, and Kuroshio (offshore Japan) Currents.

CLIMATE & CLIMATE CHANGE

Coastal upwellings also influence weather and climate. Along the northern and central California coast, upwellings lower sea-surface temperatures and increase the frequency of summer fogs. Relatively cold surface waters chill the overlying humid marine air to saturation, so that thick fog develops. Upwelling cold water inhibits formation of tropical cyclones (e.g., hurricanes) because tropical cyclones derive their energy from warm surface waters. During El Niño and La Niña, changes in sea-surface temperature patterns associated with warm- and cold-water upwelling off the northwest coast of South America and along the equator in the tropical Pacific affect the interannual distribution of precipitation around the globe. A great impact on upwelling is posed by the new installation of large wind farms.

Some upwellings are wind-driven; these new large wind farms exert a signicant disturbance on the wind speed in the vicinity of the installations. A recent study shows that the size of the wind wake is an important factor for the oceanic response to the wind farm. At certain wind speeds, sufficient upwelling may be generated that the local ecosystem will most likely be strongly inuenced by the presence of a wind farm.

Another factor that is currently affecting upwelling is climate change. In any one region, upwelling is intermittent; it can be strong in some years, and weak in others. The success of fishermen is greatly affected by this, since a weakening of an upwelling system can bring economic disaster. Upwelling brings nutrient-rich deepwaters to the surface, where algae can thrive in the sunlight, feeding the fish. Without nutrients, there will be no fish (as happens during El Niño conditions).

Scientists are trying to determine how climate change will affect upwelling and coastal productivity. The strong dependency of upwelling processes on the strength of trade winds contains one hint. Trade winds are zonal winds, which feed off the latitudinal temperature gradient; as this gradient weakens (computer simulations show that high latitudes become warmer than low ones), the zonal flow will weaken. Thus, the upwelling that derives its energy from trade winds will weaken. Off California, upwelling of cold water has become less common since 1975, and the productivity of the California Current has diminished accordingly. A similar decrease in productivity may be expected elsewhere in the eastern boundary currents, and in the eastern equatorial high production regions. Upwelling that depends on monsoon activity (as in the Arabian Sea) should be much less affected, or may even benefit from the change.

According to some scientists, wind-driven upwellings along the California coast have increased over the past 30 years, and some scientists postulate that the increase in wind-driven upwellings is largely because of increased greenhouse gas, but such an association has been speculative. A study on the effects of future climate change and upwellings in California showed negative results from the intensification of wind-driven upwelling events in that specific area.

According to a model scientists have made, the intensification of upwelling will likely have impacts on terrestrial and marine ecosystems. With intensified upwelling, enrichment can be increased that would be beneficial to organisms; however, concentration may decrease because of increased mixing, and retention may also be decreased by increased seaward transport of surface water. Overall, this could have a negative effect on marine ecosystems, as the current balance of these three factors will change with changes in upwelling.

V

Vertical Farming

—*Janine Ungvarsky*

FIELDS OF STUDY

Plant Science; Crop Science; Botany; Agronomy; Agriculture; Biochemistry; Ecology; Entomology; Genetics; Landscape Architecture; Soils Engineering.

SUMMARY

Vertical farming is a technique used for growing crops indoors in cities. The plants are grown hydroponically, in nutrient-infused water, or aeroponically, which uses mists of nutrients and water sprayed around the plants, under artificial lighting and in temperature-controlled environments. Proponents say it is a way to grow crops in urban areas or other places where the ground or climate is unsuitable for farming. They claim vertical farming may one day help with world food shortages. Opponents say that the cost of growing produce this way limits it to providing high-priced specialty produce for the rich, and they argue that such methods use too much energy.

BACKGROUND

Vertical farms are crop-growing facilities that are located inside a building. They use stacks of shelves to grow multiple layers of plants. The growing conditions are manufactured using lights and temperature and humidity controls. In some cases, where the design of the building permits it, the plants may receive natural sunlight supplemented by artificial lighting from LEDs customized to provide the ideal light needed for growth.

Most vertical farming is done with special techniques, such as hydroponics or aeroponics, which deliver nutrients to the plants without soil. The nutrients needed by that specific type of plant are mixed with water in custom solutions that provide exactly what the plant needs for optimal growth. Vertical farms can also use aquaponics. This creates an environment where fish and plants are raised together, with the fish waste providing the fertilizer for the

Lettuce grown in indoor vertical farming system. Image by Valcenteu (Own work) CC BY-SA 3.0.

plants. In some cases, soil may be used if the building design is appropriate.

People have considered the idea of vertical farms throughout most of the twentieth century, and a few prototype buildings were even designed. However, interest in vertical farming increased after promotion by Columbia University professor Dickson Despommier. In 1999, Despommier asked one of his graduate classes to figure out how to feed the population of New York City using rooftop gardens. When the class's research indicated the available acreage would fall short, Despommier and his students came up with the idea of stacked indoor gardens instead.

The plan they came up with used abandoned apartment buildings to grow crops in areas where traditional farming could not be done, such as large metropolitan areas. Despommier called his idea vertical farming and wrote about it in the 2010 book *The Vertical Farm: Feeding the World in the 21st Century*. The idea also received a great deal of interest on the Internet, and traditional media outlets soon produced stories about this alternative way of growing crops.

Microbiologist Dickson Despommier shares his ideas about how "vertical farming" can help reduce hunger by changing the way we use land for agriculture. Image by Pop!Tech CC BY 2.0.

INTRODUCTION

In the years since Despommier popularized the concept, vertical farms have been started in a number of countries. Some limited versions of the concept existed prior to the time Despommier and his students began promoting vertical farming, but most were experimental rather than commercial in nature. It is estimated that within five years of the publication of Despommier's book, there were more than two hundred vertical farms around the world. The United States, Japan, India, Nigeria, Antarctica, and a number of European countries all have one or more vertical farms. The most commonly grown crops are lettuce, kale, herbs, cucumbers, and strawberries, as these crops adapt best to the conditions found in basic vertical farms.

Vertical farms have been created in abandoned buildings, including apartment high-rises and stores. One former school in Japan, which was closed because the population of children in the area declined, was transformed into a strawberry farm. In just two rooms measuring about 1,400 square feet total, two businessmen started growing about 3,200 strawberry plants in four levels of racks made from materials that are readily available in most communities. Computers control the heat, light, water, and food, and a manager runs the entire complex with a handful of part-time helpers to pick the ripe berries.

Advocates for vertical farming say that it will be necessary to increase food production by 70 percent to feed the world's population by the year 2050. By that time, it is estimated that nearly 80 percent of the population will live in urban centers. This means that there will be more people living further away from the areas where food is produced. Vertical farming proponents say that it makes sense to begin growing food closer to where people live to save on transportation costs and shorten the time it takes to get food from the farm to the table.

Supporters also say that the optimal growing conditions provided by vertical farming techniques means crops can be grown faster. Salad greens are one of the most popular crops grown in existing vertical farms; they grow quickly and are available year-round. They have a short shelf life once harvested, so producing them closer to where they are consumed makes sense. Growing plants without soil also eliminates the possibility of many food borne diseases and prevents the presence of insects, so pesticides are not required, according to advocates. They also note that while crop variety is limited, the potential exists to grow tomatoes, rice, and many other types of plants in vertical farms.

Opponents, however, note the high cost of produce grown in vertical farms. This cost is mostly due the huge energy demands of such farms. For instance, one medium-sized strawberry produced by the Japanese farm in the former school costs upwards of $5, and the lettuce crops and herbs are often sold to high-end grocery stores that sell organic produce to wealthy clients. Such high-cost specialty foods will do nothing to alleviate potential food shortages, critics argue. The types of crops that can be grown in vertical farms are also limited. For instance, grain crops, such as wheat, would not do well in vertical farming settings.

Those who oppose vertical farming also call attention to the large amount of fuel it takes to run the electrical components necessary to keep such enterprises operational. The lights, heaters, and pumps all use energy, which increases the carbon footprint of these facilities. Solar sources are generally not sufficient to provide all the energy needed, opponents say. Nonetheless, they concede that such problems could be overcome if energy costs come down and new techniques are developed to increase crop variety.

FURTHER READING

Cho, Renee. "Vertical Farms: From Vision to Reality." *State of the Planet, Columbia University,* 12 Oct. 2011, blogs.ei.columbia.edu/2011/10/13/vertical-farms-from-vision-to-reality/. Accessed 12 Dec. 2016.

Cox, Stan. "Enough with the Vertical Farming Fantasies: There Are Still Too Many Unanswered Questions about the Trendy Practice." *Salon,* 17 Feb. 2016, www.salon.com/2016/02/17/enough_with_the_vertical_farming_partner/. Accessed 12 Dec. 2016.

Davis, Nicola. "Vertical Farming Explained: How Cities Could Be Food Producers of the Future." *The Guardian,* 6 Feb. 2014, www.theguardian.com/science/2014/feb/06/vertical-farming-explained-erik-murchie. Accessed 12 Dec. 2016.

Despommier, Dickson. *The Vertical Farm,* www.verticalfarm.com. Accessed 12 Dec. 2016.

Haspel, Tamar. "Will Indoor, Vertical Farming Help Us Feed the Planet, or Hurt It?" *The Washington Post,* 17 June 2016, www.washingtonpost.com/lifestyle/food/will-indoor-vertical-farming-help-us-feed-the-planet—or-hurt-it/2016/06/16/f1faaa98–3332–11e6–8ff7–7b6c1998b7a0_story.html?utm_term=.908a2947e215. Accessed 12 Dec. 2015.

Huber, Karina. "One More Question for Dickson Despommier: What's the Future of Vertical Farming?" *CCTV America,* 5 May 2015, www.cctv-america.com/2015/05/05/one-more-question-for-dickson-despommier-whats-the-future-of-vertical-farming. Accessed 12 Dec. 2016.

Koga, Hiroki. "How A Japanese Vertical Farm Is Growing Strawberries Using LED for the First Time." *AGFunder News,* 15 Apr. 2016, agfundernews.com/how-a-japanese-vertical-farm-is-growing-strawberries-using-led-for-the-first-time5691.html. Accessed 12 Dec. 2016.

McKirdy, Euan. "The Only Way Is Up: Vertical Farming in Kyoto." *CNN,* 19 Sept. 2016, www.cnn.com/2015/12/09/foodanddrink/kyoto-vertical-farm-spread/. Accessed 12 Dec. 2016.

Speed, Barbara. "This Is Why We Should Be Farming in Skyscrapers." *Citymetric,* 26 May 2015, www.citymetric.com/skylines/why-we-should-be-farming-skyscrapers-1029. Accessed 12 Dec. 2016.

"Vertical Farming: Does it Really Stack Up?" *The Economist,* 9 Dec. 2010, www.economist.com/node/17647627. Accessed 12 Dec. 2016.

Viticulture & Enology

—by Micah L. Issitt, BS

FIELDS OF STUDY

Culinary Science; Chemistry; Bacteriology; Botany; Horticulture; Agriculture; Genomics; Geology; Microbiology.

PRINCIPAL TERMS

Brandy: spirit liquor made by distilling grape wine or fermented fruit juice.

Cultivar: plant created through artificial selection that has been given a unique name because of its distinct characteristics.

Distillation: process used to purify a liquid by boiling and condensing vapors.

Ethyl Alcohol: alcohol produced during the fermentation of wine that is responsible for the wine's intoxicating effect.

Fermentation: process of converting sugar to alcohol, generally involving an enzymatic process occurring in response to the growth of yeast.

Finish: final phase in wine tasting, it describes the last sensory impression of a wine's flavor when swallowed. A wine's finish will take into account the wine's weight, how long the impression lasts, and the aftertaste.

Fortified Wine: wine blended with alcohol, typically brandy, to increase the sweetness and alcohol content.

Hybrid: grape variety created by blending two or more existing varieties of grapes to create unique characteristics.

Sommelier: specialist in all aspects of procuring, storing, and serving wine, generally employed in the fine dining industry.

Varietal: wine principally made from one type of grape and given the name of that grape variety.

SUMMARY

Viticulture and enology are fields involved in the cultivation of grapes and the process of winemaking. Specifically, viticulture is the science of growing and cultivating grapes, and enology is the study of wine and winemaking (viniculture refers to the science and process of growing grapes for winemaking). The international wine industry uses techniques developed in antiquity in conjunction with modern research and technology in the effort to maximize the quality and flavor of the finished product. Although many winemakers strive to maintain the traditions of their ancestors, others strive for innovation, creating blends of grapes and hybrid grape varieties with new flavors. The single botanical species used to make wine, *Vitis vinifera* (wine grape), has led to the cultivation of more than 15,000 wine varieties. In the twenty-first century, winemaking occurs on every continent except Antarctica. Worldwide, grape cultivation and winemaking are a multibillion dollar industry, and many universities offer graduate level courses in viticulture and enology.

DEFINITION & BASIC PRINCIPLES

Viticulture is the science of growing and cultivating grapes and grape vines, which includes planting grape seeds, training and maintaining grape vines, and harvesting grapes for the production of grape juice, wine, and other products. Enology is the study of wines and winemaking, which involves the fermentation of grapes and other methods used to create wine and impart specific flavors to the finished product. Viticulture and enology are the backbone of the wine industry, which also includes the commercial production, sale, and evaluation of wine products. Research in enology is used extensively in the culinary sciences, especially in regard to learning how to pair specific wines with certain foods and ingredients, and in teaching wine appreciation to enthusiasts. In addition, enologists and viticulturalists study the factors that contribute to wine flavor and the diseases that affect grapes and grape products.

BACKGROUND & HISTORY

Viticulture and enology have been part of human culture for nearly 8,000 years, when the first wines were made in the ancient Near East (modern Middle East region). The Romans laid down the foundations for grape classification and the aging process of wine. They also pioneered the use of oak fermentation chambers (wooden wine barrels) and cork for bottling. (The earliest extant prose written in Latin is, in fact, a survey on Roman viticulture.) Winemaking was nearly exclusive to Europe until the mid-nineteenth century, when it spread to the Americas.

In 1863, Emperor Napoleon III of France asked chemist Louis Pasteur to investigate the factors that led to deteriorating wine quality. Pasteur's work was a landmark in understanding fermentation and led to the development of enology. The results of his research quickly spread around the world, and in the 1880's, the University of Bordeaux in France and the University of California established the world's first institutions for the study of enology.

During the twentieth and early twenty-first centuries, most of the innovations in wine came from the development of industrial winemaking procedures.

Advancements in sanitation during this time also led to the cultivation of higher-quality, longer-lasting wines.

HOW IT WORKS

The production of wine during a single year, called a vintage, involves three basic components: cultivation, processing and fermentation, and finishing.

Cultivation. Photosynthesis starts the process of converting carbon and chlorophyll to sugar, which is stored in the grapes. There is a target ripeness level for each grape variety, depending on the flavor desired. Viticulturists use tools such as spectrometers to measure the sugar level in the grapes to decide when to harvest them. Variations in environmental factors such as climate and soil quality affect the flavor of the grapes. The ideal condition for grape cultivation consists of warm, relatively dry days alternating with cool nights. Places where these conditions occur naturally, such as Northern California's Russian River Valley, produce the highest-quality wines. Deciding when to harvest is the next variable. Harvesting directly following rain may lead to watery grapes, while harvesting following a dry spell leads to dry grapes with overly intense flavor. Although many industrial winemakers harvest grapes with combines that can pluck thousands of plants in a short time, the highest-quality wines are produced from hand-picked grapes.

Processing and Fermentation. After being harvested, the grapes are juiced. They are placed in a large container with a spout connecting to the bottom. When the container is filled to capacity, gravity will begin to crush the grapes, producing what is called free-run juice. The highest-quality wines are developed from this free-run product. Next, winemakers use air bladders or plungers to press the grapes. With each round of pressing, as liquid from stems, seeds, and skin is filtered into the juice, the quality of the juice declines.

To achieve fermentation, yeast is added, which metabolizes the sugars in the juice and creates alcohol as a by-product. Traditionally, the juice is placed into oak containers, which impart a unique flavor, for fermentation. Modern wineries may also use stainless steel or polished concrete fermentation chambers, although most conduct a portion of their fermentation in the traditional oak containers.

At this point in the process, all wine has a pale color. To make the wine red, the juice is soaked with the grape skins. This imparts a red color to the juice, a process called cold soaking. This process may also increase the complexity, or aroma, and flavors of the wine. To prevent fermentation during cold soaking, the soaking chambers must be kept at cold temperatures—typically between 4 and 15 degrees Celsius (39–50 degrees Fahrenheit). To achieve this, some winemakers cold soak in underground chambers, and others use dry ice or industrial cooling units. After soaking, the temperature is raised to start fermentation.

Finishing the Wine. Once fermented, the wine is filtered to remove sediment and other materials. High-quality wine is filtered through a relatively loose filtration system, and some sediment is left with the wine so as to continue imparting flavor and characteristics. After filtration, wine may be bottled or transferred to oak storage vessels for aging. Many fine wines are stored in oak for periods ranging from a few months to a few years, allowing the flavor to change and develop. Even if the wine is immediately bottled, the wine must rest for several weeks or months within the bottle to achieve a palatable or desired flavor.

APPLICATIONS & PRODUCTS

Cultivars. The vast majority of wine is made from grapes that are varieties of a single species, *Vitis vinifera*, the noble or Eurasian grape. Although there are many other grape species, few make satisfactory wines. Most wines are made from varieties of *V. vinifera*, including Merlot, Cabernet Sauvignon, Pinot Noir, Syrah and Shiraz, and Sangiovese, which are among the most important varieties of grapes for red wine; Chardonnay, Pinot Gris, Riesling, and Sauvignon Blanc are among the most popular varieties for white wines.

Grape Liquors. In addition to wine, grapes can be used to produce a variety of other liquors. Champagne and sparkling wine are made either by injecting carbon dioxide into already-fermented wine or by adding more yeast and sugar to the wine to induce a second fermentation process. With the latter method, the wine is then capped in a special container that traps the carbon dioxide that is released as a by-product of fermentation.

Brandy is another liquor produced by distilling grapes. To make brandy, the fermented juice is boiled until alcohol vapors rise from the liquid, at which point the vapors are captured and recondensed into a spirit. This process, called distillation, can also be used with other types of fermented fruits or vegetables to make spirits.

Fortified wines such as port and sherry have high alcohol content and are typically sweet. During fermentation, winemakers add distilled brandy to the wine. The alcohol kills the yeast in the wine and stops fermentation, thereby preserving more of the sugar in the grapes and simultaneously adding to the alcohol content.

Tartaric Acid. Tartaric acid is an organic crystalline acid found in grapes and several other fruits. During wine fermentation, deposits of tartaric acid collect in the fermentation vessels. Winemakers collect and sell tartaric acid as a food additive. It is used as a solidifying agent in cheese making and added to some foods to increase bitterness. Because of its antioxidant properties, tartaric acid has also become a popular additive for health food and drinks.

Pumice. Pumice is a mat made of the grape leaves, skins, seeds, and stems that collect during the creation of red wine. Wine pumice is collected, dried, pressed into bricks, and sold to health and wellness companies for a variety of products. Because pumice is high in vitamins, minerals, and antioxidants, it has become a popular food additive and is used in a variety of healthful beverages and supplements. Wine pumice has also become exceedingly popular in the cosmetic industry. In vinotherapy, winemaking residue is used to treat the skin. Pumice, for example, may be ground and used in the creation of skin products and makeup. The French winery Château Smith Haut Lafitte is known for its line of beauty products created with its wine pumice.

IMPACT ON INDUSTRY

The wine industry is in a state of growth. Emerging superpowers India and China have begun producing higher-quality wine in the twenty-first century, and wine is produced in all fifty states in the United States. For winegrowers, the market is generally always difficult—enologists and commercial winemakers, however, enjoyed growth in the first decade of the new millennium. Although the 2008–2009 global economic crisis negatively affected the fine wine industry, the production of affordable table wines has increased rapidly.

Wine is subject to regulations that govern alcoholic beverages and other intoxicating substances. In the United States, the Food and Drug Administration (FDA) is responsible for regulating wine, but many states have their own laws governing the production and sale of wine and liquor. In the state of Pennsylvania, for instance, wine and spirits may be distributed only by government-licensed facilities, while other states allow grocery stores to sell wine and liquor.

SOCIAL CONTEXT & FUTURE PROSPECTS

The impact of wine on culture differs widely from country to country. In France and Italy, wine is considered a staple of everyday life, and many Europeans are daily wine consumers. In 2008, Italy overtook France as the world's largest winemaker. In the United States and most of Asia, wine is considered more of a delicacy than a part of a daily routine. Though traditions differ, wine is collected and consumed in nearly every country around the world.

A major debate in the wine industry in the early twenty-first century concerns the use of cork for bottling. Cork has been used as a wine stopper for centuries because of its light and elastic nature—its compressibility allows it to form tight seals—and because it is considered impenetrable to gases and liquids. The problem with cork, however, is that it can be the source of bacteria within the wine or allow leakage. Some wine experts feel that glass or plastic corks may soon overtake natural corks as the most common type of sealant for bottled wine. Other closures gaining ground in the industry include screwcaps.

FURTHER READING

Goode, Jamie. *The Science of Wine: From Vine to Glass.* Berkeley: University of California Press, 2007. A basic instruction to winemaking and the scientific aspects of enology written for the general reader. Includes information about wine industry debates and controversial issues in the twenty-first century.

Johnson, Hugh. *The Story of Wine.* London: Mitchell Beazley, 2004. An introductory history of wine with information about wine varieties, the development of the industry, and the development of enology as a scientific discipline.

Paul, Harry W. *Science, Wine, and Vine in Modern France.* New York: Cambridge University Press, 2002. Provides information on the history, industry, and development of winemaking and scientific enology in France and around the world, as well as information about wine varieties and careers in the winemaking industry.

Pellechia, Thomas. *Wine: The Eight-Thousand-Year-Old Story of the Wine Trade.* New York: Running Press, 2007. Presents a history of the wine industry from its earliest manifestations to twenty-first-century industrial wine production, along with information on the scientific study of wine and the state of wine research.

Ribereau-Gayon, Pascal, et al., eds. *The Enology Handbook.* Vol. 1. Hoboken, N.J.: John Wiley & Sons, 2006. An introduction to the science and study of wines and winemaking, which requires some knowledge of microbiology and chemistry for full comprehension. Includes information on the history of the wine industry.

Sommers, Brian J. *The Geography of Wine.* New York: Plume, 2008. A basic introduction to the science and art of viticulture. Includes information on microclimate, soil evaluation, harvesting, and the history of grape cultivation.

American Society for Enology and Viticulture http://asev.org

Wine America: The National Association of American Wineries http://www.wineamerica.org

W

Water Conservation

—Jack B. Evett and Karen N. Kähler

FIELDS OF STUDY

Ecology; Environmental Studies.

SUMMARY

Water conservation is a comparatively simple and inexpensive means of addressing water shortages and making the most of existing water supplies. Effective conservation programs can postpone or prevent the need to construct additional water-supply infrastructure and can also reduce wastewater discharges. As the earth's population increases, conservation measures become an increasingly important tool for meeting society's water needs.

In 2015 the world population was estimated to be 7.2 billion and was projected to rise to 9.5 billion by the year 2050. Because drinking water is a basic human necessity, it follows that the amount of water needed for human existence will increase by a factor of roughly 1.4 over this forty-year period. While such an expansion in water supply is theoretically possible to achieve, the likelihood that it can or will happen is low for several reasons. First, although there is enough water available globally, it is not uniformly distributed around the world when and where it is needed. Second, the costs involved in such an expansion would be astronomical. Third, even if all needed moneys were readily available, it would be difficult to build the necessary facilities and have them operational by the time they would be needed.

DOMESTIC, INDUSTRIAL & AGRICULTURAL MEASURES

The greatest potential for water conservation in residential settings is in bathrooms. Traditional toilets use 19 to 26 liters (5 to 7 gallons) per flush. However, low-flow toilets using 13 or fewer liters (3.5 gallons) per flush—and ultra-low-flush models using as little as 6.1 liters (1.6 gallons)—are increasingly being installed in new homes and as replacements for traditional toilets in older homes. Baths and showers also use large amounts of water, and low-flow showerheads can contribute to water conservation efforts.

A great deal of water is also wasted in kitchens. Water can be conserved at kitchen sinks through simple means such as collecting food scraps for compost piles rather than putting them through garbage disposals. Low-flow sink faucets are also available. Running automatic dishwashers only for full loads is another water conservation technique; the same is

U.S. Daily Water Use, 1940-2000
(billions of gallons)

Year	Total	Per Capita (gallons)	Irrigation	Public Supply	Rural	Industrial & Misc.	Steam Electric Utilities
1940	**140**	1,027	71	10	3.1	29	23
1950	**180**	1,185	89	14	3.6	37	40
1960	**270**	1,500	110	21	3.6	38	100
1970	**370**	1,815	130	27	4.5	47	170
1980	**440**	1,953	150	34	5.6	45	210
1990	**408**	1,620	137	41	7.9	30	195
1995	**402**	1,500	134	40	8.9	29	190
2000	**408**	1,430	137	43	9.2	23	196

Source: U.S. Department of Commerce, *Statistical Abstract of the United States, 2004*, 2004.
Note: Per capita figures are gallons; all other values are in billions of gallons.

Water conservation. Image by U.S. Department of Commerce, *Statistical Abstra.ct of the united States 2004*, 2004.

true for clothes washers. Older models of both types of washers can be replaced with newer, more water-efficient ones.

Another potential source of water conservation in residential areas is found outside the homes, in plant and lawn watering. Methods of saving water include decreasing lawn sizes, landscaping with plants that require low amounts of water (known as xeriscaping), mulching to reduce moisture evaporation from the soil, minimizing the frequency of watering, and using drip systems instead of sprinklers.

Business and industrial settings use copious amounts of water. In some cases, water conservation measures may involve simple changes in the way things are done, such as substituting sweeping for the washing down of floors. In most cases, however, in these settings water conservation on a noticeable scale results from process changes of some type. For example, by converting from a water-cooled ice machine to an air-cooled one, a restaurant may improve water efficiency in its ice making by 25 to 50 percent.

The largest amounts of water are used in agriculture for irrigation. In the United States, a history of relatively cheap and readily available water led farmers to waste great amounts through improper or excessive application. Water can be and has been saved through computerized timing of water application. Modifications of irrigation procedures have also helped. For example, drip irrigation, which applies water slowly and uniformly at or below soil level adjacent to plants through mechanical water outlets, has produced significant water savings over traditional methods of simply spraying water onto the soil.

OTHER WAYS OF CONSERVING

As is true of the consumption of other commodities, water usage can be affected by pricing. Water-conserving rates exhibit increasing unit costs as volume used increases. One simple model determines a certain rate for the average amount of water a household might be expected to use and a much higher rate for all water used over that average amount.

Yet another way in which municipalities as well as home and business owners can conserve water is by finding and repairing leaks that occur anywhere in water distribution systems; water audits—careful reviews of water usage over time—can help to identify leaks. In some water distribution systems, documented leakages have been found to amount to losses of more than 50 percent of the water sent through the systems. Water reuse may be thought of as another form of water conservation. Ample opportunities for water reuse exist during industrial processes. For example, wastewaters that were being treated on-site and discharged into receiving streams might be reused as cooling water. Households might recycle water used for dishwashing or bathing as irrigation water for outdoor plants. While it is possible to treat sewage for reuse by households, most people are not yet ready to accept such a conservation measure.

The negative aspects of water conservation are generally related to monetary issues. For example, successful water conservation efforts can lead to reduced revenues for water utility companies, which may then be forced to increase their rates. Conserving water over the long term may also reduce the "slack" in the system, making short-term drought savings difficult to achieve and reducing the amount of water available for water rationing. Some water conservation efforts can be expensive to put in motion, while the actual savings from reduced water use are achieved more slowly, over a period of time. Such water conservation is not always cost-effective.

Further Reading

American Water Works Association. *Water Conservation Programs: A Planning Manual.* Denver: Author, 2006. Print.

Asano, Takashi, et al. *Water Reuse: Issues, Technologies, and Applications.* New York: McGraw-Hill, 2007. Print.

Chiras, Daniel D. "Water Resources: Preserving Our Liquid Assets and Protecting Aquatic Ecosystems." In *Environmental Science.* 8th ed. Sudbury: Jones and Bartlett, 2010. Print.

Glennon, Robert Jerome. *Unquenchable: America's Water Crisis and What to Do About It.* Washington, DC: Island, 2009. Print.

Postel, Sandra. *Last Oasis: Facing Water Scarcity.* New York: Norton, 1992. Print.

Unger, Paul W. *Soil and Water Conservation Handbook: Policies, Practices, Conditions, and Terms.* Binghamton: Haworth, 2006. Print.

United States Environmental Protection Agency. *Cases in Water Conservation: How Efficiency Programs Help Water Utilities Save Water and Avoid Costs.* Washington, DC: Author, 2002. Print.

Vickers, Amy. *Handbook of Water Use and Conservation: Homes, Landscapes, Businesses, Industries, Farms.* Amherst: Waterplow, 2001. Print.

WILDLIFE CONSERVATION

—Jerome A. Jackson, PhD

FIELDS OF STUDY

Biology; Zoology; Biochemistry; Chemistry; Physical Science; Conservation; Animal Science; Ecology.

SUMMARY

Wildlife conservation is a broad discipline that focuses on assuring a future for wildlife in this rapidly changing world. It encompasses a diversity of careers. Some are highly specialized, such as wildlife veterinarian, wildlife researcher, university professor, and environmental lawyer. Others, such as wildlife-refuge manager, wildlife-research technician, wildlife educator, and wildlife law enforcement officer, incorporate the tools and skills of many disciplines coordinated and directed to accomplish wildlife-conservation tasks. Wildlife-conservation careers all involve understanding relationships between wildlife and their environment, the behavioral interactions among species, and the impacts of human-caused and natural changes on wildlife populations.

DEFINITION & BASIC PRINCIPLES

Wildlife conservation focuses on the protection, enhancement, and management of wildlife populations in their natural habitats. Understanding ecosystems, habitats, food habits, migration, home ranges, territoriality, reproductive biology, predator-prey relationships, the nature and impact of wildlife diseases, and human impacts on all of these are essential to wildlife conservation. The development and publication of field guides and availability of quality binoculars has increased public interest in wildlife and has greatly facilitated species identification. The development of wildlife telemetry techniques and equipment, enhanced photography and sound recording capabilities, computers that can be used in the field, software such as Geographic Information Systems (GIS), Global Positioning Systems (GPS), and molecular techniques that help identify genetic differences among populations are among the many technological advances of the late twentieth century that have enhanced the quality and scope of information that wildlife conservationists can use.

Such technological advances have led to more efficient and more effective wildlife conservation and, along with need associated with declining habitat availability, have driven specialization and growth in wildlife conservation career opportunities. GIS, molecular ecology, wildlife telemetry, fire ecology, translocation, and captive breeding are among many opportunities for specialization within wildlife research and management careers. Wildlife educators, communicators, photographers, and ecotourism operators function at many levels to promote understanding and support of wildlife conservation. Zookeepers, veterinarians, and wildlife rehabilitators contribute greatly to understanding and promotion of wildlife health and to wildlife-reintroduction programs. Environmental law enforcement officers and lawyers are on the front line of wildlife conservation, promoting sound conservation laws and appropriate mitigation efforts and assuring compliance with legal mandates to protect species and their habitat.

BACKGROUND & HISTORY

As human populations grew in the eighteenth century, declines were seen in some wildlife populations, and local laws were sometimes passed to limit hunting, but wildlife conservation in its infancy was largely a voluntary effort, not yet capable of

The Siberian tiger is a subspecies of tiger that is critically endangered; three subspecies of tiger are already extinct. Image by S. Taheri, edited by Fir0002 (Own work) CC-BY-SA-2.5.

supporting careers. Employment in the field of wildlife conservation is a development of the late nineteenth and twentieth centuries. Early wildlife conservation efforts focused on human needs such as depleted game populations or fisheries.

Careers in wildlife conservation began in earnest with the establishment by Congress of the U.S. Fish Commission in 1871 to find ways to conserve declining fish populations. Concern for the welfare of birds and the scientific study of birds grew in the latter half of the nineteenth century, resulting in the founding of the American Ornithologists' Union in 1883 and other professional scientific societies. These had some conservation focus and appointed conservation committees to discuss conservation issues and support efforts to solve conservation problems. Such organizations raised public understanding and awareness of problems and were often led by government employees.

As American agriculture grew, blackbirds and other species were increasingly labeled as pest species. In 1885, the Office of Economic Ornithology was created within the United States Department of Agriculture. This supported wildlife careers in the study of bird food habitats, migration, and habitat use. Although initiated from a negative and economic perspective, the knowledge gained became foundations for later conservation efforts.

In the 1890s, feathers had become fashionable adornments of ladies' hats and other clothing and killing of birds for the millinery industry grew. Hundreds of thousands of egrets, terns, and other birds were shot at for their highly desirable and valuable breeding plumes. Chicks were left to die in the sun. As bird populations plummeted and stories of the slaughter became known, public outrage led to formation of clubs, boycotts of clothing with feathers, and public pressure for legal action to protect birds. Such actions led to formation of the Massachusetts Audubon Society in 1896, other state Audubon societies, and in 1905, to the formation of the National Audubon Society. Such nonprofit, conservation-oriented organizations established sanctuaries and hired wardens to protect wildlife. These organizations continue to be major employers in the field of wildlife conservation.

In 1905, the federal government responded to growing conservation sentiments by changing the name of the Office of Economic Ornithology to the Bureau of Biological Survey. The Biological Survey later became the U.S. Fish and Wildlife Service. Other federal agencies (U.S. Forest Service, Bureau of Land Management, Department of Defense) also developed wildlife-conservation programs associated with lands they managed.

During the twentieth century, habitat losses, species endangerment and extinctions, and pollution led to national (the Endangered Species Act of 1973) and international (International Convention on Trade in Endangered Species of Wild Fauna and Flora) legal protection for wildlife. Government responsibilities for wildlife and government career opportunities in wildlife conservation continue to grow. Nonprofit conservation organizations have also grown in number and in conservation roles, providing further wildlife-conservation career opportunities. Legal protections for wildlife and its habitats also have fostered a large service sector that provides career opportunities in wildlife conservation to assist industry in complying with conservation legal requirements. Finally, public interest in wildlife has grown to include a large, multifaceted ecotourism industry that provides many career opportunities in wildlife conservation.

HOW IT WORKS

Wildlife conservation began with the protection of animals from overhunting but quickly expanded to include the recognition that destruction and fragmentation of wildlife habitats also reduced wildlife populations. By the mid-twentieth century it had also become obvious that industrial and agricultural chemicals could harm wildlife. The latter realization led to the understanding that humans share these ecosystems with wildlife and that in a very real sense, wildlife was the barometer to the health of the world at large and an indicator of problems that humans could face. Such understanding gave new currency to careers in wildlife conservation. It was also learned that the interconnections between humans and wildlife include shared diseases. The phrase "ecosystem health" took on new meaning. Lyme disease, West Nile virus, rabies, and other diseases created wildlife-conservation career niches for veterinarians and other health professionals. The diversity of careers within the broad field of wildlife conservation continues to grow. Many wildlife conservationists work outdoors; others work at zoos, museums, universities, industries, and in the offices of government agencies and nonprofit organizations.

Refuge, Preserve, or Park Management. Local, state, regional, and national parks, preserves, and wildlife refuges provide significant habitat for wildlife and significant recreational activities for humans. Managers of these lands must deal with wildlife-conservation issues ranging from maintaining and managing habitats to managing hunting and fishing, providing for endangered and threatened species, and conservation education. Land managers must have an understanding of wildlife-habitat needs and the tools for satisfying those needs. It is the land manager's responsibility to make decisions regarding such things as the periodic use of prescribed burning, control of exotic or invasive species, closing of some areas to the public, regulation of hunting, and other recreational or research activities. All of these responsibilities must be carried out with a clear understanding of the natural landscape and the ecosystem within which it is located, the laws that govern the landscape and the wildlife found there, and the human communities of the region. Land managers always work at the human-wildlife interface, and sometimes the work on the human side of things is the most challenging.

Wildlife Telemetry. One of the most useful tools in wildlife conservation is the use of telemetry. On the local scale this can involve the attachment of a transmitter to an animal and following its signal using a handheld receiver and antenna or using an automated receiver linked to multiple large, fixed antennae to learn patterns of daily or seasonal habitat use. For larger animals, it can also involve monitoring long-distance migrations by using transmitters that communicate with the National Oceanic and Atmospheric Administration's (NOAA) Argos weather satellite. The Argos satellite system was established in 1978 has been used for wildlife telemetry as well as other applications since the late 1980s. As the satellite passes over the animal with a transmitter, it establishes uplink and stores information from the transmitter. Later, as the satellite passes over a receiving station on earth, that information, including location coordinates for the animal, is transmitted by downlink to the receiving station and from there to the researcher. The quality of telemetry equipment and software has steadily increased while at the same time the size of transmitters has decreased, allowing them to be used effectively on insects. Methods of attachment have also changed over the years. Some transmitters are affixed to an animal with a harness or collar, some are glued to a bird's feathers or to the back of a turtle, or attached to a leg or fin, and some are now implanted in or fed to the animal. Use of telemetry can help scientists understand such things as the amount of space a species needs, the habitats it uses, daily and seasonal patterns in behavior, and how the needs of males, females, and juveniles may differ. Limitations on the use of transmitters include size, battery duration, signal strength, and tolerance of the species or individual animal to the transmitter. Capabilities of all of these parameters are continually improving.

Wildlife Law Enforcement. Increasing legal protections for wildlife, along with growing and increasingly mobile human populations, have created a diversity of career opportunities. Wildlife law enforcement agents are hired at city, county, state, national, and international levels. In general, this hierarchy reflects increasing pay, increasing responsibility, and increasing specialization. At the local level, responsibilities might be associated with enforcing hunting and anticruelty laws. At the state level they can involve enforcing fish and game warden duties, helping to manage hunting, enforcing or administering efforts associated with state laws protecting threatened species and their habitats, enforcing laws regulating the keeping of wild animals in captivity, inspecting pet stores, dealing with nuisance animals, and working with veterinarians and others associated with wildlife-health issues. At the national level, wildlife law enforcement agents are often associated with managing and enforcing hunting regulations on federal lands, interstate transport of wildlife, importation of wildlife, and protection of federally protected species. Wildlife law enforcement agents are often associated with U.S. border crossings, where they assist with controlling the importation of wildlife from around the world. At any level, a wildlife law enforcement agent might have to work alone, on weekends, and at all times of day and night. At other times, wildlife law enforcement agents work with other professionals as part of a team dealing with issues that include, but may go far beyond, wildlife conservation issues.

Wildlife Rehabilitation. The care of sick, injured, or orphaned wildlife provides career opportunities at wildlife-rehabilitation centers, zoos, veterinary clinics, theme parks, animal shelters, and research centers. Sometimes this requires problem- or species-specific training and can require advanced education as a veterinarian or radiologist, or other medical specialty. In addition to career opportunities, wildlife rehabilitation provides a great deal of opportunity for volunteer efforts and internships that can provide training for individuals potentially interested in a wildlife-conservation career.

Zoological Medicine. Veterinarians who specialize in working with zoo animals or free-roaming wildlife are certified as diplomates in zoological medicine. To receive this distinction they must have at least three years of professional training beyond that required for a degree in veterinary medicine and fulfill other requirements of the American College of Zoological Medicine. Although sometimes referred to as "wildlife veterinarians," that designation is not an official title but merely a descriptor of what they do. Veterinarians with this expertise are critical to captive breeding and rehabilitation of wildlife.

APPLICATIONS & PRODUCTS

Wildlife conservation initially needed little in the way of specialized equipment and produced little in the way of products other than the game that was being protected and managed. The growth of wildlife conservation during the twentieth century, however, has produced a number of things that have been of use outside of wildlife conservation—not the least of which is a land ethic that was fostered by Aldo Leopold in his 1949 book, *A Sand County Almanac.* It helped promote the idea that not all land use should be determined by immediate economic benefit. Leopold noted that ecosystem "health is the capacity of the land for self-renewal." He defined conservation as "our effort to understand and preserve this capacity." That is what the field of wildlife conservation is all about.

Conservation and Mitigation Banking. Within a few years of the passage of the Endangered Species Act of 1973, the legal burden of providing mitigation for environmental harm done was taking up a good deal of court and industry time. In 1995, the federal and state governments approved mitigation banking as a way to provide predetermined mitigation for environmental harm to wetlands. It was a free-market process whereby an entrepreneur would restore or develop wetlands that would be approved as a mitigation bank. Then, when a court decided that mitigation needed to be done to offset damages to a wetland, the party causing the harm could buy mitigation credits from an approved mitigation bank to satisfy the obligation. The success of mitigation banking for wetlands led to the development of a similar program, called conservation banking, to provide predetermined conservation credits to mitigate harm done to an endangered species.

Bird Banding. Being able to mark a bird so that it can be identified as an individual at a later date was essential to understanding bird longevity, migration and other movements, and bird social behavior. Banding involves placing a serially numbered metal band on a bird's leg, recording details of its sex, age, and where it was banded, and having those data available in a central repository to retrieve if the bird was found by someone else, somewhere else, at a later time. Bird banding began as a hobby in the early 1900s but was quickly realized as a useful scientific tool. In 1920, the Bureau of Biological Survey took over all responsibilities for managing bird banding and banding data to ensure quality of data and so that the technique could be used in wildlife management. Later techniques were developed for adding colored plastic bands along with the serially numbered band, which allowed researchers to identify individuals by their color-band combinations without having to recapture them. These techniques continue to be vital tools in bird conservation.

Translocation Efforts. One of the major conservation problems facing wildlife is fragmentation of habitat and isolation of species in small pockets of habitat. The small number of individuals in such populations results in the population being highly vulnerable to extinction as a result of random events (such as wildfire, tornado, or predation) and also subjects the population to inbreeding, something that can result in genetic harm to the population. What was needed was a way to move individuals from one population to another and to have some assurance that they would stay in the new area and not attempt to return to their original site. The endangered red-cockaded woodpecker is an example of a species facing such problems. Studies of individually color-banded birds revealed that breeding males removed from their home would quickly return from great distances, but that young birds, especially young females, could be moved and would remain at a new site. Translocation efforts using young birds and careful timing have been improved upon through research and are now a common tool for this and other bird species. Similar efforts have also been developed and used for species such as the black-footed ferret. Radio telemetry is often used in conjunction with translocated animals to monitor their movements closely.

Photography. Whereas photography has long been used in many disciplines, it has often been as a supplement or simply for illustration rather than filling a specific need. With the advent of high-quality digital cameras, photography has taken on important roles in documenting wildlife and wildlife-conservation efforts. Researchers take photos for specific purposes and in specific ways rather than merely capturing snapshots. Wildlife photographers now play very important roles in both research and in wildlife-conservation education and communication.

SOCIAL CONTEXT & FUTURE PROSPECTS

Wildlife conservation was barely known as a career option two hundred years ago. It has since grown in importance and diversified into areas of specialization as diverse as education, law enforcement, satellite telemetry, zoological medicine, land management, wildlife photography, and the rescue of endangered species. It remains a vibrant, actively growing field that offers opportunity for travel to distant jungles and ocean depths, zoos and research facilities, parks and preserves, backyards and city streets. Any career in wildlife conservation is one from which to learn and grow continually. The challenges are great and change and innovation will be the norm for the future as well. More sophisticated transmitters, greater use of satellite telemetry, and more use of molecular biology are on the horizon.

Things like oil spills, invasive species, and habitat loss, as well as human factors such as discarded medicines and diseases, including cancers perhaps caused by factors in the environment, and climate change threaten wildlife. Finding causes and cures for them may provide hope for mankind. Challenges for wildlife conservationists in the future will be to connect isolated populations through development of habitat corridors and to find better ways of translocating individuals to preserve genetic diversity, to unlock the secrets of captive breeding of difficult species, and to find the will and the means to restore the habitats they need.

FURTHER READING

Bonar, Scott A. *The Conservation Professional's Guide to Working with People.* Washington, DC: Island, 2007. Print.

Cleva, Sandra. "Working a Beat for Wildlife." *The Wildlife Professional* (Spring, 2010): 46–48. Print.

Fanning, Odom. *Opportunities in Environmental Careers.* New York: McGraw-Hill, 2002. Print.

Jackson, Jerome A., *William E. Davis, Jr., and John Tautin, eds. Bird Banding in North America: The First Hundred Years.* Cambridge: Harvard U Nuttall Ornithological. Print.

Janovy, John, Jr. *On Becoming a Biologist.* 2d ed. Lincoln: U of Nebraska P, 2004. Print.

Leopold, Aldo. *A Sand County Almanac.* New York: Oxford U P, 1949. Print.

Mazet, J. A., G. E. Hamilton, and L. A. Dierauf. "Educating Veterinarians for Careers in Free-Ranging Wildlife Medicine and Ecosystem Health." *Journal of Veterinary Medical Education* 33.3 (2006): 352–360. Print.

Mittermeier, Cristina. "The Power of Conservation Photography." *The Wildlife Professional* (Fall, 2007): 30–31, 43. Print.

Van Riper, C., III, et al. "Career Opportunities in Ornithology." *American Birds* 43.1 (1989): 29–37. Print.

Wilson, E. O. *Naturalist.* Washington, DC: Island, 2006. Print.

TIME LINE OF MAJOR EVENTS IN AGRICULTURE

Paleolithic

30,600 BCE	Pestle used as a tool in southern Italy to grind oats.

Neolithic Revolution

8,500 BCE	Neolithic Revolution, the first agricultural revolution, begins in the ancient Near East
8,000 BCE	Domesticated wheat at pre-pottery Neolithic area A (PPNA) sites in the Levant.
7500 BCE	PPNB sites across the Fertile Crescent growing wheat, barley, chickpeas, peas, beans, flax and bitter vetch. Sheep and goat domesticated.
7000 BCE	Agriculture had reached southern Europe with evidence of emmer and einkorn wheat, barley, sheep, goats, and pigs suggest that a food producing economy is adopted in Greece and the Aegean.
7000 BCE	Cultivation of wheat, sesame, barley, and eggplant in Mehrgarh, modern day Pakistan.
7000 BCE	Domestication of cattle and chicken in Mehrgarh, modern day Pakistan.
6800 BCE	Rice domesticated in southeast Asia.
6500 BCE	Evidence of cattle domestication in Turkey. (Some sources say this happened earlier in other parts of the world.)
6001 BCE	Archaeological evidence from various sites on the Iberian peninsula suggest the domestication of plants and animals.
6000 BCE	Granary built in Mehrgarh for storage of excess food.
5500 BCE	Céide Fields in Ireland are the oldest known field systems in the world. This landscape consists of extensive tracts of land enclosed by brick walls.
4000 BCE	In Mehrgarh, the domestication of numerous crops, including peas, sesame seeds, dates, and cotton, as well as a wide range of domestic animals, including the domestic Asian water buffalo, an animal that remains essential to intensive agricultural production throughout Asia today.
4000 BCE	Egyptians discover how to make bread using yeast.
4000 BCE	Evidence for rice domestication in the Khorat Plateau area of northwestern Thailand.
4000 BCE	First use of light wooden ploughs in Mesopotamia, modern day Iraq.
3500 BCE	Irrigation was being used in Mesopotamia, modern day Iraq.
3500 BCE	First agriculture in the Americas, around Central Amazonia or Ecuador
3000 BCE	Turmeric, cardamom, pepper and mustard are harvested in the Indus Valley Civilization.
3000 BCE	Fermentation of dough, grain, and fruit juices is in practice.
3000 BCE	Sugar produced in India

Antiquity

2600 BCE	Large-scale commercial timbering of cedars in Phoenicia (Lebanon) for export to Egypt and Sumeria. Similar commercial timbering in South India.
1700 BCE	Wind-powered machine developed by the Babylonians.
1300 BCE	Creation of canal linking the Nile delta to the Red Sea.
691 BCE	First aqueduct (approx. 50 miles long) constructed to bring water to Nineveh.
530 BCE	Tunnel of Eupalinos first underground aqueduct.
500 BCE	The moldboard iron plough is invented in China.
500 BCE	Row cultivation of crops using intensive hoeing to weed and conserve moisture practiced in China.
300 BCE	Efficient trace harness for plowing invented in China.
200 BCE	Efficient collar harness for plowing invented in America.
100 BCE	Rotary winnowing fan invented in China.
100 BCE	The multi-tube seed drill is invented in China.
200 CE	The fishing reel invented in China.
600	The distillation of alcohol in China.
607	The Chinese begin constructing a massive canal system to connect the Yellow and Yangtze rivers.

Modern Technological Advances

1700	British Agricultural Revolution ends.
1701	Briton Jethro Tull invented the seed drill, an improved plough that was drawn by a horse.
1763	International "Potato Show" in Paris with corn varieties from different states.
1798	Thomas Malthus predicts impending famine as population growth outstrips food production.
1804	Vincenzo Dandolo writes several treatises of agriculture and sericulture.
1809	French confectioner Nicolas Appert invents canning.
1831	American Cyrus McCormick introduced his mechanical reaper, which was mass produced by 1847 in a Chicago factory.
1837	John Deere invents steel plough.
1863	The U.S. Agriculture Department, which forecasts crop production for major countries across the globe, publishes its first monthly crop report.
1866	Gregor Mendel publishes his paper describing Mendelian inheritance.
1866	Austrian Gregor Mendel laid the foundation of modern genetics by showing traits pass from parents to offspring.
1871	Louis Pasteur invents pasteurization.

1873	American John Deere designed the first cast steel plough.
1881	First generation of hybrid corn to increase production created.
1892	First successful gasoline engine farm tractor built by American inventor John Froelich.
1895	Refrigeration for domestic and commercial drink preservation introduced in the United States and the United Kingdom, respectively.
1913	The Haber process, also called the Haber–Bosch process, made it possible to produce ammonia, and thereby fertilize, on an industrial scale.
1923	Commercial hybrid seed corn developed by Henry Wallace, who in 1926 founded the Hi-Bred Corn Co (now Pioneer Hi-Bred International).
1934	Worst drought in U.S. history swept through the Great Plains and covered more than 75 percent of the country.
1960	First use with aerial photos in Earth sciences and agriculture.

Green Revolution

1944	Green Revolution begins in Mexico.
1944	Normal Borlaug, father of the Green Revolution to increase food production, joins Rockefeller Foundation.
1945	Beginning of the Green Revolution to increase food production through new cultivars, irrigation, fertilizer, pesticides and mechanization.
1956	Mexico becomes self-sufficient in wheat as a result of the Green Revolution.
1960	Philippines government, Ford and Rockefeller foundations establish the International Rice Research Institute in Manila.
1968	William Gaud, director of the U.S. Agency for International Development, coins the term Green Revolution. "These and other developments in the field of agriculture contain the makings of a new revolution. It is not a violent Red Revolution like that of the Soviets, nor is it a White Revolution like that of the Shah of Iran. I call it the Green Revolution."
1970	Borlaug is awarded Nobel Peace Prize for his contributions to world peace through increasing food supply.
1974	China creates the first hybrid rice.
2000	Genetically modified plants cultivated around the world.
2005	Lasers used to replace stickers by writing on food to "track and trace" and identify individual pieces of fresh fruit.

GLOSSARY

A horizon: Topsoil

acid rain: Acid rain is mainly caused by sulfur dioxide, nitrogen oxide and other pollutants when fossil fuels are burnt. Acid rain rarely falls near the source of the pollutant because smoke, fumes etc can be carried many miles by air currents. The main effects of acid rain are on wildlife. Water in lakes also becomes very clear as fish and microscopic life are killed off. It is thought that acid rain is responsible for the death of some trees, specifically conifers, which gradually lose all their leaves and die.

acid soil: A soil with an acid reaction, a pH less than 7.0.

acidification: Process of becoming more acid. pH 7 is neutral, above this is alkaline, below pH 6 is acidic.

acre: A parcel of land, containing 4,840 square yards or 43,560 square feet.

additives: Substances added to feedstuffs for preservation or flavor.

aerobic digestion: Decomposition of organic matter by micro organisms in the presence of oxygen

agriculture extension service: Cooperative (federal, state, and county) agency doing research and education for rural and urban producer and consumer groups, located in each county with specialist personnel for each particular area.

agriculture: The utilization of biological processes on farms to produce food and other products useful and necessary to man both a "way of life" and a "means of life" for the people involved in this industry.

agroforestry: Cultivated mixtures of trees, crops and/or livestock.

agronomy: The science of crop production and soil management.

alfalfa: A valuable leguminous crop for forage or hay used in livestock.

ammonification: The formation of ammonium ions from nitrogenous compounds.

anaerobic digestion: The breaking down of organic matter in oxygen free conditions.

animal unit: A unit of measurement of livestock, the equivalent of one mature cow weighing 1,000 lbs.

annuals: Plants that flower and complete their life cycle in the same year that they are raised from a seed.

antithesis: The time of flower opening in a plant.

aquaculture: The cultivation of aquatic organisms for the production of human food. (i.e. fish farming)

ark: A moveable outdoor pig shelter.

artificial insemination: The mechanical injection of male semen into the womb of the female with a special syringe-like apparatus. The process begins with the collection of semen from the male. This method is used extensively in dairy husbandry.

auger: Spiral device on a shaft used to move grain through a tube avian: Pertaining to poultry and/or fowl.

balance ration: A ration which furnishes all the necessary nutrients in the proportions and amounts needed by the animal for normal functioning and growth.

biennials: Plants that only flower in the year following that in which they are sown.

biochemical Oxygen Demand (BOD): The amount of oxygen required by micro organisms, usually in polluted water, slurry or industrial effluent for oxidation processes.

biodiversity: The term *biodiversity* was coined in 1985, (abbreviating 'biological diversity') and has been defined in many different ways. Holdgate (1991) regarded it as the 'Total sum of life's variety on Earth, expressed at the genetic, species and ecosystem level'

conservation headland: A sterile strip, minimum of 1m wide, between the crop and the field boundary which prevents weeds such as sterile brome and couch encroaching on the crop. A band along the outside of the crop is left unsprayed and receives no pesticide treatments in the spring.

conservation: Protection and preservation, in relation to: soil, herbage, species or the environment.

contract growing: Crops grown under contract to supply a specific outlet or to a factory for further process. Often grown to a detailed specification.

controlled atmosphere: Regulation of O_2 and CO_2 levels as well as temperature to improve storage of fruit and vegetables

controlled lighting: Artificial lighting of poultry housing increasing or decreasing the number of hours of light during the day will control sexual maturity, fertility, and molt.

cooperative: An organization formed for the purpose of production and marketing of goods or products owned collectively by members who share in the benefits. Most common examples in agriculture are canneries and creameries.

cost/benefit analysis: An assessment of the costs and benefits associated with a process, action, enterprise or system.

crop rotation: More or less regular recurrent succession of different crops on the same land for the purpose of maintaining good yields cwt: Hundredweight or 100 pounds.

cultivations: Working of the soil prior to and during the germination of crops. For seed bed preparation and weed control

dedicated fuel cropping: Use of resources solely towards the production of a crop destined for use as a fuel

deficiency payments: Payment to producers to make up the difference between the average market price for a commodity and the guarantee price when the market falls below the guaranteed price.

desalination: The removal of salts from sea water (generally)

desiccant: A chemical causing drying

dewpond: Small pond which forms on high chalky soils, filled by rainwater. Often man made for livestock drinking, kept full by rain.

double crop: Two different crops grown on the same area in one growing season drainage: The removal of excess surface water or excess water from within the soil by means of surface or sub-surface drains.

draft ewes: Female sheep which are sold out of a flock, usually upland to a lowland farm.

drilling: The process of opening the soil to receive the seed, planting the seed and covering it in a single operation.

dry cow: A cow that is not producing milk, the period before the next calving and lactation.

dry land farming: The practice of crop production without irrigation erosion: The wearing away of the land surface, usually by running water or wind.

economies of scale: Unit cost reductions which result from increasing total output.

ecosystems: Systems which include both living and non-living substances interacting to produce an exchange of materials between the living and the non living units

effluent: Drainage from a cattle yard, manure heap or silage clamp.

energy flow: The rate of energy transfer between elements of an ecological system

entropy: The degree of randomness of or disorder of a system.

eutrophication: The accumulation of excessive concentrations of plant nutrients in water courses leading to a bulk-up of algae.

evapotranspiration: Loss of water by evaporation and transpiration by the above ground parts of a plant. Governed by meteorological conditions, especially when water supply is not limiting.

fallow(ing): Resting land from deliberate cropping, not necessarily without cultivation or grazing, but without sowing

FAO: Food and Agriculture Organization.

feed lots: (1) dry lot feeding: Feeding process wherein cattle are confined in a small area and fed carefully mixed, high-concentrate feed to fatten them. (2) farm feed lot: Where cattle feeding is complementary with other farming enterprises (3) commercial where cattle are fed for others on a custom basis. Feed usually is purchased, labor is hired.

fertilization: The union of pollen with an egg to form an embryo.

FFA: Future Farmers of America, an organization for high school students studying vocational agriculture.

field capacity: The moisture content of soil in the field as measured two or three days after a thorough wetting of a well-drained soil by rain or irrigation water.

field capacity: The state of saturated soil when all the soil moisture that is able to freely drain away has done so

flow diagram: The diagrammatic representation of the structure of a system illustrating physical and information flows between compartments

food chain: Includes all businesses involved in the transformation of raw materials into food. It can be short (e.g., eggs purchased straight from the farm) or long and complex (e.g., transport/slaughterhouse/transport/processor/cannery/transport/retailer/consumer chain)/

forage: Crops consumed by livestock

forage: Vegetable matter, fresh or preserved, which is gathered and fed to animals as roughage (e.g., alfalfa hay, corn silage, or other hay crops).

fossil fuels: Biological materials which have been subjected to long term geological effects such as coal, oil, and natural gas.

4-H: Club for boys and girls sponsored by the agricultural extension service to foster better agriculture and homemaking. The 4 "H's" stand for head, heart, hands, and health. Members are 9 to 19 years of age.

free range: System of (mainly) poultry keeping in which hens are allowed to range over a relatively large area

GATT: General Agreement on Tariffs and Trade.

gelding: A male horse that has been castrated before having reached sexual maturity

genetic engineering: The science of modifying the genetic constitution of plants and animals directly

genome: The full set of chromosomes of an individual

geothermal energy: Energy contained in the earth's heat as steam water and hot rock

global warming: Certain gasses such as carbon dioxide, methane, nitrous oxides and chlorofluorocarbons (CFC's) absorb reflected infra-red radiation and trap it within the Earth's atmosphere. This heat which would normally be dissipated into space increases the 5temperature of the earth. An increase in the Earth temperature would have profound effects on the world's sea levels and climate. The control of theses gasses is therefore seen as a major international priority.

grade A dairy: A dairy that produces market milk (for human drinking purposes) under state approved sanitation conditions according to state controlled pooling laws. Milking barn and milk-handling equipment must meet certain state regulations.

grade B dairy: A dairy that produces milk for use in making cheese, ice cream and condensed and powdered milk sanitation requirements are not as strict as for grade a production. The milk cannot be sold for fresh market consumption.

grade: (1) quality grades for beef, based on conformation, finish, and quality: U.S. prime; U.S. commercial; U.S. choice; U.S. utility; U.S. good; U.S. cutter; U.S. standard; U.S. canner (2) yield grades for beef, based on the expected yield (curability) of trimmed, boneless major retail cuts: Y.G. 1 (best); Y.G. 4; Y.G. 2; Y.G. 5 (poorest); Y.G. 3. (3) quality grades for lamb are U.S. prime, U.S. choice, U.S. good, U.S. utility and U.S. cull. Yield grades for lamb are the same as for beef. (4) yield grades for lamb are the same as for beef. (5) USDA grades for slaughter hogs and pork carcasses combine quality and yield into one designated grade. These are U.S. 1, U.S. 2, U.S. 3, U.S. 4 and utility.

gravitational water: Water that either runs off or percolates through a soil. Not available for use by plants.

grazing: A method of feeding by herbivores

green manure: Any crop or plant grown and plowed under to improve the soil, by addition of organic matter and the subsequent release of plant nutrients, especially nitrogen.

greenhouse effect: Global warming due to build up of atmospheric CO_2

Gross Domestic Product (GDP): The total final output of goods and services produced within a country in a year by residents and non residents regardless of allocation of domestic and foreign claims

Gross National Product (GNP): The total value of goods and services produced by an economy over a particular period of time. GNP is made up of consumer and government purchases, private, domestic and foreign investments in the country and the total value of exports

growth promoters: Substances given to livestock to increase growth rates

growth regulator: A natural or chemical substance that regulates the enlargement, division or activation of plant cells

habitat: The area of an environment where an organisms lives, feeds and breeds.

haulm: The stems and leaves of a crop left after harvest.

heifer: Young (less than 3 years) female of the cattle species that has not borne a calf.

herbivore: Animals that eat plant origin feeds only.

hierarchy: A structural relationship in which each unit consists of two or more sub units, the latter been similarly divided

horizontal integration: The combining of two or more similar functions under one decision making body. A farmer who acquires and manages another farm as a separate unit and a canner that builds or acquires a cannery in another area are examples of horizontal integration.

horticulture: The cultivation of fruit, flowers, vegetables, and shrubs, also used to describe the commercial production of such crops on general farms

humus: The well decomposed, relatively stable portion of the organic matter in a soil.

hybrid: The first generation offspring of a cross between two individuals differing in one or more genes

hydrological cycle: The cycle of water evaporated from water surfaces and plants and precipitated as rain

hydroponics: Growing of plants in water containing the essential growth elements. This process is being used in "glass" houses for intensive "off-season" production of vegetables.

ICM: Integrated crop management

IFM: Integrated farm management.

immobilization: This occurs when inorganic ions are assimilated by soil organisms and are bound organically again

incubation: A process of holding eggs under controlled conditions of heat and moisture permitting the fertile eggs to hatch. Chicks require 21 days and turkeys 28 days to hatch.

infrastructure: The availability of roads, power supplies, education and health facilities, for example which all industries share and for which they do not directly pay for

integrated control: Integrated use of both biological and chemical methods for controlling pests and weeds

integration: Control by a single organization of all or some of the various stages of production.

intensive farming: A system of farming with the aim to produce the maximum number of crops in a year with a high yield from the land available and to maintain a high stocking rate of livestock.

intercropping: The growing of more than one species on the same piece of land at the same time

invertebrates: Animals without backbones

lactation period: The length of time a female gives milk following birth of offspring—usually with reference to dairy cows and milk goats.

land classification: The classification of units of land for the purpose of grouping soil of similar characteristics, in some cases showing their relative suitability for some specific use.

layer: A female chicken producing eggs regularly. A good layer should produce between 19 and 20 dozen eggs in 12 months.

leaching: The physical removal of nutrients in solution by drainage from the soil.

leaching: The process of removal of soluble materials by the passage of water through soil.

leaf area index (LAI): The area of green leaf per unit area of ground

legumes: A type of plant which has nodules formed by bacteria on its roots. The bacteria that compose these nodules take nitrogen from the air and pass it on into the plant for the plant to use. Some legumes are alfalfa, soybeans, sweet clover and peanuts.

legumes: Plants of the family leguminosae, e.g. clover, peas

litter: A group of offspring born at the same time by one sow.

livestock: Any domestic animal produced or kept primarily for farm, ranch, or market purposes, including beef and dairy cattle, hogs, sheep, goats, and horses.

lowland farming: Farming on low lying land as distinct from uplands, usually more intensive

manure: Generally, the refuse from stables and barnyards including both animal excreta and straw or other litter.

mare: Mature female horse.

marginal land: Land almost too unproductive to be farmed profitably.

mastitis: A disease of the cow's udder resulting from infection by microorganisms. The infection may be caused by improper milking procedures.

meristem: Localized region of active cell division in plants

milk (average composition): Milk contains on the average, the following: Fat–39%; albumin–7%; casein–2.5%; lactose–5.1%; mineral matter–.7%; and water–87.1%.

mineralization: The conversion of bound organic nitrogen into the mineral (ionic or inorganic) form required for plant uptake. The inorganic nitrogen is then converted in a process known as nitrification from ammonium salts to nitrites, and from nitrites to nitrates by nitrifying bacteria which obtain their energy from this oxidation process. Once free nitrate is formed, the rapid recycling process offers many options; the nitrate may be assimilated by plants, denitrified or leached.

mixed grazing: More than one type of animal grazing the same area at the same time

model: A simplified representation of a system. (May be expressed as word, diagrammatic or mathematical form)

monoculture: system of cultivation where only one crop is grown on the same piece of land over a number of years.

monogastric: Non ruminant animal having only one stomach, e.g. pig, human

nematode: Soil worms of microscopic size. These organisms may attack the root or other structures of plants and cause extensive damage.

nitrogen cycle: The sequence of transformations undergone by nitrogen in its movement from the free atmosphere into and through soils, into the plants, and eventually back these biochemical reactions are largely involved in the growth and metabolism of plants and microorganisms.

nitrogen fixation: Conversion of atmospheric nitrogen to plant compounds by micro organisms by micro organisms

nomadism: Continual movement of humans and animals with no fixed settlement, generally in search of food or water

nutrient: A chemical element or compound that is essential for normal body metabolism, growth and production includes: Carbohydrates fats, proteins, vitamins, minerals and water.

offspring: The progeny of parents omnivore: Animals that eat both animal and plant origin feeds.

omnivores: Animals that feed on material of both plant and animal origin

organic farming: Basically farming without the use of manufactured chemicals, but now conforming to very detailed production standards

organic fertilizer: Any fertilizer material containing plant nutrients in combination with carbon pasteurization: A process of treatment of milk through heat that kills all harmful bacteria, without changing its physical or chemical composition.

ozone layer: A layer of ozone found in the stratosphere, where it absorbs harmful ultra violet radiation

perennials: Plants that continue their growth from year to year

permanent pasture: An established plant community in which the dominant species are perennial grasses, there are few or no shrubs and trees are absent

permanent wilting point: That point at which a plant is dried so badly that even though put into a humid atmosphere and watered, it will no longer recover.

pest: Any organism injuring plants or plant products.

pH: A scale of measurement by which the acidity or alkalinity of soil or water is rated a pH of 6 to 7.5 is considered "ideal" for most agricultural crops. Each plant (specie-type), however, has its own "ideal" pH range.

photoperiodism: The effect of day and night length on plant flowering. Some plants are long day, requiring14–16 hours of sunlight per day to flower. Others are short day requiring only 8–9 hours. Others are day neutral and unaffected by day length. Photoperiosim also effects the breeding cycles of farm livestock

photosynthesis: The process by which carbohydrates are manufactured by the chloroplasts of plants from CO_2 and water by means of the energy of sunlight

pig: A young swine weighing less than 120 pounds.

poaching: Wet soils which are trampled and 'churned up' by cattle, they become muddy and it often leads to a breakdown in soil structure.

pollen: The male germ cells pollination: The transfer of pollen from the anther to the stigma.

pomology: The science or study of growing fruit.

poult: A young turkey.

poultry: Domestic birds raised for eggs and meat.

precooling: The process in which loads of fruit or vegetables are rapidly cooled prior to loading for shipment.

primary production: Production by plants: 'primary' in being the first use of solar radiation, the main energy source for biological processes.

productive soil: A soil in which the chemical, physical, and biological conditions are favorable for the economic production of the crops suited to a particular area.

pullet: A female chicken less than 1 year old.

pulses (grain legumes): Leguminous plants or their seeds, chiefly those plants with large seeds used for food.

relative humidity: Water vapor in air compared with the amount of water vapor held at the same temperature when saturated

respiration: The oxidative breakdown and release of energy from fuel molecules from fuel molecules with by reaction with oxygen in aerobic cells

rhizobium: Bacteria living in nodules on the roots of leguminous plants that are capable of removing nitrogen from the air and soil "fixing" it into forms that plants utilize for growth.

rhizome: An elongated underground stem, usually horizontal, capable of producing new shoots and roots at the node

roaster: A young meat bird, 12 to 16 weeks old weighing 4 to 6 pounds, used for pan roasting.

roasting pig: A milk-fat pig weighing from 60 to 100 pounds.

rooster: A mature male chicken.

rotation: A cropping system in which crops are grown in a field in a fixed annual sequence. A rotation reduces the build up of diseases and pests, aids weed control, improves soil fertility, spreads the risk of crop failure, and allows even distribution of labor requirements.

rotational grazing: The practice of imposing a regular sequence of grazing and rest on the area to be grazed

roughage: Feeds high in fiber, low in total digestible nutrients as hay and silage; the complete forage plant, including the stalk, stem, leaf, and (if mature) the seed.

ruminant: An animal that chews the cud and has a complex digestive system with a four part stomach (rumen, reticulum, omasum and abomasum) enabling bacteria to break down food. Ruminants lack upper incisor teeth and their complex stomach allows them to store and digest large amounts of bulky and fibrous food

saturate: To fill all of the openings among soil particles with liquid sheet erosion: The gradual, uniform removal by water of the earth's surface, without the formation of hills or gullies.

scrapie: Long established disease of sheep that causes intense irritation: cause unknown, thought to be neither bacteria nor virus. (See BSE)

silage: Prepared by chopping green forage (grass, legumes, field corn, etc) into an airtight chamber, where it is compressed to exclude air and undergoes and acid fermentation that retards spoilage. Contains about 65 percent moisture; 3 lbs. of silage is equal to 1 lb. of hay nutritionally.

slaughterhouse: A place where animals marketed for meat arc killed humanely.

slurry: A semi fluid mixture of feces and urine which may also include rain water and yard washings.

soil horizon: A layer of soil material approximately parallel to the land surface which differs from adjacent genetically related layers in color, structure, texture, or consistence. It also differs in biological and chemical characteristics.

soil map: A map designed to show the distribution of soil types or other soil-mapping units in relation to the prominent physical and cultural features of the earth's surface.

soil reaction: The degree of acidity or alkalinity of a soil usually expressed in terms of pH value,

soil series: A grouping of soils which have developed from a particular kind of parent material and which are similar in all characteristics except texture of the surface layer. The soil series is one of the principal units of soil classification.

soil structure: Refers to bonding together of soil particles and the resulting configuration of solid and voids,

soil survey: The systematic examination, description, classification, and mapping of soils in an area.

soil texture: Refers to the coarseness or fineness of a soil. It is determined by the relative proportion of various sized particles (sand, silt, and clay) in a soil.

soil type: A finer subdivision of a soil series it includes all soils of a series which are similar in all characteristics, including texture of the surface layer.

soil water deficit (SWD): The amount of water required to restore field capacity

soil-moisture tensiometer: An instrument which measures the tension with which water is held by soil. The instrument can be used for estimating when to irrigate land and for detecting drainage problems.

species: One kind of plant sow: Mature female hog.

stocking density: The number of animals per unit area of land at a point in time

stocking rate: The number of animals per unit area over a given period

stolon: A creeping stem above the soil surface, roots usually form at the nodes

stomata: Controllable openings (pores) in leaves, usually on the underside, through which gaseous exchanges take place

store cattle (or sheep): Animals which have been grown slowly so that their skeletal development is not impaired, but muscle tissue is slightly below the animals potential and fatty tissue is not developed.

strip cropping: Growing crops in long narrow strips across a sope approximately on a line of contour, alternating dense-growing intertilled crops. This is sometimes done with crops grown under government acreage allotments in order to increase yields per acre, since the intertilled area is not included in the allotment. It is also done in some dryland areas to conserve moisture and reduce the hazards of wind erosion.

stubble: The part of a crop left after harvesting

subsoiling: Breaking of compact subsoils without inverting them this is done with a special narrow cultivator shovel or chisel, which is pulled through the soil at a depth from 12 to 24 inches and at spacings from 2 to 5 feet.

summer fallow: Land plowed up (usually in spring) and left unseeded through the summer this is done to let the land air out and rest until fall, when it is worked up and planted to a crop of grain. May also be done to beak down organic matter or kill weeds.

sustainability: from *sustain* meaning to hold up; to bear; to support; to provide for; to maintain; to sanction; to keep going; to keep up; to prolong; to support the life of.

sustainable agriculture: The application of husbandry experience and scientific knowledge of natural processes to create integrated, resource conserving farming systems, based on respect for the people and animals involved which reduce environmental degradation, and which promote agricultural productivity and economic viability in both the short and long term.

system: A number of components linked together

terminal market: A metropolitan market which handles all agricultural commodities. The San Francisco wholesale produce and fruit market is an example.

Terms of Trade: Measure of the relative movement of export prices against those of import prices. They indicate whether a country is heavily reliant on imports or can generate income from exports

tiller: Shoot of a grass (or cereal) plant, arising from a leaf axil, normally at the base of an older tiller

top dressing: Lime, fertilizer, or manure applied after the seedbed is ready, or after the plants are up topsoil: The layer of soil used for cultivation, which usually contains more organic matter than underlying materials.

total digestible nutrients (TDN): The sum of all nutrients in a feed that are digested by the animal transportation: The loss of water vapor from the leaves and stems of living plants to the atmosphere.

trace elements: Nutrients which are required in very small amounts (plants or animals)

transhumance: Farmers have a fixed residence but send herds and flocks (with attendant) to pastures

variable: A quantity able to assume different numerical values

variety: A group of individuals within a species that differs from the rest of the species vertical integration: The combining of two or more successive steps in the production, processing and distributing processes under a single decision making body. A canner that produces some of his own raw product; a group of farmers which acquires a cannery or a cotton gin; or a feed company that owns the poultry are all examples of vertical integration.

vertebrates: Animals with backbones

viticulture: The science and practice of vine growing: Grape growing water rights (riparian rights): The rights of a person owning land containing or bordering on a water course or other body of water in or to its banks, bed, or waters.

water table: The upper limit of the part of the soil or underlying rock material that is wholly saturated with water in some places an upper or perched water table may be separated from a lower one by a dry zone.

welfare: Consideration for the comfort and well-being of farm animals.

weaning: Removal of young mammals from their source of milk

windbreak: A strip of trees or shrubs serving to reduce the force of wind; any protective shelter from the wind

zero grazing: Where forage is cut and transported to livestock

zoonoses: Diseases transferable between humans and animals

GENERAL BIBLIOGRAPHY

About Agronomy "About Agronomy." *American Society of Agronomy*, 2018, www.agronomy.org/about-agronomy. Accessed 15 Mar. 2018.

Abraham, Athol D., and Anthony J. Parsons. *Geomorphology of Desert Environments*. New York: Springer, 2009.

Abrahamson, Warren G., ed. *Plant-Animal Interactions*. New York: McGraw-Hill, 1989.

Abrams, Lindsay. "Factory Farming Is Killing the Planet: Why the Meat Industry's Future Needs to Look More Like Its Past." *Salon.com*. Salon Media Group, 25 July 2015.

Acquaah, George. *Principles of Crop Production: Theory, Techniques, and Technology*. Englewood, N.J.: Prentice Hall, 2001.

______. *Principles of Plant Genetics and Breeding*. 2nd ed. Hoboken: Wiley, 2012.

Adams, Charles R., Katherine M. Bamford, and Michael P. Early. *Principles of Horticulture*. 5th ed. Burlington, Mass.: Butterworth-Heinemann, 2008.

Adams, George F., and Jerome Wyckoff. *Landforms*. New York: Golden Press, 1971.

Adams, Jonathan S. *The Future of the Wild: Radical Conservation for a Crowded World*. Boston: Beacon Press, 2006.

Ag Gag: Safeguarding Industry Secrets by Punishing the Messenger "Ag Gag: Safeguarding Industry Secrets by Punishing the Messenger." *Food Integrity Campaign*. Food Integrity Campaign, 2015.

Agassi, Menachem, ed. *Soil Erosion, Conservation, and Rehabilitation*. New York: Marcel Dekker, 1996.

Ager, Derek V. *The Nature of the Stratigraphical Record*. 3rd ed. Hoboken, N.J. John Wiley & Sons Inc., 1993.

Agroecology Knowledge Hub, Food and Agriculture Organization of the United Nations, 2018, www.fao.org/agroecology/en. Accessed 9 July 2018.

Ahrens, C. Donald. *Meteorology Today: An Introduction to Weather, Climate and the Environment*. 8th ed. Thomson Brooks-Cole, 2007.

Akinyemi, Okoro M. *Agricultural Production: Organic and Conventional Systems*. Enfield, N.H.: Science Publishers, 2007.

Albarede, Francis. *Geochemistry: An Introduction*. 2d ed. Cambridge, Mass.: Cambridge University Press, 2009.

Alberts, Bruce, Dennis Bray, Julian Lewis, Martin Raff, Keith Roberts, and James D. Watson. *Molecular Biology of the Cell*. 3d ed. New York: Garland, 1994.

Albonico, Marco, Dirk Engels, and Lorenzo Savioli. "Monitoring Drug Efficacy and Early Detection of Drug Resistance in Human Soil-Transmitted Nematodes: A Pressing Public Health Agenda for Helminth Control." *International Journal of Parasitology* 34, no. 11 (2004): 1205–1210.

______, et al. *Preventative Chemotherapy in Human Helminthiasis*. Geneva: World Health Organization, 2006.

Alden, Peter, et al. *National Audubon Society Field Guide to African Wildife*. New York: Chanticleer Press, 1995.

Aleshire, Peter. *Deserts*. New York: Chelsea House, 2008.

Alexander, Brian. *Rapture: A Raucous Tour of Cloning, Transhumanism, and the New Era of Immortality*. Basic, 2004.

Alkon, Alison Hope. "The Socio-Nature of Local Organic Food." *Antipode* 45.3 (2013): 663–80. Print.

Allaby, Michael. *Biomes of the World*. 9 vols. Danbury, Conn.: Grolier International, 1999.

Allen, K. R. *Conservation and Management of Whales*. Seattle: University of Washington Press, 1980.

Allen, Keith C., and Derek E. Briggs, eds. *Evolution and the Fossil Record*. Washington, D.C.: Smithsonian Press, 1989.

Allen, Laura. *The Water-Wise Home: How to Conserve, Capture, and Reuse Water in Your Home and Landscape*. North Adams, MA: Storey Publishing, 2015.

Allin, Craig W. *The Politics of Wilderness Preservation*. 1982.

Allmon, Warren, and David J. Bottjer. *Evolutionary Paleoecology*. New York: Columbia University Press, 2000.

Al-Rawas, Amer Ali, and Mattheus F. A. Goosen, eds. *Expansive Soils: Recent Advances in Characterization and Treatment*. New York: Taylor & Francis, 2006.

Alspaugh, Shawn. "Farmer Ted Goes 3D." Mathematics Magazine 78, no. 3 (2005).

Ambrosioni, Juan, Daniel Lew, and Jorge Garbino. "Nocardiosis: Updated Clinical Review and Experience at a Tertiary Center." *Infection* 38, no. 2 (2010): 89–97.

American Water Works Association. *Water Conservation Programs: A Planning Manual.* Denver: Author, 2006.

Anderson, Bjørn G., and Harold W. Borns, Jr. *The Ice Age World.* Oslo: Scandinavian University Press, 1994.

Anderson, Mary P., ed. *Groundwater.* Oxford: Intl. Assn. of Hydrological Sciences, 2008.

Anisimova, Tatiana, and Parves Sultan. "The Role of Brand Communications in Consumer Purchases of Organic Foods: A Research Framework." *Journal of Food Products Marketing* 20.5 (2014): 511–32. Print.

Appendix 1: Soil Horizon Designations "Appendix 1: Soil Horizon Designations." *FAO Corporate Document Repository,* www.fao.org/docrep/W8594E/w8594e0g.htm. Accessed 8 Nov. 2016.

Archana, G. "Geomicrobiology: An Indian Perspective." *Current Science* 100, no. 4 (2011): 451–452.

Archer, David. *Global Warming: Understanding the Forecast.* Blackwell, 2007.

Archibald, J. David. *Dinosaur Extinction and the End of an Era: What the Fossils Say.* New York: Columbia UP, 2013.

Argue, D., et al. "Homo floresiensis: A Cladistic Analysis." *Jour. of Human Evolution* 57 (2009): 623–639. Print.

Asano, Takashi, et al. *Water Reuse: Issues, Technologies, and Applications.* New York: McGraw-Hill, 2007.

Ashman, M. R., and G. Puri. *Essential Soil Science: A Clear and Concise Introduction to Soil Science.* Malden, Mass.: Blackwell, 2002.

Ashton, Mark S., and Matthew J. Kelty. *The Practice of Silviculture: Applied Forest Ecology.* 10th ed. Hoboken: Wiley, 2017.

Avery, Alex, and Dennis Avery. "High-Yield Conservation: More Food and Environmental Quality through Intensive Agriculture." In *Agricultural Policy and the Environment,* edited by Roger E. Meiners and Bruce Yandle. Lanham, Md.: Rowman & Littlefield, 2003.

Avise, John C. *The Hope, Hype, and Reality of Genetic Engineering: Remarkable Stories from Agriculture, Industry, Medicine, and the Environment.* New York: Oxford UP, 2004.

Bacher, John. *Two Billion Trees and Counting: The Legacy of Edmund Zavitz.* Toronto: Dundurn Press, 2011.

Bailey, James E, and David F. Ollis. *Biochemical Engineering Fundamentals.* New York: McGraw-Hill, 2006.

Baker, Jeffrey J. W., and Garland E. Allen. *The Study of Biology.* 4th ed. Reading, Mass.: Addison-Wesley, 1982.

Bakker, Robert T. *The Dinosaur Heresies.* New York: William Morrow, 1986.

Baldridge, W. Scott. *Geology of the American Southwest: A Journey Through Two Billion Years of Plate Tectonic History.* New York: Cambridge University Press, 2004.

Balinsky, B. I. *An Introduction to Embryology.* Philadelphia: Saunders College Publishing, 1981.

Ballantyne, C. F., and C. Harris. *The Periglaciation of Great Britain.* New York: Cambridge University Press, 1994.

Barbosa-Cánovas, Gustavo, et al. *Global Issues in Food Science and Technology.* Burlington: Academic, 2009.

Bard, Jonathan. *Morphogenesis: The Cellular and Molecular Processes of Developmental Anatomy.* New York: Cambridge University Press, 1992.

Barnes, Dianna. *Coastal and Beach Erosion: Processes, Adaptation Strategies and Environmental Impacts.* New York: Nova, 2015.

Barnes, Robert D. Invertebrate Zoology. 6th ed. Philadelphia: Saunders College Publishing, 1993.

Barnum, Susan R. *Biotechnology: An Introduction.* 2nd ed. Belmont: Wadsworth, 2005.

Barrows, Geoffrey, Steven Sexton, and David Zilberman. "Agricultural Biotechnology: The Promise and Prospects of Genetically Modified Crops." *Journal of Economic Perspectives* 28, no. 1 (2014): 99–120.

Barry, W. B. N. *Growth of a Prehistoric Time Scale.* Rev. ed. Palo Alto, Calif.: Blackwell Scientific Publications, 1987.

Barth, Frederick G. *Insects and Flowers: The Biology of a Partnership.* Princeton, N.J.: Princeton University Press, 1991.

Baserga, Renato. *The Biology of Cell Reproduction.* Cambridge, Mass.: Harvard University Press, 1985.

Bashkin, Vladimir N. *Modern Biogeochemistry: Environmental Risk Assessment.* Dordrecht: Springer, 2010.

Baskin, Y. "Ecologists Dare to Ask: How Much Does Diversity Matter?" *Science* 264 (April 8, 1994): 202–203. Outlines and discusses the scientific evidence for the importance of biodiversity to ecosystem function.

Basra, A. S., ed. *Heterosis and Hybrid Seed Production in Agronomic Crops.* Binghamton, N.Y.: Food Products Press, 1999.

Bassett, Thomas J., and Donald Crummey, eds. *African Savannas: Global Narratives and Local Knowledge of Environmental Change.* Portsmouth, N.H.: Heinemann, 2003.

Batie, Sandra S. *Soil Erosion: Crisis in America's Croplands?* Washington, D.C.: Conservation Foundation, 1983.

Baumgardner, Dennis J. "Soil-Related Bacterial and Fungal Infections." *Journal of the American Board of Family Medicine,* vol. 25, no. 5, pp. 733–44.

Bazzaz, F A, and Wim G. Sombroek. *Global Climate Change and Agricultural Production: Direct and Indirect Effects of Changing Hydrological, Pedological, and Plant Physiological Processes.* Delhi: Daya Publishing House, 2005.

Bear, Firman E. *Earth: The Stuff of Life.* 2d ed. Norman: University of Oklahoma Press, 1990.

Bear, Jacob. *Dynamics of Fluids in Porous Media.* Mineola: Dover, 1988.

______. *Hydraulics of Groundwater.* Mineola: Dover, 2007.

Beckwith, Carol, and Marion Van Offelen. *Nomads of Niger.* New York: Academic Press, 1987.

Begon, Michael, John L. Harper, and Colin R. Townsend. *Ecology: Individuals, Populations, and Communities.* 3rd ed. Cambridge: Blackwell, 1996.

Bell, R. H. V. "A Grazing Ecosystem in the Serengeti." *Scientific American* 255 (July, 1971): 86–93. The great migrations of a variety of herbivores of the African savanna are illustrated and described.

Bender, Lionel M. *The Nilo-Saharan Languages: A Comparative Essay.* Munich, Germany: Lincom Europa, 1997.

Bennett, Hugh Hammond. *Elements of Soil Conservation.* 2d ed. New York: McGraw-Hill, 1955.

Bennett, Matthew M., and Neil F. Glassner, eds. *Glacial Geology: Ice Sheets and Landforms.* 2d ed. Hoboken, N.J.: Wiley-Blackwell, 2009.

Benton, M J. *Vertebrate Palaeontology.* John Wiley & Sons, 2015.

______. *The Rise of the Mammals.* London: Quantum, 2005.

______., and D. A. T. Harper. *Introduction to Paleobiology and the Fossil Record.* New York: Wiley-Blackwell, 2009.

Berk, Zeki. *Food Process Engineering and Technology.* Academic Press, 2009.

Berner, Elizabeth K., and Robert A. Berner. *The Global Water Cycle: Geochemistry and Environment.* Englewood Cliffs, N.J.: Prentice-Hall, 1987.

Bertheau, Yves. *Genetically Modified and Non-Genetically Modified Food Supply Chains: Coexistence and Traceability.* Wiley, 2012.

Bethony, Jeffrey, et al. "Soil-Transmitted Helminth Infections: Ascariasis, Trichuriasis, and Hookworm." *The Lancet* 367 (May 6, 2006): 1521–1532.

Bickis, Ian. "Agriculture Industry Betting the Farm on Innovation to Boost Yields, Profits." CBC News, CBC/Radio-Canada, 10 Feb. 2016, www.cbc.ca/news/technology/farming-technology-1.3442023.

Birchard, Bill. *Nature's Keepers: The Remarkable Story of How the Nature Conservancy Became the Largest Environmental Organization in the World.* New York: John Wiley & Sons, 2005.

Birkeland, P. W. *Soils and Geomorphology.* 3d ed. New York: Oxford University Press, 1999.

Black, Brian, and Gary J. Weisel. *Global Warming.* Greenwood Press, 2010.

Black, C. A. *Soil-Plant Relationships.* New York: John Wiley & Sons, 1988.

Blanco, Humberto, and Rattan Lal. *Principles of Soil Conservation and Management.* New York: Springer, 2008.

Bloom, Arthur. *Geomorphology: A Systematic Analysis of Late Cenozoic Landforms.* 3rd ed. Long Grove Ill.: Waveland Press, 2004.

Boardman, John, and David Favis-Mortlock, eds. *Modelling Soil Erosion by Water.* New York: Springer-Verlag, 1998.

Bohn, Heinrich L., Rick A. Myer, and George A. O'Connor. *Soil Chemistry.* 3d ed. New York: John Wiley & Sons, 2001.

Bonar, Scott A. *The Conservation Professional's Guide to Working with People.* Washington, DC: Island, 2007.

Bondareff, Joan M., Maggie Carey, and Carleen Lyden-Kluss. "Plastics in the Ocean: The Environmental Plague of Our Time." *Roger Williams University Law Review,* vol. 22, no. 2, 2017, pp. 360–383. Accessed 5 Feb. 2018.

Bonner, J. T. *Morphogenesis: An Essay on Development.* Princeton, N.J.: Princeton University Press, 1952.

______. *On Development: The Biology of Form.* Cambridge, Mass.: Harvard University Press, 1974.

Boothroyd, John C., and Richard Komuniecki, eds. *Molecular Approaches to Parasitology.* New York: Wiley-Liss, 1995.

Bouma, Jetske A., and Pieter van Beukering. *Ecosystem Services: From Concept to Practice.* Cambridge: Cambridge UP, 2015.

Bourgaize, David, Thomas R. Jewell, and Rodolfo G. Buiser. *Biotechnology: Demystifying the Concepts.* San Francisco: Benjamin/Cummings, 2000.

Bowler, Peter J. *Evolution: The History of an Idea.* 3rd ed., U of California P, 2003.

______. *Life's Splendid Drama: Evolutionary Biology and the Reconstruction of Life's Ancestry, 1860–1940.*

Bowman, David M. J. S. *Australian Rainforests: Islands of Green in a Land of Fire.* New York: Cambridge University Press, 2000.

Bradbury, Ian K. *The Biosphere.* New York: Belhaven Press, 1991.

Brady, Nyle C., and Ray R. Weil. *The Nature and Properties of Soils.* 14th ed. Upper Saddle River, N.J.: Prentice Hall, 2008.

Braimoh, Ademola K., and Paul L. G. Vlek, eds. *Land Use and Soil Resources.* London: Springer, 2008.

Brandon, Robert N. *Concepts and Methods in Evolutionary Biology.* Cambridge University Press, 2001.

Bransilver, Connie, Larry W. Richardson, Jane Goodall, and Stuart D. Strahl. *Florida's Unsung Wilderness: The Swamps.* Englewood, Colo.: Westcliffe, 2000.

Bräutigam, Amie, and Martin Jenkins. *The Red Book: The Extinction Crisis Face to Face.* Gland, Switzerland: IUCN, 2001.

Bravata, Dena M., et al. "Are Organic Foods Safer or Healthier Than Conventional Alternatives?&amp" *Annals of Internal Medicine* 157.5 (2012): 348–66. *Academic Search Complete.* Web. 2 Jan. 2013.

Brenner, R. L., and T. R. McHargue. *Integrative Stratigraphy Concepts and Applications.* Englewood Cliffs, N.J.: Prentice-Hall, 1988.

Brewer, Richard. *Conservancy: The Land Trust Movement in America.* Lebanon, N.H.: University Press of New England, 2004.

Bridgewater, Peter, et al. "Engaging with Novel Ecosystems." *Frontiers in Ecology and Environment* 9 (2011).

Briggs, Derek E. G, and Peter R. Crowther. *Palaeobiology Ii.* Malden, Mass: Blackwell, 2007.

Briggs, Derek E., and Richard A. Fortey. "Wonderful Strife: Systematics, Stem Groups, and the Phylogenetic Signal of the Cambrian Radiation." *Paleobiology* , vol. 31, 2005, pp. 94–112.

Brill, Winston. "Agricultural Microbiology." *Scientific American* 245 (September, 1981): 198.

Brody, Aaron L., and John B. Lord, eds. *Developing New Food Products for a Changing Marketplace.* 2d ed. Boca Raton, Fla.: CRC Press/Taylor & Francis, 2008.

Bronmark, Christopher, and Lars-Anders Hansson. *The Biology of Lakes and Ponds.* New York: Oxford University Press, 2018.

Brooker, Robert J., et al. *Biology.* 3d ed. New York: McGraw-Hill, 2014.

Brooker, Simon, et al. "Global Epidemiology, Ecology, and Control of Soil-Transmitted Helminth Infections." In *Global Mapping of Infectious Diseases,* edited by Simon I. Hay, Alastair Graham, and David J. Rogers. Amsterdam: Elsevier, 2006.

Brookfield, Michael E. *Principles of Stratigraphy.* Hoboken, N.J.: Wiley-Blackwell, 2004.

Brothwell, Don, and Eric Higgs, eds. *Science in Archaeology.* With a foreword by Grahame Clark. 2d ed. New York: Praeger, 1970.

Brown, David J., and Martin Lee. "From Microscopic Minerals to Global Climate Change?" *Geology Today* 23, no. 5 (2007): 172–177.

Brown, Lester R. *Plan B 3.0: Mobilizing to Save Civilization.* 3d ed. New York: W. W. Norton, 2008.

______. *State of the World, 2000.*

Brown, O. E., and R. A. Minnich. "Fire and Changes in the Creosote Bush Scrub of the Western Sonoran Desert, California." *American Midland Naturalist* 116, no. 2 (1986): 411–422. Slightly technical but very informative account of the role of fire in an ecosystem.

Brunsden, Denys, and John C. Doornkamp, eds. *The Unquiet Landscape.* Bloomington: Indiana University Press, 1974.

Bryant, Dick, et al. *The Last Frontier Forests: Ecosystems and Economies on the Edge.* Washington, DC: World Resources Institute, 1997.

Bryson, Reid A., and Thomas J. Murray. *Climates of Hunger.* University of Wisconsin Press, 1977.

Budhu, Muni. *Soil Mechanics and Foundations.* 2d ed. Hoboken, N.J.: Wiley, 2007.

Buol, S W. *Soil Genesis and Classification.* Ames: Wiley-Blackwell, 2011.

Butzer, Karl W. *Archaeology As Human Ecology: Method and Theory for a Contextual Approach.* Cambridge University Press, 2006.

Buzzi, O., S. Fityus, and D. Sheng, eds. *Unsaturated Soils.* New York: CRC Press, 2009.

Bynyard, Peter. "Pursuing the Energy Efficiency-Conservation Path." *Energy Policy.* (1988).

Cage-Free vs. Battery-Cage Eggs "Cage-Free vs. Battery-Cage Eggs." *The Humane Society of the United States.* HSUS, 7 Apr. 2015.

Cameron, Frank. *Cottrell, Samaritan of Science.* Garden City, N.Y.: Doubleday, 1952.

Campbell, Elizabeth. "Number of US Farms Fell to Six-Year Low in 2012, USDA Says." *Bloomberg,* 19 Feb. 2013, www.bloomberg.com/news/articles/2013–02–19/number-of-u-s-farms-fell-to-six-year-low-in-2012-usda-says-1-. Accessed 7 Apr. 2017.

Campbell, John R., M. Douglas Kenealy, and Karen L. Campbell. *Animal Sciences: The Biology, Care, and Production of Domestic Animals.* 4th ed. Boston: McGraw-Hill, 2003.

Campbell, Karen L., and John R. Campbell. *Companion Animals: Their Biology, Care, Health, and Management.* 2d ed. Upper Saddle River, N.J.: Pearson Prentice Hall, 2009.

Campbell, Neil A. *Biology.* 9th ed. San Francisco: Cummings, 2011.

Canfield, Don, Erik Kristensen, and Bo Thandrup. *Aquatic Geomicrobiology.* Atlanta: Academic Press/Elsevier, 2005.

Carr, Margaret, and Paul Zwick. *Smart Land-Use Analysis: The LUCIS Model.* Redlands, Calif.: ESRI Press, 2007.

Carroll, Robert L. *Vertebrate Paleontology and Evolution.* New York: W. H. Freeman, 1988.

Carroll, Scott P., and Charles W. Fox, eds. *Conservation Biology: Evolution in Action.* New York: Oxford University Press, 2008.

Carroll, Sean B., Jennifer K. Grenier, and Scott D. Weatherbee. *From DNA to Diversity: Molecular Genetics and the Evolution of Animal Design.* Malden: Blackwell Science, 2001.

Carson, Rachel. *Silent Spring.* 1962.

Carswell, Cally. "Bumblebees Being Crushed by Climate Change." *Science News.* AAAS, 9 July 2015.

Carver, Brett F., ed. *Wheat: Science and Trade.* Ames, Iowa: Wiley-Blackwell, 2009.

Casa, Kathryn. "The Changing Face of Farming in America." *National Catholic Reporter* 12 Feb. 1999: 13+. *Points of View Reference Center.* Web. 6 Nov. 2015.

Castillon, David A. *Conservation of Natural Resources: A Resource Management Approach.* 2d ed. Madison, Wis.: Brown Benchmark, 1996.

Castro, Peter, and Michael E. Huber. *Marine Biology.*, 2019.

Cathomen, Toni, Matthew Hirsch, and Matthew H. Porteus. *Genome Editing: The Next Step in Gene Therapy.* Springer, 2016.

Caullery, Maurice. *Parasitism and Symbiosis.* London: Sidgwick and Jackson, 1952.

Chaloner, William G., Peter R. Crane, and Else Marie Friis, eds. *The Origins of Angiosperms and Their Biological Consequences.* New York: Cambridge University Press, 1987.

Cheeke, Peter R. *Contemporary Issues in Animal Agriculture.* Upper Saddle River: N.J, 2004.

Chemical Revolution of Antoine-Laurent Lavoisier "The Chemical Revolution of Antoine-Laurent Lavoisier." *ACS Chemistry for Life.* American Chemical Society. Web. 16 Dec. 2014.

Chen, Ju-Yuan, Di-Ying Huang, and Chia-Wei Li. "An Early Cambrian Craniate-like Chordate." Nature 402 (December 2, 1999): 518–522.

Chernicoff, Stanley, and Dona Whitney. *Geology: An Introduction to Physical Geology*. Upper Saddle River, N.J.: Prentice Hall, 2006.

Chernicoff, Stanley, and Donna Whitney. *Geology: An Introduction to Physical Geology*. Upper Saddle River, N.J.: Prentice Hall, 2006.

Chester, Charles A. *Conservation Across Borders: Biodiversity in an Interdependent World*. Washington, D.C.: Island Press, 2006.

Chiras, Daniel D, and John P. Reganold. *Natural Resource Conservation: Management for a Sustainable Future*. Pearson Education Limited, 2014.

______. "Creating a Sustainable System of Agriculture to Feed the World's People." *Environmental Science*. 9th ed. Jones, 2013.

______. "Water Resources: Preserving Our Liquid Assets and Protecting Aquatic Ecosystems." In *Environmental Science*. 8th ed. Sudbury: Jones and Bartlett, 2010.

Cho, Renee. "Vertical Farms: From Vision to Reality." *State of the Planet, Columbia University*, 12 Oct. 2011, blogs.ei.columbia.edu/2011/10/13/vertical-farms-from-vision-to-reality/. Accessed 12 Dec. 2016.

Chorley, R. J., S. A. Schumm, and D. E. Sugden. *Geomorphology*. New York: Methuen, 1985.

Chrispeels, Maarten J, David E. Sadava, and Maarten J. Chrispeels. *Plants, Genes, and Crop Biotechnology*. Boston: Jones and Bartlett Publisher, 2003.

Clark, M. J., ed. *Advances in Periglacial Geomorphology*. New York: Wiley InterScience, 1988.

Clark, Tim W., A. Peyton Curlee, Steven C. Minta, and Peter M. Kareiva, eds. *Carnivores in Ecosystems: The Yellowstone Experience*. New Haven: Yale UP, 1999.

Clarke, G. R. *The Study of Soil in the Field*. 5th ed. Oxford, England: Clarendon Press, 1971.

Clayton, Dale H., and Janice Moore, eds. *Host-Parasite Evolution: General Principles and Avian Models*. New York: Oxford University Press, 1997.

Cleva, Sandra. "Working a Beat for Wildlife." *The Wildlife Professional* (Spring, 2010): 46–48. Print.

Climate, Drought, and Desertification. World Meteorological Organization, 1997.

Clutton-Brock, Juliet. *A Natural History of Domesticated Mammals*. 2d ed. New York: Cambridge University Press, 1999.

Cockrall-King, Jennifer. *Food and the City: Urban Agriculture and the New Food Revolution*. Prometheus Books, 2012.

Coggins, Chris. *The Tiger and the Pangolin: Nature, Culture, and Conservation in China*. Honolulu: University of Hawaii Press, 2003.

Colby, William E., ed. *John Muir's Studies in the Sierra*. San Francisco: Sierra Club Books, 1960.

Coleman, Eliot. *New Organic Grower: A Master's Manual of Tools and Techniques for the Home and Market Gardener, 30th Anniversary Edition*. Chelsea Green, 2018.

Colfer, Carol J., with Nancy Peluso and Chin See Chung. *Beyond Slash and Burn: Building on Indigenous Management of Borneo's Tropical Rain Forest*. New York: New York Botanical Garden, 1994.

Colinvaux, Paul. *Why Big Fierce Animals Are Rare: An Ecologist's Perspective*. Princeton: Princeton UP, 1978.

Colleran, Brian. "Think before You Eat: The Widespread Effects of Factory-Farmed Meat." *E: The Environmental Magazine* July–Aug. 2008: 30–31. *Points of View Reference Center*. Web. 6 Nov. 2015.

Collins, Robert O., ed. *Problems in African History: The Precolonial Centuries*. New York: Markus Wiener, 1991.

Collis, P. H., ed. *Dictionary of Ecology and the Environment*. 3d ed. Chicago: Fitzroy Dearborn, 1998.

Combes, Simon. *Great Cats: Stories and Art from a World Traveler*. Shelton, Conn.: Greenwich Workshop Press, 1998.

Commoner, Barry. *Making Peace with the Planet*. New York: Pantheon, 1990.

Communications on Energy: Household Energy Conservation "Communications on Energy: Household Energy Conservation." *Energy Policy* (1984).

Condie, Kent C. *Earth as an Evolving Planetary System*. 2d ed. Waltham, Mass.: Academic Press, 2011.

Conger, B. V., ed. *Cloning Agricultural Plants Via in Vitro Techniques*. S.l.: CRC PRESS, 2017.

Conkin, Paul K. *A Revolution Down on the Farm: The Transformation of American Agriculture Since 1929*.

Connell, Des W. "Soil Contamination." In *Basic Concepts of Environmental Chemistry*. 2d ed. Boca Raton, Fla.: CRC Press, 2005.

Conniff, R. "The Little Suckers Have Made a Comeback." *Discover* 8 (1987): 84–94. An interesting article that describes the medical uses of leeches in microsurgery.

Connor, Lawrence John, ed. *Asian Apiculture: Proceedings of the First International Conference on the Asian Honey Bees and Bee Mites*. Cheshire, Conn.: Wicwas Press, 1993.

______. *Queen Rearing Essentials*. Kalamazoo, Mich.: Wicwas Press, 2009.

Conway, Gordon R., and Edward B. Barbier. *After The Green Revolution: Sustainable Agriculture for Development*. Routledge, 2013.

______, and Vernon W. Ruttan. *The Doubly Green Revolution: Food for All in the Twenty-First Century*. Ithaca, N.Y.: Comstock, 1998.

Conway-Morris, Simon. *The Crucible of Creation: The Burgess Shale and the Rise of Animals*. Los Angeles: Getty Center, 1999.

Cook, Albert John. *Manual of the Apiary*. 3d ed. Chicago: T. G. Newman & Son, 1878.

Cook, Frederick A., L. D. Brown, and J. E. Oliver. "The Southern Appalachians and the Growth of Continents." *Scientific American* 243 (October 1980): 156–168.

Coors, J. G., and S. Pandey. *Genetics and Exploitation of Heterosis in Crops*. Madison, Wis.: American Society of Agronomy and Crop Science Society of America, and Soil Science Society of America, 1999.

Cornforth, Derek H. *Landslides in Practice*. Hoboken, N.J.: John Wiley & Sons, 2005.

Cousteau, Jacques. *The Ocean World*. New York: Harry N. Abrams, 1985.

Cowan, C W. *The Origins of Agriculture: An International Perspective*. Tuscaloosa, Ala: Univ. of Alabama Press, 2006.

Cox, George W. *Conservation Biology: Concepts and Applications*. 2d ed. Dubuque, Iowa: Wm. C. Brown, 1997.

Cox, Stan. "Enough with the Vertical Farming Fantasies: There Are Still Too Many Unanswered Questions about the Trendy Practice." *Salon,* 17 Feb. 2016, Accessed 12 Dec. 2016.

Crawley, M. J., ed. *Natural Enemies: The Population Biology of Predators, Parasites, and Diseases*. Boston: Blackwell Scientific Publications, 1992.

Crockett, Christopher. "Conservation of Linear Momentum: Newton's Third Law." *Education.com*. Education.com, Inc. Web. 16 Dec. 2014.

Crop "Crop." *National Geographic: Education*. National Geographic Society, 2014.

Crowell, Susan. "2007 Census of Agriculture: Agriculture's 'Middle' Slipping Away." *Farm and Dairy*. Farm and Dairy, 9 Feb. 2009.

Cvancara, Alan M. *Field Manual for the Amateur Geologist*. New York: John Wiley, 1995.

______. *Sleuthing Fossils: The Art of Investigating Past Life*. New York: John Wiley & Sons, 1990.

Dales, R. P. *Annelids*. 2d ed. London: Hutchinson, 1967.

Darwin, Charles R. *On the Origin of Species by Means of Natural Selection: Or, the Preservation of the Favoured Races in the Struggle for Life*. London: John Murray, 1859.

Das, Braja M. *Soil Mechanics Laboratory Manual*. 5th ed. Austin, Tex.: Engineering Press, 1997.

______., and G. V. Ramana. *Principles of Soil Dynamics*. Stamford: Cengage Learning, 2011.

Davies, Bryan, David Eagle, and Bryan Finney. *Resource Management: Soil*. Brighton, Ont.: Diamond Farm Book, 2001.

Davis, David Howard. *American Environmental Politics*. Belmont, Calif.: Wadsworth, 1998.

Davis, George H., and Evelyn M. VandenDolder, eds. *Geologic Diversity of Arizona and Its Margins: Excursions to Choice Areas*. Tucson: Arizona Bureau of Geology and Mineral Technology, 1987.

Davis, Nicola. "Vertical Farming Explained: How Cities Could Be Food Producers of the Future." *The Guardian,* 6 Feb. 2014, www.theguardian.com/science/2014/feb/06/vertical-farming-explained-erik-murchie. Accessed 12 Dec. 2016.

Dawkins, Richard. *The Blind Watchmaker: Why the Evidence of Evolution Reveals a Universe without Design*. W. W. Norton, 1988.

Day, Michael H. *Guide to Fossil Man*. 4th ed. Chicago: U of Chicago P, 1993.

Day, Robert W. *Soil Testing Manual: Procedures, Classification Data, and Sampling Practices*. New York: McGraw-Hill, 2001.

De Beer, Gavin. *Embryos and Ancestors*. Oxford, England: Clarendon Press, 1951.

De la Cruz, María Victoria, and Roger R. Markham. *Living Morphogenesis of the Heart*. Boston: Birkhäuser, 1998.

De Siqueira, Isadora Cristina, et al. " *Chromobacterium violaceum* in Siblings, Brazil." *Emerging Infectious Diseases* 11, no. 9 (2005): 1443–1445.

Dean, Warren. *With Broadax and Firebrand: The Destruction of the Brazilian Atlantic Forest.* Berkeley: University of California Press, 1997.

DeGregori, Thomas R. *Origins of the Organic Agriculture Debate.* Ames, Iowa: Blackwell, 2004.

______. *The Environment, Our Natural Resources, and Modern Technology.* New York: John Wiley & Sons, 2008.

Dennen, W. H., and B. R. Moore. *Geology and Engineering.* Dubuque, Iowa: Wm. C. Brown, 1986.

Denniston, D. *High Priorities: Conserving Mountain Ecosystems and Cultures.* Worldwatch Paper 123. Washington, D.C.: Worldwatch Institute, 1995.

Dent, David. *Soil as World Heritage.* Dordrecht: Springer Netherlands, 2014.

DeRose, Laurie Fields, Ellen Messer, and Sara Millman. *Who's Hungry? and How Do We Know? Food Shortage, Poverty, and Deprivation.* New York: United Nations University, 1998.

Despommier, Dickson. *The Vertical Farm,* www.verticalfarm.com. Accessed 12 Dec. 2016.

Development of Agriculture "The Development of Agriculture." *Genographic Project.* National Geographic Society, 2014.

Dhillon, Sarinder K., and Amandeep S. Sidhu. *Data Intensive Computing for Biodiversity.* Berlin: Springer, 2013.

Diamond, Jared. *Collapse: How Societies Choose to Fail or Succeed.* London: Penguin Group, 2005.

______. *Guns, Germs, and Steel: The Fates of Human Societies.* New York: W. W. Norton, 1997.

Dickerman, Carolyn. "Pollination: Strategies for Survival." *Ward's Natural Science Bulletin,* Summer, 1986, 1–4.

Dickinson, William R. "The Basin and Range Province as a Composite Extensional Domain." *International Geology Review* 22 (2002): 1–38.

Dimitri, C., and L. Oberholtzer. *Market-Led versus Government-Facilitated Growth: Development of the U.S. and EU Organic Agricultural Sectors.* Washington: WRS-05–05, USDA, Economic Research Service, 2005.

DiSabato-Aust, Tracy. *The Well-Tended Perennial Garden: Planting and Pruning Techniques.* 2d ed. Portland, Oreg.: Timber Press, 2006.

DiSilvestro, Roger L. *The Endangered Kingdom: The Struggle to Save America's Wildlife.* New York: Wiley, 1991.

Dixon, Lloyd S., Nancy Y. Moore, and Ellen M. Pint. *Drought Management Policies and Economic Effects in Urban Areas of California, 1987–92.* RAND, 1996.

Dixon, Richard A., and Robert A. Gonzales, eds. *Plant Cell Culture: A Practical Approach.* 2nd ed. New York: Oxford UP, 1994.

Dobrazanska, Halina, Erzsebet Jerem, and Tomasz Kalicki, eds. *The Geoarchaeology of River Valleys.* Budapest: Archaeolingua, 2004.

Dobzhansky, Theodosius G. *Genetics of the Evolutionary Process.* New York: Columbia University Press, 1970.

Dobzhansky, Theodosius, Francisco J. Ayala, G. Ledyard Stebbins, and James W. Valentine. *Evolutionary Biology.* Springer Verlag, 2014.

Dodds, John H, and Lorin W. Roberts. *Experiments in Plant Tissue Culture.* Cambridge, Mass: Cambridge University Press, 2004.

Dodds, Walter K., and Matt R. Wiles. *Freshwater Ecology: Concepts and Environmental Applications of Limnology.* 2d ed. Burlington, Mass.: Academic Press, 2010.

Dolgoff, Anatole. *Physical Geology.* Boston: Houghton Mifflin Harcourt, 1999.

Donahue, Roy L., Roy Hunter Follett, and Rodney W. Tulloch. *Our Soils and Their Management: Increasing Production Through Environmental Soil and Water Conservation and Fertility Management.* 6th ed. Danville, Ill.: Interstate, 1990.

Donovan, Steven K., ed. *Mass Extinctions: Processes and Evidence.* New York: Columbia University Press, 1989.

Doran, Pauline M. *Bioprocess Engineering Principles.* Amsterdam: Elsevier/Academic Press, 2013.

Doran, Peter T., W. Berry Lyons, and Diane M. McKnight, eds. *Life in Antarctic Deserts and Other Cold, Dry Environments.* New York: Oxford University Press, 2010.

Dott, Robert H., Jr., and Donald R. Prothero. *Evolution of the Earth.* 8th ed. New York: McGraw-Hill, 2009.

Dow, Kirstin, and Thomas E. Downing. *The Atlas of Climate Change: Mapping the World's Greatest Challenge.* University of California Press, 2006.

Dowie, Mark. *Conservation Refugees: The Hundred-Year Conflict Between Global Conservation and Native Peoples.* Cambridge, Mass.: MIT Press, 2009.

Dowswell, Paul, John Malam, Paul Mason, and Steve Parker. The Ultimate Book of Dinosaurs. Bath, England: Parragon, 2000.

Doyle, Hugh A. *Seismology.* New York: John Wiley, 1995.

Drickamer, L. C., S. H. Vessey, and D. Meikle. *Animal Behavior: Mechanisms, Ecology, and Evolution.* 4th ed. Dubuque, Iowa: Wm. C. Brown, 1996.

Droege, Sam, and Laurence Packer. Bees: An Up-Close Look at Pollinators Around the World. Minneapolis, MN: Voyageur Press, 2015.

Drury, Stephen. *Stepping Stones: Evolving the Earth and Its Life.* New York: Oxford University Press, 1999.

Du Toit, Johan, Richard Kock, and James Deutsch. *Wild Rangelands: Conserving Wildlife While Maintaining Livestock in Semi-Arid Ecosystems.* Chicester: Wiley, 2012.

Dunn-Georgiou, Elisha. *Everything You Need to Know about Organic Foods.* New York: Rosen, 2002.

Duram, Leslie A. *Good Growing: Why Organic Farming Works.* Lincoln: University of Nebraska Press, 2005.

Durbin, Kathie. *Pulp Politics and the Fight for the Alaska Rain Forest.* Corvallis: Oregon State University Press, 1999.

Edelman, Gerald M., and Jean-Paul Thiery, eds. *The Cell in Contact: Adhesions and Junctions as Morphogenetic Determinants.* New York: John Wiley & Sons, 1985.

Ehlers, J., P. L. Gibbard, and P. D. Hughes, eds. *Quaternary Glaciations; Extent and Chronology: A Closer Look.* New York: Elsevier, 2011.

Ehret, Christopher. *A Historical-Comparative Reconstruction of Nilo-Saharan.* Cologne, Germany: R. Köppe Verlag, 2001.

Ehret, Christopher. *An African Classical Age: Eastern and Southern Africa in World History, 1000 B.C. to A.D. 400.* Charlottesville: University Press of Virginia, 1998.

______. *The Civilizations of Africa: A History to 1800.*

______. *A Historical-Comparative Reconstruction of Nilo-Saharan.* Cologne, Germany: R. Köppe Verlag, 2001.

Ehrlic, Henry Lutz, and Dianne K. Newman. *Geomicrobiology.* 5th ed. Boca Raton, Fla.: CRC Press, 2008.

Ehrlich, Paul R. *The Machinery of Nature.* New York: Simon & Schuster, 1986.

Ehrlich, Paul R., and Anne H. Ehrlich. *Extinction: The Causes and Consequences of the Disappearance of Species.* New York: Random House, 1995.

Eilperin, Juliet. "Energy and Environment Interior Rescinds Climate, Conservation Policies Because They're 'Inconsistent' with Trump's Energy Goals." *The Washington Post,* 5 Jan. 2018. Accessed 4 Apr. 2018.

Elliott, John E., Christine A. Bishop, and Christy A. Morrissey, eds. *Wildlife Ecotoxicology: Forensic Approaches.* New York: Springer, 2011.

Elton, Charles. *Animal Ecology.* New York: Macmillan, 1927.

Enger, Eldon D., and Bradley F. Smith. *Environmental Science: A Study of Interrelationships.* 12th ed. Boston: McGraw-Hill Higher Education, 2010.

Ensminger, M. Eugene. *The Stockman's Handbook.* 7th ed. Danville, Ill.: Interstate, 1992.

Esch, Neal S., et al. *Soil Science Simplified.* 5th ed. Illustrated by Mary C. Bratz. Ames, Iowa: Blackwell, 2008.

Evangelou, V. P. *Environmental Soil and Water Chemistry: Principles and Applications.* New York: Wiley, 1998.

Evans, Sterling. *The Green Republic: A Conservation History of Costa Rica.* Austin: University of Texas Press, 1999.

Fabre, Jean Henri. *The Life of the Spider.* Translated by Alexander Teixeira de Mattos. New York: Dodd, Mead, 1916.

Fageria, Nand Kumar, Virupax C. Baligar, and Charles Allan Jones. *Growth and Mineral Nutrition of Field Crops.* 3rd ed. Boca Raton, Fla.: CRC Press, 2011.

Fanning, Odom. *Opportunities in Environmental Careers.* New York: McGraw-Hill, 2002.

Faulkner, Edward. *Plowman's Folly.* Norman: University of Oklahoma Press, 1943.

Fehr, Walter R. *Hybridization of Crop Plants.* Madison, Wisc: American Society of agronomy & Crop science Society of America, 1982.

Fellows, P. J. *Food Processing Technology: Principles and Practice.* 3rd ed., CRC Press, 2009.

Fennell, D. A. *Ecotourism.* 3rd ed. Abingdon, UK: Routledge, 2008.

Fenner, Janis L., Debora J. Hamberg, and John D. Nelson. *Building on Expansive Soils.* Fort Collins: Colorado State University, 1983.

Fenner, Michael, and Ken Thompson. "Soil Seed Banks." In *The Ecology of Seeds.* New York: Cambridge University Press, 2005.

Ferguson, Rafter Sass, Sarah Taylor Lovell. "Permaculture for Agroecology: Design, Movement, Practice, and Worldview: A Review." *Agronomy for Sustainable Development,* vol. 34, no. 2, 2014, pp. 251–74. Accessed 9 July 2018.

Fetter, C. W. *Contaminant Hydrogeology.* Grove, Ill.: Waveland Press, 2008.

Field, Harry L., and John B. Solie. "Erosion and Erosion Control." In *Introduction to Agricultural Engineering Technology: A Problem Solving Approach.* 3d ed. New York: Springer, 2007.

Field, Thomas G., and Robert E. Taylor. *Scientific Farm Animal Production: An Introduction to Animal Science.* 9th ed. Upper Saddle River, N.J.: Prentice Hall, 2008.

Filson, Glen C., ed. *Intensive Agriculture and Sustainability: A Farming Systems Analysis.* Vancouver: University of British Columbia Press, 2004.

Finlayson, Bill. *Changing Natures: Hunter-gatherers, First Famers and the Modern World.* Bristol, UK: Bristol Classical Press, 2010.

Fischer, David, and Thomas Moriarty. *Pesticide Risk Assessment for Pollinators.* Chichester: Wiley, 2014.

Fisher, R. J. *If Rain Doesn't Come: An Anthropological Study of Drought and Human Ecology in Western Rajasthan.* Manohar, 1997.

Five Worst Mass Extinctions "The Five Worst Mass Extinctions." *Endangered Species International.* Endangered Species Internationa, 2011.

Fletcher, Charles. *Physical Geology: The Science of Earth.* Hoboken, N.J.: John Wiley & Sons, 2011.

Food and Agriculture Organization (FAO) of the United Nations. *The State of Food and Agriculture 2008.*

______. "Forestry." 2012.

Foote, Michael Joseph, et al. *Principles of Paleontology.* New York: Macmillan, 2007.

Forrester, Glenn C. *Niagara Falls and the Glacier.* Hicksville, N.Y.: Exposition Press, 1976.

Fortier, Jean-Martin. *The Market Gardener: A Successful Grower's Handbook for Small-Scale Organic Farming.* New Society Publishers, 2014.

Fossel, Peter V. *Organic Farming: How to Raise, Certify, and Market Organic Crops and Livestock.* Voyageur Press, 2014.

Fournier, Arthur J. *Soil Erosion: Causes, Processes and Effects.* New York: Nova Science, 2011.

Fowler, Cary. "The Svalbard Global Seed Vault: Securing the Future of Agriculture." Global Crop Diversity Trust. 2008.

Fox, Stephen. *John Muir and His Legacy: The American Conservation Movement.* Boston: Little, Brown, 1981.

Fralish, James S. *The Central Hardwood Forest: Its Boundaries and Physiographic Provinces.* St. Paul, MN: U.S. Forest Service, 2003.

Francis, Charles A. *Enough Food: Achieving Food Security Through Regenerative Agriculture.* Emmaus, Pa.: Rodale Institute, 1985.

Francis, Charles A., Raymond P. Poincelot, and George W. Bird, eds. *Developing and Extending Sustainable Agriculture: A New Social Contract.* Haworth, 2006.

Frank, Stefan, et al. "Reducing Greenhouse Gas Emissions in Agriculture without Compromising Food Security?" *Environmental Research Letters,* vol. 12, no. 10, 2017, pp. 1–14. *IOP Publishing,* doi:10.1088/1748–9326/aa8c83. Accessed 9 July 2018.

Frankel, Charles. *The End of the Dinosaurs: Chicxulub Crater and Mass Extinctions.* New York: Cambridge University Press, 1999.

Frederiksen, Harald D. *Drought Planning and Water Resources Implications in Water Resources Management.* World Bank, 1992.

French, Hilary. *Vanishing Borders: Protecting the Planet in the Age of Globalization.* New York: W. W. Norton, 2000.

Freyfogle, Eric T. *Why Conservation Is Failing and How It Can Regain Ground.* New Haven, Conn.: Yale University Press, 2006.

Friedman, Matt. "Emerging onto a Tangled Bank." *Science* 324 (2009): 341–342.

Frisch, Wolfgang, Martin Meschede, and Ronald C. Blakey. *Plate Tectonics: Continental Drift and Mountain Building.* New York: Springer, 2010.

Fromartz, Samuel. *Organic Inc.: Natural Foods and How They Grew.* Orlando, Fla.: Harcourt, 2006.

Fry, Carolyn, Sue Seddon, and Gail Vines. *The Last Great Plant Hunt: The Story of Kew's Millennium Seed Bank.* Richmond, Surrey: Kew, 2011.

Fryxell, J. M. "Forage Quality and Aggregation by Large Herbivores." *American Naturalist* 138, no. 2 (August, 1991): 478–498.

Fu, Qiaomei, et al. "Genome Sequence of a 45,000-Year-Old Modern Human from Western Siberia." *Nature* 514 (2014): 445–449. Print.

Fukuyama, Francis. *Our Posthuman Future: Consequences of the Biotechnology Revolution.* Picador, 2003.

Futuyma, Douglas J. *Evolutionary Biology.* New York: W.H. Freeman, 2006.

Gamborg Oluf L., and Gregory C. Phillips, eds. *Plant Cell, Tissue and Organ Culture: Fundamental Methods.* New York: Springer, 1995.

Garner, Jerry. *Careers in Horticulture and Botany.* 2d ed. New York: McGraw-Hill, 2006.

Garrett, Roger A., and Hans-Peter Klenk, eds. *Archaea: Evolution, Physilogy, and Molecular Biology.* Malden, Mass. Wiley-Blackwell, 2007.

Garrison, Erv. *Techniques in Archaeological Geology.* New York: Springer, 2010.

Gately, Iain. *Tobacco: A Cultural History of How an Exotic Plant Seduced Civilization.* New York: Grove Press, 2003.

Gause, G. F. *The Struggle for Existence.* Reprint. New York: Dover, 1971.

Gerard, A. J. *Soils and Landforms.* London: Allen & Unwin, 1981.

Gerlach, Duane, Sally Atwater, and Judith Schnell. *Deer.* Mechanicsburg, Pa.: Stackpole Books, 1994.

Gerrard, John. "Soil Degradation." In *Fundamentals of Soils.* New York: Routledge, 2000.

Gewin, Virginia, and Patrick Monahan. "As U.S. Moves to Cut Greenhouse Emissions from Farms, New Study Finds Big Global Challenge." *Science,* 19 May 2016, doi:10.1126/science.aag0007.

Ghosh, Pallab. "'First Human' Discovered in Ethiopia." *BBD.* BBC, 4 Mar. 2015.

Gibson, Clark L. *Politicians and Poachers: The Political Economy and Wildlife Policy in Africa.* New York: Cambridge University Press, 1999.

Gibson, David J. *Grasses and Grassland Ecology.* New York: Oxford University Press, 2009.

Gilbert, Scott F, and Michael J. F. Barresi, ed. *A Conceptual History of Modern Embryology: A Conceptual History of Modern Embryology.* Springer Verlag, 2012.

______. *Developmental Biology.* Sinauer Associates, 2018.

Gilbert, Scott F., and A. M. Raunio. *Embryology: Contructing the Organism.* Sunderland, Mass.: Sinauer Associates, 1997.

Gillespie, James R., and Frank Flanders. *Modern Livestock and Poultry Production.* 8th ed. Clifton Park, N.Y.: Delmar Cengage Learning, 2009.

Givens, D.I., et al. *Health Benefits of Organic Food: Effects of the Environment.* X:CABI, 2009.

Glazer, Alexander N., and Hiroshi Nikaido. *Microbial Biotechnology: Fundamentals of Applied Microbiology.* 2d ed. New York: Cambridge University Press, 2007.

Glen, John. "Mathematical Models in Farm Planning: A Survey." Operations Research 35, no. 5 (1987).

Glennon, Robert Jerome. *Unquenchable: America's Water Crisis and What to Do About It.* Washington, DC: Island, 2009.

Gliessman, Stephen R., and Martha Rosemeyer, eds. *The Conversion to Sustainable Agriculture: Principles, Processes, and Practices.* Boca Raton, Fla.: CRC Press, 2010.

Gliessman, Stephen R., and Martha Rosemeyer, eds. *The Conversion to Sustainable Agriculture: Principles, Processes, and Practices.* Boca Raton, Fla.: CRC Press, 2010.

Global Emissions "Global Emissions." *C2ES,* Center for Climate and Energy Solutions, www.c2es.org/content/international-emissions. Accessed 9 July 2018.

Goldberg, Paul, and Richard I. MacPhail. *Practical and Theoretical Geoarchaeology.* New York: Wiley-Blackwell, 2006.

Goldsmith, Frank Barrie. *Tropical Rain Forest: A Wider Perspective.* New York: Chapman & Hall, 1998.

Goldstein, Melvyn C., and Cynthia M. Beall. *The Changing World of Mongolia's Nomads.* Berkeley: University of California Press, 1994.

Golly, Frank Benjamin. *A History of the Ecosystem Concept in Ecology: More than the Sum of the Parts.* New Haven: Yale UP, 1993.

Goode, Jamie. *The Science of Wine: From Vine to Glass.* Berkeley: University of California Press, 2007.

Goodman, Jordan. *Tobacco in History: The Cultures of Dependence.* New York: Routledge, 1994.

Gordon, Andrew M., and Steven M. Newman, eds. *Temperate Agroforestry Systems.* 2nd ed. Wallingford, England: CAB International, 2015.

Gordon, Iain J., and Herbert H. T. Prins, eds. *The Ecology of Browsing and Grazing.* Berlin: Springer, 2010.

Gordon, M. S., and E. C. Olson. *Invasions of the Land: The Transition of Organisms from the Aquatic to Terrestrial Life.* New York: Columbia University Press, 1995.

Gore, Albert, and Melcher Media. *An Inconvenient Truth: The Planetary Emergency of Global Warming and What We Can Do About It.* Rodale, 2006.

Goreau, T. F., et al. "Corals and Coral Reefs." *Scientific American* 241 (August, 1979): 124–136. Good discussions on the ecology and formation of the great reefs built by mutualistic relation between algae and coral.

Goudie, A. S., and H. A. Viles, eds. *The Earth Transformed: An Introduction to Human Impacts on the Environment.* Oxford: Blackwell, 2003.

Gould, Stephen J. *Ever Since Darwin.* Reprint. New York: W. W. Norton, 1992.

______. *The Flamingo's Smile.* New York: W. W. Norton, 1985.

______. "Ontogeny and Phylogeny—Revisited and Reunited." *BioEssays* 14, no. 4 (April, 1992): 275–280. Revisits the topic of his 1977 book.

______. *Ontogeny and Phylogeny.* Cambridge, Mass.: Harvard University Press, 1977.

______. *The Panda's Thumb.* Reprint. New York: W. W. Norton, 1992.

______. *Wonderful Life: The Burgess Shale and the Nature of History.* New York: W. W. Norton, 1989.

Graham, Joe M., ed. *The Hive and the Honey Bee: A New Book on Beekeeping Which Continues the Tradition of "Langstroth on the Hive and the Honeybee."* Rev. ed. Hamilton, Ill.: Dadant, 1992.

Grant, Gary. *The Water Sensitive City.* Chichester, UK: John Wiley & Sons, Ltd., 2016.

Grant, Susan. *Beauty and the Beast: The Coevolution of Plants and Animals.* New York: Charles Scribner's Sons, 1984.

Grant, Verne. *The Evolutionary Process: A Critical Study of Evolutionary Theory.* 2d ed. New York: Columbia University Press, 1991.

Graphic: The Relentless Rise of Carbon Dioxide "Graphic: The Relentless Rise of Carbon Dioxide." *Global Climate Change,* NASA, 26 Jan. 2018, climate.nasa.gov/climate_resources/24/. Accessed 5 Feb. 2018.

Greenland, David. *Guidelines for Modern Resource Management: Soil, Land, Water, Air.* Columbus, Ohio: C. E. Merrill, 1983.

Gregory, Kenneth J., and Andrew S. Goudie, eds. *The SAGE Handbook of Geomorphology.* Thousand Oaks, Calif.: SAGE Publications, 2011.

Greiner, Alfred, and Will Semmler. *The Global Environment, Natural Resources, and Economic Growth.* New York: Oxford University Press, 2008.

Grigg, D. B. *The Agricultural Systems of the World: An Evolutionary Approach.* New York: Cambridge University Press, 1974.

Grigg, David. *An Introduction to Agricultural Geography.* London: Routledge, 2005.

Grimm, V., et al. "Ecological Modeling for Pesticide Risk Assessment for Honey Bees and Other Pollinators." *Pesticide Risk Assessment for Pollinators* (2014): 149–62. Print.

Groh, Trauger, and Steven McFadden. *Farms of Tomorrow Revisited: Community Supported Farms, Farm Supported Communities.* Junction City, Oreg.: BioDynamic Farming and Gardening Association, 2000.

Groom, Martha J., Gary K. Meffe, and C. Ronald Carroll. *Principles of Conservation Biology.* Sunderland, Mass.: Sinauer Associates, 2006.

Groom, Martha J., Gary K. Meffe, and Carl Ronald Carroll. *Principles of Conservation Biology.* 3d ed. Sunderland, Mass.: Sinauer, 2006.

Grosse, G., et al. "Why Permafrost Is Thawing, Not Melting." *EOS* 91 (2010): 87.

Grotzinger, John, and Thomas H. Jordan. *Understanding Earth,* 7th ed. MacMillan, 2014.

Groundwater Depletion "Groundwater Depletion." *USGS Water Science School.* US Geological Survey, 9 Dec. 2015.

Guilford, Gwynn. "The New Frontier for Big Agriculture Is Aquaculture." *Quartz.* Atlantic Monthly Group, 17 Aug. 2015.

Gullan, Penny J, Peter S. Cranston, and Karina H. McInnes. *The Insects: An Outline of Entomology.* Chichester: Wiley-Blackwell, 2014.

Gunde-Cimerman, Nina, Aharon Oren, and Ana Plemenita. *Adaptation to Life at High Salt Concentrations in Archaea, Bacteria, and Eukarya.* New York: Springer, 2011.

Gustanski, Julie A., and Roderick H. Squires, eds. *Protecting the Land: Conservation Easements Past, Present, and Future.* Washington, D.C.: Island Press, 2000.

Guymon, Gary L. *Unsaturated Zone Hydrology.* Upper Saddle River, N.J.: Prentice-Hall, 1994.

Guyton, Bill. *Glaciers of California: Modern Glaciers, Ice Age Glaciers, the Origin of Yosemite Valley, and a Glacier Tour in the Sierra Nevada.* Berkeley: University of California Press, 2000.

Hagen, Joel B. *An Entangled Bank: The Origins of Ecosystem Ecology.* New Brunswick: Rutgers UP, 1992.

Hagstrom, Jerry. "Briefings Planned For 'Know Your Farmer'." *CongressDaily* 10 May 2010: 10. Print.

Hall, Brian K. *Evolutionary Developmental Biology.* 2d ed. New York: Chapman & Hall, 1998.

Halseth, Greg, Sean Markey, and David Bruce. "The Next Rural Economies: Constructing Landesa." Rural Development Institute. http://www.landesa.org.

Hambler, Clive. *Conservation.* New York: Cambridge University Press, 2004.

Hamblin, Kenneth W., and Eric H. Christiansen. *Earth's Dynamic Systems.* 10th ed. Upper Saddle River, N.J.: Prentice Hall, 2003.

Hambrey, Michael, and Jurg Alean. *Glaciers.* 2d ed. New York: Cambridge University Press, 2004.

Hamilton, Geoff. *Organic Gardening.* Rev. ed., Dorling Kindersley, 2008.

Hamilton, W. D. "Geometry for the Selfish Herd." *Journal of Theoretical Biology* 31 (1971): 294–311. Hancocks, David. *A Different Nature: The Paradoxical World of Zoos and Their Uncertain Future.* Berkeley: University of California Press, 2001.

Hand, Carol, and Chad Kruger. *Sustainable Agriculture.* Abdo, 2016.

Hand, Carol. *Depleted & Contaminated Soil and Your Food Supply.* New York: Rosen Pub, 2013.

Hanks, Sharon La Bonde. *Ecology and the Biosphere: Principles and Problems.* Delray Beach, Fla.: St. Lucie Press, 1996.

Hannam, Ian, and Ben Boer. *Legal and Institutional Frameworks for Sustainable Soils: A Preliminary Report.* Gland, Switzerland: International Union for Conservation of Nature, 2002.

Hanson, Charles, et al. "Transplantation of Human Embryonic Stem cells onto a Partially Wounded Human Cornea In Vitro." *Acta Ophthalmologica*, vol. 91, no. 2, 2013, pp. 127–30.

Hanson, Elizabeth. *Animal Attractions: Nature on Display in American Zoos.* Princeton, N.J.: Princeton University Press, 2002.

Harper, Charles L. *Environment and Society: Human Perspectives on Environmental Issues.* 4th ed. Upper Saddle River, N.J.: Pearson/Prentice Hall, 2008.

Harpstead, Milo I., Thomas J. Sauer, and William F. Bennett. *Soil Science Simplified.* 4th ed. Ames: Iowa State University Press, 2001.

Harris, David R., ed. *The Origins and Spread of Agriculture and Pastoralism in Eurasia.* Washington, D.C.: Smithsonian Institution Press, 1996.

Harris, S. A. *The Permafrost Environment.* Totowa, N.J.: Barnes & Noble, 1986.

Hart, John Fraser. *The Rural Landscape.* Baltimore: The Johns Hopkins University Press, 1998.

Hartwell, Leland H., and Ted A. Weinert. "Checkpoints: Controls That Ensure the Order of Cell Cycle Events." *Science* 246 (November 3, 1989): 629–634.

Haspel, Tamar. "Will Indoor, Vertical Farming Help Us Feed the Planet, or Hurt It?" *The Washington Post,* 17 June 2016. Accessed 12 Dec. 2015.

Hassell, Michael P. *Arthropod Predator-Prey Systems.* Princeton, N.J.: Princeton University Press, 1978.

Hauter, Wenonah. "Wal-Mart Squeezes Organics." *Progressive* 78.6 (2014): 18–20. Print.

Hayes, Denis, and Gail Boyer Hayes. *Cowed: The Hidden Impact of 93 Million Cows on America's Health, Economy, Politics, Culture, and Environment.* New York: W.W. Norton, 2015.

Hayes, Samuel P. *Conservation and the Gospel of Efficiency: The Progressive Conservation Movement, 1890–1920.*

Head, K. H. *Soil Technicians' Handbook.* New York: Halsted Press, 1989.

Head, W. S. *The California Chaparral: An Elfin Forest.* 1972.

Healy, Richard, and Bridget Scanlon. *Estimating Groundwater Recharge.* New York: Cambridge UP, 2013.

Hedberg, H. D., ed. *International Stratigraphic Guide.* New York: John Wiley & Sons, 1976.

Hegner, Robert W. Big Fleas Have Little Fleas: Or, Who's Who Among the Protozoa. New York: Dover, 1968.

Hickman, C. P., L. S. Roberts, and A. Larson. *Integrated Principles of Zoology.* Boston: McGraw-Hill, 2001.

Hidore, John, John E. Oliver, Mary Snow, and Richard Snow. *Climatology.* 3d ed. Merrill, 1984.

Higham, J., ed. *Critical Issues in Ecotourism: Understanding a Complex Tourism Phenomenon.* Oxford, UK: Butterworth-Heinemann, 2007.

Hill, Julia Butterfly. *The Legacy of Luna: The Story of a Tree, a Woman, and the Struggle to Save the Redwoods.* San Francisco: Harper, 2000.

Hill, Marquita K. "Pesticides." In *Understanding Environmental Pollution.* 3d ed. New York: Cambridge University Press, 2010.

Hilmi, Martin, Nicola Bradbear, and Danilo J. Mejía-Lorío. *Beekeeping and Sustainable Livelihoods.* Rome: Rural Infrastructure and Agro-Industries Division, Food and Agriculture Organization of the United Nations FAO, 2012.

Hinchliffe, J. R., and D. R. Johnson. *The Development of the Vertebrate Limb: An Approach Through Experiment, Genetics, and Evolution.* Oxford, England: Clarendon Press, 1980.

History of Poultry Production "History of Poultry Production." *USPoultry.org,* https://www.uspoultry.org/educationprograms/PandEP_Curriculum/Documents/PDFs/Lesson2/HistoryofPoultryProductionver3Pres.pdf. Accessed 26 May, 2017.

Hoag, Dana L. *Agricultural Crisis in America: A Reference Handbook.* ABC-CLIO, 1999.

______. *Applied Risk Management in Agriculture.* Boca Raton: CRC, 2010.

Hobbs, Richard J., et al. "Novel Ecosystems: Theoretical and Management Aspects of the New Ecological World Order." *Global Ecology and Biogeography* 15 (2006).

Hodge, Russ. *Genetic Engineering: Manipulating the Mechanisms of Life.* New York: Facts On File, 2009.

Hofrichter, Robert, ed. Amphibians: The World of Frogs, Toads, Salamanders, and Newts. Willowdale, Ontario: Firefly Books, 2000.

Holdgate, Martin W. *From Care to Action: Making a Sustainable World.* Washington, DC: Taylor, 1996.

Hollenbeck, James E. "Concentrated Animal Feeding Operations (CAFOs) as Potential Incubators for Influenza Outbreaks." *Trakia Journal of Sciences* 11.2 (2013): 205–9. *Academic Search Complete.* Web. 13 Nov. 2015.

Holloway, M. "Sustaining the Amazon." *Scientific American* 269 (July, 1993): 90–99. Describes the threats to the rain forests and possible ways to preserve tropical forests and their richness in biodiversity.

Holtz, Wesley G., and Stephen S. Hart. *Home Construction on Shrinking and Swelling Soils.* Denver: Colorado Geological Survey, 1978.

Holzer, T. L., T. L. Youd, and T. C. Hanks. "Dynamics of Liquefaction During the Superstition Hills, California, Earthquake." *Science* 244 (April 7, 1989): 56–59.

Honey, M. *Ecotourism and Sustainable Development: Who Owns Paradise?* 2nd ed. Washington, DC: Island Press, 2008.

Hooke, R. L. "On The History of Humans as Geomorphic Agents." *Geology* 28 (2000): 843–846.

Hoppe, Robert A., and Penni Korb. "Large and Small Farms: Trends and Characteristics." *Structural and Financial Characteristics of U.S. Farms.* Washington, DC: USDA Economic Research Service, 2004.

Horowitz, John, and Jessica Gottlieb. *The Role of Agriculture in Reducing Greenhouse Gas Emissions.* Washington, D.C.: U.S. Dept. of Agriculture, Economic Research Service, 2010.

Howe, Christopher. *Gene Cloning and Manipulation.* 2nd ed., Cambridge UP, 2007.

Howe, Henry F., and Lynne C. Westley. *Ecological Relationships of Plants and Animals.* Oxford Univ Pr, 2007.

Howell, David G. "Terranes." *Scientific American* 253 (November 1985): 116–125. Terranes are small blocks of crust added to a mountain belt by plate collisions. This article describes types of terranes and what happens when they collide.

Howland, John L. *The Surprising Archaea: Discovering Another Domain of Life.* New York: Oxford University Press, 2000.

Hrazdina, Geza, ed. *The Use of Agriculturally Important Genes in Biotechnology.* Washington, D.C.: NATO Scientific Affairs Division, 2000.

Huber, Karina. "One More Question for Dickson Despommier: What's the Future of Vertical Farming?" *CCTV America,* 5 May 2015, www.cctv-america.com/2015/05/05/one-more-question-for-dickson-despommier-whats-the-future-of-vertical-farming. Accessed 12 Dec. 2016.

Hudson, Norman. *Soil Conservation.* 3d ed. Ames: Iowa State University Press, 1995.

Humphrey, Mathew. *Preservation Versus the People?* New York: Oxford University Press, 2002.

Humphreys, David. *Logjam: Deforestation and the Crisis of Global Government.* Earthscan, 2006.

Hunt, C. B. *Natural Regions of the United States and Canada.* San Francisco: W. H. Freeman, 1974.

Hunt, Constance Elizabeth. *Thirsty Planet: Strategies for Sustainable Water Management.* New York: St. Martin's Press, 2004.

Hunter, Malcolm L., ed. *Maintaining Biodiversity in Forest Ecosystems.* New York: Cambridge UP, 1999.

Hutchinson, G. Evelyn. *The Ecological Theater and the Evolutionary Play.* New Haven: Yale UP, 1969.

Huxley, Peter A. *Tropical Agroforestry.* Hoboken, N.J.: Wiley-Blackwell, 1999.

Idriss, I. M., and R. W. Boulanger. *Soil Liquefaction During Earthquakes.* Earthquake Engineering Research Institute, 2008.

Illinois Endangered Species Protection Board (IESPB). *Checklist of Endangered and Threatened Animals and Plants of Illinois.* Springfield, IL: IESPB, 1999.

Industrial Livestock Production "Industrial Livestock Production." *Sustainable Table.* GRACE Communications Foundation, 2015.

Ingebritsen, S. E., Ward E. Sanford, and C. E. Neuzil. *Groundwater in Geologic Processes.* 2d ed. New York: Cambridge University Press, 2006.

Insider Info "Insider Info." *CFNC.org,* 2017, https://www1.cfnc.org/. Accessed 26 May, 2017.

International Food Information Council Foundation. "Food Biotechnology Timeline." *Food Biotechnology: A Communicator's Guide to Improving Understanding.* Food Insight, n.d. PDF. Web. 24 Sept. 2013.

International Union for Conservation of Nature, United Nations Environment Programme, and World Wide Fund for Nature. *Caring for the Earth: A Strategy for Sustainable Living.* Gland: IUCN, UNEP, WWF, 1993.

______. *World Conservation Strategy: Living Resource Conservation for Sustainable Development.* Gland: IUCN, UNEP, WWF, 1980.

Introduction to Poultry and Eggs Farming "An Introduction to Poultry and Eggs Farming." *Department of Agriculture, Environment and Rural Affairs,* 2017, https://www.daera-ni.gov.uk/articles/introduction-poultry-and-eggs-farming. Accessed 26 May, 2017.

Introduction to the Poultry Industry "Introduction to the Poultry Industry." *USPoultry.org,* https://www.uspoultry.org/educationprograms/PandEP_Curriculum/Documents/PDFs/Lesson1/IntroductiontothePoultryIndustryver2.pdf. Accessed 26 May, 2017.

ISAAA Brief 46–2013: Executive Summary "ISAAA Brief 46–2013: Executive Summary." *ISAAA,* 25 Mar. 2014, www.isaaa.org/resources/publications/briefs/46/executivesummary/. Accessed 28 July 2014.

Isager, Signe, and Jens Erik Skydsgaard. *Ancient Greek Agriculture: An Introduction.* New York: Routledge, 1995.

Jabr, Ferris. "Building Tastier Fruits & Veggies." *Scientific American* 311.1 (2014): 56–61. Print.

Jackson, Jerome A., *William E. Davis, Jr., and John Tautin, eds. Bird Banding in North America: The First Hundred Years.* Cambridge: Harvard U Nuttall Ornithological. Print.

Jackson, Wes. *New Roots for Agriculture.* 1980.

Jakab, Cheryl. *Natural Resources (Global Issues).* North Mankato, Minn.: Smart Apple Media, 2008.

James, Rosalind R., and Theresa L. Pitts-Singer, eds. *Bee Pollination in Agricultural Ecosystems.* New York: Oxford UP, 2008.

Janick, Jules. *Horticultural Science.* 4th ed. New York: W. H. Freeman, 1986.

Janovy, John, Jr. *On Becoming a Biologist.* 2d ed. Lincoln: U of Nebraska P, 2004.

Jariel, Domingo M., et al. "Plant and Soil Nutrient Concentrations in a Restored Cajun Prairie." *Proceedings of the Twenty-first North American Prairie Conference* 21 (2010): 95–107.

Jasson-Holt, Sophie. *Unfold the Chaparral.* San Francisco: San Francisco State University Chapbook, 1996.

Jeffery, Simon, and Wim H. van der Putten. "The Ecology of Soil-Borne Human Diseases." *Microbiology Today*, Nov. 2016. Accessed 17 Nov. 2016.

Jenkins, Clinton N., et al. "US Protected Lands Mismatch Biodiversity Priorities." *Proceedings of the National Academy of Sciences*, vol. 112, no. 16, 21 Apr. 2015, pp. 5081–86, doi: doi.org/10.1073/pnas.1418034112.

Jenny, H. *Factors of Soil Formation: A System of Quantitative Pedology.* New York: Dover, 1994.

______. *The Soil Resource: Origin and Behavior.* New York: Springer-Verlag, 1980.

Jensen, Eric A. *The Therapeutic Cloning Debate: Global Science and Journalism in the Public Sphere.* Ashgate, 2014.

Jochim, Candace L. *Home Landscaping and Maintenance on Swelling Soil.* Special Publication 14. Denver: Colorado Geological Survey, 1981.

Johanson, Donald C., and Maitland A. Edey. *Lucy: The Beginnings of Humankind.* New York: Touchstone, 1990.

John, Brian S. *The Ice Age: Past and Present.* London: Collins, 1977.

John, D. M., S. J. Hawkins, and J. H. Price, eds. *Plant-Animal Interactions in the Marine Benthos.* Oxford, England: Clarendon Press, 1992.

Johnson, Arthur H. "Acid Deposition: Trends, Relationships, and Effects." *Environment* 28.4 (1986): 6+. Print.

Johnson, Hugh. *The Story of Wine.* London: Mitchell Beazley, 2004.

Johnson, Rebecca L. *A Walk in the Tundra.* Minneapolis: Carolrhoda Books, 2000.

Joly, John. *The Surface-History of the Earth.* Boston: Adamant Media Corporation, 2005.

Jones, David L., A. Cox, P. Coney, and Myrl Beck. "The Growth of Western North America." *Scientific American* 247 (November 1982): 70–84.

Jose, Shibu. "Agroforestry for Ecosystem Services and Environmental Benefits: An Overview." *Agroforestry Systems* 76, no. 1 (May, 2009): 1–10.

Journal of Forestry. "Focus on . . . Private Forest Landowners: FAQs about Conservation Easements." 100, no. 3 (2002): 4.

Jurena, Remy. *Agriculture in the U.S. Free Trade Agreements: Trade with Current and Prospective Partners, Impact and Issues.* New York: Nova Science, 2008.

Kalloo, G., and J. B. Chowdhury, eds. *Distant Hybridization of Crop Plants.* Berlin: Springer-Verlag, 1992.

Kandler, Otto, and Wolfram Zillig, eds. *Archaebacteria Eighty-five: Proceedings of the EMBO Workshop on Molecular Genetics of Archaebacteria.* Forestburgh, N.Y.: Lubrecht and Cramer, 1987.

Kapur, Selim, and George Stoops, eds. *New Trends in Soil Micromorphology.* New York: Springer, 2010.

Kareiva, Peter, and Simon A. Levin. *The Importance of Species: Perspectives on Expendability and Triage.* Princeton, NJ: Princeton University Press, 2015.

Katti, R. K. *Behaviour of Saturated Expansive Soil and Control Methods.* Rev. ed. Oxford: Taylor & Francis, 2002.

Kavanagh, James. *Waterford's Discovery Guide Bees and Pollinators: A Folding Pocket Guide to Familiar Species.* Dunedin, FL: Waterford Press, 2016.

Kearey, Philip, Keith A. Klepeis, and Frederick J. Vine. *Global Tectonics.* 3rd ed. Hoboken, N.J. Wiley-Blackwell, 2009.

Kelley, Jane H., and Marsha P. Hanen. *Archaeology and the Methodology of Science.* Albuquerque: University of New Mexico Press, 1988.

Kellogg, Charles E. *Our Garden Soils.* New York: Macmillan, 1952.

______. *The Soils That Support Us.* New York: Macmillan, 1956.

Kempe, D. R. C., and Anthony P. Harvey, eds. *The Petrology of Archaeological Artifacts.* New York: Clarendon Press, 1983.

Kendall, Ronald J., et al., eds. *Wildlife Toxicology: Emerging Contaminant and Biodiversity Issues.* Boca Raton: CRC, 2010.

Kennedy, Merrit. "Report: More Pollinator Species in Jeopardy, Threatening World Food Supply." *NPR.* NPR, 26 Feb. 2016.

Keske, Catherine. *The Emerging Market for Private Land Preservation and Conservation Easements.* Saarbrücken, Germany: VDM Verlag, 2008.

Kézdi, Árpád. *Soil Testing.* Vol. 2 in *Handbook of Soil Mechanics.* New York: Elsevier Scientific, 1974–1990.

Khan, Muhammad Sarwar, Iqrar A. Khan, and Debmalya Barh. *Applied Molecular Biotechnology:*

The Next Generation of Genetic Engineering. CRC Press, 2016.

Killman, Wolf, and Lay Thong Hong. "Rubberwood: The Success of an Agricultural By-Product." *Unasylva* 51 (2000): 66–72.

Kimberley, M. M., and D. E. Grandstaff. "Profiles of Elemental Concentrations in Precambrian Paleosols on Basaltic and Granitic Parent Materials." *Precambrian Research* 32 (1986): 133–154.

King, Geoffrey, and Geoff Bailey. "Tectonics and Human Evolution." *Antiquity* 80 (2006): 265–286. Print.

Kingman, J., and P. Kingman. "The Dance of the Luminescent Threadworms." *Underwater Naturalist* 22 (1993): 36.

Kingsbury, Noel. *Hybrid: The History and Science of Plant Breeding*. Chicago: University of Chicago Press, 2009.

Kipps, M. S. *Production of Field Crops: A Textbook of Agronomy*. 6th ed. New York: McGraw-Hill, 1970.

Kirby, David. "Obama on Factory Farming: Something to Crow About." *Washington Post*. Washington Post, 26 Mar. 2010.

Kirby, Richard R. *Ocean Drifters: A Secret World Beneath the Waves*. Studio Cactus, 2010.

Kisling, Vernon N., Jr., ed. *Zoo and Aquarium History: Ancient Animal Collections to Zoological Gardens*. Boca Raton, Fla.: CRC Press, 2001.

Kline, Benjamin. *First Along the River: A Brief History of the U.S. Environmental Movement*. 3d ed. Lanham, Md.: Rowman & Littlefield, 2007.

Knight, Gregory, Ivan Raev, and Marieta Staneva, eds. *Drought in Bulgaria*. Ashgate Publishing Limited, 2004.

Knight, Richard L., and Courtney White, eds. *Conservation for a New Generation: Redefining Natural Resources Management*. Washington, D.C.: Island Press, 2009.

Koepf, Herbert H. *The Biodynamic Farm: Agriculture in the Service of the Earth and Humanity*. Hudson, N.Y.: Anthroposophic Press, 2006.

Koerner, Robert M. *Construction and Geotechnical Methods in Foundation Engineering*. New York: McGraw-Hill, 1984.

Koerth-Baker, Maggie. *Before the Lights Go Out: Conquering the Energy Crisis Before It Conquers Us*. Hoboken, NJ: John Wiley & Sons, 2012.

Koga, Hiroki. "How A Japanese Vertical Farm Is Growing Strawberries Using LED for the First Time." *AGFunder News,* 15 Apr. 2016, agfundernews.com/how-a-japanese-vertical-farm-is-growing-strawberries-using-led-for-the-first-time5691.

Kolbe, Emily A. "'Won't You Be My Neighbor?': Living with Concentrated Animal Feeding Operations." *Iowa Law Review* 99.1 (2013): 415–43. *Academic Search Complete*. Web. 6 Nov. 2015.

Komolprasert, Vanee, and Kim M. Morehouse, editors. *Irradiation of Food and Packaging: Recent Developments*. American Chemical Society, 2004.

Kondratyev, K. Y., V. F. Krapivin, and Costas A. Varotsos. *Global Carbon Cycle and Climate Change*. Springer, 2003.

Konhauser, Kurt. *Introduction to Geomicrobiology*. Hoboken, N.J.: Wiley-Blackwell, 2006.

Krebs, Charles J. *The Message of Ecology*. New York: Harper, 1987.

Krebs, J. R., and N. B. Davies, eds. *Behavioral Ecology: An Evolutionary Approach*. 4th ed. Boston: Blackwell Scientific Publications, 1997.

Kristiansen, Paul, Acram Taji, and John Reganold, editors. *Organic Agriculture: A Global Perspective*. Cornell UP, 2006.

Kruckeberg, Arthur R. *Geology and Plant Life: The Effects of Landforms and Rock Types on Plants*. Seattle, Wash. University of Washington Press, 2004.

Krupp, Fred, and Miriam Horn. *Earth: The Sequel—The Race to Reinvent Energy and Stop Global Warming*. New York: W. W. Norton, 2008.

Kyte, Lydiane, and John G. Kleyn. *Plants from Test Tubes: An Introduction to Micropropagation*. 3rd ed. Portland: Timber, 1996.

Lade, Poul V., and Jerry A. Yamamuro. "Evaluation of Static Liquefaction Potential of Silty Sand Slopes." *Canadian Geotechnical Journal* 48 (2011): 247–264. The article compares sand types and their liquefaction potential.

Lægreid, M., et al. *Agriculture, Fertilizers, and the Environment*. Cabi, 1999.

Laidlaw, Stuart. *Secret Ingredients: The Brave New World of Industrial Farming*. Toronto: McClelland & Stewart, 2004.

Laity, Julie. *Deserts and Desert Environments*. Hoboken, N.J.: Wiley-Blackwell, 2008.

Lal, Rattan, et al, eds. *Methods for Assessment of Soil Degradation.* Boca Raton, Fla.: CRC Press, 1998.

______. *Principles of Soil Physics.* New York: Marcel Dekker, 2004.

______. *Soil Degradation in the United States: Extent, Severity, and Trends.* Boca Raton, Fla.: Lewis, 2004.

______. *Soil Quality and Soil Erosion.* Boca Raton, Fla.: CRC Press, 1999.

Lampkin, Nicolas. *Organic Farming.* Rev. ed., Diamond Farm Enterprises, 2002.

Lancaster, Brad, and Joe Marshall. *Rainwater Harvesting for Drylands and Beyond.* 2 vols. Tucson, AZ: Rainsource Press, 2013.

Lancaster, N. *Geomorphology of Desert Dunes.* London: Routledge, 1995.

Landa, Edward R., and Christian Feller, eds. *Soil and Culture.* New York: Springer, 2010.

Langley, Andrew. *Is Organic Food Better?* Chicago: Heinemann, 2009.

Lanner, Ronald M. *Made for Each Other: A Symbiosis of Birds and Pines.* New York: Oxford University Press, 1996.

Lappé, Frances Moore, Joseph Collins, and Peter Rosset. *World Hunger: Twelve Myths.* 2d ed. New York: Grove Press, 1998.

Large-Scale Pastured Poultry Farming in the U.S."Large-Scale Pastured Poultry Farming in the U.S." *UW-Madison Center for Integrated Agricultural Systems,* 2003, https://www.cias.wisc.edu/large-scale-pastured-poultry-farming-in-the-us/. Accessed 26 May, 2017.

Lasca, N. P., and J. Donahue, eds. *Archaeological Geology of North America.* Boulder, Colo.: Geological Society of America, 1990.

Latrubesse, E. M., J. C. Stevaux, and R. Sinha. "Tropical Rivers." *Geomorphology* 70 (2005): 187–206. Compares the geology and geomorphology of various river basins around the globe. Discusses drainage from mountain belts to lowland plains in wet and wet-dry climates. The authors also mention anthropogenic river management issues.

Lavers, Chris. *Why Elephants Have Big Ears: Understanding Patterns of Life on Earth.* New York: St. Martin's, 2001.

Lawson, Sam and Larry MacFaul. *Illegal Logging and Related Trade: Indicators of the Global Response.* London: Chatham House, 2010.

Lawton, J. H., Richard Southwood, and D. R. Strong. *Insects on Plants.* Cambridge, Mass.: Harvard University Press, 1984.

Lévêque, C. *Ecology from Ecosystem to Biosphere.* Science, 2003.

Leakey, Richard E. *The Making of Mankind.* New York: Dutton, 1981.

Leakey, Richard E., and Roger Lewin. *The Sixth Extinction: Patterns of Life and the Future of Humankind.* New York: Doubleday, 1995.

Legget, Robert F. *Cities and Geology.* New York: McGraw-Hill, 1973.

Legionnaires' Disease Associated with Potting Soil "Legionnaires' Disease Associated with Potting Soil—California, Oregon, and Washington, May-June, 2000.

Lemaire, G., et al., eds. *Grassland Ecophysiology and Grazing Ecology.* Wallingford, UK: CABI Publishing, 2007.

Lent, C. M., and M. H. Dickinson. "The Neurobiology of Feeding in Leeches." *Scientific American* 258 (1988): 98–103. A description of how feeding behavior in leeches is controlled by the nervous secretion of serotonin.

Leopold, Aldo. *A Sand County Almanac.* New York: Oxford U P, 1949.

Lerman, A. "Geochemical Cycles." In *The Oxford Companion to the Earth.* Edited by Paul Hancock and Brian J. Skinner. New York: Oxford University Press, 2000.

Leval, Kim, et al. *The Impact and Benefits of USDA Research and Grant Programs to Enhance Midsize Farm Profitability and Rural Community Success.* Lyons: Center for Rural Affairs, 2006.

Levetin, Estelle, and Karen McMahon. *Plants and Society.* 5th ed. Boston: McGraw-Hill Higher Education, 2008.

Levin, Simon A. *Encyclopedia of Biodiversity.* Oxford: Academic Press, 2013.

Levinton, Jeffrey S. *Marine Biology: Function, Biodiversity, Ecology.* Oxford University Press, 2018.

Levy, Charles Kingston. *Evolutionary Wars: A Three-Billion Years Arms Race.* New York: Basingstoke, 2000.

Liang, George H., and Daniel Z. Skinner. *Genetically Modified Crops: Their Development, Uses, and Risks.* New York: Food Products Press, 2004.

Lieberman, Bruce S. *Paleobiogeography: Using Fossils to Study Global Change, Plate Tectonics, and Evolution.* New York: Kluwer Academic, 2000.

Limbaugh, Conrad. "Cleaning Symbiosis." *Scientific American* 205 (1961): 42–49. Print.

Lindsey, Rebecca. "Climate Change: Atmospheric Carbon Dioxide." *Climate.gov,* National Oceanic and Atmospheric Administration, 17 Oct. 2017. Accessed 5 Feb. 2018.

Lippmann, Morton, ed. *Environmental Toxicants: Human Exposures and Their Health Effects.* 3rd ed. Hoboken: Wiley, 2009.

Lipson, Elaine Marie. *The Organic Foods Sourcebook.* Chicago: Contemporary Books, 2001.

Little, C. *The Colonization of the Land.* Cambridge, England: Cambridge University Press, 1983.

Liu, Cheng, and Jack Evett. *Soil Properties: Testing, Measurement, and Evaluation.* 6th ed. Upper Saddle River, N.J.: Pearson/Prentice Hall, 2009.

Liu, George. *Cattle: Domestication, Diseases and the Environment.* New York: Nova, 2013.

Lloyd, G. B. *Don't Call It Dirt.* Ontario, Calif.: Bookworm Publishing, 1976.

Loeffe, Christian V., ed. *Conservation and Recycling of Resources: New Research.* New York: Nova Science, 2006.

Lomolino, Mark V., Dov F. Sax, and James H Brown. *Foundations of Biogeography: Classic Papers with Commentary.* Chicago: University of Chicago Press, 2004.

Long, John A. *The Rise of Fishes: Five Hundred Million Years of Evolution.* Baltimore: Johns Hopkins UP, 1996.

Look at the Poultry Industry "A Look at the Poultry Industry – How Does Chicken Get on Your Plate?" *Farmaid.org,* 2015, https://www.farmaid.org/issues/industrial-agriculture/a-look-at-the-poultry-industry-how-does-chicken-get-on-your-plate/. Accessed 26 May, 2017.

Losos, Jonathan B., Robert E. Ricklefs, and Robert H. MacArthur, eds. *The Theory of Island Geography Revisited.* Princeton, N.J. Princeton University Press, 2010.

Lovins, Amory. "Energy Strategy: The Road Not Taken?" *Foreign Affairs* (October 1976).

Lugo, Ariel E. "The Emerging Era of Novel Tropical Forests." *Biotropica* 41 (2009).

Lundgren, Lawrence. *Environmental Geology.* 2d ed. Englewood Cliffs, N.J.: Prentice-Hall, 1998.

Luoma, Jon R. *The Hidden Forest: The Biography of an Ecosystem.* Corvallis, OR: Oregon State University Press, 2006.

Lutgens, Frederick K., Edward J. Tarbuck, and Dennis Tasa. *Essentials of Geology.* 11th ed. Upper Saddle River, N.J.: Prentice Hall, 2011.

______. *The Atmosphere: An Introduction to Meteorology.* 11th ed. Prentice Hall, 2009.

Lydersen, Bjorn K., Nancy A. D'Elia, and Kim L. Nelson, eds. *Bioprocess Engineering: Systems, Equipment and Facilities.* New York: John Wiley & Sons, 1994.

Lye, Lin Heng. *Sustainability Matters: Environmental Management in Asia.* Singapore: World Scientific, 2010.

Lynch, Colum. "UN Backs Human Cloning Ban." *Washington Post,* 9 Mar. 2005, www.washingtonpost.com/wp-co/hotcontent/index.html?section=nation/specials/science/cloning. Accessed 28 July 2014.

Lynch, J. M. *Soil Biotechnology: Microbiological Factors in Crop Production.* Malden, Mass.: Blackwell, 1988.

Lyson, Thomas A. *Civic Agriculture: Reconnecting Farm, Food, and Community.* Lebanon: Tufts UP, 2004.

Mabbutt, J. A. *Desert Landforms.* Cambridge, Mass.: MIT Press, 1977.

Macdonald, David W., and Katrina Service, eds. *Key Topics in Conservation Biology.* Malden, Mass.: Blackwell, 2008.

MacDonald, James. "Family Farming in the United States." *Economic Research Service.* US Dept. of Agriculture, 4 Mar. 2014.

Mackay, Richard. *The Atlas of Endangered Species.* London Earthscan from Routledge, 2017.

Mackenzie, Fred T. *Our Changing Planet.* 4th ed. Upper Saddle River, N.J.: Prentice Hall, 2010.

Magdoff, Fred, and Brian Tokar, editors. *Agriculture and Food in Crisis: Conflict, Resistance, and Renewal.* Monthly Review Press, 2010.

Maher, Neil M. *Nature's New Deal: The Civilian Conservation Corps and the Roots of the American Environmental Movement.* New York: Oxford University Press, 2009.

Mainguet, Monique. *Aridity: Drought and Human Development.* Springer-Verlag, 2010.

Malmesheimer, R. W., et al. "Preventing GHG Emissions Through Avoided Land-Use Change." *Journal of Forestry* 106, no. 3 (April/May, 2008).

Manske, Llewellyn L. *Perpetually Sustainable Grazingland Ecosystems.* Dickinson, North Dakota: North Dakota State University, 2013.

Manske, Llewellyn, and Sheri A, Schneider. *Biologically Effective Management of Grazinglands.* Dickinson, ND: North Dakota State University, 2014.

Mapes, Kathleen. *Sweet Tyranny: Migrant Labor, Industrial Agriculture, and Imperial Politics.* Urbana: U of Illinois P, 2009.

Marchak, M. Patricia. *Logging the Globe.* Montreal: McGill-Queen's University Press, 1995.

Margulis, L., et al. *Illustrated Glossary of Protoctista.* Boston: Jones and Bartlett, 1993.

Margulis, Lynn, and Dorion Sagan. *Slanted Truths: Essays On Gaia, Symbiosis, and Evolution.* New York: Copernicus, 1997.

Margulis, Lynn, and Michael J. Chapman. *Kingdoms and Domains: An Illustrated Guide to the Phyla of Life on Earth.* 4th ed., Elsevier, 2009.

Margulis, Lynn. "Symbiosis and Evolution." *Scientific American* 225 (1971): 48–57. Print.

Marine Resources Service, Fishery Resources Division, FAO Fisheries Department. *Review of the State of World Marine Fishery Resources.* Rome, Italy: Food and Agriculture Organization of the United Nations, 2005.

Markus, Nicola. *On Our Watch: The Race to Save Australia's Environment.* Melbourne: University of Melbourne Press, 2009.

Marland, Gregg, et al. "The Climatic Impacts of Land Surface Change and Carbon Management, and the Implications for Climate-Change Mitigation Policy." *Climate Policy* 3 (2003): 149–57. Print.

Marris, Emma. "Ragamuffin Earth." *Nature* 460 (2009).

Marsh, William M. *Earthscape: A Physical Geography.* New York: John Wiley & Sons, 1987.

Martin, Anthony J. *Introduction to the Study of Dinosaurs.* New York: Wiley-Blackwell, 2006.

Martin, Paul S., and Richard G. Klein, eds. *Quaternary Extinctions: A Prehistoric Revolution.* Tucson: U of Arizona P, 1984.

Maser, Chris. *Forest Primeval: The Natural History of an Ancient Forest.* Reprint. Corvallis: Oregon State University Press, 2001.

Mather, Anne. "Arid Environments." In *Environmental Sedimentology,* edited by Chris Perry and Kevin Taylor. Malden, Mass.: Blackwell, 2007.

Matsch, Charles L. *North America and the Great Ice Age.* New York: McGraw-Hill, 1976.

Maun, Anwar M. *The Biology of Coastal Sand Dunes.* New York: Oxford University Press, 2009.

Mayr, Ernst. *The Growth of Biological Thought: Diversity, Evolution and Inheritance.* Cambridge, Mass: The Belknap Press of Harvard Univ. Press, 2003.

______. *Populations, Species, and Evolution: An Abridgment of "animal Species and Evolution".* Belknap Press of Harvard U.P., 1977.

Mays, Larry W. *Ground and Surface Water Hydrology.* Hoboken: Wiley, 2012.

Mazet, J. A., G. E. Hamilton, and L. A. Dierauf. "Educating Veterinarians for Careers in Free-Ranging Wildlife Medicine and Ecosystem Health." *Journal of Veterinary Medical Education* 33.3 (2006): 352–360. Print.

Mazoyer, Marcel, and Laurence Roudart. *A History of World Agriculture: From the Neolithic Age to the Current Crisis.* New York: Monthly Review Press, 2006.

McBrewster, John, Frederic P. Miller, and Agnes F. Vandome, eds. *Industrial Agriculture: Factory Farming, Livestock, Aquaculture, Agribusiness, Monoculture, Agroecology, Organic Farming, Urban Agriculture.* Phoenix, Ariz.: Alphascript, 2009.

McBride, Murray B. *Environmental Chemistry of Soils.* New York: Oxford University Press, 1994.

McCarthy, Colman. "Infiltrators Expose the Cruelty of Meat." *National Catholic Reporter* 27 Apr. 2012: 17. *Points of View Reference Center.* Web. 6 Nov. 2015.

McGhee, George R. *The Late Devonian Mass Extinction.* New York: Columbia UP, 1996.

McGowran, Brian. *Biostratigraphy: Microfossils and Geological Time.* New York: Cambridge University Press, 2008.

McKane, L., and J. Kandel. Microbiology: Essentials and Applications. New York: McGraw-Hill, 1996.

McKee, Gregory J., et al., editors. *Pesticide Resistance, Population Dynamics, and Invasive Species Management.* Nova Science Publishers, 2010.

McKee, Jeffrey K., Frank E. Poirier, and W. Scott McGraw. *Understanding Human Evolution.* 5th ed. Upper Saddle River: Prentice Hall, 2004.

McKirdy, Euan. "The Only Way Is Up: Vertical Farming in Kyoto." *CNN,* 19 Sept. 2016, www.cnn.com/2015/12/09/foodanddrink/kyoto-vertical-farm-spread/. Accessed 12 Dec. 2016.

McKnight, T. L., and Darrel Hess. *Physical Geography: A Landscape Appreciation.* 9th ed. Englewood Cliffs, N.J.: Prentice-Hall, 2007.

McMenamin, Mark A., and Dianna L. McMenamin. *The Emergence of Animals: The Cambrian Breakthrough.* New York: Columbia University Press, 1989.

McMenamin, Mark. *Discovering the First Complex Life: The Garden of the Ediacara.* New York: Columbia University Press, 1998.

McNamara, K. J. *Shapes of Time: The Evolution of Growth and Development.* Johns Hopkins UP, 1997.

McNaughton, S. J. "Grazing Lawns: Animals in Herds, Plant Form, and Coevolution." *American Naturalist* 124, no. 6 (December, 1984): 863–886.

McPhee, John. *Basin and Range.* New York: Farrar, Straus & Giroux, 1981.

Meeuse, Bastian, and Sean Morris. *The Sex Life of Flowers.* New York: Facts on File, 1984.

Mentz, André. *Ecological Cattle Farming: A Roadmap to Sustainable Cattle Farming.* South Africa, Northern Cape: Pula Books, 2016.

Menzies, M., ed. *Modern Glacial Environments: Processes, Dynamics, and Sediments.* Oxford, England: Butterworth-Heinemann, 1996.

Messerli, B., and J. D. Ives. *Mountains of the World: A Global Priority.* New York: Parthenon, 1997.

Metcalfe, Darrel S., and Donald M. Elkins. *Crop Production: Principles and Practices.* 4th ed. New York: Macmillan, 1980.

Middleton, Arthur Pierce. *Tobacco Coast: A Maritime History of the Chesapeake Bay in the Colonial Era.* Baltimore: Johns Hopkins University Press, 1984.

Miius, S. "When Elephants Can't Take It Anymore." *Science News* 155 (May 29, 1999): 341–342.

Miller, Fred P. "After 10,000 Years of Agriculture, Whither Agronomy?" *Agronomy Journal* 100, no. 1 (2007): 22–34. Available at https://www.agronomy.org/files/about-agronomy/future-of-agronomy.pdf.

Miller, G T, and Scott Spoolman. *Living in the Environment.* Cengage Learning, 2018.

Miller, G. Tyler, Jr. *Resource Conservation and Management.* Belmont, Calif.: Wadsworth, 1990.

Miller, G. Tyler, Jr., and Scott Spoolman. *Environmental Science: Problems, Concepts, and Solutions.* 13th ed. Belmont, Calif.: Brooks/Cole, 2010.

Miller, Naomi Frances, Katherine M. Moore, and Kathleen Ryan. *Sustainable Lifeways: Cultural Persistence in an Ever-Changing Environment.* Philadelphia: University of Pennsylvania Press, 2011.

Miller, Raymond W., and Roy L. Donahue. *Soils: An Introduction to Soils and Plant Growth.* 6th ed. Englewood Cliffs, N.J.: Prentice-Hall, 1990.

Miller, William Charles. *Trace Fossils: Concepts, Problems, Prospects.* St. Louis, Mo.: Elsevier, 2007.

Miller-Struttmann, Nicole E., et al. "Functional Mismatch in a Bumble Bee Pollination Mutualism under Climate Change." *Science* 25 Sept. 2015: 1541–544. Print.

Millot, Georges. "Clay." *Scientific American* 240 (April, 1979): 108. Emphasizes the chemistry and industrial uses of clay minerals. Offers a brief discussion of weathering as the source of clays and as part of the geologic cycling of clays. Discusses how weathering reactions vary in different climates and yield different clay minerals in the local soils. Provides several spectacular scanning electron microphotographs of representative clay types. Accessible to the general reader.

Minteer, Ben and Robert Manning, eds. *Reconstructing Conservation: Finding Common Ground.* Washington, D.C.: Island Press, 2003.

Mistry, Jayalaxshmi, and Andrea Berardi, eds. *Savannas and Dry Forests: Linking People with Nature.* Burlington, Vt.: Ashgate, 2006.

Mitchell, C. Ben, et al. *Biotechnology and the Human Good.* Georgetown UP, 2007.

Mittermeier, Cristina. "The Power of Conservation Photography." *The Wildlife Professional* (Fall, 2007): 30–31, 43. Print.

Molles, Manuel C., Jr. *Ecology: Concepts and Applications.* 7th ed. New York: McGraw, 2016.

Molnar, Peter. "The Structure of Mountain Ranges." *Scientific American* 255 (July 1986): 70–79.

Montez, Michele, and Lorraine Bodger. *The New 50 Simple Things Kids Can Do to Save the Earth.* Kansas City: Andrews McMeel Pub, 2009.

Moore, John E. *Field Hydrogeology: A Guide for Site Investigations and Report Preparation.* Boca Raton: CRC, 2012.

Moore, W. G. *Glaciers.* London: Hutchinson Educational, 1972.

Morgan, Jack A., et al. "Carbon Dioxide Enrichment Alters Plant Community Structure and Accelerates Shrub Growth in the Shortgrass Steppe." *Proceedings of the National Academy of Sciences* 104.37 (2007): 14,724–4,729. Print.

Morgan, Royston Philip Charles. *Soil Erosion and Conservation.* 3d ed. Malden, Mass.: Blackwell Science, 2005.

Morholt, E., and Paul F. Brandwein. A Sourcebook for the Biological Sciences. San Diego, Calif.: Harcourt Brace, 1980.

Morin, Peter Jay. *Community Ecology.* 2nd ed. Chichester: Wiley, 2012.

Morphet, Sally. "NGOs and the Environment." *The Conscience of the World: The Influence of Non-governmental Organizations in the U.N. System.* Ed. Pete Willetts. Washington, DC: Brookings Inst. P, 1996.

Morrissey, John F, James L. Sumich, and Deanna R. Pinkard-Meier. *Introduction to the Biology of Marine Life.* Jones and Bartlett Learning, 2018.

Morrone, Juan J. *Evolutionary Biogeography: An Integrative Approach with Case Studies.* New York: Columbia University Press, 2009.

Mortimore, Michael. *Roots in the African Dust: Sustaining the Sub-Saharan Drylands.* New York: Cambridge University Press, 1998.

Motani, Ryosuke. "Rulers of the Jurassic Seas." Scientific American 283, no. 6 (2000): 52–59. A beautifully illustrated article explains the adaptations that had to be made for reptiles to return to the sea.

Mowat, Farley. *Never Cry Wolf.* Boston: Atlantic-Little, Brown, 1963.

Mudroch, Alena, and José M. Azcue. *Manual of Aquatic Sediment Sampling.* Boca Raton, Fla.: Lewis, 1995.

Muller, Richard. *Nemesis: The Death Star.* New York: Weidenfeld & Nicolson, 1988.

Murphy, Janice. *Seed Banks: Types, Roles and Research.* Hauppauge, New York: Nova Science, 2016.

Mustafa, Saleem, and Rossita Shapawi. *Aquaculture Ecosystems: Adaptability and Sustainability.* Hoboken: Wiley, 2015.

Mysteries of the Rain Forest. Pleasantville, N.Y.: Reader's Digest, 1998.

Nash, Roderick. *Wilderness and the American Mind.* 4th ed. New Haven, Conn.: Yale University Press, 2001.

National Archives. Records of the Civilian Conservation Corps. http://www.archives.gov/research/

National Conference of State Legislatures. "State Rainwater: Graywater Harvesting Laws and Legislation." *NCSL.org*, September 1, 2013.

National, Research C. S. *Science and the Endangered Species Act.* Washington: National Academy Press, 2004.

Nelleman, Christian, Stefan Hain, and Jackie Alder, eds. *In Dead Water: Merging of Climate Change with Pollution, Over-Harvest, and Infestations in the World's Fishing Grounds.* Arendal, Norway: United Nations Environment Programme, GRID-Arendal, 2008.

Nelson, Gerald C., ed. *Genetically Modified Organisms in Agriculture: Economics and Politics.* Academic, 2001.

Nelson, John D, and Debora J. Miller. *Expansive Soils: Problems and Practice in Foundation and Pavement Engineering.* New York: John Wiley and Sons, 1997.

New, T. R. *An Introduction to Invertebrate Conservation.* New York: Oxford UP, 1995.

Newson, Malcolm. *Land, Water, and Development: Sustainable and Adaptive Management of Rivers.* 3d ed. London: Routledge, 2009.

Newton, David E. *GMO Food: A Reference Handbook.* ABC-CLIO, 2014.

Newton, Ian. *Speciation and Biogeography of Birds.* London: Academic Press, 2003.

Nicholl, Desmond S. T. *An Introduction to Genetic Engineering.* 3rd ed., Cambridge UP, 2008.

Nicholson, Sharon E. *Dryland Climatology.* New York: Cambridge University Press, 2011.

Nierenberg, Danielle. "The Commercialization of Farming: Producing Meat for a Hungry World." *USA Today Magazine* Jan. 2004: 22–24. *Academic Search Complete.* Web. 6 Nov. 2015.

Niman, Nicolette Hahn. *Defending Beef: The Case for Sustainable Meat Production.* White River Junction, Vermont: Chelsea Green, 2014.

Nisbet, E. G. *Leaving Eden: To Protect and Manage the Earth.* New York: Cambridge University Press, 1991.

Noble, Elmer, Glenn Noble, Gerhard Schad, and Austin MacGinnes. *Parasitology: The Biology of Animal Parasites.* 6th ed. Philadelphia: Lea & Febiger, 1989.

Norberg-Hodge, Helena, Todd Merrifield, and Steven Gorelick. *Bringing the Food Economy Home: Local Alternatives to Global Agribusiness.* London: Zed Books, 2002.

Norris, Joanne E., et al., eds. *Slope Stability and Erosion Control: Ecotechnological Solutions.* New York: Springer, 2010.

Norstag, Knut, and Andrew J. Meyerriecks. *Biology.* 2d ed. Columbus, Ohio: Charles E. Merrill, 1985.

Norton, Bryan G., et al., eds. *Ethics on the Ark: Zoos, Animal Welfare, and Wildlife Conservation.* Washington, D.C.: Smithsonian Institution Press, 1995.

Norton, Tomas, Brijesh K. Tiwari, and Nick Holden. *Sustainable Food Processing.* John Wiley, 2014.

O'Brien, Rachel, and Jennifer A. Roberts. "Teaching the Principles of Geomicrobiology and the Process of Experimental Research to Undergraduate and First-Year Graduate Students." *Journal of College Science Teaching* 38, no. 2 (2008): 30–38.

Obach, Brian K. *Organic Struggle: The Movement for Sustainable Agriculture in the United States.* MIT P, 2015.

Odell, Lee H. *Treatment Technologies for Groundwater.* Denver: American Water Works Assn., 2010.

Odion, Dennis C., and Frank W. Davis. "Fire, Soil Heating, and the Formation of Vegetation Patterns in Chaparral." *Ecological Monographs* 70, no. 1 (2000): 149–169. A detailed scientific account of the role of fire in the chaparral. It would be difficult reading with no background in biology.

Odling-Smee, F. John, Kevin N. Laland, and Marcus W. Feldman. "Niche Construction." *American Naturalist* 147, no. 4 (April, 1996): 641–649. Argues that niche construction regularly modifies both biotic and abiotic sources of natural selection.

Odum, Eugene P. *Ecology and Our Endangered Life-Support Systems.* Sunderland: Sinauer, 1989.

Oelhaf, Robert C. *Organic Agriculture: Economic and Ecological Comparisons with Conventional Methods.* New York: John Wiley & Sons, 1978.

Officer, Charles, and Jake Page. *The Great Dinosaur Extinction Controversy.* New York: Addison-Wesley, 1996.

Ogg, James G., Gabi Ogg, and Felix M. Gradstein. *The Concise Geologic Time Scale.* New York: Cambridge University Press. 2008.

Oldest Fossils on Earth Discovered "Oldest Fossils on Earth Discovered." *Science Daily.* Science Daily, 22 Aug. 2011.

Ollier, Cliff, and Colin Pain. *The Origin of Mountains.* London: Cambridge University Press, 2000.

Olsen, Sandra L. *Horses Through Time.* Niwot, Colo: Roberts Rinehart, 2004.

Olson, Gerald W. *Soils and the Environment: A Guide to Soil Surveys and Their Application.* New York: Chapman & Hall, 1981.

Organic Certification "Organic Certification." *United States Department of Agriculture Economic Research Service,* 19 Oct. 2016, https://www.ers.usda.gov/topics/natural-resources-environment/organic-agriculture/organic-certification/. Accessed 22 Nov. 2016.

Organic Foods: Are They Safer? More Nutritious "Organic Foods: Are They Safer? More Nutritious?" *Mayo Clinic.* Mayo Foundation for Medical Education and Research, 7 Sept. 2012.

Organic Market Overview "Organic Market Overview." *United States Department of Agriculture Economic Research Service,* 19 Oct. 2016, http://www.ers.usda.gov/topics/natural-resources-environment/organic-agriculture/organic-market-overview.aspx. Accessed 22 Nov. 2016.

Organic Trade Association. "Quick Overview: Organic Agriculture and Production." *Organic Trade Association.* Organic Trade Assn., 16 Feb. 2011.

Orians, Gordon H., and Otto T. Solbrig, eds. *Convergent Evolution in Warm Deserts.* Stroudsburg, Pa.: Dowden, Hutchinson & Ross, 1977.

Ortega-Rivas, Enrique, editor. *Processing Effects on Safety and Quality of Foods.* CRC Press, 2010.

Osman, Khan Towhid. *Soil Degradation, Conservation and Remediation.* New York: Springer, 2016.

Owen, Meran R. L., and Jan Pen, eds. *Transgenic Plants: A Production System for Industrial and Pharmaceutical Proteins.* New York: Wiley, 1996.

Owings, Lisa. *Sustainable Agriculture.* Minneapolis: ABDO, 2013.

Oz, Mehmet. "Give (Frozen) Peas A Chance And Carrots Too." *Time* 180.23 (2012): 38–46. Print.

Paarlberg, Don, and Philip Paarlberg. *The Agricultural Revolution of the 20th Century*. Iowa State UP, 2000.

Paarlberg, Robert. "Attention Whole Foods Shoppers." *Foreign Policy* 179 (2010): 1. Print.

______. "Organic and Local Food." *Food Politics: What Everyone Needs to Know*. Oxford UP, 2010.

Padian, Kevin, and Luis M. Chiappe. "The Origin of Birds and Their Flight." Scientific American 278, no. 2 (1998): 38–47. A good outline of how birds evolved from dinosaurs.

Palm, Cheryl A., et al., eds. *Slash-and-Burn Agriculture: The Search for Alternatives*. New York: Columbia University Press, 2005.

Palo, Matti, and Heidi Vanhanen, eds. *World Forests from deforestation to Transition?* Boston: Kluwer Academic, 2000.

Pankhurst, Richard. *The Ethiopians: A History*. Oxford, England: Blackwell Publishers, 2001.

Parekh, Sarad R. *The GMO Handbook: Genetically Modified Animals, Microbes, and Plants in Biotechnology*. Totowa, N.J.: Humana Press, 2010.

Park, Alice. "Scientists Report First Success in Cloning Human Stem Cells." *CNN*, 16 May 2013, www.cnn.com/2013/05/15/health/time-cloning-stem-cells. Accessed 24 Oct. 2016.

Parkes, P. A. *Current Scientific Techniques in Archaeology*. London, Routledge, 2016.

Parson, Ruben L. *Conserving American Resources*. 3d ed. Englewood Cliffs, N.J.: Prentice-Hall, 1972.

Patiny, Sebastian. *Evolution of Plant-Pollinator Relationships*. Cambridge: Cambridge University Press, 2012.

Paton, T. R., G. S. Humphreys, and P. B. Mitchell. *Soils: A New Global View*. London: CRC Press, 2018.

Paul, Eldor A., ed. *Soil Microbiology, Ecology, and Biochemistry*. 3d ed. Boston: Academic Press, 2007.

Paul, Gregory S. *Predatory Dinosaurs of the World*. New York: Simon & Schuster, 1988.

Paul, Harry W. *Science, Wine, and Vine in Modern France*. New York: Cambridge University Press, 2002.

Pauly, Philip J. *Fruits and Plains: The Horticultural Transformation of America*. Cambridge, Mass.: Harvard University Press, 2008.

Paumgarten, Nick. "Food Fighter." *New Yorker* 4 Jan. 2010: 36–47. Print.

Paustian, Keith, et al. *Agriculture's Role in Greenhouse Gas Mitigation*. Arlington: Pew Center on Global Climate Change, 2006.

Pearce, Fred. "The Greenhouse Gas Riddle: What Is Causing the Recent Rise in Methane Emissions?" *Yale Environment 360*, Yale School of Forestry and Environmental Studies, 25 Oct. 2016. Accessed 9 July 2018.

Pellechia, Thomas. *Wine: The Eight-Thousand-Year-Old Story of the Wine Trade*. New York: Running Press, 2007.

Penick, J. L., Jr. *The New Madrid Earthquakes*. Rev. ed. Columbia: University of Missouri Press, 1981.

Pennington, R. Toby, and James A. Ratter. *Neotropical Savannas and Seasonally Dry Forests: Plant Diversity, Biogeography, and Conservation*. Boca Raton, Fla.: CRC Press, 2006.

Pfeiffer, Dale A. *Eating Fossil Fuels: Oil, Food and the Coming Crisis in Agriculture*. Gabriola Island (BC: New Society Publishers, 2008.

Phillips, C. J. C. *Principles of Cattle Production*. Wallingford, UK: CABI, 2010.

Phillips, Denise, and Sharon Kingsland, eds. *New Perspectives on the History of Life Sciences and Agriculture*. Springer, 2015.

Phillips, Sarah T. *This Land, This Nation: Conservation, Rural America, and the New Deal*. New York: Cambridge University Press, 2007.

Phillips, Tom. "Largest Animal Cloning Factory Can Save Species, Says Chinese Founder." *Guardian*, 24 Nov. 2015, www.theguardian.com/world/2015/nov/24/worlds-largest-animal-cloning-factory-can-save-species-says-chinese-founder. Accessed 24 Oct. 2016.

Phillipson, David W. "The Antiquity of Cultivation and Herding in Ethiopia." In *Archaeology of Africa: Food, Metals, and Towns*, edited by Thurstan Shaw et al. New York: Routledge, 1993.

Pianka, Eric R. *Evolutionary Ecology*. 6th ed. San Francisco: Benjamin/Cummings, 2000.

Pierce, Benjamin A. *Genetics: A Conceptual Approach*. 4th ed. New York: W. H. Freeman, 2012.

Pierik, R. L. M. *In Vitro Culture of Higher Plants*. Kluwer Academic Publishers, 2000.

Pierzynski, Gary M., J. Thomas Sims, and George F. Vance. *Soils and Environmental Quality*. 3d ed. Boca Raton, Fla.: CRC Press, 2005.

Pimentel, David, and Anne Wilson. "World Population, Agriculture, and Malnutrition." *World Watch*, Sept.–Oct. 2004, pp. 22–25. Accessed 7 Apr. 2017.

Pipkin, Bernard W., and Richard J. Proctor. *Engineering Geology Practice in Southern California.* Belmont, Calif.: Star Publications, 1992.

Pisani, Donald J. *Water and American Government: The Reclamation Bureau, National Water Policy, and the West, 1902–1935.*

Plant Biotechnology Timeline "Plant Biotechnology Timeline." *National Institute of Food and Agriculture.* USDA, 18 Mar. 2009.

Plaster, Edward J. *Soil Science and Management.* 5th ed. Clifton Park, N.Y.: Delmar Cengage Learning, 2009.

Plescan, Costel, and Ancuta Rotaru. "Aspects Concerning the Improvement of Soils Against Liquefaction." *Bulletin of the Polytechnic Institute of Iasi* (2010): 39–45.

Plummer, Charles C., Diane H. Carlson, and Lisa Hammersley. *Physical Geology.* 13th ed. Boston: McGraw-Hill, 2009.

Pollinators "Pollinators." *US Fish & Wildlife Service.* Dept. of the Interior, 1 Mar. 2016.

Postel, Sandra. *Last Oasis: Facing Water Scarcity.* New York: Norton, 1992.

Postlethwait, John H, and Janet L. Hopson. *Modern Biology.*, 2012.

Poultry "Poultry." *Farms.com*, 2017, http://www.farms.com/poultry/. Accessed 26 May, 2017.

Poultry Farming "Poultry Farming." *Roy's Farm*, 2017, http://www.roysfarm.com/poultry-farming. Accessed 26 May, 2017.

Powell, Jerry A. *Biological Interrelationships of Moths and "Yucca Schottii."* Berkeley: University of California Press, 1984.

Powell, Russell. "The Evolutionary Biological Implications of Human Genetic Engineering." *Journ. of Medicine & Philosophy* 37.3 (2012): 204–25. Print.

Prado, José Rafael, et al. "Genetically Engineered Crops: From Idea to Product." *Annual Review of Plant Biology*, vol. 65, 2014, pp. 769–90.

Price, L. *Mountains and Man.* Berkeley: University of California Press, 1981.

Price, R. J. *Glacial and Fluvioglacial Landforms.* New York: Hafner Press, 1973.

Primrose, Sandy B., et al. *Principles of Gene Manipulation.* 6th ed., Blackwell, 2003.

Prothero, Donald R. *Bringing Fossils to Life.* 2nd ed., McGraw-Hill, 2004.

Prothero, Donald R., and Carl Buell. *Evolution: What the Fossils Say and Why It Matters.* Columbia UP, 2007.

Pullin, Andrew S. *Conservation Biology.* New York: Cambridge University Press, 2002.

Pye, Kenneth, and Haim Tsoar. *Aeolian Sand and Sand Dunes.* New York: Springer, 2008.

Quamman, David. *The Song of the Dodo: Island Biogeography in an Age of Extinction.* New York: Simon & Schuster, 1997.

Rackley, Stephen A. *Carbon Capture and Storage.* Butterworth-Heinemann/Elsevier, 2010.

Raff, Rudolf A., and Thomas C. Kaufman. *Embryos, Genes, and Evolution: The Developmental-Genetic Basis of Evolutionary Change.* New York: Macmillan, 1983.

Rajapakse, Ruwan. *Geotechnical Engineering Calculations and Rules of Thumb.* Burlington, Mass.: Butterworth-Heinemann, 2008.

Raman, Saroja. "Soil Degradation and Its Impact on Agriculture." In *Agricultural Sustainability: Principles, Processes, and Prospects.* New York: Food Products Press, 2006.

Randolph, John. *Environmental Land Use Planning and Management.* Washington, D.C.: Island Press, 2004.

Rapp, George, Jr., and Christopher L. Hill. *Geoarchaeology.* 2d ed. New Haven, Conn.: Yale University Press, 2006.

Rath, Sara. *The Complete Cow.* Stillwater, Minn.: Voyageur Press, 1990.

Raup, David M. *Extinction: Bad Genes or Bad Luck?* New York: Norton, 1991.

Raven, Peter H., Linda R. Berg, and David M. Hassenzahl. *Environment.* 6th ed. Hoboken, N.J.: Wiley, 2008.

Rayner, Alan D. M. *Degrees of Freedom: Living in Dynamic Boundaries.* River Edge, N.J.: World Scientific Publications, 1997.

Real, Leslie A., and James H. Brown. *Foundations of Ecology: Classic Papers with Commentaries.* Chicago: U of Chicago P, 1991.

Reed, Matthew. *Rebels for the Soil: The Rise of the Global Organic Food and Farming Movement.* London: Earthscan, 2010.

Reiley, Edward H., and Carroll L. Shry, Jr. *Introductory Horticulture.* 7th ed. Florence, Ky.: Delmar Cengage Learning, 2006.

Reinhardt, J., and W. R. Sigleo. *Paleosols and Weathering Through Geologic Time: Principals and Applications.* Special Paper 216. Boulder, Colo.: Geological Society of America, 1988.

Reisner, Marc, and Lawrie Mott. *Cadillac Desert: The American West and Its Disappearing Water.* New York, N.Y: Penguin Books, 2017.

Relyea, Rick, and Robert E. Ricklefs. *Ecology: The Economy of Nature.* New York: Macmillan Education, 2018.

Renewable Energy and Agriculture: A Natural Fit "Renewable Energy and Agriculture: A Natural Fit." *Union of Concerned Scientists,* 2003. Accessed 7 Apr. 2017.

Renewable Fuels Association. "Ethanol Facts: Agriculture." http://www.ethanolrfa.org/pages/ethanol-facts-agriculture.

Restrepo, Juan D., Sergio A. Lopez, and Juan C. Restrepo. "The Effects of Geomorphic Controls on Sediment Yield in the Andean Rivers of Colombia." *Latin American Journal of Sedimentology and Basin Analysis* 16 (2009): 79–92.

Retallack, G. J. *Soils of the Past.* 2d ed. New York: Blackwell Science, 2001.

Reynolds, J. E., and D. K. Odell. *Manatees and Dugongs.* New York: Facts on File, 1991.

Ribereau-Gayon, Pascal, et al., eds. *The Enology Handbook.* Vol. 1. Hoboken, N.J.: John Wiley & Sons, 2006.

Ricketts, Cliff, and Kristina Ricketts. *Agribusiness: Fundamentals and Applications.* 2nd ed. Clifton Park: Delmar Cengage Learning, 2009.

Ricketts, Taylor H., et al. *Terrestrial Ecoregions of North America: A Conservation Assessment.* Washington: Island, 1999.

Ricklefs, Robert E., and Gary L. Miller. *Ecology.* 4th ed., W. H. Freeman and Company, 2000.

Riffkin, Rebecca. "Forty-Five Percent of Americans Seek Out Organic Foods." *Gallup Poll Briefing* 7 Aug. 2014: 3. *Points of View Reference Center.* Web. 12 Nov. 2015.

Rissler, Jane, and Margaret Mellon. *The Ecological Risks of Engineered Crops.* MIT P, 1996.

Rissman, Adena R., et al. "Conservation Easements: Biodiversity Protection and Private Use." *Conservation Biology* 21, no. 3 (2007): 709–718.

Ritter, Dale F., R. Craig Kochel, and Jerry R. Miller. *Process Geomorphology.* 4th ed. Long Grove, Ill.: Waveland Press, 2006.

Roach, Mary. "Almost Human." *National Geographic,* April, 2008, 126–145.

Robbins, Jim. "As Fight for Water Heats Up, Prized Fish Suffer." *The New York Times,* April 1, 2008, p. F4. Scientists explain how temperature changes and diversion of water for agricultural uses combined to harm fish habitats in western U.S. rivers. Includes suggestions for restoring fish populations.

Roberts, Larry S., et al. Foundations of Parasitology. 6th ed. Boston: McGraw-Hill, 2000.

Rodrigues, Silvio Carlos, and Gelze Serrat de Souza Campos Rodrigues. "Geomorphology and Biogeography of Tropical Deserts." In *Tropical Biology and Conservation Management: Encyclopedia of Life Support Systems.* Oxford: UNESCO, Eolss Publishers, 2009.

Rogers, Heather. "Slowed Food Revolution." *American Prospect* 21.6 (2010). 12–16. Print.

Romanovsky, V., et al. "Permafrost Temperature Records: Indicators of Climate Change." *EOS, AGU Transactions* 83 (2002): 589–594. Discusses the changes to the ecosystem brought on by permafrost thawing. The article provides information on permafrost zones and the methods used to study their degradation.

Root, A. I., and E. R. Root. *The ABC and XYZ of Bee Culture: A Cyclopedia of Everything Pertaining to the Care of the Honey-Bee: Bees, Hives, Honey, Implements, Honey Plants, Etc.* 41st ed. Medina, Ohio: A. I. Root, 2007.

Rose, Michael A. *Darwin's Spectre: Evolutionary Biology in the Modern World.* Princeton, N.J.: Princeton University Press, 1998.

Rosenbaum, Walter A. *Environmental Politics and Policy.* 7th ed. Washington, D.C.: CQ Press, 2008.

Rossberg, Axel G. *Food Webs and Biodiversity: Foundations, Models, Data.* Chichester: Wiley, 2013.

Rossmiller, Ronald Leroy. *Stormwater Design for Sustainable Development.* New York: McGraw-Hill Education, 2014.

Rothstein, S. I., and S. R. Robinson, eds. *Parasitic Birds and Their Hosts: Studies in Coevolution.* New York: Oxford University Press, 1998.

Rowley, William D. *The Bureau of Reclamation: Origins and Growth to 1945, Volume 1: Reclamation—Managing Water in the West.* Denver, Colo.: Bureau of Reclamation, U.S. Department of the Interior, 2006.

Rudman, William B. "Solar-Powered Animals." *Natural History* 96 (October, 1987): 50–53.

Runnegar, Bruce, and James W. Schopf, eds. *Major Events in the History of Life.* Boston: Jones and Bartlett, 1992.

Runte, Alfred. *National Parks: The American Experience.* 4th ed. Lanham, Md.: Taylor Trade, 2010.

Ruskin, F. R., ed. *Lost Crops of Africa: Grains.* Vol. 1. Washington, D.C.: National Academy of Sciences, 1996.

Rutter, Nat. "International Year of Planet Earth 4: Utilizing Paleosols in Quaternary Climate Change Studies." *Geoscience Canada* 36 (2009): 65–77.

Ryan, William, and Walter Pitman. *Noah's Flood: The New Scientific Discoveries About the Event That Changed History.* New York: Simon & Schuster, 1998.

Ryden, Hope. *Out of the Wild: The Story of Domesticated Animals.* New York: Lodestar Books, 1995.

Salisbry, Frank B, Frank B. Salisbry, and Cleon W. Ross. *Palnt Physiology.* Thomson Wadsworth, 2007.

Samama, J. C. *Ore Fields and Continental Weathering.* New York: Van Nostrand Reinhold, 1986.

Sampat, Payal. *Deep Trouble: The Hidden Threat of Groundwater Pollution.* Washington: Worldwatch, 2000.

Santamaría, Johanna, and Gary A. Toranzos. "Enteric Pathogens and Soil: A Short Review." *International Microbiology* 6, no. 1 (2003): 5–9.

Sato, Shun, and Eisei Kurimoto. "Essays in Northeast African Studies." In *Senri Ethnological Studies.* Osaka: National Museum of Ethnology, 1996.

Sauvain, P. *Geography Detective: Mountains.* Minneapolis: Carolrhoda Books, 1996.

Savage, R. J. G. *Mammal Evolution: An Illustrated Guide.* New York: Facts on File, 1986.

Sayre, April Pulley. *Tundra.* Frederick, Md.: Twenty-first Century Books, 1995.

Schaetzl, Randall J, and Michael L. Thompson. *Soils: Genesis and Geomorphology.* New York, NY: Cambridge University Press, 2016.

Scheffer, M. *Ecology of Shallow Lakes.* Boca Raton, Fla.: Chapman and Hall, 1998.

Schery, Robert W. *Plants for Man.* 2d ed. Englewood Cliffs, N.J.: Prentice Hall, 1972.

Schiffer, Michael, ed. *Advances in Archaeological Method and Theory.* 11 vols. San Diego, Calif.: Academic Press, 1978–1987.

______. *Formation Processes of the Archaeological Record.* Salt Lake City: University of Utah Press, 2002.

Schiffman, Richard. "Amazon Rainforest under Threat as Brazil Tears Up Protections." *New Statesman,* 26 Apr. 2017, www.newscientist.com/article/2129024-amazon-rainforest-under-threat-as-brazil-tears-up-protections. Accessed 31 Jan. 2018.

Schivelbusch, Wolfgang. *Tastes of Paradise: A Social History of Spices, Stimulants, and Intoxicants.* Translated by David Jacobson. New York: Pantheon Books, 1992.

Schmink, Marianne, and Charles H. Wood. *Contested Frontiers in Amazonia.* New York: Columbia University Press, 1992.

Schnepf, Randall D. *Energy Use in Agriculture: Background and Issues.* Washington, D.C.: BiblioGov Project, 2013.

Schnepf, Randy. *U.S. Farm Income Outlook for 2015.*

Schopf, J. William, ed. *Major Events in the History of Life.* Boston: Jones and Bartlett, 1992.

Schultz, Gwen. *Glaciers and the Ice Age.* New York: Holt, Rinehart and Winston, 1963.

Schumm, Stanley A. *The Fluvial System.* New York: Wiley-Interscience, 1977.

Schuur, Edward A. G., et al. "The Effect of Permafrost Thaw on Old Carbon Release and Net Carbon Exchange from Tundra." *Nature* 459 (2009): 556–559.

Schwab, Glen, et al. *Soil and Water Conservation Engineering.* 5th ed. Clifton Park, N.Y.: Delmar Cengage Learning, 2005.

Schwartz, Jeffrey H. *Sudden Origins: Fossils, Genes, and the Emergence of Species.* New York: Wiley, 1999.

Scott, Nicky. *Reduce, Reuse, Recycle: An Easy Household Guide.* White River Junction, Vt.: Chelsea Green, 2007.

Seastedt, Timothy R., Richard J. Hobbs, and Katharine N. Sundig. "Management of Novel Ecosystems: Are Novel Approaches Required?" Ecological Society of America. http://www.esajournals.org/doi/abs/10.1890/070046.

Seeley, Thomas D. *Honeybee Democracy.* Princeton, N.J.: Princeton University Press, 2010.

Semple, Kirk. "UN to Consider Whether to Ban Cloning of Human Embryos." *The New York Times,* 3 Nov. 2003, www.nytimes.com/2003/11/03/world/un-to-consider-whether-to-ban-some-or-all-forms-of-cloning-of-human-embryos.html. Accessed 28 July 2014.

Sengar, Rakesh S., and Kalpana Sengar. *Climate Change Effect on Crop Productivity.* Boca Raton: CRC, 2015.

Seo, Sungno Niggol, and Robert Mendelsohn. "Climate Change Impacts on Animal Husbandry in Africa: A Ricardian Analysis." World Bank Policy Research Working Paper 4261.

Sereno, Paul C. "The Evolution of Dinosaurs." Science 284, no. 5423 (June 25, 1999): 2137–2146.

Shanks, Pete. *Human Genetic Engineering: A Guide for Activists, Skeptics, and the Very Perplexed.* New York: Nation, 2005.

Shapiro, Leland. *Introduction to Animal Science.* Upper Saddle River, N.J.: Prentice Hall, 2001.

Sharp, Gary D. *Future Climate Change and Regional Fisheries: A Collaborative Analysis.* Rome, Italy: Food and Agriculture Organization of the United Nations, 2003.

Shawnee National Forest. *Biological Evaluation of Regional Forester Sensitive Species.* Washington, DC: Shawnee National Forest, U.S. Forest Service, 2004.

Sheldrick, Bill, and Sally Macgill. "Local Energy Conservation Initiatives in the UK: Their Nature and Achievements." *Energy Policy* (1988).

Shepard, Mark. *Restoration Agriculture: Real-World Permaculture for Farmers.* Austin: Acres USA, 2013.

Shepherd, Donna W. *Tundra.* New York: Franklin Watts, 1997.

Sherr, Lynn. *Tall Blondes: A Book About Giraffes.* Kansas City, Mo.: Andrews McMeel, 1997.

Shipman, Pat. *Life History of a Fossil: An Introduction to Taphonomy and Paleoecology.* Cambridge, Mass.: Harvard University Press, 1993.

Shiva, Vandana. *Stolen Harvest: The Highjacking of the Global Food Supply.* Lexington, KY: University Press Of Kentucky, 2016.

Shoshani, Jeheskel, and Frank Knight. *Elephants: Majestic Creatures of the Wild.* Rev. ed. New York: Checkmark Books, 2000.

Shucksmith, Mark, David L. Brown, Sally Shortall, and Jo Vergunst. *Rural Transformations and Rural Policies in the U.S. and UK.* New York: Routledge, 2012.

Shugart, H. H. *How the Earthquake Bird Got Its Name and Other Tales of an Unbalanced Nature.* New Haven, Conn. Yale University Press, 2007.

Shugart, Herman H. *Terrestrial Ecosystems in Changing Environments.* New York: Cambridge University Press, 1998.

Silver, Donald M. *African Savanna.* New York: McGraw-Hill, 1997.

Silver, Lee. *Challenging Nature: The Clash Between Biotechnology and Spirituality.* Harper Perennial, 2006.

______. *Remaking Eden: How Genetic Engineering and Cloning Will Transform the American Family.* Harper Perennial, 2007.

Simmons, I. G. *Environmental History: A Concise Introduction.* Oxford: Blackwell, 1993.

Simms, O. Harper. *The Soil Conservation Service.* New York: Praeger, 1970.

Simonson, R. W. "Outline of a Generalized Theory of Soil Genesis." *Soil Science Society of America Proceedings* 23 (1959): 152–156.

Simonson, Roy W. "What Soils Are." In *Soil,* edited by the U.S. Department of Agriculture. Washington, D.C.: Government Printing Office, 1957.

Simpson, Beryl B, and Molly Conner-Ogorzaly. *Plants in Our World: Economic Botany.* Boston, Mass: McGraw-Hill, 2014.

Sindermann, Carl J. *Coastal Pollution: Effects on Living Resources and Humans.* Boca Raton: Taylor, 2006.

Singer, M. J., and D. N. Munns. *Soils.* 6th ed. Upper Saddle River, N.J.: Prentice Hall, 2005.

Skinner, B. J., et al. *The Dynamic Earth: An Introduction to Physical Geology.* 5th ed. New York: John Wiley & Sons, 2006.

______. *Physical Geology.* New York: John Wiley & Sons, 1987.

Slater, Adrian, Nigel W. Scott, and Mark Fowler. *Plant Biotechnology: The Genetic Manipulation of Plants.* New York: Oxford University Press, 2008.

Smith, Bruce D. *The Emergence of Agriculture.* New York: Scientific American Library, 1995.

Smith, Robert L. *Ecology and Field Biology.* 6th ed. San Francisco: Benjamin/Cummings, 2007.

Sobolevsky, Dmitry Yu. *Strength of Dilating Soil and Load-Holding Capacity of Deep Foundations: Introduction to Theory and Practical Application.* Rotterdam, Netherlands: A. A. Balkema, 1995.

Sockol, Michael D., David A. Raichlen, and Herman Pontzer. "Chimpanzee Locomotor Energetics and the Origins of Human Bipedalism" *PNAS* 104.30 (2007): 12265–12269.

Soil Genesis and Development, Lesson 4 – Soil Profile Development "Soil Genesis and Development, Lesson 4 – Soil Profile Development." *Plant & Soil Sciences eLibrary.* Accessed 8 Nov. 2016.

Soil Horizons "Soil Horizons." *K–12 Soil Science Teacher Resources,* soils4teachers.org/soil-horizons. Accessed 8 Nov. 2016.

Soil Profile "Soil Profile." *Soil Nutrient Management for Maui County, College of Tropical Agriculture and Human Resources (CTAHR),* www.ctahr.hawaii.edu/mauisoil/a_profile.aspx. Accessed 8 Nov. 2016.

Soil Science Society of America. *Glossary of Soil Science Terms.* Madison, Wis.: Author, 2001.

Soil Science Society of America. *Glossary of Soil Science Terms.* Madison, Wis.: Author,

Soils Overview "Soils Overview." *Soil Science Society of America,* www.soils.org/files/about-soils/soils-overview.pdf. Accessed 8 Nov. 2016.

Soliva, Carla Riccarda, Junichi Takahashi, and Michael Kreuzer, eds. *Greenhouse Gases and Animal Agriculture.* Boston: Elsevier, 2006.

Sommers, Brian J. *The Geography of Wine.* New York: Plume, 2008.

Soulé, Michael E., and Gordon H. Orians. *Conservation Biology: Research Priorities for the Next Decade.* Washington, D.C.: Island Press, 2001.

Southgate, Douglas, Douglas H. Graham, and Luther Tweeten. *The World Food Economy.* Malden, Mass.: Blackwell, 2007.

Spangler, M. G., and R. L. Handy. *Soil Engineering.* New York: Harper & Row, 1982.

Sparks, Donald S., ed. *Soil Physical Chemistry.* 2d ed. Boca Raton, Fla.: CRC Press, 1999.

Speed, Barbara. "This Is Why We Should Be Farming in Skyscrapers." *Citymetric,* 26 May 2015, www.citymetric.com/skylines/why-we-should-be-farming-skyscrapers-1029.

Spencer, Edgar W. *Introduction to the Structure of the Earth.* 3rd ed. New York: McGraw-Hill, 1988.

Spivak, Marla, and Gary S. Reuter. *Honey Bee Diseases and Pests.* St. Paul: University of Minnesota, 2010.

Sposito, Garrison. *The Chemistry of Soils.* 2d ed. New York: Oxford University Press, 2008.

St. Fleur, Nicholas. "Climate Change Is Shrinking Where Bumblebees Range, Research Finds." *New York Times.* New York Times, 9 July 2015.

Stallings, J. H. *Soil: Use and Improvement.* Englewood Cliffs, N.J.: Prentice-Hall, 1957.

Stanley, Stephen M. *Earth System History.* 2d ed. New York: W. H. Freeman, 2005.

______. *Exploring Earth and Life Through Time.* 2d ed. New York: W. H. Freeman, 1989.

______. *Extinction.* New York: Scientific American Library, 1987.

______. *The New Evolutionary Timetable: Fossils, Genes, and the Origin of Species.* Basic Books, 1981.

Stearns, Beverly Petersen, and Stephen C. Stearns. *Watching, from the Edge of Extinction.* New Haven: Yale UP, 1999.

Stebbins, G. L., and G. J. C. Hill. "Did Multicellular Plants Invade the Land?" *American Naturalist* 115 (1980): 342–353.

Stebbins, G. Ledyard. *Darwin to DNA, Molecules to Humanity.* W. H. Freeman, 1982.

Steinfeld, Henning, et al. *Livestock's Long Shadow: Environmental Issues and Options.* Rome: Food and Agriculture Organization of the United Nations, 2006.

Stern, Nicholas. *The Economics of Climate Change: The Stern Review.* New York: Cambridge University Press, 2007.

Stewart, Eileen. "Food Irradiation: More Pros than Cons? Part 2." *The Biologist,* vol. 51, no. 3, 2004, pp. 141–44.

Stewart, John H. *Geology of Nevada.* Reno: Nevada Bureau of Mines and Geology, 1980.

Stille, Darlene R. *The Greenhouse Effect: Warming the Planet.* Compass Point Books, 2006.

Stone, Richard. "Taking a New Look at Life Through a Functional Lens." *Science* 269, no. 5222 (July, 1995): 316–318.

Storer, Donald A. *The Chemistry of Soil Analysis.* Middletown, Ohio: Terrific Science Press, 2005.

Strassberg, Gil, Norman L. Jones, and David R Maidment. *Arc Hydro Groundwater: GIS for Hydrogeology.* Redlands: ESRI, 2011.

Street, Deborah. "Fisher's Contributions to Agricultural Statistics." Biometrics 46 (1990).

Streever, Bill. *Bringing Back the Wetlands.* Darlinghurst, Australia: Sainty & Associates, 1999.

Strickberger, Monroe W. *Evolution.* 3d ed. Sudbury, Mass.: Jones and Bartlett, 2000.

Strom, Stephanie. "F.D.A. Takes Issue with the Term 'Non-G.M.O.'" *The New York Times.* Accessed 28 Oct. 2016.

______. "Recalls of Organic Food on the Rise, Report Says." *New York Times.* New York Times, 20 Aug. 2015.

Stronach, N. *Mountains.* Minneapolis: Lerner, 1995.

Stutz, Matthew L., and Orrin H. Pilkey. "The Relative Influence of Humans on Barrier Islands: Humans versus Geomorphology." *Reviews in Engineering Geology* 16 (2005): 137–147.

Sullivan, Timothy J., et al. *Air Pollution and Freshwater Ecosystems: Sampling, Analysis, and Quality Assurance.* Boca Raton: CRC, 2015.

Sutton, Ann, and Myron Sutton. *Eastern Forests.* New York: Alfred A. Knopf, 1986.

Svejkovsky, Cathy. "Renewable Energy Opportunities on the Farm." 2006.

Sylvia, David M., Jeffrey J. Fuhrmann, Peter G. Hartel, and David A. Zuberer, eds. *Principles and Applications of Soil Microbiology.* 2d ed. Upper Saddle River, N.J.: Pearson/Prentice Hall, 2005.

Tan, Kim Howard. *Principles of Soil Chemistry.* 4th ed. Boca Raton, Fla.: CRC Press, 2011.

Tandlich, R. *Novel Approaches to Rainwater Harvesting and Sanitation in Developing Countries.* New York: Nova Publishers, 2017.

Tarbuck, Edward J., and Frederick K. Lutgens. *Earth: An Introduction to Physical Geology.* 10th ed. Upper Saddle River, N.J.: Prentice Hall, 2010.

Terborgh, J. *Diversity and the Tropical Rain Forest.* New York: Scientific American Library, 1992.

Terzaghi, K., and R. B. Peck. *Soil Mechanics in Engineering Practice.* New York: John Wiley & Sons, 1948.

Testa, Stephen M. *The Reuse and Recycling of Contaminated Soil.* Boca Raton, Fla.: Lewis, 1997.

Thieman, William J., and Michael A. Palladino. *Introduction to Biotechnology.* 3rd ed. Boston: Pearson, 2013.

Thomas, M. F. *Geomorphology in the Tropics: A Study of Weathering and Denudation in Low Latitudes.* New York: John Wiley and Sons, 1994.

Thomas, Mary Ann. "Factory Farming Backlash Spurs More Choice—and a Little Chaos." *Christian Science Monitor* 14 Sept. 2013: n.pag. *Academic Search Complete.* Web. 6 Nov. 2015.

Thompson, John N. *The Coevolutionary Process.* Chicago: University of Chicago Press, 1994.

Thompson, Stephen A. *Hydrology for Water Management.* Rotterdam: A. A. Balkema, 1999.

Thomson, Keith S. *Morphogenesis and Evolution.* New York: Oxford University Press, 1988.

Tiner, Ralph W. *In Search of Swampland: A Wetland Sourcebook and Field Guide.* New Brunswick, N.J: Rutgers University Press, 2005.

Tobias, Philip V. "Major Events in the History of Mankind." *Major Events in the History of Life.* Ed. J. William Schopf. Boston: Jones, 1992.

Toft, Catherine Ann, Andre Aeschlimann, and Liana Bolis, eds. *Parasite-Host Associations: Coexistence or Conflict?* New York: Oxford UP, 1991.

Toogood, Alan. *American Horticultural Society Plant Propagation: The Fully Illustrated Plant-by-Plant Manual of Practical Techniques.* New York: DK Adult, 1999.

Toovey, Michelle. "The Industry Handbook: The Agriculture Industry (COW, LSTK)." *Investopedia,* 10 Mar. 2016. Accessed 1 Nov. 2016.

Tranter, R.R., et al. "Consumers' Willingness-to-pay for Organic Conversion-grade Food: Evidence from Five EU Countries." *Food Policy* 34 (2009): 287–94.

Tricart, Jean. *The Landforms of the Humid Tropics, Forests, and Savannas.* Translated by Conrad J. Kiewiet de Jonge. New York: St. Martin's Press, 1972.

Trinkaus, John Philip. *Cells into Organs: The Forces That Shape the Embryo.* Englewood Cliffs, N.J.: Prentice-Hall, 1984.

Troeh, Frederick R., and Louis M. Thompson. *Soils and Soil Fertility.* 6th ed. Malden, Mass.: Blackwell, 2005.

Tudge, Colin. *The Variety of Life: The Meaning of Biodiversity.* New York: Oxford UP, 2000.

Tutelyan, Victor. *Genetically Modified Food Sources: Safety Assessment and Control.* Elsevier, 2013.

U.S. Department of Energy. "Biofuels and Agriculture: A Factsheet for Farmers." September 2001.

U.S. Soil Conservation Series. *Our American Land: Use the Land, Save the Soil.* Washington, D.C.: Government Printing Office, 1967.

Unger, Paul W. *Soil and Water Conservation Handbook: Policies, Practices, Conditions, and Terms.* Binghamton: Haworth, 2006.

Union of Concerned Scientists. *CAFOs Uncovered: The Untold Costs of Confined Animal Feeding Operations.* By Doug Gurian-Sherman. Cambridge: UCS, 2008.

United Nations Environment Programme (UNEP). Global Strategy for Plant Conservation. http://www.cbd.int/gspc/

United States Bureau of Labor Statistics. "Agricultural Workers." *Occupational Outlook Handbook.* US Dept. of Labor, 8 Jan. 2014.

United States Department of Agriculture. "World Agricultural Supply and Demand Estimates." *USDA.* US Dept. of Agriculture, 9 Oct. 2015.

______. Economic Research Service. "Special Feature: The Shift to Large Farms." *Structure and Finances of U.S. Farms: Family Farm Report.* Washington, DC: Author, 2007.

______. World Agricultural Outlook Board and U.S. Interagency Agricultural Projections Committee. *USDA Agricultural Projections to 2019.*

United States Environmental Protection Agency. *Cases in Water Conservation: How Efficiency Programs Help Water Utilities Save Water and Avoid Costs.* Washington, DC: Author, 2002.

USDA Organic "USDA Organic." *US Dept. of Agriculture,* 18 Oct. 2016, www.usda.gov/wps/portal/usda/usdahome?navid=organic-agriculture. Accessed 30 Nov. 2016.

Vallentine, John F. *Grazing Management.* Amsterdam/New York: Elsevier Science, 2016.

Vallero, Daniel A. *Environmental Biotechnology: A Biosystems Approach.* Academic, 2010.

Van Breemen, Nico, and Peter Buurman. *Soil Formation.* 2d ed. Boston: Kluwer Academic Publishers, 2002.

Van Dyke, Fred. *Conservation Biology: Foundations, Concepts, Applications.* 2d ed. New York: Springer, 2008.

Van Riper, C., III, et al. "Career Opportunities in Ornithology." *American Birds* 43.1 (1989): 29–37. Print.

Vandermeer, John H. *The Ecology of Agroecosystems.* Sudbury, Mass.: Jones and Bartlett, 2011.

Vengosh, A. "Salinization and Saline Environments." In *Environmental Geochemistry,* edited by Barbara Sherwood Lollar. Oxford: Elsevier, 2005.

Vernooy, Ronnie. *Community Seed Banks: Origins, Evolution and Prospects.* London: Routledge, Taylor & Francis, 2015.

Verstraete, Michel M., Robert J. Scholes, and Mark Stafford Smith. "Climate Change and Desertification: Looking at an Old Problem Through New Lenses." *Frontiers in Ecology and the Environment* 7, no. 8 (October 2009): 421–428.

Vertical Farming: Does it Really Stack Up "Vertical Farming: Does it Really Stack Up?" *The Economist,* 9 Dec. 2010, www.economist.com/node/17647627.

Vickers, Amy. *Handbook of Water Use and Conservation: Homes, Landscapes, Businesses, Industries, Farms.* Amherst: Waterplow, 2001.

Vidal, John. "Q&A: Rainforests." *The Guardian,* 26 July 2007, www.theguardian.com/environment/2007/jul/26/conservation.endangeredhabitats. Accessed 31 Jan. 2018.

______. "'The Sumatran Rainforest Will Mostly Disappear within 20 Years.'" *The Observer,* 26 May 2013, www.theguardian.com/world/2013/may/26/sumatra-borneo-deforestation-tigers-palm-oil. Accessed 31 Jan. 2018.

Vié, Jean-Christophe, Craig Hilton-Taylor, and Simon N. Stuart, eds. *Wildlife in a Changing World: An Analysis of the 2008 IUCN Red List of Threatened Species.* Gland: IUCN, 2009.

Vileisis, Ann. *Discovering the Unknown Landscape: A History of America's Wetlands.* Washington, D.C.: Island Press, 1999.

Vogt, Willie. "Regulations Affect Farms in Variety of Ways." *Southwest Farm Press* 19 Sept. 2013: 14. *Academic Search Complete.* Web. 6 Nov. 2015.

Walker, Sharon, and David McMahon. *Biochemistry Demystified.* McGraw-Hill, 2008.

Walker, Tom. *Caribou: Wanderer of the Tundra.* Portland, Oreg.: Graphic Arts Center, 2000.

Wallace, J. *The Rise and Fall of the Dinosaur.* New York: Gallery Books, 1987.

Walther, John Victor. *Essentials of Geochemistry.* 2d ed. Burlington, Mass.: Jones & Bartlett Publishers, 2008.

Walther, John Victor. *Essentials of Geochemistry.* 2d ed. Burlington, Mass.: Jones & Bartlett Publishers, 2008.

Ward, Peter D., and Don Brownlee. *Rare Earth: Why Complex Life is Uncommon in the Universe.* New York: Copernicus, 2000.

Watson, James D. *The Double Helix: A Personal Account of the Discovery of the Structure of DNA.* Reprint. New York: Simon, 1998.

Watson, James D., et al. *Recombinant DNA: Genes and Genomes—A Short Course.* 3rd ed., Freeman, 2007.

Watson, Paul. "New Zealand Aims for Greener Pastures: Officials Ruminate on How to Curb Methane from the Nation's Livestock, a Culprit in Global Warming." *Los Angeles Times,* 8 June 2008: A-6. Print.

Weasel, Lisa H. *Food Fray: Inside the Controversy over Genetically Modified Food.* New York: Amacom-American, 2009.

Weathering "Weathering." *National Geographic Society,* nationalgeographic.org/encyclopedia/weathering/. Accessed 8 Nov. 2016.

Weber, Gustavus Adolphus. *The Bureau of Chemistry and Soils: Its History, Activities and Organization.* Baltimore, Md.: Johns Hopkins Press, 1928.

Webster, Gerry, and B C. Goodwin. *Form and Transformation: Generative and Relational Principles in Biology.* Cambridge: Cambridge University Press, 2011.

Weide, David L., and Marianne L. Faber, eds. *This Extended Land: Geological Journeys in the Southern Basin and Range.* Las Vegas: University of Nevada Press, 1988.

Weil, Ray R, and Nyle C. Brady. *The Nature and Properties of Soils.* Harlow: Pearson Prentice Hall, 2017.

Weis, Tony. *The Global Food Economy: The Battle for the Future of Farming.* New York: Zed Books, 2007.

Weishampel, David B., and Halszka Osmolska. *The Dinosauria.* Berkeley: University of California Press, 2004.

Weiss, Michael. "Precision Farming and Spatial Economic Analysis: Research Challenges and Opportunities." American Journal of Agricultural Economics 78, no. 5 (1996).

Wessells, Norman K. *Tissue Interactions and Development.* Menlo Park, Calif.: Benjamin/Cummings, 1977.

West, Frederick H., ed. *American Beginnings.* Chicago: University of Chicago Press, 1996.

Wetzel, Robert G. *Limnology: Lake and River Ecosystems.* San Diego: Academic Press, an imprint of Elsevier, 2015.

Wheeler, Tim, and Joachim Von Braun. "Climate Change Impacts on Global Food Security." *Science,* vol. 341, no. 6145, 2013, pp. 508–13.

White, Robert E. *Principles and Practice of Soil Science: The Soil as a Natural Resource.* 4th ed. Malden, Mass.: Blackwell, 2006.

Whitefield, Philip. *The Biology of Parasitism: An Introduction to the Study of Associating Organisms.* Baltimore: University Park, 1979.

Whitten, R., and W. Pendergrass. Carolina Protozoa and Invertebrates Manual. Burlington, N.C.: Carolina Biological Supply, 1980.

Wicander, Reed, and James S. Monroe. *Historical Geology.* 8th ed. Belmont: Cengage Learning, 2015.

Wicander, Reed, and James S. Monroe. *Historical Geology.* Cengage Learning, 2016.

Wild, R. and C. McLeod, eds. *Sacred Natural Sites: Guidelines for Protected Area Managers.* Gland, Switzerland: International Union for Conservation of Nature and Natural Resources (IUCN), 2008.

Wilfert, Lena, et al. "Deformed Wing Virus Is a Recent Global Epidemic in Honeybees Driven by *Varroa* Mites." *Science* 351.6,273 (2016): 594–7. Print.

Wilhite, Donald A., and William E. Easterling, with Deborah A. Wood, eds. *Planning for Drought: Toward a Reduction of Societal Vulnerability.* Westview Press, 1987.

Wilhite, Donald A., ed. *Drought and Water Crises: Science, Technology, and Management Issues.* CRC Press, 2005.

Willett, Sean D., and Mark T. Brandon. "On Steady States in Mountain Belts." *Geology* 30 (2002): 175–178. Provides an overview of steady-state processes. Discusses the four types of steady state in orogenic systems. Accessible to academia.

Williams, John W. and Stephen T. Jackson. "Novel Climates, No-Analog Communities, and Ecological Surprises." *Frontiers in Ecology and the Environment* 5 (2007).

Williams, Michael. *Deforesting the Earth: From Pre-history to Global Crisis.* Chicago: University of Chicago Press, 2003.

Williams, P. J., and M. W. Smith. *The Frozen Earth: Fundamentals of Geology.* New York: Cambridge University Press, 1991.

Wills, Christopher, and Jeffrey Bada. *The Spark of Life: Darwin and the Primeval Soup.* Cambridge, Mass.: Perseus, 2000.

Wilmut, Ian, et al. *The Second Creation: Dolly and the Age of Biological Control.* Farrar, 2000.

Wilson, Edward O., ed______, Edward O., ed. *Half-Earth: Our Planet's Fight for Life.* New York: Liveright, 2016.

______. *Naturalist.* Washington, DC: Island, 2006.

Wilson, R. W., et al. "Contribution of Fish to the Marine Inorganic Carbon Cycle." *Science* 323, no. 5912 (January 16, 2009): 359–362.

Winchester, Simon. *The Map That Changed the World.* London: Penguin, 2001.

Woese, Carl R. "Archaebacteria." *Scientific American* 244 (June 1981): 98–122.

Wojtkowski, Paul A. *Agroecological Economics: Sustainability and Biodiversity.* Boston: Elsevier/ Academic Press, 2008.

______. *Landscape Agroecology.* Boca Raton, Fla.: CRC Press, 2003.

Wollenberg, Eva. *Climate Change Mitigation and Agriculture.* New York: Earthscan, 2012.

Woods, Andrew W. *Flow in Porous Rocks: Energy and Environmental Applications.* Cambridge: Cambridge UP, 2015.

Woods, Michael. *Rural.* New York: Routledge, 2011.

Woodward, Susan L. *Grassland Biomes.* Westport, CT: Greenwood Press, 2008.

Workman, James G. *Heart of Dryness: How the Last Bushmen Can Help Us Endure the Coming Age of Permanent Drought.* Walker Publishing Company, 2009.

World Health Organization. "Soil-Transmitted Helminthiasis: Number of Children Treated 2007–2008: Update on the 2010 Global Target." *Weekly Epidemiological Record* 85 (April 16, 2010): 141–148.

World Wildlife Federation. *Living Planet Report 2016: Risk and Resilience in a New Era.* WWF International, 2016, www.wnf.nl/custom/LPR_2016_fullreport/. Accessed 31 Jan. 2018.

Wright, Richard T., and Dorothy F. Boorse. *Environmental Science: Toward a Sustainable Future.* 11th ed. Boston: Benjamin/Cummings, 2011.

Wright, V. P., and Alfred Fischer, eds. *Paleosols: Their Recognition and Interpretation.* Princeton, N.J.: Princeton University Press, 1986.

Yadav, A. K., S. Ray Chaudhuri, and M. R. Motsara, eds. *Recent Advances in Biofertilizer Technology.* New Delhi: Society for Promotion and Utilisation of Resources and Technology, 2001.

Young, Tomme R. *Genetically Modified Organisms and Biosafety: A Background Paper for Decision-Makers and Others to Assist in Consideration of GMO Issues.* International Union for Conservation of Nature, 2004.

Yount, Lisa. *Biotechnology and Genetic Engineering.* 3d ed. New York: Facts On File, 2008.

Zann, Leon P. *Living Together in the Sea.* Neptune City, N.J.: T. F. H., 1980.

Zhu, Qiang, John Gould, Yuanghong Li, and Chengxiang Ma. *Rainwater Harvesting for Agriculture and Water Supply.* Singapore: Springer, 2016.

Zimmer, Carl. *At the Water's Edge: Macroevolution and the Transformation of Life.* New York: Free Press, 1998.

Zimmerer, Karl S. "Environmental Discourses on Soil Degradation in Bolivia: Sustainability and the Search for Socioenvironmental 'Middle Ground.'" In *Liberation Ecologies: Environment, Development, Social Movements,* edited by Richard Peet and Michael Watts. 2d ed. New York: Routledge, 2004.

Zinsser, Hans. *Rats, Lice, and History.* Reprint. New York: Bantam, 2000.

Zohary, Daniel, and Maria Hopf. *Domestication of Plants in the Old World: The Origin and Spread of Cultivated Plants in West Asia, Europe, and the Nile Valley.* 3d ed. New York: Oxford University Press, 2001.

Zwinger, Ann H., and Beatrice E. Willard. *Land Above the Trees: A Guide to American Alpine Tundra.* Boulder, Colo.: Johnson Books, 1996.

INDEX

C

D

biofuel: Fuels devised from biological materials including crops (especially trees) and animal wastes

biological control: The control of pests and parasites through the use of other organisms, often natural predators.

biological efficiency: The efficiency of an organism or biological system

biomass: The total weight of living material, of all forms

biotechnology: The development of biological processes and products, i.e. genetic modification.

bloating: Abnormal swelling of the abdomen of livestock, caused by excessive gas formation which can result in death.

boar: A breeding male hog, any age.

bovine spongiform encephalopathy (BSE): A disease of cattle (often called 'mad cow disease') caused by an agent that is neither a bacterium nor a virus. First recognized in 1986 it has a long incubation period.

brassicas: Plants of the brassica classification e.g. turnip.

break crop: A change in crop in an arable rotation in order to help control weeds and plant diseases.

breed: A group of animals descended from common ancestry and possessing certain inherited characteristics which distinguish it from any other group. When matings within the breed are made, these characteristics are transmitted to the offspring in a uniform and predictable manner.

broadcasting: Random scattering of seeds over the surface of the ground if the seed is to be covered, this is done as a separate operation, usually with a spike-tooth harrow.

broiler: A chicken of either sex about 7 weeks of age.

browsing: A method of feeding by herbivores, in which the leaves and peripheral shoots are removed from trees and shrubs.

bushel: A unit of dry measure (1 cubic foot) for grain, fruit, etc., equivalent to 8 gallons of liquid. Weight varies with the density/bulk of the commodity. Example: Oats weigh 32 lbs per bu.; barley, 46 lbs. per bu.; and corn, 56 lbs. per bu.

calf: Young (up to yearling or sexual maturity) animal of the bovine species cash crop: Any crop that is sold off the farm to yield ready cash.

carnivores: Animals that feed on other animals or material of animal origin.

carrying capacity: The number of animals an area of land can support.

cash flow: Movement of funds through a business

catch crop: Crop utilizing land betwe`en two longer term crops

certified seed: Seed grown from pure stock which meets the standards of certifying agency (usually a state government agency) certification is based on germination, freedom from weeds and disease, and trueness to variety.

chloroplasts: Discrete photosynthetic organelles within plant cells, contain chlorophyll

cloning: Producing a stock of individuals all derived asexually from one sexually produced

complete fertilizer: A fertilizer containing the three macro nutrients (nitrogen, phosphorous, and potassium) in sufficient amounts to sustain plant growth compost: Organic residues, or a mixture of organic residues and soil which have been piled, moistened, and allowed to undergo biological decomposition. Mineral fertilizers are sometimes added.

confinement: Livestock kept in "dry-lot" for maximum year-round production facilities may be partial or complete solid floored and enclosed/covered.